CW01373664

# THE FIFTH DIMENSION

## THE MISSING LINK

JOSEPH DONNELLY

authorHOUSE®

*AuthorHouse™ UK Ltd.*
*500 Avebury Boulevard*
*Central Milton Keynes, MK9 2BE*
*www.authorhouse.co.uk*
*Phone: 08001974150*

© 2011 Joseph Donnelly. All rights reserved.

No part of this book may be reproduced, stored in a retrieval system, or transmitted by any means without the written permission of the author.

First published by AuthorHouse 3/23/2011

ISBN: 978-1-4567-7241-3 (sc)

Any people depicted in stock imagery provided by Thinkstock are models, and such images are being used for illustrative purposes only.
Certain stock imagery © Thinkstock.

This book is printed on acid-free paper.

Because of the dynamic nature of the Internet, any Web addresses or links contained in this book may have changed since publication and may no longer be valid. The views expressed in this work are solely those of the author and do not necessarily reflect the views of the publisher, and the publisher hereby disclaims any responsibility for them.

To my wonderful extended
family, past, present and future

# THE FIFTH DIMENSION

By the very name of this book with its attachments, we will be able to follow monumental changes in our attitudes to each other described by those words above. They will be seen to represent us towards future consideration they will be the lead we need in the order of change. Based on the best reasoning that we are here on this our earth now, we can examine the past without the need to change it. We will be able to look at the why and how we came to be in our present situations without any recourse of blame, condemnation or accusation, but be able to carry criticism of ourselves and each other.

Already we live in a world of three dimensions to the visual aspect; by later study we have come to accept there is a fourth dimension in terms of space and time. Further examination by the medium of science in its/our quest to the understanding of all formula has led in course to the production by imagination the need of extra dimensions! Those wrapt in the mystery of science with all of its tributaries, seek that in order for future and final understanding, we might need progressive dimensions from the fourth to the fifth, sixth even up to eleven so that we would be in a position to unravel the complexities of the universe.

With this book as our production, we can deliberate without the need of full comprehension to why and how science has loomed so large in supposition that she is to be the agent of our complete understanding of all matters. Though we have given her licence on our behalf, we can claim with our human Fifth Dimensional presentation, limit to her objective by the standard of our understanding already.

Before or against any classification The Fifth Dimension in my form is not representative for any aspect of science, it will never deny any aspect of science. Where it will walk in greater clarity is to enhance the largest of scientific elements, that of religion. Without following doctrine, text, tenet or covenant the Fifth Dimension will aid us all in better comfort by the measure of human achievement on our own behalf. In link it will carry my/our first four dimensions in line of progression, but as in science to mixed circumstances as they present themselves.

Dare we question science, dare we not!

# FORWARD

Reading this book we can follow the how and the why? How is delivered from the effort of which my limitations in the skill of communication are exposed! How I deliver in promise is from what I had to start with, which is of little consiquence except to the reader?

I have not taken time to have this finished work polished, not least because of its latent importance to us all.

It is an epic work written from a heart of hopefulness, from the eternal optomist with a true belief in the honest integrity of all of us to be described as specie! My style is a marked effort for any of us to say what we think, along with showing the route of our justification for any such statements.

I deem the work to be of historic significanc, for it considers all to modernity from a time mark in the year two thousand. Without distinction my judgements are of very small events to have occurred over millenium and millenium and millinium, but the journey belongs to us all.

Why is not a bespoke reason for producing this work. In the year two thousand (which for other reasons I never expected to see). I had one small opinion of what was or what had most affected many standards of world behaviour? Of all the causes to the mass of much of our human failings, I settled on a type of what had appeared to be of religious significance of late, that is upto about the year two thousand. In taking to the one standard of cause and effect, my thinking was channelled to our two main influences at that time of our histories?

By the incidence that we are all controlled politically, that feature had to be prominent. Although never led, I had to follow the course that communism died by its own hand through its own excesses. By following the parallel, I have personally opted for our political zenith to run that democracy is it and the American Dream is in apex representatively?

Why, by my own religious understanding holds much more force in all realities, but only beholding to the expected understanding of God expressed by specie acclimation, specie acclimation!

Why, religion stands by its own force in all cases but within limits! Over the same time period allocated to politics since nineteen forty five. My study of religion started to form in condemnation of Islam for its own sake, but to make us all aware that to kill of self by self through martyrdom is not a justifiable course from any scripture!

If I had set to condemn or chastise or catalogue Islam, my emotions to the dangerous acts that some from islam might play. Had fortunately only managed to turn my head to the good of all religions connected with the God image and not Islamic shortcomings!

If I saw the one flaw in Islam it was of itself. By my own interpritation this point has been presented fairly and in relation to all great religions Eastern or Western! There is little we have to do to understand, but much to change without alteration!

Why, when in talking of God, even by my own interpritation of events. My book, our work, is aimed at all who seek any degree of understanding to the means of hope and of purpose in the future!

By centering on politics but primarily on religions, my own mantra is to the concept of God in reality of specie, even made by specie! I have little connection to listening in time to those of us who do not accept God as the entity 'he' is! For me, no belief in the concept of God is a denial of self. I have no ear for those like us who deny the future! No God no conversation!

My conclusions are ours, the key to all change is to be carried by us, even to include the three taken and one active prospect of the Fifth Dimension!

I have put this work to be an historical document, for such it has to be presented in the form so done. Because by my own best efforts the work is part of our march of time in this singular religious specie we are!

# CHAPTER 1

Eight hundred million years ago to the first forms of earth life on this our planet, hold up, not quite that far back?

More than three and a half million years ago an ape like creature stood erect and walked leaving their footprints transfixed in muddy soil. Footsteps locked in stone now when over those vast years that soft volcanic ash dried and hardened holding its memory for us to read today. Africa is the land of that discovery, being the oldest ever found anywhere to give us humans the first sign to our ancestry. To assert we were to be human from that sign is reasonable when following world wide signs to all creature histories. We have learnt to mark them all in the progressive nature they seem to follow by time, sometimes by hundreds of millions of years.

Even from our own short time span to now we do not have the full picture; assertion is not the means of proof, but will suffice in the absence of it, for us to be set in our human character, very many processes of change had happened over the ages. From an erect biped to this time of now we followed alterations to our present physical bearing, shown by us all through our differences of colour, size, habit and outlook arriving to be this wonderful collection of people we are today.

While moving through these pages it will be necessary to give mention about evolutionary concepts, the process of change determined by differences between the same specie, specifically measured in terms of progressive improvement supported by the phenomenon of natural selection. To think of the word evolution it is almost imposable not to be in mind of Charles Darwin, who some one hundred and fifty years age published his extremely well formulated theoretical work, The Origin of Species. He was very fortunate to have travelled at a time before much of the world was overrun by humanity, he found in many of the places he visited almost pristine conditions for the processes of evolution. From his meticulous studies he began to expound specie development led in part by his notice in groups of small birds by the collective name of finches. He would indicate that by association they were all of a single prior parenthood, but had developed different beak structures of shape, size and strength to accommodate them in their eating habits. That they had evolved in the cause of need, exclusive of beak structure, by deduction he made like connection about human specie development, his link there was from a group of animals with assumed similarities to us in the apes.

Unfinished as his studies were, in that connection he left a gap by inference in the ape human link assuming that the appropriate evidence would be forthcoming. That is why our erect biped theory is very interesting, for even in modern times to study apes from their genetic closeness to us, we can see they are essentially quadrupeds. They do have two defined feet but the definition of having arms also is belied by use, the upper limbs do gather, break and shell foodstuffs, which are further manipulated by the mouth. But in walking these great beasts trundle along in a semi upright position using their upper and lower limbs in continuous contact with the ground. It is therefore safe to assume that standing on their supposed legs while reaching to higher levels with their supposed arms does not constitute an upright posture. Evolution aids us to conclusion by reason of the actual structure and formation of apes leg bones and leg muscles. They are simply not designed for the purpose to be ape erectus. We are therefore left searching the missing link between apes now known and man, looking for the fossil record of any creature with bone structure approaching Homo erectus.

By good fortune the oldest time record of a biped had been found in Africa, not conclusive of our link by any means, but to be taken more seriously because of where it was found. While we are still looking throughout the world for more tangible proof of our human history, we have that fixed connection to Africa. It is that we might say we are 'Out of Africa', for other signs have been found in close accompaniment to give indication if only in part, that another animal of old could stand erect but was not quite ape or not quite man!

At no time in his life or to his realisation did Charles Darwin know of our archaeological finds in Africa they were much later than his time. His exposition of mans ancestry was a proposition by line to be followed in looking at other examples of connected specie. His revelations were scientific and heartily delivered on that basis. Where he created a great stir at publishing a great work, was overshadowed by the society of then in how they reacted to his work.

Evolution was a growing concept of time and reason, but more expressed as progression, it was a method to account for the connection of many creatures by their similarity. Other like study was also done in the lines of plant and animal husbandry to force the active change in plants or animals for seen improvement. By status in all fields and forms it was used to understand the way things were to the way they are now, accepting the time factor involved to some changes of specie being monumental. From Darwin's The Origin of Species, by indicating the scientific fact of progression evolutionary, did not take into account many societies already acceptance of all answers by the medium of creation? Evolution and creation at that time did not work together, because one idea is secular and the other from spiritual means, will ever the twain meet in harmony?

Charles Darwin the scientist published a great work with out any forethought of effect, save recognised understanding of its merit if warranted, it was. By personal status he was of European extraction, at the time vastly progressive people and generally of Christian religious persuasion by circumstance. Many to the society in Europe and religions worldwide saw the work as a direct affront to their faith. That science would follow to the possibility of mans succession from apes, when it was known as fact that man was chosen by god befitting to god's creation of all things and all things. Darwin's stir by the understanding of others still carries today, not by his fault, but from our ignorance.

# CHAPTER 2

Between three and a half million years and the publishing of The Origin of Species a lot had happened. We might now look at an imagined event of some four thousand years ago to bridge that gab by hidden association. Then ago there was innumerable well organised societies throughout the world, interacting on every level. To many a matter of self concern was their religious or god structure. Religions were a way of dealing with all that we did not know or understand historically, the management of which was left to those who could interpret such affairs, our priesthood. We revelled in the numbers of gods that we could hold, literally gods for everything. By time, doubt arose in some of our gods ability, which gave us feed to question the value of them in relation to ourselves? Like for when we lost in battle, other war gods must be better, when others prospered and we didn't, why did our gods fail? Better make a bigger offering to a different god and see what happens now!

At about those four thousand ago, from one group of men to their own guise in their own reasoning but for their very community, gave in answer to future concept! By excellence of mind through error of thought, they gave in form the settled matter of creation. By direct association they were met in thought to have been chosen in god's aim, no more to forage in the mire of godly castes. Here by choice this single people, albeit as connected tribes, could now settle in kind all of god's mysteries by the certainty of there being only one god. Not one head god, not a god of war or harvest or the weather or love, but the god of creation itself, what incredible feelings must have been felt now that the true route to the acceptance of purpose had been revealed to us! How best to react, how to translate this wonderful offering to all in tribe, unless by those most able to deal with it from our own priesthood, those most responsible for this final new discovery?

Direct contact in mind had been made upon these chosen people who were very much aware of the world they lived in and the way they lived. Society was structured in the god, priest, leader set. With the standard of this huge realisation, it was required to give the concept depth of age to validate the understanding of it for the group. Part of the excellence by these people was working out the one god's role as the ultimate creator, by selecting their own ancestor in Abraham the father of Isaac to have carried these words of enlightenment. In order to cement this link with god, records of history had to be laid down to verify fact.

This was done in best order by the skills then available but now seen in error. These fine people reacted the exact same way any other group would have then ago. They assembled from memory the previous ancestral structure of their tribes to that first link with god, by natural normal reasoning they thought in our terms of human frailty by back calculating each generational lifecycle until they reached contact with Abraham! Creating a figure in length of time by seasons to be about one thousand seven hundred years long, which was the setting the time by date for human and creation by God! To this day associated to be the Jewish year mark calendar, but additionally with time added from then to now we are in about the year five thousand seven hundred from Jewish reasoning?

Through the abstract and from my own style of presentation, the Jewish religion is to be my First Dimension, not on scientific terms. It is taken on religious terms by two main factors, the one that it is regarded as the worlds' oldest recorded religion, secondly it is the first to be truly monotheistic. Also it has very deep connections with an area of the world to be generally known as the Middle East which figures large by perspective throughout these pages. It will always remain the Jewish faith but gets my definition from these most recent of years. We will see how it could only have been from very recent times by the course of world events that I put my own code to it even completely unconnected religiously?

Even past excepted standards of conduct the founders of the Jewish faith entwined us all to the most complex of future realisations, yet to be confirmed. Where by comparison all other life in standing lived and died. We as humans were unique through the medium of the one true god to live for ever, even if our living of now was curtailed by whatever means, we of this form would live on and on and on!

## CHAPTER 3

Two thousand years ago in the same area of the Middle East, events were almost non events. Many established societies were interacting as they had done since any time past. Most of the region had been conquered and were subjugated by their new masters though left with a certain amount of autonomy. The city of Rome in far off Europe was the centre of this ruling body, in light of their success by conquest they held well to the belief in their own gods and in their organisation backed by military power. Nevertheless they watched very closely all the events and decisions made in their territories. They knew well the structure of society in these lands, knowing of one group the Jews, who had been waiting two thousand years in expectation for a triumphant unity with their single god, imminent! Although in those past two thousand years of waiting for the expected messiah, there had been many false alarms by peoples claiming some sort of divine connection in open deceit to curry favour or gain some sort of other advantage, the Jewish priesthood watched for all eventualities!

When a man of some thirty years age began to act in similar style to other false prophets the same cautious heed was given by the Jews and ever Rome, just in case? This time there was an air about the man not quite like others, he was well educated in Jewish law and scripture. Yet almost casually he claimed to be the messiah by connection he was the manifestation of god the son, God's representation to be physically man? Explained as part of the trinity, God the Father, the Holy Ghost as gods' messenger, then he, Jesus Christ on earth, God the son in the form of mortal man, three as one, remaining the one God!

For three years Jesus moved about the lands of his birth preaching new interpretations of the Jewish faith. At the very core of his own belief, although he let it be indicated that he was the expected messiah. He had no great intentions to stand over and rule these expectant people. His objective was by his Fathers sight to relieve from man his burden of original sin, gathered when the first created man disobeyed gods covenant in the realm of paradise exposed as the Garden of Eden in holy Jewish scripture. Jesus was to die as a man in sufferance for man by atonement, so said and taken by many to be so.

Many centuries before when the Jewish people reconstructed their own history to accommodate the imagination of god's time scale for creation, they had actually set the time to god's method of it. In creating the perfect human state by intention, but also by giving man the ability to self express by free will. A situation of temptation arose to ignite mans own query of knowledge, but by reacting to it mankind had affronted god and were cast in judgement, blemished for our deed. By gods good grace all were given the same chance of renewal to undo past effrontery, but at great cost. When first the Jewish tribes came to these conclusions they rightly took gods concern to have been directed to those who would seek most to correct previous misdeeds. By that very connection it was but a small step to have been chosen, enough to content Jewish thinking for almost two thousand years.

There is much to say that Jesus was indeed very well versed in Jewish law, history and endeavour, certainly to the standard of being a teacher of men. Without control of what Jesus actually said we must suppose by outcome what he might have been like. The good offices of the Jewish religion were very well established, if not a little jaded. From the realisation of a near triumphant unity, imminent, to then be in waiting for one hundred generations and still waiting, what next? It was indeed necessary to hold integrity by being prepared for some who might like to exploit this longevity. Of the many who tried history holds no mark, save perhaps in Jesus?

Although Jesus spoke in terms past Jewish thinking, at the time, his basic teaching was to and by the Jewish faith. Beyond the show of desire for personal notary, he declared in the slowest of terms that his mission was for the redemption of all mankind; all mankind through showing his father in heaven mans ultimate sacrifice. For and because of other conditions, Jesus did die by Roman authority in the lands of the Middle East. From the collection of the small band of people he had gathered about him to assist him in teaching, form the basics of his sermons by inclusion now of all men to have been chosen? After his death there was a rapid growth of this new branch of Judaism. But in the crossover, confusion between the established priesthood and Jesus own cadre of helpers ensued, one remained in line to have been originally chosen. With the new version taking to the eventual name of catholic meaning universal by connection with Jesus as the founder, but only two thousand years ago!

Because of the time difference between Judaism and Catholicism, because of their association by the geography of area, this second monotheistic religion is my Second Dimension. That dimensional heading with the first does not attempt to usurp or supplant any religious doctrine or to change it, hide it, decorate it or be of any other interference. The dimensional title is of this time now, not four or two thousand years ago but measured in mere months as forty or fifty or sixty, that recent! First or second titles are given on a time reference basis only, there ultimate connection will be realised when we are able to add our prospects as three taken and one active.

## CHAPTER 4

When Jesus gave of his ideas in the lands of his fathers, there was no particular new theme to his teaching. Anything to grow was by the good fortune of human endeavour. Although it took someone like Jesus to enhance it, in those very lands at his time, all people were conquered, all had to follow rules and laws beyond their own standards. Roman law was of the first order by concern but allowed others degrees of self control. Their vigilance was to any signs of sedition or the gathering of those against them. They knew of the Jews to be antagonistic and watched them continually. On hearing of the supposed Jewish messiah preaching in the locality, what were they disposed to do? When listening to words of unity, understanding, how were these words to lead to revolt. Jesus spoke for all men to be included in this new gathering, he even gave quarter to the Romans by including them in his plan. No defiance of Rome here when reminding all to render unto Caesar what was Caesars, then in gesture to turn the other cheek as an expression to placate confrontation.

How Jesus had conducted himself never offered a direct threat to Roman rule, where their interest might have been raised was by his very liberal approach in extending the Jewish faith for all men. As the Jews had been tribal by their own approach to their understanding by faith, this was self limiting in the threat they could offer. But this Rabbi, Messiah or king, by his inclusion of all, was this sign to gather all groups together a threat against Rome?

After three years of preaching Jesus had indeed gathered much interest, his own band of apostles had been coached to carry on with his work. His departing was at hand, at the last meal with his twelve devoted followers he turned them melancholy by once again going over the impending future. His discourse to them was of doing his fathers work, of the sacrifice that must be to save all mankind. By way of encouragement he reassured them that he would rise and sit at the right hand of the father in heaven to indicate that for all, there was to be redemption. Jesus left twelve frightened men to carry on with his favour. He did return after death, he did ascend into heaven, after their initial fear eleven of the twelve good men carried on with his work to phenomenal success.

In the Son of God as man Jesus knew of his role in that human form, his life was to be in cruel ending, his fear and desperation was fully human. There has been much debate in the cause, reason or happening to the death of the Christ Jesus. Did he contrive in his own death did he agitate those that would see it so or did he force by its crudest of standards suicide! The answer is none or no, no, no. He was a fine preacher he revelled in the fable, he told happy and educational stories, some would say without malice that he preformed many wondrous acts surpassing the skill of magic. He was a positive man liked and admired in what he did, he was also a man of his times. His biggest danger was of those times where the cut in life was severe, when harshness and brutality were of acceptance by normality. The Romans yes, they were masters but were seen not to be too sophisticated in some of the disciplines that others enjoyed. When the Jews brought before them a criminal person who had broken Jewish heresy laws, who by accusation or not had professed to be their king. Rome through the dignity of her governmental office was obliged to settle matters of high punishment by way of maintaining their grip upon the subjugated. Let the Jews have this death, their case is reasonable by the way we let them manage their law. But we will take him and show the strength of Roman law by our method of crucifixion as public as any king might want! Jesus was a victim of the times in which he lived!

After the fear of his death subsided, his eleven apostles did make their best efforts to spread the word. They scattered to the compass but were contained in distance by the methods of available travel. Success in converts followed in noticeable numbers to the notice of vigilant roman overlords. Quite soon, to these new Jews who preferred to be called catholic, by the influx of new numbers, by the physical spread of their area of influence, could they be forming to be a threat to Rome? The Romans then began to persecute these new Christian converts as enemies of the state by probability. In much better style than before and in the form of entertainment, they arranged vast theatre in their city arena creating new ways to destroy in death the catholic threat!

How the Romans ever got enough Christians to be used in this way over a very long period of time, is perhaps a statement of the times and how the word of Christ was brought to ever newer converts. Times were indeed brutal, which gave the natural feel of apathetic acceptance come what may. From the earliest Jewish teaching gods triumphant return was imminent. By linked association still hot on the lips of men, many in direct contact with Jesus preached of times just passed. Of coming to gods right hand, of suffering death in gods name and by gods expectation, then in quick succession being reborn for the coming of that unity. What a great picture, to join in gods glory very soon even if under brutal sufferance. It would not be fair to suggest that you or I would opt for death as a means of escape from Roman conquest. The connection was to be of the true god just to nullify all Roman gods and place the context of being conquered into the realms of an ungodly event. Like claiming if we lost because their logistics were better, or we lost because they had better weapons or a better battle plan. Our victory now is manifest by choosing the one true god!

For however long this persecution of the Christians went on it was the Christians who had the last word. About three hundred years after Christ had been crucified in hot and dusty lands of Roman conquest. Rome through its emperor Constantine converted the whole empire to this ever present and ever growing religion, ensuring an even greater pace of growth by connection to be now

Roman and by fit adjustment the new centre of operation of the faith was moved to the capital Rome, led now by name to be Roman Catholic. There was still vital connection to the holy land as the Middle East.

If the Romans were not yet to be fully sophisticated they were gaining ground apace. They had given all of their lands roughly the same circumstances, proof of their growing elegance was from how they had standardised much of society by law, commerce, finance, communication and latterly religion, of which they used as the final peg in holding court to the world. At that time there had not been a universally recognised year date reckoning. Rome had ruled for centuries, but from when, they had fought great wars with Carthage centuries ago but when? Rome from its own make had many great emperors, generals and of course gods, some of whom gained unending notary by having a day of the week or month of the year named after them. In the final act of order and above all the accolades given to emperors, the latest act of clerical management was to be able to refer to any day or month by a defined year reference point. This was done in all splendour by adopting the birth time of Christ as the marker year in the Roman calendar. By back calculating they arrived to the time of Christ's birth counted as the year zero. For definition they gave any year date before that time, B C before Christ. Any year after that time to be marked A D anno Domini, "in the year of the lord". They could now refer in context to say they subdued Gaul in fifty five B C and conquered Britain in A D forty three, using the standard year time mark for ever. Approximately three hundred years after the Romans had crucified Jesus Christ, they said sorry!

## CHAPTER 5

How and ever we might refer to the roman era can be now year dated by example from use of their primary skill of utilisation. Acting in self interest is usually the best way forward from those in the position to promote their own motives. Roman power and the Christian religion were just about ready to be joined at the time of its happening. For the first real time of our history the secular was to become entangled with the spiritual, alas half measured, but with the both by the best of intentions. Unseen error carried by the separate camps was the way in which this union was settled. Both companies had leaders or guides they also had government by the senate or the priesthood, working together but steered by their own particular interests full measured, thus making the half!

From the political side the feeling was that religion was an attachment of high standing, but of service, its role was to fill the needs of the populous past the supply of normal community. Priests in being fed from a different pot felt that their lead was from another plain than mere civil or political rules? Naturally they accepted the principle of managed society. Having turned a mighty empire by always offering the other cheek endorsed one of the root concepts of their faith. Endorsed a power flow of being right by standing to the son of gods very words, being right in the headed structure of their very religion, though perhaps missing an undisclosed feature of early doctrine.

We will not say or it never can be claimed that this union was the first or final between the spiritual and secular, our mark is to be taken from their particular relationship. Other Great empires throughout the world had for centuries operated in the same god/state mix but passed their signs of achievement ongoing through the medium of science from the secular. Where the roman example is somewhat different was by the set up of the very religion they had joined to be their own by name in Roman Catholicism. That society had taken the mixture between church and state one step further than ever shown before or ever known before is a mark of modernism and practicality!

Through existing politics, by the very structure of state, they had tiered representation to all peoples in their control. Although joined by a competent priesthood to administer the new religion, both acceptances were always to be misaligned, giving us today our first example of what will be referred to as our Half Measured Syndrome. A feature collected when we act to overlay/join systems of different construction but are supposed to offer the same outcome. With the secular of first concern to the civil conduct of community even unto death in force of law by it. To run concurrently with religious doctrine by concerns to the hereafter enforced by influences to have been created for the conduct of mankind but not of him, which is no bad thing, it will bear further comment.

When the main structure of the Roman Empire crumbled between the fourth and fifth centuries A D the rule of imperial/civil law was overtaken by new methods and styles fixed from the habits of the conquerors. In many cases seen to be a huge step backwards, new squabbles in demarcation disputes even worse than the pre Roman influence broke out in the emerging European countries. Peoples wanted to fully illustrate their perceived tribal or clan definition. Unfortunately this obstructed the already established civil orders of control, all roads did not now lead to Rome in the force of politics.

Even unseen then there was the vapour of the one state religion still emanating from that broken city. Given time by its own force of hidden power, it re-conquered the split image of Roman rule. Not in giving to the separating peoples of Europe the standards of law or commerce or currency or conduct previously obtained. But once again the considered care of community expressed from the perspective of the great unknown, managed in fashion by the same priesthood to the same determinations. Roman Catholicism began to become the state religion of the emerging European countries. Even so, religious determinations had a different feel Rome had indeed set standards of faith since the time of their collective conversion in the fourth century A D. For the European sector of empire the expression of religious order also from Rome had no divisive elements to it. All powerful Rome instigated an all powerful religion in all provinces, people there related to the stories of Christ through the management from Rome. They had no direct connection with any other area or with the prophets, no connection with the son of man, save through the bishop of Rome!

In the Middle East after the decline of empire the story was much the same, tentative self determination by tribal and clannish groups to match their old standing or by other influences. By religious determinations things began to change or more so found different levels by the same interpretation! Here since the death of Christ the region had maintained its own religious flavours. Judaism was still best connected with its long history, the Christ faith practiced in the same fashion until the hook finally caught its quarry to convert Rome. Following times still had direct connection to Catholicism after imperial decline, when in about the year six hundred A D; there was to be a monumental reawakening!

## CHAPTER 6

Mohammed as a man of learning, in the lands of Arabia, open land between the Eastern ends of the Mediterranean Sea and modern day Iran gave aspect to the need of change to the religious structure of first the area then for all men. His perception of religion was broad and well read by how he could refer to others by structure. We might not be able to see how he first set his mind to the task in hand, but can offer conjecture to what might have prompted him. He was aware of the Jewish faith and their connection with god through Abraham. Catholicism offered query of consequence, instigated by Jesus in claim to be the son of god as man, this religion by its own take was the broadened version of Judaism, shown by taking the prime concept of original sin on all men to be nullified by his death in sacrifice.

One acceptance we can offer is the respect Mohammed showed to both of those religions by his attendance to their relationship with god. Any deviation must have been first shown to the living religion of Roman Catholicism, which by direct association was cast in form in those very lands of the prophets. To now return headed in name by a foreign city, who had taken emphasis from lands where the very dust had brushed over the feet of holy travellers. Men who had drunk water from its wells, who had taken loaves and fish in sustenance, who had shared the same night sky in peace and harmony, who risked much by posing opposition to mighty force.

If Jesus had been the son of god incarnate, how was the faith he was responsible for moved in source to far off lands? His apostles might have misunderstood the basics of concept with the religion by brief for all men, and then what matter to its point of reference, which indeed could have been a close explanation. Might not Mohammed led by his own emotion in the situation where he himself had come close to god through the Holy Spirit, he himself could see god's intention. Then as a mere mortal man if he could read God through Allah, would not Christ also be as man, after all it could not be that god in life's blood was to be forsaken, was to have the history of human contact scattered to the wind without due deference of the obvious by intention!

From the way Mohammed was able to interpret messages revealed to him, it would seem he could only think of Jesus the same as he! Moses by all definition was a main Jewish prophet carrying god's words in first sign. The centuries dividing him from Jesus masked time enough for meaning by interpretation to have changed. When Jesus renewed old messages by refinement the gap between him and Moses was sufficient to allow new reasoning by earthly interpretation even if not then captured by the pen. Likewise the gap between Mohammed and Jesus was also sufficient to allow degrees of change, this time as the last effort, and to cover the vulnerability of man? Mohammed recorded heard doctrine to be carried for ever in the same meaning. For by revelation he saw himself to be the last of that prophetic line so long carried in those lands.

Carried in those lands, carried in those lands? By the obvious as Moses and Jesus were prophets to whom god had laid edict upon in progression. From supposed frailty, with Christ as gods newest messenger setting his own apostles who were to allow the word carried in source to other main places for representation. Was Mohammed now to suffer this last attendance to the word of god in keeping the link to remain unbroken in this area of land where god's first sign to Abraham was made!?

Mohammed created the third monotheistic religion in the same region of the world having been done twice before. His intentions were just as sincere and to mark it so his writings were to be the final sign. If Judaism and Roman Catholicism had led, then fair and just respect was shown, but from now only through this religion of peace, Islam. In the year six hundred and twenty two A D, by world standards a third religion of the one god theme was laid before all mankind. By mark of reference it was to carry forth its own date from inception to be the year zero. Mohammed died in peace as mortal man in the year ten of this third main religion to its own year mark calendar.

By my own definition the religion of Islam with its adherents standing in name to be Muslim is to be my Third Dimension. Not by connection to be now one thousand four hundred years old. It is from the same link by time of the first two dimensions, titled by these recent months, with all three bearing by new effect these twelve years past to these fifty to sixty years past, post world war two. For there was never the world climate to have arranged our connection ever before that time, by all effort we have never managed to mix as a single unit in our single specie. It might best be done by our differences with the same flags!

## CHAPTER 7

If we cannot change the past without alteration, then we can alter it without change. What ever is the first line of issue can be best illustrated through our religions better than through our politics. With three men to stand and look upon each other now, thirty years of age each, what would they see, a Jew five thousand seven hundred years long, a Christian twenty centuries old or a Muslim fourteen hundred years young? The past is the same but three thirty year olds might not be! In those declared forms looking to the exact same expectation their time measured in the future is beyond their care, for their epitaph is beyond their understanding.

If three thirty year olds by their own choice had taken any political selection, then those would be expected to deliver an ending in the form of product. Politics by the secular is to the result of the here and now even short termed, even changed, politics is of the physical, it touches it is active? Religions of the spiritual also touch many of us to deep comfort, but its feel from one to the other cannot equate to the benefit of commodity.

From cruel and wretched times these last four thousand years we had mixed the two in determined effort. From civil controls to multi god cultures we heaved in the product of living, with the overall expanding of many societies we showed our ability to succeed each time by new methods no matter how. By the Roman experience with the first real experiment on a vast scale of the new state religion cajoled or imposed upon vastly differing groups of peoples was a new chapter. Half measured by description from me is not to set the tone, but to measure the experience!

As known we had the success of empire merge with the success of endeavour. Both to continue in aim enhancing their joined though separate objectives. By some sort of protective measure they were each to aid the other, unfortunately helping our half measured syndrome effect, which was inborn by first approaches and carried to this time by other names. Since Rome and religion became one about seventeen hundred years ago, most of all politics and religions had become tainted by show and by application! By example to look at their marriage then and their separation by the death throws of one by conquest, it can be seen in how they coped with the unsuitability of the marriage in the first place. Any lessons that we have refused to learn since then have been repeated over and over past indifference, but only now to apathy.

Half measured can be the working of deed, item, idea or any principle, of any union or expectation or outcome. It does not mean half hearted which might border on the wilful. My half measured syndrome is the effect by result, where we might know of our lacking even that we operate by best intent we continue to smother the reason of effect by style. That went unnoticed at the first concerted attempt of the colossal unity between the Roman state and adopted Roman church! Both elements becoming what they were not, only to nullify their separate open intent, politics losing its cutting edge with religion losing its healing balm. The mixture of the two has remained virtually unchanged since that time spread throughout most world societies. Where it has been magnified in intensity is shown by these most modern of times from when we have made start up decisions and because steps have been taken in the first place, we usher progress to support the result unabashed. Flagging to some right but most wrongs half measured, then to consequence, which is in result of the way things carry on anyhow but not by the religious political union only. History has shown us in diversion that even by single politics or single religions we have worked the half measured syndrome.

# CHAPTER 8

To begin to understand the meaning and my intention it is for us to think in those erroneous terms set without equal consent by our first three dimensions in being religious. Naturally from their individual start they already operated with forms of civil order and politics. In all cases where the political structure had managed its own change by inevitable progression, religions although extremely well intermixed and overlapping, drew for themselves their inspirational beginnings from the same source in god. But by different interpretation allied with the current political civil standing, all designed to fit into the already operating societies but with new determined influences, giving us all the mix of the two in politics and religion.

My half measured syndrome indicated from the union of Rome and Roman religion was already set before and after. Where we might split the one or my interpretation, is for us all, that one or the other affect us all even from then until now stands. Judaism by civil order in society worked. When to that faith we realise that gods laws were imposed on man through both mediums can we see half measure working? If society knew of how civil laws worked of how the commandments worked, how could they work together? Between the dividing lines by any description when man had contrived in his own laws even to the death of man by any means, then by the hand of man received command not to kill, where was choice?

Politics had been set to always move forward by the pace of circumstance, where it could change from day to day, religions could not. New Judaism, Catholicism expressed as the second dimension, also hailed of god's commandments with the same point not to kill. These two religions kept their separate composure, operated to state fully, operated to god's commandment fully in name but chose to forget its meaning when coupled with the current political focus, they operated half measured! Was it known then?

Mohammed's course set by direction was not much different, his choices though led by best intent treated the aspect of human death in a different way than others, not to oppose them not even in the cause of ridicule. His standing was led, but also from his time and circumstances. In the need to reform society by intent, to realign purpose, to change conduct, his personal feelings were to bring mankind to gods standing as led, but by his own perception, also in the method to do this, which would ever support mans own god given free will. Respect of high regard was shown to the Jewish prophet of Moses and the Roman prophet of Jesus? Where they held in tenet not to kill of man by god but did in politics. Mohammed ruled in aspect to politics that by the character of man it was that we could kill of self, but only by determination of reason in politics always by law. Then for Islam the new religion by sighted need to defend or promote by the same protection this our faith, carried thrice and for the last time to lands of close proximity to all prophets and the last!

If any religion is taken it might be the vilest or the meanest or greatest ever of mans achievement. Those dimensional three by my definition are determined by progression time wise, one to follow one to follow one is not in cause of their excellence. It is to help our thinking if we remember their start time in relation to each other. Also that each was laid upon our political standing as and when, which has always been very fluid sometimes forcing us to think in separate ways and to different solutions. We have three dimensions linked and separate. I have titled this work The Fifth Dimension and claiming it to be the last in line. We have talked of ancient times relative to the first three yet called of their titles to have been made in a matter of recent months. How might we look to the fourth dimension, first in line of number or first in name?

## CHAPTER 9

The fourth dimension was even around at the first time of the Jews, did they know it? It was there to Christianity, did they reawaken it? It was there for Islam, was it encouraged by them? All three were aware of other influences in presence to their own acceptances, but by the essence of belief were able to contain the method of their faith controlled in line by their awareness to the concept of creation. What was to manifest in danger was how the fourth dimension took in opposition to those three and all other religions in and to the point of actual belief.

It is, was and will always remain to be of scientific standing, which in form is diametrically opposed to religious concept from whom so ever or what ever when captured in the picture of pure faith. Called as the medium of learning the fourth dimension is ageless and has always been with us, although redefined by myself these recent months to make part set of one to five dimensions, on its own account it will come in notice to us all by many interpretations. By my own determinations it will come to all of our acceptances. I have put it to be the fourth to set its line in how we deem its importance in relation to the first three and all other religions, for by its own half measured working it will seek to impress upon us its false self worth in relation to them.

Science in any form of discovery is as old as measurable time which connects directly to us as we have given it that medium, we have also set its course in the field of experiment. By repayment it follows the exactitude of formula sometimes giving us false tenet. It holds in peculiar form result, which in the theory of science is undeniable, but by the measure of faith is sometimes untenable. The fourth dimension in these terms is of my and our use to better examine the true faith of religion while containing its own realisations which we have already acted upon half measured by political terms.

It has been stated by myself that the first three dimensions represent connected religions from roughly the same geography, along with all others religions from wherever, that is so. In my terms of reference I have stated that this classification could only have been made by very recent times is so. Although practiced religions have a history from about four thousand years ago my terms stand by these recent months. With the fourth dimension it will seem obvious that the very nature of science has its own calling by many titles, not least in name by the very voice of the universe as mathematics. I will still refer to it being named by four in line after the first three and by these recent months to contain them all in proper cycle.

With The Fifth Dimension being the last in line, my objective is for us to rationalise the relationship between religion and politics, taking that we might be better disposed to look at the future with open eyes. Unhindered by half measured reasoning, unhindered by political correctness, unhindered by religious dogma or unhindered by the mere expectations of written formula, The Fifth Dimension will lead in shadow!

Recent months gives us in key and the method in which we might first and often look to correct our alignment. Coupled in expression with the adage of "that was then this is now" we might better serve our own intent if we intend to be broad of mind but narrow in thought. We have travelled great times since ancient footsteps were cast in petrified form some three and a half million years ago. Religions might explain intent but not cause while science would give all in reason without meaning? Five dimensions are all that are needed to explain all that is needed, but first taken to be recent, then for all time in all places.

## CHAPTER 10

One hundred thousand years ago an insignificant event took place with the burial of a fellow clan, family or tribesman in a place now known as Shanidar in modern day northern Iraq, such remains found were dated up to be about that age? Iraq is close enough to our definition of the Middle East for future terms of reference without compromise.

One hundred thousand years might be the start of all of our stories. When we look in sadness at any of todays burial rites of people we have known or loved, often our emotions carry us into despair. From the civil end of political organisation for such events, to the religious rites, we would each follow in sequence our own understanding. By the same means later, we might remember collectively that loved or revered passing but to our individual memory? Although many heads were to the one event, would not we each remember our own precious moments? If we went to school with them, if we lived with them, if we loved and loved them dearly, the sorrow would bear its own end to their passing!

Today if politics was to influence our reasoning, then we might suffer deep loss? If religion was our standing then after the bitterness of remorse many might hold in celebration to the fulfilment of our own and loved ones choice by god's direction? What would be the real story one hundred thousand years ago?!

After carefully picking at the bones of an ancient burial site find of some sixty to one hundred thousand years ago, what might the study of the situation reveal? Before us is seen is a body placed with reverent care, but with added amazement when found, with it was an offering of what can only be described as grave goods, even if only seen as adornment to the passing. What might be told by this find extends our reasoning of that age to be another first mark, though this time to represent us all in being of one in type by the working of our collective minds. We had used political ceremony to show religious concept!

By the very act of placing a body in death shows in us all the true measure of organisation, carried by our developed civil band of control or engagement. What we might wish to make of old dated events is to our individual taste, but only to the point of agreement. We cannot deny the sense of community displayed. As hunter gatherers of the time we had excelled in the organisation of how we coped in the routine of life. Some to one task, others to others, if the copy was taken from those in instinct then the event almost broke the evolutionary chain by the civil organisation of our attention to death.

From the very next tick of time, we completely shattered any prospect of our own single specie tie to the collective evolutionary chain? If I would say on my own account that that burial rite carried signs far past the parting of community, then I might say it carried our unarguable first sign to religion! Not from our own sophistication of four thousand years ago when we strove to understand our beginning by gods charter in Eden. One hundred thousand years ago we celebrated in death the discovery of god in life!

Civil control by the force of circumstance must have reigned, otherwise who would bother to take a corpse and lay it in peace when it was so much easier to break camp and move on. Religion as concept was not invented nor seen or sited or thought of! The obvious passion was simply transmitted from the living to the dead, but why without objective? Did religion bear fruit not from the future, but the recent past, because one of group, whom so ever, had stirred in our minds eye the memory of yesterday even from where they lay in the motionless slumber of death? Instead of usually tending the fire, or skinning the hunt, they no longer moved by the nudge or the push or the shout. But still found the motion of life by any or all of the individual group memory? Known as dead but still active? What mystery!

No claim has ever been put to any reading of ancient events simply to explain religious motivation. My suggestion is in the form of practical probability, like when we tried to justify the cause of emotion even when we never knew of its existence. Though our total active operation was in cause to propagate, we never knew why? Our driven intent was from pure instinct like all other animals and beasts in our community of circumstance. At best we brought the seasons of weather to assist us in our gathering capabilities. By the tilt of time then ago unrecognised, we foraged in the right location by our own learnt drive.

To say we are here now is the real argument, but to our own good owed self interest we must be mindful of ourselves then ago. My speculation will always be supported by our modernity, but for assurance might not we seek some answers from the past without altering its structure? Each of us can enter the portholes of enlightenment on our own account by what ever means. For me to have sited personal interest of a very, very old burial is representative of us all, that I make of it a claim of perhaps our first real religious sign is not a fact of delivery, but a proposition of discovery.

Any religious connotation is not similar to anything we might practice today the binding link comes to the movement of our condition of acceptance to any religion. Today, all are locked in the mystery of tomorrow by the avenues of heaven or paradise or nirvana or continuing or understanding. One Hundred thousand years ago the born mystery was not in tomorrow but in the past. Decay was the only sign of the future with the activity of the mind holding the past forever in motion again and again to all individuals that we buried and buried and buried, perhaps since Shanidar!

Shanidar as our oldest recognised burial site is not to be acclaimed as an appointed religious shrine for by that title we would too savagely alter the past by all steps and measures. It is far bigger than just a grave, even if claimed as our first act of civilised behaviour we would lose the full scope of its religious reference to the overlay of politics which would count half measured against us all. We should hold the two as separate, yes civil order was involved in this rite, but the value of memory brought the notice of human mystery to new levels. Allowing in the grasp of time to relate our individual memory as a source bestowed upon us!

Although we knew little of how the weather or seasons worked we followed in abeyance. Much of our food stuff was shown as ripe or ready for picking by the habit in which we had come to collect them. Not altogether a great mystery since we were readily maintaining our ability to propagate in all forms of acceptance. Having buried a member, any member of the clan or tribe, to have them trouble the make of our thought by being active in our head when we knew they would move no more grew upon us, then how?

Was it that we had come to religion, was it that we had found god or is it modern thought back transmitted by myself in order to give us a lead of direction? Without distinction it is all three of the above. Yes we had come to religion the religion of mystery. In that virtually everything by description to us was unknown, we knew nothing of anything save the total drive of instinct. All of habit to us had the same drive, we could visualise aspects of the hunt by memory, we could see in minds eye the expectation of a fruit filled tree just over the next hill. What had come without alteration was that now in as much as we carried the thought of a bountiful meal into the morrow we also carried the thought of the fire maker or the hunter or the fish catcher whom we had left dead and buried, but still walked our path though unfed. Still shared our comfort of group though un-attending, still spoke in silence, still laughed in mirth, yet took no space by the fire except with all or none of group whatever the case may be. What mysteries had been uncovered to bury kin by the name of religion?

Or had we found God separate to religion? Then ago the answer was no, but meaning yes. From the burial rite at Shanidar in northern Iraq some desperate long time ago, it would not be possible to determine that the human god kinship had been cemented. We can only run to God by the connection to religion offered in explanation to be mysterious. Life at that time was too intense to consider the aspect of self by the furrow of death. Friends were buried in the surround of mystery but not in the aim of purpose, which would give the link needed for the proper god connection. Our aspect of God by my intention is to take us all to the time when we were the one specie in being human. That single burial as the oldest found is also a mark in time by our move by some 'Out of Africa' if driven by circumstance or desire we had taken further steps than those made three and a half million years earlier.

## CHAPTER 11

We have no further connection in line with burials of times ago, save that I do link us all fully in kind to those circumstances. My connection is spread from now to then, that we all by what ever group, have shown those very same attributes in our links to our religions, or have we? By association to then, if we look to the standard time of our first monotheistic religion of only four thousand years ago, there is a real gap between the two timings. By joint intelligence those of us in the best understanding of archaeological finds in the nature of human remains from any dated time between have set their standards to what they have expected to find ex of religious/God influences. Creating in danger our aptitude to select our own excellence of character more the most meaningful to any separate group we have chosen to belong too, which by some very narrow finding might be no bad thing!

What I do say in terms of Fifth Dimensional reasoning, is that we of this single specie first trod ours steps in Africa, where some still remain by fixed choice and circumstance, while others of self have travelled throughout the our world. Holding in form that we are all 'Out of Africa' by decent which is to be our first taken prospect through my Fifth Dimension. That our small family, clan, group or tribe took long, long generational steps, by the time of Shanidar we had managed by other circumstances to have changed by small degrees, though never in the nature of our enquiry which had by then brought us all to the mysteries!

Throughout the world many human type remains have been found giving in date to the extent of our range of movement, some before our ritual burial but never of the same depth of intensity by any understanding. From times after, the next great finds of confusing dexterity throughout the world range by age to be about thirty thousand years old to the sublimely recent in the form of cave drawings or rock painting. By all standards they are wonderful in their art form if nothing else, though many bigger tales are told or to be reckoned with. Even unselected by age, cave paintings in many places seem to depict in the chronicle aspects of daily life. It is only when we stretch our thoughts to them by their own times that we might set the reasoning of answers.

Great studies in all of our names have been set to describe them in the form of influence, if we look to the two types. One in dark caverns the other on rocks in open space we might help our own hand to fill gaps. Deep dark cave paintings might hold difference of meaning than other similar works to be noted as rock carving/painting. By very modern dating methods it has been shown that the great art of our cave painting excellence finished in its form some ten thousand years ago. Rock painting went on continuing in its form until short centuries ago. The rock painting form of information was just that, information. In lots of examples there seemed to be religious connotations, first, that the places of the events were revered as sacred to most indigenous people using the chronicle form of depiction. Giving information to visiting groups locked in the same need, whereby they could keep in touch by medium while on walkabout. If the weather forces ware too severe on the depictions as they were generally exposed, then a particular tribe member might repair or alter the damage but still giving in sign the same meaning only now updated. Being a feature to show us how long a particular site had been continually used until its own extinction by the original users moving on or changing the direction of concept?

Compared to outside openly expressed rock painting, what can we find different from cave paintings? If from those that have been discovered we can date the last to have been done no later than eight thousand years B C what should be our first conclusion? Did people simply move on in more than one sense, if to move in the avenue of progression? Caves went out of fashion, no more or no less! From our studies we might classify the artistic work done by the fact that they were often attributed to different communities. We have a timescale of just over twenty thousand years to work with. There is no scale or standard we ourselves can put to the total of works so far found. If a buffalo or horse was painted in a time gap from one place to another of five thousand years the only conclusion we can draw from that is the artist was not the same person. Possibly not even the same family, clan or ever growing tribe.

For the endeavour of cave painting involved was a real act of community, for many hands and much tribal effort was required when we look at the scale of some works. Of the many conclusions we can draw relevant to all peoples involved over the whole time span of our finds, the first stands to their civil/political capabilities. Organisation on a massive scale to those communities was required, when we consider that most were hunter gatherers, if the painting acts were to continue then some would have to hunt and gather more to support the artistes'. Deep dark caves give us another clue to the extent of planning needed in terms of the provision of materials required for first painting, then in the means to light the dark in order to do any work.

Berries and ores and fats and oils were crushed, pummelled and squeezed to form paste to form colour, charcoal and chalks gave outline of shape to the elected topic of display. Was it that one person of group could carry all of these paints, in some cases a mile into deep dark recesses? Light supply was more important before minds eye or at least the provision of it if any member of group was to ever start work. It was well tried that the glow of a busy fire in times of darkness gave much comfort in the dark of night. When a burning branch was lifted and held aloft this also gave its own good feeling, but all too soon the flames would curl and die to leave the red embers at arms length offering heat only. By the extent of our growing learning capabilities, what if a small hunted animal was thrown on the fire to remove its troublesome and tasteless fur before the feast? It was noticed that if it was left too long in the flame it also began to burn, its natural fats offered better light than our held in hand flaming branch. Might not we skin those small animals and wrap the skin in tight pads about the end of trusting sticks that we could carry to offer longer light support?

Light and paint was supplied to our artists, all that we can now do is plant our thinking to be theirs at the time. How lucky were individuals to be assisted to clamber over rough terrain carrying paints and sufficient torches to be lit in consecutive need in cold caves to paint chronicles of the time! Or what were they to paint? Trying to think in their terms or those of the artist for what was to be the display of meaning, we might look by our own view, but to remember the thinking of the time only. It has to be taken in first context that this was not the work of an artist. We did not have a Rembrandt or Picasso to lend us their experiences of taste, what was given to us was in expression of the tribe, Not least how we were all involved to supply the materials needed for the tasks. Every member gave of their talent, but more importantly of their time which in some cases must have cut into the essence of tribal living, in devoting that time to decoration above the hunt or harvest. Why?

Because we are so far separated in time from those experiences, we group virtually all finds near to being of the same category. Over the twenty thousand year time span so far found in relation to cave painting, much the same activity of content has been found. Without any distinctive individual merit most studies are of animals in form. Was this the tribe's chronicle of learning, that scuffing sloughing complaining children might be dragged far into dark recesses to be shown in picture what a bison or horse looked like? When in greater likelihood they might have seen the same beast in its natural surround only days since. While the encamped tribe settled in the comfort and dryness of a cave for any period of time in terms of seasonal grace to exploit the gathering of food or the hunting pattern needed to follow migrating herds. Were paintings made in memory of where next season it was best to camp again, were they done deep in darkness so that others might not know our secrets?

It has been difficult for us to accredit the true meaning or reason for some of the works done when showing clearly human form depictions both male and female. With the former in some cases to be at the hunt or more frighteningly at some stages of battle, presumably rival groups! Women by their depiction are more set in the trend of health in perhaps showing to the welfare of community in terms of fertility, certainly a sign of wellbeing for the whole tribe. It has not always been clear how long a particular cave structure had been in continuous use when the range of painting so far found from place to place covers a period of twenty plus thousand years. None found later than about ten thousand years gives sign of how our habits had begun to change and for why?

Not the last word on such events but by my own reasoning as ours, I have found what I deem to be a more than curious aspect to many cave painting sites, we can see the blow print of human hands on walls. Taken as a form of signature this is assumed to be done by placing an open hand palm tightly against the cave wall and spitting a liquid paint colour against the back of the hand. Any over spill will hit the wall surface between the open fingers and around the shape of the hand, which when removed leaves the appearance of a hand shape on the wall, signature or not? As an effect it is certainly a mark of identity to be human, but can we guess to those times that it was of a particular individual like the artist?

If we can make the bridge of thought from now to then we must strip much of our own intellect without the cause of vanity or pride, after all such a move is in direct line with our own ancestry anyhow. Thinking most done then was of the moment, how to eat, how to protect all was carried by the base establishment of group using each to their best shown skill first, then in collective use. Having led that the enterprise of cave painting was a mighty task, I further take its intent to be of source to the mysteries! Also I would hold to my own feeling that the effort of hand printing is significant by many features?

Taking that it might have been a sign of the artist is indeed a worthy supposition, but only by modern thinking? Taking that it might have been in sign to the clan leader is more of that age, because of the shown structure of civil control above actual comparable pack behaviour. Both examples fall short if we consider that there might never have been a time when all of the clan or family or tribe were to see the finished work. If that was to have been considered then why paint in dark, dark places where by normality fear lurked. Hand mark blow painting as recognition would not have worked if never seen, and in any case the painter was known to the group as was the leader if the group was so structured.

I take cave painting in cause to be of the mysteries from the case shown by our burial rite discovery in Shanidar northern Iraq? With the obvious sign of grave goods there, by comparative terms in deep and dark caves there was also left grave goods with no need of light. For they were the preserve of those in darkness in all further use, they were the preserve of the dead! Great time and effort by the group had been used to create in memory our first tombs. By depiction showing in splendour deeds and scenes of familiarity to those departed or at least some of the departed, then to mark for whom with a blow print of their hand left in saying this is yours, meaning for the dead, "this is mine"! Dark to dark in one of the only ways such paintings and scenes could be of continuous need but for the dead!

Many of the splendid cave paintings so far found have been in the modern day countries of France and Spain. I make no direct connection between events of one hundred thousand years ago in Iraq of the Middle East and events of between thirty to ten thousand years ago in France and Spain of Europe. Save that in title I class them to have been felt mysteries in source by religious connotations. I accept the link is less than tenuous if it was not for who it applies to, our direct ancestors. In both cases there is overwhelming evidence of a trend created and followed far beyond signs given by any other specie upon this earth. What other creature has adorned their dead in expectation, what other creature has expressed a mark upon the universe beyond the self interest of procreation? Who has shown in sign such wonder when speech was barely audible then, yet proposed intent of purpose even beyond the meaning of time by the query of thought nurtured by the event of thought through the mysteries!? Are we yet this wonderful specie to have clawed life as one 'Out Of Africa'?

The Fifth Dimension is to limit our thinking by equal consent but not our thoughts.

## CHAPTER 12

That some or the oldest cave paintings were found in Europe tells the human story in part only, like findings in other parts of the world of rock painting and decoration tell the same story in different verse?. There time difference in relation to each other bears no added merit from one group to another. Connections that I make are of the progressive nature, but regressed. The reason for which is to give us all fair chance in being able to think as others did simply to see how we have changed and been made to change by forthcoming circumstances unseen. At any times before the supposed onset of what some of us call civilisation we were all exactly the same. If the start of connected civilisation began at the end of the cave painting era then we still carry the source of our mystery thinking dated back a thousand centuries to our first known burial rite?

By the strangest of paradox the more we have progressed along the road of civilisation the more we have lost in connection to our intent of purpose. Our joint intent of purpose! History of serious age can fully bear testament in what I have just written on our behalf, for by it we were directed in purpose by our own conduct far and above the leading edge of other civil control or the growing of our mysteries into our worthy god feature. No sooner than we could stand on our own two feet, we began to exploit our own good nature by half measured working? A process carried on these last ten thousand years by us all, because we were the same but allowed our physical features of skin colour or hair or bone structure tell us we were different, and still. It is that we might relish in our differences and should owing to the circumstances to have prevailed in our change from that black skinned 'Out Of Africa' biped to which we each are now. It is that we should know that by our localised climate our geography or our diet and others that we have changed in form but not type, we are as one of specie, who have been afraid to accept the obvious!

For one of the first signs of our difference by selection, we might look at the edge of civil/political conduct in the art of trade. By many of our separate groups we had used items in our vicinity to assist us in our routine. Like when all the result of the hunt was used to there best ends, what we couldn't eat we learnt of different uses, perhaps as clothing, animal fur turned into human skin. By forms of connected association by mixing with other groups, if at the height of suspicious tension, notice was given that others used slithers of flint as worthy blades to butcher prey or fixed to wooden spears that became mighty hunting tools. If we as group had no access to such magic implements, we had not seen the like on our own route of travel, how could we get them? We must guess that at some junction trade came to be of issue if the difference was fight or barter then at some time we learnt that bartering was a way to better maintain our needed supply of what was desired. In note we could also supply to others what we had in excess for their needs, the table of trade was so set!

I would make no suggestion about those times and deals that there was any sort of religious symbolism attached. Any trade of difference conducted then as long as hundreds of thousand years ago to perhaps about ten thousand years ago went to show to some of us our differences by bestowal! Meaning to my definition that we could further see between our group and others a reason for them and us in being separate, not on the condition of excellence, but either they had or we had. Even being something which might even bear to some of us today but not historically, only socially. Being a serious matter where we can use the essence of the Fifth Dimension to deter entrenchment by what ever medium and not turned by political motivation through correctness of half measured working!

With one taken prospect 'Out Of Africa' to lighten our understanding that we are all of the same single specie by our connected bearing. It is at this very early stage we can list our second taken prospect to be 'Past Gods words' holding to the Fifth Dimension to better determine our collectiveness of direction between our political overlay and religious undertones? When at first we traded politically to show some differences then buried in ceremony to tribal rites, we were simply learning of that difference and how to enhance it. Forgetting that the very shape, style, method of our living now had been determined by the factors of climate, geography, diet and even expectation in to what we are nationally now different, but morally the same!

By curios reason, politics has no bearing to our physical state in direct terms except by claim from some of its extreme forms of discipline, misguided. Politics might claim to be nationalistic of a particular set, but with no moral standing separate to our collective morality, we are each to determine its course representatively or actively, that is its thunder. 'Past Gods Words' does have a direct religious connection, because it applies to speech first, which of course can be made in representation to the written word. Why I have made this address now, is to show us all the difference between the secular and the spiritual by tone. For any person to call on political alignment is representative of human choices or our specie choice, even expressed different as we are all the same! Political discourse can only be delivered by individual endeavour even unsavoury, but wholly by human endeavour alone.

If we will ever see the aspect of the devout by religious concern in later pages we might set in memory our true religious standing now, which though very, very good is very, very divisive. Born from the magic of mystery, or not. It has intoned to all people by all collection of choice. It is and has been most assertive when in all cases or not, it gives voice in the written words of god, which on all levels is only to our good when it points in direction of our first unlearnt, untold or unimagined experiences bringing us into the thought of the mysteries and there extent, long before science was known!

One hundred thousand years ago we first operated to religion 'Past Gods Words', how ever we have tried to invoke reaction thirty thousand years ago by the decoration of our caves or by the information of our rock painting. By our own excellence of mind but error of thought, were we impatient, were we longing too much to have that final link imposed by due deference? When we would be guided to gods path by voice in that determination, not to be aware that through the repeated contact we might have claimed since it was that all time would then stand still if god was to show more than voice!

Was it that to know of god's voice alone was to know of god's reason, was to fill our purpose was to be able to see the awakening of past souls by other terms? Or was it that to have been in gods words, by the force of total community it is that we must now hear 'Past Gods Words' to select any direction submitted by these most recent of times by mere months. When at last The Fifth dimension was first able to be offered as another bridge to our intent of purpose, not as a religion or of any political connection, without law of guidance or tenet or covenant it is offered as our conscience to operate us by our own understanding and ability. We can each suffer our own thoughts to it and through it by its one and only active prospect to be 'One for the Whole'!

## CHAPTER 13

From the construction of the Fifth dimension some fifty or sixty months ago it is necessary for us to take a back seat while taking glancing looks at our histories showing its progress ten thousand years ago. Always in representation to us all, we might look to an area of the world of familiar note, the Middle East again, or to modern day Iraq. Two great rivers had cut there courses through the centre of the country, in the early times the land between them was once known as Mesopotamia by description as the land "between two rivers". By the terms of our world climate then, that area was very well suited to the permanent settlement of wandering hunter gathers. Peoples such as us who could now better see other arrangement to our existing style of living, when we had started to manage the herding on our own livestock, started to manage the seasons by planting seeds to be met by the best of weather to ripen ready for harvest and storage, saving us the wandering of before.

In Mesopotamia we had come in collection to form stable/static communities by human progress. There are other historical sites throughout the world which would agree in kind to be early places of fixed settlement. My personal interest of that area of that time bears by what was to follow there for the next nine thousand yeas, even never stopping. Although of much world interest now by other means.

I will address them for us and by us to the interest of the Fifth Dimension by its three taken prospects and one active as 'One for the Whole'.

There might be a direct link between the end of cave painting finds after about ten thousand years and the onset of our settled communities in the land between two rivers. In other places, had we now also began to stabilise by the merit of convenience. With the global ice fields receding altering local climate at places where none of the encroaching ice had ever been. The bounty of product meant that we didn't have to roam so far to meet the supply of need. Had we taken to building better shelters for protection in the long term in places where there were no caves for use at the end of our normal gathering? We had learnt to fix our location and we had delegated tasks by tribal need finding now by the formation of community we could set the need of control under civil command.

That the Middle East has remained in the foremost of much of our controlled advancement since about those times is set by many plusses and minuses. Both features have been of much benefit to us even in the open negative whereby something simply does not fit? Our first writing capability might have found root in the area of our first real settlement, Mesopotamia, with farming and city state societies, we recorded to mention the skill of hand in producing the basics for all modern comforts on and on. We now grew on our dependence on many things, where from before an individual was skilled at making torches that would hold light, they were still needed. But because of the success of community by its growth of clan we settled to sleep in smaller groups by family definition, each needing their own light supply for nightfall. Supplied now from the specialist lamp maker, one who had modelled clay dishes to hold a supply of oil soaking a wick to be lit to give us the needed light in darkness, not unlike the torches of old, but now more refined.

Much activity was spread amongst many people who had thought of themselves as different from others even as trading partners, for in the land between two rivers the success of our specie abounded. Most in the form of civil control was activated by the developed command structure howsoever run, but in general terms for the good of that particular clan or tribe. An overt rise in the separation of skill was also for the new priesthood! Sort of administrators who ranged in purpose to be the go between from the leadership set up to the general populous. Perhaps being a real consequence of our success where the function of the many had been woven to accommodate the desire of the leader, if to be the King?

From the consequence of a ritual burial in close proximity to Mesopotamia at Shanidar, even unnoticed, the time difference melts only by suggestion, because by our city state organisation we were now thinking in active religious terms, in active terms of God! Activity in Mesopotamia from starting in a settlement form at about ten thousand years ago had spread to the whole of the Middle East and of the same vein pushing into Europe via Greece. Separate trading and warring groups formed collective city states in linked tribal areas of nationality, forming our first semblance of country! At that time, among the many great histories we share, one is a beacon, that of Egypt. She gave us many wondrous things and linked with most of us her/our illuminated god structure, illustrated with the most explicit form of decretive writing, possibly the most colourful ever.

For millennium B C; our year date mark, Egypt scored her story in remembrance of her Godly Kings. Leaders who were the very essence of tribe, ordained in task to set the pace of life and the morrow in death, believing the same splendour and standard to be for all the population in that mirror of death. All living for now to purpose by achievement and skill levels having been set by normal family ties in the order of generational connection. Millers who had sons would be taught to be millers, lamp makers the same and carpenters and like and like. Administrators of city life would act in much the same way by the same tradition. Even great Pharaohs, Kings as total fathers of community had their place. But because of their godly connections, even with civil or political administrators they also needed comprehensive priesthoods to administer religious or spiritual matters determined by them through their collective Gods eyes! For instead of reaching complete understanding of the mysteries we controlled there very order by allowing some of them be managed by their own specific God. Ones who might be concerned with perhaps less interesting aspects of understanding but nevertheless needed their own attendance. Was this a sign we were loosing control of the spiritual before we had gathered its concept, had we over compensated for the magic of death and continuance!?

God culture was in no way negative to our understanding those times ago, it was a sign of our progress, it was in sign that we were linking our own lifecycles to understanding far apart to other specie thinking? It was a sign of our growing excellence of mind but error of thought by accepting the need of gods to manage that which we did not understand, but in creating them we set forces that we did not understand? If this was negative or among the minuses to our progress then in fair claim it was of that time, we had not yet become the people we think we are by today's standards. Of the many great Egyptian stories to be told to our joint development, none are more magnificent than those still standing in kingly majesty as the pyramids. Built holding sign to our human frailty our structured separation and our sheer magnificence, built as tombs, as graves, places where we might lay in corpse those worthy in society to be sent in style to the ultimate God, the ultimate of time, the sun!

Egyptian society in its advanced state then and still, is fully representative for us all now, for by display then, it showed in style how we were all to develop. Organisation on scale to surpass our cave painters was needed, from their mixing of paint and the making of light, translated into working twenty thousand workers for twenty years. Although the great pyramids were so obvious to their intent, other factors were built into their use. To be seen in the abstract, mentioned to show how one of our early manicured societies had adopted our half measured syndrome unseen, but were able to progress in the aim of purpose anyhow?

With the structure of community owing its very form to Kingly connections through the sun, on the Kings death, his body was to lead by transmission. Being the understanding of how the dead would meet with the ultimate, bringing all their worldly possessions as a sign on their importance! By status, entwined in the layout of the three main pyramids at Giza, great structure aligned in aspect of the heavens, tombs to be means of connection with the afterlife. By construction the tombs really are wondrous, but deserve investigation for what they tell us about older societies?

Often first hand, we accept that they were built by cleaver design to stop the prospect of theft, that grave goods would be stolen by whom so ever for their own needs is reflective of how our society really was. Kings, administrators, priests, all working to the perfect state of our proposed acceptances had set our course, but failed to notice that even as we moved among their style of system, we could still use our own free will. Not to say theft was an expression of that free will, but to note that we had our own ideas any way. Which questioned the methods by which we lived or more specifically the power force of our gods in influencing us, form the mighty Pharaoh through to our iconic animal icons? One thing of merited note with regard to the pyramids is to be said, although transmission was the expected course for our Kings afterlife, the tombs were over built in the error of protection. Because of those found intact with all their grave goods, even if only one, none of the goods were missing, the mummified body of the deceased was still there? If that was the result then why labour in preserving the body and storing the goods if further use was not apparent? We must address any answer in the positive, because to the good our ancestors had brought us to realisations beyond the level of instinct, by trying to grasp purpose!

Egypt's lead in that direction brought us to the three elements far from pure instinct; by various stages she gave us the god unit. Most to be the sun in open mystery, that which was reckoned to control the elements like the flood water of the river Nile, a great artery driving through the centre of the country. She brought us in link by association of the God man unity, where her kings were the main human link to spawn their own people fed from the heavenly power. She then brought that we also were carried forward by the same forces led to kings, but in our same role as now, with much more to follow!

## CHAPTER 14

By area, the Middle East from about five thousand years ago, had more interactive community divide than any where else in the world. India and China as their own areas of habitation by proof of today's population show, also had much cultured activity which is never to be unnoticed. But by the tone of my own observation, I have taken the Middle East and later connected links with Europe to pilot my observations. By structure all succeeding groups in their city state/ tribally formed units had all of modern traits, improving each as the opportunity arose. With one huge grey area, that of religious concern, even with dedicated on lookers the god syndrome was mixed and intertwined in many layers. Not least that to some people the innate the insubordinate the invented were objects to be involved in worship by us the fully human!

By indication I have already given my own insight of how the peoples of the Jewish tribes became one through their new religion. One people, one choice, one God, although it was from the locality in question, much is to be attributed to other factors, not least to the vibrant air of the whole area? Intermixed stories of great deeds and dexterity by one city or another, success and failure by human choice to their own god culture, even to better understand some of the mysteries by the natural growth of inquiry. We were channelling the mystery codes attributed to various gods, perhaps like to the god of rain. With some of us seeing the basics of seasonal weather patterns as regular deeds more than appointed events, although not yet understood. Why offer in the waste of sacrifice to events that would occur anyhow?

By the best trait of human nature many of us in our societies viewed the Jewish monotheistic religious direction to be of their concern only, for by the comings and goings of the whole area, their cut was the same as ours. By the scale of time from Mesopotamia to the Roman conquest of much of the Middle East the changes to the area had been immense. When at first we took tentative steps of fixed settlement in the land between two rivers all by show to our conduct had escalated our pace everywhere, we had made monumental advances in all fields. At the time when another success storey was unravelled in area when Roman soldiers marched on conquered streets in far of lands to their own city. By the unseen eye of the future we as part of the single human community had held there our first dimension in standing but not to be used in title for two thousand years later.

Also in movement were aspects to the fourth dimension, the one mainly of scientific value, although undated, like others its progressive steps were of clear distinction. From where our lamp maker had moulded clay dishes to hold oil, aided by further boundaries of science he might now use the metal of bronze to make and decorate his wares more pleasing to the eye. From the use of bronze a mixture of metals even used as cutting tools for great stone carving by other tradesmen. Through the natural forward flow of scientific thinking in that copper and tin had been used to form many objects of daily use by availability, weapons of the hunt or of war were also made from it. A natural link was made between mans progress by our own effort partially sighted by further acts of discovery? Like from where men had worked the forge to cast weapons of war, by first strike those to make sharper or more robust weapons could be pre-emptive. A matter which did occur whereby metal workers had come to the new metal of iron, just what the generals needed, more destructive cutting edges of swords of knife blades to give advantage over enemies!

With our progression of ideas sometimes carried by trading, it was not long in practical terms before every side had come to iron weapons and or iron implements to add to our bank of scientific achievement. By some cultural means these and any changes to society had to have bearing by the structure of those very societies. We were all moving in what we might have deemed to be a forward motion, yet had to account for change by the auspiciousness of our gods. In area and by circumstance, by condition, was it not to find the most powerful god to hold the ultimate lead? Then how to force that prospect in being viable and relevant to or amongst those prevailing circumstances. I write these next lines without authority, claim, direction or accusation?

By their/our time of history when the Jews to be led by Moses came upon their promised land, how was the future attended too? I will say without distinction it was placed in the hand of their god by the then thinking, by the then reasoning, by the then standards and by the then expectation. By reality, monotheism which was to be the blue print of following religions of at first the area, then the world in general by finally addressing quite correctly our god culture above other objectives, then how!? Without serious concern of how the excellence of mind but error of thought pattern worked amongst any people of the day, we must again travel our own minds to the thinking of then ago.

To be able and dismiss all other gods was a huge step to take, the first being that to other people we would be saying "your gods are not really up to the mark". Already active with a tidal god set up of god worship for some, this became possible, because gods were like the tidal flow coming in and out of fashion. To cite your/our own single god could be of little consequence to other groups. Perhaps even without direct intent, the surround to the Jewish appointment was done in such a way to be no direct affront to others, taken by philosophical reasoning in grouping all gods as one god, then the association was direct from man to god or taken to be from God to man. By artful creation all Jewish successes were to have been in the hands of God, all such failing were to have been in the hands of man even if god appointed! If wrongly taken to be a feature in today's world, which alas for some it is, then by the Jewish thinking then ago they/we had usurped any such idea, but was it understood?

Direct from the time of Moses, having been released from bondage as the story might run, selective Jewish culture in the purely competitive seam at the very time of their realisations beyond the error of thought, had also settled their own fate. For effort of discovery, for diligence of deed, for the acclimation of history, their outcome was settled in the imminent glorious unity with God forthcoming. They were to be heralded in god's own time which obviously was now theirs/ours, but to be imminent?

If they as we were the chosen people, it was from our time of then ago, their appointed choice was for them to turn all people of all lands to god by Judaism, which is almost what happened by two attempts since. Under the shield of Moses it was easy for peoples who had been given their freedom to accept all things in the name of god, to accept all things in the name of god's intention was a different matter? Moses carried Gods covenant in finality to be laid upon his people, then for all people, by our own complexities we followed our own nature but in kind to be of understanding from the hands of our priesthood. For almost two thousand years the Jewish religion travelled with the Jewish people in all best effort to keep gods covenant. Not least by what happened two thousand years back from now.

# CHAPTER 15

Jesus was one of many individuals highly educated in all Jewish scripture, seemed to attempt and rekindle what could have been the original Jewish mandate. By selective reasoning from writing mans total history and recording it in the bible, ancient scribes had attempted to fit all matter of time committed so far by man into the reason of logic. Hence the work was to the past in the book of genesis, and to the future in the book of revelation, one achieved, the other imminent! Unconnected, I can assume that from the style of his study, Jesus put to question the progress of the religion that was attributed to the now always acclaimed Jewish people. For two thousand years they had been waiting for the triumph of mans unity with the creator God. Cover to those who had died in the mean time since Moses, were fortunate enough to be attended in the wait by god's pleasure in near paradise.

Christ Jesus had no control over god's intent but in understanding that by choosing the Jews was the lead for all men to be brought to god by becoming Jewish. By following the record of human history in the bible this was shown, for in being written of times before the formation of this first monotheistic faith all of us were by descent from the Garden of Eden! Therefore as we were one, then we are to be one being Jewish before gods reason was met by connection, that the sin of Eden was by all men, carried by all men, then in the blessing of forgiveness it was for all men chosen?

I would not intone that the above had been a direct set of standards by the thinking of Jesus, for he was thinking by outlook of then ago. Without advantage I can think in the realms of hindsight from two thousand years later, which we all can do. I have no advantage, for all recent thinking can only be done in surround by the edifice of half measured processes. If Jesus could look and see of his/our situation then ago, would the first question be to the waiting of two thousand years of Jewish patience for god, did he or would he have considered the drastic in order to better focus the original Jewish intentions? To cut directly into what had become the norm of their acceptances would he hasten the prospect of change by the information that it was him who was to instigate the speed of unity by clearing all men in owing to our first sin as shown by the bible in Eden, our equal original sin?

By forms of connection that we have used before from within these pages, it is more difficult for us to re-enact the state of mind owing to groups of people who were now declared as racial and religious by the same acceptances. When one amongst the group told of the new beginning for all, many from within were also impressed and followed, without condemnation the whole process had been noted to have been claimed before. If the managers of religious conduct, the priesthood, had scented familiarity, then their aim was to protect the faith in the order of previous acceptances. This was done and told in many stories, the most obvious was that the Jewish preacher Jesus left the world's biggest religion, itself distorting from catholic to Christian in the natural flow of time by human standing. Although the

Jewish and Christian religions were linked by the most obvious of connections, no admission of agreement was ever forthcoming, why should it?

My own definition of having the Jewish faith as the first dimension and the Christian religions as the second dimension bears no standard of excellence. Their code is by the note of progressive thinking tied to all manner of forward thought. To imply that Jesus might have moved with impatience to bring all of mankind to god's intent is done only with hindsight, for even from these times of now with the fifth dimensional plan, it is unfettered to any claim except that of all. Both these religions by my personal secondary time link definition, will take in willing kind the three taken prospects of 'Out Of Africa', 'Past Gods Words' with the third yet to be read, and the one active 'One for the Whole'. Four prospects already in the best use by the majority of religions but alas used in their own context. By my trite title of the Fifth Dimension to include them, bears no authority, but lends to them our intent as it will be applied soon. They have not been made from our histories of neglect, but made in these times of the past sixty plus months or so, not to be disabling only to invite any discussion. Of the moment we who are all 'Out Of Africa' and so modern to be 'Past Gods Words', with the impetus of change by 'One for the Whole' to carry future intention!.

Mesopotamia by our changing life styles compressed our human specie integrity by the advances we had recorded over the preceding ten to eight thousand years B C; our now accepted common year mark. By all best considerations to the rest of the world on all continents, much in every other community was the same. We all set the forward gear, we all carried religious concept as the road map, we would all make many turns, sometimes getting lost, but never desperate for our compass always swung in the right direction, the trick was to notice it at the right time fully and not half measured.

## CHAPTER 16

All of our histories had been served well by the Middle East over the past one hundred thousand years. All peoples in all places on all continents were of our one specie, not least because we were the only to explore religion, religious concept and religious perception. Yet to be recognised is that most of us have looked in muddy waters, not that we cannot see that which we know, but we only see that what we have found and nearly understand? Religions have not been the cause to our error of thought, but our politics might have been?

Two thousand years of Judaism in the Middle East, and natural progressive changes were an ongoing process, now the influence was from other places. Europe had featured in our painted history for a long time. Rome having now placed influences upon the Middle East; the general change was to force the hand of politics. When another crucifixion was somewhere entered in the annex of recorded authority it was lost in the dust of cupboards. Later it arose to be of significance but was taken in half measured fashion, not in any deliberate style, but perhaps treated better than any of us could do today when considering the facts were reconstructed three hundred years after the event. We might try to test our self, try and think what the death by fire of someone three hundred years ago, say for heresy and piece together what their last words were? Could we be exact?

Rome the city was of Europe, the Roman Empire was of most of Europe and all lands bordering the Mediterranean Sea, plus from its Eastern shoreline moving into much of the Middle East by area. It's most extreme northerly extent was in Britain, all of which was controlled by the cold medium of economics from political leads. For hundreds of years, Rome the city state had conducted much of her affairs by political criteria in controlling market forces. She had run with her own gods in presentation and with all of their successes. When finally beaten down by, or over laid by an old/new second monotheistic religion, Rome took in style to this faith, but in half measured fashion. Politics had worked so well for her in the brutality of conquest, by it she had organised half the world to her means. If Roman gods had now become one god, then the control of the populous was to be in the hands of duty and not expectation unless one was the master of the other as normal!

By taking to one god, no fatal flaw was yet made by Rome, politics still ruled, her own understanding did not run to the attitude of her own people. By diligent control they had been linked to the careful thought of Roman law to be the first feature of life by all counts, Roman gods worked because Rome did, a new single God from Rome must also work because Rome said. Rome even introduced a year mark time calendar in total acceptance to this new single god. No factor was taken that if Rome did not work, then their new single god would not, for when the house did fall it was the servant who still held its name, Roman Catholicism!

When all roads had led to Rome for centuries, by taking a new single religious concept to be the state religion, then its focal point should be one with the past. It would have been unhealthy for Rome to have managed secular politics from the city and spiritual guidance from one of many conquered lands. Much ceremony was attributed to the places familiar to the life of Jesus, but all first decisions were from the gates of Rome. Even that the head of the spiritual wing of empire the Pope, was to speak by interpretation of religious maters, politics sanctioned all by authority. Where I might interpret this as half measured reasoning or outcome carries, because by the structure of Rome, that was a non event. Rome was the source of all information politically and religiously!

Any forced change to this situation by the demise of empire still left the authority of church with its own ruling right. Something worked very well, for Rome survived as the centre of religion to recapture spiritual fervour in lands where it began to wane once the secular hand had changed. In Europe the demise of Empire led to the dark ages, the undertow was to the firm establishment of tribal groups who had grown to begin and form national gatherings again, a process to take some time. In one sense nothing any where else really changed, survival ruled as the biggest kind of need all over, with the backup or religious redemption! Back in lands referred to as the Middle East things were not quite the same. Here where religious order had almost become a way of life running with styles of political refinement. With the structure of alien peoples government removed, life was set to continue in all the same struggles, but again reasserting the atmosphere of religious thinking.

From Mesopotamia to Egypt to Abraham, Moses to Jesus lots of religious ingredients had stirred in that general area. Types of free exchange had come gone and mixed with the thoughts of men, the central theme now running to be monotheistic by best acceptances, not least that one god was easier to turn to. If the bible had worked to give us better understanding of this in terms of human history, its two weaknesses had been shown by much of the selective interpretation it allowed and how it was added too centuries and centuries after first entry. By area in geographical terms, the Middle East after being abandoned by the Romans was like the rest of empire. Although previously fed from the great city by religious direction, the Middle East reacted her own way turning to previous main religious interpretations like local Judaism, even local Catholicism?

For many people it might have seemed unreasonable that from the very birth place of our greatest of human study, we were now spoon fed the product of our own guided mind? For one person, was this to be turned into the reality of effort in the means of better change? Roughly six hundred years after the birth of Christ, even six hundred years in the waiting, Mohammed form his own learning saw the situation for what it was in these lands that god had made his contact known. Abraham and Moses had felt the breath of god to their ears so told, resulting in the true line of belief in one God. Another great prophet Jesus had claimed to be gods own representative in the living form as an ordinary man to be Gods son incarnate, god as man to hold our link to the very emotions of man the servant. For the Romans to put in character the story of Christ dying on the cross and ascending into heaven three days later, was carried by the very strength of the vibrant religion. With the withdrawal of Rome from the Middle East the state religion had changed in so far that from the roman lead of before regarding Jesus, the base events of Christ's life and death without Roman political influence had a reduced spiritual lead, not least in that the religion created in the Middle East was being bartered to all places in Europe first, what should be done?

Mohammed's great works and efforts were studiously recorded, which in terms of the written word religiously, represent a first by result. Virtually all the greatest of Jewish and Christian writing were done post event, allowing for slightly different interpretations to be added later. Indicating how it was for us to think like then ago with our extended knowledge of now perhaps having influenced the way we might record events? Being a problem I will again illustrate by planting my thoughts to be in harmony with some thinking as it might have been one thousand four hundred years ago! With the Jews already having established the first real one god religion giving depth of purpose to our intent or least ways to appointed intent, by waiting two thousand committed years for the immediacy of tomorrow for triumphant fulfilment, how to continue with the waiting? When the new Jewish order was first instigated as just that, how was the instigator examined by the standard of the day? We must guess that Jesus did honestly believe that he was the son of God. His knowledge of the bible was such to show that he saw self sacrifice as the only means of mans redemption by clearing our original sin marked by our own history. As the real son of god he had to trace his steps as man to hold the hurt of our feelings in sufferance. To die was the most extreme way to the certainty of intent, to die in the glory of rebirth was a pretty good way to prove it. All done, was enough at the time by direct involvement, with a few good friends and growing converts lasting for three hundred years before the head of politics in Rome was turned to be religious and political. What then three centuries later?

## CHAPTER 17

Mohammed in his own lifetime created the last and first monotheistic religion, his drive was in the exact parallel to that of Jesus or so he made it so. By intent he would not have changed any thing unless change was paramount. If the Middle East had become spent of purpose, it was most certain to have been caused by Rome of olden times. Three hundred years ago they literally lifted the soul of the very people there and transported them to far off lands, when the cup was turned by the demise of empire. All that was left in the Middle East was secondary leads to or of mans obligation to the one god.

Fourteen hundred years ago Mohammed set to turn our hope by his own new religion of Islam. Virtually all aspects of the faith were new in terms of obligation, or were they, or were they the interpretations of another mere man in the matched guise of Jesus the prophet, but now as the last in line to be of god's messengers indirect? If the door of future prophesy was now finally closed, it was done in worthy style by the admission of connection to Jesus, plus by affectionate and appreciative concern to Abraham and Moses with many more additions and outlets. But the finest to have been by the style and matter in which all revelations were received through the heavenly medium of the archangel Gabriel then recorded meticulously and instantly. Always to be held for first view time and time again, but with the unalterable key of our developed traits that our thinking then could only be of then! There was one common factor where we could accept proof of all written words by any distinction even linked unintentionally and that was for all when in paradise, even heaven!

In personal terms I have taken the religion of Islam to be my third dimension by its obvious own first date stamp, having been created some six hundred years A D. It holds to be over fourteen hundred years old to its own start date, which curiously was set while Mohammed was alive? Not representative of his birth, nor could it be of his death, what that indicates for me personally is that Mohammed took personal charge to this matter unlike where Jesus had left his apostles to carry on his own verbal intentions. Here we have the expressed foresight to raise the standard, to hold the banner of matched intent, to carry in verse all written words by example, if Mohammed had erred by excellence of mind it was by themes not of his doing, but by our interpretation. From where the first sighted blur was to come was to the future that by intention all men were to come to Islam, seconded by the original Jewish concept. For the last two thousand six hundred years there did not seem to be another concerted Jewish attempt to fill this expectation since the rabbi Jesus, even if he had included all men by sight, it must have been by geography also?

My own meaning to the geography aspect of that last sentence above is of wholly human definition in how we thought then, ambiguity and the abstract were not even words of fashion. Our disposition of the future was on god's terms only determined by atonement and judgement, the obvious route in all ending was from the place of beginning. Hence our connected will with where we stood in being, was to where we had first heard. I would not say that particular trend stood with everybody like Romans of old. But it might explain the Jewish stance where some only could react within the very lands of their birth. So if all peoples were to be Jewish would they all have to come to the same area, now known as the Middle East meeting all targets! From then until now is not spaced in distance by time but changes by attitude.

For me to have coded the faith of Islam to be the third dimension has been done in scope by my own reasoning. If linked by progressive numbers with two other religions it has been done first to set in capture the fourth dimension. Where that was to kick sand into the face of faith, I have completed the list by naming our Fifth Dimension the last in line, the last needed for all forms of our controlled and available thought. Controlled because we have already reached the full limit of the fourth dimensional reaches in knowing all aspects of experiment will be fulfilled. Available because we can ponder the same exercise over and over even that hell has frozen, but by tomorrows means where to each of us individually we can all have new, fresh and unique thoughts on our own account, but now for all!

Three taken and one active prospect is the full extent of the Fifth Dimension. Without law tenet or covenant it is of this very time now to be an overlay to all styles of religion and all styles of politics. 'Out Of Africa', 'Past Gods Words' two taken prospects with the one active 'One for the Whole', is three quarters of our levy. To mention our third taken prospect now in form puts it in our term as probably the most important, for it is not to overlay anything. It is to put us once and for all by all description to the total acceptance of our single specie by the unchallenged acceptance that we are prepared to live to gods own intention seen and shown us in the most splendid of transmissions by our first three dimensions, and others and other religions.

'Natural Body Death' is the simple expression of our third taken prospect. However previously expressed it is important enough to have outweighed all of the others, for it can lead us by direct connection to our burial at Shanidar in northern Iraq pre history. From the deeper studies done at the grave site it can be argued that the body laid there was done in a tender manner. Further study could argue that the death was natural and not necessarily of a mighty hunter, but perhaps of someone who was in a weakened state. If from very early stages of our development we had aspired so, how might we have found the need to change only four thousand years ago?

## CHAPTER 18

Since then we have found other bodies all over the world some showing signs of distress in death. Whether being killed by intended prey, it had been impossible to tell, the only half truth gained is that some bodies had bone damage which could have been signs of the cause of death? From beautiful cave paintings of any age ranged over twenty thousand years we have scene depiction of what can only be taken to be humans in battle. They do not follow views of the hunt, but show spears and arrows directed from man to man! Later events in the land between two rivers would now/then by its own time date, carry the real means of battle planned, because many places were fortified, not against the elements, but each other.

For the first positive sign of admission that killing by war or want was endemic of the human character came in the very hands of Moses. He that would take the Hebrew tongue and make for his people the Jews the same religion of Judaism transmitting direct to his people the very words of god carried on stone tablets. From the time of the event this new endeavour was to benchmark many styles of proposed future living. Of the ten new rules in the deemed sufficiency of control, one of the strangest was the commandment "Not to kill". If the most blatant and obvious by statement was that, we soon smothered its application unerringly, but also unknowingly, being unable to connect it to our present situation!

Moses had led his people in flight from the history of Egypt. If god had aided their rescue then it was of harsh consequence to the loss of a generation of Egyptian first born by their suffering parents. That told, it came harsh to Egypt again when if in grief or treachery their army followed the fleeing Jews in revenge and were swept to oblivion by the mighty waters of the ocean by gods saving hand, proving the exception to have been the chosen! When in turn to the wandering at god's behest, Moses and his people meandered in time to meet the Promised Land. Was not the whole experience mixed in the mire of wonder in how god sought to direct his chosen people? Alas all worked well until by application the rules of conduct came to be mixed into or with the already standard civil codes of conduct ongoing. Without the emphasis of close examination, save on one account; the commandments were most of existing rules anyhow. Exact additions were to the one true god by means of homage. Having expressed to the mortality of man not to kill, was perhaps the most subtle of ruse to put upon any single person as we then ago or for anyone since?

"Thou shall not kill" by any translation can only refer to mankind or of mankind this human specie. We are the only one from all specie who can interpret its overt meaning its intention can only be extended to us for any form of understanding! Not to kill has perception only. If we unravel what could have been intended by god to man through Moses, what do we see? We had just left a whole generation of first born dead, "thou shall not kill". We had just left an army floundering in death. "Thou shall not kill". Even in our wandering to the extent of time by forty long years, it is in the realms of possibility that we executed in the process of executing of our civil law, "though shall not kill". If Moses was not disposed to give us all charter of gods ruling, that remains in the true realm of mystery. What must be taken is that the Promised Land by god was left to us in our own tender care without the wisdom of Moses to bear his feelings, by gods desire he was not to finish the journey with us.

From early times in our new contentment we somehow drew to the aura of god's first meaning form the commandment "thou shall not kill" in relation to us the chosen, but extended to all of mankind as we were to follow that course. Naturally we killed as needed, for the pot, for sacrifice, for protection against wild beasts, these acts were never even considered to any intention. Concerning humans the edges were blurred, we of course killed in the promotion of the law, even uncontested we could individually kill in the protection of self. God by his own good grace had killed an army in cold pursuit of our death therefore we could arm in like fashion to be protected by our own army as the times we lived in necessitated the need. Maybe this not to kill aspect really meant by intent only, it could be that in order to obtain the ultimate of religious gratification we were not to be of the mind to kill others as we. We might expect to obtain gods judgement in all of its glory so that we could answer, "yes lord I have not taken human life unless absolutely necessary". I have kept in good covenant I have honed my body to hold in triumph of your glory to be on earth imminent!

Jews/we of then ago did not act in error, only half measured, the very vibrant area of their/our community in the Middle East waned in battle and human blood letting for the two thousand years since Moses. Perhaps when all men begin to come to the one God as might be our intention, we can better address the matter of self slaughter. Unless if from amongst our number one could show a different lead? If Jesus was that one person, his manner and style were suited to the task, that of example. From studious learning he had formed for us all what was to be newer to our sense of value, where we might translate to action the standing commandment "thou shall not kill" in its final representation to man. He was from his time but past the realisation that any task on this matter was far more difficult than it had been with Moses.

In times of extreme violence, for Moses to have brought gods covenant by hand, there was a direct connection with the situation, but without him there constantly, that commandment was waived by events to unfold over the next two thousand years. Jesus almost failed at the first attempt for he had no real authority, if he was an errant rabbi, a priest with a different conscience. Reiterating what had been said daily and daily for the last two thousand years wasn't going to work when other acceptances had been reached already. To bring meaning to the expression thou shall not kill what might Jesus now say?

By my own terms, I might say this was done by two factors, if one was done that Christ might say he was there for different means and act to those, then continue to highlight all methods of future conduct. For the Son of God to be committed to his own death by human terms in our redemption from our first sin, in answer to the aspect of human death other than by law. Might not he say from now we are to be one, all unblemished by sin, all set to be in covenant by the commandments, all in answer not to kill of man? For those willing to listen to Jesus, captured by his sincerity, must they be prompted to ask in kind what if we were to suffer the indignity of lethal opposition? In answer "to turn the other cheek" was given to the change of standing acceptances more than other interpretations.

"Turn the other cheek" was not a submissive act, it was to be representative that we were actually the same? We stood shoulder to shoulder in equal prospect to outcome by other factors. It was almost a statement to remind each other that in all cases we will all be eventually turning the other cheek to the tone of judgement. It was almost in expression of "what goes around comes around", wherein if one was unduly forceful of character, if their expression was to overlay influences upon anyone, the reply was due in turning the other cheek. They would have to turn the other cheek by all judgements at a time beyond their personal control. It was not an expression of obligation it was to be the mirror of reflection for any to have received its content, "I turn the other cheek" for better judgement!

Perhaps too much was going on at such times for Jesus to remember he was only talking to us uncomplicated people. He was just like us, a normal man, but his train of thought was far to our best prospects. While delivering very moving and heartfelt stories in concern of our welfare he saw and was cheered by the numbers prepared to listen, by the swell of human feeling, did he understand that we understood? By any reference to Moses, by any reference to the covenant of commandments, by the act of expansion to this existing monotheistic community "thou shall not kill" was simply meant from man to man by intent! It was a through link because as we listened to Jesus from many different tribal backgrounds, we were now to be one by all existing religious rules. Reassertion of the expression thou shall not kill might also have been a clever way to keep Roman thinking in neutral with regard to large numbers of people having meetings?

By reflective terms without the matter of human judgement, I have stated that the Jewish, along with the Catholic religion were spawned from the same area of geography. My reference to the two in terms of the commandment "thou shall not kill" has been led directly by modern thinking, where by we associate the expression to be final and at first directed solely to mankind of mankind. We must also remember both religions have sometimes taken that stand!

## CHAPTER 19

'Natural body death' is the third of my three taken prospects of the Fifth Dimension, which is of these recent months, "thou shall not kill" has been related to four thousand years of our history. Of there similar texture by expression they cannot be related, nor have I intended it so, for it would be in the realm of half measured working to cross mix our thinking of this time to the thinking of four and then two thousand years ago. My first and second dimensional religions had held the commandment to all best uses of their respective need but with the positive stand that it was part of their covenant with god through Moses and Jesus. By later and more recent times a type of reverse or hold was put to that line of theoretical thinking. If to fill the deep lingering need in case of reaffirmed direction, what was the ingredient to have put change forward? Without my own lead, I have expressed my thinking to cover the state of conditions to have then prevailed in the Middle East.

If conditions were indeed the key then they were related to then ago, not from my or any other modern outlook back? When Mohammed first put directed thought to the written word, the moment was progressive in historical terms. With the collection of ideas already assembled by his studies he now came to relate them in conjunction to the directed text, there was no other way. For he could not have been chosen by his own direction if he was not already aware of the past through Abraham, Moses or Jesus or others and others to be able to receive the projected future. I cannot say, nor would I express the opinion of my own thinking to what Mohammed might have thought in relation to all that was revealed to him. I do submit it was a fearsome step for him to take in relation to others opinion of Christ?

For six hundred years the catholic religion of Jesus hailed now as the world religion of Roman Catholicism carried in set belief that Christ Jesus was the Son of God incarnate. Having had the mightiest of questions ask to its resolve by brutal persecution in early centuries it now settled to continue and grow. It expanded more throughout Europe simply by such circumstances carried by the needs of men, but taking on national views of some self interest if only connected by separate tribal leaders. From the Middle East the story was not quite the same, here in many ways history ruled by effect in that the core of the new Jewish concept had lost its route.

I can submit without connection that Mohammed saw and directed the need for change, his path if/although preordained, set new parameters on several different plains. Without his clarity of vision by our own understanding, his historical insight was indeed to the far past and to the present locality. Great prophets had stirred the minds of men close in that area, all revelation had been direct to the people there, it was we who had failed of its/there significance. Perhaps we were not to blame because most of what we heard was second or third hand with the added blur of later generational views being cast back to reinterpret the same story meaning differently. Also as to the supposed majority of people being Roman Catholic by connection, even they in their own piety began to home their own views to the locality where the Son of God had walked, and not from a now deposed city state!

With any collection of ideas from now, the past will not be changed, yet it was and by Mohammed, who readily established what can only be seen as religious order to the area once more! Two great steps were taken by him in order to turn the heads of his listeners. Firstly with the reaffirmation of Allah's intent through God of our conduct, holding it to be self sufficient by lead of direction from the written word. This was not an easy task, for Mohammed was directing his efforts to me and you, we who had set our own standard of conduct from how ever we had listened to stories from the past. Here and now for the first time was our own personal direction and the means to maintain it. In essence no standards were to be changed by priesthoods, who might advantageously interpret meaning to suit a particular previous political stand point! Priests of religion yes, but they also were to hold to the written word by conduct with us. Although our free will was to cut into the text from time to time, in general terms it was held as religious law compared to some religions which turned inward. Islam was now open, but the same!

By means of speculation I on my own account will forward in joint obligation what might have been a mighty step for Mohammed to take. If religions were waning by area, if the course of duty was being replaced by apathy, then a pointed object of sight might be needed to turn heads. By careful thought or direct innovation by the best possible means, could not we be focused to our own mortality in not being given gratis acceptance of unity with God in Allah! If the Jews and Catholics had expected the splendour of god's unity in heaven imminent by creed for the acceptance of their beliefs only, how was this to be forestalled by another religion? Both had worked the finest of deeds in expression not to kill of humans to god's commandment. Wherein Moses had delivered our covenant by stone tablets now displaced, Jesus of the trinity, being the Son Of God, passed the same words from his own lips. In which case in deference to status there could never be any argument opposing a total commitment of time. Unless by the unknown field of human parody where we accepted the word of Gods Son; and continued in human kind by usurping the commandment "thou shall not kill", justified by political reasoning.

For Mohammed to turn this phenomenon, even if led, then open steps must be taken for our acceptances. One by dire means of historical consequence, was to liken Jesus to himself, both to be messengers of god in the true realms of prophesy. No one of God, only by God in Allah! Perhaps one of the clearest or direct assertions made in contact on our behalf was from terms of our ability to kill of self, not by our own hand but of each other? By the two aspects this was maintained, unfortunately it further blurred into our half measured function, what was to be the meaning of it? All societies worked human death by the human hand according to civil law, some to religious law on its own account, then even some by religious armies held in commandment. I can only assert that what Mohammed was to do was not to sanction human death in any fashion, but relate to it in the manner in which it could be necessary for us to be in that position of decision?

By new means one of our taken prospects of the Fifth Dimension, 'Natural Body Death', had centuries and centuries before being aired by these recent months. By all intentions it actually was a working tenet from all times ago, the only shortfall then was by certain circumstances, we misinterpreted its intent by continual inquiry from our own free will. When Mohammed put the word in text, it was of his time but broad enough to hold fixed for future centuries. Without idiom of expression 'Natural Body Death' then, did mean that we as men could kill of each other, when Mohammed indicated it so it was by the advanced promotion or defined defence of the new faith to be. Where he could not advance his predictions was in how our own free will was to taint us all in style even aided or not. Any consequence of his cover by any direction that we could kill of self was not in condemnation of others but to restore our own human dignity of self choice. Alas once again our personal failings have done much to compromise Mohammed's heartfelt intentions.

## CHAPTER 20

For the Fifth Dimension to make comparisons that the first two dimensions had a sort of link in prophet values not to kill of self or self; and the third by endorsement of the last prophet could. Then it is for you and I to reckon by our own ability in use of our free will to quantify the thinking of mind by the three different timescales involved uncorrected, and our individual free thinking of today! Even without recourse of the fourth dimension had set half measured reasoning or not, by religious terms only. Although politics has always influenced us one way or the other it compounds our task by having its own separate and linked set of half measured standards to contend with.

As far as any seeds might have been cast by me to link the old Jewish faith with its newer version of Roman Catholicism the aim has been miss-planted. When in later chapters that link will be taken to have grown, that too in character falls on fallow ground. By dimensional definition the first three are very separate religions. They are to be treated so, yet combined by the three taken prospects and one active told by the Fifth Dimension. All religions can hold to their own text but now in prospect with the themes 'Out of Africa', 'Past Gods Words' and 'Natural Body Death' taken with one further to the possibility of change as 'One for the Whole' active. Being active means we can translate our own thought to each others religion without the meaning of doubt or inquiry, even comparison?

Mohammed from self wisdom in reference to the very geography he was born to, had his own set mark to establish his/our newer religion. First pointers are not by extension of Roman Catholicism but in link with Abraham and Moses. Where he must have viewed all previous religions to now be misaligned, it was easier for him to relate to the Jewish religion of before than to the now import of the Roman cross Catholicism, which was also an old religion? Had its founders the apostles, because of its move to Rome, lost or misplaced trust in the words of Jesus? Had they invented memories that their leader was in fact the manifestation of god the father as the son? It might be easier for Mohammed to state in honest obligation that Jesus was as he, a prophet in the manner of expectation to the Jewish belief. Not like Christ Jesus the Messiah, not so the King, but in gods word as he a prophet!

Strange pointers can be picked at will to demonstrate what the thinking of Mohammed might be without speaking of what he actually thought. When Christ had his last supper with his apostles, the stories of the future went into the darkened enclaves of each mans own imagination. To Mohammed, if they were in the presence of god the son, they would not fear the future, from that did he see that they were only in the presence of a man prophet? To bolster their own moral was it the apostles who invented the words of Jesus to hold different meaning? Unwritten words of any time have different meanings in all times.

If eleven frightened men were to carry the words of god to all men by Jewish tenet, yet had suffered the loss of their leader by Jewish law, how might they continue? Unless determined to make of that sacrifice the truer and newer meaning of the original Jewish concept. To bring all men to god by gods laws given in hand to Moses from Abraham and others and others! By prevailing circumstances, perhaps never even perceived, the nature of the Roman conquest had nothing whatsoever to do with how events were first shown or developed. Save by the thinking of the day, save by how some were allowed to precede on their own account without crossing Roman barriers of conduct. Roman conquest was simply a consequence where from later years it reasserted the theory of conquest by religion.

Mohammed by his time had nought to reckon in how old conquerors had influenced his day, save by their adopted religious concept. Their influence was by the Son of God, Jesus, whom I can take by my own standards of hindsight to have been respectively accepted as so. By time that it had taken Gods own messenger Jesus to have had a final influence to the might of Rome. Taken that it was not fishermen or tax collectors to have turned the head of conquest but the rightful ordination of gods own choice. Endorsed by events of time whereby Rome was displaced of empire but not religion. Bringing us back, linked to when and where Mohammed had now found his influences to be.

In retracing the writings of Mohammed, we cannot forget his thoughts and reasoning when being specifically led, which has its own meaning, tracing the pointed word also has its own form. Generally speaking we are genial simple people, even today. To spite how we present ourselves we are still vitally connected by our religious concerns, it has always been the case and is. By the complex structure of life in the Middle East then ago our final hope of obtainment was just as vivid. What Mohammed began to form was the open need of reappraisal to the religious standing of the area first.

Where now we were mostly monotheistic in numbers, if carried by different peoples to different meanings, then the reflection was to first contact. The first to be so was the Jewish religion set in time by Abraham and tone by Moses, when then ago it fell so that the recorded words were to the Jewish religion. Not by conduct but by area and history in terms of organisation, when certain demands were made to the correct standards of conduct for peoples even if led by the conditions of geography climate and diet. Developments to that conduct had found many accepted habits for various peoples not least the Jewish tribes who still carried those habits as religious doctrine. Circumcision was a deemed religious necessity, food preparation was also of high importance extended to dress codes and virtually into every aspect of contact but specifically of religious context.

Although by the content of the bible through Abraham all men were to come to god by Jewish influence, they had faltered in two thousand years of exercise to include gentiles to their habit. As from Jesus to have actively included all men to the Roman Catholic branch of Judaism, he faltered in progressively changing standards to suit the converts rather than the religion for what ever reason. With Abraham, Moses and Jesus no longer in direct influence to our conduct. We our selves could learn to follow the new writings of Mohammed better directed to follow the older path of the first monotheistic religion. In heeding aspects of personal hygiene if through circumcision, of diet and obligation with the endorsed focus to change the whole world to the new appointed religion of Islam by consent of God through Allah. This primary link with the Jewish faith might simply have had more to do with the area of the Middle East, a place of habit and religious culture rather than the dark edifice of Europe sending its own form of cultural interpretation to be upon other peoples.

Mohammed's first standard in seeming was linked to older practices, his defence in likening this new faith to old habits was cut in the order of how the faith was to be practiced by the individual. He was appointed to down grade the aspect of the priesthood, perhaps being a protective method so that they could not lead the flock to their own determinations as rabbis or the apostles of Jesus might have? Mohammed appointed the name of his religion while he had many years to live. Although he was mortal his expectation might have been to be able to keep the original concept on track and in note that the concept of self discipline did work well!

## CHAPTER 21

To the fact that I and others take the first three dimensional religions to be of a Western texture has its own meaning, I take the connection they are monotheistic. Plus they are further connected to god by the link of confidence through Abraham, endorsing that specific line into human connection with god first hand. By feature the three religions of area along with any others with a western flavour hold to the ultimate contact with God, not least by 'Natural Body Death'. This being part of my explanation enables us each to think in the proper terms of unity, with the direct connection through our death, but by due atonement as seen and taken by our live conduct. We earnestly revel in that unity by the venture of our connected life and death by the medium of our soul(s). Even without the clamour of scientific discovery by dull experiment, we have been fortunate enough to take to the means for our eventual understanding. Without recourse to invention we have made the fixed connection between the meaning of our burial rite at Shanidar and the means to remember events in the life of those buried!

For most Eastern religions that complication has been omitted from most text.

Human life is indeed of the utmost importance by connection where purpose is associated to the very forms of human conduct. In religious type, I do not weigh them one against the other, but as I have professed to the western clique in the connection of human life associated to the individual(s) soul in death ever standing. Then I must state that by many Eastern variety of religions closely linked in their own set, there is a total break between the individual and what might be the connected source of finality. Most very fine religions ponder in the dexterity of mans ability to reincarnate time and time again. Having spent each human form of life in the active pursuance of bettering ones conduct until the perfect state has been reached. This then is the time to be aware of the explanation of purpose? But alas lacking in the measure of how and for why any conduct was enacted by previous lifecycles and to what real end if the conduct could not be measured? How so ever conclusion had been reached I will not put to balance Eastern against Western religiously. I will however weigh the matter of Western and Eastern interactions politically and relentlessly.

I do hold by direct acceptance, that we do not know enough about Gods own intention with regard to us on any basis from the East, West, North or south. I will submit through all channels that it is for us in the excellence of our single specie unity to attempt and make ready how we might be of interest to God by gods own Terms. For that I must stand that we are all human and must set our own tenet by first aid in 'One for the whole', our one active prospect. Under, by and to any style or system of use, though through all religions first, keeping in recognition of the connected element of the soul in death tied to our individual lifecycle. If this can work politically also, then let it be so!

For most types of religion there is but one truth, that being what the holder believes. I have offered a brief explanation of how we vary between our Eastern and Western styles of religion done in the spirit of our collectiveness. If we can all now hold in open display the three taken and one active prospects of the Fifth Dimension to try and see beyond the limit of testing what they all stand for? It is that we may use them as individuals if we wish, but my own release is that they are each incorporated in the form expressed by the Fifth Dimension for all religions. I accept first that many religions have held the same features for centuries beyond my reintroduction over these recent months. Where I run too by that reintroduction, is to show they are not to replace any taken tenet or covenant, but simply endorse that we each can express the same sentiments through the three taken prospects. That we each have the ability to cross all borders through the one active prospect so far, all contained in the Fifth Dimension simply as a united term of reference.

## CHAPTER 22

'Out of Africa' represents our first and united beginning through our fathers, fathers and fathers and fathers beginning, no matter how far we might go back along the evolutionary trail, we will all stop at foot steps cast into the now petrified ash of ancient volcanoes in Africa! Some arguments might hold or prove success to the contrary that some ancestral lines were from other places? If this is so, it is not to be in notice that there is any difference to us. My own well supported first taken prospect is the united acceptance of our specie integrity. Just like Charles Darwin's finches we all had the same parentage. Unlike his suggestion to the matter of excellence by natural selection, our changes had been driven by the many different circumstances forcibly interposed on our behalf that had occurred unnoticed? Bearing in mind that today most of us are in the situation we are in, predetermined by the circumstances of our society more than to personal ability or choice. We have to do what we have to do! But most excitedly we are here now as the same people!

'Out of Africa' represents the funnel of specie development. We now have a broad top layer in our human diversity. To follow individual natural history we each would need to have had two biological parents, by the same natural link we would each have to have had four biological grandparents, eight great grandparents. Each time our parentage doubling up which is the very essence of natural selection, that we are all generationally connected to the past. If we are here now as you and I at the top of the pile then the picture produced is that of a pyramid and not a funnel.

We stand in life at the top of the pyramid while those that have supported our existence fan out below us creating the mirror of shape, which is of a true pyramid. To reveal that the position is reversed, it is the funnel we must look at, broad at the top reducing in size almost as an upside down pyramid. This shape is a truer reflection of our specie development leastways by numbers. We do have a large ever growing world population to represent us all at the top of the funnel. In order for us all to have been able to fit into it historically then without the formula of science something needs to be reviewed.

Many of us have siblings, this of course could account in some small way why I have used these shapes. With three or four brothers or sisters having the same ancestral line then the original support group would be less than the fixed numbers on the one to one basis. But that is not the answer to the shape of our single specie development. Would the true answer lie in the fact that we indeed did have siblings but not necessarily of kind, but those of us tempered by time through climate, geography, diet and other mediums?

If we better look at the bottom of the funnel first then let us see our footsteps in Africa slowly making motion to the ever widening aspect of the future before us. By the very success of our evolutionary progress in the art of propagation we moved in success, which meant growth by numbers. All of which could only have been afforded us by the area of our first breech climatically. In early times as bipeds, we moved in groups of size to accommodate the habit of collecting food with the occasional bounty of carrion. Like others our management was such to remain in small troops so that we could best take advantage of our situation. For those of us who split from the main and went our own way the answer of our birthright was lost because we no longer recognised our joined and single beginning. Unstoppably, we began to fill the funnel of human life from our united start nearest from where we left footsteps in muddy ash 'Out of Africa'. We are all here now with the same deep honest history, we are of the same specie we are of the same worth. Our funnel of human number growth stands to show that our ancient siblings had not an increasing number of grandparents but were actually brother and sister in following those before. To allow us to expand outward in all places in all times since, not by counting backwards but in growing forward to be here as we are now, always one, 'Out of Africa'

## CHAPTER 23

'Past Gods Words', sets new standards when seen that from the very time of Abraham we have been talking and acting 'Past Gods Words'. It is an extremely heavy expression if seen as a warning to be heeded. That any of us might suggest that the devout of god should work or think or act 'Past Gods Words' when they the same as we, have taken the joy and inspiration from any religious text is not an affront. My own feel for the expression has been taken by events these past sixty plus post world war two years, then added to be one of three taken prospects delivered these recent fifty or sixty months. My own structure of the prospect was set in a very broad range. By expression 'Past Gods Words' can mean almost anything. The word God is there for the sole purpose to relate this particular one of three taken prospects to the concern of our religions only. It has no political connotation unless rudely used by some who might not understand my aim or their own reasoning.

If any religion is a statement that we believe in god or purpose or the hereafter or the future then any of the codes we might deliberately follow must have the direct direction of god's words, but how? There is no doubt by the evidence of doctrine that much religious activity in the supported sense began about four thousand years ago. By all quirks they were mostly set in structure 'Past God Words' without any form of deceit. To connect to a monotheistic belief by Western reasoning those times ago, the object of group unity was to depose all redundant or inept gods. Until then by obvious assumption, the direct method of connection to ones own particular gods, was to request blessing by means of payment in homage through the attendants to the particular god's temple namely the priesthood. Even if the gods failed to deliver, there agents were not going to come to admission. There failure was represented to the believers offering by the attendant priesthood, giving suitable reason for the outcome, so much so if the general religious mood was to change then this was done in all sense 'Past gods Words', by the priesthood. No god would say in kind or could have expressed to each of there holders "I am to be made redundant we have decided to go monotheistic" unless to have been expressed representatively from agents at hand!

In grasping the concept of 'Past Gods Words' we must take care not to be alarmed by the times we now live in against the times we had lived. My own indication has been that the prospect is new but is entitled to overlay any time of our history religiously. Not so to implant modern thinking to then ago, expecting modern reaction from then ago, which would not have been normal anyhow. To look at the modern by these recent fifty or sixty months my aim of selection is for us individually. Having already had our last prophet in attendance some fourteen hundred years ago, none from these times might say upon anyone else "I carry Gods Words, this is what we/you must do". Without reaction the answer to all such questions, indicators or points of view stands on our own heels, we are now locked in our own free will of choice. Any reactive response would in the first place not be religious, nor would it or could it be sanctioned by any religious text from wherever or whenever!

I submit, but not on the same terms, 'Past Gods Words' is and was intended to also belong to the past. Though not to overlay anything said or reported to have been said by or from Abraham, Moses, Jesus, Mohammed or others in between. Bearing always we are individual, we all have our own free will which is our personal mantra of choice. Of the three main western monotheistic religions, two, Judaism and Christianity hold a strange advantage, which we must guard against holding them to be linked, other than by there obvious past religious connection. Their recorded version of the heard or appointed word from God direct, had been channelled through my first dimension by a well established priesthood of which most did not here Gods words direct. Therefore by fact, they transmitted all information unrealistically 'Past Gods Words'. From my second Catholic dimension, with the pen lacking most times when Jesus spoke, his words were heartfelt, but blemished with the hear after, however good intentions were!

From my third dimension, in the case of Mohammed the expression 'Past Gods words' has a different bearing when viewed in the time limit of the installation of his new religion. My statement to its point of reference being of these recent months stands, but still carry against what had been written then ago. His words recorded in the Quran were by consent of Gods own agent, so in a way they were 'Past Gods Words', but expressed in the exact moment as gods words. Offering no problem to my personal thinking when considered that as soon as they were written they were sent upon us individually, to be translated in guidance by our semi directed terms through our own free will. That we were supposed to learn the text vividly, hold its content and construction by intention, this was to be done under our own recognisance. Then again as the individuals we were by the experiences we had had, not least free will, we chose our own route, so much so to take in two main meanings of the holy Quran. In one group being disrespectful to the other and equally reversed, by a close time mark by both branches of Islam were fed from the same plate. Did one take to have seen god's words, while the other read by their own circumstances 'Past Gods Words'?

By the avenues chosen to direct followers of Islam in the name of Sunni or Shiah with the former of the largest number without distinction, because we all count singly anyhow. 'Past Gods Words' is forced in to contention by one or the other for one or both followed past god's words anyhow. Therefore in show of our individuality, we can speak in those terms of each others religions without the menace of offence, that gods own words have been nullified by gods own decision to our individuality by merit or our free will, expresses free will!

I have added the taken prospect not on religious terms but for religious feeling. It is indeed good that we might hold of our own opinion to operate to gods own expressed lead. A prime consideration to us all stands in view whereby if any person on religious grounds would offer his tone to have been of god's words, then this is so. This can be taken but when passed to the next and next person the words are no longer gods, but those of we individually or we mortal men. Therefore by example only, to spite the total care in which Mohammed took by the recording of his delivery. Was his work compromised in any way when in short relief after his own demise, others both Shiah and Sunni of Islam both, Muslim, gave in equal consent but by different accord to be of God's words or to have been 'Past Gods Words'? All connections of which are for our own conscience by reflection, that we might fall or wish to react to any implication is done 'Past Gods Words' on our own spoken account. Not only through Islam or Judaism or Christianity, but in all cases of religious forbearance to be above and better to religious tolerance, which can only operate half measured!

If a time spun prospect relative to our religions in first relief to those of the Western classification, monotheistic, 'Past Gods Words' still applies to any and all religions. It may not be obvious by any political reasoning because most systems work 'Past Gods Words' in any case. It is so to those who might erroneously link gods words of a religious significance to be of political importance. That avenue of deceit has been removed from use because of my introduction of the taken prospects these recent fifty or sixty months now and for future dealing. 'Out of Africa' and 'Past Gods Words' although to be two of three taken prospect are now never to be denied. We might raise or diminish either but only through the medium of our one active prospect, 'One for the Whole'.

## CHAPTER 24

'Natural Body Death' is my third of three taken prospects; and in many ways might prove to be our best choice of importance ever. It does not challenge the commandment "thou shall not kill" or the execution of secular law, nor retribution like "an eye for an eye" or other challenges to what may be seen as natural justice. First count by implication it is of mild texture if only to indicate that the best means of our eventual death is by all natural methods, at best by the event of old age. If the full secrets of Shanidar are ever revealed to us beyond them shown, will they be our best proof of original Godly concept by note that the actual burial was indeed of a comrade in tender care. That we took one of family, clan, group or tribe and laid them in rest for our conscience sake, but were spurred on by an unknown challenge to our ability of thought through memory? Although I attach a far more significant reason to that act of humanity, my mark is always for all considerations ongoing.

'Natural Body Death' by direct association is in open reference to the fact that we will all die. It is not morbid or worrisome by sight sound or intention at worst it is to be ignored, at best it is to alter all of our lives without the intrusion of effort. From religious commandment not to kill, by the very time and circumstances of its introduction we have reacted to it by situations of requirement above direction of deed. Even today it is still to be argued about who or what it refers to directly and under what criteria. For clarity, our prospect can be dissected in the three segments of its form.

'Natural' in terms of we humans, this single specie, stands to its own definition by time, natural is that which will occur to the conditions of normal expectation through any time scale. Also in our own case to include the intrusion of some very unpleasant medical conditions which is tragic and only work by the weight of linked sorrow amongst family and friends. What the word is intended to mean by my first direction is of neutral expression, whereby no human is to have intruded to the death of another by any method, reason, remit, direction of desire. War, murder, execution, ethnic or religious or political cleansing or creed is not to be the ideals in the pursuit of human death for any reason. By the greatest of good fortune these acceptances are already held by the vast, vast majority of peoples.

'Body', in like terms refers to us in our human body form. It refers to us from our walk 'Out of Africa' when in times unknown we were as one, but grew to notice our differences if first shown by cave paintings when we recorded the act of war as an act of tribal conflict. If fed by the competition of food supply then this was only a diversion, it was no sign or indication to clan excellence by divine natural selection! Our key to formality was by the/our burial rite of one hundred thousand years ago in the open unity of personal grief and wonder. Body refers to us in link from footsteps to be now cast in stone three point five million years ago it is to us being human. It refers to us collectively as any body by our continental roots to hold the same direction of start and to lead to the same end in death naturally!

'Death', by all concepts is finality, to us in human kind is to the fallen fruit or leaf, we are no more by the instance of painless decay. By nature, it is in death that fills the time of our natural existence. From our death there is no or never could be any inclination that the act should or could have been by any other means than natural cause. To die by natural means is to have been measured in the concept of reality and in the concept of God! Not so by appointed direction, but to show we are broad enough to understand Gods own requirement. For the only hearken, it is that all appointments of 'Natural Body Death' are directly without Gods lead, also death from the order of creation is to the expanse of time by human deliverance only, without human intervention!

'Natural Body Death' in prospect is to be one of the real contacts with our singularity of specie beyond the measure of evolutionary processes, leastways in respect of death being natural. Whereby in general terms many other creatures do not practice the code of inter specie killing to the same standards that we might use naturally. Some must follow instinctive traits without choice, like for example if the male line of some specie are called to fight to the death to kill those of a weak disposition in order to display prowess to mate, displaying terms of natural selection, this is done in the blindness of instinctive drive without compromise. It is done for the goodness of all, unreasoned. It is done outside the interpretation of need, intent or purpose, which only help measure our own stand apart! It is also intended that we as we are, are prepared to set for us a fixed point of reference in order to show in all manner that we have developed purpose even beyond our own ability to really understand!

'Out of Africa', 'Past Gods Words' and 'Natural Body Death' are to be three taken prospects by us. We have not needed to be selective, for the three lend their output to us all by sheer intent. We have duly learnt without compromise that if we put forward any proposition it is now done in aspect with 'One for the Whole'; the means whereby each of us can contribute using this active prospect to illuminate our own opinion for all whatever we might have to say. Without danger, we all can offer opinion one way or the other with no need to count the return, for that carries other opinion to both nullify the extreme without judgement of human fear or cynicism.

Three taken prospects from very recent times of these fifty or sixty months are real, that we now have for the first time in human history a real point of reference from which we can each join in the order of change if needed. 'One for the Whole' is the one and only active prospect for us to initiate the avenue of change but only if accepted by whom so ever if selected? Without the vision of complexity we are each here now and active. We are each here of true worth to each other if only to be included to look to tomorrow rather than yesterday without the worry of being compromised.

## CHAPTER 25

Without any lean slant or lay the Fifth Dimension in carrying three taken prospects and one active in 'One for the Whole' has set new standards creating the united point of view full measured! Although first set through our religions, it carries the same emphasis to all of our political standings, crossing all divides. Again fully measured, but allowing for different standards to be obtained by how we let our politics react politically!

For full measured intention from our religions; the mirrored reflection is only slightly blurred by our personal attitude to the concept of God or Allah. That I would suggest all religions can take and use three prospects now delivered these recent four to five years, all histories can be viewed full measured! Whereby without the cross section of area, language or continental alignment setting differences from the circumstances of climate, geography, diet or other means, we each have three reference points anew and continually full measured! Bringing us to the set conclusion of our separate religions that we each have the same goal, not in half measure tone, of what we could have been led to believe of god's intent through our own misunderstanding shown by our separate religions? Instead, now on religious terms for another first time, we can express in open declaration our human intent to gain Gods own interest in us all now, because at last/first, we now hold three prospects in tandem!

Even already held by all of our histories, 'Out of Africa', 'Past Gods Words' and 'One for the Whole' go to our religions, not to compromise them against our differing political persuasions, which can only operate in error, half measured? Having been done on the basis that no form of political control has yet been made to operate full measured, but are driven or supported through the aspect of our newer steps taken syndrome? Continually compromising the future by putting political concerns above religious requirement, if!

Politics unlike religion has been inexorably linked to the standards of the fourth dimension without consent, even as stated by myself to that burial rite in northern Iraq one hundred thousand years ago. Any organisation to bury one of clan was done by civil/political management, even if the reason was driven by other means, naturally? For the next ninety plus thousand years we had advanced or developed better to some standards, this was done by our civil/political drive only slightly supported by religious twinges? Some four thousand years ago we really set our first standard when we came to serious religion. With this new understanding, by excellence of mind but error of thought we simply laid our new direction in conjunction to/with, alongside the civil codes we already practiced? Leading today when we seem to have lost many directions we might have taken if we had not been drawn to operate each of the important features to our personal political management half measured!

If the worst or lowest point of our collective awareness had been noted by different context since the ending of world war two, then from then until now we have been in the worst situation to realise the route or cause of our many problems! Matters which have been addressed many times by many good causes but never full measured? I can state only because we are of this time that half measured working needs change? Five years since our first connection with the Fifth Dimension is an instant of time by the clock of human development. Five years to the lifecycle of an individual of our generation is a huge length of time in the comparison of other differences. Not least when for some of us the average age is only fifty plus years per group and those better connected with an average lifespan of seventy plus years; the measure of five years begs for us all, urgency.

For us to visualise what the meaning of my declaration of half measured working means, it will be better for us to begin to think in the terms of our own lifecycle is not to create worry or concern, but self assurance. When all the connections are real, it is we who work and manage our religions and politics. Having repeated that we should, must separate our opinion of the two into religions and politics, has to be done in service of both and to each other, otherwise we will find ourselves locked to always only operate half measured. Might not we now try to understand what the context really means to one another, by use of our active prospect 'One for the whole'!?

## CHAPTER 26

With our one active prospect, 'One for the Whole' by its self will promote the need for our changing some of our habits intertwined with our separate religious and political standing. From our religious definition it is easier, because one way or the other we already have my three taken prospects to be going on with. Politically things are much more difficult, for it has been by the codes of civil control that we had set our standards of political conduct. Influencing every other adaptation to have been collected for our development and setting the style so that we have only been able to have operated half measured since? In real terms we have to ask what does half measured working mean if anything; what is the meaning of the outcome in my steps taken syndrome classification?

Half measured is caused by reason of our own reasoning, when entering any contract or agreement. Human haste has always controlled our attention to detail. We accept the contract but are always prepared to iron out the details later, that is the essence or our half measured reasoning I will explain the exact relationship of it to us. If we have set our standard to be civilised as we might, then where did that standard come from? Civilisation is not a result of time or knowledge. It is not the result of understanding or acceptance. Therefore to be civilised can not be an accepted standard by any faction! In relating to our burial rite at Shanidar, which is taken by me that we were all present, therefore we are all the same can mark our similarity, but does not mark us to be civilised?

We had a living clan member now dead, but remembered by time factors unimagined. While we can actually relate to their living exploits in a somewhat different manner than other specie through our unknown of memory. That one of us was dead without any activity to the remaining group was half measured, simply because we took time to bury him, but why? Although it was through our civil control we organised the event over our natural desire to gather food or hunt, we were not sophisticated or civilised by any means. It stood in the crossover of our own reasoning, without tenure we had acted above our senses, but in doing so had given measure of the event done, that is half measured reasoning. Supplanting an event of no particular meaning to have depth in terms of how we should review it, which by circumstance is what has now happened by my own intrusion. I have erroneously put to measure something that happened one hundred thousand years ago in all significance to events of sixty plus years ago, that is half measured reasoning!

Under religious terms the standard might hold to different criteria if we can call to three Middle East religions of definite linked connection and treat them as totally separate. That is a point to half measured reasoning. By political terms the matter is much the same but changed by the ability of our politics to be timeless in relation to us individually. Our politics had and are being set to the eventual theorised concept of equal sharing, but by there lacking the ability to deliver, that is half measured working. Half measured working is substantiated by us in the assumption that our start point is always correct. Like from our different religions in the direct concern of us, but at the same time expecting us treat the same outcome differently but expressed through our individual lifecycles!

Again on the religious theme only, but without explanation, Moses did not operate his style half measured but in accord. His line of direction was from the past and remained to have first take, he had become the focal point of the new monotheistic religion. By any lead given then ago the story had been set to its expected ending. When some two thousand years later the son of man by the same connections gave his lead, this in the name of Jesus was half measured past the consideration of any connection? Jesus like the rest of us did not know of his time that two thousand years later any accusation like that could ever be laid against him for any intention!? What had happened in the time of Jesus balanced to the time of Moses was the two thousand years of our history in separating the two.

Waiting years after the pointed hope of Moses had stalled, because the imminent expectation of our chosen unity with god had been tarnished by that very time span in relation to our individual lifecycle. Within that time we had failed to turn the whole world to the one faith in the one God. I had given it that Jesus was indeed a rabbi of the Jewish faith shown by the standard of his education. Not least by the standard of his preaching of human unity by connected service from the bible and to the full extension of this new Jewish faith? Where he gave much in credit to the very structure of the faith was unread in relation to its overlong history. Apart to the structure of the faith being imminent there had been two thousand years of civil/political history ongoing in the meantime. Any religious reference would have to have considered this if only from my own thinking of now. Being impossible to relate factors then ago by thousands of advanced years is no fault owing to Jesus but has to be measured being half measured when other assumptions were made!

Six hundred years after Jesus by the connection of geography to much the same area of the world; the Middle East, Mohammed gave us our third monotheistic religion. Questions to this being done half measured are of the same significance to those in relation to the preaching and thinking Jesus had shown. He and Mohammed were led in the same drive to settle our intent of purpose. Both had set it by the same standards of their day, but in advancing time scales holding the progress of that advancement measured by changing attitudes in other fields like politics. To have operated religious concept half measured was done by the aspect of political leads from different criteria. Religions to the individual if, politics to the individual but extended beyond the range of our lifecycle, taking that so led choices by the factors of achievement displayed the conclusion of civilisation. Being the missive in which we have operated half measured long, long before religious concept.

Jesus and Mohammed are without separation to the desire of Moses. My connection of half measured working on religious terms in relation to the catholic and Islamic faiths stands because of the political conditions in the times of the separate exposition of intent to each prophet. The real depth of statement is to our civil means of political controls in being half measured to the point of influences by wholly political motives ever unread, ever misunderstood. Being matters of little historical concern until viewed by these most recent of times without assumption, but of the same desire to the discovery of or to our main points of error politically or religiously and now linked.

Religious half measured direction is not to be damning when the right motives are almost always there, our views in just audit, are for us to realign them if they had or have taken the wrong direction by the acceptance of half measured reasoning even if unknown. Political reasoning of any intent is at its most virulent when in determination between us of different nationalities. Some of its biggest points of error had been made by those determinations. If only assumed to have been right then judgements made from the wrong point of view were covered to be held by my definition of our steps taken syndrome being a direct result of half measured reasoning.

Without the direction of hindsight, for any of us to be asked when we first felt we were civilised, can only be answered from recorded history, many would agree that even post cave painting. Our first writing if recorded on clay tablets found in Mesopotamia or more openly in the Middle East, where our nomadic lifestyle first turned to settlement by forming some of our first city states. Is a good example of where the art of digital communication first began? In the land between two rivers, we had settled and become civilised now that we could show organisation and be selective of our own skills by means of choice and to our own perceived style.

## CHAPTER 27

Without any form of criticism we might try and balance the meaning of my half measured definition because of some religious interaction in connection with historical times. After the establishment of the religion of Islam becoming the main one in the lands of the Middle East, by much the same drive the Roman Catholic faith had consolidated most of Europe. Two very great religions linked by history joined by Judaism yet saw no reality in their equal quest. Although all of my observations are of this time now, by the advantage of our united histories, I on our account are at full liberty to comment to how any previous prophet, manager, seer or lord of convenience might have proposed our future. That is not half measured working or thinking. Where they might have supposed of our conduct from two outlooks the same but claimed to be different, our reactions are spun from our own interpretations to the assumption we of each group were civilised and the other not?

Jihad versus Crusade is of the same making, is of the same times is of the same result. Three hundred years of the worst related religious wars ever to have been conducted in history. Three hundred years where we of the one specie humbled ourselves to no avail, from those warring times our histories have followed the wrong course. Without any balance the final outcome of the religious wars between Islam and Christianity in the Middle East came to political impasse. If we settled to our own forms of civilised behaviour after such events it was done half measured, not to be challenged until some two hundred years ago by slightly different reasoning, but this time between Europe and the Middle East!.

Three hundred years of our holy wars had failed in the overall support to our two separate religious concepts. With both standing to the behaviour of the individual by the show of our earthly deeds from which we expected just reward in death, nothing was to have changed. Three hundred years of battling in the name of Allah and Christ God belied the reason of time involved, unless to have now become politically motivated. Reflectively, initially we might have taken up arms to counter heresy claiming that each side was right so therefore the other must be wrong? All points were moved when we killed each other in those three hundred years, from the first set of battles in the care of each society the dead were due the equal reward by effort in God and Allah's eyes. After that the emphasis was promised the same, but had changed through its own longevity. No more could we fight in Allah's or God's reason when we now carried the motives of revenge or hate or fear. All now textured by previous war battles leaving the taste that our fathers even our fathers, fathers had perished at the hands of the infidel!

Any divisive feature between the connection of half measured working from religious and political influences are to be taken in context only, are to be taken in direct reference to the area where conflict was almost a way of life. My intentions are not to prove what half measured reasoning is, but to show how it can be disguised by not being noticed? From the times of our religious wars of Jihad and Crusade, in asking what the outcome was, the answer is always in the negative. If the theory of both actions were correct on religious terms, when both sides had only ever thought from their own religious perspective. Post wars, in quieter periods our standards although still religious were to become more politically inclined under changing circumstances?

After the death of Mohammed, things in the Middle East internationally slowed down and returned to normal life and the pursuit of happiness. Likewise in much of Europe the same applied but from very different circumstances. Without comparison, when the Islamic religion had split into two main factions early after Mohammed's natural death. The Catholic religion of Rome rumbled into the theory of dispute for centuries in the growing cauldron of questioning displayed from the European format of nationalism. All over Europe even with the same religion, different peoples strove to be more sympathetic to there own national standing. The first real demand for change was by Martin Luther. His main query was to the language of use in the presentation of Roman Catholic theology in relation to how the holy sacraments through the mass were delivered. Latin was the standard, an old language of Rome but throughout Europe its use was less and less desirable in the name of nationality, hence the need of change?

Church and state were of bonded unity, but in the reality of nationality with the same language being spoken through the mass, some people sought there own solution? Martin Luther's most famous call was that the voice of the church should be in the language of the people, in his case Germanic, which was local to the tribes of his vicinity, being a simple idea at least to political reasoning, because although his ideal was self motivated it suited all other nationals also. We also wanted our own language used? With the language of the church to be national, politics would carry new weight, by some sections of each society there were those both for and against the stand, which in many cases was only half sorted by protracted argument over many cruel and brutal years. By first outcome this era of time in terms of the Roman Catholic religion came upon itself later to be known as the reformation. Loosely a time of reform to aspects of the faith through protesting, first against doctrine then clergy as might be needed in order to rationalise the mix of many nationals with one faith.

With Europe having split more religiously and nationally over the following centuries since Martin Luther through reformation and protesting? Our Middle East over much the same time period generally remained with the religion of Islam unchanged, politics to Middle East determinations remained of a lesser entity as did science, but was of no less importance!

## CHAPTER 28

At the time when Martin Luther made his stand, science or the expanding aspect of study was to fly in the face of accepting what we did not know given to be in Gods realm only by being the holder of the mysteries? European standing was also beginning to expand on its own account, that is not to say new revelation or the way they were treated held any specific importance in Europe than any where else. I mention the relationship in those terms to highlight the probability of ensuing circumstances to have made that feature in relation to Europe having its own type of effect. With the political structure of Europe being monarchical much of the civil controls were overseen in part by the well established priesthood(s) taking their lead from the Holy See in Rome and or political entities?

Questions from within had been put to many separate national religious Roman Catholic orders, although led directly from the Popes council, some countries used their own authority for local influence. Open opposition to the ideals of Luther were counterbalanced strongly by the Spanish branch of Catholicism who set through church law the manner of the inquisition. A method then to determine the right order of commitment to the holy mother church, a fearsome method to punish those who opinionated against any aspect of church law, a legal method to set in heresy any declaration unaccepted. A brutal treatment for some men judged by others to have erred in the standard line of acceptances as ordained?

From the last quarter of the fifteenth century the Spanish inquisition set to standardise in a backward direction the core of belief in order to account for the wind of change by the breeze of consent which begin to settle over Europe. When science began to reveal new truths, began to move boundaries, began to reveal some of the hidden mysteries of accredited belief previously given by consent of the church. By example when new lands were found, some say led by the Italian astronomer Galileo, who made of the world to be round? The Catholic Churches having set to accept the creation theory as per the full bible of gods own work in the making of the universe to have taken six days. Worked to the assumption that the earth was flat instead of round tallied in what was and could be seen. It looked flat, it was flat, naturally god made it of the first importance because it was created for all men in his service. It did not matter to be known or was suppressed in learning that Egyptian mathematicians some thousands of years earlier, had proved by the movement of shadows cast by the sun that the earth was indeed spherical and of its approximate circumference to be of today's accuracy?

While the church protected its own standing to its obvious connection with mankind set in the rules of creation. Close to the time of Martin Luther and the Spanish inquisition in the year fourteen ninety two, irretrievable journeys were made. Christopher Columbus an adventurer of his day had devised a plan to shorten the journey time for trade routes between the spices and silks of India and China. His intention was to sail by boat from the Western landfall of Europe, until he came upon the spice lands of India and the east. Lands where the normal accusation of these very fine goods was obtained from land caravans moving eastwards, yet he was prepared to sail west in proof that at least he thought the world was round. In effect his proposal was to sail west and arrive east! In principal an action that could only be done if the world was round. With much trepidation and deep fear held by many of his crew he proved in finality that the world was round again, and still is?

Christopher Columbus from his travels found many more questions than answers, in show to his beliefs that he had achieved his goal reaching the spice lands of India. He gave to his first landfall to be the West Indies and coupled in the general confusion that the peoples he first met from various tribes were ever known to be Indians! His realisation did not extend to the fact that by sailing west he could not fail to discover completely new lands of such magnitude. Lands that extended from the northern ice cap unbroken in length reaching far into the southern ocean, creating a natural barrier between the world's two largest oceans. By size and shape the land form before the realisation of national boundaries within, was given to be two vast continents of a north and southern complexion held by a central isthmus, where on its own account much hurried interest was to fall. Columbus to the European establishment then ago proved that the world was round at least not to fall off its own flatness? He had not found India but vast continents to be later known as North and South America, with his own West Indies to the central portion of the land mass.

From my own observations, Christopher Columbus forced many of our hands long centuries after his own death. All of which is to be carried by our own conduct this time later in the matter of reflection by implanting our learning up until now to the balance of how we operated after his discoveries then. We will see how we had managed to operate our half measured reasoning by religious means and more fully by political means and to the latter endorsing our steps taken syndrome!

Columbus and the Spanish inquisition are only connected by the loose band of there respective timing, with the one being a driven adventurer holding to the spirit in the possibility of true discovery. Inquisition holding to the stamp of fixed time taking that all that was known was all explained by Gods good grace, but only by mans own interpretation half measured in one context. With stories of new lands, of strange people of strange habits and of much wealth and by returning with great stories of discovery, messages reaching at first the lands who best supported his journeys at the extreme western land fall of Europe, Portugal and Spain set there own course.

## CHAPTER 29

Christopher Columbus in fair recognition should be given great credit for his achievement, not fully counted from his own time, for then it was almost a simple act of expansion. People connected were looking to the possibility of investment for future gain; the circumstances that prevailed in Europe at about his time were vibrant. The known world at that time of recorded history accepted the single landmass having Europe at on side with china at the other. India of its own mixed history almost set the union as the intermediary with its own input, whereby to the best conditions for all of our circumstances we were able to link some political interaction like trade for example. India and China, vast countries at the time connected by tribal to be racial links were fully able to determine their own conduct and form their own council. Through that political link of trade they reacted to each other and Europe as necessary through many intermediate mediums. That they made their own judgement of individual political management and religious determinations apart to established trade partners set their own mark. At such times the Middle East, India and China had extremely sophisticated forms of civil/political management along with definitive religious concept. Therefore what need to hurry change?

For many inhabited areas of Europe and Asia, being landlocked set certain habits for personal development. To others areas with coastlines interest in the sea gave a different outlook if not only by the conditions and circumstances to have occurred by other means. The open point of that observation by me stands on any urgency generated by those prevailing conditions that had come to set our standards. At the time of Columbus the standard of shipbuilding in Europe was what it was, which was to follow improving designs. With European taste being set by the better use of the wind to fill more sails replacing the oar, making for more efficient travel? This helped the building of larger ships to traverse new uncharted waters helping success in long journeys. Also Europeans were able to copy and gather information from each other for continuing improvement.

Unknown to most of Europe, at the same time and earlier the Chinese had also used the sail to good effect, they had also developed larger and just as seaworthy vessels. Any difference to outcome had been sealed by their first connection to maintain their own standard. They were content to develop on their own terms by self reasoning because there society was to its required need. From what they might have gleaned by strange travellers coming from western lands past India could have helped set their opinion to remain fixed to their own affairs. Although Chinese ships were truly sea worthy, in their contentment they were only used to sail the areas of settled familiarity, local waters? Even when facing the vast Pacific Ocean which had shown its own cruel wrath many times to all affected and known lands of the area. They saw no urgency in sailing eastward to come to Europe by faster means than they might have travelled landward westward! Beyond my own capacity to know and of any proof to show; the Chinese might have made many journeys into the vastness of the Pacific Ocean. Where Columbus with his frightened band of sailors were blown for thirty days into the same wilderness of heaving waters until the welcome hail of land ahoy! Chinese sailors of the same determination might have sailed and sailed for thirty or forty or sixty days into the swelling mass of the ocean and ocean and ocean without any sign of its ending, so much for the value of Europe seaward east and not landward west?

Of the adventures taken by Columbus and my Chinese sailors, both are worthy of recognition, but tell us little of what we understand about our different cultures which by the circumstances of circumstance had remained in mystery at those times. If china had become self entrenched and kept a closed society, then that was by her own desire and from her own reasoning. If India and the Middle East mediated between dealings from the extremities of our single continental landmass, that was to all of our input! Even that we knew of our differences by colour by style and habit by politics by religion, we had failed to see the cause; climate, geography even diet but all undistinguished by the time in which we all then lived Columbus from the eventual discovery of America north and south began a trek for us all to have lasted for almost five hundred years until about the end of the second world war in nineteen forty five. Fortunately his travels have now led to bring to us the ability of acceptance that we are of the same beginning through 'Out of Africa' and of the same ending by 'Natural Body Death', unconfined, but truly linked by the Fifth Dimension.

## CHAPTER 30

I have indicated what I consider to be a great debt owed by us all to Columbus for his travels, by them he revealed a huge anomaly to all human philosophies. He on our behalf had found for the first time again our own lost brotherhood, already cemented with us from our burial rite at Shanidar in northern Iraq one hundred thousand years ago. By my reading of events I Have taken that by some unconnected memory, Shanidar was the spark to all religions. With the proof carried by what we revealed in the two new continents owing to Christopher Columbus travels. Before we might look more closely at the extent of or differences of these new peoples, leastways to us of Europe first, we might try to measure Europe's own actions or reactions to those discoveries?

Tales of great wealth was the dynamo of response for those connected to the knowledge of new lands. Portugal along with Spain took active charge to explore these new challenges. At the time both would carry the blessing of respective mother churches being fully Roman Catholic pre the influences of protestation and reformation. Without the claim of connected association, it cannot be said that any new travels to the new world were only church driven. For at the time all political acts even of discovery were sanctioned and assisted to carry church doctrine only as part of other objectives. Coupled with the fact that the Spanish catholic church of Rome through the inquisition was re-examining the text of faith in counter to proposed alteration by any means, then even if some were motivated by profit, they had two masters in attendance, national politics first, Roman Catholicism second?

From the junction of two societies of old Europe and new worlds, nothing could be seen the same, which by dour consequences bore heaver on the new more than the old. If from the spirit of adventure the Spanish conquistador came to reap the rewards of gain, they were assisted by the un-imaginable. They found these new peoples well able to cope in the management of civil and political community, that they had established cities far better organised than any others in Europe at the time. By extent of population the new world outnumbered the whole of Europe, and like Europe had better developed different national structures. By the level of civic pride belonging to most of the new societies they outclassed Europe having the world's largest active building structures. Built like the stepped pyramids of ancient Egypt though unconnected by adoption or use, they did not have metal tools to work stone or the wheel to move it or beasts of burden to further assist. Above all of the fantastic things to meet the conquistador at ever juncture of discovery, was how these peoples had managed to work the precious metals of silver and gold, which now became the future motivation of all decisions just to acquire those element?.

For the Spanish conquerors to focus on mere metal was set by standards of developed trends for individuals from political leads in Europe, whereby self gain in conjunction with the arms of progress were leading us above spiritual requirement. On the front of spiritual concern by inquisitional trends, the Spanish had no difficult decisions to make by religious tenet if considered by some European reasoning. By no connection with the God of Christ, these new people who practised the barbaric act of human sacrifice on a monumental scale were open to forced conversion or destruction, which ever was the easier? Unfortunately it was mostly destruction that followed and for centuries latter in the new world by the embodiment of European exchange in what we brought with us to the new world and other places. In the first exchanges we brought the marvel of steel weapons, of the horse, mighty beasts that tarried to our every whim. Of the unseen, we carried in our own bodies the destructive force of continents. We carried in minutia, viral diseases which decimated millions of people in the new world without the built in ability to fight such pestilence.

By the standards of the day, the early fifteen hundreds, Spanish conduct was not very much different than any of the other societies in which we had previous contacts of trade or war or religion. Without attempt I will not soften the matter of overt cruelty and brutality by how the conquistador were our representatives, for similarities are always almost the same. By other reasons and reasoning the unexplored mystery of the new world whenever, was truly astounding with many separate aspects taken out of context or incorrectly overlaid by standards and styles completely unconnected by measure at that time. In order for myself to visualise the thinking of the day, I hold that we have the representative presentation of how Europe and the Middle East had so far run. I hold how the connected separate religions could each force reaction from one another. I hold how the structure of politics and society could have maintained self to be the drive to all change that we have now obtained in its own progressive forceful standing. Then even further to my own taste I hold that through our connections by the Middle East, India, China, Africa and Europe all can reach Shanidar and still, all can connect to the future by the past!

When Spain who were generally first, came to the new world, how could they have related what they found to what was actually there? Any half measured approach then could only be counted that we only knew half as much at that time than from now where we already operate half measured knowing the all. If Shanidar could be knitted into our pasts then it would have the same connection to most of the people of the new world.

I have indicated that through our first mystery from tribal death one hundred thousand years ago, we marvelled in query of our own ability to still relish the comfort of our lost companion. This would not have needed to change even by the sacrifice of our own person unless our lead had been inspired by other broad factors. Not least to have breeched one of our three taken prospects from the Fifth Dimension being contrary to 'Natural Body Death'. If bloody acts induce bloody reaction and counter claim, then no balance can be put against any party, leastways not five hundred years ago when the old world met the new world for at least the second time!?

## CHAPTER 31

Our own fourth dimension though its avenues of learning from which some have been able to hold scorn to some religious opinion, might relent and assist us to other outrageous observations. We do know by processes of natural migration owing to our own specie behaviour that our conduct by unrelated moves had surpassed other specie. In most cases they followed or moved to seasonal determinations, berry or nut harvesting from the best availability of supply. Over very long periods of time, but parallel with many who maintained the seasonal habit, our own forbearers when faced with the split of the clan or family group were set to move, which we did; and on and on. Breaking the rut and finding ourselves always having to continue and move even eventually to break the habits of seasonal feeding. Which gave us the open ability to move and adapt how and when we needed outside the requisites of evolutionary demand.

All travelling was done without written finality, but accepted because in a trickle we did pour out of Africa by the only means of travel we had for several hundreds of thousands of years, walking and learning? Without known histories, time is sometimes irrelevant, but not now when we know of its mark at Shanidar Northern Iraq. Since the burial had occurred we still had time to make our way the furthest East by land that we could, then even further by the then land ice bridge to North America. Once there, from any time span between forty to twenty thousand years ago, we continued to percolate southward until the two unknown continents had human habitation. Much later our best recognised settlements in modern day Mexico and others extending south through the isthmus joining the two great continents. Came to be of European interest from about five hundred years ago, from where we could exploit the different tribes in us, but not seen until now through the Fifth Dimension!

Checking comparisons reveals how we were prone to operate half measured from then ago leading into modern times, when we half tried to correct some of our own blundering. In the first instant when countering those who practiced human sacrifice, we assumed all people on these new continents were of a low standing compared to our own high European standards! With that fixed focus we were unable to see or understand that there were two distinctive religious outlooks by the vastly different peoples there, though all were from the same issue as we, 'Out of Africa'.

Looking to the Northern part of the northern continent, most of the land north of modern day Mexico was given to a huge number of tribal groups first misnamed as red Indians, but who interacted in the most intricate of ways. Perhaps their greatest mark on all histories is that they had developed their own deep religious concept beyond the recognition of those in Europe. Were they monotheistic that they gave most reason to nature and the elements through the medium of the Great Spirit? People who had associated with their surroundings to such an extent that they were connected with the eventual unity of all people confirmed by 'Natural Body Death', but under their own terms!

By our own reflective notice of now, to cast our eye to the style of some religions including the Great Spirit doctrine, if I have mentioned my declared difference between our Eastern and Western religions, one monotheistic and the other perpetual. Then they are somewhat straddled by any concept to take in that we are all part of the whole, affecting unity through death with little challenge because it is so final. Questions I have asked now are not fairly relevant to our own good ancestors, for we can think in hindsight. But it is that we must make the connection between our misnamed North American Indians and the burial rite at Shanidar. It is important to make the association these peoples of our same single specie were disposed to enhance that experience between man and God, also through death, but not hailed in the mystery of the unknown when the conclusion is that we were to be returned in self form just like from creation carried by them in the Great Spirit?

Hindsight can bear for us the above reasoning, but from any complexity it cannot change old circumstances to be modern, yet all above fits in connection through the Fifth Dimension. 'Out of Africa', 'Past Gods Words' and 'Natural body death' three taken prospects with one active to be worked 'One for the Whole' hold to all people in the same standard regardless. At this time without conclusion I can state that no religious concepts would not be able and fully adopt the stated prospects, because the prospects are naturally inclusive. Religious human sacrifice is broken by two prospects and therefore supported by the other two. 'Natural Body Death' defies human sacrifice and 'One for the Whole' destroys the concept even in thought. Therefore without the mixture of the past and the present it is that we must again look to five hundred years ago when the conquistador first found the practice of human sacrifice but without the weight of hindsight.

When the Spanish came to the Aztec empire and were stunned in greed by what they saw, any implication of human sacrifice was to be no different than what they already carried with them through the inquisition in any case. Any defining feature was locked into the actual philosophy of the people entwined by fatalistic outcome, outcome, not outlook. The one biding question was not even offered in comparison to the standard of one creed from another. No connection was made then or had been successfully worked since to how the Aztec Empire and others had copied human sacrifice while in primary worship of the sun in transit by day and the stars in movement at night?

By association and connection we might piece together that because of the ancient land ice bridge between the continents of Asia and the Americas the one human flow followed. That is not to say we all came as one group to split into our own organisation when we came to be aware that as we moved further south into the vastness the conditions of climate and geography set new acceptances? In the mould of hunter gatherer other factors in the name of ready supply aided our clannish and or tribal splitting until by the span of those twenty thousand years we were presentable to the new travellers arriving from Europe five hundred years ago, the rest is not history. Because it's making was done not from human considerations but by the prospect of gain, wealth and more gain on all counts by European thinking and for centuries after. By the recording of our history from one stand the measure of balance was lost by the other, but not in changing history only ignoring it.

Europeans sitting on our new gold and silver mountains could easily extract the justification of our actions on those who had worked the metals generally within terms of their religious beliefs. Human sacrifice seemed to be the corner stone of all directed worship to the mighty life giving Sun God. Although we were never there then, we can work between some of the written text in the diaries and notes of adventurers, explorers and exploiters from those times ago. First associations were always to human sacrifice to have always been human slaughter. Erroneously used to support the claim that we conquerors were in fact liberators, hence any such action or reactive response by the new arrivals was justified by the higher divide of Mother Church, if! Fortuitously missing the connection that if the liberty of the people was assured in the stopping of this barbaric rite, if we were to spare our brother humans, how could we set them into the liberty of slavery, usury and abuse? And still not reason of their condition in the prime instant!

To my own personal reasoning, plus for our sake, I have laid the ground that there were two styles of religious acceptance in the Americas at the time of the conquistador. I would also state that by first contact and to its conditions; the fate of the two continents were sealed by we of Europe without being charged, for we were locked into our own blind driven ignorance of greed without the expansive knowledge of hindsight to then ago.

Looking first to the Aztecs and trolling the isthmus into other South American empires, no real connection was made to the belief of the Sun worship culture and of its history. Especially when through some of the writings by early travellers we can read of these people being trusting and expectant, how was this so, if not from the depth of reasoning already accrued? From then to now tantalising theories have emerged of which I will view on my own account. Not in the method of fact, but of the possible, not least to account for how the same group of people from Asia had gathered two diametrically opposed religious philosophies connected by the same migration from Africa to Shanidar to the Americas.

## CHAPTER 32

Sacrifice is not the leading word or the sun or the two together, unless added to the night sky, which aid to complete the cycle in the two latter, sun and night sky, producing the former, sacrifice! Sun rise indicated a beginning night fall foretold an ending, by association the two led to the third by medium that any sacrifice, which was an offering to the ultimate of belief religiously, was allowed. Our rising sun was an open sign to the meaning of life in general when every day it gave witness to that fact? Having suffered what the night sky from the predicted movement of the stars indicated the ending to our society, if? What else to do unless offer the best sacrifice that we might to the life giving power of the sun? It gave life in every aspect including us, therefore by return might not we reverse what was told by the night sky and pay the sun its due, even by human sacrifice!

I have indicated on our behalf that there were two religious attitudes between most of the peoples in the new world, where the Great Spirit theme was near monotheistic and had plausible links with most other religions elsewhere, even if through Shanidar? My next indication stands to another link which could have been from any time over the last five thousand years. Even unproved, I utter without authority or sentiment, what I speak of is in defiance to the fourth dimension. I do not need the proof of experiment to mitigate my thoughts, we are all at liberty to suppose, which is a medium we could better use to share our own hopes fears and understanding.

I have used Shanidar which is the burial place in northern Iraq, first to show signs of aftercare in human death, simply to indicate our collective unity and as the focal point to be the nerve centre of our religious thinking. Close by, but some ninety thousand years later in Mesopotamia, by historical fact, I stand that measurable human organisation began to steer us to the development of society as we might know it even today. Of the same area by geographical connection from the river Nile delta in North Africa the Middle East; likewise for us; the Egyptian nation showed on our behalf massive signs of advancement of our tribal standing. Including with intricate designed aspect to purpose in religious terms!

By varying degrees, through the study of astrology they began to form in structure the image of the sun being crucial to the meaning of life. Under time scales to have eclipsed the normal span of any individual lifecycle, the general theme was to the King the Pharaoh, being the father of the nation linked with the sun itself. That the physical form of the sun in the heavens was also to be the direct giver of all life, even human life. It controlled the seasonal flush of the mighty Nile River, bringing regeneration to all by the bounty of the harvest. Which further induced the trend that the whole Middle Eastern area was in much religious activity in the sense that all were in the search for total reason like the Egyptians!

It can be accepted that the Egyptians had come to the monotheistic plan first, but reasoned of it with an object with no human characteristics? In the normal error of human impatience this plan worked well at its time, but in some cases its biggest weakness was from its management by its own priesthood. Working that the Pharaoh was to be the life giver of his people encapsulating the power and force of the sun into the authority of his being was acceptable until the style began to compromise other thoughts of those in the management of the plan, the priesthood? Who might not complain of the plan in working, but were ready to support others who might be King by the advent of normal timescales even politically?

If I would infer that by regal descent the ruling houses of Egypt gave the extreme picture that from the King all said was all that was to be done no matter what the consequences, for here stands the giver of life to the nation in the form of the sun god incarnate through the blessing of the sun. If this was the case, it was to its own result, for when we have looked into the past we automatically crown the events of then with interpretations of today. Following, if there was a Pharaoh who so upset his people that the next in line to the thrown set to obliterate his memory from all history. His images were to be systematically eradicated from all monuments, writings and recording from human history, what of that errant Pharaohs priesthood? This procedure failed, for we do have a cloudy record that there was indeed splits in the royal Egyptian house of old, what is not certain is the timescale, which if between three and five thousand years ago allows more for the possibility that at least parts of an errant priesthood might have seen the need to move or flee to maintain there own worthiness and or belief in the sun god theory!

I do not suppose that you are to draw some of the conclusions that I make, for I am simply tying to repair the shattered glass jar of our joined histories into a reasonable shape that we can all recognise. Because of the activity in the Middle East from ten thousand to fourteen hundred years ago we can all now come to a core belief of how we had or might have been able to perceive our religious structures full measured. Eastern, Western, Monotheistic or perpetual all run in connection to Shanidar. Egyptian sun worship through astronomical connection can also be steered in that direction when seen connected to the same monotheistic flow. Where some difficulty has fallen to remain upon my own shoulder is asked by events in the new world because of the anomalies found there?

Given to most acceptances the continents of North and South America had only been populated by this our single human specie from any time between forty to twenty thousand years ago. Then the people also can be associated to events at Shanidar, particularly when connected to some of the burial rites shown in the new world. It would be hard to see how the Great Spirit philosophy could be compromised by other beliefs if that was the root of the only belief to have travelled across the ice land bridge between two great continents! Right or wrong to imagine how we did develop the aspect of human sacrifice from the Great Spirit philosophy needs important study outside what the fourth dimension can deliver?

One of many failings we took with us to the new world at first contact again, has been left open because of uncertainty over our direct historical link over any period of time from five thousand years until five hundred years ago? We had been unable to associate how we had more than cultural similarities between two continents, Europe of old, and this vast land which was to become two continents yet to be named? Did we notice the general structure of these different societies, not only different to us but each other also, yet all remained the same in that we all aspired? Did we not make it fortunate that because of our own standards of acceptance from our current European drive politically, which was commercially, and religiously which was styled by reformation? Our judgement became the only one that mattered, because at the same time of our efforts of discovery we were met by the strangest of all mysteries that beyond all reasoning, we the visitor, were somehow expected!?

Without forfeit my own submission of now, but not yet in relation to human sacrifice, is that five hundred years does not represent our first contact again with the new world. Also without forfeit I would not make argument against earlier Viking voyages reaching Newfoundland or other landfall elsewhere on the northern continent. Where I delve in the possible, is of travelling done at any period from three to five thousand years ago especially by discontented Egyptians priests!

What If there had been circumstances whereby there came about violent realignment at the leading end of Egyptian society with the royal faction in dispute, with there own following taking sides, if to the victor the spoils, what of the vanquished? Perhaps in calling on what always had been part of human instinct, in order to survive, the deposed group might up root, even flee. Perhaps in keeping with already practiced trends, if that group was to remain priestly led, might they look for divine inspiration of what to do next?

Looking to the heavens for Godly inspiration, is it possible for those who would flee that the calling was to always follow the setting sun ever westward? Even if it meant following the North African coast until at its extremity west, all that could be seen is the setting sun dipping into a mighty ocean. If to follow the sun on crafted boats brought them to eventual landfall at the isthmus of the Americas, what now for the future? Even if to have taken many, many generations of travelling almost like wandering in the wilderness. If and when these deposed people had come to new lands why would they not try and impose there own standards not least because they in their own eyes had been delivered, it was time to make sacrifice to the giver of all life our guiding sun, but now, in who's land, but ours!

If taken that the American continents were first populated by us through the ice land link between the continents of Asia and North America as late as twenty thousand years ago. Then at the other side of the world when Egypt took to the sun God philosophy some five thousand years ago there is no reasonable connection between the two definitions. Unless the Vikings and Christopher Columbus were pre-empted by a collection of disgruntled threatened or banished people? Peoples living in the face of changed society who might have been eradicated from the history of that society. People to be cast aside in the shame of their adopted belief, even forced by that banishment or flight to follow the remnants of daylight to the setting sun westward!?

What ever the story is I am committed to make reference to certain matters ex to the structured reasoning of the fourth dimension, which is constricted by formula or experiment to the proof of substance. My theorising is about the shape of human conduct adopting half measured working or theory and turning it into our steps taken syndrome. Not unlike how we might have worked Judaism, Christianity or Islam under the same misinterpretation. Or allowing how Hinduism and Buddhism have taken to their own discovery in some cases by unconnected means. Our simple reasoning of now stands to the suggestion, not of cross Atlantic Ocean travels by ancient Egyptians. But of how two peoples worlds apart came to the same understanding of the astounding notion that it was the sun to have been the direct giver of human life? Not the supporter or supplying too, but the embodiment of the human form, and further, that by its resolve through the same daily pattern, pulled the night sky into view so that it might depict in prophesy mans determination?

Along with other theories of sun worship, either as the head god similar to the monotheistic religion of Egypt long centuries past, who were able to promote their worth shown by their prowess in battle in the fringe areas of the Middle East. Being lands where our first settlements began to abound, lands from where the collection of ideas were percolating to other tribes and clans in proximity also with expanding outlooks. To become not unconnected from how any tribe or tribes could study the success of the Egyptians, even thought their religious concept was considered wanting if by confusion? With all beliefs generally led through the Pharaoh the King being responsible for his people, even to their births in conjunction with the gods and the sun God in particular with its own mysteries!

From the mixture of the sun God and the King combination, little imagination in the hotbed of human query was required for the connection that the king and sun God were one and the same. Not by Egyptians for their own standing was virtually unmoved by their obvious successes reminding them at all times by their trophies of excellence through their fine building, monuments and battling? Others might have been able to recognise that as the king was actually mortal, then he was human. Perhaps creating in that query why the sun god might pick a mere man against other creatures to carry its light, unless by Egyptian excellence of mind but error of thought they misaligned primary thinking? Unless they did misalign primary thinking to the detriment of the king, where he was man the sun god was representative to man, therefore the God was as man. All representations made by the sun God to Pharaohs in line by any determination as man were made not by a hot circular disc in the sky, but God in the image of man to man!?

It could be half measured if we use the above paragraph to form firm convictions about the possible. What has been written has not been done in deference to fact but again to show how modern thinking might be used to blur the past when we have the advantage of hindsight. I do make the point that from the extended area of Mesopotamia into the Middle East from ten thousand to four thousand years ago, any thought processes were possible and happened in compliance to connected thinking. Adding that with further connection we were vibrant and competitive not only by battle and commerce, but in collective thought, that is how we are by being human and of this single specie!

For myself and without announcing claim to any earlier reasoning, on turning to the middle Americas some five hundred years ago, Europeans had met a culture who practiced human sacrifice, or so it would have seemed? Men as we, ravaged by greed saw this abomination and determined its eradication, if not first by the sword then by church sanction through inquisition beyond the aspect of tribulation as the word guilty spoke in the names of pagan and barbaric. I speak in those terms because of the after effect to have settled upon all colonising Europeans since then in altering attitudes even up to very recent times. Alas the unnoticed was perhaps the most important in helping to cement future European conduct.

Of the many cultures in the American continents, by our best understanding they were from one people the same, who had travelled the land bridge from Asia. Having split by tribe and group during our respective settlement over the whole forty thousand years we seemed to adopt two religious outlooks. If the one was to the Great Spirit in connection to us all by the avenue of nature and natural progression, then of its own shape it was also monotheistic in type. What ever range of thinking we might have to the act of human sacrifice then ago, it should not stand on deed but to expectation. If we can think by that time we must also supply reason. I have shown the possibility of connection from Egyptian sea travels some thousands of years before Christopher Columbus. By events which can be settled by choice past the exactitude of proof when reasoned that both societies, ancient Egyptians and some of the new world worked to the acceptance of the sun in divinity. Although ritual human sacrifice on mass was never carried out by the ancient Egyptians. Why then in question did people carry out human sacrifice on the new side of the Atlantic Ocean only?

For me to have considered supposed travels by disaffected Egyptian priests and group, then we must clear further ground to make all matters viable. The first point we have to reckon is that the isthmus extending north and south to the two continents were already inhabited. Although the flow of migration that had led some of us 'Out of Africa' originally took us hundreds of thousands of years to pace. Now, then in our well organised tribal and clan societies we could make faster and deliberate decisions to move when ever in need or desire and by the same natural progression we settled to our own desire accommodated by the circumstances our climate and geography gave to us.

At any time on the great new continents up until three or four thousand years ago, would have seen most of us settled by the convenience of our situations and interacting with each other like peoples everywhere. Although not quite on the same level because internal competition was belayed by the very vastness of areas we now inhabited. There was less of the competitive drive that might have occurred in the lands between two rivers being smaller in size. Here without the understanding of the need to change or with less to learn from each other because we were much the same in any case. What then if strange new peoples arrived as bedraggled and exhausted travellers from where the sun would rise daily? If to have naturally neutralised the possibility of any threat, might not the existing peoples there listen to amazing stories from strange lands and of strange beliefs?

What we might never know is not of the travellers but of the peoples they first encountered. What say would they have by giving aid and sustenance to these arrivals already knowing they were not of any other tribes to the locality that they had encountered? Who also by style presented themselves in the flamboyance of expected outcome under the protection of their ultimate God in the sun!

Without grounds or grains of experimental proof I now offer my own opinion of how human sacrifice was to come about, not to be a final supposition. Not for sensationalism, for my own tenet is fed by the Fifth Dimension through its three taken prospects and one active in 'One for the Whole'. If in the Americas all peoples had followed the open philosophy of the Great Spirit tied to the form of natural occurrences. What could be more natural than the sun, especially if told of its leading importance in what it was ever responsible for, then why not honour it in its own right, similar to how we regarded the Great Spirit?

A meeting of differing philosophies in the Americas of only five hundred years ago would have to have followed certain steps of initiation, on the comparative grounds of how others on the continent presented themselves to us European outsiders some short time after the Spanish conquistador. When other European explorers came to the Americas their portion of discovery by the English, Dutch and French was mainly on the northern continent first. Separate discoveries to large portions of the southern continent were added to by the Portuguese. All of us travellers looking for discovery to our own best advantage. Finding placid accepting people on our initial drive ready to welcome us clouded our objectives, yet it was we who closed all portholes of insight into the how and the why of their condition. We saw acceptance we saw human sacrifice we saw gain, profit and enrichment, and worst of all we fully missed the connection that they were as we in being of the one specie. But what matter if plunder was to compensate for our own shortcomings.

To another time scale, if displaced peoples in the form of a deposed priesthood after harrowing journeys due to banishment or from flight, came upon people who had affected their own philosophies to suit where they now lived. Much the same could have been expected of these travellers from the Middle East, who having made journeys over long periods of time and vast distances, had the impetus to maintain their status quo of belief.

If to find that this welcoming body of people had a God culture also, then in similar style to the exchange of ideas that had occurred in the Middle East previously, why not use the same reasoning to influence such people? But now with the refinements we had considered while in those long years of wandering. Having reasoned that all societies must have forms of leadership just like our own, but to belay any further leadership or kingly disputes which again could lead to the persecution we were to know. Accept this society by endorsing their leaders, but only to the working element of having and maintaining a vibrant priesthood as primary administrators!

For me to have suggested that the ancient Egyptians might have sailed the Atlantic Ocean three or four thousand years ago is truly incredible. From first indication I had made connection with referral to a deposed priesthood with other referrals to sun worship. If to indicate that this was how the peoples of the new world were to come to the God sun and all attendant aspects of it. Other questions without the review of human sacrifice stand for the same consideration. Perhaps not so obvious in terms of the written word is that the Egyptians had a very well developed vocabulary of the pen the brush or reed, depicted through the written word of hieroglyphics. Also they had there own dress style, of the two features, comparisons can readily be made between the old world and the new? It cannot be said that everything was replicated from one to the other because there are considerable anomalies, not least to the end result of how or why human sacrifice came to be a major feature!

## CHAPTER 33

El conquistador showed us by terms of semi recent history that most peoples of the new world were trusting and in the amazement to what these travellers had brought in the aspect of differences to what they had so far encountered or had developed in their own pace of need. Some were fuelled with the incredible expectation in the reality of someone, some form of leadership, or deliverance was to be expected from where the sun continuously rose in the eastern sky? As I have already made that connection by the Egyptian story of its own time, I can set justification in how human sacrifice came about but only by my own study in the abstract.

When exhausted travellers landed long pre Columbus and were met by tentative reception, for the host the time to certain realisations would not be long in coming after all gratitude was exchanged. By any period of time of from months to years these Egyptian travellers had moved continually westward along the North African coast, if not pursed then in fear. The reason was on religious grounds to save the true relationship between the God King and the Sun God. After the rigours of monumental sea voyages, they found in semi isolation peoples who were open to god but by the style of acceptance to the Great Spirit? Where all standards were of the general aim to live part of the whole then from death enter to the understanding of it, but now, how long might the meeting between two religious philosophies last before one is to supplant the other on the grounds of fortitude? Given that one seemed somewhat inactive even laid back, while the other had been endorsed by the reason of deliverance enhancing all base structure of their faith and new ideas!

Even from the abstract, examination can be fair, explaining that the end justifies the means. I do not use that example for my own reasoning, but I take the stand to the reality of now because we are all here. Talking of this, those possible reasons to any aspect of our history will be seen to show how some events or happenings would later cause reaction and reaction from the very same outlook, but by different interpretations? From which I submit have been caused by ourselves in transplanting our methods of thinking of and from various time slots of history onto today's plan, leading to blame each other because of our natural trait with the developed ability to know understanding of our selves! Not least by example, although we had worked and moved by half measured reasoning to cause the Second World War, our best reconstruction plans by the authorisation of individual human rights politically to make amends, were half measured once more!

Four thousand years ago in the central Americas was really too early for us to be half measured religiously, for none of us had yet reached the broad network of understanding that we were all of the same being 'Out of Africa'. Although if we did eventually have two different religious philosophies interacting, how could we have been taken from our Great Spirit belief in acceptance, to one of human sacrifice in action? If never to know, even after prolonged European colonisation and immigration, because too much of what could have been evidence of Egyptian travel had ever been lost or discarded?

Organisation would not have been the key how Egyptian travellers were to change attitudes at journeys end, experience was the pointer. Egyptian priests had been the product of prolonged interaction between many different tribal or city states or nationals from lands now known as the Middle East. With a very broad base if interactive experiences from whatever means, their own desire and resolve would or could be cemented to their invulnerability as perceived. They did not arrive at their destination by chance, when every day the sun rose behind them and set to the direction of where they must aim. By quick succession, countering helpful relief on meeting these new peoples, everything by normal interaction was restored. We might now have been back in the Middle East living by the edge of competitiveness to our own best advantage? Taking and working our own ideas with the best product of other peoples and mixing them to fix our own interpretation as the most viable? From our own station on religious terms, we had many channels to filter because of the number of tribes with their own special gods. Although we held them by the best service they had shown to us, we were apt to gather some of what others had selected also to our own advantage maintaining always the groundwork in searching for the ultimate god.

For mixed reasons if Egyptian travellers had made such journeys then they had travelled with the final belief of there selection to the ultimate force, now endorsed by their safe arrival and further so by the people there to greet them? Those people who had not had the experience of complex multi god societies interacting by religious means, but readily accepted what was the obvious by determination into the realm of the Great Spirit. People content by their own standard and to the irregular interaction with other tribes elsewhere on the continent. Who also seemed of the same religious determination, there had been no drive to continuously search for religious fulfilment, which was already there to all in similar likeness through the Great Spirit image, unless now others might submit another plan?

Hailing to the Great Spirit in concept is a real form of divinity, more so when coupled to the total meaning of all things intertwined to be of the whole. If there was a simmering doubt then it could come from the imagination, perhaps not understanding that in taking the whole to be all creatures, all life forms, where did the idea come from? If to the Egyptian priesthood that enterprise was likely to compromise their position eventually, then it might be worth directing full focus to the essence of what the Great Spirit really is? By direct connection in association from their travels, the Great Spirit was in fact the sun. That which by daily sign gave charge to our very existence by it vital support?

In order for any change to be made without direct conflict, which might not have been an option, forms of coercion could have been used but by very well devised methods to sway people entrenched in their own beliefs. From first contact, if old travellers to new worlds were of mind to notice, that to those who essentially were their saviours and held comfort to the full story of life through the Great Spirit. Then in just opportunity this was the meaning of their own struggle from the time of banishment or flight. Now was the selected time of their payment in reward for their suffering. By the very character of these people in open handed welcome, our own good fortune of destiny is to be there from our determination, led by the fact of our deliverance, the Sun God?

It would of course be now necessary for us to lay the real truth before our hosts including the subtitles of change we had made on our travels to this new beginning. It was we as priests who were the real determiners of interpretation. Although we had a great king, it was he who drew embittered resistance to our plan by his own delivery, taking in that through the giver of all life the sun, by its own daily cycle, it was of his service! He took his own power in translation of his service to his people, forgetting we as the serving priests were for the harmony of the Sun God and the God King. We were appointed servants to the will of the day star and were there to translate the night sky.

For any method or style to any transposing that might have been done, care was taken not to alienate these peoples, done first by heeding and then leading or moulding the civil end of met communities to the explicit need of our redefined priesthood code. Priests, now able to carry as much care as normal or existing civil administrators, but with the addition to serve all with our own extra requirements, if not yet fully shown. On that basis their leader was always to be the leader, only form now his mantra was to be run directly from and by the priesthood. No more for the king to destroy the link of magical aura between priestly men and the vital Sun God. To have a king and king yes but also no longer to cast upon the memory of his people the value of his greatness over the sun God or serving priests!

What ever way to promote these or any plans would have been driven from the real circumstances already in place, which would have unfolded at a reasonable pace by our awareness of the situation. Meeting peoples with the God culture in the Great Spirit, yet without written text in explanation of conduct, how should this situation best be met? Like with like might be first option, not forcing upon these peoples our own hieroglyphics as a subject of the brush or pen. In order that we might mould these peoples by inclusion then let it be in the guidance of experience instead of how we had recorded our own fate to be removed by the whim of others. In relying on the written word dangers were, if the only learning was from them, then if lost to the mishap of fire or flood or pestilence or whim, where to turn next? Why not better construct the style of management to our own Sun God culture in the care of the generational weave? Holding in that it was for the priests to interpret all signs and indicators in relay from the Sun or its own agent in the night sky, to relay to all others for direct attendance howsoever, but always to be maintained by the priesthood!?

I readily accept of my own lead above, I do remind us that the effort is run from abstract reasoning, not by evasive means but to hamper those who might hold sanction in order of the fourth dimension who demand the avenues of proof. With the success of any adventure, the results might not represent the outcome. If I have set it for our Egyptian travellers to take the lead for the spiritual welfare of all, then most factors were in place for this to have been fact. In the world of the ambiguous Great Spirit, in the name of tomorrow, action might be turned by the plausible of today, to see the sun only? If we are beginning to connect to linked societies of ancient times then it should be seen that in as much as the one could offer comparisons, the other was in balance. Meaning both new world tribes and Egyptian travellers were locked into their own acceptances being quite similar in that both endings were of the innate, holding to something outside of the human experience? Without the collective exchanges known to have grown in the

Middle East, in the new world it would be less difficult here for new priesthoods to challenge old habits in order to change?

By my own synopsis, stating that the travellers of lore were better at imposing their will by the hidden design of organisation, keeping to the general structure of society through the royal or leadership house and cajoling others to also accept this. By overlaying the Great Spirit with the Sun God, by comparing both were to the natural attendance for the welfare of all, then further being able to attend to the sceptical by any means to apply. All actions of change only needed authority from the royal house, but now managed by priestly sanction? I further state my case in the abstract, that by coincidence, this meeting was done between closed societies where active reasoning had no opportunity to flourish. All said has not yet brought us to the aspect of human sacrifice, until by accepting the sun, now for the cost!

For the existing people of the new world, death was the result of time by all means, even through rare human conflict owing to the vastness of the landscape. If new travellers had formed their own death culture image it might have been pointed by some remembered events. They themselves had been forced into the self sacrifice of position of station of stature in holding to their belief of divinity. All features since had implanted the sacrifice theme in attendance to their own salvation by two counts. First through the very bounty that the sun actually generated, with the other in how it told of its intention by shifting the night sky over ages and ages to foretell expectation. All of which could be taken in line that from the past, many offerings were to be set against the sun for its bounty, for the continuance of tribe and clan and priesthood and society and all and all. Then what for the best sacrifice?

If intrepid travellers had undergone many hardships to find sanctuary, then to the total giver of life by existence, the sun, why not make best heed in best gratitude? Which by show to the connected bounty, in returning some of those that owed all to its existence by the ultimate sacrifice in man? Then continuing in style to further enhance the deed by more offerings, until the business was almost an accepted art form! I will say without condemnation the possible or probable or any other expression to describe the act of human sacrifice, must be connected to the time and circumstances of occurrence. It was to have been the mentor, in that the actions were under the permission of factors not realised to have been by our specie drive in the form of our own free will, which its self was not always taken on that individual platform, until after the Second World War!

Three or four thousand years ago if we came to accept the offering of human sacrifice, then we carried the act on mass in the new world until about five hundred years ago when we were forced to review the situation. By that review and stopping under force, nothing had become clear because those who led the change, visitors from Europe, sacrificed us or made us join in their catholic religion. Which told of their god in our human form who self sacrificed for us, but only if we were to be like Europeans religiously, with our, their God?

Had any others in the name of Islam first rediscovered America and found the same forms of human sacrifice, I venture that the total practice would also have been eliminated by many firm and final means deeds or actions, just like how the Catholics had? I leave it that the overall result, if carried on until today would have seen the eradication of human sacrifice in those far of lands, but I will not speculate in any fashion of how the world would now stand politically. My connection in relation to an Islamic overview to human sacrifice stands in parallel to the Christian aim, because both had violent connected histories on religious terms and each ended in supporting their own aims for life. From the wasteful endeavours of human sacrifice there could only be seen an ending with no beginning, plus other reasoning from the Christians who did eradicate it and the Muslims who would also have!

My collective speculation of how some acts were introduced in foreign lands like the new world by alien people like Europeans stands proved. If we look at the unknown history of our specie, tribal and clan movements throughout the world any time over the past forty thousand years. It is generally accepted that the flow was one way through the land ice bridge between north East Asia and North America if over several millennium. I had indicated that by circumstances these tribes altered and changed habit by their immediate situations, making the further assumption that of these small changes there was not any urgency in self promotion. Especially when there is the real possibility that each tribe carried their own view of the Great Spirit in all endings? Even if carried since Shanidar, though the meaning had long since been lost, the mystery was always vivid.

Unconnected by all options we have to assume that at some time in any period from forty thousand years until two or three thousand years ago. Tribes, who would eventually inhabit Mexico, the isthmus, and much of South America, could only have picked their way south from present day Alaska and Canada by driven intent. Past, but mixed with ideas of other tribes and clans holding the same basic philosophy owed to the Great Spirit, but in mystery, at the same time carrying the means to promote belief in the sun as an emblem of creation? Many pictures can be created from abstract thought, even to alter history without change. For anything we might not be sure of, it is that any of us might propose ideas in order to broker discussion but only in these past sixty plus years. Covered by taken prospects and one active in 'One for the Whole'. My own speculation of Egyptian travel is full measured; also it disallows any factor in the defence or promotion of the steps taken syndrome.

## CHAPTER 34

Christopher Columbus turned many keys and opened many doors by his travels, some good, bad, right or wrong. By first notice he broadly proved that the world was round. Many following observations were not quite as direct, even wrong, but best intentions always prevailed. When landing in the Caribbean thinking it to be part of India, the first peoples he encountered he named Indians, without intent, making a half measured judgement furthering the error to the steps taken syndrome. Later travellers compounded the error but not always for told reasons. In being Spanish, Portuguese, Dutch, French or English without proper navigational aids, each were prepared to maintain their sovereignty in the name of discovery. They did not confide in each other of specific discoveries even now branching to the north and south continents proper. But even so, half measured reasoning was continued because they still referred to all the inhabitants encountered on the continents as Indians from Columbus title, supporting the steps taken syndrome!

Any difference between Egyptian or European travellers is made because we had developed from the prevailing circumstances and the time difference. Those from the Middle East carried in desperation the culmination of religious fervour maintained never to change, by unknowingly supporting the steps taken syndrome even when unreasoned. In finding the new world holding to the Great Spirit, the Egyptian forced philosophy of the Sun God was never to be compromised in any time scale, until Columbus first, El conquistador second and Europe finally. It is to be remembered all three factions brought concept half measured, half run and only half done but to devastating effect following steps taken. These new influences had come from the zenith of religious interaction even if at first being Roman Catholic of Spain, they had originally come from places who had self developed three basic religious philosophies of the same and linked balance. Linked further by the factor that the Middle East was the core and original area for the meeting of new thinking religiously to eventually become monotheistic? But in the new world when found, even by the Spanish inquisition, there never had been awareness to previous Egyptian religious concepts having being carried elsewhere, which means that five hundred years ago no connection was made between sun god worship number one from Egypt and sun god worship number two in the New World!

Any exploitation of the new world since Columbus in historical terms, had now moved the focal point of our historical base from the land between two rivers and the covering area of the Middle East to any where in Europe where the action was? Any where in Europe by the next new deed of discovery, by the next deed of usury, by the next new deed of centralisation, misaligned our focus and funnelled concept to be political of Europe rather than 'Out of Africa'. A feature balanced by who was to record history, rather than who was to suffer the consequences of it, helped by being underwritten by the still dormant but rising feature of the fourth dimension and scientific discoveries?

It is fair to say that Christopher Columbus was the kingpin of European future exploration but not to be responsible for all future results. He was an adventure who took in tow the whole of the old world of Europe who followed his deeds and climbed on the band wagon of exploitation. All of us had set targets from the European perspective come what may, not only from direct gain where Columbus had been, but in our drive throughout the world prompted by revelations from the fourth dimension? From Galileo through to Martin Luther, of whom one sought to say that the world was round, and the other that God was more rounded? By all determination I now say that from about the year fifteen hundred, it was from being European that trends were set undetermined, but by determination. From within Europe with one borrowing from the other to prove any advancement; the trend of politics was growing, for that included science and commercialism which should have been the case, because our Roman religion was splitting to become national to the same God!

All the circumstances from the past were to be the pivot of European advancement. In place we had become the most competitive of any collection of tribal groups throughout the world, not least by the unseen conditions of the real circumstances like climate and geography. Other connections were set by the actual size of Europe itself. She was compact enough for the original city state of Rome to control most tribal groups. But not too large for the same tribal groups to split apart and never meet again even after the fall of the Roman Empire. All of which kept us competing even when nationalism was breaking out, each group by size power or for other reasons set to reclaim standards of historical connection pre Rome, but remained linked by Roman Catholicism. It was therefore by that guise that world influences began to be formed from the European temperament and to the desire of individual European states, but by unknown half measured reasoning?

All future interaction in the name of politics and religion was to flow in the same mix after the fall of Rome, even to future conquests made on behalf of different European states. Later on religious reformation was part of the engine to drive Europe on, but the fuel used was politics mixed with the fourth dimension. With the open war of Christianity taking place on the home continent itself, most results were carried to the new worlds of increasing discovery and exploitation. Over various time spans, half measured reasoning was the standard of delivery from all new lands treated from the respective standing of separate European states. Some to be self confessed catholic in denomination with others to be reformed protesters as new Christians, interwoven political and religious regimes working uncontrolled to our steps taken syndrome, unknowingly!

Because of the Catholic reformation in Europe the steps taken syndrome was in full promotion from about the time of Martin Luther, yet in parody he was unconnected. His aims from the abstract were of high political reasoning. If he wanted to say I am German and catholic it was done. What he inadvertently promoted was a deep cut into political reasoning when all people could now be catholic, and or protestant and national by that division, which brought us to use politics only for fuel to drive change topped up by the fourth dimension. Science by one of its veins in the study of improvement came to prominence in the mix of European national, religious and political reforming. From necessity it was to assist one or another of the split national factions, but not all religiously, because religious change was not driven by modernity? While showing how we could better forge steel or build ships to sail the oceans we spiralled the aspect of discovery in the art of science to better drive the discovery of new resources in general.

By the ongoing timescale mixed to some of the above events, situations in parallel to many countries of Europe were that some had split their Catholic Doctrine of Rome to be with the same Christ the Son of god in man, but now to be Christians nationalistically. Even if in due time the ruling body of any nation took to revert to previous religions on various grounds. This resulted in bloody internal religious and political upheaval, very often of a bestial nature by murder and mayhem, supported or rebuked by our steps taken syndrome. Being a matter where we follow first choices of change or adjustment doggedly, without going back to see if the very minor changes needed to implement that chance could not have been done by other means before following steps taken. From being half measured by approach, Christian by division inclusive of our steps taken syndrome, we Europeans took our blemishes to each other and carried them in practice throughout the world. Not yet to be challenged until we had perfected the art of human slaughter by the deaths of some fifty five million people of this our single specie, from direct European commitment to the Second World War ending some sixty plus years ago.

## CHAPTER 35

In this new era of change some way supported by the fourth dimension, from what might be termed expansionism, European nations in the ascendancy conducted all manner of change by internal national means deemed necessary. I have indicated that much of it was done on the religious front, even if to decimate the internal workings of some countries by there enclosed strife. Outward appearances allowed that for the reason of conquest or colonisation or exploitation the matter of internal carnage was of no concern to those to be subjugated. By no more than the prevailing circumstances Europeans were coming to another point of our united histories. Though the greatest of dangers now, was that we were beginning to lead into the aspect of total specie separation in terms not yet to have been written about, we were taking to be naturally selected and selective. Politics was showing this and by other means our Catholic religion had survived in fracture to be active and Christian and progressive and futuristic, blessed with directed aid by the sciences!

Although science has always been, every aspect of discovery was for everybody, but generally taken on more quickly by some over others? Science gave in revelation when we understood its method, which was always there to be used, harnessed or left in abeyance until we understood how to use it. We had already learnt much, but an historical event is due reflection for us to remember Archimedes. He was of Greek extract from the times of city states some two hundred and eighty years B C, Greece had always delivered much in our education. By background Archimedes was in the study of form and reason connected to scientific fields. By some effort the turning over of stones gives its own results. His action at one time was in the field of volume and mass, he might have sought to understand the mass weight relationship of an odd shaped piece of stone or almost any object? As the story might be relayed, while getting into what we can only assume to be a type of portable bathing tub which was amply filled with water. He noticed that as one limb then another and another began to be immersed in the water, the liquid began to overspill from the tub. In his rampant excitement he began rushing hither and tether shouting "Eureka", "I've got it". What he had got was an incredible insight into how mass and volume could be forever calculated by his own lucky chance of spilling his bath water.

Fortunately Archimedes is also remembered for many incredible inventions and rightly so, but Eureka stands to us all showing the use of chance, for from his discovery then, many new avenues emerged to the recording of science. Greece was close enough to the Middle East to be included as the first European country in our sphere of natural development, for in her own right she offered much to world opinion in any case. Even that Archimedes studied much in Alexandria, a very fine city of Egypt, detracts nothing because he was one of us any way. Eureka has now become irreversibly linked with the expression, "yes, I've got it". "I understand", but I urge caution stating that it must only be tied to scientific matters, if used religiously in could belie our taken prospect 'Past Gods Words'.

By fact, the story of Archimedes and the incident of eureka are true, my relaying of it might be wanting, but I have recounted it to show how we may come to various discoveries on our own account. They do not have to be mind blowing, but levelled to our own control by our own judgement and driven by our own free will without plan. By his time of history Archimedes if linked to the Middle East by my own spurious connection, it was done because he was of his own time long before the second and third dimensional religions of the Middle East. Yet his eureka was to all for all times and still by scientific criteria. My take on that finds better recognition by how the Christian second dimension and the Islamic third dimensional religions related to science and its discoveries in style, but from there own different timescale and to there own judgement?

Before we look to cause and reason, positive acceptances must be acknowledged of the differences between European attitudes nationally and religiously in balance to the attitudes to those of the Middle East. Taken from the start of the fifteenth century at the pivotal time of new European discovery, this was done by people in the high state of competition. Whereas Middle East interaction was taken in all normalities by religious consensus first, illustrating without direct comparison how the two separate groups of our same specie had now taken to behave at a different pace of concern. Europeans might have been able to speed the pace in the movement of scientific discoveries to suit individual needs if to gain advantage at first over new discoveries and then each other. While those of us in the Middle East also had been relentless, but in being more settled by our own representation saw less hurry to advance the same scientific discoveries at the same pace as others, perhaps our overview had been better settled to our religious reasoning?

Another feature from Islamic thinking, is that the religion although newer, it kept to the same standards from the first writings by Mohammed. Formed to be the last addition to Catholicism from Judaism, by locality the religion of Islam itself might have been enough to settle its people as we into a fixed stupor. Maintaining us to have the final say on religious doctrine and added to whatever effect that comfort brought us, setting us to be naturally laid back in most attitudes for the time being? In the same speed that most new European scientific discoveries were made, we of the Middle East had also been handling some for centuries in any case. What had not affected us yet, was the speed that the Europeans had gathered that information while embroiled in religious reformation. Gathered and used it for or from its own complexities into forced advancement, but why!?

Archimedes had reached the Middle East before Christ and long before Islam, his eureka was pointed to all people and heartily taken by all, save at their own pace. With Islam the speed of acceptance was ever to be used, that it was also of science in general was expansive, therefore to wait was to be included. All of which means that at times we held different attitudes if led by our religions or politics. We all knew of innovation, but our characters had developed beyond our control through circumstances, we did manage them to our need and desire. Perhaps for the whole fourteen hundred years of the Islamic religion the attitude of the faithful was able to remain the same, because from within, they had never had serious forms of the reformation that the Roman Catholics had had for some to be now Christian.

Islam's reforming was done close to the time of its founder prophets death, when by two divisions it was settled to be Sunni or Shiah. Although the tragedy of conflict by each separate branch had played its own part in this relationship over the years, Muslims had almost nine hundred years to settle into running standards. Which I have indicated might help us accept that people of our single specie had become more settled in attitude or were more laid back in comparison with Christians fighting for the advantage on new land discovery, while infighting on religious grounds. Both communities in First and Second dimensional terms could always meet by equal religious attitude. Both by the fifteen hundreds were in equal balance by all means of measurement. Where the Christian move gained momentum was to the open fact of there own internal European nationalistic strife. Not on the grounds to be better or above any other religion, merely on political grounds in the name of discovery under the umbrella of gain and exploitation. Europe in the midst of self discovery used science to self advantage by hurrying the processes of competitive discovery. Islam at its own pace also accepted the new ideas from science, but was unable to make the competitive change between Sunni or Shiah on wholly religious grounds. Taken that both were full in the avenue of their own understanding, division was accepted by the terms of the split nine hundred years earlier, so there was no double shock wave to be spread in upsetting style like in Europe at its reformation? Islam was not looking for further reform just then, if any observation was taken of conduct elsewhere, then there pace was such to allow things to develop in the way they always had in life as before!

By most reasoning from times past, we have allowed ourselves to think we are fundamentally different, we are not. Yet another capital feature to enhance the actuality of some of our difference came upon us historically some two hundred years ago, brining us to the start of this modern era. Two European nations, France and England, found themselves fighting a navel battle collectively known as the battle of the Nile being near the delta of the river Nile in the Middle East.

Two Christian nations although at sea and at war, had taken conflict to lands of completely differing religious philosophies to their own, not on terms of religion, but politics, which I now use to illustrate how we had become disaffected to others of this our same specie without knowing the why, except through superiority.

That Europeans had taken self judgement to be separate from other peoples on different continents, then this brought us to the full pointed direction of half measured working endorsing the steps taken syndrome. This was to be our total charge that in this truly modern era after the Second World War ramifications, we continue to so interact. From any patch work we might have done in the name of human rights or to the aims of political correctness or religious tolerance since, we have almost irreversibly signed the contract of intractability with the only outlet of change by negative means. All of which point in certain fields to how some of our best errors had been made since that war by sometimes discarding religious intent with overbearing political formula. Leading me to connect our half measured approach we had come to adopt unknowingly, which by various undertones the most forthright is the fourth dimension. If from my own reasoning then it is to be recognised as the science medium over and above religious concerns, which endorses all other human attributes save religions proper. Science which runs to politics unspiritual, also by means of civil control in government, sociology, trade, commerce, study, invention and the despoiler of the mysteries. Can be now run in title to be connected to fourth dimensional reasoning, which we actually thrive on and by it, where by intention I have put it in that field, is so that we all have a working platform of the same recognition, but not to affect us if not desired.

## CHAPTER 36

Having put the three main Middle East religions in numerical order and including most others of the same intent, I had introduced the infamous fourth dimension by showing of its conceit to the aspect of open faith and belief. By the same indication I had put it to be linked to other aspects of our joint importance, though not necessarily for all. With the fourth dimension standing for science primarily, it serves us all politically, that has been the route of my connection. My last Dimension the Fifth is just that, the last, in order to contain further and further expansionism when all the ingredient for perfect human harmony are already here now, and have been since Shanidar.

What we have inadvertently done since Shanidar, including the misuse of the natural drive that has propelled us to the greatest of our realisations since, is overcomplicate everything in half measured fashion. Automatically employing the steps taken syndrome to account for previous half measured steps. I can warn for our consumption, that the essence of my meaning of both terms has been mentioned for how we have or had worked in the style of half measured reasoning? It had not been done in deliberate fashion but emerged because of all that we knew at a particular time therefore anything started could only be half measured until we knew more. But to support that move we argued that the steps taken were right!

In principle half measured was an attempt to do something, although our pace had quickened, unfortunately we always entrenched to what was done right or wrong? The measure of which will be better seen when we can relate to many decisions made in this modern era of the last two hundred years. Steps taken is a very dangerous precedent that we readily adopt in support to the condition of accepting the half measured style unnoticed. Another way to consider the negative influence of both syndrome and system is from the realisation that the very questions of change in need are being asked at this very time since the end of the Second World War? By the best of intentions our form of the human rights mandate giving equality to all, should have been in place if not centuries ago! This remedial act created in aftermath set the total confusion of intent that when in charter, we assumed all men were equal, the idea made from the settlement of victory, erred in concept? Not least when from the war zones of two continents, those who had taken their stance being superior over others of the same specie, like the Germans and Japanese claimed, were treated by political management only. In the assumption that now, under new political designs we could all mark the same standard. Even from the first draught of the concept of human equality maintained by human rights the method of our actual political outlook at the time virtually compromised the whole issue, which was purposeful half measured working, but not known?

Having adopted the first principal of human rights even under serious question, it was compromised in draught because it did not recognise the fact that although we were of the same specie, we were not all the same. Which was enforced by the circumstances of climate, geography and history by delivery? Therefore by updating our human rights mandate we executed it to our steps taken syndrome mode, being not quite right, but adopted because half measured steps had been taken in the first place. Therefore we must now run with the principle although wanting. Compromises of the concept ran in open confusion to how our thinking had actually run. For at the same time when we were introducing these new principles of fair conduct through political reasoning, we were marking the nature of the world in commercial terms to be of first, second or third world standing economically. Being an open invitation for some to hold to the view of being better than others, which are steps taken from half measured reasoning? Further steps taken to supportively hold us all to the same standard of human rights had been added by strange entities of no particular definition!

Political correctness, combined with the principal of religious tolerance, have been added in charter compiled in principle by the new standing United Nations, which is a grand gesture in expression of our humanity on solely political terms. I personally point in the critical, that those with the biggest say on all matters belonging to it are those of high economic standing and or those most powerful militarily and or those of the first world. All of which at this very time is fair and acceptable when the alternative might have been differently aligned and we would not have the same open forum!?

The Fifth dimension by the simplest of reasoning is to be our bridge between the world of two hundred years ago and the post war era up to fifty or sixty months ago, when it was devised to hold three taken prospects and one active in 'One for the whole'. Under all open invitations it is for all religions to take the prospects in concept, along with all political persuasions even from different perspectives. By suggesting that any political system should take the three taken prospects including the one 'Past Gods Words' is done so that no separation can be made. Even if a state is communistic or atheistic, by holding to the three taken prospects they can come in the correct uses of the one active in 'One for the Whole'! For a godless state by self admission to be asked and take a prospect of godly reference is not half measured reasoning, is not half measured by suggestion. If any state is of that inclination the idea is to be one of full inclusion, they would not ever have to make reference to God or Allah, for they could utter all 'Past Gods Words'.

To have connected in the realm of human conduct by political means only creates for us all the dilemma of definition in respect to human rights. For from within recent time, running from about the year nineteen hundred we stood in mood by another medium of our thinking, progression? With due regard I have mentioned the work of Charles Darwin in supporting the natural development of specie, I had also indicated that without intention he caused controversy, this I personally endorse on political grounds anew, but only politically. That Darwin was not responsible for how we interpreted his work will always carry with chastisement, for we have the personal choice of free will.

## CHAPTER 37

French and English war ships in sea battle of the coast of Egypt some two hundred plus years ago have been used by my self to indicate the endorsement of changing attitudes by some of us. We had two European nations in the continuing process of war even after their respective religious reformations, conducting matters of primarily political concern by their own judgement, without any consideration for where battles were fought except from the European perspective. By following all the assumptions that our development was primarily because we were or had accepted our own standard for being civilised, what can be judged by this conduct? I add to remind us, that looking back to particular incidents carrying the baggage of our present understanding and acceptances is not balanced and is half measured. When the thinking of the day had been set by the obtained and claimed standards of civilisation from which we set all criteria even religions. The master of future decisions in the name of civilisation, although to be claimed, is not of or for the whole!

English and French conflict of two hundred years ago was indicative of the behaviour for most of the different nationalities in Europe, who now carried on working politically. All done, even under support of the same religious concept albeit splintered from the Catholic image of universality to the Christian one of unity. Indeed now, or more precisely then, the show of political influence as the first lead in determination began to collect the momentum of order. The Christian religions of Europe were still extremely important, but had now taken second place behind some means of pre thinking from the political sphere. Which from design was to the processes of planning in the national interest from political perspectives while maintaining suitable deference to our religious doctrine? But could only work half measured when one was to continually compromise the essence of the other, then be sanctioned by the steps taken syndrome. It might be that a clearer study of events was instigated by Charles Darwin on the publication of a biological reference book?

Without any complication to his good name the memory of his great work has been carried by two interpretations, even if not pointedly linked. Evolution without being a set theme follows in following. This was beginning to be a principle accepted by other studies and in proof by the actual efforts of mankind in general. Previously expressed by farming methods, of careful seed selection and in the field of animal husbandry, it was seen in the course of time, better crop yields were being obtained. Healthier and bigger stock animals were also producing more milk or more meat on the hoof. Even by other means in the interest of sport, dogs, horses or birds of prey could be seen to improve by slight change in what was gaining note in being driven by evolutionary means!

When this process of change was pointed out by Darwin, it was read and understood how small birds could change their beak structure from need, also in principle most creatures could change from need. Until the pen in print began to refer to the same progressive nature for mankind, our link then without knowledge of the genetic code was to have been from the apes, based on the similarity of appearance and some known likened traits. If Darwin had been one hundred percent right in his assumptions, which are palatable, then the wave of concern that was created still holds ramifications, until we see creation is part of evolution!

Mankind from apes, how could this association be made when god had created all life to the style of his own desire and by particular choice to have created man in his own image? Not quarter was given for the reality to the scientific nature of Darwin's studies of the natural world, no understanding was taken by the point of reference that all life was subject to the same conditions of survival counted by various timescales. Coupled with his other observation in the same vein referring to natural selection, he created in the form of a smouldering theme a focal point open to any type of query. Natural selection referred to the specifics of outcome, that the end was the result of the means by misinterpretation, that the end was an endorsement of the chosen by their own effort in the form of being the end product. Some of us could relate to this by connection of our own ties of belonging to this our group without religious sanction, without sought or claimed unity by the natural link from tribal ties through the medium of God. Some of us over later periods of time could take it that evolution was in fact the means by which we are here now, no God, no creation and no purpose, but by natural selection which was the result by forming our particular group over the right length of time? Darwin cannot be rebuked for our shortcoming, although not finite, his work was purposeful in offering to us all cause and reason. Any religious reaction to his work was in the spirit of evolutionary inquiry in that our human forte has always been to ask questions from the answers given.

Darwin in real terms, had delivered to most of us, his "eureka", taken in the scientific context, his studies laid out formula and were of experimental terms to proved his theories in the most part. Any doubts can be squeezed between the lines to fit in the overall explanation of evolution by time forces well past religious acceptance. At the very time of his publication, England was his home base in the continent of Europe, which was at its most expansive time in practically all forward fields. Industry and enterprise was the driving force to commitment and on the growing of manufacturing capabilities, markets of trade needed protection. If this was by military means then it was from Europe at the same time that the complimentary armament industry began to hold it own force. Almost like evolutionary processes whereby driven need was the driving force to satisfy the end product of natural selection too/for and from the secular only.

I had stated and reiterate from my brand of abstract evaluation that there was indeed much concern to the suggestion that we of mankind had developed from the apes. Much vehement opposition was unbalanced but in the most part delivered from sincere concern at the open effrontery to God the Father as God the creator. In those real terms Darwin only spoke of timescales to cover the full expanse of other specie that we share this world with. It is to be paradoxical that the depth of his studies were done at the other side of the world in newly discovered lands of Australia, and in passing most other continents. By result from his voyages it was from Europe that most criticism was given. Again to show us how the importance of Europe was in full hold to most worlds affairs.

Darwin's ape theory unsettled on two counts, from the ridicules concept of the actual line, apes into men, nonsense, why it's like calling apples pears, well they do look similar, but everyone knows they are from different trees, they have different roots! More serious the challenge was to God. It was obvious we were here to God's design he made us just as he made the apes. If they wanted to swing in trees that was there concern, if other creatures changed their beaks or fur or teeth, that was their concern, for God made them all, God made all. Most importantly he made us to serve and honour him by charter, tenet and covenant for all time until his appointed time of judgement. Even at or since the time of the Catholic, Christian reformation in Europe centuries ago, no one has dared doubt god. It was only how the story had been told by man, that doubt was generated.

From all the condemnation Darwin might have drawn by his indication of the ape man linage, along with that possibility and along with mention of separate specie natural selection, the resound was of monumental proportions. From where evolution was to stall the directness of the god man relationship through creation, natural selection by study of content could literally dispense with God altogether.

I have made the connection between Darwin and Europe in its base form but not of coincidence. His opportunity at the time before he published in eighteen fifty nine was solely appointed from the European connection and fitting circumstances. Europe at the time was active and in the forefront of innovative creativity on practicably all levels. For a long time her peoples had wore their religions on the cuff only, politics had raised newer and bigger issues by its immediacy.

Although religions did hold the attention of all the political parties, they by conduct and from other obligations, set to expand the civil management of the ever growing populous. When times were hard through political shortfalls, religions were encouraged more so that we could draw comfort when other systems seemed to fail or falter, leastways for when we die! Where politics operated from its need in being run by its own priesthood, politicians, it set its standards to accommodate the long established acceptance in the formed structure of society and therefore helped diminish the importance of our religions. For Darwin to have indicated our succession from apes was a big dent to religious thinking and acceptances. For the connection in/of natural selection to apply, it was a big double plus to politics, which could not only tip the scales from the spiritual to the secular, but cloud our real genuine first insights, as it did?

Natural selection by inference was just that, the best or better or fittest of the pack, heard, pride or flock would be able to maintain best standards in need of the group. When mixed in metaphor by extension and taken to mean by implication it was from the strongest and fittest who were owed the success of all, different pictures emerge. If religions turned that all were equal even if only by death, now politics could say it is not chance or appointment to be the decider, but being classed by the distinction that the strongest survive by natural selection, meant that we had no need to touch any codes of any religion?

## CHAPTER 38

At the same period of time that Charles Darwin was collecting his data and formulating his style for publication, the world was active in terms of human progression everywhere. From some of which is due balanced recognition later, not to be side stepped but held in abeyance until it fits the frame? My connection with Europe remain pivotal because of the sense of urgency injected mid nineteenth century between her and the emerging north American country of America now known as the United States of America.

America by construction was little Europe with the capacity to show bigger ideas. She had been formed mainly by settlement from the countries of the old world to eventually be able and commit her own approach to all decisions. Specifically at the precise time of Darwin's publication America and mainly England had traded in the commodity of cotton which was of serious concern to both countries. Because America's civil war was fought on political terms between the same Christian factions in her northern and southern states. Amongst other trade the raw cotton commodity was stopped from being exported to England, with the immediate effect of closing down all the cotton mills where the raw material was turned into the finished product of woven cloth. In purely political terms this simply highlighted how the secular had lost reason of control by not being able to support the very people it represented. When cotton workers whom had no work or national support were in extreme difficulty, trouble with many heads exposed the weakness in some mediums of developed trade and management politically. Notwithstanding that there was always the spiritual to provide solace in the form of compensation!

My references above are generally of factual reasoning, my statement of America with the little Europe adage stands also to factual reasoning. Because all of her influences since Columbus were from European statistics. Over the centuries to befall the native inhabitants, we Europeans flooded the new world making it our own, assuming in the positive we were taking updates to religious attitudes and political reasoning. That we had not always been able to see clearly of our full intent will be answered from later times. Top line and still ongoing is the open reality that America is now irreversibly tied into the structure of European thinking, even that she leads by a different style. She is to be representative in the positive of total human development so far, not only to Europe but the rest of the world. By all the codes of progression in relation to the human condition politically, she has showed the most forward motion, but in caution still needing interactive input by others to be listened too and sometimes heeded.

Darwin and the American civil war were concurrent events, for a further overlay in the same time bracket and not unconnected for my own reasoning I have looked into the works of Friedrich Engels and Karl Marx. My abstract reasoning is again to be employed, for me to suggest their motivation even if never addressed or acknowledged is done to open pages and not fill gaps. It would have been of no particular significance if either of them had abandoned all forms of religious sufferance. With the fourth dimension dressed as science in the attitude of explanation, ready to lay bare most mysteries in those very vibrant times, from which the idea that all reason was to be explained. If religion had been abandoned by reason, concept or because of its own imbalanced presentation, the only thing left of human influence was politics. Even by its structure it had collected its own definition, our political format in all cases old and new was exactly the same. No matter what promises or what any system might offer, even with religious connections or of regal semblance through kings or emperors, the one theme ran to all systems. Under all conditions all politics had been constructed to maintain support for the top tiers?

By presentation, Engels and Marx saw the existing structure of most political systems to favour reward for those as they saw the inactive few, from the effort of the many. From times of extreme hardship perhaps in comparison to the English cotton mill workers having no cotton to process. They could further note the unfair working of developed society by such occurrences, with the church and state in support of their people, this was read to be of pious platitude to trust in God so that the hardship and suffering now would be paid for by reward later? I am to remind us that my thinking is in the abstract about how two very accomplished philosophers might have arrived at some conclusions?

If Churches had been condemned for being inactive or offered as a placebo, then value to the community in the active pursuit of day to day living was to be utterly diminished. With both men working from slightly different thinking grounds, their formation of ideas in loose connection, turned them from religious forbearance to the practical resolution of mans plight by political means only. What had been started then by the influx of emerging and developing ideas, by the mixing and examination of winged philosophies, by the practicalities of implementation by the positive deed. Was the foundation of the ever desired complete compendium of human requirement through the taking of old tunes and making wholly political music in the form of communism but sold as Marxism!

From where the principle of communism might have worked in some city states in the time of ancient Greece some three thousand years before, any application now, was to be complicated in the extreme by the new rendering from Marx and Engels. As they seemed to want to represent the proletariat on their fair deal basis on the all for one and one for all style of sharing, their calculations in the terms of the sheer volume of the population were amiss. Ancient Greeks fairly gave up the system of communism when other styles of civil management had emerged, some able to cope more readily when joined city states had come together forming larger groups for self control, civil and religiously.

In Europe proper, at the same time, there were very many countries with relatively large populations most accomplished in their own style of political management and set Christian doctrinaire. Without being deterred Marx and Engels published much to the proposed betterment of mankind in general and the workers in particular. From their attitude and by the display of conviction they turned many from the need of religious solace being any way necessary for fulfilment. Yet unwittingly left in prospect a completely new form of understanding to the nature of purpose through country, almost like the sun worship!

I do submit, but not for the first time by anybody, that both were fully aware to the situation surrounding them revealed in the name of pure science and to the prospect of evolution in the cause of mankind's development. Here for the first time, real doubt was cast upon the acceptance of creation as the means to the only appearance of mankind. Almost pure scientific evidence that we had evolved by the avenue of vast timescales needed to develop as we had were revealed by Darwin. Forgetting, or not being able to connect too, Marx and Engels already understood that greater time was involved for our existence, from examples of our emerging fossil record. Belying the five thousand seven hundred years of human existence explained by the Jewish calendar from the bible, religiously?

If discounting the priesthood of religion to make better contact with the proletariat by disassociation, steps taken would account for the next move to undermine the social hierarchy if needed to gain forward advantage. In general terms Engels and Marx were set to correct the unfairness of developed society, first in Europe, then quite naturally throughout the world. Their plans were made from all the remnants of history, their thinking had been held by the past, not least that they would always have been unable to assemble future prospects from it without altering parts of it? Their thinking did attend to the past without leading to the future, by placing the formation of the unobtainable in objective, not being able to fully express the real desires and needs of man, even the proletariat, they left the book open!

Communism as a modern political medium was translated to Marxism in order to be acceptable, palatable and obtainable. In general Marx held certain scorn for the church; he almost saw it as an agent of the higher end of society, who in protecting their own profit margins used its spectre in pacifying disgruntled people on the political front. His indications were of no particular concern to it except scepticism. Before he and Engels could have considered all the aspects in the need to change most of the existing societies in Europe and of the American style also, time intervened. Unfortunately the shortfall of their work and studies was superseded by Marx's death in eighteen eighty three in London and by the death of Engels in the year eighteen ninety five again in London. They had done enough to set the basics of some needed change to the present structure of society, but not enough to guard against how their works might have been usurped by any subsequent application. With the masters dead and gone the questions we can ask our selves by reflection is, was the baton taken in honest fortitude, for the running was done on our behalf, did the runners take to what was intended or did they compromise the ideals because of their own shortcomings?

In picking at the bones of our history by my own particular style I have not set out to be intrusive or corrective, just informative. For in as much as I have been selective to my own subjects it will be seen that any where I have touched or will again are of concern to us all. My task and aims are made easy, because I know some of the answers already without even setting the questions. Without direction, on reading these pages we will be drawn together, because writing these words are more than my opinions carried as our experiences. The Fifth Dimension carries much of our understanding and from being without law, tenet or covenant is of a flexible nature to be able to form reaction by its one active prospect in 'One for the whole'. I have previously added caution for us not to relate our thinking of today in overlay of the decisions of yesteryear. It is that we will have to hold that opinion until we actually write of the times we individually were involved in, but only using the three taken prospects of the Fifth Dimension to temper self opinion!

At the ending of the nineteenth century, bringing us to about one hundred years ago and the next phase of the modern era leading to the final phases of human self discovery, we were still collectively able to demonstrate our ability not to understand the obvious. In the far of lands of South Africa by means of colonisation European peoples were once more engaged in conflict big enough to be another war with much the same intention of gain. Originally people of Dutch decent had settled the land to be their own, to what had become open plan exploitation of the vast continent of Africa. Europeans in similar fashion to the colonisation of the Americas had brought their own set of standards to most of the continent. In this later case we had set the plan of various European countries having their own sections of the African continent for self management rather than to one grabbing all, which really meant that the continent was too large to be managed by any single European nation!

With much of the southern part of South Africa of British interest in parallel to Dutch settlers who had formed the structure of society based on farming settlements. When gold and diamonds were found in various areas of interest to many parties it seemed that the British stand was to be clarified to the effect that they had fully intended to administer the whole area allowing the Dutch settlers carry on farming, but by British mandate. By more than odd circumstances one country of Europe was at war with another country of Europe in far of Africa, both Christian in religious terms, but forgetting how or from when or why they had become Christian from Catholicism! They had become national rather than remain universal by concept, which through the unfolding of events is to be seen how half measured we had now become. We just tripped various switches to suit led thinking by what ever method or style was of particular interest to us at the time. All of which is a statement to how we had developed with the thinking of then, which seemed to have been centred on how the thinking in Europe was at that time.

With the closing of the century marked by the terms of political deeds the die was cast that all settlement was now to be of political influences. If Darwin had allowed questions to be put upon creation in a well balanced society set by inquiry, then from the diminishing of religious substance by some of us if from other argument. The next unseen phase to be answerable was also to be from Europe, if not made by Engels and Marx, then to have been stirred in the cauldron of proposition without the ingredient of religion.

All of which is to say religion had not diminished in any sense except politically; the whole world was locked into the comfort, tranquillity and expectation of religious service. Many other societies with vast populations had seen their own conduct over the long centuries meet and serve all religious commitment. I had indicated that by natural processes like from climate or geography and other circumstances, had been the unseen hand to some of our physical differences. It is to be taken that they were to have been the mother of our attitudes also, of how we have been able to address issues from different perspective than how Europeans had. As the pace of life does run in other countries and on other continents, we of Europe were unaware that many concerns elsewhere still remained religious first and political second!

# CHAPTER 39

I have laid trace how some of us might have been able to translate parts of Darwin's work to suit or endorse attitudes that we had already adopted, but were not quite sure of how we might call such opinion while holding to the essence of religious doctrine? Half measured reasoning in disguise had always helped that we might have known that something did not quite fit, but we wore it anyhow. For Darwin to have cast doubt on the time scale of creation in relation to mankind by connection with all other life forms, then it was by half measured readings we took umbrage to the suggestion. Based on my partly made argument, that if we had found the time line connection between apes and mankind offensive to ourselves and an effrontery to God in Christ. Some of us were only too quick to make the connection between natural selection and the strong surviving, to have already been our blue print for human conduct come what may, but from our own doing?.

By failed reasoning we were not able to connect to the actuality of all events up until Darwin's revelations. We had already broken from the collar of evolution by uncounted millennium; we were not of instinct, we were not embroiled by all efforts to meet the single requirement of procreation. Any ape, man, inference was harmful to some of us, to be judged so, was to ignore our own good effort to the ability of understanding. Over suitable time spans from each human society we strove to exceed all the abilities of all other specie, carried first by our religious intentions however perceived. In that sense, evolution was a little cold because it was ever to lack purpose, unless natural selection was of the ability to mix those successes amongst other specie to the same standard claimed by some of us?

To have often stated that some of our differences happened because of the circumstances of climate, geography, diet and our following attitude from any of the above, I will be more specific in the context of Europe. She by reaction was most like our beginning over again, creating the biggest future changes by chance only, extending until these very recent times of the last five or six year from where we had found new beginnings in the Fifth Dimension.

Mesopotamia having been one of our first places of settlement with separate city states trading together, succeeded because through our close proximity we were able to exchanged ideas as well as commerce. In most cases the process was generally successful, but sometimes that success strained relations and conflict ensued, which escalated the processes of gaining advantage in conflict. Which further generated the same type of progress, but now the line of exchanges were made through the avenues of battle aided by how science released the art of weapon making? In the near nine thousand years from first real settlement in the land between two rivers, Europe now stood in the best position to exploit the ever increasing avenues of science by simple equation. Climate, geography, diet and other complimentary factors were included. If attitude was the driving force to the accelerated changes in Europe, then without the study of natural selection, the endeavour was to be that the first is right. Leading to the adage that might is right, but curiously comforted by the same religion which was inevitably right at all times, which only needed death to prove that fact?

Climate, geography and allied circumstances were to remain the driving force for Europe to gain the most in terms of scientific advances since the time of the discoveries of gain from the new world. Coincidence through the medium of the Roman Catholic Church first and the Christian churches following was the prime contributor to the same scientific advances. My own take on other religions is indeed that they were also in the lead with all discoveries, but without the significance of the pace to have been injected into the European standard, therefore into the American standard to be.

Previously, from Roman inclination Europe and the Middle East were set to the same control for a considerable time, by the same connection all areas of Roman influence had been given a year date reference by the adoption of the monotheistic catholic religion. Originally, Christ's religion had no first hand written record kept about its formation, even when becoming the adopted religion to be now Roman Catholicism the story was the same. Even so the religion was to remain and grow managed ably by its associated priesthood who had connected with the written word of their own style and with what Rome might have agreed? In particular to all matters about the translation of the mysteries by God's terms alone! This feature expanded on its own account when the Roman Empire collapsed form the multi cultural society it had been. By good chance Catholicism survived in most of the European countries that it had been taken to and remained on the basis of being the only standard seat of learning throughout. Uncounted for the future it remained to its learning forte till the present time. In reflection the Middle East on the demise of Roman rule felt somewhat displaced in tending to the words of the same religion still seated in Rome. When Mohammed introduced his religion supported by well written text, new mosques were also to be the main seats of learning, but now from a different perspective. With the religion defined as the complete entity in itself there was no need for comparative anecdotes to contain the matter of scientific mysteries. Allah ordained this religion, science and all its attachments were in complete mix with the final plan in what ever style. Science and all discoveries were only necessary to mankind by discovery when exposed and used when required?

With Islamic thinking introduced some time after catholic thinking, divided by six hundred years, then the difference is that for the first religion by time, God maintained all the mysteries and from the second religion by their time, Allah maintained all the mysteries? In subsequent relief early Catholic thinking was that mysteries were of god's domain, they were needed to be in the control of god's first servants the priesthood? God could reveal any of the mysteries by his grace, but only the priesthood could deliver them of meaning; hence the first houses of learning from Catholic thinking were not to open expression, but containment. In order to be able to shout down scientific exposés if they were to cause doubt to the sanctity of what mysteries really are; the managers of learning had to be aware of what science had revealed, which inevitably meant as we knew, we grew, to spite hiding answers!

Moslems were not quite so stirred by the mystery syndrome, what did it matter if the sky rattled in anger, of course Allah is responsible, when anything is revealed to us it is by holy desire, when needed, we will be told? Our aim is to keep to our religious studies and honour all that we have read so that we might be rewarded in our destiny. Time by any extent can produce mixed meaning, that the Catholic and Islamic religions were separated by centuries is of great significance to how they both have different outlooks upon our humanity. Time by the same extent can be affected by the circumstances of climate, geography, diet and other factors no more than in this case, when we the same people, formally of the same religion, were split in outlook by those very separate connections with God and Allah!

The Middle East and Europe were once of the same people, if loosely connected by the same appointment from the one God the memory was to be lost in the course of the future, because it is we who have divided his intentions by our own standard of politics. Although my example is specific to the one main area of the Middle East we have to take into account the whole world measured from there. Mesopotamia is ten thousand years old counted in time by the setting of standards. India, China, America north and south, Australia and the great northern islands above her on the map, Indonesia, New Guinea and the Philippines are also of the same mix by two or three or four times ten thousand years. They are also to have worked at first from religious perspectives before allowing the standing level of civil controls hold sway. They had all been connected to religious determinations by burial rites in all examples, if not through, Shanidar then by their/our ancestors!

I might not have included all of our singal tribes, clans or groups by name, because all our connection had been made at Shanidar, for we are all of the same worth. If I have claimed representation on our behalf, I have looked to set the scales in fair balance for now against all of our histories, because of the urgency of the situation we are now in, which works to draw us apart. We are always losing the view of reality that we are very small people in a huge universe, which loses in us the importance of our individual ending. In as much as we might see the picture of the almighty to be the aim of purpose, from my own interpretations over recent months, questions are always to be open. I use the factors or circumstances of climate, geography diet and other reasons why we differ, providing we do not have to accept that because of those reasons does not mean that we have to have separate gods!

## CHAPTER 40

Five hundred years ago from the first new foot in America, the European angle on religion and politics came into prominence, by all the means of el conquistador in Mexico and leading south. Whatever was found of human conduct did not meet with what Europeans had been used to. If for any other reasons this was as excuse to promote remedial consequence, then it was done by the valour of the day in connection to that one standard only. From unfortunate circumstances in dealing with what was found in Mexico; the isthmus and further into South America, we Europeans used our own standards in excuse for brutal repression. Not even looking to the political structure in the new world against our own, for just at about that time, Europe had awakened to the possibility of her own flawed religious countenance. Left to deal with people considered savages by reason of the found religious practice of human sacrifice, while in Europe the sacrifice of heretics were merely acts of purification in preparing them for the glory in god by this atonement!

Connected but not readily seen, was how to these peoples of the new world, Europeans might have been their saviours fitting to the told stories of ancient visitors from the east reappearing to further sanctify them? The tragedy of time expressed outside the importance of the individuals own lifecycle, was lost amongst the mixing of stories between the day sun god and the night sky predictions read by the inhabitants of the new world then? When first led to this philosophy of the sun God; the only other active belief was to the Great Spirit, which although fulfilling, it still had shortcomings. Most in the sense of being un-guiding, for it led to the fixation that all things were ordained, almost holding the people of mixed tribes in suspended animation. Setting opinion when there was no real method or history whereby separate tribes could have exchanged ideas to promote progress if required. Until Egyptian travellers arriving from the direction where the sun always rose, who could have laid new opinion to the point where all could work to the ability of planning our own route to the future? Even forestalling the need to keep written records for all members of the tribe, when religious affairs could be managed by an able priesthood, allowing all to know the validity of belief from the certainty of the sun rising daily and daily and daily?

In my own connection, I have loosely attributed that in the Americas at the time of rediscovery we found two basic religious concepts, one, the Great Spirit, two, of sun worship. I had set it that at one time, there might only have been the first theory of the Great Spirit, with the other being added later by direct connection with peoples from Egypt? Without provable justification, I would suppose that the reason the later theory extended southward only from what was to have been first contact in Mexico long before Columbus. Stems because of nothing more than human attitudes bestowed by the unknown input of climate and geography along with some features unrecognised at the time and even unseen since!

Mexico, the isthmus and further south into the full southern continent proper by area, fall into a beaten zone by the sun. With the same area straddling the Earths equator, in general terms the sun has very measurable time patterns than further north, from perhaps Mexico and further south than Peru. Unknown at the time when the earth tilts on axis the greatest range of time differences between night and day increase the further North or south one might travel. By regular patterns near the equator the day night measurement is reasonably even, therefore if desired or inspired, any measurements taken of the sun in day and of the sky at night could be pictures into fable, If this was to be captivating in any way, why not spread the good news and new views?

Climate and geography, if giving longer day and night patterns, offer natural exchange variations in temperature and seasonal swelling. Which means the further north and south the earth tilts induces natural seasonal cycles unlike nearer to the equator, which if followed can create different thought patterns to local inhabitants if encouraged? With the comfort of the Great Spirit in attendance, it is easier to connect to seasonal cycles under control of such a deity providing all the needs to the natural flow of events? Being the standard to have captivated a common union for all peoples of the Americas living by sparse contact amongst vast lands, unless by the introduction of change? For any change, if it was put that the Great Spirit was in name the sun. Then the resound by some could have been why does it not have the power to banish short days or long nights or reversed, and or hot summers and cold winters everywhere? Unless of course the Great Spirit was in test, so that we might show our reason in the understanding to it! Belonging to the Sun-God or the means of the Great Spirit was not enough to quell how Europe settled upon the new world!

If to say one group of many tribes could not connect to the sun god because of how the seasons were more attuned to what was expected from the Great Spirit. At the same time imply that others could connect because it fitted that the sun was in control of all events anyway. Therefore if told, any further telling would turn the tale to fact!

## CHAPTER 41

In our name, my Fifth Dimension carries the baggage of change in order to review our histories without change. My three taken prospects and one active in 'One for the whole', are an existing extension for all religions and all political systems from now on determined by new viewpoints from these past five to six years of revision? Where religions stand to their own concept confirmed, is not so for politics. Religions by the mystic of there formation are extremely flexible because in substance they have been constructed by the medium of mans thoughts to gods desire. Politics on all levels is to the art of tying our shoe laces or putting clothes on or preparing our next meal to family desires by extended connection, or in the liberty of making fat profits from commodity deals. It even hails to our welfare under the wings of human rights and to be politically correct, which is only adding the image of its own importance. Although politics runs half measured inducing steps taken, from now, with our active prospect of 'One for the Whole' the reflection will be realistic?

When Charles Darwin published his Origin of Specie, religion and politics were almost set in concrete by acceptance, that is until the printing press went into production on his behalf. The whole world had mixed a thousand political solutions and a thousand religious settlements into the prospect of interaction.

From then to now, how Europe had come to any sort of prominence was from the trial and error basis coupled with the outfall of the conditions of climate, geography and circumstances. No body there was special, except from the drive of vanity, until from the written text it came possible for some to ascribe natural selection to a single group of people like any single group of animals who were able to show selected prowess by the edifice of power! If natural selection was used to be accountable for the situation of being naturally selected, and be implanted to the European outlook first. Then although not to be blamed, Darwin on publishing when done, fitted to the situations of how and at the time where we were developing fastest, but to what?

Europe to America since Columbus was brutal and uncaring by distinction, without cause the church was used in the forefront of excuse to that overbearing force. But not to account for the reality of the abstract, whereby people of the old world transmitted new devastating human diseases to the peoples of the new world. Resulting in mass murder by contamination, but in context to Gods Desire by the rules of inquisition and by the political rules of circumstance that Europe was now in America anyhow? As to how Europe had self promoted in America until the time of Darwin's publication, when he was perceived to have filled gaps, not by intention but convenience. By the mid eighteen hundreds when Darwin produced print from the pen, he inadvertently gave credence to all for Europe over world matters politically that the strong are destined to survive by natural selection? He had not intended to deviate that this was a political or European or religious matter by selection, but in terms to the fact of reason whereby our representation carried our new impetus by natural selection onto another plain.

All of which has alerted us to our own ability of misunderstanding the written text viewed, but not read? Of all the best features from the past to hold us fixated, let us not look to our own duties whereby we always attend to the views of the few above the majority for convenience. From the mid eighteen hundreds we Europeans had now the appointed reason to our methods for any thing we might do throughout the world under political determinations. On the religious front, deeper complications were set into the natural selection of the strong having first choice, because in theory the idea was not to gain from surviving but dying?

When Christian churches had been ready to accept the concept that all men were created equal. Then in regard, that message was imparted to all others contained by the umbrella of European thinking politically, that those in convert to the Christian religions were to follow all religious examples by the overriding political combine. From all considerations the Darwin ape to man theory and the more poignant readily accepted concept of the strong surviving better by natural selection, carried this single human specie from when first argued up to these recent five to six years. In that time we have now developed a frame work to set the standards of behaviour through the Fifth Dimension with the three taken prospects to be in overlay on all religions and all political outlooks. Then to be used to guide needed alteration if required, but through the one active prospect of 'One for the

Whole'. But not to be carried half measured like from where Marx and Engels might have earnestly tried to operate 'One for the Whole'!

Both men dead and gone before the turn of the twentieth century had set new political standards to the general idea of government by the people for the people, not there to clarify the best methods or style in obtaining their objectives. Others to have taken up the mantle travelled in the first choice of revolution as the most direct or only possible way to change things to the advantage of most people. By connection people known as the proletariat, a class in most European societies who were to be the power source in the engine room of industry. In positive terms they were a focal point to which it was best to direct the necessity of change, because as well as their physical status they represented us all. The rest is not history for it is to be attended to how the effect of communism was to affect the whole world since its first real success in the early part of the twentieth century?

From first successes aided by the bloody mayhem of the Great War in Europe between nineteen fourteen and nineteen eighteen, Communism turned into a cold master. That Karl Marx had referred to religion being the opium of the masses was a point of reference, his inference was that it was used as a means to placate the masses in terms of expected reward, while his own ideas were to the positive of now. Others in the drive to change the country of Russia from the autocratic ruling house it had become, used the new form of reconstructed communism by all of its worst standards. At first, complicating it without the parallel function of religion or of what could have been the restraining hand of shared opinion. New communism by the revamped ideas of the ruling political group got it so wrong to blight us all for more reasons than ever to have been seen or connected to it.

I can reconsider my opinion of new communism, but make the statement without trying to think in terms of then ago. Its new masters who sought to oppose monarchy and church by there replacement with politics, the shortfall was in their lack of understanding the need in ordinary people. From habit many of the proletariat wanted the additional security of religion, even as the new dictate was being implemented by revolution, religious fervour remained. Unfortunately so as not to allow the possibility of opposition by religious overtones, the connection to church and or monarchy were promoted to be counter revolutionary and all means to eradicate that prospect was pursued by the most severe measures. When Marx first promoted his ideas of economic sharing it was done to spread the working wealth of any country in the political spectrum. Not understanding the concept, those in the revolutionary cause used all ideas mixed to achieve their own ends, that choices had been made to depose the ethic of religions, the reflected result was to ultimately serve ones country in place of God! Mother country as the essence of purpose, mother country the provider of all bounty, belief and faith! Without religious sufferance those in power set us to lose human self respect in a worst manner than ever before in our histories. When the state alone could take its creators in man and use them in all forms of abuse to its own heady ends. Mother country or fatherland rang there own death knell!

Russia in revolution in nineteen seventeen was in my title to have occurred in the next modern era, unknown by any of us then or even translated now, her lack of religious support was to instigate her total down fall through political means? Of the era in question the influences of Darwin's bequest held no form of recognition to the communistic system of politics, for its drive was to the innate. Evolution had no real stir to non religious concepts and by the same token, if the state meant owing all to country, then the element of natural selection was compromised, for countries are just that, countries! With the substitution of religion turned to the deity of nation, all being done was to sew the spirit of god into the directives of politics, which proved to be unmanageable?

**CHAPTER 42**

Had Karl Marx and or Friedrich Engels discussed the prospects of the Fifth Dimension mixed with the debates taken up after Darwin had published. Had those who took up the basics of there works misguidedly and moved to curtail the aspect of religion in the cause of revolution also had the prospects of the Fifth Dimension. Communism in Russia might still be in command but with religious fixtures. From the supposition above the connection between politics, religion and the Fifth Dimension, then unknown, is vital in connection to the same two aspects now, and will be shown so.

At the very time of Darwin's publication, across the Atlantic Ocean in the new Europe of the United States of America, political war was taking place or more precisely civil war but defined by the interpretation of political ideals. Which like in Europe then and since, the war was between the same peoples of the same Christian religions where religion was never an issue, politics was the driving force as economics was the vital cause in the difference between the separate factions. If the American civil war was to tell us any thing then it should be measured from all of our interpretations, for in some respects it was an apex in future human determinations. By the coincidence of circumstances one of the offshoots from America to Europe was to affect the cotton workers of England having no raw material to process. Perhaps drawing interest from Marx and Engels of how the proletariat were left to suffer by there own devises being unsupported at that time, while previously making the wealth of others to flaunt it at their own discretion?

At the time of the American civil war, without connection, the Fifth Dimension did not come into the position of use politically or religiously but some of its prospects were shuffled! With war between the southern Confederate states and the northern Union states the disposition of politics under different determinations were beginning to direct the American form of Europeanism to be new Americanism, but from political perspectives. Like many other examples following and in context the changing world leading to the twentieth century was done on a trial and error basis. With the overspill to most interactions taking us to the situation where many or our responses to any situation was to be filled by applying our steps taken syndrome. Although I had attributed that phenomenon to have come from our half measured actions, the steps taken were to be serious when the deeds it applied too, had far reaching consequences?

In as much I have allowed myself hop skip and jump to various examples of our conduct, my aim has been to funnel our attention into how we have come to interact by this very time slot of now, just past to the start of the third millennium! If we do look to the statistic that we have aimed to achieve like the existing standards of human rights, endorsed by the practice of political correctness, touched on by natural justice, then it is that we must dissect the reason for our present situation. Our reasoning can only be illustrated in the abstract and to most matters of history without alteration. Said again in context, it is now to be taken that we use my base, having been constructed these recent five, six and seven years in the unmitigated form of the Fifth Dimension, to be our collective conscience!

The Fifth Dimension has shown already how I have conducted our research to lead where we are now. I continually stand by Shanidar as our first reference point past the three taken prospects with one active in 'One for the Whole' to be our illustrator of the importance of time to us. When at first we might have stood to the belief in god, we had fixed that connection to our understanding of our own life span. Although far and above how creatures of evolution could ever think, by our sense of awareness then, we were unable to mix evolution and creation? Even when God intervened the fact only closed our general thinking!

By our own reasoning we accounted to our existence by the hand of god, but forgot it was by gods own timescale. To us, we did account that the same hand was to have created all things, fixed by the extent of eternity explained by religious concept. When Darwin questioned the theory of creation by the fact of evolution, was it from our conceit that we felt that this was an effrontery to God? Not least from the assertion that god by concept had directed all issues through mankind. Therefore the theory of evolution was challenging to faith, belief and of purpose. When now we can see the open view that evolution was in fact to the spirit of gods own time scale; and from it creation was in terms of our understanding to fit mans own time scale by the very same connection through God!

Darwin made no error, God made no error, which leaves only us. It is only done by mankind in the doubting of evolution but worse and still we compound and doubled the error at concept by holding that creation is not, is not part of evolution. Evolution is of gods time span which includes the three or four hundred million years to some specie still living by extension; or dead and gone by events. Evolution is not measured in days months or years or millennium, it is measured in the event of ideas. Of any ideas to bring reason and thought to the aims of purpose through intent. Evolution covers the extent in which this earth was to cool in the circle of space and slowly bear upon this planet the eventuality of any life form whatsoever! Evolution by that same extent is only to be questioned and from the product of time through God in whose timescale we have no judgement of. When by season's decree we came of age to be human from ape's if/and/or, measurement to all events was still by Gods connection. We erred by the questions and assumed answers about creation by our own realisations that creating creation by our own judgement to time and in connection to God on our own behalf, we chose to ignore God and only apply our own thoughts of his desire!

Evolution is still evolving, it is called the natural progression of time, creation had evolved from it, but not in context to all life forms for none had switched to the limit of existence through the complexity of a single being or single lifecycle! When through our religions we were in the certain discovery of god's intention we erred to assume it was of gods intent that we were in the holding of that mystery as it had occurred? Our error in exuberance was that we needed to have the endorsement of discovery in context and so created Gods own delivery to us above the reality of our connection to God by the knowledge of the need in creation apart to the actuality. We were disposed to make the event suit the occasion in the only way possible to us then ago. That was in deed a half measured action, but not fringed in the aspect of deceit of desperation, for we had put a deed to action without the cause of understanding. Further expressing how we had surpassed all of evolution into creation when now we set solution.

My own feeling to any of the above is that by best conduct we had taken our half measured action and lent it to the certainty to become managed in the steps taken syndrome style, creating for us the complexity of now and how? Darwin is not to be the catalyst in the order of change, but taken as the fulcrum from which change can be tilted. His time by date with Marx, Engels the American civil war and the general situation in Europe being Christian by religious overtones, set in my next era of time the final placing of events which must induce future change without altering our histories, unless only by attitude!

## CHAPTER 43

Half measured working is not a disease or blight, in some cases it might lead to the only action that can be taken for any reason. If in turn it leads to the implementation of the/my steps taken syndrome then that on several counts and in many directions need looking at. For far to long in any situation we have ambled along, even struggled, yet have had to make huge decisions for our own welfare. If that has been so, let us cut again into the skill of our reasoning if even half measured, which will be determined from where we start. To look at some events of these last eighty or so years, but with the added ability to always visit the framework of our societies before then. Being a place from where we might have seen change in many forms, though always to remember our thinking must be of the past and of past standards not to overlay any meaning of it from today's standard of thinking?

That we had unavoidably used any means or methods and styles for self development in practical terms, the scars left have to be reckoned with the same pragmatic approach forward looking, to make any new decision?. From the Fifth Dimension, yet to be released into the domain of self examination by this publication, are three taken prospects, 'Out of Africa', 'Past Gods Words', 'Natural

Body Death' and one active in 'One for the Whole. To be our best means to review the past, not in sympathetic fashion but to set in the future our ability to change if needed? The Fifth Dimension strips the singularity of our religions and politics allowing any thoughts, propositions, full voice suggestion or inclination or ideas that may occur to any individual, but from now, reviewed by the value of our active prospect in 'One for the Whole' through religious or political channels?

All of which might sound comfortable until the immensity of what we must sift through is seen the same to have occurred to the same people of the same specie. Allowing for the slight time warp to have affected some of us differently, if not by the circumstances of climate and geography, then attitude could have played its part in creating of forcing reactive attitudes. For our better sense of reasoning I will proceed with my own format in relating events by my understanding. I have used the image of the funnel to be of use in explaining or pointing to how some occurrences from the mixture of some of our tribal or national interactions have blended to be poured from the one spout at the same time.

Any concentration of Europe, America from about eighty odd years ago until now, is a result of the previous five hundred years when Europe only went to the new world and entered the funnel. It is also representative in the time sense of when we were at Shanidar and split to our needs, desires or driven requirements resulting in change. America before our recent European mix of the full five hundred years, has to be seen on two counts religiously! By whatever means el conquistador ravaged the marked lands of Mexico the isthmus and further into the southern continent proper. All was done to the catholic codes of religion at first delivered by Spain added to in quick succession by Portugal. Any forms of conversion were done in and to the confines of inquisitional codes, but kept largely to be of Roman Catholic design, being a matter we might review time and time again.

Northern most of Mexico the tone might have been slightly different but the style was much the same. Here in these lands no great cities were found, no fashion of national politics was discovered, no real blend of existing tribal unity was displayed, except through the calming relief in the Great Spirit? Although like elsewhere on both continents Europeans flooded there for the prospect of gain, on the northern continent the religion brought in the most part was Christian by protestation through reformation. People of many different nationals carrying their religious scars from the turmoil created after Martin Luther's reforming demands, people wanting to get away from state, church persecution. Beyond the romance of the idea for gain, many wanted to be able and express religious sentiments in their own light, even if tied to former political systems? Which made for a different mind set in the northern continent than in Mexico and the southern continent?

I would not wish to lessen the destructive force with which both continents were brutalised, being ravaged in the context of the people already there? The space of one sentence will not be used to side step obvious points of concern. My open view is to us all, that we in the practice of human sacrifice or in atonement to the Great Spirit and were led to change by a fractured second dimensional religion of the Middle East having turned the circle without notice. In no time than before we mixed past the recognition of our sameness, we were at all losses to ever be in notice of specie connection, never to connect in reality that we were one, but for the circumstances of climate, geography and or other reasons?

From much of the mayhem we had come to in Europe, some was transmitted to the new world. While in the full ability to literally roll over the native people in the Americas we conducted our own European conflicts there, until under the final act unfinished, our colonists of the full Christian brotherhood took to their own determination. After fought for and claimed independence, the northern continent moved at a different pace than did the south by other reasons, but nevertheless as we in all cases, but ever by European influences?

America had become European again by competitive steps. From first settlement the original takeover of the aboriginal people classed as red Indians by Europeans after Columbus, had exemplified European conduct in the growing concern of self. By the war of independence the split from Europe was made on political terms only, adding although to be a Christian society there was to be differences in approach! North Americas own civil war of colossal cost in human life to both sides, changed her standing in the swing to futurism by political thinking more so. Even that the American language was to remain English it was by the same connection how Catholicism had become Roman Catholicism, due to the ready and easiest prevailing circumstances.

I had earlier mentioned how I felt that the North American model was a direction we might all aspire to. Without retraction I can add that it could, but if only looked at through some of the avenues of political reasoning? My view is laid by historical connection from the event of the existing European exploitation started about five hundred years ago. By all ideas, propositions or suppositions this great European drive, was representative to us all and of us all fuelled by the processes of climate, geography, diet and other influences under the umbrella of circumstances like gain. My repetition is on my terms only, but for us all and not even at this time screened by the prospects of the Fifth Dimension. Further inference is in relation to how at the time, Europe began to advance the pool of our scientific knowledge if by no more than attitude.

From which, because of the effect those actions had with having pace like natural selection injected into the European model, some could command that the lead was directed? Which we know it was not, we simply carried on expanding like in Mesopotamia! My suggestion that Europe and European standards were changing holds, because we are directly connected in attitude the same when influenced by life's standards set in the land between two rivers. Now, when the American model is simply an extension of that situation is still representative to us all, which might be questioned by us all, but only through the pen or from the heart or even religions!

My assertions of the American model being forthright stands from the progressive terms of politics only, again not necessarily in the presented form of democracy, my line of association stands to commercialism in its branch form added to the line of political influence. Mesopotamia gave us linked styles of interaction from which I accept that connection is not to be directly added to all conducts throughout the world. Many great cultural experiences happened on their own account in Africa, India, China the rest of Asia, the Americas, Australia even parts of Europe. If we turn the pages of political history, all events have been funnelled into these most recent of times from which America stands on point, because of world events over these recent eighty plus years, politically**.** To be further highlighted by events in Europe when Russia became revolutionary!

## CHAPTER 44

By my own abstract lead, in this chapter I intend to set the scenery of today's era where the character of the Fifth Dimension is to be first played. The first quarter of the twentieth century, driven by well trodden circumstances, saw that we are at liberty to use for any reasons of cross referral acts performed, initially against the backdrop of scenery having been set by European standards. If the second modern era had shown its colours, then the mixture was to form where all of our ingredients were best suited to come together.

Without the definitive study of religious concept and by working on the principle that collective Christianity was ok, Europe had managed her political affairs to her own taste. America had refined the spectre of politics and had been able to sidestep some of Europe's political traditions relative to its royal houses. By untried reasoning through the bloody conflict of the Great War and the Russian revolution showing its own bloody hand to halt the carnage and support its people to the quality of true sharing, cast shadows? That it was to the proletariat all aims of restraint were put on by the old systems, the motivation of change was only to be political. Refined forms of communism were taken to what suited any particular situation at the time. Any type of political reformation this brought, if past the intent of Marx or Engels, then those in promotion of new ideals, set us all into our most difficult times yet ahead? To this new political regime in Russia the first chance of concerted change was lost to be hidden for almost twenty years. In taking politics only for the refinement of society and in suppressing the human spirit of religious choice through religious repression, the masters of revolution acting on all of our behalf failed us in the concept of the future?

I will remind us all that although many actions from the past were stunted, they were from us in trial, I make no assertions that any of us are of a better standard than then, but put in place that we cannot but help and gain from what ever was done then. Meaning in abstract thought that it is not necessary that we learn from our mistakes, but we learn by knowing there are better options. From the first attempt of national communism in Russia; the new picture in Europe was set, but now to set it communistically world wide? From the winter palace in the city of St Petersburg Russia, European politics of a new kind to American democracy had set standards whereby the two new aims were the same. Government for the people and by the people politically, but swayed with one being set with ancient religious standards and one being set to the concept of the mechanics of the earth itself. Setting surreal standards to both camps, if one would say let us operate by all political selection in physical human service so that we might reach God in all accomplishments religiously? With the other to say by direction of political expedience, let us compare our worth in material value to all others of different standing in the total support of statehood and mother country?

Our new era beckoned by now, on our timescale, we are close to the start and the closing of the Second World War. I have mentioned how some European nations had been reacting to each other nationally over the last two hundred years since the battle of the Nile off the coast of Egypt. With the same amount of activity continuing right up to the year of nineteen thirty nine, accepting that alliances were changing at furious paces by political determinations. The Great War of Europe having been set in national tones was fought by Christian communities for political ends. After conflict the situation was left to promote the certain reason for further strife. An additional factor to the forthcoming conflict that must be, was in design by how much of Europe from the early nineteen twenties took to new kinds of political experiment. Dictatorial fascism began to emerge, not to any particular defined code, but in the possibility of connected association to the ancient Roman era of empire!

In the upsurge of political feeling post world war one, new politics called for authoritarian leads, but to comply with the modern ethic; the initial method to those ends was through the ballot box, even so, many new governments leant heavily toward dictatorship. It is to remind ourselves again these acts and actions were of our doing. Darwin of the mid nineteenth century advocated the concept of natural selection being the result of the strong of specie in having survived by best standards. Marx and Engels as contemporise advocated the measure of change in society, if the ballot box could not be used because of an autocratic monarchy, then by what power, but by force if revolution was the call?

In the cauldron of mixes in the heartland of Europe proper, Germany as the main defeated nation of the Great War sought solace by any means. On her own behalf some of her people by political strands sought to implement change and also on there behalf, sought to correct the past? Germany with the rest of Europe by the progressive charter I had previously mentioned, was awash with the philosophy of ideas. Some who could focus in their own direction might have been able to translate the full meaning of Darwin's text in self explanation. From where Darwin might have catalogued the propensity of stronger males setting certain standards to the success of specie, read Nation!

Some might see the obvious that selection was not an option, but a command, that it was not a case of the strong surviving for the success for the specie, but that they could determine that the week must not survive, for the success of Nation. By caressing such notions the simple link is made that only the strong can make such judgement, only the strong deserve to survive. Judgements which can be double turned, even using the best traditions of Christian behaviour, if measured that religion in service will be deserving of its place once all the political wrongs are righted. From the avenues of thoughts by reactionary philosophy in touching on what Darwin had written about natural selection and converting it in sphere to be superior selection. How far is that to carry in the human choice of superior humans? How far might we not go to proving the structure of such philosophy accepted in the statement of concept when already driven to make wrongs right by force if necessary!

Counted in nineteen thirty nine, Germany the most organised nation of Europe was to lead the world by her own superiority, into devastating war! Sadly by my own crucial time span since, we have been able to learn little from the horrendous experience, except of our own frailties in countering the possibility of repeats and repeats by the very same set of politically thought superiority? Without explanation, it would seem imposable that any European country could have come to that standing philosophy from religious determinations of its people being superior to others when having entered the era of catholic reformation some four hundred and fifty years ago. Martin Luther when setting the order of change, was done on all of our behalf, he wanted local language expressed in religious terms, not for Germans but Catholics. That we turned religious standards to political ends over the short centuries of European colonisation, climaxed when a single group in Europe claimed racial superiority through the politics of fascism over all others, when all had thought the same any how! In the most bizarre of circumstances by what was to have emerged out of the funnel at wars end. Was that although god was on all sides, politics was the winner in taking us to the last part of that era entering us to the first part of the next and last era?

Nineteen forty five saw the end of the war in Europe, which included the destruction of one form of politics in the overbearing form of state fascism. Two other types of politics one democratic the other communistic, both counted to be right in the words of victory. All three types which are an example of how politics had finally taken control of the worlds rational, because in effect politics was the overall cause of conflict in the final place? Although gone, we can connect fascism with democracy to Christianity to religion, then communism of no religion. We might finally see how up to and from the last sixty years we have tried in failure to re-reconcile all with that disaster, we have set it so that we are loosing the ethic of politics by meaning, and the soul of religious understanding?

After that blood lust, again unashamedly, we settled to redress the past from the prospects of the future, but managed to set the wrong criteria from the wrong beginning from the right ending? In too quick a movement the spoils of war were shuffled with one hand only, that of politics. From which alliances were made to be democratic or communistic in form, working by the prevailing circumstances, even that the abhorrence of state fascism had been removed the original reason for war was lost? But again replaced by further political alliances when being met with new situations where democracy came to be right wing and communism to be left wing. Both types using their different standards of government for or by the people for the people, matters of which might not have been of any concern, if our religious will had not been ignored from both sources? I make no criticism on our behalf, for at that time, because of the aspect of war much was seen in straight forward fashion to the need for direct change, but still to be worked out to our best advantage.

With the democratic front against the fascist front, who were each Christian and against each other in the war, with the democratic Christian armies joined with the irreligious communistic armies against the fascist Christian armies, then much concern needed to be given for the final settlement? Having posted notice of where we might begin to move, my point is to state that the future of politics was less commanding without religious influence as we will see. From the ending of full conflict in Europe, we were to set the stage of our future standing, determined by events in the familiar lands of the Middle East!?

## CHAPTER 45

Early nineteen forty five with German fascism defeated, all heads were turned to Asia and to the defeat of Japanese imperialism. Amounting to be fascist by the same twist employed from the ruling political party of Germany. Both had gathered the mantle to be superior by national selection. Without merit, from when the European fascists took the helm of racial superiority at wars start in nineteen thirty nine. Imperial Japan had been conducting her own forms of superior specie ideology in China from ten years before. Her drive was in the same coincidence to European fascism, but tied by serious religious codes, which nevertheless were eclipsed by the political drive of commerce and gain.

War done and settled in both theatres of Europe and Asia, saw the first real modern era of true politics. I had mentioned of Americas move to the forefront of most political agendas to be a process of natural progression owed to us all. How ever that view is to be taken, it can only be done now, because of these five to six years reversed. Where I had indicated that Americas lead now was political, is not half measured, but can be misused to work the steps taken syndrome! Settlement of the wars conduct was treated separately in Europe and Asia. Americas influence was pure and simple by political division only, but allowing the use of religious sentiment creep in to discolour the clarity of opinions or views to the product of future events that could have only been imagined?

European settlement was done from political reasoning, with one of the victorious sides holding to Christianity in the same way that many defeated fascists had, with the other victorious side over fascist Christianity ungodly by its root of Marxist communism. Without distraction by the ready mix of political overtones, much of the peace settlement done in Europe was earnestly degenerate and degrading to all concerned. Having at first curtailed and then destroyed racial fascism by the course of events and done by political means, allowed us forget our underlying drive into the cause of purpose religiously. From the new division of Europe having borders redrawn and reallocated, only by half measured reasoning, we committed vast human tragedy knowingly, but under the auspiciousness of the steps taken syndrome, unknowingly. In the aim of permanent European peace we moved vast numbers of displaced peoples back to former countries, even that borders had been redesigned, although this was done to the best of intentions for political expediency. New concerns for the American, British and commonwealth side, arose when by sending many people back to former or new communist countries, and knowing of there fate, that they were to be murdered out of hand! Helped set new criteria for future political references, not least because by wars end our objectives were no longer clear except to form our bloc culture divide!

Settlement against Japan also carried many twists, but the handling of affairs was done to a different style under unseen comparisons. Most of the direct settlement from the stature of victory was done by American ideals, which due to there input of American blood and money, was balanced! Where the whole structure of the Asian conflict was of a different standard, is that most of the conquered peoples by imperial Japanese aggression had already been conquered by European empire builders, which was to complicate the American peace plan with Japan. Although the Chinese people had taken the brunt of prolonged Japanese aggression, there renewed political reasoning was to follow communism, with guidance from the Russian form of Marxism!

America from the political stand only, proceeded to implement her own style of settlement with connection to her European allies, now reinstated to their original levels of colonial government. But not at first before setting to promote national independence for some countries by the earliest means, if only to open trade markets that had originally been managed from Europe. In order for this to happen, and in light of the political agendas emerging throughout the world, her ideal was to create world market economies. Her reasoning in some instances was to make us too busy for future mass war, which could destroy those levels of comfort forecasted about free markets politically? Without direct intent, the form of consumerism was evolving as the banner to the best and most successful form of government, which then was only guessed at, but now has become so!

With the American people in all extracts the same as all of us, yet unlike most of Europe they carry there religious concepts more openly, yet since wars end they have run most political agendas from the aspect of commercialism? I had put America to be representative of us all to our aims of obtainment which stands, but now we must all better reckon our own religious standards? My contention is for us to see the aspect of religion and politics un-criticised over these last sixty plus years. Having stated several times that we have been led by America commercially, bears fruit by result. In taking us three and a half million years to reach the standard of the Second World War, might be a means to question the standard of our civilisation. Taking to Shanidar one hundred thousand years ago and to reassert our religious obligation through the much later first, second and third dimensional religions of the Middle East, along with other religions. How we had allowed wars come to political settlement only has to be always challenged, not least to include religious understanding? Now it can be done in taking to the overlay of the Fifth

Dimension of three taken and one active prospect, which does not endorse that any particular group are civilised?

## CHAPTER 46

From the endings of world conflict in nineteen forty five, politics by self promotion saw its own solutions to most questions? From what ever angle or method it was operated by, it worked in falsehood by assuming that by the style adopted, we were right. That we alone, politically, operated in the appointed standard of those we represent, that we by the process of leadership, were entitled to maintain that representation by the means of civilised community. Being able to operate strait faced in that manner, we produced our half measured reasoning which automatically led to the production of our steps taken syndrome. By thinking that we are civilised, has been enough to cover any actions from what ever era of time we were in. By using the developed forms of politics suitable to particular circumstances and addressing them in terms of our own interpretation, has been proof to some of us that we are civilised! Even that without any form of religious input to the recent settlement after our world war, we acted in a civilised manner? Which only stands until we grasp the nettle to say we are political we aspire to be religious, we are not civilised?!

Most of the above paragraph has been written in the positive, my particular point in expression is to be able and check our own credentials politically and religiously. If there are key words to be amongst us all, then along with the best to be love, family, hope, joy, friendship, help, sharing and caring, civilisation also makes its mark, but has been vulnerable. Because when uttered by any body or group, clan, tribe or country as a form of status, it immediately sets the statement of the uncivilised in cross referral, it instigates the mood of half measured working. Not from all terms of malicious intent, but in the attitude of the bespoke. Holding that the style, system and manner of all previous conduct is correct, because of effect taken in acceptance that what was done accounted to be civilised behaviour? In order to complete the circle of circumstances to our being civilised, it might be that we should remember when we first could be accredited so by settlement in the land between two rivers!

Better that we remember then ago, we barley had systems of civil control working to any level, we did not have politics anything like today, we had no religions anything like today, but we did hold to the mysteries of life in there abundance. If I have leant it that civilisation was born and slowly moulded from the minimal of social combinations, then we might hold to that supposition as our undercoat before final decoration. For many societies between now and the long past, politics had been woven as one road, religions as broad avenues, with the scientific exposition of the mysteries as great boulevards. This was felt at different times from different circumstances to different interpretations, but holding that now we are civilised by self choice, if only seen through political eyes up to now.

Until timed by these five to six years of the Fifth Dimensional roundabout, from which we can pass roads avenues or boulevards to take the correct route!

Any lessens to be further learnt are to be from the making of the future, even if using no other harness than the way we educate our young. All of which can be done by looking at the past and our histories, but not to change them only by altering our expectation of it? From the reading of this book so far, we can begin to think futuristically in terms of what should be and what will be. By all positive trends it is to be seen in all counts that our religions are vital to the discoveries that we are yet to make, even for the atheist or agnostic? Since wars end, by the exact political standards of today our religions are in decline because of political servitude, now it is that we must look with balanced hope and desire to set the picture without admonishment.

For me to have reasoned from the abstract; and translated my points of reference through these pages is a product of us all, just so when Darwin published his Origin of Specie, but he had far more style and worked from the positive. Points to be taken are that by some of his evolutionary lineal solutions, humankind had evolved from apes. After initial very strong reactive protestations and slanderous comments about the man himself, matters of politics and the world continued apace. Leaving the duplicated religious community to enter the theory of theology, explaining things as they are meant to be thought of by ordination, in most cases to Christian standing, simply because where the publication first took place, in Europe. That aside, being a real hard knock to the establishment on religious terms, the political element was functioning at its frantic pace and faster in all political attachments like trade, commerce, colonisation and ever empire building. Assisted by the productivity forthcoming from the fourth dimension further promoting the irreligious sciences, lending to compress the events in the following twentieth century, ongoing. But turning in a spin to reach the last half of that new century taking us all up to about now!

From how I have pin pointed various aspects to our welfare, I have made my own view clear by showing the importance of what the Fifth Dimension will deliver. I have set the scene from obvious events to have occurred after world wars two settlements on the separate continents. Again I make it clear that when I mention any person or group, East, West, North or South, is done in representation to us all, always. Specifics to the Middle East are to us all, Specifics to any national of any country is to us all, my theme is to us all, but by best reasoning from religious undertones first, politics second!

## CHAPTER 47

From all the lessons we have learnt we have perhaps missed the most important one, not by subject or title, but of time. What ever is to have been or was, is done, which means it cannot be returned upon by the two reasons I had previously mentioned. One, at the time of there doing, was from our capabilities of that time.

Be it our burial at Shanidar, through the fact of cave painting or rock marking or from settlement in Mesopotamia or ancient Egyptian sea travels to the new world, all were of the days thinking then ago. For us in our own vanity, it is almost impossible to think of the past by any age without adding our own input. From which I submit has been the cause of how we have continued to offer solution half measured, likewise we have allowed some decisions be covered by our steps taken syndrome!

Time is of the essence? By looking into the future is to use the aspect of time fully. In doing so there is but one solution of no fixed abode. 'Natural Body Death' is to be our closing event, which by all standards is the most noble of aims and must be set in all standards of achievement, including from politics and the more informed, our religions. That taken prospect alone supports evolution and creation, and by good choice has been with us since Shanidar without question, not half measured! Our record of it has only just been released these five to six years in conjunction with the other strands of direction to help us set the motive of time correctly. Without the concern of worry, it means that we will all be able to centre on the correct use of our politics led by our very many different religions.

On occasions I have moved that we can learn or teach or interact by our own historical standards without alteration, on the grounds we cannot alter the time of it, it was the past, and actual events are not entitled to be carried forward to cloud the future! By the biggest of margins we must all use our collective and personal histories in this time now, but only to alter the future by perception through 'Natural Body Death'. By that consent I viewed the past by abstract thinking in the clarity of representation, allowing on the one hand that the past present and future are vitally important politically, but the future is only to be important religiously. Which in turn will become self descriptive by adoption of the Fifth Dimension when overlaid upon the element of our politics and religions.

Post World War Two has allowed us enter the most intellectual human movement of mind on international levels, which give nothing to where we had come from or why, but all to how we think and have been able to. Colonialism and imperialism in general terms were good casualties of war even if from dubious and mixed means of change. What was being exercised by different peoples due to the circumstances of climate, geography, diet and other reasons was self determination? But now in the togetherness of global unity, if not being exploited politically, then at least by theory we could all now be of the same voice if not opinion. From the offices of the new United Nations generated to be progressive, countries had an arena from which we could be seen in true nationality. Although many of us got our independence at different times, changes had already been backed or supported by different outlooks, which had led to the support of our bloc cultures? This from source has immeasurably helped us be able and expand our religious perception, because when the bloc culture failed, our religions are all that is left, providing we realise they are our key!

Without prior understanding the political forces at work in this last modern era were underpinning religious fervour yet to be determined, but to be most obvious in the new Middle East. Without comparison yet, much change was going on throughout the world at the same time. We know of the political camps set in Europe, with Russia being ungodly and communistic, standing face on to American Christian democracy. Across to Asia among the ashes of the same world war, China also former allies of America, had now turned in full coat to also become communist, even if taking the works of Marx under a different interpretation? All of which was poured at slightly different timescales into the finished product of the United Nations emerging from the funnel politically, but not yet fully working. A measure of which, was how from the two camps evolving the consensual agreement, was that religions were to be of their own concerns?

All of our histories have been united by this last modern era, taken by real time to be the start of the third millennium. Looking back to any war will be seen from where our biggest crime to humanity has been committed. From the ruin of conflict, ordinary people have been laid waste, which even from abstract thought, can not be justified by any measure claimed from civilised society. One difference only is always allowed, if one was to defender a family member or close friend or even a child from being mortally attacked to the support of crime, the question of restraint does not need to be answered. If the attacker was felled in death, circumstances already laid by the first attack are covered by our instinct! This is in cause that we might not yet be civilised, it might become fact, but applies to the role of our taken prospect of 'Natural Body Death'. For which we are civilised!

War and die, even that the self desire is to kill or be killed dominates, then each of us on that basis is covered by 'Natural Body Death'. Apart to war, if through personal defence from the most extreme form of provocation, death results one way or the other then the loss is to 'Natural Body Death'. Our taken prospect is not to sanction forms of death but show us how we can encounter the processes of evolution and draw them into the spectre of creation. Also to embarrass our political standing, where if by national desire we go to war with another country and lament at the loss of human life, lament also when the political quota of retribution is extracted in human death after peace. I would further set to embarrass our political agenda, who when side stepping the aspect of our religious reasoning by standardising the concept of the spiritual to be concerned only with our welfare after death?

From our runaway or run ahead politics, it has been easy to shift emphasis to its own importance. It works to the born assumption that we are civilised and therefore demands direction from that base. It purports to lead in the avenue of support upon all those connected, but does not, because it only knows how to operate by half measured reasoning, like from the example of war, it separates us by the steps taken syndrome. Since the end of World War Two, with the slaughter of fifty five million souls, we still continue in the belief we are civilised people. We have run to hundreds and hundreds of separate conflicts with a large proportion of them having ended in the status quo, but all errant because we still kill of self, which has been by political trends in line with commerce and trade and gain? From which we are not to side step religiously, because in balance by the scales of natural justice, politics has tended to lead us in the prospects of a timescale to the supposed welfare of generations of people. Religions are to the moment of now, to their own balance in relevance to the individual's lifecycle. No larger picture can be painted to support my own reasoning than the one painted in the Middle East, from where was led the second formation after Europe, of our political bloc culture standoff, to taint future political reactions and the re-emergence of religious fortitude? Hardly noticed when for the first time Islam was to come to the fore in religious terms on the world stage, simply by the reasoning that we have pretended to listen!

For an active example of half measured reasoning, which was to carry the steps taken syndrome, we might turn again to the Middle East from just about wars end, but first take in all of the settlements in Europe at about the same time. I have stated by numerous forms of reference of the general state of European harmony with its own brutal edge. Even by much longer timescales than expressed by my own reference point of two hundred years of conflict in the first new era of modernisation. When the boundaries of Europe were redrawn at world war twos end by tracing lines on a map; the differences were absorbed by normal reference in the European style of horse trading areas of one country to another, but settled there in haste because of the formation of political blocs. Eastern or Western by definition according to the political standards held, but only by the form of politics being democratic or communistic, but not then to speak or offer religions as standards, until new standards were claimed!

By the same paradox, our European form of communism, was aiming to create world imperial domination supposedly through the ballot box, but were never able to suppress our ingrained concern of purpose through religion, which stood well for us all individually. American influence had now, then, become paramount to the reforming of European political alliances. Not least from the complete successes of her economic muscle gained from world conflict, which has to be commended, for she literally was the rainy day nation when needed to regenerate most other economies laid waste by the cost of conflict. In so far as she carried this means of assistance over other matters, by long quoted reference, she was to push for a plan of self determination by other nations, even if former colonies of European countries. For reasons of political morality, this aim was to promote self awareness by us all and open all markets to the world base of business, which seemed to be a good political solution to help America and thwart communism?

One of the most prevalent forms of national realignment was to be done in the Middle East, but by the most extraordinary of circumstances still not realised? In the first modern era of de-colonial internationalism, the eastern end of the Mediterranean Sea landward, had been mandated to Great Britain and France by treaty after the end of the first world war. Under several and various mismatches to the countries in care they were holding mandate over, Britain and France found the cost prohibitive. Bankrupt from the Second World War just gone, they almost allowed what happened in the Middle East, happen?

In the case of the French, matters were addressed by their own particular style of management. From the British perspective far more complicated instances had several bearings. Although in the whole of the Middle East there had been and was persistent forms of political management one way or another. From indignation there had been no religious substance given to the feel of the area by the power of mandate. Britain operating by her own style of management in the matter of form that she had worked the control of her depleting colonies anyhow, continued in the area generally known as Palestine just the same. But like us all, was unable to see the consequences to the future or later, the far future, bringing us to the turn of this millennium? Still not knowing how actions then could have stimulated reactions now with religious fervour used on both or all sides!

## CHAPTER 48

Mandate to Great Britain of Palestine then, was simply a political task of running the matter of control upon the people already there. From the colonial perspective nothing was new, from the political perspective in the form that things were expected to have progressed, nothing was new. Here in Palestine after taking over control from the Ottoman Empire post world war one, the situation in hand was to control the management of the area, nothing more. That there was considerable disquiet already there between peoples of different religious alignments seemed of little consequence to the politics of political mandate. From the Christian perspective through the form of government in the form of management, what consideration needed have been given to those of area practicing in the religions of Judaism and of Islam, none, if one religion was pre Christian and the other post Christian?

Between the First World War ending in nineteen eighteen and the Second World War ending in nineteen forty five, in the area of Palestine particularly, there was an increased force to the aim of self determination on religious grounds. Much of which was uncounted if not unnoticed, what was or could have been or is or was the effect there has had a bearing on the whole world ever since? Even now to be still unseen, because all matters of attention then were half measured and in all cases every further political move taken carried the steps taken syndrome, unknown. By natural development in the area of British mandate, there were two main religious groups operating to their own faith of Judaism and Islam, overseen by those of a Christian influenced political standing? From the European perspective the matter in any case was almost only driven by one of the connected wing of politics, through the arm of commercialism that steered choices to bloc culture satisfaction?

After the time of the Ottoman Empire, even while being ruled by mandate, peoples in Palestine were awakened to the broader idea of self determination, seen by the way things were turning out in Europe. By the natural habit of the time, in order to set the best case first, means of force were adopted against each other to establish demarcation. Unfortunately, in these or under those circumstances, the matters turned to the forum of power by suggestion? Even that most of the criminality was directed against those of the different religious standing being Jewish and Muslim. When the opportunity of forcing any particular case was put forward to enhance a proposition, shortly after the close of the Second World War, this was done by the Jewish community in the matter of political terrorism!

On wholly political grounds the Jewish community in Palestine began to make the certain case that they were ready for self determination by the bomb! To counter the dither of political time wasting, they proceeded at pace to take advantage of the political situation in Europe. Now moving into the bloc culture that was to bear upon us all politically for some time, also to take advantage of the situation of which European countries were to make active plans to hand their colonies back to self determination. In so far as these decisions and opinions in the lands of Palestine were ongoing, all views were opaque, for at the demise of the Ottoman Empire the Arab communities were fore-promised their own determination, unfulfilled? Where the second European world war came in juncture from several and various political decisions, even if half measured, new opinions were created to lay standards for future world, bloc, American or European conduct!

From one hundred thousand years ago at a burial site in northern Iraq we were to refocus to events in the Middle East. From where we had created in name three of the world's foremost religions the latest only some fourteen hundred years ago, we were to attend to such matters by political determinations only. Half measured working has been the bane of modern society by more measures than one, and I believe I can show all of its destructiveness even unintended, unless connected to the steps taken syndrome? I also take advantage of the abstract to cover what I am prepared to say, I have already made fair representations how we have tackled most of our affairs with the political coat, which has been the driving force to the steps taken debacle wherever found. My format of presentation is to be taken in full measure of myself holding understanding to the principles of the Fifth Dimension and its attachments. Where I wish to illustrate my own ability to pass on to others, is from how abstract reasoning works, already holding three taken prospects and one active, without overlaying them to the political thinking before they were announced.

Because of recent events and the expected forecast of future events, for best reasons in the bloc culture mentality, actions above intent were to drive some of the decisions made about the Middle East to be half measured. With the follow up right, but mostly wrong, in full promotion of the already steps taken syndrome? Effect then, could not have been counted to events now. If no more than by the pace of events and even to be of cultural recognition, when the Jewish community of Palestine enforced a terrorist regime on the British mandate, at the same time gaining some British and American support to their ends? It was very much easier to cede to their demands simply by the force of their political organisation above all else. For the Arab community, less well organised and in smaller politically designed groups, although from a larger population base, was less productive and organised to their objective? For the British it remained best to conduct matters by half measured reasoning, just to get the job done and hand over self determination to the symptoms of organisation politically? Then allowing all future events be conducted by the peoples there themselves from where they actually lived to what ever ends, but only politically!

Although throughout the world there had been various moves by everybody everywhere to the political form of self determination, religious conduct in many cases was closely intertwined within many societies, with none more so than in the Middle East. By the indecent haste of half measured reasoning, the events there close to wars end were overwhelmingly half measured. Features to drive the misdirected haste were long standing and driven by political motivation, which again was fully half measured by not including the religious aspect or mood of the area and more so to have not included the full religions of all the people there. Political features by the reason of trade and commerce were in some respects tied to it by the fact of the Suez cannel and the growing awareness to the bonanza of oil reserves in the local area even if not in Palestine directly!

If due to the bloc culture divide evolving between Westernised American democracy and Eastern communistic imperialism at point of reference from Russia, then the Middle East decisions were made to fully cover that format. For on those political terms, when Palestinian mandate was foreshortened, the new Jewish state of Israel was to emerge, backed in quick succession by America nationally, being the first to recognise a newborn country! If by countermeasure, Russia on her own political agenda went to support those of the same area in direct opposition to Israeli desire, then the second phase of the block culture was completed, extending attitudes in Europe to the new Middle East.

I have made much of our ability to make decisions half measured, forms of which vary considerably. If at first, half measured means for expediency, then things might not always be too serious, because we are just trying to get things done for the best of reasons. If errors have been made, then to untangle them and by correcting them to deeds done is not to be difficult? For me or any of us to label things so, can further be done by inclusion of our one active prospect of the Fifth Dimension in 'One for the Whole'. That medium invites us all to consider all aspects to any decision, yet through the one ideal of accounting to the whole!

When Israel first raised her flag, it was without that one active prospect, when America hastened to endorse Israel as a sovereign state politically, it was without that one active prospect, but was half measured. Half measured because the Palestine question was prevalent and needed long term settlement, America and many countries of Europe showed unsettling bias to support Israel against others of the area, country or people or religion?

Nationality and self determination had been a promoted factor from American politics since world war one. If market forces were the main cause, then even that carried its own morality of judgement, for it implied that each nation was given the option of choice. De- colonisation was a requisite to these ends, and although this was done over the next forty years after Israel, most was done half measured. Where the end has been set like self determination or nationhood, then in order to proceed at the right pace full measured, all considerations must be given to all points of reference, by all the people involved! With the first argument to be displaced, that it is not the strongest or fittest or fastest who make the choice because of natural selection, but the broadest, those of us who already see where the line is to be drawn and from where to where!?

From Israel's time, the factors not in place were unavoidably set to be carried forward as normal conduct, then later to most of Africa, then later still to most of Asia. I had set the conditions to be in the hands of those involved; and of the preceding time from where there was an upwelling of Arab distrust to the possibility of things going against their desires by the attitude of those in mandate and others! It was known that there was much opposition to Israel by a vastly superior population in numbers, virtually the whole of the Middle East, not just from Palestine. Through the whole matter, what was unseen, even misunderstood, was that the Arabs although volatile people generally, were by other circumstance to be of a laid back nature; and in case, allowed what was happening, happen?

If from the bloc culture mentality developing in Europe, now to be more under American influences militarily, then to extend or have a politically aligned country in the Middle East became a necessity before the considerations or morality. That is how half measured reasoning has blighted us by many divisions since and continuingly. By the appointed time, Israel declared her sovereignty on political grounds and to instant conflict with the surrounding countries. When to be the first in supporting this new country by recognition, America endorsed the half measured reasoning by the more damaging steps taken syndrome, America supported Israel politically, forgetting the other religions of area!

In process, to dissect my half measured reasoning to look at the above, we must take away the vitals in respect that it was Israel who declared her own independence. That many of us will bargain to gain our objectives was done, that any form of opposition anywhere world wide by the same example, was against any so proposal, then half measured steps were taken for expediency. By the haste of declaration, Israel did not even define her own borders, but kept close in line to some of the proposed areas if the new state was to be? One point of reference is to be of note, that at the time, in nineteen forty eight the new countries capital city was Tel Aviv. The only real appointed concern to the parties who originally drafted the proposed new borders because of pending partition, was to state directly that Jerusalem was to be of shared influence because of the tripartite religious connections to it!?

Half measured means the taking of actions without the full knowledge of circumstances. By the cut and thrust of politics, if Israel acted by her own best interests at that time, it was done without scope? By unconnected circumstances the making of the country was mismatched in tone to plant a European country in the middle of lands holding to a different cultural standing. It was done with the biggest support from America who in her supportive role of Europe proper against communism, the ungodly, was hastened to maintain the prospect of Israel's European standing. Because her overwhelming forms of direct support had come from both Europe and America by peoples of the connected religion of Judaism.

Who for generation upon generation running into centuries and centuries had never been to the Middle East, who by their own motivation never saw to consider other peoples objectives which were exactly the same by life's drive but not necessarily by political ends only. By half measured mandate it was only required to focus on the lowest common denominator, in this case, to nationality and then to self, which again was its proof of half measured working by not considering others as self also!

By the same consideration, it is that we must not drive on what has just been written above in respect of Israel, for reactions to that statement would be half measured. For me to have indicated that the essence of decision taking was done by European political thinking gives only part image of intent. Indeed there had been an underlying intention of eventual Jewish statehood; if the only or best way was to lobby the European holders of mandate, then this was done by the need of European thinking? Not least because certain direct offers to have been made to the Jewish Palestinian representatives in the name of the European powers were that they might be offered a permanent homeland cut from the East African colony of Uganda? Which by half measured European thinking, was a way to solve the rising conflict between the Arab and Jewish communities in Palestine at the close of world war one!

Half measured thinking now, has the name to be vacant thought, a means by which words are expressed to accommodate any situation as it arises uninterrupted. Many people like me will comment on the total effect on world affairs of actions or judgements taken at the end of the Second World War. If change was imminent then in some cases it had to be forced. Unfortunately the half measure approach of offering one people part of another's land space, had to be obliterated in substance, which might have been done through the use of terror tactics! If the outfall was to change political thinking, then it was too slow when the medium of our religions were already there and had surpassed political thinking by the age of centuries. From which the only key to understand this was in formula?

Un-bespoke in the channels of history, is that little chapter, whereby for a time some European thinking was that it would be alright, even correct to set the standard so that a new state of Israel could be set up in the African colony of Uganda. Fortunately the arrogance of such a plan saw its own demise, but was not easily opposed. Until the far reaching consequences of religious sufferance was truly considered even if yet not fully understood, but only one way and only half measured. At that time, for Europeans to have considered thinking in those terms is a good indication of how far we have advanced since then, but we have to update new actions so as not to allow steps taken endorse bad original half measured notions! By the same expression that Israel took to her independence away from Africa, does not always count as correct against that original European plan, because along with American support all now had a foothold in the Middle East politically?

# CHAPTER 49

Religion is a force held by us all without the realisation of its importance. Many of us hold in the comfort of its fortitude, solitude it's bracing strength and the peace of will it can bestow if left to flow over us by its very structure and meaning. But all it ever gives in the reality of time is expectation to for and of the human character only. Religions are of humans only; religions belong to the aspect of creation drawn from the time force of evolution. Where evolution is the span of gods thinking, creation is to the knowledge of gods time! For the mighty crocodile to have been upon this earth for some two hundred and fifth million years is to the value of evolution. For the next same length of time more crocodiles will propagate and propagate which bears only to evolution and not creation!

That we humans have been here for only three and a half million years until one hundred thousand years ago, when we buried a clansman in ceremony, we were then of creation, for we had broken the boundaries of propagation. As the fourth dimension has raked and parried the substance of belief by the mediocrity of unwinding the jumble of scientific experiment into formula. We had cast the idea to the force of all learning if needed, which is to be encompassed by a single lifecycle in the not so sophisticated form of believing in God, setting us to many, many religions. But to alter the form of our future gods by the delivery of the conditions and circumstances to have fallen upon us by climate, geography and other reasons, expanded from when we had found god one hundred thousand years ago? But did not quite know what to do next, in some cases until more than four thousand years ago when we associated what purpose was, but forever to be attached to our future?

Where our drive was to be in the direction of hope and purpose, we had no learning to our own ability, because the mixture of god and time and mankind did not quite fit. Not least on the principle that god is everlasting by his role, yet we are of a lifecycle to be of our own making of fifty plus or seventy plus years, stamped by the circumstances of climate and geography afore mentioned. While all was going on in the ranges of history; the policy of human management grew from the civil control of long, long times past to the modern era, where we had come to be super civilised in the pursuit of political ideals only, because from some forms of politics we had lost the ability to focus better by religious intent in putting the innate to be the aim of purpose above the unobtainable? Even from now where we have the Fifth Dimension and its three taken prospects with one active, in 'One for the Whole', we have to rest with that situation. Because we cannot overlay the past with them, but only imagine how we might have cut our decisions better equipped in the balance of our human singularity had we had such help previously?

My intentions by reference to the Middle East, has been done on two fronts, to show us all how any variety of reasons can be acted on from perspectives of self interest. From which the areas importance is raised on every level by its unique qualities off tenure. No where else in the world since any time, had there been the compressed images of three self emitting religions. Nor had we the convergence of two colossal political ideologies of the same magnitude any where else in the world. Here in a small portion of the world from forms of alignment to either member of the political bloc, communist or capitalistic, was a collection of countries predating the concept of self determination by intention. Kingdoms, Princedoms, autocratic manifestations or plain dictatorships were plying in the craft of unified modernism without political consequence. Many overlaid by serious and direct religious determinations, not least Israel, who might have pulled America in the forced action of employing the steps taken syndrome supportively, but for other reasons!

From her own determination Israel was to take her actions for that end even if having misled others? Whether of not her cards in hand were only partly shown to America, the consequences were open handed. If the political suit was all that was seen, then the cover was much the cause of our present world situation, for the hand that the emerging Israeli politicians held was held to what the Americans had thought they saw. By abstract reasoning it can be given that the American support for Israel was to political determinations only. Then in sidestepping the fact that Palestinians were already there, American and European emigrants going for the beginning of the Israeli state, were Jewish by religion!

America might have seen the installation of an allied state in a sensitive area of the world to be westernised, from that the political drive was prevalent. She knew also of the religious significance to some of the Jewish thinking of the day if only through Zionism. But sought in thought to place these issues in context of political importance only, unfortunately showing by representation to us all for us all and by us all, how we had let the medium of some of our religions become less valued than others by default. Israel set in clear motive of her intention to return to the lands of her birthright that might have been or had been ordained, by whomsoever? Steps taken from half measured reasoning, has since forced America to unstintingly support anything Israel might now wish to do on the world stage. Even in force of Israeli religious determinations by whatever standard or just by claim politically. The steps taken syndrome can put any of us to the embarrassment of actions unread, unseen and misunderstood, by having followed half measured reasoning if even for expediency.

By the unfolding of events in the Middle East from wars end until now, from the Israeli point of view, they always acted in their own best interest, which could not have been questioned religiously or politically until these past five to six years. It is not that any finger is to be pointed at them as people or Jews or in being Israeli. Fingers from us are at the mirror, and in reflection above each image is the text in all languages, 'One for the whole'.

All of the above events, even my own interpretation are linked to there own value of truth, where the political divide could cut and preen to all circumstances, our religious shoulders were to carry the head of reason, but tied to its own conviction. Until these recent five to six years, religions had been forced to take in the boundaries of ridicule imposed by the fourth dimension, unopposed. From where all religions had felt the swoon of discord by scientific impositions and revelations, some carried on regardless, which entrenched staid attitudes of righteousness into open aggressive postures, but only to reaffirm political standings taken already. That the Middle East is to be of much significance to our future reasoning carries weight from the pictures we had painted in other parts of the world over the same period of time but not so intense, unless religiously!

Politics of no significant flavour was enacted in other parts of the world without being underpinned by religious standards. With America having set there control to most commercial enterprises in the former Asian theatre of war, the picture there, although not of the same colour, had bearing to the Middle East. In Asia the divide was more directly political, almost set to the two colours grey, owing to American westernised democracy with a Christian background, and the newer form of Chinese communism aligned to Russian Marxism, ungodly? Each side came into dispute over the former Japanese colony of Korea. By no more than the reasons of expectation, the peoples of Korea wanted their own sovereignty, which had been usurped by soviet Russia, China and American demarcation at the end of the Second World War.

From the desires of expansionism by result, China exerted military rights to the whole of Korea, from the particular time set, the responsive reaction was from the emerging new world order in the form of the United Nations to be a body of control in future world affairs. This sad action in the form of a localised war was to have created or endorsed the sanctity of war by the means of political standards only. For the first time in world history, because of our bloc culture stand off situation in Asia, it was possible to have vile and costly bloody conflict while holding to the agreed stand off culture elsewhere, but only by political determinations. By the same evidence, at the cost of about three million human lives only; the Korean War returned to the status quo having no further input to our education since, for political actions are nought but burnt gas!

Beyond being critical for reasoning, the war on the Korean peninsular was doubly wasteful, for although it was by political motivation only. With eastern communistic values against western political commercial values, with the final cost in human life weighed against which side was best able to supply the most shoes or hats or gloves, whose politics was to be the most satisfactory commercially? Communism having turned to the philosophy, not of provision, but commitment only, had run all for the sanctity of the motherland endlessly and to human life ending! For the United Nations who essentially was to be the voice on all political occasions, did not allow China any opinion, which was of the time, but was to give the impression that religion was to hold its own weight, if because America was Christian?

America, in the overwhelming role of the worlds mentor, not least by her financial support, found that she was in the unenviable place of being the first line of defence to alien types of others political choices. When met with such cases as the Korean War, from her international obligation in the form of leadership, she was prepared to give of her own sons in payment for the world which at the time, it was better for us to strive for? With or without her own obligation, but thinking of the possibility of more communistic wars, America could quote in chapter and verse her own religious sentiments in being Christian and in being right by Gods ordinance? In some cases with God on her side as an acclimation held only by her own feelings, which could become a matter out of focus if dealing with the politics of communism by treaty, reason or War!

Not far from the human blight of the Korean political ideological conflict, where from result there were only the dead to count. In the Middle East, troubles between the Arabs and Israelis was turning its own course but taken in the triangular, from a flat political understanding. From any number of reasons that America was to support the Israeli state ever growing in population from European extraction politically, further incentive was added by the counterbalance used by the Arab people. Who accepted in ready numbers any aid supplied by any other countries even if communistic, Russian led and ungodly by declaration!

America's western foot in the Middle East was Israel, yet by the familiarity of having co existed with forms of religious forbearance from her constitutional beginning, she quite naturally took to the fixed political partnership there. For any informal treaty between the two countries on western terms, neither side saw any mismatch, even if the combination was to become sinister politically? With further communistic support to some Arab countries in the Middle East, bloc culture was being defined under different criteria than in Europe. Even that the ungodly communists were in face off across the iron curtain; the aspect there was to the political structure of defiance. Whereby the Arab and Israeli stand off was to become primarily religious, no hint of understanding to the emotion of that fact was seen by either America or Russia from western democracy, Christianity or eastern communism!

However pointed that last sentence rings in any ears; the full situation to any understanding of outcome is now covered by the Fifth Dimension. Even when from the first instant of declaration the Jewish nation of Israel was born, it was openly done by the code that they as people were reinstating conditions to Gods appointment, returning to the land of their fathers?! That they did not or had not since defined their appointed borders, is indicative that the whole event was to the full emotion of modern politics only. A longing desire was met whereby peoples who had been stateless or were integral to many, many others states if most were from Europe with many from America, also found religious unity politically in the new Israel!?

Because of the politics of communism, in a twist of the macabre, the worlds other super power of Russia was also posturing, but by the same un-recognition as America might. Russia of her own lead in support to many Arab states, done so from her own ungodly stand, yet to peoples of the most active of the dimensional Middle East religions. America and Russia looked across the political divide at each other and both failed to see the significance of our religious fervour from any hand or from any God or from any voice! Perhaps the biggest of our errors might be seen how we had allowed the arm of our civil latitude become the voice of purpose instead of listening to all, but hearing nothing!

To compound the style and depth of the wrong turns we had made by using the wrong maps, when not less than fifteen years after the Korean War a pinnacle was reached in the Middle East. Although of short duration, there was to be direct conflict in the stylised form of war conducted between some new Arab nations and Israel. This conflict had more than one facet but was only measured in terms of the bloc culture standing. For in the clouded indifference of political intrigue the line drawn under it was to the measurable ability of eastern or western weapon systems. America and Russia supporting separate sides by the means of finance, also laid great weapon arsenals in the area to maintain the status quo. For what they had supported politically, when battle commenced even if in the background, both superpowers were guardians to their own philosophies. On that phase of events half measured, at that particular time, the steps taken syndrome was self perpetuating, because once again when the dead were counted, the stand off was executed in the same manner!

I make these next observations by my own reasoning from standards developing these five to six years ongoing, but not in hindsight. That democracy and communism were standing toe to toe at several places throughout the world in the confrontational role of being weapon masters to many countries of new nationality. Of all the posturing done, it was in the style, backed up by the whole political spectrum with all attachments. Alas some features of very great personal concerns were glossed over being understood anyhow or not being understood anyway. By intent, superpowers nose to nose on the primary concern of political matters, simply accepted gratis that those whom they supported precipitated in like manner. Therefore they carried on and on the political avenues in the overall conviction that all was well by both camps at that time?

Without back thinking to then ago and not carrying some of todays reasoning, I can make further observations in relation to cause and effect but not in statement. From that first real war between some Arab states and Israel, victory was measured in the small matter of survival and in the gain of territories by the Israelis. One such gain was for them to have overrun the city of Jerusalem, which was also of immense importance to other religious groups. That this historical city was to become the new capital city of Israel tells many stories, but to different audiences by effect or result or notice or concern! Not least when considering such matters it is to note, that by not more than twenty years before Israel took her place in the matter of nationhood in accord if overseen by the United Nations? Formed to the theory of natural justice, how then was our United Nations to support natural justice?

Although Israel did not mark her borders from the instant of partition into nationality, she had complied in form to leave Jerusalem under the mist of agreement an open city. By no form of judgement can we relate todays thinking, even if clouded by other events to then ago? But other questions raised without that form of judgement, stand to that very time ago on religious terms only. If Jerusalem was part of the bounty of war, that it was of religious importance, it had to be rated by the concept of religious fortitude under all religions by desire only, so that the matter of religious fortitude could not be ruled over, only questioned. Not sided by the tone of political expediency, which compromised the whole structure of the prevailing situation first in area then later to the world! Communism was not concerned to the matter one way or the other, counterbalanced, democracy was not in the form of reaction to religious fervour if not expressed by a better communistic weapons systems?

Half measured application and our steps taken syndrome are just as they are by certain features I have been describing. If Israel had indicated at any time before, that it was her intention to incorporate Jerusalem into her divide, even through religious means, then it was now done half measured. For America not to prepare for such consequences then, it followed to be of the steps taken syndrome, which means for a primary action half measured, steps taken is negative and detrimental to our wellbeing. Because we are continuing in the vein of endorsing the wrong decision by indifference while following our own track habitually. Which can close all the avenues of expansion owed to this our single specie. All of which can be better measured how we are to review certain religious fixtures, particularly in the Middle East by looking at the whole of this last modern era added to in the last five to six years when we could propose for ourselves the Fifth Dimension to the status of our conscience!

## CHAPTER 50

From the Middle East once again without assertion we must look to the future not because of twenty, forty or sixty years ago but more to the reality of fourteen hundred years ago. For that is the mark in time when all of our histories were to converge by unmeasured or uncounted circumstances through the religion of Islam, my third dimensional faith from the one and same area of the Middle East. Judaism into Christianity into Islam is the actuality of progression, that is why that same area belongs to the whole world. Microcosm or supernova is to be all the ways to describe our reactive sentiments for the area and also because of recent complexities of world politics.

Communism and democracy are world political systems, that both are for the people by the people or by the people and for the people shows there complexity? That I would mix them in small contempt stands from my own aspect to be political first and religious first. That politics might not see this is of our failing, and where religions might not, is of disastrous consequences. No where else in the world can this be seen than in those lands of our first monotheistic religions? Having created a giant of human understanding to be the holder of all mysteries, we sought in the field of discovery to endorse that what we had, was sufficient to what we needed, for by first reference we began to believe in the sanctity of god the creator. Four thousand years later we began to believe in our own doubt unless carried by the iron gauntlet of political management. From the small example above we have come to the same conclusion from three different standards of reasoning through three dimensional religions from one area and from one people!

At the time of the first Israeli Arab condition of actual war, in addition to those two military combatants, their religious codes were one to be Jewish and the other to Islam. One side supported by a massive political connection being mostly Christian by belief, and to the other side equally with massive support being atheistic. All of which was then and is now impossible to unwind into the sense of reasonable compatibility, because over these recent years of this last modern era, it has been found to be politically expedient is to ignore all else unless of that political hue, unless expressed religiously, but by whom? By reason of all half measured actions unconsidered the patch work of the steps taken syndrome has been unmanageable by the first count that both are done from the same source. Where victory in any case had been counted in the strong surviving by natural selection, then from it, to pronounce joint equality is unfounded if made so by the victors. Because of such examples, the patchwork of repair after the Second World War, was and is now mismatched.

How ever the vital repairs of this matter were done it was unseen, but they were done by religious determinations. Where they have double failed is that at first intention any remedy was political, but when turned to be of religious significance, because the western bloc was coated with two religious attitudes Judaism and Christianity. We broke the code of uniformity or any thing that the United Nations was supposed to support us in, because we were or pretended to be faced with ungodly communistic support of the other side as it was. No realisations had then or since been given to the fact that the opposition was not political or Arab, but only Islam, of whom no voice had yet been heard!

In as far as any side was to succeed by the total means of political realignment it was to have been the western political bloc who raised the final flag. For the communistic banner was lowered by political determinations leaving countries who had unfortunately hoped for better aims short fallen by several counts politically and in self need of realignment. Matters which should or could or might have been attended by such societies as the United Nations, who were wrapt in the concern of washing hands of such things by political expediency, but so far could not count what any true or forceful religious reaction might be. Because we had not been aware of how strong religious feelings were on all sides, except from Christianity!

How I present those previous paragraphs and the following few, stand to the Middle East only along with its three native religions. In notice of the new country of Israel being first dimensional by my own religious classification is only for reference. She is, and my connection is purely historical. In the same note that my second dimensional religious code is for Christianity, also originally from the same area by extension, but now supporting a religion that was originally opposed because this was its replacement? Although all objectives were always clear to all sides politically. When my own classified third dimensional religion of Islam reacted to any deals or results from whatever motives, then the die of modern expletives was cast not least to my own sited explanation of our total half measured reasoning and our steps taken syndromes.

Christian America and Europe supported the country of Israel at birth and still, in the Middle East by political terms only? This action was endorsed by two factors, one that the very style of the thinking in political terms by the Israeli people at that time and still, complimented western thinking. Indeed the founding fathers of the new state of Israel had all been briefed by American and European influences, which is no bad thing, but gives insight to how they were to lobby support for intention, leastways by political determinations. Factor two is still political reasoning that this western aligned Israeli state could be a buffer to other Middle Eastern countries aligned to the rising communistic influence in the area, but now to have waned? So is the picture of some political intrigues?

I venture to say from the minute the Israeli state was created in the Middle East there are peoples to recognise those facts in the above paragraph, yet will be incapable to rationalise my following comments. I will further point to the reasoning of that last sentence. Politics then post war, was for the first time a new business, with the intention of new nationalities, with the intention of commercialism, business globalism leading to total materialism, politics was set to produce the golden carrot. As the prime considered form of our human sophistication, it was to be the medium to rationalise all the purposes of mankind futuristically. It is perhaps very unfortunate that at that particular time of history, two great political enterprises who had contrived to destroy the worst end of the political spectrum in fascism, which ran to the master race theory. Were themselves now faced with having to offer reasonable representation on the physical level to their own people by the separate codes of democracy or communism. Neither fully able to deliver our requirement because we were now being forced to react to political concerns by the fear factor incorporated in the rising competition of ideals.

## CHAPTER 51

When the Israeli flag was raised in its capital Tel Aviv on Independence Day, there was undoubted joy cast over the occasion. Unfortunately by a feature unseen although clearly displayed, this great occasion for some, was to blight many others leading to unresolved changes in the whole structure of community world wild. Self determination by nationality on political terms is a great objective for everybody, save by our common factor! When done and in entering the global network of countries, all of our histories have to be contrived and adjusted to suit any circumstances. This has been done by virtually every country since the time of Mesopotamia. From where this picture unfolds from different creases has been led by religious factors considered, declared or imposed by what ever means and from whatever past, but not carried politically!

Having developed the concept of the Fifth Dimension as a product from us all, I speak only by my own reasoning with it, but without referral to any of its three taken prospects or one active in 'One for the Whole'. Much what I have laid by opinion is in and of the abstract, which in cases is the only way any balanced statement can be made. Mentioning again about the savage and tragic events of mutual murder between Arab and Jew in Palestine under British mandate, is done to highlight political indifference to the obvious. Although enflamed by religious reasoning, that factor was to be ignored by outside political thinking, who were already planning their own regional strategies and were set to remain indifferent to any religious lobby, unless self imposed by those involved!

Where I lead without direction stands, because the side of support given first to Israel being Jewish was political; the Israeli format was western. Even if the support was Christian and in like manner western, the Jewish sector moved by their emotion religiously, with the Christian sector moving politically only. Even that some support was offered to Arab communities outside of Palestine, no religious communication was entered into accepting these people were Islamic by religion. Without gained sophistication, our Middle East at that time had turned history without alteration, but by acceptance of irrational behaviour in all cases. Therefore because tomorrow is the start of new history, therefore we can repair the past without alteration?

I still have to offer my own reasoning to those events of recent times, which have been done on religious terms only, because of the particular area. Israel in first order in the name of the Jewish religion declared a homecoming to the lands of their fathers where Moses had led them lost millennium ago? Sentiments well expressed that they were justly carried by faith, belief and expectation, because gods hand was involved to this great occasion? By those involved first hand in this new adventure, when thinking in such lofty terms in did not matter that Christian countries were closely involved to that fulfilment for political reasons. Did it not matter also that in the very same area, later religions had developed, and the latest held sway in area in an unbroken line for fourteen hundred years through the religion of Islam? From the same bloody contest from the same reason the religions of Judaism and Islam were brought into the new contest of political competition, involving different factors through the political aspiration of some Christian countries?

By my own linked association, we can see from now at this time of reading this very book whenever. It will be taken in situation that the world's first monotheistic religion seen to be Jewish by many, sought in need the blessing of the new Jewish religion of Christianity to regress in habit and form, ignoring some parts of its own real history. Then in line we will see more clearly the reactive forces of the third balanced religion of the same area opposing any such perceived union to a political factor that confounded it religiously! That is only from now, because at the very time of these realisations by the first Israeli Arab war, no such thinking was ever, ever available to balance or counter balance any situation? For by the adage of some thirty to forty years the Fifth Dimension had not yet been ready or formed on any account, but only by stages until I set down its standards!

Even said now, at the time of reading, my motives are ours to compensate us all by reason of how history must not be changed by todays thinking. If it was for the Arabs to lose such a war to political determinations, where would the image of religion stand to such an outcome? Even if fully supported again by communistic means without a God force fixation, where/how could the measure of political, if economic defeat, be taken religiously? When all three Middle East religions were to their own format, connected but disassociated. How then to read or react or make the future balance by the supposed new era of human advancement and reconciliation post war, continuing! From where ever we begin to posture by one set of events big or small, we begin to set our own personal objectives covered in full measure by half measured token, with reactionary fillips to accommodate the outward gratification of doing the right thing at the right time which are sometimes proved,wrong again!

## CHAPTER 52

Some forty plus years ago from the tiny Arab Israeli war in the Middle East, we had come to the apex of some of our histories, leading now to the four cornered triangle. Simply to the fact that we were growing to be extremely advanced technologically, we began to lose our judgement and have been unable to compensate our opinions by the rational since? If our gains were under the shadow of fourth dimensional exposes moving the physical essence of the sciences forward or ever onwards and upwards. This feature began to cloud our best of intentions, being those through our religions on the scale of self examination, and by comparison, even if incorrectly matched to other religions and some political situations?

From my own thoughts I began to notice from a European perspective, that the order of church involvement with community and self, was turning or stood at the acute angle of change, but for the direction to be still decided? Change being instigated by the overbearing success of some or our political aspects, not least in the form of supply and materialism, spread to some people personally and to some countries nationally. If from theory there came a new awareness that politics was indeed to be of the foremost in concern to our direct wellbeing. Especially from the western pattern, with the Christian churches still promoting the benefits of life through death, yet the commerce of political industry was actually supplying demand, even if self promoted, that was cause?

By that same western pattern, it was taken that success in any quarter was deserving of reward outstripping the quality of effort. Allowing it to be altered or hid that the individual in such progressive mediums had used all of our histories. That we had put iconic reference to some of our peers because of their achievements was half measured and therefore carried to the acceptance of our steps taken syndrome. Progressively allowing us to react to any situation whereby notice given to an event was of individual excellence. Ruefully forgetting that apart from celebrity, thousands of others in all fields had improved our lot over thousands of centuries. Apart to religions which some began to admonish, we continued to promote the driving force of politics as the medium of our deliverance, which might have been, but never at the expense of our religions, except to ignore parts of doctrine?

That a trend or settlement to the promotion of gain was being set by a section of our world community, which happened to be European, American, western and Christian, which at the time seemed to be winning in the bloc culture confrontational epic, only from chance? Where politics was in the ascendency, and as I had indicated from my personal standpoint, religions were in soft decline. Automatic drive simply cut in to promote the continuing roll of this standard of development politically? Political choices of one sort were beginning to set all standards to all people, even done by all people, when the exact properties of success to one individual, eventually become the properties of all through business. Making the situation to generate human conduct past the limits of considered religious management of equality already foretold! Politics was seen to supply by result, only because of our shared advancement of technology!

Where the situation was really turned, had come from unintentional means rooted within political activity, wars end had brought bloc culture confrontation, which from its own drive had encouraged the mechanics of scientific advancement. With both sides promoting their own philosophical political demeanour across the divide, compromise had to be entered into through the medium the United Nations. This being done by the learning and standard of the then thinking, was set in the obscurity and obscenity of human classification by an accepted standard from the top down, almost in evolutionary terms? With the strongest where they were by natural selection, not from or because of the circumstances of climate, diet or geography or other reasons, but because they claimed it!

Through the promotion of stated human equality we sought to standardise events when self determination was being introduced throughout the world in the name of progress with all new countries welcomed into the fold to that equality. Alas in the reality of events from political determinations one way or the other, the measure of national equality was determined by the factors of, to and how independence was achieved!

In the new post war world, our standard of equality was set on the supposed measure of economic development or capability, equality being measured by political means only, from half measured thinking? Equality was pushed into acceptance by the full packet of political commercialism. In order to quantify status on one level only, efforts were compromised to and by the steps taken syndrome. We were to be classified into being first, second or third world countries by the standard of our economy. From a fact of politics then ago, our wellbeing was measured from or by the strength of our currency, illuminating the worst excesses of our steps taken syndrome. Having had time to be able and set up the model of the United Nations in part from the ashes of the defunct League of Nations. Where it would have been repeated and repeated that we must learn from our mistakes again and again. Much was done in honest pursuit, but from the exact same standing and by the same method to have occurred before. With the chair of the United Nations leading on its own account, being naturally first, then all subordinates were second or third world?

With western Christians having the biggest say in such affairs as the restraining force of ungodly communists, also in the bloc culture mode, little voice was offered on those events from any system to deny the principle of three worlds, which was to under-write steps taken? By other means and almost at the same time, our human rights charter was being redrawn to state categorically that we were all the same in equality by first second or third world standards. But by compensatory measures, we gave religious tolerance to all, even those that claimed no religious connection, only if to maintain false harmony by silence! All of which was to show us the absolute need of those recent times where we had tried again to set fair standards from where we were, automatically to include the attitudes of the day. Not knowing at the time, but for some of us even then the plan was flawed, not by intent, but from half measured construction. Without the force of opinion, I can set this question which can be left unanswered without any penalty. Had we set to revise the standards of the United Nations now, how could we start without having to compromise the standing between the eastern and western bloc culture, which is gone without victory!?

We could not pose the same question in any different time slot from wars end until very recent times like now, because we were motivated differently, certainly in the Middle East forty plus years ago. In the land of vibrant religiousness the history of the worlds two largest monotheistic religions, was set to crescendo from interaction with the first one god religion. In mandate the area had been in turmoil between its own inhabitants the Jews and Muslims with daily murder almost the norm. With the new state of Israel first to take full nationality, even ably assisted by America and Britain, although always first hand from and through political management, most standards of political uniformity were met. Even when from result Israel showed self religious gratitude with no inflammatory intentions. To have rejoiced so was the dangerous display of vilification over past events, yet worse, but without malice, to have regressed to a history situation of four thousand years ago, was uncounted!

With the concept of religious tolerance placed as a political concept for world consumption, this steps taken syndrome was to become the heat that generated the fires of some forms of religious fundamentalism. As far as the nation of Israel had entwined itself in alliance with vastly populated Christian countries holding to their different religions, because their conduct was close politically, somehow avoided the obvious, but that did not cover other perspectives and other results or what might emerge from other religions? With religious reaction set to react, it was done from current events then, but with a reduced concern for politics. With an almost two hundred year history of involvement with European conflict between the French and English in the Middle East, Christians as they were then, although there by conquest and mandate, in theory it was by the quicker use of weapon technology even then. Throughout occupation, in the whole time Muslims remained Muslim, reaction was not unsettled, until the new world order was set to allow maintaining the status quo! Being to some, the point of no return by the best effort of the religious doctrinaire in now first operating to religious requirements even if misperceived?

I believe and state now, but by reflection, that because of the standard and type of reaction to some matters in the Middle East, was born the mixed ideals of religious fundamentalism, not least by first reaction to political outcomes. Not to be a flat or dead statement, but the comparative note now, even with the Fifth Dimension and prospects, stands that the Islamic faith increasingly shows that finite edge by responding to the political outpourings of others!

## CHAPTER 53

Islam by my referral is its own religion, which is the same for all religions, from which none will be able to ignore the listings of the Fifth Dimension proper from future study. I look at Islam from my own perspective in not having joined in at the specified time of our most important political underwriting if through the United Nations, my understanding and reasoning are my own. One of the biggest complications of its very structure is from the very heart of this fine religion, although set to the working capacity of the individual, lead is given by the standards of some other individuals from their own point of view. Being no bad thing, but showing perhaps the need for a more defined priesthood structure, not necessarily of the Jewish or Christian pattern, but to be that one voice only might prosper Mohammed's intentions, that it is to the individual to express their own grace! From which the matter of being Sunni or Shiah, different aspects to the same faith, might fall to the better ability to recognise self and follow in unity of all and others!

I had previously mentioned the importance of fourteen hundred years ago with the start of the third monotheistic religion by association to the same area of the Middle East along with the first two. Taken from the conditions there then, the impact of its introduction overshadows many natural occurrences, which were not necessarily religious. With Islam now being the definitive religion of Abraham through Moses and Christ, then its picture was set in the conduct of the individual by doctrine, for which was to be the overriding influence to the living and lifestyle of self first religiously. Where the previous two religions of the same area were unable to maintain direct contact with adherents in the place where they also first came to Gods will. Islam through the teachings of the written word set to maintain this new religious continuity, because unlike others, the same written word was always at hand.

Without any veil of deceit, without the force of implementation, this religion began to hold the area as its own and has done so these fourteen hundred years, even if misaligned politically by recent events only? From how I have laid the supposition that we are the same single specie, has been supported by the reason and methods for our obvious differences of character. Geography, climate, diet and other reasons are to compliment my own and other people's theories to that fact. I have suggested that some peoples as we, perhaps directly from the Middle East some time ago, were or could be classified amongst other differences, to be laid back. I know for some of us we will first see that many people from the Middle East are very volatile people, but I labour the point using that very modern term "to be laid back", is to be recognised as a state of mind above emotional responses. From the wars of jihad and crusade between the young religion of Islam and the comparatively young religion of Roman Catholicism support of my theory might not seem possible, which is to be taken in both modern and older reconstruction.

When understood that Mohammed worked on two counts, firstly to the order of control by organisation to his specific areas of interest, secondly by religious fortitude. From the latter, by presentation of the final step to our glory in Allah, any conduct depicted was only set to improve the lot of all with the presentation of our own personal conduct for fulfilment. An extremely important point to this new religion, was how it had set to take in all aspects of normal living of the time including with people from city states or larger tribal groups, the combination of civil control also touched all by natural interactive cooperation. Islam was not to deny this, but include all to its banner in conduct and in spirit, allowing for the natural flow of civil to political controls take there own course without overbearing upon this new religion. Given perhaps that for the first time to a monotheistic religion the different point of focus was how the religion was structured on its own base without recourse to civil amenities. All of which relayed to new adherents of these new teachings, that civil order was only an amenity whereas Islam was an answer? Why should not any body feel content and laid back now, with their own piece of mind cast from religion, at last!?

I had referred to the Middle East to have been the Muddle east in connection to the last fourteen hundred years of our united histories. My reference was not in condemnation of all examples, but allied to religious concept at a time then, when nothing was really sure or happening in a positive manner? Even that the area was settled to all civil concepts of control and leadership of any type, expressed by any medium through the local tribal or city state communities. For Mohammed to have made his assertions so clearly for the conduct of the individual by the individual, this in concept changed attitudes to authority. Although generally taken that there was the overall need for government by whatever means. People in the growing awareness of the teachings by this new religions holy orders, were prepared to accept almost any form of leadership to provide the basics to the standard of controls already given. When from here and now, no matter what ever leadership structure there was, it was possible to accept its shortcomings because irrespective, our faith in Allah, was self assured by self practice. Giving in addition to what might have already been there, the further ability to take a laid back standing in conjunction with the living of the now to the overall outcome in the future. What ever politics was to give, our religion was to fill all of our needs!

I have not attempted to have set a standard of conduct for anybody by whatever style we might have come to accept, prompted by the circumstances of climate and geography. My direction is through my own classification of the three Middle East dimensional religions! Given that all three in following had set by intention to bring all men to God or Allah from their own doctrine. In the event that the link brought us to the comfort of being laid back by character has to be measured how we differ and from what cause? To have sited that Islam by linked association created for its adherents that fixation of character, even if the same as how Judaism or Christianity might? Then we must look to the changes taken and the reason for them, which I had touched upon, but only in the ambiguous.

When the eternal city of Rome took to hold the banner of its new Catholic religion some three hundred years after the death of Christ, it erroneously changed the emphasis of that same proclaimed religion. Rome in its own political structure of then ago already held in unity, countries of Europe and of the whole landward surround of the Mediterranean Sea. By the time of its demise, spread amongst the emerging countries reverting back to tribal independence, most held to the same religion centred on Rome. Differences to occur were because from that new nationality, types of political structure followed forgotten pasts? Yet by them, in some cases and places the political voice came of its own age following better self interest allowing change to be political first and religions not!

Coupled with other circumstance emerging, European countries moved at the same pace as everybody everywhere, dark ages of history bestowed itself upon us, but only in course to the fact that on political levels we were going backwards. By about the time of Islam the picture in the Middle East developed at a different pace than in other areas, here a new religion brought some new ideas, but many influences making religion more important than civil control? Many secondary influences from the Rome religion of Catholicism were brushed aside in area. In very quick time the new religion of Islam took in sweeping control to the lands on the southern shore of the Mediterranean crossing it at its western extremity into Spain, being Europe proper.

At the same time in other parts of Europe, words fell that new religious conquests had happened by others, which was always possible anyhow, but this time such changes to be forced were for religious objectives. If the fibre of the Roman religion was a force of unity to this threat, then it was promoted in self interest on or by two counts, from the priesthood in general and the competitive edge of nationality politically? I do suggest this was a partial source of fuel to European competiveness which did not really manifest its self for centuries running until the next sweeping change to our society, instigated by Christopher Columbus? Then in quick succession up to about two hundred years ago back to the Middle East by European interference. Then to Charles Darwin and his first publication, then to natural selection and the strong surviving, but not quite to the changing reality of now!

From my own reflections to times past and with no connected weight, I have been leaning to show how Islam now stands on the world stage, by mixing semi recent events and reviewing them from the same angle differently. Being a valuable asset in self assertion from all of our perspectives, carried in abstract thought until certain cross referrals occur, but without changing history. I have given much mention of the Israeli Arab war of some forty years ago when the main result was to increase the land space of the sovereign state of Israel. That world reaction was unheeded by Israel, had been put that the overall outcome had been of a religious settlement, which was to break existing United Nations codes of religious tolerance, only because our bloc confrontational surges were still very dangerous politically!

That Arab/Israeli event just mentioned although turned to have been of significant relevance religiously shows how our bloc culture mentality clouded many issues? Like when the whole world was almost plunged into unstoppable nuclear war, instigated by small events in the sunny island of Cuba in the Caribbean Sea, which is in close proximity to the United States of America. Highlighting the dangerous proposition of our developed bloc culture settlement by political means only, for in religious terms Cuba was Roman Catholic historically. But who had recently turned to the political aspect of communism for measures to improve the general lot of most of her people, stirred by the general world aspect of opinion towards better self determination, yet her standard religion was to stay without normal Marxist communistic oppression. Mixing all the metaphors of interaction we had allowed develop accepting half measured working, was also to bear in Cuba. Although Cuba took communistic self determination it was to be bound to the overall influence of the actual area of the world in which she was, not by religious acceptances but political. Even when America of Christian standing, set upon Cuba of Christian standing, the demand of political ends to thwart any supposed advantage taken by the bloc culture mentality drew its own solution. In at first trying to correlate all world sentiments at the same time without qualifying events to have occurred in Cuba, by almost having to go to world war following our steps taken syndrome supporting half measured reasoning, we almost reached Armageddon!

Short years after what had become known as the Cuban missal crises, how was the Arab voice to charge the ear of world attention to matters of competitive political management? Although some declared to the total destruction of the state of Israel, even if measured from ancient times, when we did have the violent edge of competition in the land between two rivers, this tack now was not viable! Not because either side was a separate tribe or clan, but that of this last new modern era, all of our aspirations have to be set from the one perspective always. For this is the time of the Fifth Dimension proper, where we will soon measure our differences by three taken prospects and one active as 'One for the Whole'. Before this equality of voice is raised, few alternatives were left to some of the Arab peoples by the feature of political defeat which was not a challenge to religious countenance. Leastways if the looser was only from the active strength of political ideology!

Of any victory, with the winners magnanimity dispensed with further deeds of repression, what might be used to raise attention to further misdeeds? Historically, balance done in the Middle East from political determinations, can be counterbalanced by religious overtones. Did this happen is a question in the rational, that I now classify actions into deeds by religious ferment is done in the pursuit of my own and our understanding. From that time, and following the outcome to most political affairs in the Middle East due to military excellence, any connection between Israel and America on those terms seemed to have been done in plan, whereby the most modern of weapon systems were to be tested in area under the true conditions of war. If the outcome was to always improve the aspect of democracy over communism, what voice might the sound of religion now make? If first voiced in that medium through Islam, perhaps for the first time in this recent modern era, we must take care by reference, not to treat the matter half measured!

## CHAPTER 54

The religion of Islam had not really changed for about twelve hundred years of its total history, that there had been different interpretations of some of it, is to be accepted by our own individuality. Although the internal factions of Shiah or Sunni both being into the heart of the same religion, were maintained to that religion only, which was not realised internally until other religions were involved? My mark in history to the first Arab Israeli war and beyond the reactive response was of a purely political hue. If it might have been seen by others, who thought Islam first, and military expeditions of force had failed them by the standard and type or application of weapon systems, against an arm of western democracy, through Christianity and Judaism. What better force than to rearm in the form of Islam. If Islamic fundamentalism was to prosper then by what measurement and from whose direction was this epitaph set? If fundamentalism was an aspect then, has my own timing been right to judge it so? Because from this work is offered strait forward complexities that might be added too by supposition or possibility, save by reflection in thinking now with our Fifth Dimensional prospects to offer change, but not to alter the written structure of history!

I have indicated on behalf of us all that our religions hold much more importance for many of us, but by varying degrees. I have and will use this last modern era to gain matters of clarification of how we should begin to react better amongst ourselves. I have made much of my first, second and third dimensional religions, which are to be marked in reference by there three different timescales. In consideration, at various places I had added to them the maxim of being in line, but also intermixed with all world religions. This is still my opinion, not by judgement, but in open acceptance that we are all the same single specie. Any hidden or covert reading of my intention is done erroneously, which can be established first by my own tone in not leading that we are different. Secondly, from how I and others had previously and still class all of our religions into two distinctive groups, not politically, but to be eastern or western by character. From the eastern sect inclusive of the main being Hindi and Buddhist has been given my collective other title of being pre progressive to the unfinished or incomplete life cycle relative to the physical aspect of us all. When by transmigration through the collective expression of timescales yet unmanaged we will find the same edge of complete understanding owed all religions, which can only be described as a warm and comforting fixation. Which means that we can now confirm that all religions set to the aspect of understanding God, have made the matter of the future to have been confirmed?

Without recrimination, from the same aspect of geography, climate, diet and other factors to have influenced us all one way or the other, my reasoning has had its own common base structure set. Because of the span of time for collective reasoning, that any of us might focus more to western religion relates only to our birth circumstances. With Christians and Muslims holding almost half of the worlds population between them, but to there respective religions, any standard was simply due to the progress of time! But in the placing of time and looking to the three religions to have come out of the Middle East proper, all have set the same standard, but from the different thinking of the time they were originally set? What we are to achieve now is to show our understanding of what is meant by a single lifecycle. Which is the same for Jew, gentile, chosen or appointed, Catholic, Christian, infidel or of Islam, Sunni or Shiah, all now to the importance to the value of a single lifecycle?

Although the Middle East has the historical connection to always be of interest to the whole world, geographically it is of the exact same importance to anywhere else measured to the individual and linked through our single lifecycle. My concern and others shown are to its importance on the religious level which without the realisation of political dealings had on one side taken to be religious instead of following the confrontational aspect of the recent bloc culture situation. Israel was Jewish by cause, but political by application!

When for many different reasons and in its own cause, Islamic religious culture began to re-emerge from any time since the first Israeli Arab war, stretching on to the demise of our East West form of bloc confrontation. From the Middle East of these recent times with several different themes, the mixture of Islamic assertion even if through fundamentalism took its own view and was considered to confront the political structure of the western style of rule and or government or so it might have seemed to some. What was happening in the fact of reality was that we through our single specie mantra, were beginning to look again in measurement of our histories and how they had been managed. Even without the notice that all relevant histories were and are of these very recent times marked by the end of the Second World War. We have so far been unable to equate them to each other and in what form or from what source. Until we can begin to reconcile our differences, which are the same in any case, how might we now look to our future histories when we have not really learnt what any position was at wars end, because we were prepared only to think first, second or third world!

I have made much of our bloc culture world wide, this of course was ever only by political terms, of which the two main types then, began to bat and ball each other. From where one would serve a wide the other might return a googly, from where one would hit a screamer the other would return a lob. When we tried to promote general world unity for the first time, not least because of the wanton waste of human life that had just been, there were many of us who actually did not believe in specie unity. Because of the cultural mixture of all combatants, from each step in any direction, actions were sometimes balanced from who was writing the report!? Some would evaluate the cause of success or failure by there own particular standing in relation to who did what and when and how? If there is to be any denial to this, then let it be explained how we were able to begin to set the supposed standing of human equality by reason of our economic come political standing to be first, second or third world countries? Yet from within each separate economic classification there were peoples from the three Middle East religions.

Our attempts were openly valid, because there was real belief in our ability to change for the better, but using the same tool of politics as before, we shored change up with the wrong mixture of material? The United Nations does earnestly stand to that end, but leans over when agreeing our differences? First, second and third world economic countries with equality of religious stature, fail when left as so and is compounded further to fail by that method of half measured implementation. With other supports hastily added to maintain our balance, our road was simply in the course of happening to be of our steps taken syndrome? Half measured, not sure why, but best keep going, steps taken syndrome! By action those conditions are felt when we know we are slightly ajar. In order that we must continue, in the haste of time we lay plans of thought without thinking! Like how to cover what might be an economic political anomaly, how can we justify our difference by finance yet be same specie by religion, but different by politics, yet the same from the United Nations?

I had hinted that we might not yet be civilised from the reasoning of abstract thought, I have appeared to make much of our differences although we are the same. From wars end we attempted to prove this or at least tried to. But for our laxity we must suffer our own embarrassment by all means. For us to lay our charter in open view with it's built in anomalies then paper over the cracks by thoughts of idealism, fell short on two main counts. By then forming concepts of human rights and civil liberties being the paste for the paper to hide cracks, the cycle was almost complete. Civil liberties and human rights have their own value which is not the same how ever we might try and deliver them, because they are both from our political edge. They pertain to conduct, expectation even outcome, they are to us all in equal measure by that reasoning yet fall on stony ground by application. For the simple reason that they cannot account for what we are, they have no ability to measure our differences by the mammoth overlay of our character bestowed by the standard of geography, climate, diet and other factors. Civil liberties and human rights are like chalk and cheese, because at times they are used to paper over the cracks they cause!

Even amongst us, from some of our religions those very expressions have no meaning, if from base they are individual by conduct or other means. Then our actions to the spiritual outcome we suppose, does not carry the edict of political control only the essence of civil management. Even if we depict it so that we stand in fairness to religious tolerance by our human rights standing politically, this also falls short in method to force our unity by some factors and not others. Civilisation is not set to stand or fall on the few lines of script from any writings, but it has to be seen again, that all that has ever been needed to say had already been said and written. Our connection by today is not to ignore anything, but set the situation to the very best time that we might begin again to change nothing more than our attitude to each other.

To be uncivilised is no bar to the future, if that is our lot, my own connection to the expression is not judgemental, but kept to the standard of query. For the word offers a barrier to our understanding if tied to the matter of fact whereby we can classify ourselves by the status of national economic value. Being a means how some could marry that classification on a scale to be civilised or not by natural selection where the strong are forced to survive by that ability only! Therefore any judgements made to the matter of individuality are quashed, not by result, but outcome. Given that the strong automatically survive, thence the strong also decide, eluding the matter of fact that our connection is not from the generational pull of self or personal contact, but more so from the likes of Shanidar? We have only changed a little bit on our way here, not least how at last we are beginning to realise that there are to be better means of communication above the methods of half measured choices by steps taken syndromes. Then even further from where we might ably connect our thoughts by the medium of our Fifth Dimensional reasoning carried ably by three taken prospect and one active in 'One for the Whole'.

## CHAPTER 55

My measurement of sixty plus years ago, almost equates to some of our reasoning of long centuries past from our time in the meeting of the old world and the new world in the isthmus of the America. Unlike the model of recent time in the Middle East, when from the Islamic perspective, there is the opportunity to advance our reasoning collectively; the meeting between the new world and Europe of old had its own sad outcome? Even if only balanced by the thinking of then ago to each separate side, but not to be carried forward in historical terms, because then, history had not been made. From what was found by the style of life and the standard of living in the new worlds of the Americas some four hundred years ago had a real bearing on our thinking since, seen or unseen. Caution must be taken to avoid giving opinion about matters of which we knew very little. I take licence because I have been fortunate enough to be able to measure a moment of eureka even if unconnected, I can evaluate what might have been in the same timescale of then ago without tarnish from todays thinking.

When we first travelled to the new world five hundred years ago we assumed we were civilised and in charge, being the nature or who we were and measured by many other accomplishments? On finding two distinctive spiritual aspects shared by the people there, nothing was seen for us to consider changing self opinion or motivation? Without counting the consequences then or even knowing how to project them to the future, we were obliged to follow the norm by our own distinctive patterns of behaviour. For one group of people there to hold to the Great Spirit concept was enough for us to be assertive in our own judgement, for we had been directly appointed to our task by the Son of God direct! Against other groups who had come to human sacrifice as part of their religious code, to take human life in tribute of the inanimate such as the sun in the sky. All styles, reasons or aims were employed by any means howsoever, to convert this unholy abomination to the human psyche. Of the methods used it did not matter that human murder and slaughter were the norm, not for the first time in our united histories, it was done for their own good!

It can only be said by imagination that any features, any events in the new world of the Americas was to be part of our future funnelled outpouring of human development, if only measured by the theory of progress. If the total effect of the European intrusion there was to set the objective of standards, much was done under the supposed influence of religious fortitude. I can suggest, but recognise it for its ambiguity, European countries moved at pace on religious terms, but mainly for political financial motives, being a trend used throughout the world from then until the supposed end of imperialism or colonialism. What better reason for expansion than at first to say we are here to stop these people offering human sacrifice, and in comic book fashion promote our correctional theory of murder to eradicate the practice? All of which from various time scales and mind sets have helped some of us realise our own appointment even if led, but using the feature of natural selection to show that the strong always lead. Then natural selection coded by time to be correct, all was set to progress by this manner by self reasoning, though we were not able to reconcile it with evolution against creation?

Even now, looking back to the rediscovery of the new world, we tend to paper over its impact upon us all. Without exacting any similarity, I hold that by my own reflective view of then ago we will be able to see where religious fundamentalism might find support if not total justification! We can now think today on our own terms without altering the past, but to always notice it for what it was without the overlay of modern thinking. From the many different tribal factions in the Americas then ago, there were two basic religious concepts, the Great Spirit and Sun worship. How they had operated and coexisted for what had been centuries was the complete norm. I had offered by indication previously that geography, climate, diet and other influences were involved in this, which is always the case?

Up until foreign intervention, there had been no historical signs to show us how the two new world religions had come to their own particular standing or how they interacted on the broad scale. I had intoned of the incredible link, between the incredible possibilities of journeys by Egyptian travellers from any tine up to three thousand years prior. The thrust of that supposition was relative to such travellers, if when they got there, they found only one basic religious philosophy embracing the Great Spirit theory. Having been preset before any planed route was used to populate the American continents, then the collective behaviour of the tribes there, were linked more probably to our specie unity through the burial at Shanidar some forty or fifty thousand years before. These clans, tribes or family units had travelled with that accepted recognition unknown, but because all group reasoning was similar, there was no interactive reason to change theory as all were of the same thinking.

Without comparison, if we look across the world at the continent of Australia; the aboriginal peoples there from ever or about forty thousand years ago from accredited first settlement, did not have other influences like Egyptians! Yet these peoples maintained the unbroken Great Spirit concept without any conceived form of connection or collusion with the population of the Americas. By direct comparison the inhabitants of the new world of some four hundred years ago and the aboriginal people of Australia shared roughly the same transit time in reaching their final settlements without any perceived contact to each other. By direct comparison the aboriginal people of Australia along with the inhabitants of the new world who had not shared the cave culture in the same manner indicated in other places of the world, maintained like spiritual outlook in or through the Great Spirit concept?

What can be said about the two American continents and the continent of Australia is that all are of the size to allow any populating be done in the manner of minimal contact with others because of there immensity. To the acceptance of the hunter gatherer lifestyle, the atmosphere of contentment is conducive to absorb the essence of ones surroundings as part of the medium to or for life and the living or ending of it. In balance, by similar settlement of others as we in the land between two rivers, producing its own competitive edge previously described, which further induced the style of natural competitiveness at all levels even religiously. Although they did not take the accepted form of most religions today, in area they fell into position to influence each other and to be altered from anybodies suggestions, but did not in Australia or the Americas.

My point of query in relation to our religious standards can almost be offset by how some have recently been reactivated to be fundamental or basic or searching. My connection is to link all religious fortitude collectively, not in the reason of comfort, but in an act to the reality of life today. When studied by what ever age I will continue to prove that our religions have been the driving force to the discovery of more than we can recognise, my motives are ours. Much of the delivery in account is to my own style because I have been fortunate to have been in the right place at the right time representatively.

What I deliver is in the face of perfect management, which is of no criticism to any of us personally, but is given to stand in support of my style, I have been making politics my and our enemy by name, because of how we spread conceit through it. We seem to relish operating it by auto drive, which is not wrong until we examine it by what it delivers and to whom! By example of semi recent times again, I returned to the central Americas and reiterated about our practice of human sacrifice religiously. My connection to be, was that it was not our religions that abominated that tragic practice, but politics only. Politics only was the engine most times, even when at the time from about the year fifteen hundred on the European calendar, we practiced inquisition religiously, from which we could judge the theory of human sacrifice was deemed pagan. But politics only could abuse others, while religion dressed the renewal in conversion, if anything was left after secular treatment!

## CHAPTER 56

From that very medium of politics only, is to be balanced how we can look at what was considered settlement in the Americas, particularly at first, rather than later settlements in other places of the world by Europeans also. In the new world by example we had another case of the funnel format, whereby lots of different ingredients were poured into the funnel, but the mix emerging from the spout did not relay what had been put in.

Since Columbus, over the next two hundred years the Americas were settled by Europeans with the baggage of intermixed national, political and religious wrangling which had not yet been settled? With political intrigue driving some of this migration, in many cases religious fortitude was the direction. Shown in the way that to the southern American continent and as far north as Mexico, the religious influence is mainly Catholic of the Roman kind if readily introduced by Spain and Portugal. Each giving to the people there the legacy of Spanish and Portuguese languages, but tied in concept to the one second dimensional Middle East religion of Roman Catholicism which generally remains today!

By line of connection, the modern country of America, bordered by Mexico in the south and Canada in the north can show much in support of my own interpretation of events. Although the language of America is English, to all states there are areas with very large other European communities, who broke with Europe wishing to improve their lot in the new world from the opportunities offered. Further north in Canada the language situation is somewhat different in respect that in some places there is a duel complexity, whereby the languages of French and English dominate by historical settlement. Indicating from the language situation on the whole of the two continents, the intricate way they were settled and from whom and by the mixture of standards set by individual European countries, of which I have used to show our differences are the same!

Not for the first time will this American example be used, my intention, difficult that it is, will help show how we can side step some issues by the normal use of half measured reasoning bringing us again to this very modern era. I had mentioned before, that the American experiment is the example that we all should be aware of, even that we should aspire to copy it in the abstract, where the mix can be immeasurable? My observation to that end is to be collectively taken by our own example, we had continuously progressed much of our political thinking in one direction. My own drive is done because the reason that we have to meet on common ground is not to be political but religious, being daunted by very recent events of deep concern, so that for some of us the best and only answer seems to be to politicize religion!

The picture I would like to see having emerged from the American example is the correct one. Any status given is not for European or American excellence one way or the other, but simple observations of history. I have not ignored the open, wanton or deliberate brutality by any side, medium or conduct. In the open statement that we are here now bears all to my intent that we must progress from here and now in form, not to continue as we might have by example of half measured reasoning or from the employment of our steps taken syndrome. By being here and now in world terms, our position before we enact extra half measured reasoning supported by any steps taken? Has to be measured in the reality of confrontation leading the standards of conduct by actually being here and now, but from the perspectives of this last modern era? Then we can better begin to understand the coming together of the tide of politics and the tide of religion renewed, where we must see all with our own eyes and using our own judgement from this very time.

From all of our yesterdays, tomorrow is unknown, therefore we must begin to focus all to the aim of discovery not least if the interactive face of conduct between politics on one side and religion on the other stand by my own definition. America by the circumstances of geography, climate, diet and other reasons I say represents politics? With the other side expressing forms of fundamentalist religious fervour through my third dimensional religion of Islam, showing distain in the only way possible by the meagre avenues of communication between religion and politics? All of which even with other referrals to both and each other again and again, will be done again and again, but now if taken in time that not only are we here and now, but more precisely at this exact moment of reading this book. From this time of now any matter means or type of politics is to be overlaid with my Fifth Dimensional Concept past the reaches of nationality only. Taking our selected and made forms of self control to have been from the civil order of politics and the creative edge of our religions, to be both tempered with the three taken prospects and one active in 'One for the Whole' from the adage of the Fifth Dimension to be the last. Working in the future with new and flexible tools one way and always forward!

## CHAPTER 57

Negativity is a positive word but an empty promise; it is the long way to say no, on the positive front the answer is yes. That is why I might dare suggest that we should overlay any and all of our religions with my maxim of the Fifth Dimension. To remain in the positive I have also relayed that we should overlay all political systems with the same adage. From that suggestion I do not regard them of the same importance to us on my own account. I have raised the standing of our means to secular control out of the respect we should have to it for help in leading us to our achieved heady heights to the spiritual through our religions.

Although accorded similar gratitude by me, my underlying motive to cover them both in the same distinction is forestalled because the actual overlay of the Fifth Dimension to our civil, political means of control is to keep in check the very real medium of the fourth dimension. Being the back door way to admonish the wealth and strength of our religious fortitude, gained by our stealth of sidestepping reason and cause to the reality of time, that we alone of this single specie gave cause to the very nerve of existing in the taken form of creation in the name of purpose. Even that our style has been very erratic, it is wrong that we allowed the doubt of our own existence be challenged by the feature of politics of our own character that we nurtured by best intention to answer the same questions?

At the very time that we professed to our God culture by determined means, we also offered our own denial by the same feature when we accepted theory into fact. My coded fourth dimensional standard is proposed to be of scientific matters to explain cause and reason, exemplified by experiment and recorded as formula. How could the many different and some strange religious beliefs of faith compete with the show of result that politics could actually deliver? From that standard of proof, almost at the first inset, religions were challenged by the aspect of science if at first expressed to be done by peoples of a different clan, tribe or national. Leaving others to religious reasoning from where and from whom, parried by the edge of discovery in doubt of others as we, but not yet able to recognise our similarity, only our differences for what ever reason!

Who ever was to carry the fourth dimension forward by us, was done under the unrecorded or unknown standards of many of our histories. With the greatest of fortitude we each in our separate groups used any of its expositions to the best advantage that we saw at the time. Even that these acts of discovery were done from the same outcome, we managed to respond to them to suit our own developed style to or at what ever time we felt best to use them, depending on the way our characters were developing influenced by climate, geography or lifestyle or other reasons. It is not to say we were aware of the fourth dimension at any time, but it has to be accepted that its presence was well used. I have tidied up my perception of its scientific bearing in order that we might see it in the comparative way we should think of it in relation to both our religious and or political reasoning.

By description the fourth dimension is a stage of our development, it has no call or say or opinion except from our misuse. When Archimedes yelled in abandonment "eureka", he was expressing the medium of discovery in and by the realms of science. From that time his foresight was for us all in that form, and for all time until other features of science exposed themselves to us by its own means, which could alter anybodies original thesis scientifically? Another more defined feature of the fourth dimension is that it is constructed in layers, so that all steps must be taken in chronological order until the direction of change is pointed to by the turntable of study. I have classified science in its own category to highlight how we have been draw to misuse its well intentioned expositions, but not to challenge religions?

If science as a subject is to explain habitual conduct by formula progressively then there is no fault to my definition of it collectively, even in linking it to or with the subject matter of politics. Why I had sought to contain it in a submissive quarter is to its own reflective mode by our misuse to its description of religious matters by contemptuous indifference. Where a body as a group might stand in the pure belief of faith, is an option by the un-measurable progress of the human mind in relation to how we had formed opinion or from what incident? If our thinking strays in query to a mighty clap of thunder having a divine route, then that counts first that we had been open enough to consider what the sound was? From where we saw a mighty jagged fork of light shatter to the ground or cut its route across the sky, that we tied the following roar of noise in connection, how could this consideration be thwarted as mediocre? No other creature had our perception, yet as soon as we showed pace, others from amongst us decried our wonderful efforts into the negative of no, rather than the positive yes, we are moving!

Although the fourth dimension is of us all, it has different bearings to many of us because of the different times each group might have used its revelations and how or why! Many of us in open acknowledgement of it fully recognise its scope, but for some of us we still fail to see the damage it can cause by how we use it? Those of us most locked into the method and style it presents the mysteries of the universe to us, have allied it to the positive aspect of our political motivation, not least because that to in some forms is almost equally progressive. Where my own issue of concern rises from, is that some scientists individually have used the offices of the fourth dimension unknown, to deliberately undermine religious expectation. Not from the overall splendour of the concept, but whether or not thunder has a bearing to the essence of a particular faith?!

The fourth dimension in line by my own coding follows the first three by number, yet was here in design long before any others. My particular style in classing it so, follows by these recent five to six years only. In looking to code our religious practices and set them in there adjudged equality, I have put the three dimensional Middle East religions by time but with their own reservations, to be inclusive of all and others religions. I have seriously included the fourth dimension in line to the first three religions accepting that its theory far pre-empted them, my particular note has been forwarded. My dimensional standard is to be raised as a means to show our acceptance of the obvious and to lead in the acknowledgement of our most recent errors by the staging of political excellence over religious fortitude. By this style I raise no negative view, even locked to our histories by four, two or one thousand four hundred years, because now in answer to how we have slipped since world war two ending, we have the control to all of our futures through the Fifth Dimension!.

Without increasing the pace of delivery our tempo is on the up by the natural flow of time. Not to forget that my own rational is to focus our attention to the single life span of any individual, presents the enormity of any task entered into.

Especially when looking at the way we have become civilised by any measurement, when the very nature of acceptance to that end is its own contradiction. For it is notice of the manner of which we behave and not the motivation, which is to mean that most of us measure our own conduct by it in not what we do, but in how others behave? Without the avenue of sight we can each look by reflection and see the opposite in each other. Not very unlike how we have come to accept our standing to matters such as some of our human rights codes. Where we cede them or the concept to all and sundry, yet from one band to the other or reversed, who would or can call the level style or manner of civilised behaviour accorded to it, them or us?

In my previous references to our half measured standing and our steps taken syndrome approach, we will have to include our civilised prospect in part from now on. The purpose of which is to re-measure the direct effects of our conduct in this very modern era by the turn of this third millennium. I will and can only try to keep entering my own view that the issues in the world today are a direct result of the post World War Two years. At no other time had our political nerve structure been tested to such degrees, not even from the realisation when we all met among the devastation of the impact of slaughtering fifty five million souls. Then, because of the enormity of that impact, where before we could walk on political legs, we failed to make religious judgement, now! In many of the past pages you will have noticed that my style has been conducive to our concerns, after all, I have only used our emotions from knowing of our differences and knowing better of our singularity. Where I make no separation is from our religion and our politics, but the Fifth Dimension will!

## CHAPTER 58

Nothing has changed, but we are in the depth of our last modern era, for those of us not to understand what is going on, it means that minute change is necessary, but without alteration? Much of what we are to do is to be through our young, not by the form of formal education, but in the general terms of the future as opposed to the past. Where the learning is to be done, little change is to be affected, for the future is to carry its own impetus. The only tools ever to be used are in the learning of our differences and the whys and wherefores. In as far that we might travel into the past we will see in most circumstances that much of what we have had and did, was sufficient to the dominant society at any one time, which has to be looked at for referral if necessary, but not in condemnation.

By all initial advancing by whomsoever, we began to add more and more to our baggage of consent, which to best advantage carried us further into the realm of creation past evolution. Alas what we might have done, is take on too much, not in preparation for our future expectation, but just for the future, forgetting motivation if entered some four thousand years ago. If our result is what we are now, that is all very well, excepting how we openly classify each other on the one hand being different, yet embellish our equality through those differences. By all and most of our assumed civilised standards we have truly tried the new beginning since wars end, but our luggage must be reweighed?

What we did not have was not there in any case, because most of the realisations I have put forward had not been monitored in a like manner to that which I had since been able to visualise. Therefore not to be praiseworthy only to the individual, if for no other reasoning than I have never been able to value some of the laws of copyright. I have always seen the product of the individual as part of the whole, nothing is new to the individual we all speak in the words of those that have gone before. I do not deny credit for innovation, but question collective effort claimed as a product of any individual, even religiously? From which can be made the point of issue whereby from political means we can and have been able to separate our standards, but for different objectives, even from dictatorial leadership!

In as much as we have tried to establish political equilibrium if not somewhat led by events in the Middle East, simply because that area has been active to all of our concerns since World War Two ending. By the classic application of my noted half measured reasoning we have moved by political intrigues there, and from that, plied our steps taken syndrome politically. Where is the call back facility to those religious outpourings or discontent or questioning, given by the accepted standards that some of us have been more civilised than others? If politics had been too multi faceted yet accepts in concept the feature of religious tolerance to cover all religions representatively, how is that equated by today's standards? When any religion driven or felt driven to its own course of action, carries out that action in any case even if using politics standards, which automatically falls to our steps taken syndromes?

Without full examination, which is now opened to be given or taken without the extent of copyright, I have proposed the maxim of less is more through the Fifth Dimension, not least because from our political perspective, whereby we have the simple straight forward classification of equality. We classify our equality through our first, second or third world countries, of which we cannot see, that by our measure of equality, we classify people by colour and legislate so? We classify ourselves and others by the difference of religious tolerance even in knowing that most religions were born from one to create another yet diminish purpose of intent. Most of which had been built in form long before World War Two but at wars end had not been demolished, only added to by the legislation of half measured working into the steps taken syndrome?

Of our best intentions we are truly open to try and try even if a little of what goes wrong has been because our very nature of enquiry, which has cast us to be progressive in the competitive sense? Being the very trait that brought us out of the regularity of evolution, we should not be too hard on ourselves. Nevertheless it is imperative that we begin to focus on a different bearing. How that is to be done in the first instant is not half measured. It is not necessary that we trace our own carbon foot print unless seen by other means. Our first realisation is to the individual of now with the consideration of a single lifespan, which has its own implication. I make that statement to thwart politics in its excuse to comply to the whole community, I also make the distinction that the individual is not valued above the community, or majority, but cited individually!

Three taken prospect and one active in 'One for the whole' is the essence of the Fifth Dimension in focus, without law tenet of covenant it is to be a wedge to force the changes needed in our attitudes and aspirations.

## CHAPTER 59

Today in the third millennium as a date mark mocks all political enterprise towards our supposed equality, yet I expect the Fifth Dimension to be received by all as our main means of communication. All political accounts from now on are to be held by it in overlay. I myself have used it frequently in describing this last modern era having its own time mark from the end of the Second World War, which is counted to be nineteen forty five by all histories. Also I will continue to use it non half measured but in consequence to the importance of us all individually represented by our own lifecycle. I do not follow any steps taken syndrome, because my intentions are openly honest in representation for us all. My justification for review is now supplied by the Fifth Dimension of which we must learn at what level is it to overlay politics and religions differently?

Before five or six years ago there could not have been any attempt to alter our political or religious attitudes, simply because we had not been able to unravel the essence of our histories. Although we are indeed proud to be progressive and forward looking because of the singularity of our specie being totally unique, we have surpassed the boundaries of evolution. In setting our forward path we have erred by little steps when looking backward and not remembering where we have come from nor why? That is why I have frequently referred to the past in the context of alteration without change, whereby we can change the past without alteration! My style of description will become clearer as we progress in time, but only if we accept small principles which have already been clarified. I use the term this last modern era to cover the actual time of now as we are in the time of the Fifth Dimension, which is the only way to change without alteration.

It will come to many of you from publication, which is to be naturally expected by the normal course of events. That I have used the frequent adage, this specific time, is the only time that any such proposal to include anything like the Fifth Dimension could have been presented. Running in formation these past five to six years also stands by historical means, even though there is really nothing new to the three taken prospects and one active in 'One for the Whole'. There never has been any intention on my part to create four commandments instead of the ten or more held by some religions, nor have I tried to create four pillars of wisdom instead of five attended to by some religions. There is no shape or style to create four principles instead of three or seven held by other religions, nor to set the wheel of life with four spokes only. For immediate notice, and once again, my four prospects as ours to overlay upon all religions and all political systems from now, even allowing that the concept has been remoulded on our behalf only these past five to six years.

This quarter of my last modern era from about the year Two Thousand A. D, is representative to us, it goes hand in hand with a curious factor not disassociated with the Middle East. Although I have made much in connection of our total world histories through the Middle East in the same way that I would not wish to indicate any diminishment of that fact. I take the A D year time reference in an off handed manner by its illustration in terms of association or disassociation? Civilisation, political correctness, human rights are all mediums of consent by our use, selection or appointment. By our very acceptance to the terms illustrated in the form of the A D year mark from that appointment some seventeen hundred years ago by the Roman emperor Constantine, standards were set to unknowingly create by endorsement half measured reasoning, adoption or application!?

Half measured reasoning from which again we must look at reflectively, is my illustration to many of our misjudgements leading full bore to our steps taken syndrome. My style of cross referral is in order that we remain alert to an open failing that we have unfortunately not yet been able to control. Where I might sight one example of the many, has been from the Middle East by how we have only looked in part focus on the factors involved to the broad situation there. Through delivery of the Fifth Dimension it will be better seen how we can alter the outcome of always following those episodes by the steps taken syndrome, now we might react to intent, with all the right answers?

From the term A D in translation to be anno Domini, "in the year of our lord" we take open acceptance to the significance of the term year date wise as the now international year time mark? That term is and has been applied by the course of circumstances if linked by geography, climate and other factors unknowingly, then apart it is to accommodate such a need of that measurement scientifically. It was not the birth of Christ that introduced it, nor was it from the realms of thoughts to the sanctity of Abraham that introduced it. From what we know it was not marked to the specifics of our burial at Shanidar, nor cave painting or rock drawings. Our Egyptian, Chinese or Indian cultures had no such mark, which further illuminate my scientific termination to it. Further still my European direction of it in situ gives or sets no pointed form of excellence to its implementation. Although in poetic terms I might have stated that three hundred years after the death of Christ the Romans said sorry. Our taken meaning of the event is from our own terms of history which we might address to change but without alteration. The Roman Empire needed a year date time mark!

By about the same time difference of roughly three hundred years when Mohamed created his new religion watching it in its own infancy, although to Muslims throughout the world a new year mark was set to the year zero in due deference. Even now the Islamic religion in its own fourteenth century year time mark, expresses it so for their religious purposes, yet use what has become known as the Christian calendar politically. All of which shows the complexity of the mix between the two forces, complicated by area from ancient and very recent times by different interpretations! It is to be accepted that the Islamic code was not introduced to thwart the means of acceptance by European standards, for their intention was for world conversion eastward to the furthest landfall. At the same time pushing Westward as far as the land of China, which was the total world then, the Islamic year time mark would have been appropriate then, but relative to that religion only.

From history falling to its own pattern by events in these very recent times if stirred by ultra modern occurrences, from the proposition of supposed acceptances in balance. We from our half measured standing move to our set standards of equality in acceptance by our steps taken syndrome. When using full employment of one of our first principles in bestowal of our human rights set standard to all, we compromise it by political correctness. All of which is to be shown in its truer light by the later course of events, but only with new attachments or at least broader outlooks if only delivered by less in the means of control. All of which is to mean in the order of control by any level that less is more. We are not to look in examination that the more we do is the order of correctness, but that the least we can do correctly will suffice by any interpretation.

Without the obvious connection of any of the major historical events in all of our histories, like when it was mixed through Abraham, Moses, Jesus or Mohammed, from each of whom set to us the most direct of directions. Whereby we were to have been fulfilled by the concept of understanding to or by the intention of purpose, which at first in all cases had to be wrestled from the course of life by normal living, their leads were to be taken in different context. Any lead given was of the times then, it was we who added further decoration that by each step taken we added more to the understanding and reasoning of that original intent or purpose no matter what. To be spread later by the means of the existing standards of civil control expanding to the matter of political management. Leading us to our present situations of multi faceted controlling elements, with most not really understood, but again used under loose interpretation by political means only, half measured, steps taken!

If how we have progressed is to be represented in the number of rules, laws, tenets or controls we employ under the headings of being civilised to the standards of unity by living in first, second or third world countries? Then the matter of control is to be levied by the same ingredients but now based from our one active prospect 'One for the Whole', which uncontrolled in any way is one of four rules we might need to recommence? Not that we would be able to simply select any given number of short rules to replace the existing plethora of controls we readily accept. My overall intention is to have us look at a shorter route to suit us all irrespective of direct creed or politics, with full notice that we all start on this level playing field!

Accepting that civilisation is a real human condition, I would not suppose that we are not just so. Any doubt offered is not to the prospect of civilisation, but only to the proposition of it. For it is a single word to embrace all of us in context to the positive or negative aspects of the state of being civilised? Although we are of this last modern era, with two people as you or I having gone under the same time span, there can never be any guarantee who is more civilised by judgement of what we might think from reading the same book? When our own definition may be turned from one to the other by where we might stand politically or from our individual religious holding if used one against the other?

Civilisation in all situations has to be recognised as a true cornerstone to our united understanding, but taken in caution from our best reasoning. We are not to condemn it wantonly from appearance, but through the caution I now offer. Yes we are civilised by advancement, which adds to our progressive mantra carried these one hundred thousand years, and that is where I would point our direction of judgement. From what ever level we would like to think we can set our own standard to, it was done before our realisations when it applied to the simplest form of self recognition by the most obvious. At Shanidar when we buried our dead, we were civilised, because we did that from amongst the very tasks we do today? If about half of the world's population have to concentrate on one task only each day then we are only civilised because that energy used is to the direct requirement of the supply of personal sustenance. Is to supply us in our need; yet allow us concern ourselves in the course of religious sufferance beyond the call of all other creatures. To supply self need in food and spiritual need in the expectation of guidance is the full matter of civilisation!?

That we have collected far more in the attendance of human rights or political correctness does not make us more or better civilised, it just bears that we are more complicated in being civilised. Being well fed does not make civilisation better, being well thought does not make it better, with the line to be drawn at being less, to make it viable in all best reasoning is a step to be seen to be taken and understood?

## CHAPTER 60

'Out of Africa', 'Past Gods Words', 'Natural Body Death', are three taken prospects with one active in 'One for the Whole'. All of which by my direction to be of the Fifth Dimension installed these recent five to six years past for interpretation, but yet to be fully implemented. If they are suggestions without law, tenet or covenant, what might there use be and what might my overall intention be, unless to promote our singularity of specie!

Who ever might have already assembled single prospects of the Fifth Dimension had done so when done. Our difference now is that newer circumstances have lent different interpretations than from my own four presented together. Five to six years has been my time frame to have assembled the broad concept of the Fifth Dimension as the focal point for the means to change. I have stated in direct challenge that they are to overlay our politics and religion in equal measure and indicated that they are to have different affects on both.

What happens now is without president, because the story in page really starts here from your reading and absorption of this offering. I will continue to write from my own level and standard such as it is, but without false modesty I am writing for us all. I am writing beyond copyright laws beyond tenet or covenant, I write in the spirit of eureka so that the scales of the unfortunate medium of ignorance and fear have fallen from my own eyes first and others now. My fourth dimension in the realms of scientific formula and experimental discovery are ever around us all, but not now in the order of challenge which might have been considered? How ever we will need to refer to it in the negative at times, is to stretch our own ability to read certain situations in there real form, what is to be pointed by that remark stands by the limit of scientific studies known. Meaning that for those that we do not know, the fourth dimension has opted to call on other unknown dimensions perhaps up to eleven in number in order that in time we might be able to understand ourselves and the instances of the universe scientifically!

My own view to that is not one of rejection, but management, because we are the driven specie by whatever means, we are looking for answers in the realms of purpose through discovery. Therefore, I see it that we might not actually have the self ability to know what we discover, which separates politics and religions when science of the fourth dimension falls to be political by show or is taken to be of political importance over religious standards. If any result gives that impression from recent times, then we have worked the political oracle more steps taken, that raises my point that although we treat all religions by the same tolerance we do not acknowledge that fact, because we cloud the theory politically!

I have indicated that it is from this time and these paragraphs that we should be prepared to be more involved in our own destiny. What is apparently happening is just that, but shown to be politically motivated through or from or by events in the Middle East. From where political intrigue will bulldoze its route by one set of standards to one set of goals. Religions cannot set that trend because we have not been able to consider them in the full benefit they offer for us all. Some thing would or is needed to be able and reflect on all matters uncommitted, hence the Fifth Dimension in overlay to both factors.

Without listing all of our religious and political problems in area; the small turn we have to make is to the realisation of our own status in dealing with each other by different standards. With some of us to be first or second world politically, cannot be done when they as we are all the same religiously! Cannot be done when recent political thinking is set to overbear on the political standards of others, which translates that a first world country cannot threaten the conduct of others, if by first world standards? Which when striped of all political referral would mean that one religious standard was imposing its will to be upon other religious beliefs. To creating a situation to be worsened when seen that former religions which had been surpassed in tenet and covenant by the later religious ideals, even if from the same camp, were to be diminished by political activity only. Making the need in response for any sort of reactive behaviour to establish the religious status quo accepted by the natural flow of time in all cases, even if that reaction was to be basic or fundamental or course or unread, but needed by reaction!

Religious fundamentalism or political correctness might never have been intended to fall into the same sentence, if I have steered it so, then it is not from my motivation that we look to the why or how. It is for us which heartily includes all, so that we can view the two attitudes above within the perspective of this latest new era politically. From when with the ammunition of change by our own consciences given by the standard of the Fifth Dimension, we can overlay all religions and all political affiliations, keeping them ever linked?

Three taken prospects and one active so far is the matter of form to my conscience concept of the Fifth Dimension. No law tenet or covenant forms its shape, all of its substance has been done over and over before by many of us, yet in the most unfortunate of reasoning we had been unable to ever see the reason of intent. Mainly because of the stance we had been forced to take by our then reasoning induced by the prevailing circumstances whatever!

'Out of Africa' in application to us all are three small words that are balanced by all the powers of scientific study to fit within the concept of creation. Three and a half million years is the scale of time to strengthen that prospect in its own validity for in the terms of evolution it cements us to that single specie that we are. Accepting that there has been many archaeological finds through the world, giving us sign to our progress and movement, does nothing to contradict that prospect. If human type remains have been carefully timed and documented else where, all still show in abeyance to the scale of three and a half million years. All still focus to Shanidar, when we finally broke through the barrier of evolution. For in the finding of any remains was accompanied by the registration of our ability to use tools above the requirement of accidental advantage. Giving us the linked ability to bury such tools or emblems as grave goods before time began and after our footsteps cast in mud had now become stone. Making it so that by direct connection we were to thwart evolution, we then were the single specie to be on the road to creation, 'Out of Africa'!

'Past Gods Words' might seem a curious prospect, not least in this context as one of three to be taken. My inclusion is to accommodate our religious fervour and more particularly to indicate our secular standing by and to the aspects of evolution in leading to creation. What I ask of us all is to consider the timescale of humanity in relation to other specie. Not by the judgement of comparison but in the ability of total time. On page one I had made reference to the first life forms occurring some eight hundred million years ago. Even in recent times that time band might be questioned, it remains in reference by scientific reasons. My own reference in terms of it happening long, long ago was an indication for us to think in a different timescale than suited to our own existence. When life forms first formed if that is the correct word, then the situation of all life was formed in the prefix of evolution.

From any such time when we as specie became aware of evolution if assuredly connected to Darwin's publication, then we have been fraught with the compounding of some basic errors pertaining to the importance of our own existence. From that time we have continually parried in debate to the standing of evolution or creation? I had indicated if we look to creation as the product of evolution, then that would stand as one line that we might position our own personal viewpoint on. I had further indicated that in the worlds of time, the total span of evolutionary events was in fact aligned to Gods own time scale, not even to be recognised by us, because it is so vast? We in fact needed to create creation to accommodate our own lifecycle which in real terms brings us all closer to our ability to understand the concept. When first seen that over these three and a half million years of our own development we have been drawn to ability beyond our understanding and now beyond the understanding of the sciences. We pre-empted cause by matching us alone to be past evolution not really understanding it, but nevertheless to be of creation by linking us alone to the concept of God.

Once done, by every means, history followed a different course, not least religiously and politically which in cases and at various times has forced us to introduce traffic lights of thought for our management. One of which is these prospects, one of which is my second to be put forward in 'Past Gods Words'. Allowing us review our relationship first with each other and then with ourselves.

It is more than important that we are all to be involved in these exercises simply because we are involved. I do not seek to curtail any accepted, devout or real connection that many of us might have with involvement to our own perception of God religiously, that emotion stands by the whole tone of this reading. One aim is for us to understand the different time factors in relation to the three religions from the Middle East first, and then apply the same interpretation to them taken from normal human intercourse.

Judaism, Christianity and Islam have each compromised their own standing from within, but not founded to be damaging, only if misused. My prospect is not necessarily to be used as a bar or ban to the prescribed thinking of any of these three or associated religious orders. God's very words are wholly transmitted through the core of each religion, one to one to one is without objection save in the crossover appraisal which might be indicated at by a group or an individual within the single religion. From the standing of Judaism in claim to be the first and appointed religion direct by god, then in open review it is operated by the direct link to be of gods words supposed. On its own account, looking into the future by Jewish terms, then operating the religion required no particular control, because the verse was self maintained by being self directed.

By the later times even when operating for two thousand years to gods own time scale for imminent deliverance, Judaism had not faltered. When from amongst their own number, new verse was given to the same religion, had they adopted my prospect to maintain operation past these newer gods' words? I make no slight of any religion except in the course of self reflection. Jesus in hand offered gods words for digestion, his motives were simply to enliven the aspect of his own religion by the medium of its own promised expansion. If the Jewish religion had faltered past its own understanding, then this was done by our incapacity to control or even think in the terms of real time properly. Which is yet to be answered by any religion after publication of this work.

I have given that some of the motivation owing to Mohammed's decision making about the formation of the last of three religions to have emerged from the Middle East, was for reasons that the area in part was disassociated by the church of Rome from the direct contact made by Gods own prophets? I have made that remark with balanced reasoning and from no other motive. 'Past Gods Words' is my own expression, but all of one of our taken prospects. Without distinction it is not any more critical to Islam than any other pre or post religion, Islam from where the written text was recorded in direct connection to the time of delivery, makes it seem more volatile to the tone of criticism. Not from the time, but in reference to the content from where the text was self structuring to be definitive, but always allowed the grace of opinion leastways for some!

Mohammed acted in his own good grace not unlike Moses or Jesus, his standards were no more exacting his approach was no more demanding, all three Middle East religions were to bring man to the expected triumphant unity with god. Although at the time of the introduction of this new faith there ranged bloody conflict, nothing was amiss to the harsh and brutal standards of life in general to all peoples at that time some fourteen hundred years ago. What was in continual discord through any age then, was how none of our religions embraced all the people at the same time. If by direct connection it was seen that there was to be forms of opposition then the standard of the brutal age was employed to accommodate success by any means. That we could not recognise then and even now, that our differences were by reason of the circumstances of geography, climate or other means then to employ the full edict of religious doctrine was not viable for natural change as it should!

From where we might not look to change history can be seen the similar trend of where Islam has brought us also. Each of three religions from the same area of the world locked in the spirit of truth had tried by guidance of our own understanding to set us in life's harmony. Each if undirected has drawn the same result in conclusion but yet unseen, each are open to accommodate our needs but now on the broad scale. Four, two or one thousand four hundred years of continuous history in the collective words of God direct, has not been enough time for any of us to equate to the single lifespan of the individual. It is from that basis I have added the taken prospect 'Past Gods Words'. Meaning in all fullness, that the individual in personal harmony may take all comfort and joy from the very real experience, but none from any level or inclination, may so cross any religious barrier save by other consents. None can use the prefix of speaking god's words by means to promote a course of action or support a deed. All of us from now in this last modern era can only operate 'Past Gods Words' by the reason of our acceptance to the importance of our single lifecycle!

My third taken prospect 'Natural Body Death' is headed last because of 'Past Gods Words'. From which there is much in the mix of cross referral, therefore I want the prospect of 'Natural Body Death' hold its own ground and hold a new force and create new self respect for ourselves and others. I had previously indicated the intent and first meaning of the expression to belay our own proposition in the art of human killing. I make more serious warnings about it whereby some might try and refine it to purpose in or covered by our half measured methods of achievement. By other means I have uncomplicated the three words for direct digestion. If we take the first word 'Natural' then that is its definition, even linked with the other two they are representative of this our single specie only!

When we die it is to be by 'Natural Body Death' even to encompass all the misery and sadness that might induce it is a further unifying concept to our sorry needs. By a means of expression it is a strait forward statement, but somewhat complicated by how it can be interpreted. From the previous prospect I have asked that we should give up the warming pleasure from the feeling of being firsthand amongst gods own words. So that some might not have the powerbase to direct others to their will rather than intended. This case offers almost the same ambiguity from the mixed means of how some of us individually bring value to its overall intention. So that we might not suffer the imposition of half measured reasoning, we can reset some opinions by means of our one active prospect.

'Natural Body Death' by first intention is read to mean that we die of old age. All other meanings or interpretations are designed to move us from any consideration that it is right or correct that we can kill of each other. By its sheer simplicity that core objective has to be questioned by many other means not least based from some of our differences promoted by driven or changed circumstances. By the heaviest caution I must warn all that the very study of the prospect questions our levels of civilisation and the behaviour of? Most questions will fall to the adage of "what if" or "what about", which is fair charge of how we react in most cases even from amongst some of the circumstances of our changed character.

Politics will reign very large to some of our queries, because from practically all of any political baseline throughout any history, we have used the standard of human death to set up or change our political opinions. We continue to do this in the name of good practice or from how we had managed the tumultuous transmission in the handover period post World War in ceding self determination to the desire of national groups. Then to correct any seen wrongs, we again papered over the cracks so that we might carry on in our own civilised manner of desire. Politics is very important in one other way by admission, she is the master of our civil obedience plan, and she is also the controller of our legal systems. Some of which take to the matter of human death in the form of sacrifice to the crimes of the perpetrator, which have a multitude of meanings but are not to be addressed on the political emotional level, the emotions of effect come from our religions.

Religion is to be our divine achievement, why do we complicate it so? 'Natural Body Death' in a way has no religious eyes, for most of our religions are set to their own ending or beginning with our death, we are each to receive full recompense in that final act, again maintaining our break from evolution? Nevertheless many of our religions have taken to the human mantle or as I submit have been given our interpretation of how we are best to behave by the hidden expression of our own free will. Some of us have compromised our own understanding by operating on the two different levels as if they were the same. We behave to the best edge of political standards where they touch in comfort to religious determinations. Yet prepare for the eventuality of life continuing through and because of religious meanings only? So mush so, that by varying degrees we allow, encourage and practice various and vile means of murder to each other in the best name of some of our religions. Some, to mean in translation by first reflection, our three Middle East religions, mainly because they are now reacting together howsoever and by whatever means at this crucial time of history expressed through this recent modern era of our united histories?

'Natural Body Death' in voice is easier to follow in concept than the other two taken prospects from the Fifth Dimension. So that all three can be rationalised; the one active prospect 'One for the Whole' is there so that any one of us might offer adjustment to any proposition. But now, not half measured, which would nullify the possibility of following any lead by the steps taken syndrome. From the above it is not to be taken that there is a disguise to any aspect of the future when we are entering it with new insights. I have given notice in recognition of how this work is essentially a collection of other deeds, I do however assert that it is the way I have assembled some of its form that deserves its attention. Our collective effort has been born from the need for small levels of reorganisation to achieve some growing aims. In as much that through publication it will truly be our work, it is that there will always be the means for any and every bodies input through the one active prospect of query and suggestion in 'One for the Whole'.

## CHAPTER 61

The Fifth Dimension in form is not a means to an end, it is the latest and best means so far, where we can acknowledge our specie singularity in the proposed management of our political structures and lead to the better understanding of religion? No law is in its name no rule is of its structure. In the conscious sense to be associated to our conscience it is brought to meaning by the abstract where it is to constantly cut into the manner of how we have come to operate by political and religious means. From where I have put it to overlay both features I favour the religious end of our concerns more seriously. For in as far as we have been able to develop, it is from that medium that all of our outstanding achievements have been made. Even at this point I make no apology or excuse for that direct statement because from politics we have empowered the course of our own destruction over and over, yet it is religion to revise that situation?

'One for the Whole' has more than one application, not least that it is to overlay any and all political structures housed in the secular. At the same time overlaying all of our religions from where it is to be better used in the form of a reflective zone and in cross referral of or from one to the other in the cause of balance. Its further and better use is from the event that in use it can be employed by any individual to examine their own conscience without charge. A person without the use of any sort of religious commitment can employ its service on religious matters of any sort if required, even switching to political concerns and not having to listen to what others might have suggested about a political or religious venue or both?

I have no cautionary note about any use in employment of our one active prospect 'One for the Whole', simply because it embraces what some of us consider to be the two biggest matters of concern in our lives. Also taken that it is so recent by time its establishment will control its own output. In light of that fact and without the matter of successful contradiction, I have stated as much that any proposal I have put forward in these pages is only part of what has already been done. My particular connection is in the actual time of the doing by these very recent five to six years of this modern era. From that very reflective mode the application of any proposal through it, is set to be mirrored, so that the recipient of the idea or observation receives the same as the proposer? By that means even if a religious idea or suggestion was offered, a political answer might be the reflection, from which the same might be reversed!

By outcome 'One for the Whole' is to be one of our best mediums of change even if only used by political means or separately from religious examples simply because it is not affiliated to either, therefore it is the weighing scale of our collective judgement singly. From the picture of the Fifth Dimension it will not be difficult to see my objectives unless further over complicated by some of our negativity. I have placed it with out the force of acknowledgement in our midst, at the same time making demands of its importance to us all. Which will be done, not by its structure or set up, but more in keeping to its absolute need at this particular time of our histories, it is the means for us to be accountable? 'One for the Whole' is all that is needed in the active vein to mount the winning rostrum, when looking back is the answer to the question posed by it? To be able and look forward with it, sets the scene for better understanding of each other, for no other reason than the Fifth Dimension is prepared to second place our political standards to the importance of our religious ones.

So that we might make better contact with my own feelings in our representation; the Fifth Dimension having its four prospects, give us all equal voice on the same terms, we are all of the direction to talk for each other that is what it does. Without law, tenet or covenant it opens all gates and closes all barriers. Through it we are not nationalistic, racial, political or of selected, specific or chosen religions. Nor are we a babbling uncontrolled rampaging mob of the disenchanted, by all leads we are in the direction to alter our biggest difference from amongst the same groups throughout the world. Those on the one side that have an expected life span of some fifty years plus and those of us on the other who might now expect to live for seventy plus years in normality?

To bring us to the reality of the importance of time to us individually, it has to be seen that no time of any length, has any worth without direction of one kind or another. This has to be reasserted into our thinking to rekindle acceptance of our single start point by unity and our equal finishing point by the exact same reality. Although our religions have been doing this for centuries and centuries, it has been because of their similar differences that we have managed to forget the significance of our single lifecycle. In locking ourselves into their directed tone we have forestalled the reality of the significance of time by any measurement. In counting to the expected outcome of life through them, we have stretched time beyond our own capability in relation to our religions length of time already!

Unbalanced as it must remain, from where we have settled for two core religious beliefs recognised to be Eastern or Western, we work them both by assumption, but from the Eastern network there can be better ranges to suit how we have developed? With most adhering to the spectre of life continuing in cycle to the object of eventual fulfilment, in loose terms unrelated to the specific lifecycle accepted to be Western thinking. Hinduism through to Buddhism and others all call in mixture to many life forms interwoven to allow the time slot whereby the ultimate of desire is eventually achieved to heavenly understanding. Without any selective motive, these very fine philosophies, have unerringly produced answers to many of the begging questions of mankind, like the understanding a different time span required?

In starting with the prime object mankind, and looking to the prime objective understanding, we had formed the other best method in the human ability to understand the mysteries. Some Eastern religions by type were also formed about four thousand years ago. It can only be conjecture of the widest girth to imagine that the factors of circumstances by climate, geography or diet and other reasons had bearing to our imagination. By all the most obvious of signs the peoples to have formed these philosophies were affected by the locality of generous contact. Their influences ex to the development of earlier success in the land between two rivers developed of its own accord in the settled modern day lands of India and China along with associated regions. Without negative distinction Eastern religious reasoning bore well on our imagination because in a like manner to Western ones we each sought the same ending in or of the eventual ability to be in the position of our appointment in being able to accept Gods own opinion?

I have almost submitted my own acceptance of this, but add in our own caution and in relay that such an opinion can be offered in terms through our 'One for the Whole' active prospect. Open for all, but by cautionary note to the factor of our developed differences from or through our relationship with geography, climate, diet and other reasons, not least displayed by our differences in any case. In the point or realisation to this factor it is that we must be able and continually review the relationship of time relative to both set of philosophies in order to reduce there self competitive edge?

My own viewpoint to the Eastern religious network has its own format by that element of time which I feel is of great importance to us for mixed reasons, even if tinged by example of abstract thought self imposed. My views can be offered for our consumption but only taken that they are open and not competitive by any means. That many Eastern religious reasoning readily accepts the concept of Nirvana or truth or understanding by the same reality of 'Natural Body Death' then our different code is narrowed. Where standards might never meet is because of my own troubled time factor, where we can see the different fitting to the range of Western religions where time is to a single lifecycle. Cementing the standard of reasoning that Western religions are singular, in the sense that any thing to be determined by the death of any individual is to be measured that while alive, all judgements are cast against fifty plus or seventy plus years of living!

For our eastern philosophies, all of the same gates are to be opened, but with the one difference through the time scale of reasoning, taken from the realms of our differences not least by climate, geography, diet and other means to the actual development of our character. Given that many Eastern religions by my reasoning to Islam are laid back by style, or taken from self development are fatalistic by character, giving to them the necessity of working to reason without any particular timescale? Setting the die, where people can look to the outcome of the future without the consideration of the specifics in actuality of a thirty, forty, fifty or seventy plus year lifecycle. Being how from those philosophies we can prosper to challenge the interference imposed by the response of our fatalistic outlook imposed by the harshness of geography or climate at least? I would not suppose that we are all that much different, for without any explanation we are all connected to our burial rite at Shanidar, which geographically happened to be at the centre of our east west religious divide?

## CHAPTER 62

Without any sort of lead we are not to think in terms of our differences any other way, except that we are different. Without directed direction, there is no route barred so that we might not be able to improve our situation? What most of this our work is set to explain is where we are now at this precise time in all relationships, but to have been set in real time since world wars two ending? I had sighted that one method to the aspect of change can be best achieved through our young, that always stands, but changes complexion when considering both types of our religious philosophies.

Where one of my prime objectives was set in the past for the realisation of our young to view it without alteration, will for ever be essential to that approach. We are not to invest the sins of our fathers upon the responsibility of our young, we are not even to invest fortitude in the same manner. Our young are to be given the view of open competition to our own imagination of purpose through all religious and political thinking via the Fifth Dimension. From its simplest definition we are to ably inform them that this actual time of history measured in these days now, is open to the realisation of us being at the time of this modern era. So not to complicate the issue we might simply make them aware, that in the past many mistakes had been made by many and many people while we were in the act of learning the art of humanity. Therefore by first acknowledgement it is through them that in the direction we propose, we can set them the standard of change with us all on the same starting point, but now to listen to the thinking of innocence!

Have I just suggested that we might forget about the past and the aspect of measured brutality that many of us have felt, have I offered an excuse that by reconciliation the slate might be wiped clean? Has this referral only been directed to those of the Western religious or political philosophies, because much of this effort has been set from a Middle East and European perspective? Nothing I have so far written has been offensive which it cannot be, because I like all others have used my own wit and our cunning and our adaptations to express opinion. All of which can be drawn from how on our behalf I have proposed the Fifth Dimension as our united means to alter the future without reference to the past if necessary? We cannot be separated on that opinion, therefore if by no other means any of us can offer in suggestion any route map, but only looking forward expressed by our young, because as of yet they do not have a soiled past! One shared concern recently raised by the growing concept of this our work stems form these recent paragraphs from where I had raised the issue of our young and the future. Then seemed to walk past that concern by referral only to the Western kind of philosophies, which is a matter to be best addressed from the abstract, again!

Without starting at a particular age, our young means those of us not yet versed with particular points of our histories, especially those of deep emotional concern to the affected living? I will not lessen the power of any emotional hurt or upset, nor disguise the effect of physical abuse or torture in any recent manner. It cannot be dismissed that those responsible were cast into that position by the cruellest of circumstances against other circumstances. Although we have tried to measure the responsibility involved in many recent cases of open brutality to each other shown in open review, if through such edifice as the United Nations in representation, then we have so far failed half measured, and promoted our steps taken syndrome. Again this does not diminish my attention but is a wedge of diversion if taken to oppose our direction and even if self induced to make a point!

From the pristine nature of our young in innocence, could be the best way we can self appraise what we might have done or been involved in to the good or the bad from the past. We can look at the Middle East Western pattern differently than worldwide, because by religious interaction from there, the structure of human continuance is to be immediate in connection through human life continuing after human death? Eastern philosophies follow the same pattern to final fulfilment, which are in no way contradictory to other styles, but offer different propositions of experiment? From which my intention by the use of that word is to associate our thinking to the fourth dimension in terms of formula?

From the supposition that by character some of us have a related and direct style of historical connection through our own ancestors and or peoples we have come to associate with, like our tribal groups. From whatever angle our connection is made by our western religious style to the future, expectation can be steered by our young from how we have aided them to review the past. In relation to the Eastern religious philosophies this on one level is open but then contained by the hidden nature of the very source of planed, adopted or declared reasoning and the conditions thereof. All of which is to mean how can we overlay my reasoning to the benefit of the young nurtured to the reasoning of these Eastern religions, when driven from the core of unconnected historical backgrounds lent through reincarnation?

By all views, we all have connected histories by the reasoning that we are all here now. Eastern and Western religious standards have the same linked connection if at first by political interaction, if at first through the medium of trade and dealing by whatever means. We have to accept this beyond any standard how history might have been recorded, which could afford its own self appraisal to some interpretations, but not from the past only. By normal connection in any case, by our own good human dogged determination, we have gathered the aspect of our young automatically to be of our first concern futuristically by this very last modern era!

Again it is for me to reiterate, this book could not have been written at any other time of our collective histories. Simply because we have not had this ability of approach laid open to us by any other set of circumstance until forced change was introduced collectively since the ending of the Second World War. In as mush that it is difficult for us to think in new realms, my approach is covered by our three taken and one active prospect of the Fifth Dimension. That is why I can assume and make forward proposals to change some of our methods of conduct no matter from whatever previous source, but always now holding to our very modernity.

For me or any of us to make the suggestion that the future progress of our human endeavour will pivot on how we might direct our young is paramount to those ends and totally reflective of why at this time before others that offer can be put. Without individual praise to any governmental rule or clause, but in line with the best advances we have made even if somewhat misguided. Taking any particular result of efforts from the world war period unconnected to human morality, because then ago we began to shake of some of our technological restraints in the field of international communication. From the requirements of deed to the repugnant, we turned our advances to best advantage for all. Information became a by word to the realisation of our actual similarity by all counts, which if worked half measured at second attempt through the United Nations was fairly tried.

From running all ideas forward sixty years after the war and not side stepping other peoples reaction we can now suppose on the same level because we are the same people, even seen through our differences? Not by our religious or political standards of now, but owing to the circumstances that our geography in casting our climate or in determining our habits of diet, shelter and lifestyle. Had also contrived to change us of appearance to each other to give some of us the doubt to our single heritage, alas allowing that factor bear upon us by other means that some of us have sought to promote. Resulting even now that some of us by self appointment through natural selection take to be better than others, forgetting we had all contributed in the past to our own futures. Making us the same, not from where we suppose to be now, dressed in all manner of sophistication, but back to our first realisations that we were not as other specie, we were not appointed by evolution?

## CHAPTER 63

Our young have always had a special bearing with us, not least from their total dependence from the first few years of living. This can best be seen in any of the comparisons we might make in nature by the study of group, troop or pack behaviour. We all supply the direct needs to basic survival in the name of feeding, protection and in the matter of teaching, which is the mark of the human animal. In the case of most comparative groups to us, all will generally follow predetermined plans in the name if instinctive behaviour. Other creatures or insects only follow instinct, even if not ever involved with procreation, many are only part of the whole, like in a swarm of bees or a nest of ants, only assisting in the product of allowing another self group do all the procreation necessary for continuance. Almost like some plants where the decisions of the future are carried on the wind, unlike from our young who not only are involved with what we do, but like adults, they too can expand in mind for all!

Heeding our young as an objective or point of reference can work right off without the need of political management or any overbearing from the scholarship of political rules. It can also work through the Western styled religions to have first emanated from the Middle East. All by trend look to our finality in the understanding gained through the matter of human death, supposed by one of our taken prospects in 'Natural Body Death'. By that image and matters leading to it, it is possible that we could instigate my laid proposal but not to be fixed in the bindings of pure legality, only simple morality. By attitude we can fully introduce our young to all aspects of the past by our own reasoning, not for them to react on our behalf or by our standards, but lay them open to our reasoning. So they can temper their own judgement to now, from where future reaction is to be dealt in the fair mood of our equality expressed by relearning the past without alteration.

In the open name of our equality we have to reconcile where we can apply the same points of reference to our young who might be marked with Eastern religious philosophies. Already belonging to us all, yet steeped in as ancient traditions from separate religious reasoning than others, which is not the asking of the question of who is right or who is wrong, save that our young are our young? The implication past assumption is that the Fifth Dimension is to be spread across all of our political reasoning and religious standards irrespective, even to cover a single lifecycle or the aspect of reincarnation, if necessary!

In open conversation I have set the proposition of standards in the making from the development of my own views as a feature of human advancement from no more than need. I had long realised that marking this time as our last modern era is not sufficient to cover the full range or our achievements. By suggestion that we can overlay all of our political systems with the four prospects of the Fifth Dimension works on all levels, because for the first real time there base line has been equalised. I have indicated that by the choices we have ,we also overlay our religions to the same extent but to the conclusion that this matter is of more importance.

By making that distinction my reasoning has been cast from the importance of the time spans in relation to either. With politics covering all times by means of civil control over our full extent of interaction generationally sets the lessened standard, whereby any political reference can be made fit any mixture of individuals at any particular time.. On the religious score at least to the designated Western styled religions, there influences are to the individual measured from deed by their own lifecycle. Making the connection between the Fifth Dimension and religion more important gives acknowledgement to the reality of the importance that should be attached to us individually.

In bringing our young to be the masters of change to existing attitudes is to be a token to show a united effort from all and to be of the same intensity for all, remembering always the measures of control installed by our three taken prospects and one active 'One for the Whole'. Any doubts festering by religious standards in cross referral to such proposals of using our young in this matter can be dispelled by more than the obvious no matter what religion we belong to?

Eastern as a term used to religious definition generally applies to the way that I have used the expression, referring to the style of approach more than the outcome. In some respects by some quarters, as a collective set of religions they are less different to any other block religions than might at first have been imagined. Both Eastern and Western fall into the definite belief in mans destiny by the exact same final expectation. Both accept our human inability to be of full enlightenment in our present state of awareness when showing serious failings by other means. Both accept the full range, even of reduced mysteries in proof of our link to divinity because of our asking before any other awareness? Eastern styled religions had assembled their own particular base of study by a different application, like not to worry about mysteries in the same way that western religions might, no where does it have to be seen that understanding all mysteries is the key? Although we are talking of styles the maxim applies across all religious codes from Western to Western or Eastern to Eastern or others no longer used, all mysteries remain until we decide they have a use!

Although I have referred to Eastern religions and left it to assumption that the basic difference between them and Western is through life's conduct individually, I mark it so that there differences are somewhat further apart. Not in the critical, but as further indication how due to circumstances, what ever difference we in our locality interact of each other from, and by local thinking first? If not honestly endorsed by my interpretation of matters to have occurred at times past in the central Americas, whereby very limited exchanges might have accounted for the situation there, because of the relationship of the tribes there, that is until newer input by new arrivals? Hinduism in general terms is a monotheistic religion, but with colourful additions to its own liking and bore from its own deference. The other main Eastern religion of Buddhism follows a different path in that it purports to the term of enlightenment as the symbol of any supposition to be monotheistic. Enlightenment being the aimed at goal to determine our final obtainment, perhaps deliberately falling short of claiming to the acceptance to the God figure by lowly human frailty or in pure humility?

Both of these select majority Eastern religious philosophies follow to the perpetual cycle of life, partly expressed by the wheel in symbol or the tree from where enlightenment might bloom. By other incorporations they openly follow in the aspect of new incarnations to extend the human time of reasoning giving us a better overall time band to be able to understand Gods intention, if to be delivered at or in a time that we might be able to understand! Looking upon the broad scope of our actual ability, Eastern of western religious philosophies are well balanced to our best human standards. If we might exercise our natural human desire taken from first acceptances to speed our unity with God on his own terms, then we have the essence of westernised thinking to our religious attitude first delivered to us from the lands of the Middle East.

If we might exercise our natural human desire taken by other circumstances from the area of India now spreading eastward incorporating similar thinking patterns, then by reasons of sited humility, might not some of us approach matters to our own reasoning? From adoption can not we by measure, offer the agent of time to the matter of our enlightenment, then from that base of self reasoning incorporate the expected time needed, by imposing the reality of reincarnation to absorb the full time period needed for the total understanding required? Then by the study of human nature, offer it so under licence that these matters are not to be related to our own wisdom. Even that we might take it so to use the spirit of other creatures in our exploits that we were all bound by the style of Gods own purpose in the creation of his own subjects in what ever form. Only allowing the factor that we might, might be the object of reason above other meanings, kept until told, until known!

It is not easy or fair that we can say in sentence we are of God, which is why we must labour to our commitment, which does not include destroying the myth of our achieved excellence if done through politics or from other bases. We can however change political direction at the merest of whims which cannot be done by some of us under any circumstances. That is why I have gathered to put it in focus that we can best use our young for these ends in changing the direction of the future only for the better. If when first said there was worry or concern how we might bridge either political or religious gaps, that is now done but not fixed in stone. It is not done yet in using our one active prospect from the Fifth Dimension 'One for the Whole', it is not a specific tool in the order of direct change, but the forum of consideration expanding on all levels. When taken that the Fifth Dimension is to be our conscience then it can be taken by us individually how best we use that conscience and how meaningful.

Taking that we might use the unmarked outlook of our young is to be always representative, I have used this cautionary note to avoid the implementation of any half measured reasoning. For that reason, it in concept applies to all of our young, even to those beginning to be brought up in the best traditions of any and all of the eastern style of religious standards. I would not try and steer anyone so that we might all be embodied into the western style of monotheism, therefore the same standard is to apply to us all young and old. By the terms we have come to accept the range of mysteries holdings accredited to God, Allah or Brahman being the Hindi one god link through the sway of Eastern religious thinking. It bears nothing to the fixed standard, of my Fifth Dimensional prospects, which because of its own unfixed standing is perfectly acceptable for any Hindu or Buddhist or other eastern religious code, the same it always is for all political determinations?

From any standard that we are to adopt or utilise, holding that first change, might be gathered by the proper use of new thinking from the uncluttered mind of most or our young so that the best of possible situations falls even to all. With the projected line of progress from those in the Western religious sphere, it will be easier to move into the one position to carry the same or similar lines of thought of how we might be received in the event of 'Natural Body Death'! Never to challenge young or old even of an Eastern religion which might have incarnated to several life forms ranging from animal to human to animal and human again. Even so there is no compromise from the Fifth Dimension per se. It brings no challenge to the theory of eastern religious thought, against the theory of western religious thought, under any circumstances.

To visualise the exposed hidden working of Hinduism, if the prospects of the Fifth Dimension were fully open to any number of individuals in the real time of now. How might there standing be if they were still unsure of the understanding of their standard of projected obtainment? If it is to become that the charge is to be put upon Eastern Religions, in respect of there being no link between the number of reincarnations in all forms, then that only elongates the time of realisation, which is not part of any challenge to any other style of religion whatsoever. No question ever should arise of who is first or who is longest to any conclusion which will be seen more evenly when we express that style through our young.

I make no case to bend any rules that could have been formed by any standard of thought set by the medium of different circumstances. I make no case for the setting of half measured reasoning to accommodate what might be interpreted from that apparent separate standing, most of all in open view. I can declare without the obvious, that the essence of our eastern or western religious concepts are to be seen the same through the balance of the taken prospect 'Natural body death'. For all indications are that all religious bases are linked in tenet or covenant to that key, therefore in using our young to settle some of our differences is small measure in picking the lock of understanding.

In open conversation it is not easy to state that by form our religious bases are the same if led to the deep understanding that we are different through them in most cases. There is to be no convenience to my supposition that we might connect the two main styles in such close order. If western style religions run from frail human life directly to the means of fulfilment after a short span, then eastern religions do the same in the step from human mortality to enlightenment. To have made sight of that multi transitional means expressed by eastern thinking, then dismiss it of virtue to fit or suit the western expressed system is done by no other means than to show our similarity. For no matter what rule we think under or by, the exact same conclusions have been reached from two different camps. Both have sought to manage their changing and growing attitudes to the advent of the final discovery to answer to the status of our mysteries? Both camps working to the obvious outfall of the successful conclusion to our joint search in the singular, but expressed to suit the particular trigger of initial enlightenment. Both monotheistic in means to surpass our own image by any classification yet twisted by our own imaginations fed from different quarters!

Without distinction, I have mixed the structure of our eastern and western religious codes to suit the line of advance in the matter of how we might all change again for the first time using the same ingredient of consideration. No person has done this, yet it is only now for another first time that this matter has again been tackled. If such circumstances had not yet prevailed as I might claim by our exact time of history now, where until now all the circumstances had not welded in form to promote the thinking of this time, then in part our job is beginning. Not least that now we can confound all matters of scientific discovery formulated from whatever experiment, but always through the fourth dimension. Death into life is completely inexpressible by any scientific reasoning no matter how many dimensional surges are needed to that end. It is we ourselves who have arrived at the beginning of conclusion, if not in full order then it has been from the matter of mayhem we have combined to set our own standard by our own efforts, which makes mayhem now undone?

I make no excuse for how I have seemed to juggle the order of compliance to the standard of placing our reflective trust in thinking how our young might behave. If I set a unique set of standards or at the very least encouraged in a fresh perspective, I have made much of how our religious thinking has always been similar even spread across our eastern and western divide. We all do know that there are questionable religious sects, which are ever to be under review now by the Fifth Dimension, not in condemnation but by inclusion. No longer able to promote their own standards when we are all to be connected by the more modern thinking in both our religious fields and political management inclusive of three taken prospects and one active 'One for the Whole'!

From where we now talk in discussion of any religion by any standard, we are fully entitled to call its doctrine to our own level and understanding by our own terms, but without obstructive intentions, unless the religion itself is self limiting by its own form. For me to have implanted the image that eastern and western styled religious concepts are similar is of no judgement. My expression is to show my view to the importance of how we have structured them to suit certain specifics to the time limit of eventual understanding by other means than those generally accepted to be in our capability. At this time we might only question them in the adverse if their objective in any way shape or form encourages the shortening of our accepted lifecycles by any means, by any means against the open support to the prospect of 'Natural Body Death'!

## CHAPTER 64

One of the two greatest influences upon us all for an even longer time than our religions has been our politics, measured at first from the organisation of our small bands of hunter gatherers wandering in the midst of necessity. Finding that the best results were always obtained by our inborn ability to think ahead, and listening in thought to those prepared to literally take the chance in the discovery of our next meal? From the groups that didn't quite make it there is no story to follow, which is representative to us all because we did make it. From any group clan or tribal decisions in the making by what ever means, we displayed organisation of a different comparison to pack or from heard culture creatures? Not least bestowed upon us by the other circumstances of climate and geography whereby from our own ability of what seemed to be natural organisation, we were able to delegate separate tasks amongst ourselves to suit our circumstances, not like others to the style of instinctive leadership won to the sole duty of propagation.

Our standards drew there own conclusion in the cementing of group singularity, when at any time we recognised, not our similarity, but our differences within that same group? No great moment of eureka here, but it came into notice by whatever unmade means that one of us became better at maintaining the fire, of opening seeds or shells, one of us was swift enough to catch small prey unassisted? Even leading that one of us over the open standing of pack leadership took control of certain group activities by best choice. From where a supposed subordinate could direct us to a better choice, our pack organisation was led to group control, which led to the skill of civil control by newer means?

Although most forms of human organisation have traceable histories extending into the hundreds of thousand of years ago, illustrated by designated tool use for example, our best pointer to civil control is met by the conditions at Shanidar some one hundred thousand years ago. I have made the suggestive remark that from then ago it was also an indicator to our first religious concept, which is to belay the statement that Shanidar was allowed happen because of civil control first? Updating us to the ending of the Second World War, with religion diminished and our politics was now forced to take its last turn to change for many of us. In order to set the rudiments of equality, of means, of intent, not least how we have concentrated to make politics balance by adding any and all means of title to placate the idea. From where we might think in terms of political correctness or to the matter of human rights has not been the same to the real discovery of our ability to cope in separation to all other specie by political terms only led by human civilisation!

From the same medium of pure politics, since the beginning of world organisation in these last sixty plus years, much has been achieved, yet in the final analysis, we are still wanting by intention. Unfortunately the first moves were thrust upon us in fair latitude by the victors in conflict, who began to redesign our future from the habits, styles and conditions of the past. In all good faith we had only recent events to work with and we did by the best assumptions that evil had been removed for the last time, therefore a new world was awakening. We proceeded in best fashion but only arrived at our conditions of today in needing a newer approach to rekindle our thinking to what our conditions of choice were at Shanidar? For better reasoning it could be that we might only look to the joint times of our decisions on religious term marked by four, two and one thousand four hundred years ago. Even that we might have erred on three counts then, expressed by the gap between the three choices above, when each is set to the conduct and outcome of the individuals own efforts, but on Gods terms, which is no bad thing. But who calls Gods own rules, three times?!

Politics was the arbiter at wars end and acted to its brief in the best way that it could from what other examples? If we can compress those recent sixty years in to a single word of actual achievement, allowing that my lead is purely representative I would pick the word consumerism to fit the bill on some levels. Simply that it refers to the concept of supply in how we relate to each other and fits into the operational manner how we can delegate our similarity by definition to be first, second or third world countries. Consumerism fits to the level or standard of organisation through the management of politics. Most other factors of alignment roll around it, it hails to supply and demand, monetarism by reward, and is set to the mark in protection and defence, albeit profit margins. Then in a confusing tone it has been offered to have been by our direction that it is a process of force to our own benefit even in open acceptance of how we might differ by a fifty plus or a seventy plus year lifecycle which had been delivered by politics only!

For all the reasons that we might try and offer cause to any situation, there will always be division, my own take towards our political rule is to the aim of self recognition in respect of us as individuals. Rather than serving from the staid controlling features of manufactured political thinking by following standards thought to be meaningful. If we are to be civilised, if we are to be democratic and show it by being politically correct or setting the core standards of human rights. If we legislate in law to drive the measure of our equality under level control by stating our singularity irrespective of colour or race, then we show open awareness. When we would also offer the control of reaction to each other by the coat of our religious standing, questions the manner of our judgement by terms of reflection if first set again in these last sixty years or so? Because instead of always looking to say yes politically, we have to sometimes say what?

There is to be no doubt we need matters of control to our conduct, in full duty to that statement we ourselves have arrived at the best means intertwined in the medium of our politics and religions. Much of what I have offered is to the recognition to that fact, but as seen my real effort has been to cut into our actual style of operation for a better pointed outcome? From where we had the chance to commit to new emphasis at the close of conflict in the year nineteen forty five, we were not actually ready to do so. Not because we lacked the will, we abounded in good intentions, but failed because we knew too much and were unable to see the cut from where our learning had taken us? From our total sophistication and being civilised, any decisions were to be the correct ones not least when all structures of our standards had been formed by intention and motivated by our own free will!

At the close of conflict we were unable to note the obvious that it was caused by political reasoning all around. Driven first by economics to the support of eventual market bases, the drive was not to share equality, but manage its proportion by the taken acceptances that some of us saw ourselves to be superior to others. Fortunately for our own benefit the outcome was what it was, for any other consequences would have seriously altered all histories. Unfortunately much of our repair done was to the same reasoning and to the same ends of market forces tinged with the fact that the winning sides were of diametrically opposed political ideologies. With one side leading that democracy was government by the people for the people and the other being communistic that government was for the people by the people or similar, with further complications that half of the winning side was religious and the other half was not!

When first draughts of peace were drawn post world war, most of the rational was of political reasoning only, by first reference the standard of dictatorial fascism which caused the whole war for stated reasons, was at last shown to be a vile and empty proposition in this modern new world. We still had to acknowledge the nature of fascism adopted particularly in Europe pre war being able to run with religious conviction, if not wholly devout to the specifics of those religions being mostly Christian, then for some keeping to nationalism! Giving the aura of international acceptances if set to oppose other non religious political philosophies. Also in part being able to rationalise the standing of superiority held from the base standing in being first nationalistic and further to have been naturally selected!

If delusion was the manager of any such events pre war, alas the first post war settlements steered clear from the importance of our religious acceptances almost matter of fact. I make the case again that we as victors holding the two main philosophies of communism and democracy concentrated more to stop the future formation of more fascist regimes henceforth. Not least by setting further political agendas compliant to the route of commercialism, like driving a wedge to belay the need of further war, by opening new open trade markets. With those first considerations and because of the dreadful way repatriation had been handled, political decision making in haste took precedence over other considerations, mostly religious, which we have now found to our cost?

Most choices on that political level were supposed to be of equal balance to the two main political regimes, but began to change by the adopting of our bloc culture mentality. From where we began to forget our own sense of purpose, generated by opposing the excesses of fascism, some realigned European and Asian countries turned to communism as a means to address the future. Not least because under certain circumstances many were breaking from autocratic monarchs, dictatorships or former colonial rule which was enough motivation for change. Once done by the very reasons we had fought in war, by the expression of our choosing differently now, we were setting future standards to compete on the very same level. This time due to the advances we had made by the former exercise of conflict, the dangers of total war contained our effort to prepare for conflict in the cold war era? The importance of which was to focus our modern emphasis to the political arena first, not least by some accord to the changing processes in the Middle East.

Even beyond political reasoning I had previously attached specific importance to that area in terms of how we had set reactions by our cold war bloc culture outlook beginning then. Much of what I had indicated interest of, was to our three monotheistic religions all of which had been closely connected to the area. However I had also set the tone that it was from there that our new modern era was to come of age and the reasons thereof, connected to how we began to react by religious differences up to this very modern time of now. I with much mention of the possible active backlash between the Arab and Jewish influences of the area known as Palestine, had set indication to show some of our modern reasons, though not yet finished. I would set it so that the main matter of alteration was for reasons of religious importance if seen by one people or another of the same standing? Then also how was the method of reaction between the two factions of Judaism and Islam if acting in the time of our bloc culture political situation?

Having the Fifth Dimension now to overlay all aspects of politics, what might we gain? I have set it so that the three taken prospect and one active are to be put in place to all political reasoning henceforth, so that we might set our judgements by different standards. To be reminded, we are to take our findings on political terms only although we know the importance of interaction, my objective is for us to set a range of thought in this modern era from wholly political leads. Not that we might pretend that we suffer religious tolerance, we are to act and react by all reasons political, but only because political vein having no Jewish, Christian, Islamic, Hindu, Buddhist or any other direct religious input!

Of course religious thinking cannot be excluded, for this is no party trick I am offering. Where my direction leads without study, is how we had formed our previous reactions and what to do about them? Politics has always been a great machine to our full service, it is we who have tarnished its image by using some of it one way at various times then in reverse at other various times? Consequently it had produced its own rules which worked to best standard by the need to question at a particular time! In coming to the turn of this century and millennium, once again with new impetus, intention and realisations, we can use the Fifth Dimension to set a new course.

Unfortunately in maintaining the core of old standards, we sought to correct our situation by adding the means of change without alteration. We added new political laws and legislation for what we adjudged to be harmonious settlement to some of our perceived and uncovered separate attitudes. Trying at all times to be all things to all men until we had to set terms of separation because of our failed reasoning, from where we cannot now determine who is right or what is wrong. Not least because from that very same area of the Middle East over the past sixty years we have determined that at different times and to the same circumstances.

Some people who might have been terrorists then or now, are or were national hero's to the welfare of their new nation, while and if others who might act in the same matter and to the same reasoning are merely terrorists because they are counted with different heads!?

We can make no judgements from the above, for the arguments have been crafted by my self in order to show how any line of thought can develop from moulded circumstances, even when the sculpture is never to be known, save we? Much in design to those situations are from our long past, which cannot be changed nor should it. That is why I have been able to put trust in our young to change it without alteration, done where we can learn from their best expectations. That is why by using the idea of their input we can focus our own intentions to the meagre future which in total is only fifth plus or seventy plus years long in all cases yet casts a shadow over all religions. Politics is in need of the same lessons but in due acknowledgement from a different perspective to its age. Not least when shown that we had indeed manufactured it to serve us and in setting the limits of that service, we gave it the wrong grace to its own immortality!

All that I now suggest is for us to overlay any and all avenues of political reasoning differently with the three taken prospect and one active from the Fifth Dimension. Even if used no more than a back drop so that we can reason direction by the correct medium if that is best? So that we can limit motivation to the direct care of who it is supposed to serve, remembering politics is part of evolution. For it is the true formation of natural instinct, it has been a means to check our progress in so far that it has always been with us. In saying it had been with us by those terms, is born out by when we used and traded anything, but if only proved by flint findings, that is enough to qualify my statement.

At Shanidar we had the organisation of civil order well before the meaning or understanding or religious input, which I am not embarrassed to mention, because religions were keyed there. In turning the step to advancement, is why I have been able to set the two apart because although one may have a connection to evolutionary trends and the other to the aspect of creation. Both are products of our development outside any possibility of any connection to any other creature. We are quite simply to be unique in the universe, and owe it to ourselves to exercise all thoughts, ideas or imaginings to that reasoning. My aim at this precise time is for a general reorganisation to some of our best ideals; although politics might be first lead I make our religions to be our first priority. In order to balance that statement as an observation, is the reason that I have offered that we overlay both features with the Fifth Dimension on there own terms. This to be done in that manner to cover the range of our political differences, whereby some are closely linked to religious matters and some are not, some in the pure medium of communism who would deny any religious connection, are to be put to the same terms. In order to show that politics has drawn us to the matter of too many rules clouding intent, until we test the ideal by giving all systems the same consideration!

## CHAPTER 65

Every time we might look at one of the four prospects of the Fifth Dimension, personal feelings at the time could determine what we might think. That is why they can overlay either of the two biggest features to our living standards in our religions and politics now at this time! Like other considerations the four prospects are not to replace any guidance from the secular, although there place will be seen to be different than when we overlay our religions by the same measure. 'Out of Africa' almost by definition will always be our first mentioned taken prospect, precisely because Africa has shown the first incline of human type movement encapsulated in stone. By far better judgement than my own, others amongst us will make a better case to our beginning; my own input is to our continual span from the one time to have been three and a half million years old.

Even that I readily accept our start like being to those of Darwin's finches with changes following the necessity to fill need. I also submit that there was no definable time to class us being directly Homo sapiens by what was found on our specie trail? Our age is of time, but most assuredly 'Out of Africa' from that single moment some three and a half million years ago? Accepting other signs to our separation from evolution by early tool use, it was not until Shanidar proper that we were at first to creation? Without definition a factor of that burial in note is that from all realisations it was carried out by us first hand but in scientific standing to have been done by a group as finches classed as being Neanderthal man? Further classed in being of a primitive nature pre to our scientific standing in being the last order of finches, above ape erectus, above biped erectus, above Neanderthal to be Homo erectus, being us now at Homo sapiens!

I have stated more than the obvious that at the time of Shanidar, we were blessed with civil control, which in total justification allows that I have made the correct judgements to our special definition. There can be no argument from any quarter to the effect that there is a superior vein to the human condition, there most definitely are standards to our difference and types, but these cannot be expressed by skin or hair colour ever. Any explanation has always been hidden by the nature of our mysteries, that is why I would lay our three taken and one active prospect over the separate reasoning of political studies, which are obliged to lead us to the consideration that our mysteries are our business and to be dealt with as we wish? Having used the fourth dimension as the scales of balance, politics had begun to set her standards to be ours, being no bad thing, but leaves the matter to be checked religiously!

Knowing better can only happen from politics when determined by mankind, it is a feature always to have been with us, our unfortunate errors have not being able to see the wrong implications at the right time! By example, if we take separate groups of people living at this very time, and make comparisons, the differences shown would be negligible, but monumental politically. Taken from the collection of people to express there activities from the full call of nature, like Bushmen of Africa, tribesmen of the South American rain forest or clansmen in the vast

Indonesian island network north of Australia. Who herself has peoples to their own aboriginal habits; and compare any to a highly westernised modern outlook.
We could only notice with distinction that all such groups were in full possession to their own spiritual accomplishment, taking them to the equality of humanity to be seen later when we over lay their societies with the four prospects of the Fifth Dimension proper?

Unfortunately, also at this time and under political reasoning, from our sophistication to be numerically expressed by our joint involvement through the United Nations, who only count small communities in block representation and protect there integrity under the maxim of total human rights and or treat them in reference politically correct. Which simply sand storms there importance, but allows us to claim the legislation of natural justice for all in the terms of conversation, providing evidence first hand of our rue how we work to half measured reasoning? I do not say, nor can I expect that we have the energy to divert the standards of vast groups of people to accommodate some very small and intricately poised people living to such adventurous ideals. My own observation is from the political perspective only that we might already deem to have included such groups under the guise of religious tolerance politically but still cannot balance how all are to be treated?

Closing the circle to why I suggest that we overlay our two main influences in life with the same prospects of the Fifth Dimension, but maintain the effect they might have by the different conclusions we could draw? Not forgetting how religion and politics operate on there own time scale, nor there real influences, not forgetting our staged manner of interaction by best intentions. We are to look at how we might repair the storm damage from the incidence of life and the living of without following half measured tracks, without being committed to trace them to be steps taken by the simple matter that we must follow them through, because we started in the first place?

Politics on its own account is to reason of its first overlaid prospect 'Out of Africa'. By our own curious standard of political reasoning, this prospect has been with us almost in direct form since wars end some sixty plus years ago. Reviewed by scientific means there was no real argument against the standard, except our complacency which has been tempered by how we have actually addressed the issue. If we have chosen the name of our own political representative to be the United Nations to express the meaning of the term, then our floundering has been done in these last sixty plus years. From every continent from every indigenous group of people, no first sign of our collective unity has been promoted because we have not been able to equate our separateness by purely political means. By the previous manner how some of us migrated to other places of desire driven by what ever means. We purposely failed to recognise our similarity to those already there, which was only a judgement by political determinations, political determinations only!

Unfortunately in the non admission of our differences shielded by the clouds of circumstances determined by geography, climate or diet and other reasons, some of us could feel selected or of selection naturally or for other reasons, being a human feature not to be surrendered by any means. I have to make that point associated to our political standing only, examples will show us to have acted in some terms by religious attitude, but my rational is only political determinations for now, but still allowing those religious insights. 'Out of Africa' is the progressive indicator of our specie development but not in progressive stages, which is the core of all the errors we have made and still do politically. Most ably expressed from that political perspective when we look to how we can class the same people by the standard of economics to be first, second or third world?

From that I do not even question definition in title, for in some respects it can be used to best advantage if applied to the standard of requirement of available need? Where we falter politically by attitude is from our attitudes becoming increasingly confined by half measured reasoning and the excuse of politically correct offering to the standard of imposed human rights. I speak not in the critic, only to attitude, further and most deeply I also speak in the terms of change, how we are to look at our histories without alteration. Sixty years ago as a new political beginning had been marred by the first standards we applied to our collective attempt at moral unity. By setting the first picture of the United Nations on mainly political terms, complicated the matter with our two main victors having opposing political philosophies. By that very reason the charge was not in the cause of unity but in the chase of superiority in open support of the method and style of politics rather than the supposed concern of human sufferance!

'Out of Africa' was pinned to the mast gathering dust and foiled into indifference by the pressing drive of political determinations, it is from that context I have set to reassert its importance to our religious acceptances anyway. But first needing the separation I have accorded to it now politically, to re-burnish it in meaning and to reshape our attitudes to it. Even that my reading of our economic division by class to be of first, second and third world standards will be rebuked, there is no political standard that can be used to that end. Therefore without malice, I openly caution those who would oppose my interpretation in the support of that opinion, do not proceed. Remember politics is our tool only, it falters in our service to it by over complicating self issues and running in guidance by the fourth dimension. If having the power of uncovering the mysteries by holding a multitude of taken prospects, then my further caution also runs to the expression of less is more. Showing from where with politics we might have over exerted our means of control or acceptance to a single and simple prospect to be taken, because it is not complicated enough politically!

Leading where we have to stand by our steps taken syndrome, because we approached the matter first in half measured fashion, 'Out of Africa' is a good example of how we can mix the same theme differently when entered to be those three words and in sole connection to this our single specie. Problems occur when we are led to assume the meaning by other interpretations by the same people, but with only slightly different outlooks politically. I have intoned that it is one of four prospects that we should adhere to in acceptance in these very modern times and only because of these very modern times. That is from where we have to make choices or at the very least put ourselves in the correct position to enhance our own judgement. My own directed split in overlaying our religions and politics with the Fifth Dimension is to give us the feel of reading the same form from two different directions and allowing us to come to the one conclusion.

By the nature of politics anybody anywhere from any division could and might say we are naturally out of Africa by the self same connection that has always been put to that theory. It is only when we realise that we are different but the same that even politics will begin to make sense of my intentions in the persistent pursuit of review to that simple expression. Although we might all appear to agree that it is self explanatory by the timescale and historical flow to our first archaeological discoveries to the first images of what we might have been like and when. Political reasoning only, has allowed that we might indeed have first developed in Africa but now it was personal dexterity that brought some of us to greater heights of achievement. Leastways until this very review taken from our total achievement by any passage of time up to now and from this very modern era which gives to us licence to change aspects of the past without alteration. 'Out of Africa' is not only a question it is a solution?

## CHAPTER 66

By the same review and to the same political standards and reasoning, I put to that edge another of our taken prospects in 'Past Gods Words' to be worked by political reasoning first hand. By this very modern era coinciding with the introduction of the Fifth Dimension, my second prospect in use hardly seems relevant to modern politics. Strangely I see it so because of the very way we now use political strategy in order to make the point of one system being better informed or better appointed than another.

During the Second World War, particularly in Europe, God was on the side of the fascists and democracies when in open opposition. With the bloody development of that conflict, when it spread by aggressive intention against the mass of Russia in terms of land, population and an ungodly communistic culture? Without determination and from the early shock of catastrophic defeat upon communist Russia; without any formal declaration the Russian people gathered to the old habits of their former religious acceptances, taking to the former standards of Russian Orthodox Christianity. Even that the regime of communism was fully operational to its own bloody ends in Russia. It was forced to cede to the will, desire and need of the people as a means to abate all the reverses of war through the comfort of god's support! Therefore almost in the time of actual conflict in Europe, we had all the antagonists with god on there side, justifying in argument my claim to use one of our three taken prospects 'Past Gods Words' to overlay political reasoning.

Although in the case of Russia, taking in god by the open re-adoption of its old religious form was permitted in those times of need, shortly after the war the harsh and brutal communistic anti religious standard was reintroduced. If leaving us once again to the situation in Europe whereby there was a renewed political stand off, then also at the same time the Middle East shared the spread of our bloc culture standard. From where political ideals had now become braced on either side by defined religious cultures in ways to connect my three dimensional religions yet maintain there separation by the rules of politics, is why I set if for the prospect of 'Past Gods Words' to be laid on political reasoning also. It is to be understood this first view stems from the rules of reflection prompted by these very recent times of the last five to six years of considering and modernisation!

When the world war first closed, in line with some of what I had noted, we began to set collective standards for all, alas measured from one perspective to be sufficient in the formation of all future standards. Victory unfortunately did not tell us the meaning of how to do what we should do and or how to do it properly. Working to our best assumptions and if taking the best of leads, most lines of decision making come from political determining. Putting us to the situation of change without alteration, yet having to cope with broadened expectations from the understanding of the new needs for human standards?

In view of the standing of the two main stations to victory, leastways in Europe, with one being Christian and democratic and the other to communism, after already depleting any religious fervour to have arisen in times of early defeat and desperation, a question beckoned, how are we to proceed? In the best way that we could, our undeclared ruling was to put our faith in the decisions of politics wholly. How else to continue if larger and larger groups of people were adopting Marxist communistic traits counting politics to be the only means of communication with others who might be Christian but dealt in the politics of democracy? I have to cite the first part of this paragraph to be an unending example to modern half measured working. Fortunately a key was left so that we all might better understand how that behaviour can become the engine for our steps taken syndrome?

By our world bloc culture standing on political grounds, any collective decisions made through the United Nations were always at first to have been by political consensus only. This method was generally accepted by all players to the open standard of appeasement while preparing in the cold war our total destruction, which was no bad thing? Any political standards or actions then ago were at the cutting edge of our reasoning where all players were in a new game at the same time unseen. We began to set for repair all previous errors! For while we studied in the art of conflict by extension using the best gains of world war two through the field of communication, we were beginning to learn how to communicate. In getting more dangerous our world was getting smaller, in getting smaller our understanding was growing to the realisation of our specie singularity. Even if not justly notified when we made such rulings in the order of human rights equal and for all, cementing it in core by political correctness, being a means to be aware of our differences, but standing on them in open similarity. All was and could only be done by political awareness to each other, at least until another trigger point of our history driven from the Middle East?

'Out of Africa' overlaid to political reasoning is core to any adaptation, 'Past Gods Words' by political reasoning might not have any particular standing, save that is how politics had recently thought it was best working by. Amongst all of these early intrigues in this modern era political moves dominated almost all of our decisions, until the spanner to hit the works was our religious countenance, pointed at in no other area more than the Middle East. I had previously led us to consider aspects of how the Arab Israeli conflicts in the lands of Palestine were brought to what might become our nemesis, of which the reality is afar!

If politics had been the reality of how we had gotten to our bloc culture stand off there, then religion has no say except by expression! I had also mentioned that sided to the general support of the Arab side for various reasons, was the communist lead from its political perspective. Although the vast majority of people in the Middle East then and still are Islamic by religion, politics was the coin of settlement one way or another, without direct motivation no indication of this was sympathetically relayed to communistic thinking. There quota was political and in political opposition to other parties on political terms only, without acknowledgement to that opposition's religious standing of any inclination.

As told, democracy supported Israel in the Middle East by the exact same terms that communism supported those to Islam nationally? That is why I have set it so that we can overlay politics with the taken prospect of 'Past Gods Words' under those circumstances. It however can only be done now because of these five to six years in the time of the Fifth Dimension. Not to use modern thinking in replacement of how we might have thought any time before, but only compliment our own standards of the day in the duty of change we tried to carry out when still not fully aware that we are indeed the same people.

Unfortunately if looking back, our methods and styles from any time have a different reflective value at other times? How we best remind ourselves of this now, is tempered by all we know from the past, not in terms of what we have learnt through the portholes of science and discovery. Our intention is in the art of preparation for the future to other standards by being aware of any time limit that we ourselves hold to our own lifecycle. If a question is in the asking, it might be how was It that by the open code of politics, it has taken us unmeasured years to have set our time span to be fifty plus or seventy plus years long or short, and had to have been pointed out from a religious perspective?

## CHAPTER 67

Politics greatest difference needs to be shown when overlaid with my last taken prospect, 'Natural Body Death'. For this one prospect more than others question the full value of all political reasoning. By implication it even questions any standard of politics to have so far reached the core of civilisation? 'Natural Body

Death' also questions us to the meaning of standards when set against any natural human lifecycle to be fifty plus or seventy plus years long. Because politics has no rule to our individuality, therefore we can temper it with the overlay of all four prospects of the Fifth Dimension, even if the very outcome is to ask that question, from where does civilisation come, how important is a lifecycle!?

Flint tools, Shanidar, cave paintings or rock drawing, even in Mesopotamia when we penned our first verse in the record of ledgers to take us from the rigors of recording the past by other means, we had moved unrecognisably in the line of advancement. Although by further refinements to the nature of our differences it was civil controls that brought us further into the means of understanding. Any and all forms of control in places of livelihood, like in the land between two rivers, brought us all to copycat most systems of the forms of civil control growing into government. Creating over a span of a couple of thousand of years perhaps, the core methods of how we would continually interact until open realisations began to bloom post world war two?

Fifty five million broken corpses gave full reason that we did not really understand the matters of purpose. By the very same number of dead in the name of those that we loved, cared for, cried for and longed for. We had condemned them in the means past natural body death, because we by half to half wished to have destroyed those that had killed or were killed by our own hand! Which was to be proved in the deceit of our renewed structure of political standards, emphasised by our bloc culture representation of how we were to prepare more in the reason of cold war slaughter? By preparing the amendment of cultural ambiguity through the police of the United Nations, to be the level platform of all change expressed from all angles, but only heeded by the few. Or at least those that might have set the rules to be politically correct, laced with singular human rights, passed in the chamber of liberty in expression of our status to be first, second or of third world prospects nationally or economically. Done by the best assumption that some are more civilised, taken that we all have the same standard of self value, all of which was at first attempted to make our political plan the same. So that when one spoke, all would at first hear and secondly understand in the same manner, even if third second or first world!

On those terms it could only be easy to overlay political thinking with the taken prospect of 'Natural Body Death' but to questioning that we are truly civilised, then by what standard or more precisely by what moment of time? When taken it is from these past five or six years it has only been possible for such a proposal, our thoughts have to be projected to our similarity, but in allowance of our differences. Politics has always allowed for our similarity but never our differences. That is why I have separated the two controlling aspects to our government, our religions and the political medium and show if we overlay them with the same features of the Fifth Dimension we will be better able to judge our self worth?

If from now, thinking on the political plain only, our objective is to openly promote that taken prospect, how might we think past the law, allowing that by the unwilling death of any person, we are open to 'Natural Body Death' as a reality? I would not state the reason we are not yet civilised, where I do direct query to, is how is human death measured? Like for example in the case of political crime before or after religious crime? On the political side to cover how criminals might act, in the case of murder, then victims die covered by 'Natural Body Death'. Whereas to die by any religious hand or motive allows that the victims are, but those who commit murders are not, because religions cannot deny the future?

Can it be read that because we offer choice we make the right decisions even if structured by the means of steps to be taken following our unseen standards of half measured working? Put in the open, it is that we assume we have the correct means of control to all matters because the picture is seen. Our deviation is only ever noticed when we come to the limit of some of our abilities by the questioning of living morality, which in all cases is an aspect of expression by our religions in full sway? Being the matter of full discussion now presented in part by this book, but without intention or drive. All that might be expected to happen is that we can now review the same subjects, topics or inclinations from the same start time in all of our view, because of these recent five to six years of extra study!

Looking at the range of changes to be made, we have only two avenues to traverse, one from our religions and the other from our political determinations, which we are actually working on. 'Natural Body Death' in political terms can be seen to be a prime objective even dressed to be one of our three taken prospects, which on its own account, expressed by this modern era, we can make the direct statement of the total effort to all forms of civil control running for well over one hundred thousand years. Our only achievements are to have placed ourselves in the difference of either living for fifty plus years or seventy plus years determined by our own lifecycle, or in the order to be first, second or third world economically. Both of which have allowed us to ignore the very value of human life across the spectrum of human life, weighed to the better expectation of our intent by other political activities. Weighed also by how we have been unable to control the matter that we have addressed our known differences by half measured reasoning and from our steps taken syndrome, when only looked at commercially?

By overlaying our political standards with three taken prospects and also our religions to the same standard, we might better see how we can improve our direction while on the move. Politically, 'Out of Africa' is almost a true statement of where forms of civil control took root and were carried by us all when we were naturally connected in form like the first group of finches? Politically, 'Past Gods Words' is the actual measure of how we have behaved, highlighted in this modern era through how we conducted the matter of Palestinian settlement post war by reason of our bloc culture standing? We assumed to only deal on the political while holding religions in abeyance to standards of accepted religious tolerance, yet from the developing force of human rights we supported that aim half measured?

'Natural Body Death' hangs politics in dilemma by several counts, but by the tone of this examination, my taken prospect overlays it for our separate examination, we can look deeper into its proposal? From the law of politics we can and do kill of each other laid down in honest pursuit of all forms of proposed justice, even natural? Fortunately Shanidar gives us all first sign of tender care to our dead without any signs of that burial having been done to anyone ritually slain? If we are wakened to the possibility of our first intended means of murder from the depiction in some of our cave painting sites, then even if I was to assume some were actual tombs, all reasoning would have been of civil tenure, we acted only by political reasoning?

To carry our will forward even into Mesopotamia when we began to mix the study of religion and politics by the very nature that we carried the aspect of human death in the hand of human responsibility. We almost tarnished our religious reasoning first time around, which was only saved by looking to the nature of study in the type of supplementary gods of religion that we actually carried. From the best example of about that time in the land of the pharaohs, Egypt, of the many gods carried, the main one even if expressed through the king, was in the heavens shown as the sun, to be the provider of all? Although the sun was first option, the heavens of night also played there part to the thinking of the time. It was naturally quite difficult to bring into the same focus the means in the relationship between man and the innate and work them to mans own point of reasoning developed from the organisation of growing civil controls? Allowing for any combination of events, even to transport such a philosophy and any ensuing theories to its structure in development to far off lands or new worlds and further developed them in theme without additional input from the possibility of intermixed interaction, like that which was open in the land between two rivers?

If political drive was to have flourished in some places mixed with the accepted core religious feelings aimed at the forces of nature through the image of the sun or the hoped imaginings of natural extension through the Great Spirit of nature itself? Then death in human kind by the advent of political reasoning could have been accepted without the challenge of query about if the means to an end was already foretold? By those circumstances if; and if from further development on those lines was to have proceeded, politics would have needed no answer to balance the prospect of 'Natural Body Death', because it already worked so!

If that situation had remained as the best standard of how we had and do continually develop, questions in the asking would never have been put in query of how politics was to act anyhow. However all was to be brought into open view about some four thousand years ago, back in area from the lands of the Middle East. When at first we made the direct connection between we as mankind and the creator, all mirrors were broken, we could no longer look to the past only the future. All tin gods were now dead, or at last were to reside in our minds and hearts under cover to the reality of the eureka of creation! Even if by closed theory then ago, we had come in conclusion to what was to be our expectation in all cases, we erred at another first take?

We made the obvious mistake in the assumption that because of our discovery to the essence of God, it was intended for us specifically? Once again I take personal licence to the structure of that previous remark; naturally we were delighted that at last we began to see the means to the purpose of all things. Our reasoning to how we had developed to the edge of competitiveness under those circumstances also erroneously turned us to think of ourselves as being different, and some of us to have being appointed. Which was fully reasonable from all the factions of how by example we had interacted while living almost together again in the land between two rivers in our city states? All of which is to bring us to the open realisations of having two guides to the absolute necessity of our conduct by what ever means, but ever positive if looking back and keeping our eye on the future.

Politics, our politics has a lot of work to do in order to better represent us all, 'Natural Body Death' is one of three guides even if taken to be used as a cornerstone of change if necessary? From that prospect against the other two taken, any conclusions could be drawn simply because some of our political reasoning is based not in the service of man, but to the aspect of ideals. As we are actually run or controlled by civil order expressed politically, then for the other additional means of control to change or enlightenment we must choose, is best expressed by our one active prospect so far 'One for the Whole?

## CHAPTER 68

'One for the Whole' is to be a giant in the movement of our political ideals, which is to be seen when we overlay it upon our religions also. By first notice it is an example of how we might manage the future better with fewer rules, laws, edicts or sentiments. By the same reasoning we can use it in form to be one additional rule, law, edict or sentiment, then simply allow its own scope flow over us for application if desired or not, but to have been readily relayed for equal consumption politically. If not already in notice, then by the exercise we must take in comfort, this one active prospect can be done first politically without any bearing of any religious substance. Making that if entered into, any comment put forward can hold religious intent but stands on political terms only by solution.

I have put it so that we might reconsider the situation in the Middle East of some sixty plus years ago. When in cover of the political bloc situation we allowed religious factions have precedent in some of the politically motivated moves carried by the new development of unilateral considerations to have been made. In having to support any such actions taken if by compensation with the underwriting mandate of human rights, when almost least practiced, or readily practiced half measured? When caught in the headlights of reaction between the Arab and Jewish peoples by actual war, most during the early years post mandate to the area, on religious grounds! By the circumstances of the situations we were developing from our political cultural standing, we further reacted only on political terms, with one group being communistic and the other being democratic! Only to be realised these past five to six years from first notice of how politics has allowed us alter the past without alteration, but accepting that form of desired change to have been the intended route anyhow, which has proved to be a false economy! With no alteration being necessary, as the past could be rewritten in the order of who was the recorder and on what terms, even that the cloud of cover included such reasoning of being politically correct and of the one standard of human rights, the past could not be changed by looking forward!

What that means effectively is that by result of whatever events the justification has been made to suit the methods, which is the meaning to the standard of our steps taken syndrome. In now working our one active prospect on political terms only, even if mixed with some religious inclination, we will begin to see how less difficult it is to reconcile our assumed differences. 'One for the Whole' is to be seen in outcome by the reflective means I have already inclined. On first count it is our only active prospect so far, therefore it is of equal standing to us all, but by reflection and in this political medium only to be used to the same effect or proposed outcome for us all.

In best light it will be seen with the ability to curtail some of our political reasoning, which has only managed to confuse some very real issues over these past sixty years. Without prompt, it recognises the route of our present standing, without condemnation, it has led in direction to how we might best begin to alter things for the better. I on my own account have laid doubt to how we might have set the first standard of civilisation that we honourably carry. For our case note we are to be reminded that the outcome of our efforts in the craft of being civilised has openly resulted in a two aged life expectancy for our first second and third world economically based societies? Therefore what are the standards those differences are readily accepted by, or from where?

'One for the Whole' is already used by our best reasoning which will be seen more clearly by our religious definition when looking into the one active prospect so far. Of political ideas it is expressed more than a guide when asked to rationalise our actual political pronouncements. From example, where the United Nations might wish to express further grants or aid to our liberty, it is now to be done in the form of understanding carried by one of our first 'One for the Whole' examples. If I personally have shown signs of my doubt how we police ourselves through the maxims of politically correct behaviour or more poignant our human rights mandate. Then by political reasoning only at this time of now, I can generously offer that we might best conduct our human research through the maxim of common decency, allowing for future reference that we could be C D, in understanding rather than continuing to be P C politically correct. Which intones to be political first which is half measured being a divisive expression now that we are beginning to understand its implications?

In its closest example P C by forms of unity let us down by that very statement that is why it is half measured, when considered it is expected to be carried by all of our political rationales? I would question the ideal again by looking to the standards of political management in different countries almost anywhere in the world. Some countries are dictatorial and have only one rule of law by a single party state arrangement. Some countries have multi political parties offering a range of choices or methods of control or conduct until the winning of power. Some countries profess to be managed by the root core of a particular form of government, yet across the range of politics, deliver different messages by their own national or regional standards. Yet in the spread of supposed equality we are all to behave in the managed form of our declared unity by being politically correct?

Along with many other adages, symbols or tokens, some taken by personal control to separate countries of similar standing in any case, again if through the United Nations we conform to one standard of human rights, what are we trying to achieve? If the United Nations is to be the political voice of our reasoning, even if only erected in this modern era then we are all at liberty to now challenge its standing from our events of these past five to six years? First eyes are to be set not against good intentions, but upon the possibility of manipulated motivation, instigated from the political world standing at the close of total world conflict in nineteen forty five. From which I could enter that the idea of human rights to all was set in blanket fashion, first to accommodate the nature of our own consciences in all cases? Other motives could emerge, but will be confined to that particular time of our united history even if not taken in even choice?

In or from the total confusion at the end of the greatest to have been the meanest of human conflicts, our means of political management was shown to be wanting, not least because we had entered war in the first place by those same political leads. Further confusion was taken and given from the same set of circumstances whereby the victors were not really able to focus on the next direction that we should collectively take. By first token in banning the trait of fascism as a further exclusive form of political management, we set the aspect of political management ongoing. When from victory, the two main political ideologies by any division were communistic and capitalistic while at the same time showing to the aim in purpose of going to war on moral grounds or otherwise. Each side wanted to set there accepted morality to the overlay of all decisions to be made from the radiant heat of recent conflict without seeming to show the aspect of intransigence.

Although in the case of Europe early settlement in the run for peace was conducted in the haste of thought. When shown of the entanglement of humanity brought about by how the German fascists had manipulated the manpower of Europe to her own needs of desire. By trying to unravel the identity of peoples having gone through imposed hardship, large groups of forced labour refugees were sent on mass to the lands of there supposed origin. Because of the total nature of the European conflict, because of the political structure there, ending and in reforming, many of those repatriated were simply led from there trains of transport to the place of their execution? In full dress regalia of civilisation, after the worst of all human conflict, in the name of peace and satisfaction we allowed hundreds of thousands of people be sacrificed in the name of politics!

In the full probability of our reaching the aspect of regained humanity as soon as possible to those sad and terrible events, we set the charter of human rights in our own attempts of admonishment, but as yet unprepared. Human rights was to be the controlling beacon to our group management of all and each of us individually, or at least taken to be the route of examination for the sake of our consciences. By concept it was supposed to be our guide of accepted standards transcending all political leads whereby each person was to their own merit equally. Unfortunately, although set for political treaties, in minds eye it was directed through religious reasoning taken from one aspect only. That is why we will see different reasoning in the overlay of all religions with the Fifth Dimension and read them the same way as we read the political picture, but in finding the punch line also?

In the forced haste of our human rights mandate, even if managed by our new United Nations we had set it in sight to curtail any excesses of all political regimes, all political regimes! Without defined control without the bracing cover of our 'One for the Whole' active prospect, it as a means fell into its own corruption politically by the very structure of its aim by bestowing the mandate upon all, no questions asked? In as much as we had tried internationally to equalise a standing of human rights. Our biggest unseen error was at the very time of introduction through the capital vein of the United Nations, who by intention set this plan in open gesture to all of all nations ongoing. When alas from within each and every well established country and to all those emerging from the confines of colonialism and other shackles, all of us held to our own bias. Not of the same standing to who we were or from where our roots had first been noticed, but much deeper and darker to any consequences yet admitted, many of us held that we were not actually the same!

By much the same entitlement, I will use Europe as my own example in explanation to my own acceptance for an offering of 'One for the Whole' and how it works. For Europe read the world, in as far as all nations are made up of several different tribal or clannish groups. We naturally compete if only in relief to our carried instinctive traits involuntary, those instincts once ignited by the advent of life move to support us in that drive. Other outstanding features of our obtainment have been cut from the passage of time in our lonely path of our separatism to other specie in our total quest, only recently recognised these past four thousand years ongoing. In use of that time span, I relate to one of our first religious pronouncements, though vigorously point that in being made by choice or in being selected, peoples then were totally different by attitude. Not least that to the chosen, they were themselves from twelve tribes anyway, different with differences?

Europe carried on that tradition in best example by how she split to be different nations with some countries having developed thousands of years ago and with some new national borders re-emerging creating new nationalities by wars end and even later? Nevertheless within all countries there was the structure of tribal enclaves by the way the countries had been managed up until when? From these localised areas within the same national boundaries peoples did not think of themselves as the same, which is no bad thing? However we all did manage to pull in the same direction when needs must, but were still aware of those achieved differences even if of slight character traits, where for instance one group might be thrifty in comparison. Where one group might have a more liberal sense of humour or another be more politically minded. Where from other matters some rang a different bell by a physical trait of being generally blond or red headed or dark haired. Being European based by the above comparisons has been my own rule of demonstration to some of our differences. Europeans are also classed by the colour of skin in being Scandinavian, pale skinned or of a Mediterranean hue in being darker in tone by complexion?

All features in relation to our open difference to be European are carried by the same circumstances of geography, climate, diet and other reasons shared in mixture and comparison to the rest of the world. Again to the argument of description, I take a quick step to the new Europe of America, where in recent history another chapter has been added to our complexity of decisions in the making. By development and from other statistics, America was forced into a different settlement both in creation and from very recent times, which is why I can sometimes incline to be sympathetic to her overall intention being ours but under duress. Because she presents the full picture of all of our hopes!

I had offered in brief description that her making was from the partial breakdown of old European standards of how to progress politically and religiously which stands, but in legacy. Building her nationality, she inherited a bad aspect of humanity through the climate of forced labour generally carried by early European opinion to the new world. That slavery was forcibly imported from the continent of Africa, which meant African peoples were the cargo of trade? Peoples who in many cases through their continual lines of descent had remained black of skin as were we all once! Where many of us changed through the circumstances of geography, climate, diet and such in any time scale over the last three and a half million years, there we used those commodities to maintain our original standing?

While Europeans in America displaced the aboriginal people there by the force of supposed progress, in using the black skinned peoples of Africa as slaves we had set the tone to create the actuality of our differences by specie terms again unseen. Post world war two, by purely political leads and long after the official legacy of slavery had been abandoned there. America in victory, as well as working with others, through the need to finally introduce equality for all and from the prompt of conscience, accepted the drive to the standard of blanket human rights even if not fully practiced at home yet?

Although this matter was hugely complicated and made further so, it was done under the cover of political correct behaviour. If not from perspectives of being democratic and religiously tolerant, the attempt was to thwart the excesses of being communistic and nationally driven only. Whereby the state was to be the religion of the people that all purpose was to the ideal of the motherland even if named to be Marxist, which in theory was nothing less than idolatry. Allowing the flow of communism focus not on God but on men like Karl Marx and Friedrich Engels, further extending to the politburo members!

Although at the first time of the United Nations renewed, the world's largest newest communist country of China was excluded from representation. Russia was the focal point of deliberation in or to the terms of international human rights. I had stated that our conscience could have been prompted to such ends but reiterate to the complexity of the situation without recourse to personalities. At the particular time of the Russian zenith of imperial communism and the Chinese introduction to Marxism the heads of state in both cases had lost the plot relative to any terms of humanity. So desperate and despicable were these people as individuals, new political blanket decisions were hastened to curtail some of our excesses, not least from the intended standard of human rights to be introduced?

From what ever route, communism working to the one edge of politics only was destined to fail. One unfortunate outcome is that it has taken us far to long to see that the way we had operated it was more than lacking by being criminal in ignoring our real humanity and not understanding our separate specie standing as a reason in cause to our separate aspirations. In other words communisms biggest mistake was to ignore or try and curtail our religious aspirations? From the Russian example we can trace through the actions of one leader the total wasting of our own commodity of time. His vision was stunted, his focus was not of human progress but national stability, mistaking that nationality was a factor and not a reason. In the drive to eliminate the church; the mother country was offered as the alternative icon, almost in parallel to ancient Egyptian sun worship? By similarity all was to be given in the desire of need to the mother country, we were the sacrifice to those ends by no more than inclination. Shown readily by how displaced peoples were treated returning to communist countries after the Second World War!

'One for the Whole' to all political regimes has been put in place by me in token to show how we might operate it in activity. With it only running from these past five to six years it could not have had any bearing upon events just described nor to the situation in the world's biggest democracy at the time. America in true mixture had peoples of virtually all nations to her own banner, with the vast majority originally from Europe and her old habits. There was also to be counted from her minorities, the largest to have been black of skin, of which of these peoples originally transported as slaves, many had longer fixed ancestries in America proper, than later European immigrants?

With all that was going on in Europe and Asia, at wars end the matter of human rights to be addressed then, simply forgot to take into consideration just such factors. The standards being set was to compromise other political outlooks if nothing more. Without a byword such as the Fifth Dimension then ago, we nevertheless worked to the best plan that we could get away with. Our largest democracy did not work alone, but in setting standards for what ever reasons we erred in the haste of settlement for the best of intentions. In terms of world history these past sixty plus years have been and will continue to be our best, greatest and most marvellous years. Not only for what we have achieved but for how we have been able to still leave our choices open for all decisions. We are almost at the last turn, but will need to finish the questions just put to the situation of the instigation of our human rights standards and all ramifications since being first put, but how?

Our triangle is set by these three corners, America's black skinned population, representative to us all, communisms anti God/religion culture, then thirdly the Middle East religious standing to its own triangle of Judaism, Christianity and

Islam? All factors yet to be fully read, but now overlaid with our four prospects of the Fifth Dimension!

## CHAPTER 69

'Out of Africa' by religious terms is to its own reasoning, which we shall see more clearly as we progress upon time after reading this work while carrying the same burdens from the past and the new means to any obstruction put up from it, now tackled with the Fifth Dimension. Abstract thought and solutions have been some of my methodology to align our thinking to indicate that we humans are of the one specie. I have always offered in reminding us of our own free will, that we are also very individual. Amongst the other methods and styles of thought I have tried to impart and implant has been to the proof of that fact we are all connected by being 'Out of Africa'.

So not to confuse issues, from my own political references in the above chapter I had stated of Americas black skinned population, I remind us all that my purpose was to replicate the situation of some sixty years ago at wars end when we were renewing or vows of intent under very difficult circumstances. Although America was to the fore in promoting the reintroduction of our United Nations, in the fore to promote national self determination from whatever reasoning, in the fore to refinance world economies. Her internal matters of conduct had been left on the back burner for too many years after her own civil war ranging for five years in the mid nineteenth century. By self arrangement after that war and running into settlement by our recent travesty, she had neglected some of her internal affairs which were to have a new relevance tied to several other factors.

To have been so prominent on our behalf at that time, was to rebound upon herself at more or less the same time that the Middle East came into new prominence and the democratic versus communism war in Vietnam hit in the nineteen sixties! I hasten to point out that America is not on trial by me or others for we were all involved, it is just that many of us were quicker to wash our hands of matters that we did not agree with or understand. In some cases we still don't, but I will split the statement of the three occasions to focus us to the religious end of my presentation in discussion about our taken prospect 'Out of Africa'.

At that time black skinned Americans were quite simply not treated or regarded in the same manner as were the former European, generally, white population. By the open spark of enlightenment in terms to the aspect of eureka; the biggest internal process of any nation was started to address this imbalance if by attitude or from the indifference of ignorance. The human rights issue was raised in the most part by black skinned Americans. In its most generous form it simply asked in question of the unnameable rights of man, that all men were to have been born equal? I have mentioned it now to highlight how in thinking in terms of black and white the answer is not to react in grey. Further suggestion is in meeting how we can take any particular set of circumstances and misread them.

To the matter of this book and of its time by these five to six years, the American black or white skin phenomenon is not an issue, save by the misdirected inclusion of the European African slave trade making a modern inroad to an ancient established prospect?

'Out of Africa' is an historical observation, but not connected to modern reasoning, because of the twist of what natural selection really is! From first signs of our biped ancestors some three and a half million years is the sign of its meaning. At the very time by the prevailing circumstances of geography, climate even diet and other factors, we would have been dark skinned, in using the modern term we all would have been black of skin. Remember we are amongst the circumstances of geography, climate even diet and other factors, like all other creatures we were to continually adapt to our circumstances by the natural flow of movement drawn from new territorial use. Hence in show, we like other specie changed in form for those un-understood reasons from when we found different foodstuffs or took to different habits in combating the rigors of the weather.

Without the concept of realisation, from any time between the facts of our ancestral footsteps in volcanic ash some three and a half million years ago and our arrival at Shanidar, we had physically changed by small differences bringing us up to date by our present circumstances. From our developed status of query, much of our wanted understanding has been made in the haste of wishing to know. No more highlighted than by how many of us misinterpreted what Charles Darwin had to say. He and we settled to the main factor of our quest from the wrong imprint of what he had indicated. Remembering he had no knowledge in date to the discovery of our biped footprints in Africa, they were found almost one hundred years after his publication. By indication, when he implied that in our own human chain of development, we like finches had the same beginning and sequence of change. His impatience of putting the theory that we were not just conceived, but had developed in line from a similar group as we, is still being unravelled?

Unfortunately when his theories and some misinterpretation of them found debate, some of us centred our arguments to the religious belief in theme of Gods creation, clouding the issue even until today from where we still attempt to rationalise a one hundred and fifty year old publication. By his own reasoning and from a new method of factual study he made the connection for how we might have developed and from when, but not dated. By implication that we were to have expanded from similar ape communities as a comparative base, was covered in space, reason and time to the expected future discovery of our real link physically, or the missing link theoretically?

From eighteen fifty nine after publication and for some of us right up to now, we hold in opinion one way or the other to what Darwin intended or meant and how we have self interpreted anything he had said or had to say? All to be further complicated by such features as our own footsteps in Africa some three and a half million years ago, when taken readily by myself to have been a factual point of indication of our arrival? Others and this also includes me, had read his meaning in terms of the missing link between ape and man of paramount importance.

Allowing that when such footsteps were discovered and dated then the time of the missing link was to have been before then? All of which is only half a sentence when in the interim many of us took more readily to his indicative conclusions that for some, our status was stamped in the cause of natural selection producing a superior class of the same specie!

Alas by any translation, that interpretation was to lead us to readily accept those terms politically shown from very recent times of how we have been able to classify our collective standing to be first, second or third world? By the related factor to the manner of such classification, Africa as a continent holds most countries to be in that third world bias, has allowed some of us to continually rate the people therein so! Alas not being a matter which could be corrected by our human rights mandate supposedly addressing all people to be the same or of the same standing. If by inclination it is to be seen that there is a gap between the same nationals of the same country, like in my example of America. Then the form of separation is spread almost in even fashion throughout the world. My reference is in good quarter to all those black and white Americans who unknowingly at the time had given us so much ammunition in our real quest to understand the human condition for and by the best terms of humanity.

My own theory of explanation that we are all the same has to be justified against that particular condition of affairs in America up to the time of change in the mid sixties, for which much of this book is intended to do. I bear no favour above the need to others; my situation has been to marry some of our past conditions to the reflective sense of our future expectations. Although human rights was contrived to thwart the excesses of some political regimes, at that particular time of our histories, we simply did not think of us all being equal. If some theories incorrectly drawn from Darwin's writings were accepted by reasoning that peoples under whatever circumstances implemented their own determinations by being naturally selective, that was self limiting.

My American experiment is not against anybody save us all, for by other reasons it shows how we can neglect the importance in time of our own individual life cycle by adopting political issues to lead over the spiritual. Secular determinations can be taken by the means of any scientific come political delivery, because they carry no time scale or plan, therefore by them it would have been possible to introduce human rights for all selectively? 'One for the Whole' in political terms will not compliment our taken prospect of being 'Out of Africa' when considered by the wrong time scale that we can have black and white Americans. 'Out of Africa' is to all expressions that we as a single specie originated in Africa and our conditions of now worldwide, are a product of that origination in all cases by our single specie connection, from first issue in Africa!

By the mid nineteen sixties even when black America as we, moved it so for equal recognition in all matters, we carried our own errors that had got us into that divisive situation anyhow? Human rights then was not an issue of equality, but remained an issue of conscience under the sole representation of political determining. By unfortunate adjustment by calling that all men were created equal was of a religious acclamation looking for a political solution. That much was eventually done to redress some forms of imbalance since, has not correctly served either our consciences or our intentions, which will find bigger notice on addressing our religious concepts separately but in unison overall.

By that specific time ago and in the measure of some results in the form of our better understanding, our religions are yet to have there day, again by religious determinations only, along with the relationship of the balance between communism and democracy in the fields of Vietnam by political headings? Even so our trilogy of events in all having some sort of resolution of outcome by time, were lacking of our collective ability to manage by our active prospect 'One for the Whole'. Not least ably illustrated how we could not resolve one case to be our American anomaly misdirected? I hasten for input to state again that at such times of our trilogy of events, the Fifth Dimension was not at hand to lay the foundations of how we could best resolve what some of us saw for the reasons to our differences?

In America when the question of black equality was raised, there was a positive move made to call all to the virtue of that equality, but it was done only in charge to the political status of the day. It was done to a new understanding of the human rights mandate even unread in conjunction with the proposition to be politically correct, alas half measured and running into our steps taken syndrome! Resulting in two not so very clear resolutions, one whereby in American society the percentage of population split at first count to be either black or white, was to be balanced in opportunity by that division? Secondly that by division there was now to be either black or white Americans!? I would not suggest that a trend was started by any such settlement, but in many other countries mainly from Europe, steps were taken in law at about the same time to mirror that example of equality?

We were all measuring our populations on political terms whether we were black or white at our extremities and defining in law, how best to treat each other equally? Perhaps missing the point by not attracting ourselves to the fact that we were all the same once the circumstances of change promoted by geography, climate, diet or other features had been removed. By connection and in error, we took to the moment and tried to create equality from the understanding that we were indeed different. Perhaps not unreasonable when considered most of the emerging countries had gained their independence from colonial rule and needed a measure of reassurance to our political equality at least first hand. I do however submit that in classifying peoples to be black or white Americans, although well intentioned, entrenched us all the more to be different and not alike. Even now we can measure the outcome by the balance of international law, whereby if one applies for any job or measure of employment, questions are asked of our racial background?!

With particular concern to the country of America, even by my own reasoning, it is that in factual terms so far, our migratory habit, which started 'Out of Africa' ended in America, even if uncertainly by European means? By sheer coincidences of history taken in time from some three and a half million years ago when our first embryo was to emerge developed, it was black of skin only to be changed in the interim by the circumstances of climate, geography, diet and other reasons. After so long a time, and by any measurement, we were queried into half measured reasoning, allowing us to accept that our finches were in fact different, because we had no other rational? If we have it now, it is through our Fifth Dimension and associated prospects, which by first examination allow us to rethink the past without alteration, and travel other routes better seen through our religious perceptions?

## CHAPTER 70

To overlay our religions with the first prospect 'Out of Africa', what might we find to interpret things differently from any measurement other than by purely political reasoning? Using the picture of evolution to be of Gods own timescale, the term in reference to this our single specie could enter to be on the first page of the bible? Which by reasoned and understandable coincidence has assumedly always depicted our beginning in Eden by forbearers of similar standing to those who first made the connection by what ever reasoning? By our status then we were unable to account for our differences, not being able to see through our available wisdom the reason for some of them, except by birthright! From where the clock had been set to enter the space of evolution and turn into creation forever, tying the matter of time to our own concern, yet by Gods terms un-assisted! It was not Darwin that gave us evolution but time itself, creation in all terms of reference is the rational of that image carried when we first accounted to it!

Four thousand years ago our own sophistication surpassed our abilities of understanding, that is why by relation we accounted life's span to be of gods standing, but on our own terms. If by collection then ago we had answered the only question in abeyance, why were we suited to stand all that we did not know and count it to be Gods own mysteries? Was it the splendour of our own impatience that hurried us into conclusion charged with euphoria that our imaginings were the spoils of eureka, or had we just guessed right!?

Indeed we were readily able to think as God would have directed us by any link of concern when we already lived in the true culture of gods and gods in all living memories. We had gods for everything, one way or the other we could account for all events or happenings through our own gods and those of others by the natural tribal or clan competition ongoing. At that time, post Mesopotamia we had many well defined cultures that in general terms practiced life by the prevailing standards, but connected in the final quest to understand our purpose, hence the use of many gods to be our mystery holders? When realisations came to a group of tribes in making our God culture monotheistic, a new era of errors ensued, but now for the first time in human history we moved with the key to that purpose in hand.

By measurement the scale of our erroneous thinking was not damaging then, because we had accounted for the past through Judaism by reason to settle for the expectations of the future? Such minor problems to have later occurred were also best intentioned at there own time mark to have been two thousand, then fourteen hundred years old, and in also being monotheistic. Making three religions almost the same if entwined, but to only account for the future. Raising without the measure of religious conflict our thinking to the past having been put to question those and all other religions by political examination through the experimental fourth dimension. From where the mysteries were readily exposed sometimes showing our religious sentiments to be lacking or misdirected, not least that by them we focused on the future and not the past, because the fourth dimension is only forever!

'Out of Africa' is a question upon all of our religions, but only from the political perspective, because past any reasoning from the fourth dimensional charter of formula, we as individuals are here and now reading this our work. Mitigating political thinking or projections and focusing us to the actuality of future prospects religiously or otherwise, but on those terms. How ever we might have envisaged our sense of purpose to have been prompted by creation above and apart from evolution, stands no matter how looked at. Evolution is gods own time scale in relation to all creatures great and small, it is we who are the only specie to translate it in time to have been the forbearer of creation by open recognition. Our great crocodiles of two hundred and fifty million years ago and still going have not been able to relate to the concept of God. We have done so from just such the same evolutionary beginning but have been able to rationalise our evolutionary changes bringing us all to the same sense by creation and through our religions!

While running with 'Out of Africa' as a taken prospect, 'Past Gods Words' sounds to be of a purely religious nature, but when seen in review it lets in better light than first noticed. By suggestion there is not much given away in terms of definition to it, especially when I had examined it under political standards. By those terms the similarity is much the same when seen that it is used to curb political fervour so that we can still be open to the application of our own free will? By religious note the edge is both sharper and deeper, because it carries the word God for definition? My intention of use is not to lay it upon individual religions and cross examine their spiritual integrity. It is to be used in close quarter as necessary amongst any and all religions that would question or attempt to limit our free will of choice or inclination, which is to be more fairly challenged by our one active prospect. By further expression 'Past Gods Words' is not an adage to any form of control, it is simply an addition to balance how we can weigh our separate thinking to be of our one specie selection only? Therefore it is to overlay our religious concepts and not underpin them. Against the political overlay, by religious standards, any person even again and again who has had in all certainty to them what could only be described to have been a wonderful experience in the hearing of god's words? That they must shelter that feeling of sheer delight in abeyance to the fact that we have all been bestowed with gods mandate of free will allowing by the charter of creation we are all past gods words unless by self choice? Which is not a reflection of this or any future time, but the only way we are all the same, 'Past Gods Words'!

## CHAPTER 71

Our third taken prospect 'Natural Body Death' is poignant by description because it uses a dark word which should be taken so by political reasoning but not religiously. Death, had always applied to an ending in moral terms. From all concepts no sooner than four thousand years ago it has applied to us all religiously in terms of beginning. Even to be examined by our burial right at Shanidar when by the subtleties of care we rested a fellow tribesman in expectation. Later recorded again when we marked further burial rites to the same fellow tribesmen in caves of darkness depicted with the memories of yesterday for the uses of tomorrow? From which I would draw the partial conclusion that our preoccupation with death in very ancient times has been turned from the religious, leading to further life expectancy, into the political view of termination?

I can draw from those conclusions that our politics is not the best able factor to deal with our concerns of death, when all views to it have been worked in half measured fashion these past sixty plus years. If to draw us to the steps taken syndrome, then the measure of change required is to be done most by political reasoning, simply because our religions have been most alert to consider the aspect of our death properly. Where most of our religions have not been able to judge this factor have occurred in the very aspect of individual death if spoken in terms from any particular religion? To that, I have already offered the open conclusion of how we have set the error of some religions by assuming their direction is paramount! Not to allow any measure of our chartered free will is that error. Because if directed against free will into religion only in token, to be appointed so, is a break with gods time scale of evolution, not to be followed by creation, but plunged into the void. From which it will be seen in beacon why we must for our own integrity operate in terms 'Past Gods Words' in open appraisal of our taken prospect 'Natural body Death' religiously, separate than from political determinations!

One very important feature which cannot be explained in political terms against religious reasoning, runs past the explanation of experimental formula. When taken that God is Allah in any case. Even by fourth dimensional depiction we might be able and travel to the expanses of the universe through worm holes if lucky enough not to be forever sucked into a black hole? Which are simple topics we can all run with in some generalities of normal scientific conversation because of theory? 'Natural Body Death' is normal conversation of the same magnitude, but spreads in meaning by the two features of politics and religion. Where we must deal of it now is politically, but under religious instruction. Without any political edge the score is to be settled by those terms because no other method is apt to run better, because the religious influences are to be the most subtle.

When running with the prospect of 'Natural Body Death' we have been unable to rationalise it in term from political determination. Even being civilised when offering the proposals of human rights or of politically correct management, all are shown to be short of breath when explored of there long winded application under all the wrong headings. Almost in fashion, we have been too clever for our own good in exposing godly things under political management, resulting in this or these to be our present standards? In reality on the religious side only, death is not closure, it is advancement! In the form to complicate matters unintentionally in accepting the one reading, we have compromised the other by maintaining our option possibilities for other reasons or from basic ignorance?

I would assert it is essential for our religions to ferment the point of how we are to accept or react to the taken prospect of 'Natural Body Death' by better means than the expression of god's service as the purpose of life for deaths reward? My reasons for direction in the matter of religious dominance are and have been to temper political actions and curtail the whimsical approach of the fourth dimension to the core of human purpose! When from the necessity of civil dominance in clan or tribal control we took to the rod of human death by the hand of our own made law. It was done in the far reaching concern of our general welfare to maintain the ability for our particular tribes, clans of groups to continue in all the means of instinctive behaviour, but improved in the human form only with better intentions. Politics was not the best master of such ideals because at any time ago or since we have used it in compromise, working it half measured even to our different religious determinations, but used lit up politically, making the standard by human rights, we could kill politically!

By other and many examples used to explain the need for us to look at this particular taken prospect with one political and one religious eye, has been presented by events in this very modern era of these past sixty years plus. Not least directed about events in the Middle East in its broadest area, extending from the eastern shore line of the Mediterranean Sea further eastward past Iran running into Pakistan, but contained at the Indian border to its furthest extent? By all the supposed codes of political management through civilised behaviour and from other leads, countries from outside the area came upon part of it in open conflict of war. If one sentence is to personify the meaning of war then it is to mutual death of humans by humans always in outcome. Even if sanctioned by our United Nations, when America and European countries went to war with Iraq carrying much death and destruction as usual, all was done on political grounds only, but at what cost?

Of the particular incident above that I mention it is placed at the height of all of our civilisations gathered in voice through the United Nations. My own view is indeed fully ours and therefore we are all culpable having in the first place gone to war and later washed our hands of all the bloody and dirty ramifications associated with it! By result, with the new or reinstated sovereign government of Iraq given lead to set their own standards of justice, if passing the death sentence to those responsible for past vile murderous deeds, justice was done if so measured, but by whose hand? When from some of the leading countries who had instigated that dreadful prospect of war, where even there own sons were to die, then at the time of the national execution in law of mass murderers. Some American and European politicians who had readily supported there own countries involvement in the war, proclaimed that in our/their national terms, we as a nation do not execute criminals or mass murderers! But under those sovereign terms that we had helped create, others might do their own will, how half measured might that be to accountable reasoning?!

By the reality of the above situation and countless others, it should not be very difficult for us to keep check on our separate valuation to our taken prospect of 'Natural Body Death'. Politics is not an obstruction to our thinking, it just has to be up dated so that we can change our opinions if we do look at the same statement from different backgrounds. The conflict referred to above does have serious religious ramifications checked from religion to religion, but not through religious concepts. For the war that it was, was by political reasoning only, even at this very late date by any time scale we have compromised the political version of our aspect to 'Natural Body Death', in that we do kill by political means only, even while being very civilised?

By those events in any context our skill of observation has to amend such results, which is to be done from our religious standards, for none of them have a mandate of open conflict. 'Natural Body Death' is already embodied in all religions, nothing more nothing less. That is why we are to empower them to voice our opinions in better accord to each other and from each other by those religious differences to make our political understanding carry reason to be nearer the same! A little caution to be used stems in part by the use of our one active prospect in 'One for the Whole', it extends to the same value in political and religious use, but is not yet definitive unless expressed by us all?

## CHAPTER 72

Human rights, political correctness or political management have so far been able to support our separateness taken by the most or best perceived standard of care through any medium of any political ideology! From any of the means we have tried to rationalise our approach, unfortunately we had failed to set the correct objectives, because we have always tackled our supposed problems from the standard of being too complete! Where doubt had ranged, we created more and more channels and turns in order to clarify matters? When our religions of recent times came to bear more fully by what ever means or under whatever circumstances, we have simply cast upon them political directives. Resulting in the present mix of situations where many of us have lost our sense of values, yet we all have to run with at least one leg of our controlling influences of our religions or politics!

My own task has been to set the compass to the shortest route that we might take for future progress. Of my biggest concerns about our present situation, I have worked to offer the lowest number of corrective measures we might take if needs must? Encapsulated by the Fifth Dimension in boxed maxim form are my three taken prospects and the one active so far in 'One for the Whole', which is to work in or for all purposes for us to compliment our single specie standing. On our behalf I fully understand that some of these proposals offered are to note our accepted differences, but to be a matter of personal conscience how we individually look at some of them? Although set from now, our similarity study in these most recent of times post World War Two, is timed from these past five to six years, with no past or pre alteration necessary!

Some of our less obvious problematic occurrences had been generally laid down by our own best seat of international law if offered through the United Nations, portrayed to be our forum of mutual expressive relief? Where I have cast doubt to its genuine approach has been aimed at the chair in point of reference when making the same deliberation to the same people, but under classification to be first, second or third world! Also when all are to carry the core standard of human rights, of conduct to be politically correct along with all manner of religious tolerances to prove we are the same? Therefore allowing misinterpretation on the much larger scale to the divisive nature that we have set basic civil law internationally, taking it so that laws have been made only to recognise many of our differences. So much so to set one group against the other without justice of consideration to the cause of those differences having been from the circumstances of our geography, climate, diet and other reasons?

'One for the Whole' by equal overlay to the theory of our management in terms of politics and religion, will always be to individual choice. It is not to supplant the measure of control that we already operate by. Its main point of reference for us all is that it directly relates to the individual at its point of contact religiously or politically, but not in the form of law tenet or covenant? It is our own wall of retribution from where we can measure our inclination to others by how we would address them. It is to be our new barometer of success by application when it represents our own sentiments to the mirror of reflection by the nature and from what source we offer query or solution by the expected reply, if due?

Apart to the most obvious of examples already put through the process which we will further enjoy, by that individual request, which has no limit of number. Our first observation in the political and religious field has to be fed from the ready adoption of our three taken prospects. I have stated by the obvious that when proposing any maxim the outcome is to be balanced by the effect of the question or direct suggestion. From the total mixture of any religious supplement in all extremities, questions have to bear within the expected outcome of that or any particular religion. Therefore any individual offering a suggestion by the standard of their own religion from the standard of their own belief in it, must allow for the circumstances of others?

I have not aimed to embarrass religion in general, but labour the point that most religions suffer their own failing through impatience? I have tried to give notice that we should curtail the excesses of some of our extreme forms of worship and what they might contain with personal mixed feelings or misinterpretations? Where some who would lay for open consideration that we should all follow doctrine that they already are involved with, this falls into the type of query to be under scrutiny? Although unclassified how we are best to use our one active prospect 'One for the Whole', when others suggest if only under religious terms, that we should be content in the preparation of human sacrifice to continue! Reaction in the first instant by reflective terms is set to counter that type of proposal, simply because the suggestion is for the whole, in being for or to all of us, therefore is to be for all!?

Any proposals in the terms of human sacrifice, fortunately an out dated concept to have reached its height of practice almost five hundred years ago in the new world, cannot now be repeated. If suggested so then our active prospect is there to offset the sublime that by application it is not from who the idea comes from how it is to be worked, but directly from all of us. By that involvement on those terms it's built in safety feature is there to nullify those that would offer in the same context that their suggestion carry Gods words of instruction. 'One for the Whole' by its activity only, operates 'Past Gods Words'. It operates within the confines of 'Out of Africa' by relationship of our specie integrity so that proposals are not put by those who deem it so that we are of a continual collective form of natural service by the terms of evolution! Therefore in consequence any proposed application has to abide by our terms of 'Natural Body Death' relative to this one specie we are.

Making it so that we do not have any human, however self engrossed who would offer any suggestion that they were representative of the past or from the dead even tied with the evolutionary calendar! We cannot accommodate anybody who claims representation of the crocodiles two hundred and fifty million years of history, relative to our collective lifecycle of fifty plus to seventy plus years. We all are to be inclined to 'Natural Body Death', therefore even from our religions most in practice, no suggestion through our active prospect leaves us in any other choice. None may offer through the challenge that we are to be represented in human murder or slaughter represented by any of our accepted religions, save by example in lead, but then only through the forum of 'One for the Whole'.

In as far as we look at the prospects offered from the Fifth Dimension there number is to explain there need. I have intoned that less is more, which works to a whole range of subjects and topics. Without the cause of worry or chastisement my offer in the first instant is to be representative, not against what we already do, but to supplement it. We have been in continual flux since wars end applying old medicines and new potions to much the same problems. Alas in swallowing the pill some of us have been in better positions from better circumstances to apply the suggestion for ourselves and in suggestion to each other from the basic set up just pre world war. Even that my Fifth Dimension quiz is of four small parts it will at first be recognised to have been a product of us all by these past five to six years. Releasing its medicine upon us all in antidote, so that when we do look into any mirror we will each see the features of geography of climate of diet and of other circumstances that make us this one and the same specie!

## CHAPTER 73

Without the full mark of dated time, I now offer what could be seen in mixture to some of our present conditions. I have given much mention of the Middle East covering a broad range of countries there and to the area of Palestine in particular? By indication I had attached much importance to a time slot some twenty years after settlement to the creation of the state of Israel and the type of conflict between the Jews and the Arabs already in Palestine? My references are to bring us all to the same focus in this modern era, for how we should begin to think, because of new introductions these last five to six years!?

In the mid nineteen sixties with the progressive reformation of most national ideologies keeping the political perspective to the fore, whereby nationality in the form of self determination was progressed by deliberation. If the United Nations was party to the basics of change and or political regulation of sorts, then without charge she as we missed the meaning of what was happening in the Middle East in the name of political expediency? If trying to regulate our similarity by the mixture of our broad based political standards through communism or democracy, we shortened our address by the means of human rights or similar political opinions only. Not least so that we could sanction each other to be of like overall outlook in the name of that unity for purposes to show progress!

By the main condition of our different political setups we kept progress by those or any political terms so that we still had that meeting point. Even if strained by the shadow of our world cold war standing, we could still run small wars, like between Arab and Jew and or democracy and communism in Vietnam! By the intention of keeping things on a primary political level, we though in terms of nurturing that means of future control giving us the same and equal means of representation. Without direction or open perception in the same terms that we applied our concerns and abilities to political reasoning, we headed religious influences by equal tolerance in supposition? In remembering that by the European conflict of world war, most sides were Christian or non religious communists and the mixture in law to the United Nations was set by that group, religious tolerance was a recognised byword and seemed sufficient to accommodate further reckoning?

Against some reasoning to the time of events in Palestine in tandem, from when it was to become clear in closed recognition that the Jewish Arab conflict was not effectively political. Then new lines had to be drawn which was done in the effective flow of political progression only. Making the situation to find other solutions of a religious nature, or accepted means to those involved who were at liberty to draw there own conclusions, but in compliant mood to the prevailing standards of our proved political expedient measures?

In overall mention to how the state of Israel was formed I had intoned of there own Jewish standing by religion. Each time I had made reference to a time of our history being four thousand years old was to my first dimensional religion of Judaism. Further mention to the religion of Islam being fourteen hundred years old is covered in general terms by cross referral at times to the Arabs and my third dimensional religion. By the realms of the unknown and to most political failings without condemnation, both of these religions if not portrayed by myself for having direct religious input to the vast European world conflict, is a matter to be now settled in harmony, not least of these recent times?

Even if unseen, when or if the state of Israel drew on independence and was assisted by America and European states although done to political ideals in all probabilities. When in later times with Israel still in further conflict with the Arabs, much the same as in the days of mandated Palestine! New conflict of any kind now, had to be accounted for through formal and for legal reasons, which was more readily done from the Israelis, because the outcome was generally in their favour? If and so, they, even in closed style celebrated the cause of there successes from being appointed following religious circumstances in the style of old forgotten/remembered promises, how was that to be rationalised by religious tolerance?

Without reasoning, if such events have to be considered to balance with any similar events, but not unfairly through another example of Vietnam occurring across the world at much the same time. That protracted conflict was horrendous upon the peoples of Vietnam in the first instance, but was fraught with purely political dogma that they were to measure their victory by, but only for political reasons! America by her involvement for world stability also had very serious wounds to lick, in both cases this was done in compliance to certain criteria embodied by the United Nations but always political motivated, leaving religious matters to the concern of tolerance? Accounting for further steps taken from half measured pronouncement, whereby it was alright for some to be guided by political reasoning and take any measure of their perception of religious conduct to any level. Even if ignoring the religious standing of others from the results of war by outcome or from the given political support to carry out war in the first place?

In the very same fashion and style that we as individual human beings are here right now, I am reading the events of the past without justification, but informed of our own lifecycles when all doors must be opened? From however any international intentions were to have attempted to rationalise our self understanding, if by assumed consensus or planned objective, all was done half measured. Although we were to be tied by political acceptances and most procedures were of that ilk, the format was by our steps taken syndrome in carrying on after half measured commitment. No more readily seen than in the conflict of war when and if some actions were claimed to be by religious determinations, right or wrong and could usurp the aspect of supported religious tolerance?

Without condemnation, but in compromise, I can offer in supposition that if the Fifth Dimension was ready at the crucial time on our behalf, then the Arab Israeli conflicts post world war would have had a different view. Then our balance would have been better suited to future prospects without altering the past of any Middle East country politically or religiously? Instead we had the situation whereby I and many others have speculated into reason or cause. We have troubled our own standards to accommodate the factor of blame from the aspect of result. Even in counter or recognition that the end always justifies the means by political determining, our only error is we allowed the result to be ongoing by what ever means or deeds. Always pushing religious holdings in to the classification of political determining whereby might is right?

From any of the above by drawing opinions, which I make on my own account now, I can put it so that if reactions were forthcoming to any so described events or occurrences then the reason was by those events? If the Arab/communistic bloc culture was to find second place by military/political engagement ongoing, from the cultural terms of spiritual engagement, how and by what standards were or had these been set? If in or by ungracious recognition to the open fact of political manoeuvring, Israel for instance took in religious celebration to outcome? How was that to be balanced by religious tolerance in the growing accepted standard of legislation to cover one aspect of our supposed likeness in allowing us worship to the same God by different methods?

Where I can turn to the reasoned standards of my three taken prospects and one active 'One for the Whole' is done only by these past five to six years. Arab people and nationals, who by general terms the vast majority are to the religion of Islam, did not have my quarter of the Fifth Dimension at those and subsequent times. If political failure was not of prime concern to them or any particular group, then religious fervour was paramount like to those of the Jewish religion, but not regarded internationally or politically so? After all, religious tolerance was to be the manager of all religious concerns in that context. Therefore if one group added acclamation by it to political deeds it was a matter of tolerance without upsetting the standard of balance in maintenance by the United Nations!

In consequence or not, if the other side was shown to be supported by a wholly irreligious political outlook and was to show disdain and make ready and prepare its own defensive or protective posturing to the interpretation of any outcome? How might that/their religious standard be seen against tolerant conduct? Half measured reasoning and our steps taken syndrome take further hold over the outfall of such described situations if that was not always contained to the specifics of any one particular incident or one particular time slot. Some facts do not need the bearing of truth to be, so they become so by being done! It is from the basics of that small statement I had proceeded to manage my own thinking on or about such matters to the understanding of the ends justifying the means, if ever political results could be balanced by religious reasoning?

That climate, geography, diet or other matters have helped us to our differences of style they have also helped us to the difference of our attitudes. By indication I had put it forward that some of us are dynamic, while others might be of a laid back character. This had been done to broaden our perception to cover that we have very many traits while being the same specie. Where by indication Darwin gave of his finches with their beak structures to have evolved by the rigours of time and to the ends of need, this was just like other creatures who changed completely by the evolutionary timescale. We from change have not changed our condition in the full terms of creation, which has simply broadened! Whereby from one side to the other in our character, from the working need of aboriginal Bushmen, peoples still acting to their direct needs on all of our continents, to the richest of merchants in the dynamics of business, nevertheless we are the same specie. Much of which is expressed how by those very extents of our range we all have religious convictions expressed by our own codes of introduction. Shanidar first showed us when, and it had taken us ninety plus thousand years to discover that from our slight changes we were able to promote our religious perception. From any time of four thousand years ago to one thousand four hundred years ago or between or since, up until the mid nineteen sixties, we have promoted our differences by religion suiting our conditions. When by the restudied element of political posturing results or outcomes were not as expected, what might be our religious questioning to the matters of political expediency through religious tolerance?

## CHAPTER 74

By all the processes of change we had introduced since nineteen forty five on the political level, including the concept of religious tolerance. We have only just managed to avoid the catastrophe of total war, but are still capable of it by political terms only. By the crudest of example I now use without fixation, I intone beyond the scope of fettered proof the standard of our half measured reasoning is becoming more responsible in keeping us in this perilous situation. When seen that it evokes our steps taken syndrome then the management of the affair is perhaps beyond the running of political matters even if by the/our United Nations?

By other standards I have mentioned the fact that we are all here and now, therefore in the style of our living and in the purpose of it we have religious countenance though yet to be fully measured? From the same perspective I have directed our line of thought in or to the two terms of our most important issues in and for the purposes of our existence. If all I am saying is that our religions should be given more value to our prospective planning than our politics, then it is from this very recent time of these past sixty plus years that we must review matters. By the best of good grace under any circumstances, it has been our religions on there own account that had been responsible in all past histories that have enabled our progress, if progress is what we have been doing all along?

By the value of human emotion in all terms of its form, we have always worked better by the guide of religious instruction than from political expediency? From the same position where I have laid claim that we should overlay both our politics and religion with the four prospects of the Fifth Dimension is still to be directed, but to the conditions quoted. That the overlay is to respond to each medium separately allows for the factors of the many branches of politics like commerce, trade, and other specifics? Our religious overlay is to be of the same fashion, but in the open acceptance that there are other elements to or by our religious thinking. Not least from the form of our two main branches one group being of an Eastern hue tied to inclusion of several processes for completion and the other group directed as Western, generally monotheistic and of self completion through a single lifecycle?

By any connection or support we might find for the matters to be done from the above directed format, back in the mid nineteen sixties we were not of the same opinion to both aspects. We ruled by political incline allowing the statistic of religious tolerance cover our concerns and our controls to religious emotion. Which was to bear in the aftermath by hindsight when we saw we missed our biggest opportunity of specie understanding?

From the factor of Israeli celebration for victory in the Arab conflict of the time with particular concern to the town of Jerusalem, when the dust of grief in our own human terms had settled. It seemed to go unnoticed leastways to the United Nations the significance of the act of conquest? Israeli emotions took control to the full meaning of that victory, when by declaration on religious terms they made that small town their new capital city politically! Misdirected or not, to proclaim that this was further sight to there homecoming by appointment, which could only be read to have meant by religious means. From where it can be accepted in all terms to religious gratitude it was never to be understood by political pronouncement unless half measured!

That any Jewish, Israeli victory was to have been by religious terms, then in point the aspect of Gods involvement was clear? If and so by deed of recognition no matter of mention was made by referral to this/these or those circumstances save in political terms to religious tolerance. Where was the balance to come from in either explanation or acceptance or understanding to include all of us, unless clouded by our political bloc culture reading of events then ago?

Half measured working would certainly have allowed for the fact in explanation that the Israeli victory was fed by better preparation, better planning a better intelligence network and better weapons. From that basis by result the steps taken syndrome would simply ensue by show that the forces of selection to our assumed matter of continual international control worked so far. Leastways for how future war could be contained to anybodies own particular area of concern by political conduct, but encapsulating religious tolerance? Alas by referral to the bones of the last World War conflict where religious tolerance did not really matter, because most sides had Christian Gods, where was balance to be found?

Although religious tolerance was added in context to the list of political influences upon our conduct, it was left to its own manner of policing by the natural assumption of Christian thinking! Without accusation and to the point of direct alignment it was through the offices of Europe and America from where the new rules of order were laid down from the year nineteen forty five. In the mid sixties the fruit of that misdirection in terms of religious tolerance was to bear to the need of urgent change even if to compromise our human rights mandate and politically correct reasoning!

If I have laid direction that the ideals of religious tolerance was based on assumption of Christian thinking only, then by what manner of voice were other religious groups to have; and by what conditions? Acting by result only and from direct indication, religious tolerance has no bearing if misdirected? When from Jewish reasoning they had had a great victory by their own religious lead, forsaking the political aid in terms of democratic fortitude against the standing of communism! Ignoring in all consequence that democracy in triumph was of a Christian bearing against the irreligious communists who by supporting Islamic fortitude with the vast abundance of followers to that particular area of geography being the Middle East, how were the numbers equated but through religious tolerance?

If to the Jews of Israel a circle had been turned, then by all consequences and from their very own reasoning, there concerns should have included the intended matter of religious tolerance. From the case that they didn't give it full recognition, shows that the feature is not workable by modern reasoning. Religious tolerance had been in better management for hundreds of years by my third dimensional religion of Islam, even if stated to have replaced the essence of Judaism and Christianity. That is why there were only six hundred thousand Jews at the time of partition at the end of the political mandate by the religiously Christian British, at a time when Islam was more religiously tolerant than she might have been?

For the Jews to have thought to their own religious concerns is assuredly of their own business, but in sidestepping the Christian influence to there aid, in ignoring the Islamic influence upon the area in general, curtails in use any further reasoning through religious tolerance! Not least for the reasons or events to have occurred since then and later to have been misclassified under the wrong heading promoting into perpetuity our misdirected steps taken syndrome? Without qualification it was quite natural for some factional Islamic groups wanting to react to the growing situation of change by the etiquette of political dealing in at first concern to the Middle East. Then further misguidedly spread to be upon Muslims in general if displayed through their religion of Islam, misunderstood! Unfortunately the concept of religious tolerance has in part been responsible for some of our actions and reactions because we had failed to put any standard to it except politically?

When I had indicated above that it was from Islam that the concept first came, expressed at introduction by Mohammed, although the specifics were different because religious intention was to come more to the fore in the time of Mohammed than civil or political controls, although prominent? Then ago I would give the situation to have been one of religious forbearance, similar in terms to allow the avenue of change develop amongst the influence of better judged religious concepts to be? Rather than the broadened base form of religious tolerance we later adopted, which in loose general terms allow that any form of a deemed religion might be practiced, even the unheard of religions of mass human sacrifice? Unfortunately the real failing of religious tolerance stands to its blanket coverage if explained, where by political determining one way or the other some claim of its use unnoticed, yet make pronouncements of its use upon others if required?

If noticed in the mid nineteen sixties, when with the medals of victory, Jewish pronouncement lacked the relevance of total political involvement even with democratic aid, then in tone their victory was not over communism but Islam!? Usurping the concept of religious tolerance yet invoking its intention if even originally set to thwart the matter of communism by the embarrassing tone of the supposed generality of religious sufferance. If our intentions were good or to the best means of understanding then ago, it is that by first referral now, we are at liberty to mitigate on each others behalf. Therefore in what range might we compare Jewish behaviour then and since, in balance to Islamic behaviour from then and since? When both religions and through there own political connections, will and do bear upon us all in terms past the ending of our political bloc culture stand off?

My own observations although carried by my own version of abstract reasoning, thought and supposition, intentionally holds that we do have a purely religious from of divisive settlement to revalue. Done through the area of the Middle East is set to focus us to the main nature of division even from our progressive religious histories tied to the area. Our modern rational does not clear the matter of cause and effect even by accepting some of our own features untried. Remember half measured reasoning and our steps taken syndrome? Well intentioned that we were, under the extreme conditions from nineteen forty five, we did not have the correct set of aids to propagate progressive settlement as it should have been, after all that is our nature, progression? Then, we did not have the Fifth Dimension in place for discussion or for argument. Even that it is claimed to have been of very recent introduction by my self over these past five to six years. We did already have the mixture of needs for progression by the progress of time, but did not know how to label them in there present form as three taken prospects and one active in 'One for the Whole'.

By our own best good fortune and to the route of our progressive nature, my rendering of the Fifth Dimension now is to be upon the separate cases of our politics and our religion. Not of a competitive nature but because it can read either or both by there own standards, even when we might have intended them for other uses. Our best appliance of the Fifth Dimension will be upon our religions, because in every case they already hold that we are 'Out of Africa' that we do operate 'Past Gods Words' that we are bound by 'Natural Body Death' and to our free will we can aspire by 'One for the Whole' in application? From the same measure of political reasoning we might examine how from such determinations we have been able to suffer each other to death by religious standards or because of religious reasoning if taken to extremes?

# CHAPTER 75

Through and from my third dimensional religion of Islam, I had referred to its formation by date to be inclusive to the feature of human death by the same human hand as we. My comparison was in balance to the terms of the first two monotheistic religions of the Middle East because they both carried god's commandment not to kill. Past the justification of argument and the edge of reasoning the simple term not to kill was a religious determination in regard to us humans as a single people. If from the first interpretation by the Jews this was so, then the ruling of the time was inclusive of the chosen tribes as perceived, that we as Jews are not to kill of our own tribesmen by god's determination to us? Even by the nature of our studies in the reconstruction of history to the beginning of time where we saw death all around us, the limit of death by the means to we of the Jewish religion was by the control of civil law or human battle through civil war?

Four thousand years ago we determined in Gods name not to kill of each other by religious standards delivered by the hand of Moses in the name of God, yet at instigation would readily kill of each other by legal requisites. On any personal level I make no application against the Jews of then ago or in link now, save by those in our present circumstances of living today that we are here now! On the same scale and through the same mirror of reflection the new Jewish religion now termed Christianity carried that self same verse, also not to kill. This time it was done two thousand years after Moses though taken in the same strength of meaning, but carrying a different agenda. Then ago with the feature of the extra two thousand years of our histories, although virtually nothing had changed except small measures of progress, we were given an extraordinary option that all men were now chosen to this new Judaism? Unobtrusively taking the tribal version of religion away and making it universal in the same medium, even if relaxing some abiding rules. Now the meaning of "thou shall not kill" had a direct bearing by intention to all men by god's terms for all men to each other?

Any reasons of success to this newer form of Judaism was compromised by the actual time of delivery, because then ago our usual approach was that we had always sought better reasons to be separate than together. Nevertheless the term of Monotheism by religious codes had been established forever which through the porthole of our reflective image from the mirror of life was further extended by Islam? Having cast doubt to our reckoning in the commandment not to kill as delivered to our first two monotheistic religions of the Middle East. Much in explanation of how we did not really understand the intention of that specific commandment was brought into a more defined form of acceptance even until now?

Previously I had put my own mark to the different standing of my first three dimensional religions of the Middle East to the aspect of not to kill? I do maintain that the concept was too large for either the Jews of Christians to really grasp, not least from the standard of normal conduct in daily life at their respective times of our history at their instigation. Islam from and through Mohammed gave better judgement to this whole matter if only through or from or guided by what we really were and how we really behaved?

Six hundred years after the death of Christ, by indication from myself, but not necessarily to the feelings of the day or area. Religious fortitude might have lost some of its impetus, because the core direction of the main faith of Christianity had ceded its point of direction to Rome, a landed capital city from outside the area of the religions first activity? Although many important local symbols and specific areas of worship were aligned in the direct memory of Christ the son of God, like places of his travels, his death and resurrection, like in Jerusalem? Overall concerns to the direction of the religion had almost forever been compromised since Christ's actual parting and from the adoption of the Roman Empire to the religions uptake and the manner of that situation? By similar connection, how the empire had reasoned to take on the religion in the form of unity incorporating many of its standing rules, tenets and covenant. While at the same time operating its civil code by the policing of state, including the real matter of human death where necessary by civil consent from civil law, had the religion been compromised by this situation?

Even without cause, effect or reason, by the simple matter of locality and proximity, any form of study by inclination to review the matter of open religious fortitude could only take effect in relation to that specific area where prophets had walked. Without any form of claim or connection save in our single specie link, by any suggestion that I might make in connection to how Mohammed might have thought by means of direction, stands by result, not least to how the application of human death had been offered in core function by or from the direct thinking of the day, even applied to this new religion to be?

By direct input, any mention or inclination to the matter of humans to kill of each other were so cast to be of religious significance also? They were maintained to the specifics of our individual lifecycle but curtailed by the thinking of the day by not being able to encompass the maxim of 'Natural Body Death' in reasoning. Where from the rational in the collective thinking of expediency the matter of human death in religious terms was cast to be for the direct defence or promotion of that, this or the new religion forthcoming? For at the time of his revelations, visitations and exultations Mohammed had not readily entered the overall concept of the religion shown in expression by how the largest core of the Islamic rational was by former Jewish thinking and planning.

It should never be suggested that Mohammed was playing at politics from his lines of reasoning if interpreted in error by others as myself, make no mistake all of his hearing and writings were through the accumulation of knowledge accredited to the hearing of God in Allah's desire. At initial presentation the matter of our individual free will was accepted by all terms, it was through our own endeavour that the religion was to be fulfilled in concept. Any protective edge to the base structure of the religion was also installed to the rigors of our individual free will, our individual free will? All of which is to show in concept of how careful

Mohammed was at the instigation processes of the religion by how due deference was given to Abraham, Moses and Jesus historically, linking us all by religion!

Unfortunately his task then ago was monumental, simply because of the people he was dealing with, who by all explanation are all of us, where some had held to the traditional and fully tried proved and attested religions of old. To be offered by demand a new status was naturally confrontational, which to the emotions of the day would need direct reaction. Even that the two main religions of the direct area held to the concept of God's own commandments including one not to kill. Now, by religious terms the standardisation of spiritual law into civil concept was carried by the laws of religious hear say for other means of self protection? Set to the supposed threat of outside reasoning, managed through the concept of fear in misunderstanding led by able priesthoods, if nothing more than wanting to protect the image of their existing religions against what was proposed?

By all comparisons, from where Mohammed saw or felt or reasoned that Jesus was only a man of human form like himself and was inclined just like him to have had a similar experience. With purpose renewed, Mohammed was now to read upon all men the one true route? Following that course either past god's words or void to the concept of us all being out of Africa, if by his own misinterpretation of those acting to the other two main religions of area, he was to strengthen the concept of religion. Then in like error his application of defensive or protective measures to that new faith by means of human death by any circumstances, compromised the ethic of 'Natural Body Death' then unseen!

By our own good grace all three of our monotheistic religions always have been open to the last and one active prospect of the Fifth Dimension in 'One for the whole' by todays uses and understanding? By the mid nineteen sixties this matter had not been settled and consequently opened all the queries of supposition to any political events that had recently occurred. Subsequently from the religious standards, which were reasoned to the acceptance of tolerance, we were then obliged to act or react in or on self denoted religious terms only. From the measure of one, two or three, our Middle East Religions, now of our recent histories had gained the reactive prosecution of religious retribution to political outcomes? Leaving them as the dish of spaghetti to the centre table of world politics intertwined through the United Nations in open display, where the meal was cooked and from the recipe of half measured reasoning served with the dessert of our steps taken syndrome!

From the mixture of our recent histories, some individuals to the religion of Islam have set to accommodate the matters of concern to the justification, whereby some of those to Islam could impose the will of Allah by religious terms and in compliance to the religious medium direct! I make the personal note but only in hindsight that at the time of such happenings I am now proposing to give explanation to cause but again only from my own reasoning. Where I denote the difference and cut of point if relevant, stands to the time of early conflict between the Palestinians the Israelis and several Arab states whom by direction were all to Islam by religion except the Jews of Israel.

## CHAPTER 76

By the very great art of those closest to any situation the recording of the event or reasons thereof are generally given first sway to those in connection to whatever situation we might arrive at. By best example I chose Archimedes to express the point already done by his acclimation eureka! By formal documentation his works have since been carried beyond the realms of doubt, because most were covered by the result of scientific experiment to be successfully expressed as formula. A method to hold the original concept in whatever field to the bound limitations contained to the working of any theory through the realms of science in that specific proof!

By most formal reasoning all scientific experiments have a basis in fact by result, that all fields of science have been measured by there interactive properties are due to such times when individual people set the trend in the must follow lines of experiment. That is why we can set the past to the aspect of evolution because of the measure of time involved to allow some experiments mature. Even from its bespoke formality science still does have areas of controversy, but these in good grace are discussed among the scientific community and released to us in the certainty of the possibility of exactitude? By its association science is to the realm of politics, leastways in the habit of delivering its products for our comfort of use through the mediums of commerce and trade and technological advances. By our worst uses some of us have been able to acclaim its merit by its delivering to us better weapon systems, allowing the claim of victory in the darkness of conflict by war!

If by the result of conflict, technology was the best aid to the victors, then all attachments to those and victory could be written in the extended terms to be covered by whatever was written. Actual reasons could indeed be attributed to such things like a groups religious standing, even without mentioning any religious affiliation to those on the other side in conflict or to those in the application of aid? For instance in reference to the Middle East with all three monotheistic religions involved, but not included by religious persuasion. Like when the state of Israel had most gain and through that factor claimed it so by religious determinations historically or not! What matter of reaction was to be expected by those who most supported them or those in opposition by their own religious terms? Unless already established by the new and gaining standards of rationalisation even if through the United Nations where religious tolerance was granted to all evenly, even to some to be of a first second or third world national rating by other factors?

If the situation was to be raised even without the aspect of planning, that a westernised first world country such as Israel, who by what ever circumstances even covered by our early bloc culture standoff posturing, was now able to promote her own religious aspect to cause and reason, but by using religious tolerance one way only? And was able maintain that aspect in compliance to political reasoning being westernised like her main contributing ally of America, who by the same tone was also first world and Christian to tolerant means? Who by her own grace, allowed it so that her religious tone was met by the same tolerance from Israel, but not returned? Or had the edge of religious tolerance been forever compromised by the style and method of half measured reasoning applied to maintain the aspect by its first real test, endorsing the mixture using our steps taken syndrome?

I had indicated in previous pages of how I read it so that America and Europe had best supported Israel from its creation, but by political reaction, balanced by the development of our growing bloc culture mentality. That reality was set because balanced in that the rated opposition by unspecified terms was communistic politically linked first in measure to others of the very real affiliation to the religion of Islam. Unfortunately the ensuing years have not been able to rationalise the instigation of religious tolerance at any time from where it had allowed some promote their own style of reference to it and questioned the approach of others? Not least from the time in question when political victory brought religious reaction by the victors yet shunned the counterbalance of religious reaction by defeat!

Without changing anybodies method or style to have recorded either their own history and in part that of others if even from the linked perspective only. My line of direction has been to openly account for how the Islamic religion had allowed fundamentalism take sway to set future direction by that religious tone? From where I myself am prepared to lead some of our thought is fully spread to the Middle East, not by result but by intention. My biggest concerns are not from the past, because that has been constructed in so many different ways anyhow, although in trying to understand the recent past, we may better alter how it might be finally recorded, again without judgement!

From where I have concentrated many references to the mid nineteen sixties as a time barrier, I have delayed what I have considered to be vastly important to the reactive capabilities from some in the Islamic religion un-led? That fixation has to remain, not for any other reason, but from our own capabilities to perform the act of self gratification by our own religious determinations irrespective of our political standing. Ably demonstrated from how I had laid the relationship between our religions and our political determinations to have been? When from the result of that particular conflict or one of the classified wars continuing in zone, acclaimed victors if like Israeli, could openly call to alignment by Gods desire in fulfilment by or on their own account! At the same time give no satisfaction to the Christian religions of their allies, no concerns to the Islamic religions of those that were second at this/that particular time through siding with communistic ideals?

Without defined definition such events by the cover of actual history are continually altered or ignored or changed by suitable circumstances, that is why my own abstract approach has been used to challenge any amassed collective histories especially those taking the most active response. I have cited religious tolerance was almost an after thought to earlier political reasoning in some cases to thwart the standing of communism. With that so it is by other results that we must look to our quest for reason and cause, but not only by the political route, because that was under new construction to be fully half measured?

By my own judgement and ably supported by this book in terms of true religious sufferance, there are no losers in the cause of faith or belief when overlaid with the three taken and one active prospect of the Fifth Dimension. In not having it in its regulated format of now then ago, beyond the scope of political defeat, there was no other avenue of expression that could have been used at any previous time by way of lament unless through the religious holding of any particular group. After all if religious praise was encountered or religious indifference enforced through any particular reading in translation to the terms of tolerance, what was left of religious fortitude?

From my own terms of abstract imagination, aids in the progressive charting of such events to have occurred, can by built from the mixture of events as they might have happened or were reported to have occurred by those most affected. This can be done without the apportioning of blame or the demand of retribution, because we are cataloguing the manner and matter of how we have affected our own treatment by the manner and matter we had conducted our own forms of reactive bias? First caution has to be expressed by myself, even if we have moved to consider the balance of any form of religious reactive or counteractive responses. Political determinations in the aspect of civil law will and can justifiably override religious intention by our coded national status and from that, our attempted united front on that basis. Wherein our political aims are set to overall unity and our religious aims already express that unity in all forms, from now we have to first take the political lead but by the religious ear. Not least because of this very real and new time we are now in, where from times levy we are all here and have to better coexist, even against the excesses of one human to another or reversed?

If by indication, intention or tone I have led us to the situation where our Islamic reaction to any form of circumstances to have occurred in terms of battle, war, defeat or victory in the Middle East from recent times, how is that reaction to be measured or accounted for? In stating that we have different opinions by our own religious leads and political determinations, complicates the matter when our original plans of betterment by national standards were managed through the offices of the United Nations in theory? Politics was to rule with the manicured ideology of religious tolerance to cover the aspect of our religious differences. What then if and how any particular group were to react if indentured to a particular religious holding by their terms of reference, even if they were far outside of the expected returns of how religious tolerance was to affect us all!

What if any group reacted within their perceived field by means which were considered to be both unwarranted and unethical by the supposition that they indeed met the requirement of religious tolerance? Not least by the committed exercise of human sacrifice to be the means of expression in the terms of protest or through the simple exercise of turning ones own body to be the means of death and destruction in the name of our religious reactive licence!? By any calling too or through the Fifth Dimension of now, we might be able to see into the question so that we might be able and balance by these exact times a means to answer such events in understanding? By what ever means we might wish to recon of any so deeds done in the past through these last forty years or so, it must now be our direct intention to alter our histories without change!

I have made no argument in my presentation of this work to suggest that there are specific reasons for any reactive processes involving the deliberate or systematic planning of anybody associated with the Islamic religions to commit murder or create the mayhem of human death, through the good offices of the religion. Where the trend has been, is to show the real difference between the three monotheistic religions of the Middle East, if not by creed then by application, then by there relationship with each other. Although cast from ancient times in difference to be four, two or one thousand four hundred years old, the total core of there separation is not that they are Jew, Christian or Muslim, but in the education of our religions to the same meaning and for the same reasons? All of which has never been apparent to us until these past five to six years when by the drive of our new learning we can better understand at first our very recent past. Following that it had been cast by the living circumstances that we had previously tried to live under, even misdirected by our best efforts of change half measured and followed to be steps taken!

## CHAPTER 77

Politics has its core, religions have there core, in as far as we manage both, our direction has been taken at the wrong turning from the nearby ending of hostilities closing World War Two. Without condemnation the actions taken were of the utmost importance, but directed from the smallest of apertures. We exercised tunnel vision in the assumption that our judgements had been right so far in attaching the greatest amount of importance to our political decision making above our religious needs. Any other format could not have operated to kick start our programs of fated human unity when at the time there was no real evidence of it. Our first real blunder into the unknown was from our accepted political determinations and in trying to generalise future settlement using old formula.

By the political actions of the day and of unspecified times, but close to the cessation of hostilities we began to err in concept, not least in barring the aspect of fascism as a means of political control. No case whatsoever to the good can be made for the vile means of control that it can become, because it is structured to be state found. By the factor that under the scheme there is or can be maintained structured religion, does not lighten its concept, because those same religions can be challenged to fully co-operate with the dictatorial demands of the state first hand. Even so with fascism that gone, by indication the two other main forms of government in democracy and communism continued to barter in consent for there own particular lines of concern ideologically. The significance of which has been by my own referral, that by there status with one being mainly Christian and the other being ungodly by manufacture if through Marxism, the results were to be volatile.

This thinking had allowed the gap in our rational form, because when instigating the standard of religious tolerance by after thought, if also to accommodate other featured religions being eastern or western? If all leads were to have been fed from the United Nations, then in some cases, even that concept was compromised when some national states under whatever determination, took to their own solutions and reasons in cause anyway? Bringing us to any such times like now, whereby it is religious fervour to take licence of voice and action in application, just as others might have done or as others might have claimed licence from? Our religions had now come in full circle again, but this time more vitally honed to there separate clauses than we should expect of each other.

Much of my intention through this effort has been to raise points and offer solutions to what can only be seen as reasons to some of our differences while at the same time we perceive it so that our successes are most likely from current thinking. I have laid ground to the aspect of our religions and by reference given my own view of how I have absorbed the image of there creation! From what ever I have said and indicated, nothing has passed my lips other than from my own ability of understanding or from my own level of education which is wanting. Nevertheless I have proceeded in this endeavour by the basic principle that we are all individually involved in our own future and that we are here now!

Being here now, is the key to my own explanation of how some of those to the Islamic religion react or have reacted by determination to political events, recent or ongoing? From what I have laid to be the foundation of change through the Fifth Dimension now, without bias we will be able to unravel the cause of events from the course of events, but by my own study. I take the stand that I do not represent any separate force or combination connected too, or affiliated to any political party or religious philosophy. I accept from normal or general query that my motives will always be suspect by the imagination of some of my contemporise, simply because of how we have separately developed and forced our own progress. In fair and from open reasoning, I remind all that my reasoning has been done from within the confines of our three taken and one active prospect to the Fifth Dimension. That which I give full credence of has always been with us and in one form or another expressed throughout the ages by people far more versed in the cause of humanity than might appear from myself. In credit due, I have taken my own stand from how we may not have been able to understand other intentions because we had received their wisdom under different headings and with holding different viewpoints?

Without fanfare, though ably supported by my addition and by my own inclination as ours in its own form, I have levelled the Fifth Dimension by way of contact, but by the specific timescale of introduction to be of these recent five to six years. So that we should be in full notice of its relevance to all and how we are able to look to the future with all recollections of the past, not only by historical deeds but in full remembrance of our own developed thinking up to then? We can now plan the measure of our future prospects equally, but first play has to go to our political reasoning and our religious determining in reverse order? The how and why we can overlay both aspects is considered by how we have already developed, not to be measured from what range of politics or religion we use, but how they should be used?

Although I have tried to set our focus to events as they should develop from now, even if it means walking past the complexity of our combined long past. Nothing is beyond our consideration, but now all is contained by the Fifth Dimension! From my overlay proposal to our political standards the results will bear there own outcome, because the same prospects have a different application of intensity in overlay upon all religions! Expressed in there own time span where religion in more ways than one relates to our single lifecycles. That is why I have found the division amongst them where from the same source some profess not to kill of mankind and some of us do kill of self by self?

## CHAPTER 78

Islam by my own interpretation has been classified to hold the aspect of human death by our human hand within its own religious boundaries, but tied to that interpretation is the feature of acting to defend the faith or in line with inbound tenet, to promote the faith? I have set it so that this reasoning has never been taken as any sort of action to the rational of the religion by our political shortfall in working half measured. Points of which are to be now laid open by our recent ability to look at religious and or political determinations with the same eyes to a different past?

In this context no comparison is made against or with any other religion to be first or second dimensional. Nor to be of an Eastern texture by their own determinations of fulfilment through regenerative cycles in life forms connected by the unseen thread of our singularity with the broad band of all life's evolutionary processes? By my own specifics which can be varied by any connection historically, but maintained to be of these close years by the ending of World War two. I draw attention to any reactions by fundamental religious outpouring and the possibility of how and why. When in determination and I say this in full context, to have laid the equality of religious tolerance upon us all was so misaligned that it alone has helped aid those who already had disposed of its principle. Such in case is perhaps the real nature and cause of much of the penance related to many of our religions by that theme and not through political determinations only. Which we will see better with the overlay of the Fifth Dimension, which is intended for us to be able and rationalise our conscience!

Tolerance as a by word for religious acceptance is an empty shell, where this is most seen is by our political awareness to it in display of recent times, and can best be taken in view of the relationship between our three Middle East religions? I set no standard of blame, retribution or recompense to any of our past recent histories, save in our ability to maintain single standards to collective decisions! Expressed through half measured working inducing our steps taken syndrome, and when once done all measures are deemed to be complete or in contentment. Supposedly shown in our standards of human rights, of being politically correct, being religiously tolerant, unless seen by one or the other of our benchmarks in supposition, acting outside of our limits of proposal, yet supporting that action by the same reactions half measured and following by steps taken? In working from or by those standards I do submit that any leads percolating from the top down if directed through our United Nations by intent, fall short in working to the concept of our steps taken syndrome!

From about the mid nineteen sixties, if Islam was to raise a new voice against the outcome of political ideals, then the matter of those concerns if expresses so have to be reviewed. Connections made are not from the aspect of defeat but the attitudes of victory, because both sides with political determinations acted and reacted by religious fervour! Of the side in most victory to be the Jews of Israel, if and to their own settlement they took good heart to that outcome by religious terms, then in all cases there is no query by their own terms to what they might have done? In taking and holding to any religious determinations by the full capture of Jerusalem and exploiting that gain was in full terms of religious tolerance? By token, if to the defeated on political terms associated with communism because of technology, then by the same standard, political wellbeing was not compromised, because democracy and communism had the same voice through the United Nations again?

My own concern by religious reaction is a two sided sword, but yet to be balanced, not necessarily by the two main religions involved. By indication to how the Israelis might have celebrated on religious terms is to there own terms then ago, but is now open to review by the Fifth Dimension representatively, but by determination through our one active prospect in 'One for the Whole'! In aftermath past political representation, all that was left for Islam with the bitter taste of political humiliation expressed by a huge religious downturn, was the need to respond religiously!? In as much as any bloody political battles are won or lost, in this case the broken and dead bodies on the crimson field was either Jew or Muslim? If democracy smiled in the glee of victory and communism licked the wounds of humiliation, and the Israelis took to religious celebration, which has already been suggested, might not Islam review her own defensive or promotional steps to be taken?

Defensive or promotional steps to be taken, is my own descriptive term of how the religion of Islam has allowed the fixture be maintained, whereby human life might be taken by self upon self to the final ends of the religion to all determinations! Although in first step to the religion of Islam; and under my own heading, whereby I had given a more readable term from present day thinking to it, it in term as the third religion of the Middle East more so than to the other two before? Must be read to be more progressive, owing to its formation, which had added times of actual human progressive thinking than the two earlier! But the concept of defensive or promotional steps taken has now to be checked. Not religiously, for by tolerance Islam can cut its own course, our full check on all religions, not just from the Middle East, is now made with three taken prospects and on active in 'One for the Whole'!

# CHAPTER 79

By other expressive terms in world politics of that time pre and post the mid nineteen sixties, our world in the full form of adjustment was taken by all manner of methods to accommodate some monumental changes instigated by the cessation of world hostilities? Although in the throws of change beyond the scope of religious determinations, one of the first calls in the effort of international peace by the new United Nations was to police the political war in Korea? Another war, sadly by distinction, perhaps the last of international importance which was conducted in the format of standardised battle plans, whereby armies faced each other to gain advantage on the field of battle primarily? I do not endorse the style or method of conduct, but make the point to differentiate between later conflicts gathered under many different headings but were dirtier by outcome, which only went to narrow our human outlook instead of endorsing our natural progression.

Closely followed and continuing through the Vietnam War the emphasis had switched to the outcome of self interest only throughout the world, splitting the seed of intention if to have been laid by the United Nations. From Vietnam and including some of its outcome, the end did not justify the means, because by political determinations the same people of the one country were treated differently, which was not a singular feature. Nor was it fated to be an aspect of communism, because in that case communism was ultimately to be the victor. In following the political ideological route in the outcome of such a fraught war the opening not yet realised, was in the final failing of such an ideology. Bore by mixture whereby the society has switched from government for the people by the people to government by the people for the people? Alas turning it so to be measured by the feature of personal gain through the arms of political reasoning to be in the coat of commerce and materialism by what ever means, even if communism was promoted the moral cost to area was monumental!.

By the worst of realisations from places throughout the world since, from Africa to Asia to Europe to the central and south Americas, our attempts at social unity on political grounds have continually remained unfulfilled. Unfortunately expressed by the medium of personal greed in relation to gain even from the effort of others, whereby we de humanise each other if not on political grounds then by social means even expressed by religious tolerance. In taking that to be the situation by the unnamed destinations on all of our inhabited continents, we might be under self query of our achievements on social and religious terms if expressing our standards to operate by any of them morally. When one had been designed to cover our plan of universality encrypted by the United Nations and the other was to be of tolerant conduct in abeyance to the former?

In walking amongst the human tragedies to have occurred post war and post the mid nineteen sixties, the only clear picture of our intentions have been set by political determinations. By my own ability to first understand that the reflected image of what we see in a mirror is not the reality of cause, I have set it so that in order for us to formulate our future prospects we might again look into the same mirror. This time though with the same product of our own imaginations in lead with the Fifth Dimension and its four prospects to be our guide. In that case, we will see the separation between our longevity expressed by political reasoning and our intention expressed by our single lifecycles religiously. Understanding that the split is of two different outcomes but for the same objectives in us, because we are here now and still want to know the answers!

Being here now is only part of the key in our general understanding, but is never to be used as an excuse. My insight is to make us aware that by that interpretation, we can and do operate to standards set by the one or the other medium of control from our politics and or religion. In setting that the mediator of discontent by or for whatever reasons was to be the United Nations, then the seal was set to operate by political or civil means first. Our afterthought of religious tolerance failed its first test by some measures of reaction to what was to come from a sound religious footing in the religion of Islam. I hasten to add that there is or was no specific incident of control where actions were made or instigated in cause to any other action politically or religiously, save post the mid nineteen sixties?

Occurrences forthcoming at that time if cited to have been by a Palestinian hand in recognition, then the link was through Arab nationalism to have been Islamic by religious definition. My implication is that reactions of the political medium were openly expressed to have a religious basis. We are both of us now to think in terms passed my own lead of how we have managed the first outfall of events and how they have changed to be self defined by our own errors of working half measured in following induced steps taken? My first serious intention is to bring us all to the realisation that we are indeed the same by the supposition of politics yet have been allowed to express our differences religiously by the idiom of tolerance!

From my own stand point now, I cannot follow the rational behaviour of times past when we are in the knowledge of our four prospect of the Fifth Dimension. That is why we can take the measure of our histories and change them without alteration. What ever direction I might give to the execution of Palestinian deeds post the mid nineteen sixties, if expressed by my own view, then they are for and by whatever means they occurred! That is to say they were what they were? If peoples were openly prepared to kill and murder at the expense of there own lives, then this/that or those acts were recordable on whose terms, which is exactly what happened? Allowing in our first turn around whereby we have given cause of the need for further open review to our political and religious interactive means, which until now have been lacking. Simply because until now we had followed the dated hall mark of formula presented at wars end over six decades ago?

Using my own ability of collecting and storing the value of our histories, I have set it so that my personal reaction is based on how I think. Which is in the open form allowed and bestowed to us all in the aim of free will if required, for there is no requisite by it that we all have to think or act or react to these or those or any other terms led. My lead is that we are here now and do actually think and react differently, where by indication I have tried to show cause and reason has been attempted in the manner of a no reprisal mandate owing to the future. In recently having cited the Palestinian's to have taken a set of reactive steps to the political outcome expressed in the Arab Israeli conflict of some forty plus years ago. By my direction that the Israelis took to the outfall having religious connotations and expressed their will so, had been done to show our ineptitude in recognition of our religious differences by promoting the spirit of religious tolerance.

By the very same token of one to the other, I make no direct suggestion that the reactive force of guerrilla warfare if, by the means expressed from Arab and or religious fundamentalism was religiously led by the Islamic religion from doctrine! It has only been my determination that the aspect of human death by our human hand was deciphered in the outlets of defence or protection? Unfortunately my explanation holds no character when overlaid by religious tolerance misread? Without my own cross examination, much of which has been laid down in these pages, by all terms of religious difference, what ever specific Muslim groups might do by their own terms of reference? Can be carried out by our expressed terms of religious tolerance politically, which from those terms cannot be laid by other political reasoning, therefore whatever Islamic responses are, is religiously based!

If we have laid the matter of change to the aspect of the future by political terms then the next religiously progressive step we have to take and must, is to relinquish religious tolerance in its present form! Then reasonably, either operate wholly politically or allow religious determinations determine direction but always allowing the aspect of progressive change from both mediums using the same starting blocks of the Fifth Dimension. History is to be the measurer of how we conduct the changes required, because of its age and through its representative time slots in which we can always learn more. If not at first from the events at Shanidar, then by our reactive studies dated some four, two or one thousand four hundred years ago. When we gave three examples to the importance of mankind and us all individually expressed by our own conduct through our own lifecycle?

In reasonable terms our personal consciences can be held within the borders of the Fifth Dimension yet without confinement. We are free thinkers we are also free minded so that we can maintain the rational beyond the limit of the study of science held in focus by the fourth dimension. From our best studies we will be able to see better how we can proceed in the careful consideration of maintaining progressive steps religiously when we can almost see an end to our progressive steps scientifically. In as far as virtually all religions have done the course by imagination and ended to the prospect of heaven or paradise or nirvana or fulfilment. By those terms we are still open to learning, best seen when we look to our young and see how by no matter what we teach in or of or to them, at various stages in their lives, some without distinction, will show the measure and excellence of our skills in question? Not least because from amongst them in us all having the same urgency of life thrust upon us, there is no measure of how it should be read, save allowing that in many cases it is the pupil to become the teacher!

# CHAPTER 80

From any picture I have tried to frame most of my reasoning has been of our long historical associations squeezed into a short period of time from the ending of world war in nineteen forty five. My urgency has been prompted by what that time span has come to represent in terms to myself and us all from and since then? In notice of how I had mentioned several times of our oneness, although separated by our personal life spans to be either fifty plus or seventy plus years long and counting, some of us born at that particular time whose living is now gone. Even if wrapt in any sort of fulfilment however imagined or promised, the feature was only carried by our religions pasts only, nothing political! To have already died by religious terms the end does justify the means, sacrosanct, some of us are in the state of fulfilment without the need of justification!

For others like myself in terms of age and from the good fortune of belonging to the group who exercise the average lifespan in the seventy plus year bracket. Our expectations have been blighted by what we have come to learn and from how it has been applied in our life spans and from our relationship with each other. In our form of living by today's means, the political matter of control displaces any length of time that any of us might live, because it works to the feature of supply and demand only. By that it works on the level of replacement to the needs of the individual as they arise irrespective of age. It has done this from the times of our first instigation of civil control by means of conduct and to the flow of our own expectations, but extended not to include us individually?

Even when learning from the growing style of our religions, politics changed in part to accommodate the need for change, but has now come in full sweep over all religions. In doing so, it as a medium of control has taken the main bread off table and handed to us our best form of communion, by operating religions through the political enterprise being appraised by tolerance! All of which is routinely shown to us by how we conduct or have conducted political matters, even at such times when some amongst us have been prepared to take what we considered to be religious reactions by what ever means? From which I submit on my own personal terms that some of our very recent reactions have had total religious perception to be the cause and route of reaction by means to infringe the aspect of religious tolerance. Making or creating the wrong further reactions or supposed countermeasures in forms of political measures in counterbalance, but again marring the intended reverence of religious tolerance?

I have made the statement and reiterate in the same loose fashion that from the application of living by politics and religions in parallel as we do. Instead of making the blend of harmony, we have allowed the two flourish on there own behalves, but measured there control by the means of politics only. Although relative to this exact time of now, when instigated in concept of our human unity sixty odd years ago, we at first assumed that what was suitable for one group was to be parallel for others. Our reminders came, when from the Islamic bloc a new arm of religious reprisals as a means of voice was used for what ever say. This, if for remedial reaction brought us up to date to our new last modern era if measured by myself to be from these five to six years, now we have reason, reaction, counteraction and countermeasure all unmeasured. Before we did not have access to the Fifth Dimension and its three taken and one active prospect in 'One for the Whole', therefore it is from now, matters will be better seen and understood?

In naming Islam by bloc association, I name religion, but from whatever context religions have been driven, all is through we? It is us who are now past all of our prophets, seers, and enlighteners, even our well directed priesthoods; it is from us that we ferment in our own juices of realisations and understanding always balanced by our own free will? My open declaration by the above sentence is to the matter of concern of how we read our own religions and therefore relate our own consciences to god by that? It is also to us alone collectively and individually that the matter of our actions can or might be measured even from outside of the concept of our own religious set up! Such in the meaning of that measurement will be carried by others if not linked by our own religious doctrine then to the specifics of what might be their religion or ours? Expressed from another view by our free will, but in determination of the true specifics to our religious doctrine, magnified and amplified beyond what we can actually understand, yet in the vein of our own religion!? 'One for the Whole' is the balancing pole that we might use while traversing the chasm of doubt around our own certainties while asking different questions to get the same answers?

What the Fifth Dimension and our four prospects allowing one active is to achieve, is for us to make the obvious association, realisation of our specie singularity. We are not ten or twenty different kinds of finches, but one of mankind, who from the recollection of history cast our footsteps in volcanic ash which hardened into the memory of stone some three and a half million years ago! Which since and running has raised the most wonderful of answers and questions that could ever be thrown upon the face of any universe? Of those questions it had taken us less time than imagination to seek answers posted by our query in death by our burial at Shanidar? From which if nothing else can ever be gained, it is but the one realisation that since then all of our lives have ever changed, that unlike the throws of evolution we evaluated concepts beyond the requisites of simple procreation, breaking that chain for the essence of creation!

## CHAPTER 81

By the mid nineteen sixties in response to sited provocation, if that is how the situation was read, then in the same breath of exultation that evoked any religious reaction by political determinations. Small groups of Palestinian activists took matters into their own frustrated hands, but climbing bigger barriers than wounded pride. I make no connection in how or by what fortitude any so acts were originally done, because in that exercise we have to traverse many different recorded histories and the means thereof, even if written with rose tinted glasses. I make no claim for any side, group or connected interest, which sadly denies to myself and others the reasoning of truth and acceptability to any action whatever. I make no claim to the understanding of what might be considered the active fulcrum in the order of balance to the matter of action or reaction?

The only leaver that I use is in the form of the religion of Islam reawakening our religious complexities in the study of three different religions having the same reason from the same area, but being led to have different perspectives? Beyond its own scope of realisation some apostles to the religion as a separate entity since our Palestinian first counter action if, steered the religion of Islam to new heights or plunged it into new depths of reactive confinement by self study, but in the wrong context on two counts. Either by not realising our own potential on the political front or through the religious process having been tackled over the last four thousand years in the same vein continuing, but past basic understanding of knowing of other religions!?

Before the blinking of reasonable consideration the Arab Islamic voice in the matter of world affairs lost its own political edge, because of how we had allowed our own self reasoning if through the United Nations, classify ourselves into the categories of being first, second or third world people? A small and simple means to down grade some of our religious holdings because the implication politically was that there was standards of excellence to behold that grade, and consequently the same standards ran to religions in being first, second or third by character. Allowing for us to maintain that grading in response to any form of claimed religious reaction by other means politically to our other own determinations, compounding any errors that might have been started in error in the first place?

Although it was not from any deliberate standard setting us by separate or coded reference to be first, second or third world. That reference was in overhang from how the world was to have been set when we professed to be in the League of Nations post the First World War of nineteen fourteen to nineteen eighteen. Being another political attempt to rationalise the then supposition of natural and advancing modernity, but professed from European and American European standards after our first European world war! My linked observation about then was how the European side of our development hung to and practiced the Darwinian indicator to natural selection. Whereby the strong were in the best order of survival and therefore in the best position to indicate any strategy by the then reading of standards if not only by events and of outcome?

My diversion to the First World War is best to be related by European thinking then and since, even if some is imposed by my own view dressed in hindsight by abstract thought? At that time we were centuries away from any consideration of the true aspect of our specie singularity. We didn't even think in terms of human classification being inclined of a first, second or third world grade by any link because for most countries throughout the world then, sovereignty was a matter for the desire of empire by others? By that very means and to that fixation, the League of Nations carried its spores even after being down graded, into the new leadership aspect of the United Nations, but continued covering the process of advancement by misdirection?

I had put it so that by any new rules in the offing many were aimed to create a different attitude in the form of unity. Although tried in concept the cooking of recipes did not bring the taste required. Not least because we did not only have three different classification, first or second or third world politically or religiously, but dozens from all angles! In stating what I consider to be the real situation has been done to show how and why we or in this case, I can talk or observe matters, yet bring different results by the same conclusion. My case stands from whereby I have been able to evaluate the reason of or for our main differences to have been caused by geography, climate, diet and other reasons. Keeping them in context, whereby judgements made to the whys and wherefores, are from the basis of us being the one people in the first place?

When by open knowledge we can see the way of our differences in any case, then from legislation or tenet or covenant, matters of control by the means of natural justice have only to show our understanding to this situation? My observation, is that with this being done through the imposed or theorised lead of the United Nations, we have fallen short in the object of our commitment to the expected result? Not least in running our concerns from the civil aspect through its means of legality, yet allowing other civil projects such as trade or commerce or business, determine forms of policy. Covering the aspect of our separate if different religions being managed by their own defined doctrine under the umbrella of tolerant behaviour, yet by expectation to conform politically!?

Having left the First European World war far behind us and out of range to any living person to have had any say or sway to our future means of conduct and behaviour, even if through what the League of Nations was supposed to represent. Looking to the connected relief of introducing the United Nations post the Second European World War, part of the same story in respect of those who first instigated our mandate and means of expression, are now either no longer working in that field because of the time aspect. Or by the events of time in any case, our attitudes have expanded to broaden our original outlook and output to approach problems from a more open base even if using half measured processes and endorsing them by our steps taken syndrome? All of which has covered our best of intentions by future reference, because without malice, we have tried to conduct our rules of conduct not by either of the above, but for general relief? It has only been when we began to slip beyond sighted expectation that we were automatically pushed into some of our other measures in the panic of desperation when we began to realise some of our errors by some of us, yet were left with the old format of solutions?

For me, the individual that I am, to promote some of my own rational in the form of solution to what could be problematic is not done save by our commitment. Although not in context, we each can speak as we find on all matters, but the requirement of what we say to be heeded or not for each other is done in open competition without conflict. From how ever I would hope to show influence, must be done for what we each can bring to be offered in the conversation of normal thought, not least to indicate that it is we who are doing the thinking differently, looking at the suggestions of conduct from different perspectives. That I offer a different approach is balanced from our past conduct and from the way we have accepted how it leads to future conduct. All of which is ensured by how we react to others deeds by reaction or other means, when not considering the basis of how other people are prone to act from within their own developed style, system or with mannerisms set by the range of circumstances to our differences! No defence is to be given to acts that confound the core and basic principles of our personal humanity, even if led by our political or understood by our separate religions from their terms?

## CHAPTER 82

I only speak from the centre of our different approaches to the matter of our politics and deeper into our own religions. I have been fortunate enough on our behalf to have prised open the door of discontent and looked behind without being caught, all of which has no real value or balance if employed to re-relate to what we already do in the name of self, even if not fully understood?

By all and any means for us to have instigated the charter of human rights in working half measured even unseen, for us to have policed the shape of some of our political directives in the code to be politically correct is only steps taken. In following both and many more laws, edicts and determinations by first insights to political leads, we had compromised the more important of our values gained from our religions. Unknowing at the time of instigation if considered, we used the only available outlook that we had to the future, which was how self comfort to the means of life was to be considered from our bloc culture state of affairs? Ongoing we used that political edge always first, when questioned by the one religion of Islam these more recent post World War years. Our reactions anew showed old reasoning to be brought to the fore, but wrongly giving higher status to that political coat, as that was the first means in which we were to build this brave new world!

By all the methods of innuendo, subterfuge, direction or from the very pointed leads I have indicated to the specific time of the mid nineteen sixties and the Arab/Israeli war of that time ongoing. Only one feature has put the full body of reaction to view, though not to the specifics of that problem. I had stated that the main reaction was expressed by the means of military action through all the keyholes of association with it to it or by it from political determinations. Whether to be recognised as gorilla action, being off key to full battle, or types of action to include self sacrifice in similar terms like fighting to the last man? I had also indicated that many actions were Palestinian led by self determined factions, though by the connection that most operatives to these or supposed deeds were from that area in the first place and Muslims also?

However I might have seemed to relate to the conditions and or the reasons for some past actions if to have been politically motivated and religiously led, then that in statement is representative to us all. The matter of me relating to a specific incident of the past has no real bearing when by the very core of what I have been driving at throughout this our work can be related in test to whom so ever and by how so ever we wish to record it? Also how we would record our reception of another's ideas as a matter of our history to be carried without alteration, because in first sight it was by others perceived perspective of the time, or of a particular time relative to us all? Beauty is in the eye of the beholder bears nothing, when both eyes see only into darkness?

Muslims, Palestinians and others from the same area, in any sort of connection by my suggestion is an accurate mark when making a point across the void of political understanding only? In acting as I would wish through the perception of our children by my own references and interpretations, what I say or suggest is of this very modern era first. Through them there is to be none of the affected hurt or damage relayed to us in fair reference. That is why I have drawn the long route to make direct points and to cover us all by any tone from where we might react. I have labelled much of this work to be modern, not least by direct reference to these post world war years, staging some of its importance to my interpretation of events since the mid nineteen sixties influenced from the Middle East?

Muslims also, is the term to denote at first Palestinian dissent to some Jewish claims by some sections of the Israeli community throughout the result of our minor wars in a bloc culture setting, covered overall by the shadow of the cold war standoff. If all matters were to be conducted through the United Nations, then by the time of judgement to one event passed, another was set to be judged on similar terms, never obtaining a goal! With religious fervour locked in the vale of tolerance, what actions were left to voice opinion in equal sound when the political arena was closed to innovation, except what had been already made to thwart another's means of political control like communism, but in subterfuge?

For me to have indicated that some Palestinian reactions were to be of methods of gorilla warfare tactically or in the same terms of resistance to a sworn enemy: is not being divisive of defensive to any faction. My implication is that serious, harmful, dour, deceitful and murderous obligations were instigated as the only means of communication from one or others in group to prove dissent? I follow in quick succession, no consent of mine can ever be attached to such rulings, I do not represent any group of people affiliated to any religion or of any political connection, save us all as this single specie that we are. My collective points are to be read that in so far that I have set different camps to one deal or another, is done from how we react under particular circumstances by the unseen. Like when one side in any conflict only represents their own views without considering others, then respond to be happy to settle matters by political judgements only!

From political intentions we thought we had at last carved the best path so far in genuinely taking treatment to our other biggest means of difficulty by setting religious tolerance? It is only from these late times when seen with the expansive profile of the development of what was actually intended, we had seen the strength of our misdirection by applying religious outfall to political mandate? If at first some Palestinian reactions were headed to be political then that view was made grow by the intention of completed affairs if the perpetrators of such acts were to fall by religious intention? By my own brand of abstract thinking that last remark can pull thoughts to the image of acclaimed martyrdom, which was not active then, but has followed from untimed incidents since. Of which the most remarkable issue is that by the very lines of direction that we have chosen since the Second World War ended in nineteen forty five. We have ably manufactured the very causes of our continuing strife, by not listening when we should, or not speaking when we must!

If not being able to envisage by the secondary motion of adding religious tolerance by means to be collective, we directed our separation? Where others like from Islam at instigation in times past hailed to forbear with those of a different creed, the connection was of change and the ability to perform it so at least by example. Now in our haste we are tolerant, which by the hope of intention allows us the liberty of self choice even without understanding? Which from my own insight to some of our religions where human sacrifice was ably carried out, by today's standards that would still be so and blessed politically?

By just the same methods of abstract meandering that I have been lucky enough to allow pour over me, in making the connection between the three Middle East religions with Gods own law "though shall not kill". My note to the first two by date and to the third of later times, expresses the call of that commandment in relation to all three religions. My direct note that the implication in all cases of the kill aspect, bears only to man by man stands in all contexts, we as man of this our single specie are not bound to kill of each other by any means even if directed by gods own hand in declaration, even delivered by gods own prophets?! By further ploy my note contained the civil aspect of the kill injected by our basic codes of group conduct at first, then delivered to the broader aspects of civil community growing to become political directives?

Without direction, but from gainful observation, I on our behalf like many others have related how or why the matter of this very important human concern has a different reading by the religion of Islam if set to the same goals as in other religions? Outside of the religious experience by looking to the darkness of normal life by the throws of evolution, there already were natural times when driven by that code, we would kill of self without compunction. Driven in the total need to propagate, we were set by the cast of survival in all of its brutal forms. Killing of self was sometimes necessary and often done, which could be drawn by some studies of our cave painted tombs giving light to the art of battle and self death? In many ways fourteen hundred years ago did not represent much difference to our actual behaviour then, than at any time to twenty or thirty thousand years earlier. Only this time we had the language and written word to pontificate above the realms of creation from the aspect of evolution, not yet understood?

It is not that Islam through any directive from Mohammed had set the act for us to kill of self for religious determinations. Any inclination by my seed of thought was done through the defensive or protective posture of the faith to suit the very times of which we lived in then ago! Alas by quick succession to the core of the religion and in keeping with our own ability to use our own free will, some of us in close proximity and linked by the same religion, used the aspect to kill of self by self even being Shiah or Sunni to the one new faith? Showing in context that to kill of self was not whimsical but directed to the perceived core of the religion itself, above the civil standing of the time, because that was befitting to the thinking structure of any enterprise fourteen hundred years ago?

Falling back to that habit of recent times, any deeds were not done in the style or method of enlightened intention, but in answer to the overbearing of political directive of all descriptions, or not? Unfortunately the open sore of reaction to kill of self by self outside of Islam, but by some Muslims, was soon taken past Mohammed's ideal or intentions or recorded instructions, especially when implanted to the different age of now run by political reasoning only! Making it so that by impetus, such acts of taking human life for religious means are or were made redundant by the natural progression of all religions, not least because it has only been by these very modern times we have at last seen for the first time what should be really important to us all. Questions set have to be made ready to receive appropriate answers, but answers given can only fall within our actual life span standing to be fifty plus or seventy plus years long? From which we will understand that all meaning will be revealed after 'Natural Body Death' only!

It is to be noted that by some of my own reasoning, by some of my own laboured thought and experiments in the spirit of humanity, by using the broad aspect of abstract awareness on our behalf. It is we who might sit at the table of discontent and parry in the ideas of recorded history by the cause of some of our reactions from any deed recent or close in time or distant in memory! For the one reason that we are to promote escape from the inevitability or counter reaction in a time slot unsuited to the nature or the deeds, least way with the new air bags of safety that we have procured these recent five to six years.

However we might wish to delve into the realms of our own particular records of history in publication, there is never any value to any of them unless suited to us all! Because these past months as opposed to decades of other forms of interaction were not aimed at us all, but in us all by division, and suited to that limit by how we had actually valued our historical records. In the space of one of our longer lifecycles, seventy plus years, we had gathered the nerve to tolerate our political differences through the measurement of commerce. Being the only medium in which we could all be engaged in, but far stretched beyond the limit of equal grace to us all. Because we were to hold to that equality in the full range of our religious differences by religious tolerance, allowing us carry the same measure of spiritual wellbeing even if disposed to be uneven religiously? Yet ignoring the accepted standards of political classification by first or second or third world classifications economically, we being looked upon as being even from the bottom up and the top down?

No change has yet been brought to accommodate our true equal wellbeing unless from religious insights. That is why we must look again and again at the pictures we have painted in best order from the two differing aspects of them being Eastern and Western by style. I have previously pointed in short order that to do this for best advantage we now have better means whereby we can judge them to experiment for two reasons. By employing the Fifth Dimension to overlay politics and religions with the three taken and one active prospect, our vision will better see the real relationship we have between the two along with the proper relationship between all religions!

## CHAPTER 83

In all equality by any definition we of this single specie are locked by that very singularity to the prospect of 'Natural Body Death'. A fate owing to us humans in complete separation to any other life form upon this earth or in the universe! Instinct also being carried by us helps keep that break from other specie, being trapped by evolution they seek the single aim of propagation, we seek progress? By the intricate and varied way other creatures exist or had existed many, many of there life styles compromised the meaning in aspect of 'Natural Body Death'. All that the evolutionary code extracted from them was that one conduct in the order of static reproduction and reproduction, whereas we sought purpose even beyond our understanding?

From that most direct of requisites, there development by all ranges showed full compliance to that method of purpose, but translated to be survival. When in order for actual life to normally continue by the forms of different development owed to some groups by the same means of geography, climate and other circumstances similar to us. Part of there food chain was in fact there own young, being a feature spread throughout the range of creatures from the reptiles, mammals or birds. By brief study in the name of the evolutionary process, looking to the depths of the worlds oceans at the living of innumerable types of fish shows the real scale of this feature.

Many types of fish and sea creatures have the reproductive capacity to produce in the hundreds to thousands of eggs per reproductive cycle. In sufficient cases to simply deploy fertilised eggs to add to the total infrastructure of oceanic life forms. Some in developed stages of minutia, float at first range to be plankton, filling the vastness of the seas and drifting on the currents of water flow determined by the feature of the weather. Then by the same cycle of habit to be followed and taken in sweep by the hungry mouths of any sea giants that might have developed to feed so.

Other abandoned eggs in there own cycle developed in the time set by previous millennium and formed to there own protective collective schools in the urgency of growth. These fish by many different types and groups when feeding on the bloom of plankton would grow by stages, but still keeping to there own habits by time and size. Unfortunately by the true nature of natural events some numbers to be a large proportion of any particular specie, found there own size to be complimentary to the eating habits of bigger fish or other sea creatures. Clouded by the complexity of these real events the picture in reality was almost designed by the requisites of evolution allowing life be maintained by the supply of food to enable procreation and the cycle to always continue. The obvious was always maintained in living equilibrium shown by the able historical value in time to countless specie with a deeper time draught than us, living by there means over and over. Almost always stopping short of not entirely eating all of there own young because the cycle was continued and continued until first recorded by one of the other specie who did not feed by the accidents of availability. Who also did not naturally break the fixture to be made in us allowing the development of 'Natural Body Death' be a part of our cycle, because of the importance we could and would attach to our young?

There are no real comparisons to be made between the exploits of fish and other sea creatures against us humans of this our single specie. The connection that we all die stands to the advent of time in all cases; any distinction I have been making is to the cut off point between different or from all specie? Where natural body death fits in all cases is recognised to the time of death, but how? From this section where we discuss the eating habits of the oceans to form comparisons is done for the vitality of all life. From the fish line, if a bigger and a bigger and bigger fish was to eat in the same fashion, then without the format of connections if made by human religions. Fish would act in all matters by natural body death in that all habits of evolution were to be maintained under whatever circumstances but by maintaining the ability to reproduce.

Our tenure is far in excess than the habits of biology. In not only looking to the processes of reproduction we have been able to liberate the vastness of the universe into the reality of substance by open acceptance to the care and welfare of the future, if at first recognised seen through the portholes of heaven, paradise, nirvana or fulfilment. But better to the desire in care that we are prepared to offer to our young by accepting that they are a future element to our own wellbeing on all counts. From one of our taken prospects in 'Natural Body Death' a measure of care by intention is already to be realised. But only from the set relationship between us and the next generation who in some matters are to carry more in understanding because they are already implanted with our thoughts and attitudes without the time worn feature of having to experience our trials and tribulations?

I realise in coaxing a moral stand to better enlightenment of our young, has been misted over by my referral to the habits of creatures locked into the realm of evolution and in eating there own young in many cases. But I am charged at this time and in this section of the book to make more defined and open statements of intent losing the aspect of supposition, innuendo the probable and the possible. I am further obliged to talk in the positive by my own tongue, charged by our own multi habits of doubt, suspicion, distrust and sheer bloody mindedness, which can be translated into serious detestable types of unjustifiable action or reaction!

For any objectives I might hold, 'Natural Body Death' is an unshakable cornerstone to be used by any of us, but in the fixation of its relevance to us of this our single human specie only. Without my determined list of direct application to it, there is no extra factor other than it is not bound by the standard of human rights nor is it to be politically correct nor is it determined to be a feature of civilised behaviour in compliance. It is a cornerstone as lasting and immovable as those at Giza in Egypt belonging to a structure that rose mans own expectations skyward in the name of future expectations?

By the same value from my own emotions, the Fifth Dimension is now presented as a direct feature, which will use the same pointed direction about the other two taken prospects 'Out of Africa' and 'Past Gods Words', but offer respite through our one active in 'One for the Whole'. Along with structure, the Fifth Dimension is the backbone of our religions and the rib cage of our politics. Holding and or supporting these vital features of life and drawing different experiences from them using the same overlay.

'Out of Africa' is another most vital of our connections. Although readily expressed by many praiseworthy and extremely honourable people in past reference, it by expression through them had only delivered the shell with the nut gone? All of us, all of us living today, tomorrow or next year had our direct ancestry in Africa. At the time and continuing we would have been black of skin, that is it, black of skin, our direct connection in all cases is that we in all of our present forms had that same multicultural beginning to be black Africans!?

I have made that statement in honest pursuit and stand without comment to its validity. Red necked bigots did not leave their footsteps cast in stone three and a half million years ago, nor so fair skinned people working the frozen ice fields fringing the north pole of this planet. Nor the yellow skinned or brown skinned highly sophisticated cultures belonging to the vast continent of Asia were responsible on their own behalf to have given us our first lead. Nor either from the continents of Europe or America did we have our first lead, simply because the time mark of our development there has been viewed from the evidence of our development there? From which like those of us in Asia had already changed enough due to the circumstances of geography, climate, diet and other reasons to have altered enough not to recognise us as the same in the first place! Without successful argument or disapproval we were all first 'Out of Africa'.

By intent 'Out of Africa' is not to be an inflammatory statement, there are only very few of us who can trace a direct ancestry of say, more than a thousand years, yet we are all linked by direct connection through mum and dad and mum and dad and mum and dad. In all the comforts that our pasts might bring, and for some of us that is extremely important, nevertheless the future is the total intention of what the past gave us from all angles. That is why it is important to establish for acceptance that all of our pasts are vitally linked simply because we are here now?

Without looking to establish the total links in line with the examples of microscopic life by whatever adventure it first formed. Without the pains taking examination of the spread of evolutionary processes to all of the classified groups we now have inhabiting the land sea and air. One of our taken realisations if by being 'Out of Africa', is to openly conclude that the connection is only to us humans through the avenue of creation? It was by the living and struggling to do so on land its self that aided our gigantic forward looking and progressive steps to advancement. Not least expressed in our daily habits to have been established whereby we first recorded the incidences or life or to living before we came to be aware of our own nationality today, there is enough evidence to show that we did indeed have a collective but singular beginning.

I have been fortunate by other means to site the connection of our unity through the events to have occurred at Shanidar? Beyond the splitting of hairs or the crossing of T's or the doting of I's that situation in occurrence to have been a burial rite, encapsulated our first habit of organisation by civil codes. On that principal it was also indicative of our connection with Africa that it was to show the habit of organisation by that standard and level, eclipsed all other creatures except those who had cast footsteps while in an upright posture? Along with my political prognosis I have put it so that by the signs of ritual attributed to the skill of such a burial, we addressed matters far beyond the capabilities of other creatures locked into the instinct of evolution. By further suggestion that the act contained some sort of religious significance has only been done in the preparation of how from any start point we managed progression? In note by the organisation of such an event and to its complexity of time and consideration donated by hunter gatherers who themselves lived on the very edge of existence. For me to have indicated that there was a specific religious point in it and that it had since been carried by all of us in time by our first direct linage from Africa, stands from our own manner of further existence symbolically?

In the two cases, that up to one hundred thousand years ago we in the collective sense buried one of kind in the tender preparation of comfort in death had been our first sign to the consideration of the future. By civil conduct to expectant preparation and in the nature of what that preparation was for, we gave our first sign to the break from the unknown idea of evolution. It may have taken us a further ninety plus thousand years to clarify the meaning of any such events, which was done again from the same area of the world the Middle East. But on others spheres, nothing had happened to take us from our unperceived drive or to detract from the fact that our spread now throughout the world originated 'Out of Africa' from where we had left our first mark?

Shanidar did not bring us too, or take us to be 'Past Gods words' for although we had laid one of clan or tribe to be part of the future, we did not understand the why or wherefore of it. In form, that particular expression fills the gap between the reality of evolution and the supposition of creation. Taken that we had just simply developed in the first place like all other creatures, our standing was very frail by many comparisons. Even that our organisation was similar to any other particular species like the crocodiles that had inhabited all the continents in one form or another. For them, keeping to the nature of instinct and befitting to there own longevity by group status, they only maintained the processes of habit. Our balance although done over a compressed timescale, had been formed by whatever to enable us carry the same progressive drive on our different continents and through our developed changes, but with a much more open approach!

From my own pointer given from the events of Shanidar some possible one hundred thousand years ago, the taken attitude has to conclude that we already had been moving to a different set of standards from any other creature by any comparisons. By the same scale, evolution represents the code of civil conduct only, that some groups separate to those that had abandoned their offspring yet unborn from deserted fertilised eggs, to those that gathered in pack or pod or herd communities. Conduct was only to have been led by civil order through group determinations, but by better inclination to there young? Quite naturally we acted the same by child care in our tribe or clan set up. Our differences if unseen by the score of hundreds of thousand of years, was that through an unseen pole of enlightenment, we in our communities recognised each other as being the same in type without real understanding. When brought against similar groups to ourselves we saw the same cunning and guile met with the same deeds of feared confrontation? From the exact time of when we might have first encountered the same prospect of understanding, what could have been the additional spark used or taken or found to stretch us into new behavioural patterns, unless chance?

Perhaps anything from using animal skins as clothing or fire in the use of heat or for light or purging the remnants of food from bones or closed shells, were all tradable commodities when meeting a similarly inclined other group in locality rather than trading in confrontation. Or wildly taken, in order to best protect our young to be able and maintain our driven need of procreation, it was only with similar groups it was better to jaw, jaw rather than war, war! Promoting by extension the aspect of adopted conduct of civil order, into the art of exchange by trade or even tribute? In connection I do not make the statement to the order of those events by factual reality, but to the edge of the possible?

Based to the fact that we are here now allows abstract thought bond in connection that there is sufficient evidence, like from the use of flint tools that there has been a considerable time slot where we actually traded with it, and with similar tribes? We might have killed, skinned and cooked many of other different specie by using flint tools or blades or weapons already seen. But there is no evidence that we traded with crocodiles or tigers or sea lions or even apes or gorillas or orang-utans by civil determinations. In counting the core of our history to have been long, by not finding any specific receipts written in stone to the actual time of when flint was first used or how we conquered the control of fire or dressed in animal skins in some environments. No sign had yet been given that we were anything different than other specie in following instinct connected to evolution proper? Save that we had managed to do it in a more comfortable style than some others, until perhaps we showed an interest to our own dead in justification to our thoughts of our young for the same future.

At that time, Shanidar is perhaps our best example by age of ritual burial on any of our continents, whatever it tells us now was proposed when it first occurred. That is not to say it was done with the intention of modern discovery and in the form of an enigma to stimulate our perceived speculation of its exact meaning. A clan or tribe member had died and was buried to accommodate the expectation of continuing in prospect? There was no great fanfare of deliberation, for in most realities the clan were probably on the move soon after for the necessities of the living. Unlike from other times and by the same connection owed to our own intellect, while collecting our goods and chattels to move on for our food gathering role, we passed the laying of the dead for the last time. Unburdened, we also carried them in minds eye by our memories without the complexity of supplying them with food and camp comforts because that was to be supplied evermore and elsewhere?

By any imagination and mine included, events disclosed at Shanidar have given us first insight to a part of our character seen no where else in the world at the time, at the time, be it ranged between sixty or one hundred thousand years ago. Where I have cloaked it to be versed in the spectre of religion or of religious significance stands to the ability of us as the single human specie we were then, and projects us evenly to this time of now. I hand the event all attributes to be of our first shown spiritual event in terms of human conduct. I do not hail it as anything more than a small accomplishment by our natural progression invoked by the rigors of life's existence anyhow. My acclimation is that it was done 'Past Gods Words' to stand as so and be lost in the void of the misunderstood by our very haste in finding all to be understood by homage to life's mysteries?

Shanidar is the record of how we can see the future on our own account and was ably followed by other fantastic burial rites in all places and at all times since, but to the same desire and in mixture by some of us in calling to Gods own blessing as a right instead of an expectation! Burial rites to have been a means of representation to us stand to the certain proof of our singularity, because of representations throughout the world. If I might suggest of Shanidar in significance, but maintain its quiet relevance is done to keep it in context of what it really was by suggestion. If our wounded heart laid here in death, is to see tomorrow, which it is to be by expectation and not demand. It is to be fed from the desire of hope beyond understanding and made in conjunction with the nature of things by the feature of seasonal perpetuity noted at the rising of the sun and the turning of the seasons? It is to be done outside of the scope of our reasoning but carried beyond forces such as evolution unseen, but closer to creation, imagined!

'Past Gods Words' was some time from being an active statement then ago, but to have reached the border of creation at such an early time stands testament to our worth as one of earths specie! By our unconnected lines of tribal or clannish interaction, save through the means of trade by the rule of civil order to separate groups. In not being able to converse almost in as much as our sophistication today limits some of our best points to each other. Then ago by the same unconnected mist we carried the prospect of the future to be of huge significance by each and every group, again displayed in our cave paintings and rock drawing ranged in time from hundreds to some thirty thousand years ago all carried 'Past Gods Words'. Although the phrase cannot be projected to the understanding of the past or for past societies, it is measured by today's standards that we must also operate to the term. To allow our own free minded ability produce the same understanding to the unknown by the same aspect of our free will as those at Shanidar with 'Past Gods Words' also expresses our unity!

## CHAPTER 84

Three taken prospects just seen will be seen again, although almost self explanatory they can be mixed and stirred by our own attitudes from what ever religious or political standing we use. However, I still impose them for forthright use, not least to be our most modern method of communication. Our one active prospect so far in 'One for the Whole' is to cover them in query, simply because all three are derived from the one source. With 'One for the Whole' in static representation of our ability to forward our own ideas in representation we are all voiced across the total range of methods in which we communicate. Far in excess of how we already sometimes agree in terms of collective decision making if expressed by bodies as the United Nations, or politically enforced ideas by whomsoever?

Four prospects offered by the Fifth Dimension in presentation pivot on that one active, which already shows how the principle of representation actually works. Even that we have had and operate in global conduct by many means, not least with political or religious influences first. We had done so by compressing natural errors by not being able to relate to our selves equally. All of our functions had been honed to interact on the world stage, but were continually compromised by the need of hasty action expressed in the short time since the end of World War Two. Forcing us without full awareness to operate in the haste of half measured reasoning, but not knowing that by implementation this process endorsed the right steps taken syndrome?

Only by first relating to these past five to six years just gone, can the picture of necessity be seen whereby the factor of the Fifth Dimension is to be realised, not as a means of condemnation but one of direction and correction. Its form has simply been to examine what we already do and how to be able and do it better. By me speaking in terms of half measured working, of our steps taken syndrome, of our policing methods whereby we have set to cover fair inclusion by our human rights mandate or with our politically correct code or our attempted understanding to religious tolerance. Has not been carried to show the error of our ways, but to be a means of enlightenment to show how some of our listed properties just given are open to the ready abuse by any of us under our own misdirected reading of any said reasoning?

Another feature not to have helped matters is to be primarily read from the overall standing of our political conditions of acceptance, in cases the very style or type has been used to promote success, and therefore set the drive to continue on par. Of all our misdirected applications, although politics in the theory of conduct is paramount and from the theory of success is demanding, its control of direction must be set correctly? I have stated in reference of how we had been able to classify people from the time of our world based League of Nations. To have attempted to update the version through the United Nations and have been able to accept some commercial political standards for some of us to be first, second or third world economically. Allowing that through the intellect of our religions we are all the same by tolerant means has not been the medicine required to dress the wound!

By such standards our means of study to theses matters ongoing have only been political and commercial, allowing us reclassify other forms of conduct to be anti corporate. Therefore, by connected entitlement we can be dismissed or fired without the use of committee, with no voice or standard, reactive measures to present ones own case after half measured judgement must follow! To have been presented by any half measured result under political terms can be well documented by many of us over some very minor issues. Even if self induced by not adhering to a particular misdirected issue, it is only by our act of deed under accusation, that if the steps taken syndrome comes into effect the picture is irreversibly changed. Inducing countermeasure half measured, escalating into further steps taken, changing the matter of concern not to the original misdemeanour, but for not beholding to the further matters of chastisement, being made guilty without trial!

On the world scale, the above format can generally be duplicated by the same reaction of faulted assumption in some cases. If done from the political medium to matters of religious concern, then the projected solutions or countermeasures, are subject to the control of our standards of and to religious tolerance? Which if employed can or might seem to be half measured and lead to our religious steps taken syndrome, complimenting the matter of tolerant liberty in that region. But in placing the medium of politics own countermeasures we fall to compromise our built standards since the ending of the Second World War in nineteen forty five, in separating religion and politics at the wrong join?

Without any specific desire and falling on no head to be blamed, all matters to our progress need total management, which from first reading sounds pretty intimidating. By long and ancient processes we have arrived at this particular juncture more so because of our political management than to our best abilities from religious instruction. This in outcome is not to be taken in a competitive way it simply belongs to our progressive nature induced by driven need. By first resulting in the attention of control to be our best provider by the means of organisation, we naturally found it fitted to our requirement by open success even if not read or fully understood? From that suggestion if to have lasted for hundreds of thousands of years then the outcome is obvious, we are here now?

In balance and from my own observation at this time, when first finding a religious inference to our real desires for total understanding from those human remains at Shanidar, we expanded. When first endorsing that new aspect four thousand years ago there was to be no release of our pointed objective to break sight from the mediocrity of life into the splendour of life through death? Therefore without that specific realisation that I in part had been pointing too, we have had no other incentive or objective than try conduct our affairs in the open or closed opportunities presented or invented?

Many of our objectives in this modern reality, have had to be bent or reformed to suit the changes that further study or understanding we have been exposed to by working to the aim of discovery, are to be taken in? In so far as the occasions of eureka fall upon certain individuals and their own specifics have been revealed to us by the medium of science, we have been drawn to lean to the aspect of politics as the main means to our standard bearing. Scientific awareness in peeling off our mysteries over these past four thousand years at least, have helped force the divide between our two divisions of control. Allowing the situation arise whereby any objection to a particular decision in relation to one or the other is dealt with by means of political legislation or in the guise of religious tolerance to accommodate both factions. All of which can and has worked by some methods of mediation in the running of new ideas, but only until the time when the form of rules in law are un-transmittable to a particular aspect of one or the other of our choices.

If the above has not confirmed to us that we have travelled in a certain direction and remain wanting, then the obvious has to be seen to be declared by these pages that by our present standards, politics and religion do not really work!? Do not really work, has to be extended with the adage of, by our best efforts of recent times by how I have assumed them to operate on the same plane even if politically led? Stated is the presentation of facts in open confinement so that we all might see how we actually operate by each others standards. In coming to that realisation as many of us had been doing for years and years, we have tried to push the advantage to the well being for all and each other. But in not denying our system and measures already taken, we from the best of intentions have muddied our own waters by operating from those or other previous proposals?

Unless allowing the overlay of more political legislation fall to be upon our mandates of collectivism expressed by the standards of human rights or political correctness, our first, second or third world standing, along with a host of other set standards and ending to be religiously tolerant? In being the only operational options that we can seize upon, then by further adoptions within those standards can only work in finality from our steps taken following the extra introduction of half measures working? Where another extra suggestion comes in its own strait forward form in 'One for the Whole' induces its own examination, but this time it is one of the first ideas brought in these times to compliment our thinking since nineteen forty five. This time within these five to six years it has been able to view our condition by all aspects and associations from all angles in who we really are, because it is further balanced by presentation through the Fifth Dimension?

Three taken prospects and one active is a small proposal to overlay our political and/or religious determinations and expect further choice. I have set the task so in order that we can openly assert our standing by religion and politics in the vitality of our own need to operate by both, but in the best order of there own understanding of each other. Our varied and magnificent histories tell only part of any story; there best charge is indicative of how we have continually improved in the matter of life but not in meaning? By reasonable examination they also show how to maintain the order of that supposed improvement we have inadvertently directed our first decisions to all choices from or to or because of our political objectives. Which is not to be fully condemned save by result, for in the general supposition of meaningful progress we have let it lead by its own basics.

What I might wish to determine is that those clauses were taken in first issue from the time of settlement of the Second World War, but raised from the conditions of political strife anyhow. Not from defeating world fascism in one theatre or regal imperialism in the other, but by how the victors saw future prospects politically. From the positions of one side holding to have a mixed and vibrant well established religious background and the other in open determination to excommunicate any religious standards so far achieved in replacement to politics only? Perhaps making at one time the view by others to endorse the aspect of religious tolerance, just in case!?

## CHAPTER 85

Within the cause of observation I have been able to determine from my own aspect of religious instruction first administered to me in my own youth, to be able and relate how for the sake of time in a very restricted time slot, our young are to be of prime concern to that direction of enlightenment? With myself living at this time so far by historical connection to the total time since the close of hostilities in nineteen forty five gives me a connected view. From some instances that I have lived through, my own captured vision of any particular events were naturally to have been formed form my own religious perspective and from the prevailing political situation to have been most prevalent? By that association and by direct connection to events that I have referred to as the Israeli/Arab war or the mid nineteen sixties, in not to note my observations then, is done now from the space of time measured in these recent five to six years, which fortunately is open time to better thinking by all.

Not one single result of the events then has changed except that now they are history and carry all interpretations to have been mixed by the recorders of history and from whose perspective? I have often stated in the same relation that we can change the matter of our histories without alteration. An example of which has been given but from my outlook of change without employing the need of acknowledgement by any other source? To that time I have made pointed remarks to the political activities of some Palestinian groups leading to the form of political activities but cased in religious fervour. Form the same mixture to have occurred long before the ending of the Second World War when the Arab/Jewish forms of reactive antagonism led to the balanced form of murder in both instances. While both sides were posturing to the eventual legitimisation of nationality and associated claims, they only acknowledged their religious differences.

Nationality sent across our best judgement, changed the aspect of direct concerns if to be only labelled of political importance because of the development of our bloc culture forms of political posturing. With religions first and finally sidestepped in the future course of internationalism, what or from where was any voice of dissent to come from if accepted internally or not and not counted externally? Just like from when the Israelis were given standard religious tolerant bearings in the conduct of their own faith from the offset within a Christian based political ideology. And others might have been covered by tolerant means when supported by a non religious motivated form of justifiable communism if used in support of Islamic ideologies. But when put together or of the same balance, they did not match by any dimensions, from what or from where was any voice of dissent to come from if by tolerant means only?

If, and if I can see it now with the additions of my own inclination to the Fifth Dimension, then history is to be changed without alteration. It is to be put into the realms of intention before result without change and can be done in this very age. Because we can now begin and recognise that the past was the anvil of the future and the future is now encapsulated by our own single lifecycles without religious compromise of any tolerance! Set by the standards in collection from the Fifth Dimension, but buried in the essence of all of our pasts in the single aim of enhancing all of our futures?

'One for the Whole' to be our one active prospect so far will turn locks and open doors in many guises. It is only a small expression and so should it be, for we are only small in the nature of time it's self, but from our own good fortune have been able to value its concept in being able to rationalise evolution into the openness of creation. From which we are able to be self evaluating of our own picture of how creation is now relevant to us at the expense of evolution. 'One for the Whole' by asking questions or in offering its own proposals for open approval is bound in acceptance of our three taken prospects, not least that they are our very first 'One for the Whole' acquisitions. By any reasoning and or reasons to so offer that we are 'Out of Africa' where the intention is to include the meaning to us all that we are one specie, no contest, our one active prospect is operational. By the same measure 'Natural Body Death' in self awareness to the same single specie, no contest?

The third taken prospect being 'Past Gods Words' is as easy as any to say, but tests the openness of our one active prospect from or by the method of presentation made to it in that particular context. 'Past Gods Words' by direct and obvious implication brings into full focus the acceptance in understanding to there being God. Therefore in any suggested proposition having been put to the active prospect in a collective manner for appraisal if so, could be misconstrued to be a form of impeachment, taking that the matter was already concluded? Likewise if just and fair proposals were put from anyone inclined to communistic political leanings in also being ungodly, then they might feel excluded to the common good? Making in error what we all have achieved since nineteen forty five by less obvious inclusions or omissions in working by assumptions created by the implementation of our steps taken syndrome, but not to accept that 'Past Gods Words' is of a political origin!

'Past Gods Words' is not a restrictive measure nor is it to be in containment to any likened declaration. It works in the open and unashamed assumption of a sacred entity, but in the most pleasant of matters denies political denial of such a form or force if included. Nothing has been decided against communism being unreligious by any specific claim or charge from what it has not already shown of its own construction or operation or actions. By the same token and in balance 'Past Gods Words' is not judgemental on any religion from there core or from there proliferates. It is not mentioned so to encourage religious sentiment either, but to bless a basic standard incorporated in all religions based on us having free will! By not denying the concept of religion through God, the standard is to be through our own consciences in any case. So that we as individuals in entering any proposals do not usurp the intention of this taken prospect when the very expression is to remind us all that any so offering by 'One for the Whole' is only humanly directed!

By compression our three taken prospects in result of our one active stands within the confines of the Fifth Dimension made from the course of my own observations over these five to six years and in being done from within the confines of this last modern era. If formed in part from the dust of abstract construction and now hardened as footprints of memory, all are to be directed in guide only of our intentions not to compromise our own free will by religious determinations. Accepting that the line of our political direction is a little askew from the spiritual, the Fifth Dimension is another means of our forced unity without penalty, when any and all expressions of desire on any personal level can now be filtered through our active prospect in 'One for the Whole'.

From any expression in political terms, like in contrast if a communist by example was to support any idea of there own political grandeur in relation to personal management, then in reflection there own proposal has to be worthy of compliance by themselves? Then by extended means the offering is to include us all by consent without the ready force of judgement. My use of singular political definition, that in theory is of no fixed religious determination, has been to separate us by division representatively in balance, using a proposal from a declared westernised democrat of no religious affiliation, the same criteria stands? Political suggestions from whatever field can be forth coming and are welcome in the aim of communication, but all carry the same levy of reflective recall, that if one of each was to propose of there own determinations being better for all, then in or from both proposals the only settlement is where both sides switch normal allegiance for all!

That feature has to be counted again by the means of a non religious political philosophy to recount the method of most styles of government showing allegiances to an object rather than the supposition of a source. When by some dictatorial means the object matter of the leader or head of state is transferred to be the embodiment of state or of the mother country or the fatherland. And in showing the fruit of desire through our one active prospect in submitting that for the good of state or property it is necessary or will be necessary to purge the background of all elements to that objection by permanent removal. Or it will serve the need of expediency that these objectors will be murdered or put to death by the means of exhaustion or degradation in compliance with political policy! That edict by means of our 'One for the Whole' prospect refers to all adherents of that running political philosophy in service of state or land or the earth, even the head of state!

I had put the argument before in the same context, but by using an arm of religion in example. Although drawn from our long past, the comparative tone relates to some actions of modern breeding. Human sacrifice as practiced on massive scales some four hundred years ago in the lands of the Americas and the joining isthmus of the two great American continents. Fortunately has not carried its broad concept to this time or the time of asking or suggesting to our one active prospect for endorsement to the practice as a means of unity. For in object our 'One for the Whole' addition is to be a reflective means of self evaluation so that we might pause and gather the consequences of our offering. By placing the sun or any object to be the centre of adulation in service and operating by means to placate its needs? Could have been called, but the tone and ending would be the same for the priests as to those that were to be sacrificed, they too must die, in other words priests are to be sacrificed first to show the meaning of the implication?

So stands one small part of the intention of our Fifth Dimension in cover to the tone of our politics and of our religions. Placed in that way it has been offered for our perusal in relation to the three taken and one active prospect so far. It cannot be stressed enough that my overall intended application of it in use is only open to the critic by our misunderstanding of the concept. That is why these last five to six years of our first study to it, are to be contained as an opening, not in any form of judgement, because we have far more serious work to do with the idea before consigning it to the rudimentary feeling of old habits. Beyond range the Fifth Dimension permits that with only three taken prospects and one active so far, it will work in full view enough to be our conscience. When seen that we are able to carry it into the realms of our recent histories from the close of hostilities world wide in nineteen forty five it will give us all the same reference point to be recorded howsoever?.

## CHAPTER 86

A very important addition carried within the range of our four prospects owing to the Fifth Dimension is by how it can overlay our two guiding influences of politics and religion and produce different standards which are the same? I site the reason for this end because of the way I had been fortunate enough to formulate its first construction. My first clue came from the outside of any supposed or proposed influence I might have hoped to put upon our duel influences. For like others I too can lament at the sorry state which we had allowed develop between us. Not only counted by the rule of internationalism, but in our wanting from our inability when dealing with our very neighbours. We had let our conduct form or be channelled by the considered dexterity of the United Nations. Alas it worked best by half measured reasoning, where it could best understand its own manufacture from the foundations on which it was built upon, being the failure of the League of Nations. With both entities at first giving the standing that they were the lead in the means to progress and if in the need of control we had been left short delivered it was because of our arrogance?

By this stage of our enquiry it is not fitting that we blame such an organisation as the United Nations on its own behalf. My direction is to show us example of how when given a lead, it is easy to follow the trend by running with the hounds. It as a body was of our construction and given our assent to do our bidding, which is the way that many other projects have been started. My dissection of the situation is not to be seen as a damning condemnation of that which was our intent, but from the way it had worked to its best standard having to employ half measured means. Not least from where it is supposed to represent us all equally in or to the first measure of politics, yet conceded that we are of three different bearings by the aspect of political commercialism?

Then further from the same complexity, determine our religious standards to be of our sought singularity and evenness through the leveller of accepted religious tolerance! From which I have been able to set the examination of both factors by now overlaying them with the three taken and one active prospect of the Fifth Dimension being the benchmark of our true intentions.

By setting a question, it is not always clear if the answer is fully known, like from the example of the chicken or the egg, which came first? As old as that maxim is, it will always set the slide rules sliding or the calculators humming and buzzing in the effort of concluding to an answer? Even looked at a second or third time that small question will still draw a different number of reply's depending on circumstances relating to us in any context from our own particular standpoint. In first consideration it is a very complicated question and shrouded in the aura of the evolutionary cycle, because it implies to that process, but at the same time is simple in reading as referring to procreation? Half measure working is of the same mark to that problem, not from the asking but from the answers. Because by result the outcome is seen from different eyes, because by working in that vein the steps taken in solution further separate two separate answers or many, many more if offered.

Steps taken as the working of half measured actions, are naturally supported by all means, because from there application they have nothing else to do but support the original half measured proposal. I am almost lost for examples to highlight the offered prognosis but in general terms they are most obvious in the civil list of affairs that we attend too, and some have been most obvious since nineteen forty five. Taken on the individual national scale by choice, from most countries even determined by being first, second or third world, if from the governments of any country decisions were made. Then in most cases the matter of fruition in the end product or result became paramount, became steps taken, it was our duty to support them in support of the original plan. For instance, to supply better housing or better medical care or improve the infrastructure. In the course of the plan in operation and ongoing, if it was pointed out that there was a better or cheaper or faster way to proceed, little was done because of the original steps taken!

Counted from its less damaging outfall if the only result was the loss of money or time then the matter might not hold to much inconvenience from one group to another nationally. Unless the matter of money loaned by one to another, even if by world banking if there is such good intension, then the results of repayment might bear heavy in cost to the people the original plan was supposed to help and support! On a much more harmful level, if from ill advised information one country was prepared to have a small acceptable type of war with the people or politics of those of another nation. Then in leading to that direction and taking the advice half measured, the same situation by our steps taken working would and does still apply. Plans in the making by huge effort become the driving force of initiative to succeed in success, steps taken! So much so that even in the case of house building where cheaper material was found which could be used to shorten the time taken to do the job, original steps taken still ruled. Like even in the case of going to war, after all that planning and effort in preparation, steps taken, war was to be the only course for all concerned!

I do not state that everything we do or have done falls into that category, but both aspects by being half measured and steps taken, are from my own observation of how we sometimes work on all levels. Without condemnation of a conclusive nature, those methods have been used since time immemorial. My own input has been to modernise our intent, not just in the shape of natural progress, because that in process has brought us to the need of better modernisation in any case. What I do not want is for us to continue in the way we have been acting in the assumption that those actions enable better understanding are only tied to the political lead. Which has done less than expected to promote the sought harmony our religions have professed to deliver?

From any delivery we have received since nineteen forty five, to look and acclaim our near two hundred different independent countries is a fine mark. To look at some, if mostly classed to be second or third world commercially politically, and note of there conduct is a shame to be upon us all. Not for being second or third world by any imagination, but through the standing from where we have not given freely of our determinations to each other from the process of neglect which was naturally endemic, bore from natural separatism? From the way we had developed in Europe and America up to wars end in nineteen forty five, to move in the haste of self sovereignty for others, no time was given to the consideration of us being actually of this single specie that we are. Though I had criticised the standing of the United Nations, it is we who can better claim its failing, because from the aura of best intentions and in applying the medium of equality by other standards. Our application has been lacking on the account of assumption that some of these new countries were not second or third world but on the same standing of us all, except by the political criteria of personal wealth, viewed differently for some new European countries?

Leaving many cases world wide, that to the inhabitants of each country, any standards obtained or devised were due to there own effort, again compromising the collectiveness of our singularity of specie, but not seen? When considering that pre nationhood, in many cases no study to the effect of becoming independent was seen in how to operate, save by standards being set by American European influences. Our new nations of Africa by the biggest majority, and of Asia being proffered to the bloc culture standoff, even of Europe in the same bloc culture vein, were almost left to there own devices to obtain set goals by international standards. Promoting political terms we are not the same, but should have the same objectives in the pursuit of wealth at least politically, which is not to say that this outlook prompted any eventual outcome. But seemed inclined to be an official fixation by direct intention from the writers of intended mandate through the United Nations and in lead of every body every where politically. Then in covering the probability of reactive forces by the position of religious tolerance to endorse our unseen singularity?

Examples in shadow cover the value of that first political outlook over and above the religious standards we have come to collect and revaluate by these times if mostly done after the mid nineteen sixties. From whatever source and by whatever inclination, again to the land of all our fathers, Africa, who from its vastness complicated the measurement of future national boundaries? From that very feature of size there might have been errors in the settlement of national tenure. Mixing tribal factions by drawing boundaries though the middle of former tribal lands, expecting all concerned to accept others good intentions? Setting new countries from the old standard of European colonisation without notice of effect save to comply with the overall political control of possible bloc culture developments!

In equal fashion Africa by all determinations was plunged into sovereignty on its own account, but as a spilled jigsaw. From which a large part of our own political diplomacy has been lost in allowing the dreadful situation in some places go unchecked. If the only answer used was by the means of internal genocide to decide the ruling factions in new countries if not by European standards? Allowing that the mixture of unmanaged commercial dealings to have gone on and in pursuit, were done in the cause of commercial politics over the staid morality of civil unity enhanced by representation of us all through the United Nations, European based! From Africa earring by inclination, any deliberate acts of brutality to the human condition count as our failing, not in relation to a specific tribe or colour but because it was we who were all originally 'Out of Africa' who always prepared the conditions?

From any events hinted at or that I have implied, I again intone in full measure to show the expediency of half measured working. Even with the situation that a country could be rated to be second or third world commercially, is a standard first set, therefore in pursuance to those ends, outcome is to have been steps taken! From which the wrong conclusions could be drawn by the process of replacement in taking to the assumption that being right by the steps taken syndrome, others are wrong by all counts and at all times?

Although I have contrived to simplify a method of study to this phenomena there is to be no let off or excuse to the action of such deeds as genocide or murder or unbound brutality in conclusion to having been left on ones own to now continue after previous times of others national management? And in cross referral to suggest that there were similar cases to such deeds over a period of hundreds of years gains no quarter because most happened outside of our new era thinking since wars end in nineteen forty five. Even from where our focus is to be by these recent five to six years, it is not that we steer future obligations in Africa but that we can now offer better guidance and reversed, but only from now on!

## CHAPTER 87

Religious tolerance has been the by word of our political steps taken syndrome in any format, simply because it implies an ending without any judgement to the beginning. It also represents the rules of legislation allowing that there is a named condition for every thought or deed. But like from what I have mentioned in taste before, our ruling is done from the reflective nature of looking into a mirror. Where in the standard of normal life when a person holds up their right hand to brush the brow, the reflection moves the opposite hand which when mimicked would be the left hand?

Religious tolerance is of that ilk, whereby the rules or codes set are not to the intention implied, unless because we would not wish to hurt the feelings of the religiously inclined, we might give in blessing to the idea of reopening the practice of human sacrifice! But this time sanction it in awareness not to the sun or the mother country or the fatherland; instead sanction it to God in Allah or the Great Spirit or any representation to have been an entity for the living in place! By such cases politics has lent its own covered hand to the possibility of such deeds by stamping worthy acceptance to the idea of religious tolerance. For our own fair judgement now, in this new era of open enlightenment we are disposed to listen to all save those in contradiction to our half measured working and our steps taken syndrome. That is why the Fifth Dimension employed to overlay politics works from two mediums!

By starting with listed mysteries of the unknown and by the ability to think or imagine in the sphere of the future, from that very asking, produced the answers in the abundance of not knowing, which produced in form all of our mysteries? Before we were of the ability to write or talk in a legible manner, all levels of reasoning were given to the self maintained opening of any perceived particular subject. Some mystery busting was done when we began to understand animal or heard migration, we pieced together the why, which was an event of understanding. Mysteries were further broken by normal acceptance, like when others of tribe by association with fire tending, lost the awe of being singed and burnt through familiarity, and for themselves showed no query to what they alone understood on our behalf. Then discovering small and linked applications such as flint napping, order was being placed around daily necessities allowing a better fixation to what we were doing. In these acts of discovery, still along way from the tidy nest of scientific formula, we began to pile the true nature of the other mysteries, save to purpose?

By sign, again from the same reflection, in looking back to Shanidar, although the image is not necessarily what it might appear to be, there is enough in our general store of knowledge now to think in a range of ways to the same topic. I have taken it to be our first encounter with religion, because of the fact that there seemed to be grave goods left for the purpose of continuance. My picture is not final, but to be suitable in connection with the matter that we were different than any other specie and that we had shown signs to the beginning and to the purpose of creation, which is to settle the aura of all mysteries? By using that connection I had jumped in time to four and two and one thousand four hundred years ago in continuing that theme.

I had also referred to the land between two rivers in much the same way, but by implication that it was our first place to have shown a better form of organised civil control to some of our already different tribes and clans. In being one of the first places in the world where we were able to show forms of collective behaviour amongst our different groups, standards and ideas in the form of mystery breaking were in part self maintained. Enough to make us think in different ways about the same subjects not least religion, shown by the number of different gods we had amongst us. Further endorsed by some of the ways we adopted others god's in the mixture of wanting to understand more and gain equal favour if applicable? Possibly creating a kind of mental backlog where we sort of applied a little reverence to all gods in a tolerant manner.

From that very early time the Middle East was one area to give us all a boost in the race to accommodate all the mysteries showing and of there release? By that time and in conjunction the matter of them or that which we did not know began to be tackled by some of us who had developed our own skill to the method of organised thought. By means of academia through the pen or scratching stick, we began to record the methods of daily life on tablets of clay, reed leafs or other pliable materials. We were simply logging the events of now with an eye on the future, in terms of supply and demand, we correlated the harvest or our livestock to accommodate need in covering unseasoned months of production, because now we were static societies. From the same balance being able to record one matter, all were now open to the book. Even from this fixture there were societies of the same dexterity of equal outlook who took to there own use of similar methods and in some different uses if driven by different impetus.

What this feature really brought to us all fortuitously, although not realised in its true form then ago by anybody, was by one channel of our true internationalism and singularity, from which we were drawn to the same conclusions, but as of then unseen. Writing or the ability to record matters of concern to any particular group brought us all in parallel awareness of our singularity, though far from being seen then or in cases even today. What was happening at the exact same time and in at least one area of the world was the beginning of us to manage all matters even our mysteries? When from the past we nibbled at knowledge out of sheer necessity and taken in relief the ability to control fire or use animal skins as our own. By other means in trying to understand the nature of rainfall we learnt to be near shelter if familiar signs began to show that it was ready to rain soon? From this generality matters began to make there own pace to all events.

With our ability of being able to record matters of necessity, two lines were drawn but not yet faced off against each other. From one corner where the mysteries were piled so high and in such numbers, our unravelling was done in the sophistication and in the name of our godly societies. In so far that all of the unknown was to be in the knowledge of one or other or shared amongst all of our gods for there delivery, but by a strange coincidence at or in or after our own death? From the other corner the pile is perhaps even higher, because by method and style when one such feature is exposed like being able to control fire. All subjects related to fire explode in profusion, by asking all associated questions to it, what will burn, how will it burn etcetera, etcetera and finally why dose anything burn?

From the two answers given and now in being able to be recorded for posterity, in one case they remained as mysteries to our Gods becoming our one God or divinity? In the other case by the complicated means of exposition to general knowledge or now basic understanding or even to the collective names of science linked first place to astrological sightings in the heavens. Physical things like solar phenomena in parallel cases, came to be of two different meanings, for the religiously inclined to be inspirational and for the more mundane of us, to be scientific? Not to be unfair to either source, the connection is made to show our separation by two mediums not even considered then. Because from within our separate societies defined by our clans or tribes, some peoples from each group do not necessarily hold to the standing of the majority in house. Being a wonderful example of our singularity whereby we shared the same likeness but were able to mix across group by thought and study and learning and wanting to understand, not just for self, but all.

## CHAPTER 88

From my own cross referral and back dating, but without definition, I would refer to those people as we, belonging to the path of my own fourth dimensional prefix. By token that my referral covers all aspects of science, then by modern standards I have cut to the value of their own importance, because some are dismissive of our best religious judgements? There is to be no denial to the value of study, save by invention and false opinion if used as the means to blind comprehension in the search for general understanding.

For the same mention, but to a different application, I again relate to the name of this very book being an open study of some of our ills linked to some of our best intentions? I had indicated early on that the name had been drawn from the hidden meaning of science calling for extra dimensions! By direct connection, science needs those extra dimensions for formula to be expressive also we have had to look into the darker side of reason, not for sinister intention, but for discovery to know! By any style, no particular scientist, and I take this study to the very early days of its opening, sought to oppose religious perception. In fact many worked within the confines of all religions, albeit in tone to the particular religion by affiliation. By the double mix of that religious connection and the routed course of experimental exposition, a natural barrier to the means of understanding grew between the differences of science piled in one corner and religious mysteries piled in the other?

Outcome did not always agree, for instance over simple things like from where the church by any coat might tremble its adherents to the wrath of god expressed by the mighty clap of thunder or the flash of lightning as a form of judgement. When science from its own early study could now translate the event, of not just from being a condition of the weather, but caused by the nature of ions or neutrons or protons reacting positively or negatively. Creating superheated flashes of light burn great streaks across the sky, expanding moist clouds at a fantastic rate of speed culminating in the crashing of god's wrath as thunder? Any first studies if commissioned by the church, were expected to show compliance to religious doctrine, not from the standards of dogma, but simple explanation for religious reasons? One unfortunate drawback of this magnified study was to the church, science even from the early days of writing, could only write in the truth!

Results of which by some of our greatest times of change to all societies, from about the time of the discovery of the new world in the fifteen hundreds give or take a few years either way, we had become global for the first time. Bringing us to be protective of what some had gained by us becoming global, which brought us to use science for self advantage? In some ways to be suppressive of some discoveries while being expansive to the openness of gained learning by all mediums of the sciences, but now inducing open competition. If some of our religions were to combat this expansive flush of discovery by the means of their own reformation in covering all aspects of dissent? Some of which Included the condemnation of scientists, who in general openness were simply working to the art of study through the fourth dimension then unknown, even in having been commissioned by the church to curtail others expositions?

Without direction, but by the advent of time, with the churches in general being involved through there own religious character by natural progression in part related to Charles Darwin's publication of The Origin of Specie, we were set new opinions? Not only was science to decry some held mysteries that were sacrosanct by acclamation, but now the question of creation was to be exposed to more general and open study by all and any means. In taking Darwin's inclination of evolution to be the provider, was to have a huge impact on all future outlooks. Not least that his ideas were the last nail that some scientists needed to lay most religious concepts in peace. By indication, now that science was almost in sight to understand the meaning of all formula exposed by the simplistic means of experiment, her own voice grew.

Although evolution was a lively and moving concept scientifically, the idea was also linked to some aspects of political reasoning since Darwin's publication; and gathered pace when science moved to the field of physics and super physics. There was no holding the medium and all the branches of science now, which by close order was of or to and for a political definition, because it also worked on the same level to the matter of conduct and interactive behaviour. Science in good and true order by its own impassioned commission is drawn to the answer of all mysteries like our religions? The difference has to be exposed by influences outside of its own field, not least to have been mentioned in full explanation by myself and others by the cases in question of what formula is? By being nothing more than the total explanation of all study, given that with a set number of objects under a set number of circumstances will produce a set number of answers, that is the essence of the history of science beyond the inclusion of the human form!

Even looking to our mighty crocodiles, their own standards have been to the above criteria, but have ended the experiment least ways for the last two hundred and fifty million years, because they do not think on? We in our own singularity of our adopted human form had broken all casts from the time of Shanidar, being now free to compromise all features in the quest for the mystery in what happens after life and more vividly what happens in death? In perfect balance to any such theory it is from the same we, that questions are asked in the name of science on the same behalf where answers are expected.

From where most of our religious postures have come in conclusion, that the answers will drop to our understanding when we die, then that is unquestionable. I say that in complete comfort as a worthy statement, not from committed dogma, religious excellence through supposition, or sheer bloody mindedness. I say it from my own first realisation that we are a unique force in the whole universe yet to be challenged from without? Not least because it is from us that our own thought has provoked the study for the meaning of all things and by better cases we have done it from the only possible two different perspectives that we could have, in our religions and from our study of our mysteries through our sciences, which later can only compliment each other!

In closing this paragraph without pointed assertion, recognising the factor of science and its own tributaries has not come cheaply to any of us. Without refinement by obtaining the status that we hold now, some of my own reasoning accepts the concept of our varied histories but allows for our full contribution collectively. We cannot leave our religious standards as they are, because all are yet unfinished and furthermore, delay promotes doubt, from which we can again nurture our apathy. Which can turn us to the haste of reactive posturing by all of our standards of acceptance and in the open necessity to react, when our religions are yet to be active and active and active, not to each other, but perhaps to our twinned concept of science? From which in the full order of its own will to know, we have entered the study of knowledge in the production of answers. Yet for those answers we are owed in due, we have sought or seen fit to manipulate religious acceptances to obtain that true scientific ending. Like following the field of physics and theoretical physics, the emerging picture is of a blank wall, almost as death? But by scientific dexterity and political manipulation we have contrived to create the correct criteria to proceed.

By the cleaver means of scientific pontificating we have been able to extend our known and working structure of our four dimensional world through our discoveries already related to the nature of the universe as it is. Because in that quest other reasons and means of query were naturally raised outside of the compass of our four dimensional world as it would be? Our method of study in the one form of theoretical physics, has induced us to theorise to there being more and more dimensions to fill our expected requirement to eventual understanding? More, simply to explain such instances like black holes in space, or of a different line like wormholes, being the theorised means in which we might traverse our own galaxy to green pastures in someone else's domain? Also quantum physics where the parallel exists with or without substance belonging to a time warp unmanaged? From where we know of the simplicity of the speed of light or that a vacuum carries no sound but transmits heat but might hide the missing dark matter that is stringed in waviest profusion, all thought is still open?

Of all or any lists we might make to the welfare of scientific knowledge so far in open discovery, it is unfair by that stream we have compromised our first better means of study in our religions. I have said in most cases we have been driven by the course of circumstances, this is a case in question when science in the medium of politics has set the scale to be unbalanced against our religious expectations. From where science has moved close to politics and in doing so she has had no cause to openly declare against the aspect of religious doctrine. But in making the observation that in order for the study in science to reach fair conclusions, we might need the effect of additional dimensions by whatever expression, needing extra numbers to accommodate the understanding of the whole universe from beginning to end. One for instance might be called the distance dimension, employing a set of hitherto unimagined formula definitions to account for the continual expanding universe. How ever perceived, some scientists envisage up to eleven dimensions in number to accommodate our ability to take everything in, with perhaps the last being the random dimension, to cover all eventualities?

It has been no coincidence that I have related my fourth dimension to the offices of our scientific query, although I cede in or from most standards it operated well before our religious concepts. My take in relating it as the fourth was to deliberately follow our religious concepts, for in open declaration I perceive them to have been a better driving force to the state we are now in than the mundane pattern of scientific revelations alone. Three religions from the Middle East in direct linkage are for me representative of all our religions, but allowing certain features to differentiate because we have been subject to different aspects of change stimulated by geography, climate, diet and other factors.

My Fifth Dimension title was also added in a small way to compromise the extended needs of scientific explanation if necessary, to put a limit on the need of complete understanding? Because it is not necessary that we carry complete understanding, by our best nature we had complicated the aspect of religion from first base, save at Shanidar! In connecting God to the nature of understanding we gave charter that our levy was to understand the nature of the deity itself. By the trick of association and in being foot printed to that ideal, we set our own standard to the nature of total discovery, which our scientists have been pursuing since and before?

By token my introduction of the Fifth Dimension is no more than a line to be followed and for us to make our own connection at each juncture. By open display and through much of this our work in study, from where we stand is to be of the same complexity of how we interact now. We each might walk past the format of our first three dimensions in being a Jew a Christian a Muslim or of an eastern inclination by religious means, in stopping and looking there is no charge. Not even on any competitive grounds for all three are the exact same to a Jew a Christian or a Muslim, the cost is to recognise that aspect of our religions, they are ours and remain so by our choice. There is no competitive edge added, although some of us might think in the misconception that our object is to only see God by one view, when the feature God is eternal and broad by all readings and able to carry our consciences for us on our behalf, from which we know we are on the right path!

By the same indication where I have set the fourth dimension in a later context to the first three in line, has been done with the same commitment but to enjoin science and politics in venture by balance to our three and all religions. From wherever my own picture has been drawn, from and by whatever ideas I would hope for us to be able and relate amongst the first four dimensions. Then it is from the Fifth Dimension proper that we can use the ready made paste to cement in cautious rigidity our own new perceptions if so desired. Only now in the use of our three taken prospects of 'Out of Africa' of 'Past Gods Words' and of 'Natural Body Death' all supported in change or query by the one active in 'One for the Whole'. Taking us to a corner for turning, but not with the liability having to know the meaning of the expression of science, when in standard we can connect to the one importance of our own lifecycle by being fifty plus years long or in the luck of being seventy plus years, but in the object of turning both to be eighty plus years long. Making a better and more honest expectation first hand rather than half measured?

Showing by all implications to the matter of the three gods in being Jewish or being Christian or of Islam and on our own joint accounts, our objectives of now are in relation to the future by human terms in promoting human unity before Gods own judgement? Not just from the aspect of death by doing very little in life, if seen by cross referral. Because on and by our separate religious determining from one to the other, two Middle East religions each can only see two sets of failing by any time scale. Because four, two or one thousand four hundred years is only counted by a fifty plus or a seventy plus year lifecycle!

# CHAPTER 89

Many of us in good heart wake each day to the best normality of life, even if this action applies to where across our range of activities some of us will hurry to the rubbish tip in order to get first call in the chance of redemption? Not spiritual or political redemption but the best chance to rummage through the city trash that is daily dumped in massive land fill sites. From where some of us in the form of our own children are disposed to salvage whatever we can in the light of supplying the meagre means of salvation in order to exist. Even from those few sentences can be seen there are wrongs in our society, which of course would be so if they were true? True or not as conditions of existence they are not the worst to greet some others of us elsewhere by any location. All of which have been caused by our selves in the service of our religions and more so in being served by our politics.

By each token in any situation the balance to our standards have been set from the one difference we hold than all other specie, we alone operate through the aspect of creation without limitation. That has been written in formula since the time of Shanidar which across all the questions related to the whys and wherefores that any individual as me would or does ask, be aired in open display. From any time in my own lifespan, to turn away from the desperation which greets some people daily can be done by two excuses, one political and the big one of religious significance? From that tart observation it is fair to discharge some of the direct blame in cause by some terms, because only recently we have genuinely improved the lot of a lot more people than allowing certain acceptances continue by whatever reasoning? My own display in showing our good and bad sides cannot be balanced by any moral plane, but opens the argument in relation to our dual mentors of politics and religion by any devised opinion that any of us individually might wish to use.

In so far that I to am individual, my collection of ideas by opinion is truly owing to the spirit of this our single specie expressed by historical reference from our peers. Some to be giants in the names of science, humanity, to the spirit of generosity in aid of charity, in help, even politically attuned to believe in the standards of equality. That we work to the best efforts of others before us, and that the rudiments of open change are to be there, in many respects we have been beaten by the advent of time. It has elongated the processes of needed change by the flowing methods in the way we have to operate and by whose agreement? So much so that by the way we operate politically had been set half measured leading to steps taken, which compressed the outcome of results.

All of which can become complicated at first by political standards and systems in use today, of which we must look at open eyed, but not in denial, which by various degrees of measurement will highlight different aspects of politics through its own channels. Politics in being set to rule in service is complicated by how it has to behave for us. Although we believe it to encompass all of our needs and desires, in true operation it never has been able to supply all demands. That is why in many first world classed countries throughout the world today, there are to be scenes described above. Because of the wanton nature of the commercial success generated by listed countries, the throwaway value of the same society generates without care the advantage for some of the collective population to rummaging through the rubbish in a competitive manner. In other words by political determinations we are happy to see this process of gainful employment in comparison to the act of begging for aid, in city streets, where financial institutes struggle to maintain there profit margins?

It is to be seen that turning from the ready possibility of global war again, we have been working some of our political strings to this end, but only in the success of maintaining our standards of human rights, of being politically correct and well intentioned. All from our funnelled devices of using half measured workings counted to produce our steps taken syndrome style of management. Then through the biggest crossover, I do hasten without devious manipulation, we have been cleaver enough to include our religious determinations across the board by the adage of us all being in or to the liberty of religious tolerance? It is to be stated in positive terms, because in open recognition when tolerance of our religious inclination was applied to our running, when done even if half measured or steps taken, it was done in the bluster and from the properties of being right in following the other standards set for political reasoning.

I had used the term "without devious manipulation" to imply that religious tolerance was set to cover all religions kept in the wings until some clear political outcome had been decided from the operation of our bloc culture divide one way or the other. What I had also referred to within the same time period, was of our Arab/Israeli war of the mid nineteen sixties having the effect of at first raising some Palestinian objections of serious political standing to any outcome of the recent and associated conflicts?

My own implication of now is being able to backtrack and look through any recent events since world war two because they all occurred in my time! It is possible for anyone to have been here then, to still experience the emotions and feelings of such events. Unlike having to create the same emotions spread through family or national ties going generations into the past when from even then our objectives were the same religiously any how? Which leads to why I have included our young and how we should feed them the past, but in respecting there take of it by the good job we will have done in relating it honestly both politically and religiously! Unfortunately emotion being carried in openness can set its own standards if half measured then the following creates the future which is now?

From the stated conflict of some forty plus years ago, although we were in the throws of many changes on international levels politically, some changes were also occurring in the religions of the world. From the extent in limit for some, they either began to open their view or contained it in the standardisation of fundamentalism done across the board by all, but maintained to reaffirm original concepts by other examples. From where I submit we all got it a little wrong in some ways bore out by how we have used or taken on board the concept of religious tolerance. Some of us have used it to imply that instead of the possible supposed intention whereby we were to respect each others religion, has been turned to the picture that each religion is of its own standing, unless determined politically!?

From any lead made in reference on religious terms only, my past referral to the Middle East by historical connections has been coincidental by the means of our conduct emanating from our times in Mesopotamia long ago. Incidental from our religious connection in relation to my first, second and third dimensional religions brought up to date by interaction these years since the close of other hostilities in nineteen forty five. Without charge incidents being of a religious nature are to be worthily sensitive, not in personal contact, but in there relationship, because of there direct connection to each other by any described means and now by any described interaction by action or reaction.

At this juncture I am making the same statement that I had made several times before in order to reconnect for us the state of affairs today as they should be by linkage, also to highlight half measure working politically or religiously from any particular times of the recent past when it had been operated. With the overall intention to put in a clearer light how the events to have occurred actually did occur, which can only be done from my tool of abstract reasoning with other determinations also allowing that we can now look at our religious and political standards separately. But from here on in with the three taken and one active prospect of the Fifth Dimension. Because when we now record any political decisions they are done of this instant, even if with bias, also some holding to certain particulars of restraint if through religious tolerance can be seen by their separate outlook?

On the basis that we are here now, allows any claimed set of circumstances be responsible to that event, which accounts for evolution. In the same trend, politics as the track we must follow is also of evolution, even from its smallest beginning of civil control to this modern day giant of multi functional activities that it is. From where it has been forced to reel in some of our religious activities, if for instance like being part responsible for the ending of the religious practice of human sacrifice. It has taken by assault the lead over any other of our religious activities by its growth into modernity represented by how its records are up to the minuet and able to change by the next minuet to its own ends if necessary? Against the recording of many of our religious standards in cases to be four, two or one thousand four hundred years old, which have to stand the test of time imposed then ago, even amid changing circumstances of now, politics must change, religions cant!

What has happened in this last modern era has been the effective crossover of our religious and political standings and in cases standards? When said that we operated a bloc culture in the Middle East as well as in other places, it was from there that most of our telling changes of interaction began to take place. In the position I personally would like to promote with regard to how our young would be expected to take issue, is to first relate all things to the future. Which becomes a difficult thing for them to do in the respect that from our political standing we are supposedly up to the minuet, while at the same time from our religious perspectives operate to timeless tenets and covenants introduced and changing over the past four thousand years? Being matters it is possible to highlight from the events of the Arab/Israeli conflict or furthermore of the separate religious determinations brought to bear from about the same time in the nineteen sixties but now recorded differently!

From where it is very difficult for anyone even as my self to translate some past events using the approach of abstract thought, is misted by who and by what division history had been recorded. All of the difficulties associated to such a task have to bear in mind that particular factor more than any other in case our or my delivery is half measured. By approach my lead has been and will remain to our religious standards, but in countenance of political reading from who the recording of events or the time will be different because the standards were and still are dressed to suit opinion?.

By time, I had indicated, that from new Jewish sovereignty by the instigation of the state of Israel the standards set were to have been politically motivated. In actuality this would have to have been so because even before our present standard of religious tolerance, there were accepted standards of religious separation by respectful tones of forbearance. Allowing the full balance to our separate religious practices to continue in relation to each other without overt intrusion or by the same standard of open acclimation apart to others but never to be over others!?

No questions in the asking of maintained standards then ago will reach an open answer, because the reactions now although supposedly linked to those times, have no relation to the religious spectrum that it is now, even without tolerance? Any statements made by innuendo are not worthy of acceptance unless offered in the construction of future reference for the ability of realignment. That is why abstract thinking has aided my drive to correlate the true relationship between politics and our religions, like when from examples that I have alluded too. About some kinds of Palestinian reaction without direct title, I have avoided making half measured pronouncement. Not for social effect but to avoid another example of steps taken or inducing steps to be further taken even if not sure of why or in what relationship the original steps were taken?

In naming the case in point, the object has yet to be covered not only for the future, but to realign the past in the required forbearance for us all. Palestinians as a group of people are as worthy as we, because without examination they like us all are connected to footsteps taken in volcanic ash all those years ago. In the reality of our own nature if not ably displayed by the Israelis themselves who acted to their own best interest. Fed from the rush of national status which over spilled to half the world at that particular time, from what went unnoticed at the time, has been our haunting, which in part has extended from some Palestinian thinking. Who not by half measured connection had been cast to represent the religion of Islam as a political means to express dissent or opinion!

Taking that I have come to associate the religion of Islam to be practiced by most Palestinian, is done to make clear of Islamic involvement in this modern world. Surprisingly up until after Israel in the form of a Jewish state was formed and even up to the results of the Israeli/Arab conflicts of recent times. That statement is lodged in fact, but by the worst of reasoning being attended to in manner to have been steps taken under consideration, we have loaded some of our own religious tolerance against us more than half measured. From where I had intertwined our religions and politics in reasonable harmony supposed since nineteen forty five. Direct inference to have been shown from my construction of events should have allowed us draw the self same conclusions, but by proceeding in fashion half measured some factors were overlooked until breaking there own ice and involving all of us to different responses?

From the European world war in the most part between combatants of Christian religious backgrounds, even Russian communists who allowed religious forbearance ride at particular times of worry to the people first and the state second. At the end of the conflict the attitudes to have changed were created in the new standoff determined by political reasoning only. From the altered situation between the main victors where only political standards came to the fore, because again the Russian leadership at that time reverted to a non religious culture by internal repressive means as required? To such times, that political bloc culture was being built in places other than in Europe like the Middle East. By the direction of alignment one side to the other, with the communist Russians in support of aid to many Arab countries for what ever reason, this standing brought a new emphasis to all reactive thinking since that time but unrecognised!

Unrecognised but noted, the matter of dealing with the situation was only political, because in the active process to the birth of nations the drive was to maintain or promote the political persuasion to have been Eastern or Western. Making the case for Israelis to be politically western, while opposing countries of any other religion to be classed as of a Eastern bloc culture politically only! Effectively giving no classified determination to any country by religions status unless only through tolerance. From where the United Nations had been involved in this or these matters at the time, and cases were to come to the fore in the matter of religious concerns, there was and still is no machinery for any cases to be heard, because of religious tolerance. No matter what judgements could be made if cutting into the aspired spirit of any religion, this was nullified by religious tolerance, which unfortunately has been unable to manage its own concept by any direction or from any quarter!

From the example of when by the first open lead in connection, the United Nations gave acceptable conditions for the formation of the state of Israel in the former mandated lands of Palestine. Direct mention was made to the partition of the city of Jerusalem in terms of remaining an open city, which was to be maintained. Not least because of its known significance to the three Middle East religions of Judaism, Christianity and Islam. When from the time of any Arab/Israeli/Palestinian conflicts of the mid sixties when the city was to become the Israeli capital, replacing Tel Aviv, no changes were made to compromise Eastern or Western political standards. Although in prompt the United Nations could comment on any so situations, they were inept at carrying out there own forms of supposed legal alignment to there own proffered appointments in the first case. Again allowing religious tolerance play on by who held the hand! Perhaps allowing the opening to some of todays reactions politically or religiously by other factors??

## CHAPTER 90

Whatever the case for or reasons to violent action or reaction, we have created the situation from the blind acceptance of working half measured allowing our steps taken induce the reflected reaction from that same production. That sentence is in rough report of the way we now openly accept violence upon the human form by religious determinations and without forfeit because of religious tolerance miscued at a vital time of specie self discovery!

Politics always failed because the most up to date application of its drive is how we can go to war by the most peaceful of means. We can discuss the matter in open debate as though we were preparing the means of our next daily chore, divorcing the outcome to be a manageable or permitted decline of standards self induced by those we are about to massacre? Whatever is to be the expected outcome and is not, all factors are covered by our steps taken syndrome, which in a number of cases already work in expectation from our half measured reasoning. I would site without application that much or our modern thinking has been due to the supposed success of the westernised approach of how we deal with matters politically? If our bloc culture standoff has been curtailed on all the Russian fronts so far, likewise we are also depressing bloc culture on the Chinese front by encouraging commercialism through the window of the Western political systems. Then some measure can be given to the rule of political determinations in so far that most of our main decisions are made on those terms. By working off half measured reasoning, the quality of war might be even discussed in the terms that I had intoned, which questions the basis of cultural civilisation. By putting the divider of that possibility in the position of tolerance has done little to aid any of our accomplishments. That standard needs us to employ our Fifth Dimension remedies by three taken prospects and one active in 'One for the Whole', but to be applied in cross reference by separate intentions to our religions and politics.

## CHAPTER 91

I had previously mentioned of the effect that any of the prospects might have on any of the priorities in our drive to be involved in our own destiny one way or another. I have ceded that from the four mentioned in the Fifth Dimension all have been done before by others, but only qualify by being separate and applied at different times. My own approach is cast from the time scale that I will continually use, guided by these past five to six years. We have never yet had such a feature like the Fifth Dimension to be the core of our collective consciences by any description. It is to be the most demanding of all of our tests because it is completely open to each of us individually on all levels without any measurement of concern to it. Meaning that it is not required for us all to understand its intent because it automatically will operate to our individual aid through the study of our expectations by politics or religion, but not forced?

I had previously covered some of its reasoning to matters of political concern which are naturally linked to all the features of our existence. In now applying three of the four prospects by religious reference we might see how they are to be of our prime importance in preparation for come what may. Quite apart, I will offer my own judgements closer to the end of this rendition to our driven intent by the mediums of religion and or politics or reversed. I make the study in two halves because of the way we have allowed our main influences become different and have treated them so by the production of some of there results so far. By contrast in virtually all spheres but derived at core, the one difference between them by most measuring and from most angles contrived by most societies is so! Politics strives to serve us in terms of being the daily provider to all of our needs in life, continued in generation upon generation which sounds evolutionary but in decidedly not. That is the most direct of our considerations to it and also covers why and how many people adhere to that standing as a finished philosophy. Although being in many forms, they are in the minority to those of us who are involved one way or another with or amongst the many hundreds of our religions throughout the world.

Apart to those considerations most of our religious evaluations can be achieved by representation from using our three Middle East religions in equal service to us all leastways by their own timed presentation. By purely religious tones 'Out of Africa' in respect of our singularity is complete; all Religions already state the sentiment fully and definitely and in finite acceptance.

'Past Gods Words' in open context complicates many different cases unless wrapt in junction with the scope of the Fifth Dimension delivered these five to six years recent. It is not a barrier or condemnation, nor is it a tool in which to limit or curtail already said written text. I use it specifically in this example case so far, to overlay our three Middle East religions, especially when all speak to the very words of God one way or another, I enjoin them of the same ability to influence our perceived objectives designated and our desired expectations in hope. By that very core, 'Past Gods Words' has little to do in the belief or desire of when Moses or Jesus or from when Mohammed relayed words to us. The implication by our religious funnel is from how we ourselves have poured our determinations out of the spout in the difference to how it was wholly delivered in honest piety for discovery in use, from our common ability with the use our own free will!

'Natural Body Death' in all terms of the human condition is to be our guide by all religious objectives in how any doctrine is to be accomplished, 'Natural Body Death'! From any picture painted over these four thousand years by three channels and all other religions, it is they who have induced our spirit of inquiry, have been the first to conquer all of our mysteries? By three examples whereby we came to the condition of eureka, we have continually tried and walk the road to the future already knowing where it was leading. From the least time involved, it has taken us one thousand four hundred years to reach the already known promontories of all intentions, but from the picture of four thousand years of memory!

Using that simple range it is easier for us to see some of our errors in the case of how we have perceived the intention of the mysteries we did not quite understand. Our selection in the terms of God had always been misaligned in human terms only. Even from that, we had shown the first interest ever in the whole universe by our ability to rationalise the concept of creation, was just exactly the right medicine in the call for the understanding we needed to have in order to be of understanding in any case. With the aspect of creation being the sign of our or any break from evolution, then this is our case. From the study of two hundred and fifty million years in the span of time owing to the mighty crocodile, that solitary time has been given to procreation, which fits well for the crocodile, but offers little to the prospect of God. By any reflective means that we may now look, it is only we of this single specie who have been able to relate the totality of time in the space of fifty plus or seventy plus years into the aspect of God and from that very realisation, sets our case? Without the fanfare of political blustering, that simple realisation has been managed by many religious concepts, but relayed to us by political determinations until perhaps the mid nineteen sixties when some choices were remade?

By the feature to overlay our religions with three taken prospect and one active from the concept of our Fifth Dimension is a levelling for all of our religions and an effort to compromise our political directives in cases. In all suggestions, pointers or offers of direction by any medium the one and single objective is to us as the individual. Not a democratic or communist individual nor a Jewish or Christian or Muslim or Hindu or Buddhist individual, nothing more than to be a fifty year plus or seventy year plus individual. From which our politics are to be rationalised in the one drive to make us all eighty years old and then plus some? While at much the same time our religions in more haste are to now present their case in the one objective which has been done for almost fourteen hundred years anyhow, but only changed by the political perspective. This time again when looking in the mirror of change, we are to be better aware to the fact that when the right hand is raised the reflection shows movement of the left, which is not to be complicated by the fact of our three Middle Eastern Religions when we only have two arms.

In looking into the soul of all of our religions has to be done from who we are as men, individually cast in our one form of the same specie? In all cases that review has to be done considerate of our free will, undirected, but past the need to compromise any of our heartfelt feeling of religious devotion! Our open examination if not through the portholes of blind and obedient service in open faith, is done in the full acknowledgement of our single specie acceptance but to the imagination of Gods desire complimentary to our own free will, God bestowed, if? We should dispel the order of unwarranted reward by the use and acceptance of matters to have been led, if inclined to harm the very substance of our existence being of this our single specie. Being able in all future cases, disarm the flush of political concerns to matters of less concern than to our wellbeing as the single specie we are yet to expose!

# CHAPTER 92

Religion is like sitting down to a great feast and leaving the table fully refreshed and full of all of the delights of eating without having eaten any food, by that picture it is an emotional event before being physical. From any other similar description giving insights to how they can excite some of us in small ways and leading others of us in rapturous acclaim is spread amongst them all. Politics does hardly enough to raise a hand unless in the disguise of replacing the concept of God with the insight for us to hail the very earth that we walk upon in mother country or fatherland. In positive exchange there is too much difference between the two for us to have given them there same status at any time of our histories, then allowing the less emotionally charged lead us to every decision created by its own tentacles.

Although readily driven by events of equal concern at the ending of world war two, our main lessons had to be in the name of politics because of the real nature of how things had occurred or had panned out and the solutions we arrived at outside of plan. In giving our religions the equal status of tolerance in the aim of inclusion to our duty, was alas proof of how we thought and worked half measured to create steps taken, but uncharted even up until this very minute? Half measured and steps taken in any combination can deliver forms of excellence or of catastrophe by political standards, which are generally bland. It is when that same flat standard has been assumed of our religions by emotional reaction that our thinking is to be made and generate proper solutions to said actions or reactions! The nature of which are not to be made in or from the based assumption that religions are tolerant entities and any reactions as might be are in their own right.

No matter how much further we might delve into what our real standing by religious reaction might be and to avoid the mixture of half measured devices knowing what they can produce. Our first objection of firm standing is to be done in the theatre of our religious battle front within all new experiences, but now realised and set to be reviewed by standards that can only first be used ever, with the key of the Fifth Dimension unscientific! A key implanted by these very pages which in form of representation is a method whereby we can overlay both of our challenges in politics and religion extracting different results, but again being un-judgemental? This quarter I have given to our religious overlay only, but caution the inclusion of political references to be used only by comparison. So that we might consider our religion from their own perspective and not tied to the subterfuge of their standing of tolerance to other religions, having in the first place been set by themselves and now reinterpreted by the forced rule of recent political standards. By any stages or intentions, I have set the picture that we can each be in the best support of our own religion, but in doing so, we can only act religiously within political boundaries to curb religious excess?

By religious perception, 'Out of Africa' is just such an expression to cover all of our needs by accepting our own religious standards to have also been a product of us all in joint query, but tainted by the inclusion of our circumstances of geography, climate, diet and other means. With all religions a product of us all, because we are here now, then by direct association our conduct in setting some of their standards has been due to our changed circumstances unseen. Although the same cover can be added to politics, that concept is less defined because of its separate referral?

If ever any formula is to be challenged, it is from my own prospect of 'Past Gods Words', because in expression I too have been drawn to include something past the rational of the requirements of full scientific experiment. God in form; is taken to be the core of proof and is offered to thwart the understanding of science by the occasion of gazumping it, by not accepting is own requirement of total proof suggested by formula and endorsed by experiment. The only danger to the belief of what we might describe as the entity through God, Allah, Nirvana or the Great Spirit or others and others, but not the mother country or the fatherland. Is from where we have set the answer to be known? By that or those inclusions we have moved the concept of any religion to be a wholly human concept above the matter of science and connected politics. To be represented in most comparisons by using the three religions to have formed from the one area of Geography to be the Middle East, in using their own times of conception by all comparative means. Not to highlight in the critic how they worked, but in notice of how we must now see them in three, and all others for the same reason to be 'Past Gods Words'!

By the range of four, two or one thousand four hundred years is the method in recognition of how we can justly tackle this our God phenomenon, whereby we are apt to acknowledge our own concept, while at the same time borrowing in source the political edict of religious tolerance differently than religious forbearance! Even through the course of history by the use of time, when we changed in many cases some of our first taken religious concepts, in course of those particular timescales and from whom in this single specie only? Our changes by fine detail were always interconnected from one to the other just like from my most prominent example using the three religions of the Middle East representatively. Without charges of being deliberately misled by any individual with particular reference to our three case religions, there would always be no proven case that this never happened by any description. Even by using what Jesus of two thousand years ago was to have reportedly said or by alluding to what Mohammed had recorded on his own behalf in direct connection to the events of that time some fourteen hundred years ago. All were connected to there own adherents lifecycles, all was done within how we actually lived, then?

If I would wish to make any kind of case that the burial at Shanidar had any religious significance, then it is easy for us all to connect in fair judgement that any events by any depiction there and then ago, because we were honestly 'Past Gods Words'. If for no other reason than the fact that we were unable to relate any significance to the burial, save that we wished to remember our dead. Then that is only proof that we could not communicate by any means except through the mystery of death in the proposition of it holding some means of light to our unsanctified human query! In wandering in the same manner for the next sixty to thirty thousand years we could not make any statement of our emotions until somehow one of kind wandered into the back of some deep dark cave and created history. They began to chronicle the standing of human life to the concept of death and the future more so than to life and tomorrow, which still carried us 'Past Gods Words'!

I with my own choice of possible explanation had alluded to the elders of some of our separate communities in times past of about four thousand years ago, who in the course of their reasoning saw in image, concept? Who saw the rational, bearing the means of delivery in understanding covering all the raised questions of the past centuries of our tribal interaction through the relationship of our multiplex God inclined adaptations in all processes of change where necessary, even if led! Without the possibility of having made a mistake we, progressed in image of the future being accountable to the past as it simply must have been? By ensuing error of thought, but in excellence of mind the mistake we then made and which has been carried since, was to our selves without the chance of escape by the feature of it lasting four thousand years?

In carrying that error for two thousand years, then without charter it was compounded by Jesus in the same style and manner, but this time making for the earlier error to be compounded if even not being understood for a further two thousand years? I had stated of Jesus' credentials which were certainly not altered by any indication that I might have given. His intentions by style were of total human dignity, in making his own particular error in duty to compare, was by his own personal belief that we were ready to accept the format of our specie singularity, which proved to be a sentiment expressed rather more than a little before his time! Even by the time gap of six hundred years when Mohammed again looked at the mixture of beliefs ushered in the lands from where our first direct link with God had been given and indeed taken. His expressions were of the same tone that had gone before and in the same justifiable meaning, but encompassing the added two thousand six hundred years since Judaism and the six hundred years since Christianity of our natural progression. New insights of thought applied in there best form, were not enough to separate his delivery than had gone before, because his style of delivery was connected to be the same as some of what had gone before, by looking back through Jesus, Moses and Abraham and of their time?

From Jewish elders, if expressed through Moses about that formed religion, and likewise for Christianity and Islam, all were carried in line of succession with the true belief of duty to their respective religions in or with or from direct connection to the word of God. Yet on our behalf, I have proposed that we each should offer solace and comfort through our respective religions by taking all of our learning in the pleasure of our own faith to be 'Past Gods Words'. But in union with the other two taken prospects of the Fifth Dimension, which is in manufacture by us in this and these times of the past five to six years only, the past five to six years only, the past five to six years only! Once again said in the time now to the core of our three Middle East religions, to draw them in duty to operate in open compliance to our specie singularity religiously first, and now politically second, because of how we can show honest religious fortitude'.

Why any of us in the picture of joined independence might suggest that we are entitled to ask of our religions to present to us together, our own doctrine 'Past Gods Words' has been done to highlight our God Given derived aspect of our own free will! Free will is choice and choice is what we humans have in the name of creation, which in part is why we can ask of ourselves the matters I have raised. In as far as I have laboured the point and will further, is done from the small realisation owing to my and our particular time of now with our shared histories. Expressed by the image of funnelling, and from where I myself have used the remark to highlight points that I have considered important. They have been done on our behalf in the collective sense that my thoughts have been generated by all historical events compounded in time since the ending of world war two? Like funnelling political propositions for my own idea of solution, I have made plenty of my own errors based on assumption from my own meagre outlook and past knowledge. Although not having to enter all the combinations for any experiments of funnelling, when applied to something like our political differences and entering the mixture of democracy in all forms with communism in all of its forms at the cup end. When funnelled we can observe of democracy that it also allows the support and blessing of religions, yet communism does not and is therefore anti God as opposed to be anti religious. Not making the full statement but alluding to something that catches the eye and can lead to further misjudgements!

Funnelling in terms of religion has allowed us pour any mixture of them, if three religions from the Middle East and all others are poured into the cup to meet at the exit in religious tolerance. Being a standard where we apply religious sufferance under political acceptance if for no other reason than to compliment our half measured working by taking to concerns that in some cases are not really understood? Allowing that politics can operate past Gods words, but by its own standards in not giving credit to the true worth of most religions and can set religious tolerance, which is supposed to operate 'Past Gods Words', but being religious does not yet know how?

In the time lapse between the introduction of the Islamic religion and that of Christianity, although I had indicated there had been many changes made in our growing and moving advancement. One partial regressive step was to accommodate what might have been misconstrued by events more than the proposition whereby Jesus gave first consideration alluding to our specie singularity if even from separate tribes? By statement to that direct consideration and by endorsement for us to turn the other cheek in recognition, unfortunately by progressive actions and reactions since, both intent and application were lost to reality. When Christianity was adopted by the Roman Empire in the interim, they through the political medium still conducted all affairs from that lead. In still warring at the boundaries of empire with ferocious tribes, it was difficult to imagine specie unity as indicated by there adopted religion of Christ. If by alluding to one ideal and living by another and if this in part was to lead to the desire in need to rationalise Gods true intent from looking at the outside of the Roman laws of government and now their religion of centuries. What motivation apart to other ideals was needed to better the prospect of the future?

Undirected but in being led by one set of circumstances, for Mohammed wishing to extend our driven aims by all other circumstances it was to be done in taking all existing attitudes and styles to the fore. Without the specifics of pure direction in taking to the human character which was ever forming, no open sign could be honestly seen from where the Church of Rome had practiced to the words of Jesus not to kill. Its most difficult reasoning was by political reaction that in self promotion it was necessary to kill for that end. That the former religious aspect was overshadowed had a bearing to those times, we were not yet ready to that conception of humanity. From that structure I have been able to given opinion from where and when I first indicated that within the bounds of this new religion of Islam, a case had been made whereby we could kill of self politically or for religious determinations in being defensive or protective. What was not made clear then, was by what forms of motivation to any aspect by any reasoning were permisable?

How and why the picture has changed without alteration is not to be set from any description of Mohammed's own thinking apart or separate to what had been settled by the pen from his own time. My previous input with regard to the Islamic religion was for us to lay open comparison to the difference of attitude in why the Jewish and catholic religion professed not to kill of self, yet allowed the practice if through civil sanction? I had submitted to my own reasoning that Mohammed had taken the full situation in context and saw in display of what we were really like by continuing conduct. Virtually nothing recognisable had changed since the time of Christ, even with him being the prophet Mohammed saw him to be, although executed by Jewish religious law and in compliance to Roman civil law. If from no other base it was felt emotionally, seen directly or encouraged by self awareness to the standing of society at that real time. When Mohammed gave in portion that we by individual need in expediency, could at first defend the word and in token also promote the same written text. If and how or why any human death occurred then it was due to the days or from the circumstances of the times in which we lived, but in making the single point of defending or promoting our one true religion. Endorsed by full acceptance to all other religions not least that the trail was from the two preceding religions from within the very same area of geography, allowing the true link by association of God's final intent!

Fourteen hundred years ago with the situation of normal conduct rated by siege standards from one city state to another, even extending to federations by political coexistence. Under or by or because of loosely defined codes of conduct, others to the drive of envy and in the assumed ability of strength owing to there political format, might through desire invade what was considered a soft option if expressed by some pious means to turn the other cheek. Even amongst there own communities within the same structure, allowing that to the heart of society there was the well defined split of civil control by whatever means and in direct relation to the living of the day by any standard, with religious acceptances under there own administration. In having the twined standards, the situation was not twice as strong in running, but half weakened!

Without offering additional status to our political means of rule, in order to overlay the one with the other style, then by changing a standard previously delivered in commandment form, some core remedies would need changing without alteration! "Though shall not kill" could only have meant by specie distinction through religious interpretation from the good or pious or indeed of the working priesthood, but maintained in the spiritual format of description yet unseen? Where from within the Jewish structure of their religion, it was maintained to the one class of people to be of there own tribal mix or to be amongst the twelve chosen tribes? The combination worked by accepted standards, but as I have indicated and reiterate, was not open in any sort of acknowledgement to become defined by any other means even from how Jesus studied the situation or from when Mohammed was later to recall any further situations!

Killing of specie by any means or in any circumstances was open to all societies at the time of Mohammed. That by the hand of Moses the ruling of not to kill was delivered from the voice of God to this prophet, further endorsed by Jesus to spite two thousand years of the Jewish faith in all compliances to daily life in that very real time? When entered into consideration by whatever factors, but in first order to curtail the growing political overlay of politics to religion as Mohammed might have seen things? How could it not be so to interpret the same method of delivery to he, as was to Moses and in supposition to Jesus also. Not to kill was only an option of choice in the aspect of relevance to cases whereby God himself had interposed to free Moses own people from bondage, and he had dispelled those that would harm his then chosen? Making the case of Gods renewed interest in the form of realignment to our considerations only by our own religious conduct in meeting all of the situations to have occurred in the following two thousand six hundred years since his previous own intervention by what was delivered to Moses?

If and when delivered into our own hands through Mohammed in the same status as Moses and Jesus if understood so. Then the message was to humanise in open selection all men by his/our own choice and to our personal conduct in making the grounds of that choosing, but now through the religion of Islam. Without the raw ability for any of us to make our own choices within the confines of any religion, rules in having been laid were forever there in the option to be taken up at any time by us individually. But in not allowing for fourteen hundred extra years without the felt proof of God on hand, how could we be sure those choices can be correct now? When at instigation all three Middle East religions were constructed to the time limit of our own lifecycles not even fifty or seventy years long but in the instant of time from birth to death!?

# CHAPTER 93

'Past Gods Words' allows all the leeway we need to construct time to its best need for us to understand creation, which fully allows us to consider other options in the realms of evolution also. Where we have struggled in compliance to any reasoning had always been pressed by the medium of science and its exposition of our mysteries, although now to be referred to in the collective name of the fourth dimension. Leastways by my own study, so that we can differentiate between my first three dimensions relative to three and all other religions, but always in a comparative way. Not to their specific functions, but to our own errors of judgement, if indeed we have made any at all? Said before, and to be taken again, beyond any appliance of the fourth dimension by formula and or experiment, how can we have three different religions plus, and apply them to the one God?

By the same code and in fair delivery, even that all signs made to us were or could have been delivered to us direct through agents of the lord or one to one by choice. By the method of further delivery how we were relayed such conduct, was always graciously delivered in tone to our own excellence of character and to our level of understanding in our own ability to ever look for answers by our own specie terms. Being a mood to feature our own excellence of mind to the process of creation by evolution in setting time to gods own scale and pattern if desired. Not in compliance to us as mankind but to the continuing aspects of changing evolution accounting in time by other changes to all other specie until one kind might utter of Gods own words heard! Allowing that as we did see the flash of lightning and hear the crash of thunder all was heeded 'Past Gods Words'.

Putting to be upon us all, who having struggled in the shortest of time spans by comparison to some other specie, from where we first moved in query, and in carrying the same quest to the four corners of the earth. Again to be separate to other specie by that very fact of habitation tied by the bonds of settlement to be lonely in our continuing search to the basics of creation, above levels that might be limited through the medium of the fourth dimension. Not competitively, but in the course of imagination and by using features of self, far in excess of what the processes of actions taken to be met by equal and opposite reactions scientifically. Allowing that we might choose to study in the fields where science might not dare tread risking the apprehension of ridicule being laid against our discoveries, because the records of experiment were not adhered to correctly by sciences own excellence of mind but error of thought? 'Past Gods Words' has always been there to accommodate any means of science, but has to take judgement by how science is operated, bearing in mind it is by us of this single specie who in context study both religiously and scientifically, 'Past Gods Words'.

'Natural Body Death' by open description although alluding to death carries many traps to its understanding, when in reality it bears nothing more than to a specific moment of time? My deep concern of it will fortunately be supported by how we as individuals will be able and express our own will by show in using our one active prospect in 'One for the Whole' to dispel its negativity. By definition I have constructed it to hold one direct meaning, but in the full realisation of our own and personal query induced by other influences, it bears its correct standing when taken by time only.

On the vast scale of politics by historical standing in close proximity to the end of World War Two, we actually gave up the counting of the dead when we reached fifty five million. From that terrible extremity in human history the meaning of death is almost too complicated to reason, because we operated beyond the scope of any rational. From any examination in the final analysis, we didn't really understand why we were at war and even how 'Natural Body Death' was to have any bearing by the study of attitudes in the future? Because human death was an open necessity by all given standards following into the first half of the twentieth century just gone! 'Natural Body Death' is of these recent five to six years, not to account for past happenings, but to overlay our religions first so that by their influence we will all aspire to it past any political control or suggestion.

By inclusion of our three taken prospects it is of vital importance that we value the full intention, for it spreads across all of our divisions if indeed we have any unnoticed? I have no personal desire in treating the study in open format excepting that if untold we will saunter in our normal fashion in compliance to what we already do prior to the recent months. Allowing our usual standards come to the fore by half measured reasoning, with the following steps taken to endorse our standards of human rights being managed politically correct. All of which are aimed far before the process of death, but without concern of how it actually occurs and in blindness of how it does occurs! Shown by all the examples post World War Two from where we have better fed the ability to kill human beings by our arms race and bloc culture standing. Even practicing the art of going to war for the flimsiest of reasons endorsed by our United Nations concept, but in practice of human rights, which is again to the individual status in a first, second or third world standing, unstated. By mixing the just said ingredients the result compromises some religious standards supposed, and embarrass some political standards who will not kill to some degree in upholding civil law nationally, but will internationally, in upholding to the supposition of terms by natural justice!

'Natural Body Death' applies to the standard that already exists in relation to our split in terms of personal longevity, in some of us having that span to be fifty plus years against others with a general span to be above seventy years, all applications are the same, save by natural justice. Geography, climate even diet and other reasons are the causes of those differences by the average sway. That they exist perhaps shows that the first world nations of our planet are not as successful as they assume to be, even religiously? Nevertheless the objective now is there from which we can never turn, being encapsulated by our three taken prospects and one active of the Fifth Dimension.

'Out of Africa' by implication has been set in stone to our singularity of specie even now dressed as finches. 'Past Gods words' endorses that fact into fact by our own developed different character whereby we show our singularity in comparison to all other specie, through the conduct to our religions and even from not holding to a religion. By each case we show the ability of concern to matters that are of no concern to any other creature. Which is not to be taken in doubt, but fortunately seen by there own contribution to the full splendour of creation separate to their own base of evolution. How ever any of us individually will look to study any creature, be it the monstrous blue whale or buzzing bees or elegant gazelles even hunting tigers. Of all their excellence if seen so, no picture can be drawn of any other intent save procreation. From which can only be spun the picture that from the development they are beholding to, none by any example had ever shown outward signs of aspiring. Even from the great apes of all forests, having been set by science most to be of our inclination, there had been no signs of their understanding to the matters of creation? Because all apes and we operate 'Past Gods Words' only, aids us humans to express choice when they cannot?

'Natural Body Death' is almost the perfect headed note in description to all other creatures, it alludes to procreation and in most cases is the standard. When on the broad base all animal and creature deaths occur in the very processes of each creature type by suited circumstances achieving there objective, then by case that is to the fulfilment of evolution only. When by further example animals and creatures openly kill of self or others to that very end in the purposes of evolutionary drives, the standard on there account can be suited to 'Natural Body Death'. Can ours be the same?

My comparison above is of a general nature, but set to show the separate path between us and all other creatures, most ably to illustrate the method of that difference being through our religions and not others? With religions taking the high ground, is done from a separate score than by our politics doing the same. Both have seen to be different than for the creatures of evolution, but both have also given us a distasteful acceptance to the factor that we alone as specie wantonly kill of self, expressed by religious intentions of promotion or defence to all religions? Then from the owing of the political tentacles of greed forced in some extent by market forces and from profit and gain or by the charge of emotional distress in cause by the two factors of action and or reaction, depending on whose lead? Making for a very difficult scenario when worked that such judgements are made to be of human rights but not tendered the same!

Forcing the imbalance between natural justice and the spectre of human rights, which on those terms compromise our ability of making civilised decisions using the generalised standards we have so far accepted by our first, second or third world rating commercially. If to comply with one of our taken prospect in 'Natural Body Death' then although that in form is of a final and equal settlement it has to be made so by constant effort, which was tried and denied post world war two. Simply because we did not have the sophistication in the character of any civilised nation or bloc to aspire additionally to the/our/an ending, because we never thought so? Not thinking so is the measure of how we have buried the importance of religious settlement from most political perspectives, because of the way politics has been forever busy in self determination.

To some methods and styles of politics, like when the influence of communism was of equal prominence to defeat the political system of fascism held in central Europe at the time. By their own charge the German system to have been fascist and the Russian system by there own charge to be communistic, at that particular time of our joint histories in the ideological battles that ensued. Even past the concept of military action, both sides inflicted upon their respective populations even in defeat, the most horrendous examples of human to human brutality. No manner of blood letting was forsaken; no manner of bestiality was limited although hidden in the vagaries of war itself. Our actions were close to the lowest ebb in the distinction of any other similar actions taken in the name of civilisation. Unfortunately in the other theatre of war as mentioned by the Asian conflict, Japanese conduct upon the Chinese nation was of equal brutality in much the same tone by the political opinion of German Nazis in there claim to racial superiority!

Later in having our smaller wars to the exact same political reasoning, instead of rekindling our intended political form of controls to also encompass religious standards. We pursued the consideration of the new habit of politics only by action, with religions by tolerance, which was inclusive, but has allowed us diminish the aspect of purpose more so than intent? Any weakness to that particular mindset when exposed, if from religious quarters, was ill timed and to the wrong balance, even if the intention was to drive religious perspectives to the fore instead of being subservient! No story of equal bearing was given in offer to redress wrongs and or misdeeds politically. But unfortunately if the deeds in first recognition by myself and others only taken in hindsight, were to have been first exposed to recognition by myself without the blame of inclusion of others. That I could take such deeds and align them to have been Palestinian reaction to whatever the outcome of local wars had been, and put a religious shade to be upon them right or wrong. That image has to be examined in the light of events which have been translated into the actions of now, but first by religious attitudes in close proximity to actual political rule?

Any inferring of those incidents above by inclination to some acts of Palestinian reaction to at first specifically Israeli impositions, is not done to sanitise either case, but rationalise how we first began to behave in this new last era. By showing in full relief how we have not been able to really work the situation correctly in light that we have now supposed profound differences. Not by the economics of being first, second or third world, but from our religions in being first, second or third dimensional? By my own awareness to some situations if even spread over a time span of forty years and growing by re cutting and reviewing such events even if of a distasteful nature. There had been no avenues whereby we could look to the methods to communicate our case one to the other except by direct action in reaction, blessed or not?

What could be seen in open review was the once again profile of might in the force of being right, not unlike when the fascist Nazi party of Germany and the superior imperial military war lords of Japan took intent to promote their own standards. This time though the forces of opposition to each other were communist Russians and democratic Christians no longer deposing those odious features of political fascism, but now each other, away from the long forgotten victory over tyranny in remembered history. By the very tone of the bloc culture standoff mentality, even with both sides taking to the aspect of religious tolerance, no grace was given for the voice of religious opinion, either way!? European, American democracy allowed there cohorts the intoned liberty of religious freedom through the inclination of the said tolerance. Whereas the opposing side in a non religious standing of pure communism alluded to the same, but in support to the mainly Islamic ideological enclave were unable to rationalise Muslim thinking expressed religiously. With both democracy and communism not being able to rationalise some religious Islamic standards and standing even if unread. We were all disposed by purely political reasoning to accept in unreligious forms the condemnation of some Islamic/Muslim religious reactions outside of the supposed ideal of what religious tolerance was or is supposed to have meant!

All of which is changing now to the developed standard of assumed excellence from the dust of resettlement to Russian communism, by de-communisation to have now embraced some forms of commercial democracy at the same time re-welcoming Russian orthodox Christianity in acceptance. Giving further support to the wrong measurement of religious tolerance that we are supposed to maintain equal standards to each other, even if not to have mentioned some ancient acts of religious human sacrifice! Then to ignore what could be a method of voice in claim whereby the only means of voice was again by the bullet or bomb which were features employed almost since the close of hostilities in nineteen forty five. But differently enhanced if some could only see such matters to defend or promote a faith being challenged more by the indifference of political dealings than in the need to respond in such an empty fashion. All of which again is to be reconsidered through a mirrored image that was not there at instigation but is readily here and now carried by direct association in one of our taken prospects from the Fifth Dimension, in 'Natural Body Death'!

Before we might use any new form of conduct our one active prospect in 'One for the Whole' through more defined reasoning has to be put to, against or for whatever reasons that religious actions have been turned into political ones. By any study yet to be done is in relation to the fact that we are of this single specie always, and have ever been so, save by our own failings aligned to some of our own personalities. Such failings to be exposed are not to be measured against the individual in respect of each other, because we can balance the result of our own collective drive these last four thousand years, or at leastways from our collective histories since Shanidar, to have been in the right direction religiously?

## CHAPTER 94

Before we enhance our total religious attitude to be jointly expressed through our named dimensional religions, to spite what has been generously and courteously stated of our singularity even by Moses, Jesus and Mohammed and of course many others. Although at the time of each introduction to our new standards of practice as foreseen or foretold, no real picture was taken to our separate singularity? From the wonderful acceptance that creation was afoot we saw in profusion the number of splendid life forms that there were and only partly adjudged them to fall in plan to our own acceptances taken to be understood. This I say is a relative feature of our three Middle East religions because in that broad image of Gods own creation it was as easy to see that human tribes were as different as the lions and the tigers. Therefore it was as easy for any separate tribe to expect and go there own way religiously if the desire fit the circumstances. On this I say further, that is what we have done since Moses, Jesus and Mohammed, gone our own way, which is not to be a judgement, only an opportunity whereby we can change direction without altering the past and its rules. From where we can alter our histories without change, because of who we are and have always been.

One of the many keys we need to turn in respect of that minor acceptance falls to the realisation of some of our unaccountable traits making the case of our differences apart to our natural unity. By the broad case of using the title of an evil connection to a single person, has helped reinforce our separate standing by allowing the application of that title be representative of a particular persons own traits and in conjunction of who they are nationally, politically or religiously? Said so that we have another means to be judgemental of our own standards, which are to be naturally broadened by using the simple feature of our one active prospect in 'One for the Whole'.

In statement that evil is not an entity leaves the expression weightless as it should be, especially when there is no evil force in operation any where throughout the universe. By any specie trait do we have evil fish or birds or insects, do we have evil lions or tigers or cats, dogs, horses, and do we have evil whales or butterflies? By reasonable study the answer is unwanted because in our own acceptances and through concept set by most religions, other creatures are to gods service in their own right expressed by evolution. We are the exception because on our own account, but in vision, we accelerated the scope of time by finding creation. In doing so, because we were not then able to equate the flood of mysteries we began to see we labelled all things and all other things to be unravelled by the fourth dimension or if not to be of Gods design, and to account for one of our own emotions, we acquainted evil by Gods own appointment? From set standards or in having them set, by the plan of geniality working that God only levelled his desires of us to the good, anything untoward was labelled outside of that province, hence, we created evil, and not God!

By direct association on human terms in reply, anything so was of a human failing and associated with the same disgrace that God may have shown or will show at a later time in settlement of our understanding! In our God religions, and by human forms of understanding, the standards being set were unobtainable then ago, because all of our actions were to have been driven along a direct path, which should have only produced pious and godly people, by not allowing for the spirit or our free will of inquiry! When some of us might offer opinion of kind to be different, if taken in effrontery by the simple range of not understanding what suggestions might have been made, then by various standards pre ordained, rebuttal might have come in the same source to consider the suggestions in the offensive and respond to acclaim evil as the source of introduction! Accounting in part for what could have given us the status of the entity to be against our own religious standards in there growing form, and from where anybody informed within the religion to suggest past the accepted was displaying forces from outside of the godly concept, was in the course of evil!

When drawn in parallel to the political situation the picture is not quite the same, keeping more in line with a true mirrored reflection, where if the real right hand moves the reflection is of the left? Politics in any of its true forms does not hold the concept because although it can rule instead of serving, although by civil tone it was requisite to our first advances by organisation. It holds no personal drive except through us which alas also lends to the assumption that anything we desire even if placed by our selves is often the only way and the right way. Becoming much more complicated when worked half measured to some situations early in the first half of the twentieth century.

With religions having only two sides, with God on one and us on the other, anything in a contrary fashion to God's intention or desire understood in portrayal by us, when balanced on our own terms to our instructions, any incompliance was blameful to our invented entity of evil? Effectively anything not to God's direction was evil and in being against, was to have been created by God for its own failing, to have made in picture a place of evil for evil doers. To have been set in mark for our memory to the open balance of compliance, being to the good, was in reward, say in heaven or in paradise or in the timeless Garden of Eden. Being bad was countered to be away from the good and to account for the measure that we could be bad in Gods light, had to be produced in Gods own cast off by an errant angle if only supposed by we as humans, but by God's own idea!

In true fashion and by our ability to reason half measured on our own account, but then ago based in the realms of ignorance that we worked from, we were bound to make the connection between good and evil almost in equal measure from God. By further thought in reasonable deliberation, our connection that God was responsible for everything prompted us in the line of thinking to assume in error of the balanced domains of heaven and hell? On the very edge of supposition and to rationalise our own know deeds the two entities fitted to our own ability of understanding. So much so that in back casting to the ignition of all history it was caused by that very damned agent of God in the devil, who was responsible separate to Gods own lead in making us to be cast from the realms of paradise and Eden!

Subsequently every evil deed done since supposition carried an outside force in the blessing of those deeds by whatever rule or excuse. From the religious perspective in which all of my comparisons have been of or too, suggestions have had to contain the devil God relationship on that basis, which has ever allowed us express our own free will. Further expressing the duel concept on political terms only was gained from our histories when at times we battled in type over national matters without the rule of God in equal measure to both sides! From where one could claim of the other or reversed that influences were in force to outcome but never in agreement. From such division the religious perspective was able to be managed because of the final claim of good to always be over evil and judgements could always be made later?

Even from early times and in context to civil matters the feature of evil was seen to operate against the rule of normal activity, least ways by conduct. If a person was to operate outside of the style befitting to small struggling communities, not carrying there share or taking more than, force in recognition had to be classified by some means of chastisement. If any pointer was given at very early times of our discovering parts or our natural histories to the division of our own communities, unfortunately we have little evidence of how some groups operated. In taking Shanidar and our cave paintings as measure, what might we find in collective recognition of how evil might have been a feature of our trend over those past ninety thousand years? Although a weighted question relating to part of our character, asking it has set parameters to enable us view different perspectives in the same light. If Shanidar gives one picture of our first intentions to religion only, then it could be from our cave painting that the spirit of evil entered our vocabulary, but only from a very imaginative base that free will permits?

To probe the matter of evil even at this time is done in recognition that we are here now, and from what we still do in the form of interaction one to one. I am leading the matter in such fashion to neutralise the entity of evil if that is what we still have, my studies do not even cut into our religious perspectives of what evil is by any tone. Not least because my lead is that the matter of concern to it had been shown pre recorded history and outside the first concepts of any religious perception!

When viewed separately, Shanidar, no sign of evil yet, cave paintings in spanning twenty thousand plus years of activity tells many pictures? From those paintings without the display of any particular trend even if taken to have been tombs by my own calculations, some picture depiction having been said before, are in display of human conflict. By any mixture I have now put to statement, that part of my own judgement to the purpose of the effort of drawing in dark places, was our first ledger by at least two counts, showing tombs, and for information? To indicate that evil was afoot by the same meaning is given to compliment the picture of the Fifth Dimension managed by these five to six years only. Without self criticism my pointer of recorded evil if, is there to alert us to the force of political miscalculation. Human conflict by any description is indeed a basis for evil deeds, my take on it at this time can only be worked through our one active prospect in 'One for the Whole' futuristically. That is where we are now moving in all forms of progression and because of that I have found it necessary to curtail any half measured approach in suggestion to our three taken prospects in becoming four or more at this time.

My standard and standing to the concept of evil is negative on all fronts, there is no concept in any form nor can there be claimed so to be. By direct comparison through our taken prospect 'Past Gods Words', evil although being displayed mostly in religious fable has no voice of recognition in any true direction. It is a non expression used wholly by political definition, yet no person can now claim of its appointment by political terms either to have been chosen or to lead that will cause human suffering of any measure unless warranted, but only by moral determination? In use it implies to the matter of us being uncivilised, uncivilised, and uncivilised and in the wrong charge of our own destiny in the name of creation only. Unfortunately used as an excuse since nineteen forty five it has clouded our judgement to many matters we had assumed to be in the right direction with our conduct, but now needs the revision I am offering for better self understanding, not least because we are all here now! Evil will always raise its head, that is what we have made it do, leastways now we will have no need to duck when shown, because we cannot see it for what it is supposed to be, we now have the Fifth Dimension, which can only smother any evil intention!

## CHAPTER 95

With a non concept like evil questioning the very root of civilisation, we have yet to seek better answers to baying questions that have been scattered upon the floor of all political institutions since the ending of World War Two. In as much as we have been trying to progress our standards, they have been naturally set by half measured action, which in itself propagates by use. In core one of the best examples existing without pointed condemnation is our own United Nations. By the supposition that it is in unbiased service to us all through our national tendencies, yet has set standards from its own classification from a or a group of first world countries, require further study and enquiry?

All questions of query put to that condition from the same political standing and directly in this time of now, is to be done for enlightenment only, because without deliberate intent. Our political judgements have moved to take full control of all guided movement without comprehension, which moves to the fact that we have to operate in accordance with our steps taken syndrome, because we set the picture half measured. Also in containment of our religious rights by planting the order of religious tolerance the field had been further clouded. Whereby no questions were to be asked from any quarter as the matter had been reviewed politically. Meaning if religions were to raise opinion because of other misdeeds, they by means of classification could be set to work blessed in the conduct of evil!

Without any formal or considered thoughts to the matter of geography, climate, diet or other features in making some of our differences, from the one first world political property, we have carried on the imagination of evil being the master to some of our differences? By which we can override the possible inclusion of religious excellence in classing some of us who might operate in the highest standards of our own belief by tolerant means, yet are confronted in all avenues?

By no simple means is this picture to change because some steps taken are difficult to retrace unless captured in stone, some are made in haste without looking to the outcome of delivery or of what reaction might be by whatever means. Creating in cycle supportive half measured solutions,, which once again are delivered as steps taken including the pronounced accolade of evil to others to account for justifiable reaction or corrective action one way only being political?

Again in lining up the chase, I have used fair measure of my own abstract reasoning not to create mixture, but leads to the solving of some of our freely taken political attitudes, which fortunately are now showing there own cracks. From where we stand in any matter, politics is our main ruling body both nationally and internationally. That is the nature of things in this our world, that is the state of affairs at this very second of reading these pages, even from the day of publication or in one week or one year further on. An instant of time is an instant of time, by our own good charge we can now recognise that feature but through our religious perceptions only. All of which is to be more noticeable by the way we round the outcome of the human rational, which is to be better expressed by our religions than ever our politics. And from now is to be better led by our religions than our politics, but in the closest of allies, not just to check and recheck each other in progress.

From all of most histories we have moved ever forward, which again is a very great example owing to us all. My own simple exercise is to have placed a mirror in front of our progress so that we might check the journey just travelled and the prospect of arrival in the course of our individual lifecycles to be fifty plus or seventy plus years long? With the method of travel as progress then the fuel is to be the Fifth Dimension shared on the two levels of our conduct. Although I have given vast differences between politics and religion my clear aim has been to the inclusion of all of us in the future, of all of us in the future, by political leads and religious determinations. Even that many of us in extension of our own free will might not wish for any participation, the exercise is vast in concept because of religious expectation and automatically includes everybody. As have all deeds since our ancestors first stood up and walked leaving memory encased in stone!

Personal choice is not forfeit, because of how and why this whole universe had developed, without direction my total inclusion of everybody equal is not form a religious standpoint, nor so from any political affiliation. My standard is from the nature of our united histories from where and how even without the why, but from where all of our past deeds by the smallest, mightiest, blackest, brownest, yellowiest or whitest have continually had bearing to each other. In having spread throughout the world by tribal or clan affiliations, in having settled in every corner of the globe to be of any mixture of the above by colour or creed, was done in the same specie link that had carried this small planet to be the voice upon the whole universe. Making sound reverberate to the very heavens in all of our names which also is not a religious statement but one of humanity, therefore increasing my tendency to hold in self respect by our specie link without compromising our thriving free will!?

# CHAPTER 96

Holding in line to our young is of its own importance, but to be used in true caution by not creating any means to blame them for our own actions and at the same time to recognise in them our creation! From initiation it is we to represent ourselves by how our young should react in the education that we are set to bring them, not by our own examples which have been wanting since nineteen forty five and again in the mid nineteen sixties and again in the turn of this recent millennium. By all counts from as far back as we might wish to go, we had been unable to express reactive thought save by our own emotional feeling to deeds already done and from what purpose or drive or intention was, until these last five to six years.

From whatever motive we had set future plans, they can only be done in reflection of the Fifth Dimension via its three taken prospects and one active. Dare I make such a claim is done from our joint pasts; it is even done to account for the unaccountable drive we as this single specie hold. Now it is done to set the matter of choice without compromise to our free will, without compromising the past and most certain not to compromise the future. That is why we can overlay the two most influential features to be upon our lives in our politics and religions with the same feature and get different answers?

In order for joint understanding of my intentions and ours, this is not a test or competition we are not given games to play. Although our sensibilities will be stirred it is not a trick or lottery in the testing of any individual by and determination, but in our overlay feature it covers everybody uncounted. Even those amongst us who have either a very strong political affiliation in all fields through communism or democracy or to regal dictatorial systems, we are included past personal choice. Not so in the manner of force, but because representatively we were all together at the beginning in the times of un-gathered control which were filtered by natural instinct in taking us to the order of group civil control for the basics of any form of success. Therefore because of our collective beginning no person in individual terms can turn away from our personal responsibilities to self and each other. It is to be true we are still each entitled to follow our own path by whatever style, but not in terms of the supposition of personal human rights. They remain a topic that needs our entire input if by no other means than with use of our one active prospect in 'One for the Whole', that alone will show the vital necessity of group involvement, because in cases, that will be our only guide to the matter of change!

Civilisation, human rights, politically correct behaviour is more linked to our political endeavours than other leads to our study of purpose through religion? My view of our past conduct in relation to how we may have behaved since nineteen forty five, can be taken in true reflection of what I have been saying about our political attitude to each other since then? In holding to the necessity of human conflict by war is the measure of how much improvement we still need, but in reverse, to look how we came through the tunnel of darkness by preparing for total destruction from our cold war attitude, lay hope upon us all. In that we still operate by our desired political standards and have primarily looked to them to fill some gaps to the open recognition to our singularity. In turning most political systems to now operate past their prime claim and in use of their joint subsidiaries like commerce, trade, business dealing and market forces and profit and profit. A different complexion has been put on our thinking because of that drive?

Unfortunately from some small measures of success we have allowed expediency rule our head based on the assumption that we were right in the first place, by holding first order to human rights under political direction and religious tolerance to the same standard. But in or over our spiritual needs or intentions, we assumed we had done enough to maintain balance. Unfortunately it had never become clear that there was a difference between our spiritual and or secular needs. Not even highlighted by our political bloc culture standoff, where one side was to be democratic in its own field with religious outlooks to be Christian and or Jewish. Against a political anti religious motherland fixation of communism in support to peoples who readily follow the line of Judaism and Christianity continued into the last religion of Islam! Leastways in an area of geography which always had a huge influence on world thinking and action over the past ten thousand years?

In pointing by association that the land between two rivers bore our first signs of what has been given in accepted behaviour for us to be civilised first, then it is through that avenue we can study reaction. Not in covering the full ten thousand years day by day, just the time since nineteen forty five will help, but although in cross referral by mentioning some religions and the spiritual, our cover is to be mainly political. In that context, civilisation only applies to that aspect of our concern, it is a matter to have formed by making our interaction more formal into our political sophistication, stemming from the early controls of civil order and earlier simplified forms of organisation! Of all the contributing features in our aim to be fully civilised, most are done in assumption that one naturally follows the other and the next, which is reasonable until we notice that the core of direction or intention is carried half measured in cases. On the political scale, and only by this example, where a tenet is pronounced beyond proposal but in assumption of full adherence, how might it be received if offered to be of religious tolerance to true communists not even thinking in that field? Unless to account for the discrepancy of untendered half measured working to conclude by our steps taken syndrome and in balance we generally include human rights to all, which automatically includes religious tolerance wanted or not!

Unfortunately stemming from our worst directed carnage in progress of the Second World War, it was almost from shock we hastily gathered our thoughts to rebuild the future on moral grounds. Some specifics of the carnage were so shocking to be left unattended in review but covered by imagination to forestall the same kind of operation this time from amongst the victors if similar was intended. Through our political manoeuvring and noticed through the way, reason and result of repatriation of the displaced people of Europe and in far too many cases their murder as a levy to victory. The displayed openness of our conduct showed no signs of collective civilisation in any shape way or form save by assumption and from different quarters, but to conclude that might is always right!

As our total management at that time was wholly political, any continuing failings were of that ilk and so can be blamed to have caused our un-civilisation again, in respect that instead of adding to our sophistication, we were peeling it quicker than we could an orange? By my own use of hindsight the given picture has not been cast in stone but I use my take of the situation to highlight the factor of our deemed assumption to be civilised and therefore from wherever that theory came from, politically that was the case for all by assumption?

For the same reason that I have made our differences in cause of geography, climate, diet and other features, no recognisable view of this understanding had been seen of this by any indication at the time in question. Read from the fact that in Europe alone, our smallest continent, we held the direct opinion that we were vastly different peoples yet generally accepted that we were of the same racial colour. Leading to the open and obvious results of the mountain of confusion we faced then and continued to face even today? Unfortunately one of the biggest leads for us to continue in this vein was tempered by European thinking at the time in the way that the different nationals involved, tried to justify the political standing that had just been and was now to be reconstructed? Fascism and democracy in loose association with there own Christian religious culture which had developed, set the first levels of later confusion? Complicated multi fold, when the mixture was to include the new form of communistic political control into the broad European equation.

The political structure of the world having been set in the year nineteen-nineteen at the close of the Great War, as it was known then or World War One it had become since. How ever it is thought that the League of Nations was to have operated, all was done in the cloud or progress and progression? By Indication earlier I had given notice of America's first involvement to world affairs proper on the political level by her inclusion in the Great War. From which during, her fourteen point proposal was put forward both for the cessation of hostilities and for points of reference into the future. Although done to the best of intentions for all concerned it/they could only be taken in the form of unrequested proposals in view of the fact that it was by the European victors that all deals were subsequently done.

It was Great Britain and France, who were the arbiters of peace at the great wars end; both after there own blood letting had become blind to the prospect of permanent settlement or so it might seem in the coat of hindsight. Looking back with today's knowledge, learning and understanding, it is hard to realise of our total difference in attitude than most of us have today? Then with an eye to be in control of matters of concern on the European continent, all bartering done was to maintain there own place of influence there, from which both could run there respective empires of previous colonial success? France and England at the time in open willingness to borrow financial support from America to rebuild from the ravages of battle, set the future from the structure of the past?

Although for a time they had listened to America through aid, many of our unread chapters of history in the making were done at the ending of the nineteen fourteen eighteen war. In direct preparation for the ending of the nineteen thirty nine forty five war, generally with the same victors of democracy, but now complicated by the addition of communistic Russia. Who were also for a time allies to England and France in the fourteen eighteen war and of the same Christian background then, but by the end of the war in nineteen forty five had reverted to her non Christian Marxist ideology. To the matter of personal choice in the avenue of theory, Russians could adopt any manner of government they so wished but much of the difference to outcome in this world conflict was early realised by political means. Whereby if America who in open honesty was to have been the major player by prevailing world standards politically expressed by its tentacles of economics, trade and other commercial interests. From new prospective she was certainly now going to be of voice in world matters of peace, not least in being goaded by an unchristian communistic Russian and later Chinese Marxist states.

My own take and understanding of the above events are done in the ignorance of knowing the results only, without being part of the real experiment, without being involved in the setting of criteria in making those expected results. I have used the expression we are all here now, in the concept of expediency to make relevant points? This I have done in fair and open pursuit to draw attention to events so that we do not have to troll the day by day record of any of our histories. Not in a malicious manner, but in taking the initiative my self, has been done to dispel some or all of our arguments, which will be tendered in the term of "what about" and continuing in lists of comparative narratives for our genuine study.

One example I will quote and I believe to be fair, reaches what I had said just a few lines before about America being a major player by prevailing world standards. With Russia having lost almost twenty five million claimed war dead out of a supposed total of some fifty five million world wide deaths, expectation would seem to lean to a large Russian input to any world standards to be agreed. Unfortunately her own pursuits without any defined religious background determined all public standards to be unremittingly centred to the mother country and not self. Which although not seen at the time and since is a total regressive step to the realm of pure evolution in taking the matter of human life to the single importance of procreation!

My point in stating the adage, not seen then and since is enough to dispel similar requests of "what about" scenarios. I make that statement un-weighed because although we all have a serious voice and in many cases quality reassurances to deliver of our selves, I have directed most of my propositions to the future and only in reflection of the past? By the same token in early chapters I had indicated that on or by some political determinations we should all aspire to the American example? Although drawn from a brutal life history in similar fashion to other countries, over their thousands of years of continual inhabitancy, the nature of new integration in America brought us up to date! Her place in history is representative to us all in modern terms and in direct connection by today, but not always read at the same speed. The best example of her very existence is how she almost compromised one of my taken prospects in 'Out of Africa' fitted to these very recent times which gives good focus to her self importance in all of our concerns by, to and from us all. For any doubts that I have raised in memory, we are to look at the situation in terms of our politics and religions separately, but always in close proximity, for they are both on the one road!

Four hundred years ago in making our renewed connection to the new world and therefore making the world global in its first modern context, was a feature to bring us to our most serious of consequences. Although we were now truly on a round planet instead of the flat earth of some thinking even if of an advanced state, much in happening from those events were to set us on the final path to judgements upon this last era. By review, unbalanced and uncharged although of considerable complexity, I have taken the most direct and what I perceive to be the most likely route for some of our interactive processes to have manifest themselves. By delivery, my opinion extends beyond the probable, because on our behalf I have used a measure of our own cunning derived from the reaches of faith and belief in an object or an ideal rather than the proof of formula?

In full acknowledgement to the matter of our differences by the feature of geography, climate, diet and other reasons, throughout the world emanating from Europe on its own account, but to the measure of progression. I had cited the event of the discovery of the new world from the European perspective and although driven by the political intentions of expansion and gain, there was also a huge religious overlay to all moves. By the nature of the discoveries there the picture was like nothing else that had been experienced anywhere in the known world then. Although in places and by leads in the field of political conduct many existing societies there had more advanced methods of supposed civilised behaviour in the field of civil organisation, alas the noticeable lacking of religious standards was poured upon in excuse by the fated European intrusion.

From all the mixing that had been done between Europe, the Middle East and Asia in the past, proper reaction by most means seemed to remain at the general status quo. Yet here in the new world to the Europeans only at a time, the difference was unimagined, so too was reaction. Here by the two camps under study with one group defined by different tribes holding to the natural habit of survival overseen by the Great Spirit and the other in well organised city nation states tied to sun worship and the habit of human sacrifice in placation thereof. Because there was no other standard from which to measure such situations even judged against the worst kind of reactions in the past with the different peoples of Asia and the Middle East even religiously. Here all fields were considered neutral and any desire of conduct was pursued with the desire of political objectives and the desire of religious blessings, but from European thinking only?

From that edge added to by all the means of time over the next two hundred and fifty years even with the Europeans fighting amongst themselves on new American continents, by that time when nationhood was taken by the former settlers of European colonisation to become American at least on part of the northern continent. It was a progressive step to have been taken by us all representatively, which is no form of excuse to the good, because at that time the American conduct to self was no shining example. But I add for a point of reference, Americanism was to become futurism to our good, which should only be seen by the first order of political motivation. My other term of reference to her on religious matters is of her continued acceptance of the former European Christian religions maintained as Christian but in cases by her own reformation and still maintained in that standard?

Without the ability to walk past the dark shadows of her young history by the same representation of or for us all, she is not a nation of only two hundred years, but our own long centuries tied by our religious perceptions. That is why without the warrant of direct praise I can say on my own behalf and in our representation her station is maintained in the positive and progressive motion that we have all been locked into for centuries. If laid in parallel to the political order of any irreligious nation, then America maintains our united drive by her own religious connection along with all of our own. From now though, she is obliged to suffer upon herself in parity to us all the matter of the Fifth Dimension, for self clarification by equal compass?

Although I lead that America is to be representative of us all politically and inclusive of our Fifth Dimension, it is for us individually to take in understanding the measure of how we must separate our religious intentions from our political ones? But at all times now in review of these years since nineteen forty five for the most part, but also that there is no charge that we might forget what is precious to us from our own deep pasts. The picture to be drawn in that representation on the political level only is to the nature of our own drive for human self sovereignty. By the same parallel in how politics has led, we of very recent times still search for the true intention of all purpose expressed through us as you and I. In using the example of communism as a means to express our individuality, the open failure of that method of politics has to be seen in cause by the rejection of religious values wholeheartedly. Without the desire to give undue credit to Americanism linked with Europeanism our emphasis is to show the modernity of our political effort when looked to how China had adopted the Marxist form of communism since nineteen forty eight and is set to reject many of its principles some sixty short years later.

Again with pointed observation it is not that we are challenging the style of politics that any country might select, what our notice is to be of is how that adoption is translated for others by example, because of this modern era we are in, is where we now interact. Whereas communism by general acceptance holds no religious properties then by our very modern terms not least shown by these past sixty years, our very competitiveness barters the two mentioned political ideologies in open competition and in active reactive formation, but ever in flux without clear direction either way. I here give notice only of the complicated style and manner of such interaction not least because the balance is upon us all while we look to our political standing under the Fifth Dimension. Nevertheless while on this drive and in our time since nineteen forty five, the biggest, although not ceded to be the brightest form of our political reasoning since then has to be given to the factor of democracy even expressed as Americanism. Because even unguarded, it complies in acceptance to religious tolerance assuming of its containment, which is no bad thing because our religions are at least included in the prominence or our future and from communism they are not?

I have to remind myself that my direction now is of our political determinations first hand, but not in any competitive vein. Since the time of the Greeks some two and a half thousand years ago we had experimented in democracy and communism with other forms of political/civil controls. By our modern comparisons the difference is highlighted by two main reasons than in those former times. Politics by style now covers continents and hundreds of millions of people and has been generated by conduct in need over hundreds of years of social interaction through specific personal traits. Also, since early Greek culture, although they were informed of the Jewish religion in lands afar at that time, there had been no additional religions to the area of the world we now give in name as the Middle East. Christianity nor Islam had no bearing in how the Greeks thought or might have thought of the two new religions in relation to there own religious plan of human instigation by the will of there own king god Zeus, which fitted there thinking bound by multi god references? Nevertheless there working in terms of pure politics gave a finish in style to how we have been able to value our own political standing separate to our religious needs and requirements. If from the lands of the two rivers some of our first modern interactive fables began to form, then it was by the Greeks that we were set in part to be able and maintain a religious outlook and political experiments. In other words it was from them that the two important styles of human management was modernised, to be separate in politics first and gods second, which can be seen only from futuristic hindsight!?

## CHAPTER 97

One of our biggest problems and one that will always remain amongst us, is that very few of us understand some of our own basic character traits, which fortunately happen to draw us further from evolution and keep us into the reaches of creation? From the first time when I indicated the means of some of our differences to have been caused by geography, climate, diet and other reasons, although met in acceptance by many of us long before my own method of explanation through our taken prospect 'Out of Africa'. By me indicating that we should follow America's example to be a guiding light for us all is done in line that we should all attempt to understand. Not, not for Americas sake, but ours, when the picture to be drawn is not, not of Americas success, but ours? Without condemnation, but in transfer of our general thinking exposed by these recent years as late as the first time when Islam came to the fore in real world terms by making us look into our divine side again, but now in future prospect first religiously, but by looking back!

Naivety expressed in most terms stands to the lack of experience, which is generally gained over expanses of time, yet I have asked that we might follow America's experiences to be ours in looking to the future. From the terms that I have used by collective consideration with words like ours, brings the American population in parity with us all as has always been my intention any and all ways. I had previously stated that her mixture of peoples as the one fifty state nation comprises those from almost every other country of the world, but now as Americans. My reiteration now, is to expressively show how the aspects of geography, climate, diet and other features work on a time basis in terms of evolution but not necessarily nationally. In as much as peoples of the world had poured into the funnel of America bringing all the mixtures of humanity to the one core, the outfall at the funnel spout belies intent, because results are sometimes misread.

Insofar that in cases through some of her endeavours on our behalf, have led to us adopting a religious standard of tolerance only, poses the questions we would like to have answered. Never loaded in deceit, but in assumption that we can manage our religious codes separate to those of commerce and trade and civil law and the social ethic, has all been shrouded in the political expediency of cover by introducing extra and more and more concepts to deal with situations that are not really there? Human rights by description are perhaps one of her/our smoke screens, but carried only to the political perspective, for it has always been attended to through most of our religions. Other features are only products of doubt, like using the feature of being politically correct, the picture given is on what levels are we supposed to operate its hidden intent? Are we to assume that everything thought of or learnt, bear the same meaning of acceptance over and above the features of change that might have been directed upon us by climate, geography, diet and other factors?

I do count myself as I see all others to have an American connection by our joint and directed forms of natural progression in all cases. At the same time I put the breaks on any further progressive suggestion or steps to be taken or imagined simply to advance the standing state of political conduct we have now reached. When seen in reality it is not the head of politics that rules in any fashion only the tentacles of trade, commerce, gain, profit industry or law and other self motivated idioms, ex to our own human excellence. In support of my view we can look to the connection of our religions in relation to politic? Both have been open to monumental changes at different times to our wellbeing and have had different receptions from essentially the same people. But in doing this exercise even by all the criteria of formula we have to always consider the climate?

Even that some of us might wish to avoid any connection that I would lay upon us all by our association to America, the standard is unarguable if taken with the overlay of my Fifth Dimension three taken and one active prospect in 'One for the Whole'. Although a product of these past five to six years only, it is timeless in open concept, but has varied readings when used under the different attitudes that we can readily adopt. By comparison to other life forms, but in the process of living, as the one specie to have escaped the restraints of evolution, even by show at Shanidar, when more conclusive evidence of that break was given. We have been connected not by our political drive shown or read as our means of civil order. But out stripping all other creatures not least by our studies of them from our shown work in our cave paintings and rock drawing over a period of some thirty thousand years collectively, when we depicted them and none showed awareness of us!

Even allowing for my own judgement that some caves were to have been tombs, another closer look at what was painted is designed to show the animals depicted were indeed subservient to us or to a form that we deemed to recognise past our ability to record the issue in any other way than we actually did. From virtually all examples, the beasts shown were recorded in there own majesty, while we depicted ourselves in or of a lesser standard by the abstract observation that although they were mighty and strong, it was we who were masters!? Making the double point that the effort required to do the work far outstripped the meaning to our tribe, clan or group, but alluded to a driven direction over civil control. Then later by all channels in the true abstract form, were to lead us all at the mouth of the funnel, to Americanism politically!?

## CHAPTER 98

All religions are only fifty plus or seventy plus years long, but cater well for longevity by questioning all the reaches of the scientific universe, leastways that is part of my study in this our work. Acclimations are given in the same context that we are all involved in our political outfall and behave accordingly, but have not yet learnt to behave religiously in the same manner, because religions have no status politically? Or that is how some matters of religious construction have settled over these past recent years in only reacting religiously when the real control is to be through our politics under religious instruction?

All religions are only fifty plus or seventy plus years long, when reviewed in study of our three Middle East religions of four, two and one thousand four hundred years by there own time spans. No matter how we might try and manipulate time by any source, the one binding feature of my first three dimensional religions so classed by my personal standard of presentation. Comes from there own tenets and covenants directed for instigation from God himself or through his chosen prophets or messengers? From each and in constant recognition is the one and linked source of how we are to conduct our own affairs within the limit of time owed to us by whatever source or from whatever circumstances. Few arguments can be given to our conduct because in all cases from our three Middle East religions the same standard had always applied. It has been without direction how we have faltered by our supposed parallel, constructed in the order of civil controls when we first found that we had both features to contend with?

From the same scale and movement of time, it has been our own uncertainty that has cast self doubt to be amongst us all and in all the same places of particular concern to our religious wellbeing, that of our own single lifecycles? So much so that in open confusion we have on to many occasions laid religious plans to have political solutions. Leading to our present situation of using the same religious tolerance by political directive to accommodate our religious differences which are not there, but only assumed to be so? Alas it is from another religious feature that small confusion has worked to blanket some of our best intention religiously? Although covered in the realms of political tolerance our Eastern religious attitudes have set the possibility of some query generated by the source of political thinking by stunted reference to all features outside of its own controlling domain!

When stating in number of my first three defined dimensional religions, I had always added the note of other and others, meaning in obvious display to include virtually all religions? I had made that connection so that all are included in the same force when we meet the proposed objective of showing our preparedness in our own presentation to God's desire expressed through our own free will. From my own staid and determined drive, by consent, I have frequently spread the difference to the monotheistic religious style of being by our direct association to God in direction or supposition, being from three different time bands? Also by their classification in being Western, monotheistic, with Eastern religions being perpetual, the divide only allows us use our western models in the closure of most religious experiments that we might wish to use? In holding that our representative Middle East religions have only two structures makes the study easier when realised that the split is of our own lifecycle guided by our own free will in self judgement. Followed in our death and direct association to God by all previous and fully open events to our own choices that had been!

Eastern religions formed virtually at the same time as all others by the study of self, have simply generated a different core which can be claimed was driven by the same understanding of creation, but by a different reading! Changing acceptances, because our attitudes had formed differently from the means of our circumstances set by our personal conditions of geography, climate, diet and other reasons? For my own acceptance, the notice is to where most Eastern religions in the form of Hinduism and Buddhism had the connection to human conduct also? But taking the view that the difference between monotheistic religions and now, was not to associate man alone to Gods image, but also accept that all other creatures were to Gods own purpose! Therefore it was for us to include all life forms in the process, whatever that might be, but now turned by more than humanly generated intentions? Making for a broader inclusion of life forms, interacting to a formed plan from the same study into the future? By forming the wheel of thought, whereby without a starting point or an ending, several forms of life in generation by could regenerate and regenerate to the aim of full understanding? Which is as good or fine as any religion can get save that by feature the end is not always known to have been justified by the means. In so that by some processes, the recipient in line to the benefit of all connected by the one wheel, might do as much in the balance of evil deeds than others to good. Because alas in formation by its finest structure there was no log or record of who was to be connected with who or what other creature in all making a kind of progress for betterment, but, if, uncounted!?

From a not unreasonable connection, Eastern religions sprang from the same well of time that our earlier first talked about Western ones recorded notes, unbalanced, both types were of our human connection even that in time they seemed different. In aiming for the matter of final understanding by the monotheistic route being Western, has allowed us build up a separate end story line. Of Eastern and Western philosophies our measure of possible enlightenment was to be readily unseen by the opted for standards, so that in the intention of our quests to fulfilment by understanding we would show our own disposition to what Gods own intentions might be. By self preparation from both, the note was ultimately to be the question, if there is any body there, are we now ready to receive your wisdom?

Without the need of microscopic dissection my loose directory with regard to all of our religions has received classification so that we might be able to put the correct amount of our own judgement to them. No matter what other choices made, the Fifth Dimension can speak for our intentions? I have offered objective doubt to some kinds of our study for the deliberate reason to expose my own ignorance of some matters, mostly religiously, while at the same time knowing some of the answers in there entirety. By that example the three taken prospects and one active in 'One for the Whole', are to be enough to demonstrate that point, if directly laid upon the form of conduct to any type of religious sufferance, religions will self contain them, so that none are able to contaminate our own learnt wisdom?

Across all religions, 'Out of Africa' speaks of our human singularity, our human singularity. Across all religions 'Past Gods Words' is indicative to our own judgements and ability to self pronounce Gods existence, therefore in keeping with reverent dignity our actions are to be measured by our own deeds. With acceptance of the expression 'Past Gods Words', we openly show the fitness of our standing to exercise our own free will, from which without, Shanidar might never have been measured? By the oddest of events, from any angle our or for definition, my three Middle East religions are in virtual proof of our need to operate 'Past Gods Words', not least by the measured events of there own structure. Four thousand years ago without the ability to record concept, by our own excellence of mind but error of thought we envisaged what the meaning of life might be, exposed by the prospect of our own death! Unknown and without connection we had done that at Shanidar some considerable time earlier. Our difference of now was drawn from our own sophistication in the belief and understanding that because we carried the written word, varied speech, civil to be political means of control. Because some of us could make judgements claiming to be civilised apart to others of a similar disposition, we were disposed to know all things and all mysteries of which we had allowed be managed by our collection of gods?

Where excellence of mind but error of thought might have been a prime mover to any supposed situation of any time past, even as long as four thousand years ago. In consideration of how we had several Gods to manage our different mysteries, in consideration to how by the study of self some of these mysteries were to by allocated to the unseen fourth dimension as mysteries? By the rolling process of natural human progression, not specie selection, but natural human progression, from which by instinctive traits un-evolutionary, particular members of clan or tribe handed on their own skills in the family line but to benefit all. Just as so from some societies the God minders in the form of priests might have influenced choice beyond all the reaches of deceit, but incompliance to the good of their community first, because that is how things were done?

From where a carpenter or a tailor or a fisherman or a tax collector might pass there own tricks of the trade to family or an apprentice, simply because developing society then worked on a need to know basis, being a mark of civilised behaviour? Not everybody had to be proficient at all tasks, interdependence then has become the same feature today which is a measure of civilisation unless having become tarnished. By the same study the multi god tenders as priesthoods, had there own reference points of self study or service. They acted in very much the same way as we all, but from being involved with the keepers of the mysteries, there own expositions seemed to be heralded as a form of deliverance, not now by many gods to be mystery keepers, but from some of us, one God only, over all things!?

If not blinded by glory in joy, then features to have clouded our interpretation of events some four thousand years ago, were taken by the full excellence of mind but error of thought from our priesthood only, included, or not? It is to be without doubt that Moses in bringing upon us all Gods own desire through the Ten Commandments, in most ways cemented our faith, but at the same time, created for us our biggest doubters from the politically motivated fourth dimension, our sciences! By at first tackling the smaller of our mysteries, science had cut into the aspect of blind faith and the belief in hope, the rest is being formed through history but not yet fully told. When Moses first gave us our covenant it was to be from priestly scribes who were to form the record at first giving us the concept to heed Gods own words, and at the same time, promoting what was to become one of our three taken prospects in 'Past Gods Words'!

Excellence of mind is to be a worthy accolade to that time of our history; yet we can loose my adage "error of thought", but only through that historical medium of referral to a specific event. By supposition and collective reasoning and now actual written work in the form of the Dead Sea scrolls we can post our notice of the importance of our own free will to be carried 'Past Gods Words' without any form of condemnation. This study in a relaxed manner is to be very important in relation to the following religions of Christianity and Islam from the same area of the world where Judaism sprang, and Moses walked?

Much thought is to be given to the writing of Mohammed, mainly expressed through the Quran but because of the particular time reference it was set in? From the same inclination I have intoned to reduce the standard of Christian writings in direct relationship to the time span of the life of Jesus and what he actually said there and then? By processes of similar texture taking that the Jewish work of the Old Testament of the bible or in their own terms the Bible in being our formal history book, has been supported by the Jewish calendar to be of stories of some five thousand seven hundred years old. Giving for us a total human lifespan in relation to Gods own input and desire to that length in time, so much so in belief, that when Charles Darwin first objectively questioned it's referred to dating, it was by the first flush of Christian opposition to a scientific prognosis that had been put. Being a matter of small concern, but allied to the fact that Darwin seemed to oppose Gods plan by our judgements, by how they had been interpreted and from when?

Although the bible in setting the Jewish and subsequently all of our connected histories by what was supposed to be humanities first year reference calendar, the records of its formation from original dead sea scrolls were almost fifteen hundred years after the time of Moses. In context they were 'Past Gods Words', which from outside of the critic is to be noted by the double reference of why the Jews as we, would need to record history in such a told fashion? When they as we had lived the days since Abraham by all the direct connections either before Moses and certainly after him, 'Past Gods Words'? All of which is to be seen in representation of how we had reacted since the times when the Jews might have acted in error of thought but by excellence of mind. To be repeated from that same stem by us as Christians and Muslims who also acted by excellence of mind but in error of thought, evoking Gods words to us by assumption that they were to be for all and all others!?

'Natural Body Death' by what ever meaning or under any description, but only to the human spirit, is a huge challenge to all religions? Particularly from where I have put the maxim to stand shoulder to shoulder with our acclimation of not to kill of specie in "thou shall not kill". The total tragedy of which, is not from death, but how the question was asked? By the same mixture of fair and balanced reasoning, the vast majority of our religions have only aided personal dilemma to that obvious event. My challenge through these pages has been for our religions as separate as they are; even so, it is for them to allay all of any concerns to this matter. It is from all of their records charged in representation of us all, but to be treated individually in tackling what could be our misdirected sentiments.

Religions in all cases have deep and sensitive histories, yet by the oddest of circumstances the vast majority have continually worked to the good and benefit of all but with kinder regard to their own faithful, which has never been a bad thing, unless stated that we are more different than we are the same? In all events our levelling feature by all religions far more so than by politics or any other influence to be upon us, like free will or personality, is balanced to the unavoidable necessity of 'Natural Body Death'? By concept in our own human terms the taken prospect is set without the requirement of factual explanation or study in examination. There are no real negative forms to the implication, when noted it is to be extended to each of us within the period of our specific lifecycle to be fifty plus or seventy plus years long. From now on, force of this realisation is to be put upon us all by our very religions and in case of some four hundred years ago when one group of us might have delved in and with true respect of human sacrifice the realisation in error can be made by our statement of 'Natural Body Death'.

Our concepts then, although readily made and recognised as performing acts pertaining to the essence of our belief in placation of the/our gods? We cannot promote the acts to be repeated by other means of live sacrifice other than human, because in relation to the Fifth Dimension no consent could be found through our active prospect of 'One for the Whole'. Although from historical connections we have bound some of our religious practices with the inclusion of animal – human forms of unity, these are all neutralised by our agreed concept of creation. Taking it so to be our own time reference in gods time span in having waited for its understanding to be drawn from the reaches of timeless evolution in full example by consideration to the time span of crocodiles? Creature although changing by that very process within there scale of some two hundred and fifty million years, had not changed enough to prove the reality of evolution against our span of three and a half million years proving creation?

For that length of time crocodiles by there separate continental marking and style and shape had indeed evolved to be slightly different by the needs imposed through there own geography, climate, diet and other reasons. Yet from any time within there total span of two hundred and fifty million years, were we to overlay upon it our span of only three and a half million years, no time could be seen to have moved so fast in the order of change without alteration. Our three and a half million years became so magnified out of all proportion from about the time of Shanidar when by my own indication we first got the concept of religion, which was not to fit within the realm of evolution and procreation? But through a stir to be upon the universe into the study of death and therefore creation, it was we alone as this single specie that carried the change to the meaning of time from where a single lifecycle was to be the key in opening that door of expected understanding?

Taking the above general idea forward as it can only be moved, any magnification of any item belonging to the past must be carried in the realm of its own creation. It also has to be carried in the spirit of its own form until shown that the theory if portrayed does not work? In this case our religions have been the theory and have always managed to self police there own magnificence even to be self destructive, that is why we can now and only now overlay them all with the Fifth Dimension and its taken and active prospects. By these very recent five to six years we can now lay in harmony the basis of much of our dividing features blamed to be from our religious attitudes, but having that blame amplified by political perspectives or the consideration of. Fortunately from the same mix it will be seen that it has not been either our politics or religions that had seriously failed but from the mixture of our own attitudes. Applied in context from the assumption that we are at first civilised or at least some of us in our comparison that the orders of control we apply to that supposition are fully maintained by other declarations, that some of us are not!

Human rights and religious tolerance do not carry our banners of desired conformity, that is why things in the world today are as they are, especially when adding conduct in fitting to be politically correct. All of which evoke our differences and not our similarity or our singularity by the growing mistake from within the above framework of adopted studies. We continually add the measure of our differences by pretending that we are the same through political valuations only? Where through conversation by that aura of righteousness we might suppose that the best way to aid peoples of the second or third world are only from the prospective of those in the first world? Continuing in line our half measured banishment and leading to our steps taken syndrome, which induces more of the same. Even under correction by inclusion of assumption that religious tolerance offered a balancing guide, falls short when it is already assumed by religions that they are tolerant because most do not have a structured form like in politics. To be a first second or third line Jew, Christian or Muslim cannot be stated when all are covered by religious tolerance to be of the same value within each religion. Even that in some eastern religions, if including the lifestyle/cycle of other creatures for the eventual obtainment included in Gods benevolent view, we have to stretch time past the understanding of its need relevant to a single human lifecycle?

Without the separation of the two banks of religious philosophy being Eastern and or Western, with both kinds to the adopted settlement of ideals to include time scales past our own ability of imagination, past even all the portholes of scientific discovery yet known. Our religions are to be our guide to the eventuality of study whereby our final direction might be better agreed upon. And in doing that chore if using the prospects of the Fifth Dimension we can all be better equipped to see our similarity from the one beginning by what ever time, that is time well spent, because nothing is changed!

Without charge I had set our first record from when a biped left footsteps in soft volcanic ash some three and a half million years ago, and in doing so, left the first sign of a life form that could have been eventually human. What I had not made clear in my initial remark and since, was that the footprints if, were in tandem of size to indicate that two life forms had moved across the ash field. By fair comparison on or to any specific specie type, judgements could have been made due to the style of the petrified footprints, that one was from an adult and the other was from a child, how fine a thought is that?

By religious terms firstly, from the position of the footprints the picture could well be imagined that the couple who had cast our die, were close enough to be holding hands in the comfort of living. For any of us holding that picture is not a blemish on reason or thought or the sciences, but an open realisation that in order to progress out of the ties of evolutionary conduct, a new kind of emotional reckoning had to have occurred! For at least to have first awakened Gods own interest to what could only have been the antics of another developed specie from within the realm of evolution, perhaps another to remain in cycle for twenty or thirty uncounted millions of years before eventual extinction? Unless in the space of only three plus million years Gods own time span of evolution was turned into mans realisation of our own time span of creation in Gods name!?

# CHAPTER 99

One of the smallest objectives of the Fifth Dimension is to bring religions back into unified equality by religious forbearance, always allowing for change if needed but without the necessity of alteration to our histories by whatever measure. All histories although cast in the soft volcanic dust of occurrence, automatically mature into the hardened rock of change by the advent of time, because the attitudes of the new readers are sometimes different to those of the original writers. With some of this reasoning in mind and to update some past considerations, to alter our general acceptance of recent events to have occurred since the ending of World War Two. It is for our religions to now step forward in all ways shapes and forms to announce in unity, their understanding and acceptances of what religious tolerance is to be offered as, better translated to religious forbearance?

These next few lines or paragraphs are not of my hand, but in the form of guidance by us all, I have used that expression so that we might not confuse in issue that this delivery is by any other than the likes of you or I as we. Simply because from how it is presented it is not for any of us to say what is to be said, but to comply with the prospects of the Fifth Dimension, what needs to be said is from the base of acceptance of the three taken prospects and one active yet to be used fully. By reasonable representation I take the Christian religions first but not in first order. In its two thousand year history, much has been passed over by our first acceptances of Catholicism. Even when added too in change by its own representation to become the named religions of Christianity through several reformations, the structure of all or the vast majority of linked Christian religions have kept in line of form to the one God!

Without real visual effect but in acceptance to the loyal martyrdom that many Christian religions were to suffer in the name of individuals, this feature was highly complicated to the irreverent memory to some martyrs cast to be heretics from the same core religious belief? This could only have been done from our other controlling influence in the name of the same humanity, but to its own lead by civil demand in and from our politics! Which having taken hold was like a terrier shaking the life out of a lesser animal or what could have been considered, that the more open influence to be upon our lives was to have the primary say. Not unlike when the Roman Empire of some three hundred plus years after the persecution of Catholics did succumb to that very religion, and by association made it truly universal within empire. Even to the extent of installing by the supposition of direct linkage a continued linage of the papacy still emanating from Rome, from whom the first was connected directly with Christ Jesus through his original apostles?

From the time of Jesus death and resurrection three days later, a new code had been introduced to this new Judaism in becoming catholic. He had made arrangements with his apostles that after his death one of there number in Peter was to be his earthly representative in terms of doctrine or the meaning of it? If doubt was to be raised about any form from its original instigation, Peter was to be the first in line as Gods open representative on this our earth, leastways to the consideration of religious doctrine by replacement or interpretation, but only for the intended new religious form and format! While given that status to be the sole interpreter to the faiths real intentions, his role was that of infallibility in those matters being a tradition held today by the direct line of descendants even if now popes are chosen by ballot?

By features of history in brief description, it has been told that as the new Jewish faith in its universal form of Catholicism came to be the adopted religion of the Roman Empire. Then quite naturally its new seat of office would also be from Rome for reasons which were to be obvious, but not necessarily to have been the right thing to have done even then. Nevertheless without raising retrospective doubt bore from hindsight, by taking that action the understanding of Christ's representative as the Pope still held to infallibility over matters of doctrine, but now by political scrutiny? Although the matter of state and religion was interwoven by the increasing complexities of the way we were now interacting, mistakes of a very simple nature were later to bear fruit of a bitter kind. From where eventually the state was to reverse the condition of having being conquered by religion and reinstate the supposed vital importance of political judgements along with all of its tributaries? Perhaps until the Fifth Dimension was able to raise its own right by the assertion that it is now for our religions to take first lead of our conduct irrespective of any proposed religious settlement or settlements!?

No battle plans are being drawn no religious strategies are to be formed by the proposal of alliances especially within the Christian sector. Although I had set the Christian religions first mention, it has been for general enlightenment to show there normal structure without distinction. Having the Pope at the head of the biggest Christian sect, Roman Catholicism, aided by the seat of cardinals and other devoted tiers of a well structured priesthood. My point is not of condemnation but to show the parity with other Christian sects, instigated by pieces of our own free will expressed through our long religious histories up to the point of reformation proper. From which time historically, the church had been worn down to be a better servant of state than of there claim to serve God?

In using the offices of the Pope, there is no aim to create tensions in the name of clarity, where our line is drawn is to mark a set of time, in this case to already be two thousand years long? By that division in relation to all Christian religions and in call to all branch leaders in the reverence of cardinals, bishops and priests, it is for them to take us all in command to the ethics and essence of their own sects in relation to each others as Christian, but above all to lead over politics. We are not in any sort of competition to the needs of man expressed by what politics can and might deliver; our pointers now are to stand religious tolerance to its own acceptance from the ability we now have in overlaying both of our different influences with the same medium of the Fifth Dimension!

For our Christian religions to speak in our name and for all men by its original intention is done from the same instant when Jesus was onboard and in life, which is only a measure where we can each reconcile how we would or could listen in the timescale of then with our thinking of now? This being a unique situation, for in this case and in the same similarity to the other main two Middle East Religions, death had a different concept? All religions were set to the inclusion of human or tribal or group conduct expressed through the individual. None of the Middle East religions were meant to travel in time, they were to have been instant that was there construction! Leastways that is the impression which had been give us all by the word of God from which we cannot comment upon, because now our religious energy is to be 'Past Gods Words'? Which is not to deny God in concept or reality, which is not to imply that God in Allah is of another intention or force, even reduced?

Our move of acceptance listening to modern theory by this recent era at first through the Christian religions is aided by self examination of our relationship not with God but each other. Since Jesus in Christ was a real person as we, his insights were tied to his own understanding and more importantly to his direct surroundings. In general terms with such a brutal world by all normalities, virtually all societies had turned to the future by concept in all considerations. By that role we looked to the comfort of death to have its meaning for most of our earthly deeds. Jesus was no different and from the fact that he carried his, this new Jewish religion to that conclusion by our own conduct was not a new theory but contained us to our found monotheistic footing. He generally endorsed our life span as a role in life to be endorsed from death, which was no bad thing, except even for Jesus his manly thinking of then ago was mostly of then ago. All fitted in the comfort of reason, save in our generational conduct of living since that original time and from the fact of the one hundred or so extra generations as we, had been subject to certain changes of attitude over these past two thousand years?

What I allude too, with aid from the structure of the Fifth Dimension, is endorsement of some of our changes. Not to be granted separate status, but to account better for such attacks that might have been laid against all religions through our political setting and or the fourth dimensional bonanza? Matters which have been able to contrive and place all religions to a standing of religious tolerance, but fortunately not yet to have been able and stifle our religious intent. That is why in this case relative to Christianity in first mention, it can be put by any of us individually that we are aware of modernity, that we are aware that times have changed, in terms of attitude, of situations, but not by outcome? That is why I as we can make the statement that our religious leaders by any Christian church from the Pope to Archbishops to Cardinals and priests and reverends can still make our representations.

Not least on two fronts, by first count that although God in all wisdom gave us the finest of insight through his son incarnate Jesus Christ? When he spoke we were not yet ready to comprehend his open intention and therefore sanctioned his delivery to allow us come to self conclusions to assume we were still on the right path! Perhaps setting the standard whereby we would continually react from assumption, questioning the original line of delivery and of its meaning to us? Sustained for centuries even under the severest of persecutions, which must have been so because we maintained our catholic structure in its original form even to have been filtered somewhat, until our very attitude changed the thinking of our imperial overlords?

I had indicated that with some of the uncertain future of the catholic religion now secured by the very working of its own priesthood headed by the Pope in Rome, which by a range of misalignment set for us some wrong signals to what would eventually become modern thinking. With attitudes changing in the field of politics, particularly in Europe over the ensuing centuries, with the concept of nationality allowing us to think of our separate status, yet encompassed by the one religion. We began to miss the range of the changes that we were making to most of the previous accepted standards of our existence. Our mixture through nationality was working more on the political level not least recognised by the first serious catholic reformation instigated by Martin Luther in Germany over five hundred years ago. Being a matter no Christian body has ever been able to counter in conformity, because essentially his objections if, were not of the spiritual plane but secular, loading the gun of politics to intimidate standard religious acceptances?

What we know of the events over these past five hundred years when politics was aided by the expositions of the fourth dimension proper, driving most religions into there own spiritual zone and there realm of silence, becoming a matter of fact. Shown in translation by most Christian churches of Europe who suffered two world wars of the last century in silence, while being correspondent to all sides by association? Once again without the critic, but in terms of our political associations, our manner of conduct now, since the ending of the Second World War, has been ably manufactured by American political associations which also happen to be Christian at base. By direction our standards have been set in the form of communication by political needs and deeds, which is no bad thing until we address its own method of style in coping with most religious settlements by the false standard of religious tolerance?

That is why in part we must address any form of procedure that redresses opinion to the understanding of times no longer here and in some ways dated by recognition, but not intention. That is why our religious leaders have to associate our modernity with past intentions in the open translation of submitting in plan the further objective of political reasoning by religious leads, any religious leads! That is why it is for our Pope, Archbishops, Cardinals, priests and reverends to lay upon the political surface, objectives for both fields. They in we are to second place political standards, not in Gods name, not in Gods name, but in cause of our own religious visions!

At this time it is impossible to issue change by decree unless forewarned, but to utter intent is not to arm those we would wish to influence as some in politics might perceive in the art of divide and rule. Without belligerence it is for the act of change to be foretold and in reason thereof? What we might utter is all and only all from human emotions, the conditions of which have been in the making these past four thousand years at least, nevertheless when the light has been switched on for the first time we are all apt to blink a little. I will still keep to my original plan in calling for Christian representation first, but again I say not by importance, but because by direction the religions of Christ have been spun from the fact of Gods own intervention to our wellbeing, but wholly instigated by ourselves in time lost to any history, because of our real need!

Beyond all imaginings, when we buried at Shanidar the key was truly turned. When Christ stood in name to be Gods son as man, the door was further ajar, when told death was salvation, the questions answered opened the book of human query? By first call how and for why was that move to be made, answers were immediately collected from the circumstances and conditions in which we then lived? By required examination the rest is almost history by our self conduct, simply because from that vital time in Christ's own breath, we were unable to understand that any relationship of then and in Gods name was in Gods time span! We were still in an evolutionary state having found creation but unable to comprehend its broadness and relationship to either us or all other creatures. In our open pursuance of knowledge for the greater good, we hastened in acceptance of what we were genuinely told, but not of its core?

For a religion big enough to conquer the worlds mightiest empire in relation to the conditions of existence three hundred years and fifteen to twenty generations of people since the time of Christ. We were unable to equate the meaning in anything of what Christ had said? When using his best examples we were forced to change direction by political overtones over our most recent five hundred years certainly in Europe, but mainly by those changing political perspectives even if generated by some and other religious experiences we were to have encountered until we could apply religious tolerance as a by word for quiet contemplation but now only by political determinations.

From where it can be seen that the Fifth Dimensional overlay can assist our Pope, Archbishops, Cardinals, Priests, Rabbis, Imams and other reverends, is to be taken by what is our first say representatively to be a new era in recognition, not of what God in Christ said, but in what was meant! Both in conclusion to the separate hundreds of generations since Jesus and to curtail the indifference of political reasoning from where some branches might impose there will without any religious consideration. Our first voice in objection is to show our own open understanding revealed to us by Jesus himself, is that Jesus in his style then ago was speaking to us as a man. Although carrying the stated words of God the Father, he relayed them to us in fashion that a Rabbi might, or even a carpenter? It was we who did not really mix the solution, but when we did, we related all outcomes to the splendour of death, of our own death, which fitted to the thinking then when there must be something better than this dour existence? That is perhaps why so many people continually converted to the faith even knowing full well, that part of the cause of empire then, was to eradicate this new catholic menace by the most intricate of cruelty?

Even that the above could be addressed as fanciful, our needs nevertheless are to indicate that our religions are not to be submissive to any type of political overtures. To disarm the blocking feature of the fourth dimension in any state of operation our Christian religious leaders in all good faith can prepare the status quo? In being that our total connection to God by intent is from and by how we might show our ability to receive direction. It is that by indication we are prepared as the only specie in the history of all galaxies to show any inkling of our relationship through creation, above the staid standard of procreation only! In doing so our limit is to be endless in relation to what we might need to do to alter the course of our own political standards and or standing, not by the methods of anarchy but cased in the order of clemency. Our Middle East religions are to be prepared to lead direction, but past the requisites of standing political policy.

After all the energy used in the production of a finite means of political control charged to produce total welfare for all, care for all, comfort and security for all, we are wanting. Even from the time of Christ, although standards had been obtained, we lacked so much, which was to keep our society divided so that we counted the individual story of life by death. Today after our achievements of organised society the best we can count is the difference by fact that some of us will live to be fifty plus years of age and others will manage to reach the same average of seventy plus years of life! That is not a charge to be aimed in a political direction, because we have kept our religions too quiet so far!

My own take on the situation has been that until now, we have been content to run with what we have so far designed, being no bad thing, unless taken that is all we can do under the restraints of our Human Rights charter or politically correct motivation or other political guides? Each imposed so that politics was our leading horse and all referrals needed her assent, especially when religions were quietened by the addition of our religious tolerance standard. Even while carried in open view working half measured, choices leading to our steps taken syndrome were tendered in honest and blind content, perhaps doubly backed up by the premise that we were also truly civilised by the comparison with others as self?

Showing the real matter of our disconcerted misalignment and ushering the need for and of our religious leaders to set the right objectives in projection of the proper outcome whatever? But being ready to attempt and understand Gods intentions from any time, conveyed to us as if we should have known by self expression, is all we need to realign political intent by preparing for the future in such a way to compliment us all? Notice is to be given that the essence of organisation in future prospect is to be given in line of religious fortitude but spread from the equality of all religions in that by each we are to show our preparedness to accommodate any intentions of the future through a single lifecycle!

## CHAPTER 100

From a history of four thousand years the Jewish religion is to be charged with the same care as the task in hand for the new Judaism or the collective Christian churches more readily known as catholic and now Christian? Historically there is little difference between Judaism and Christianity, save in the most obvious that the Jewish religion being almost twice as old as the latter forms, has the thinking and reactive tendencies of an extra one hundred generations to contend with. What we have to adjust on there and our behalf by most standards is a feature in relation to our selves and how cause might be given for how we react to each other now?

By carrying tenets and Gods own covenant for two thousand years until the time of Christ, fair and balanced consideration can only be drawn from the abstract because of our own nature. The Jews in us had lived since the time of Moses in the swirling vortex of actual events encompassed by the following two thousand years, from the proposition of expected triumphant unity with God to the righteous deliverance of those that had been chosen first hand. All that had been was the additional one hundred generations of the same people in faith, but carried by the slight changes to the natural progressive steps taken by society or at least perceived so? In changing by attitude even in very small ways, when addressed with a new religious alternative, even if expressed by one of self, by any degrees of conflict for some, which might have followed, such effrontery might only have endorsed reasonable entrenchment to original ideals, which is never to be a bad thing given free will!

My own points to be made by the above direction are not made by any forms of religious comparison, but are to be taken by historical associations in context without criticism. My aim is of enlightenment which can be viewed later if required by our active prospect in 'One for the Whole' offered in examination by any quarter. We have one hundred generations of self, who had lived in good conduct as declared Jews for two thousand years. We have one hundred generations of self, living as declared Catholics but now intertwined as Christians living for two thousand years until now. Intermixed with the second one hundred generations of Jews as self, each group mixing in the turmoil of political and scientific changes to have occurred differently over the last two thousand since Christ to now, as did all in the first two thousand years from Moses to Christ!?

By small degrees of reality, the above is indicative of how we are dealing in these very modern times expressed in term since the ending of the Second World War, but now we carry the taken and active prospects of the Fifth Dimension of recent association from these past five to six years. In not being designed to force separation we each are at desired liberty to overlay our own specific religion with the said three taken prospects and use the one active in mirror form for self assessment without outside influences, but not to ignore natural or timed change. Which does not cede to the need of change, only the balance of how some changes had occurred or had needed to occur?

Without the prospect of desired change or even the possibility of it, I charge the Jewish religion in the exact same way method and style that has been put to the Christian religion, singularly. From the chief Rabbi, even if directly from the one Jewish state in Israel, is to make self imposed representations of religious intent above the needs of political imagination, if it has been imposed upon us all to be religiously tolerant! Then Jews through the Israeli, American or European Rabbi Structure along with others in our representation, also set the same standards of obtainment in putting religious choices first for human determinations above political! Not to be competitive or obstructive but in making the same order of concern from religions to politics and not reversed by religious tolerance.

Our purpose of which might not readily be seen when considering some of our finest religions if to have been classed of an Eastern flavour, which likewise are to face the same recommendation. When and why Christianity and Judaism are to take my suggestion is done because they are two in the same by connection being monotheistic. From which in normal comparison likens our human link with or too God through the event of our death, which has no consideration whatsoever from any branch of politics!? In placing them so, I have desired to keep their method of proposal to be separate if only to represent their own religion, but in each case to be set in opposing political rulings or assumptions? Perhaps better expressed by how politics has covered all religions by the expedient management of religious tolerance aimed to touch all religions in the same way and manner, even those in support of human sacrifice, if being done or ever to return?

Religious tolerance is a method of contempt from which politics uses to express her own view by what ever means and by what ever regime? We can see its first flaw from expression when laying it on all religions even to be Eastern, the expression is self compromising. When on the one hand it purports to all religions, even those that include multi life forms of all and any creatures also as self, to eventually gain the ultimate objective of purpose, how does tolerance stand? Once again without the critic our Eastern religions in operation are attuned to us individually and some of our monotheistic claims naturally. For all religions offer in acceptance the style of creation being of Gods own reason of how we all of this single specie see the advent of time. By eventually relating the shortest of evolutionary moments expressed to be fifty plus or seventy plus years long against at first a three and a half million lifespan over others that were only to see evolution?

Eastern religions unblemished by any count can recognise the same fixture that we all suffer for, but make it more difficult to challenge what could be our nemesis in politics. Which in being able to control religions that essentially until now have not challenged, because both groups have been aligned to a single human lifespan in theory? From the Eastern view, if other creatures are mixed in cycle with we humans, politics can use theory through the fourth dimension to diminish its flow in being religious. Because in or from those regenerations there has been no claimed or amiable concern to how each reincarnation has passed its own merit forward or backward whatever the case might be?

Not to be any sort of test to be put on such emotional thinking, for both Eastern and Western religions are virtually of the same mix and time scale, both rely on the unobtainable by achievement through living, except in eventual 'Natural Body Death', when they signal our difference in understanding. With one group being monotheistic, there general stand of recognition to that connection is by human death from which we are to enter a realm far in excess to what political aims could ever be? With another group simply stretching that fixture for the same ends, whereby we are prepared to include gods own creatures in mixture to our living standard by obtainment, stretches only imagination? Which might be like mixing the living and dead in the same objective, but without the need of clarity until that state is made by such mixing, being just another way of using Gods own imagination, we can readily show our preparedness for his desire. In our own case most of anything readily goes by religious need except perhaps the expression of human sacrifice, from which never can it be shown expressively, that those to have been murdered were done so by gods own hand!

By opposing politics in the way I have indicated through our religions of Christianity and Judaism, all Eastern religions are of the same choice, but undoubtedly not of the same need or desire. From their mix, it is not necessary to settle the account with any type of civil or political drive by the events of recent years. Because from our very character in that loop such peoples as we, by our own force of attitude generated by the circumstances of geography, climate, diet and other means, had never felt such a desire even related to these very recent times since the close of hostilities in nineteen forty five? Nevertheless on our own account collectively from the way things have panned out, Eastern religious acceptances will continue to work within all the frameworks of change generated by the moment, even if expressed politically and or wrongly read?

From the same wheel as the monotheistic one of now, it is perhaps that Eastern branches of our religions have always questioned our civil and political ruling, when from there own pace of development there never had been a direct need to challenge anything except change. It is only from now and more directly related to our American political stanza that we have come in renewed challenge. Without question, since wars end and I make the assertion qualified by the fact that we have not proven our correct intention by the muse we seem to be in. By basing our assumptions to be correct and modern and supported by our management in human rights and religious tolerance has not been the answer or solution we have been looking for. In carrying the proposition of new deeds leading to new deals has not been able to address how we are drifting apart leastways religiously. When counting that our politics is prepared to operate half measured full blown, and reacting to the consequence of the difference between steps taken in following or in the steps taken syndrome. Instead of marrying our two methods of our overall control by aim, we have been aimless in taking one only as politics half measured and ignoring the other as religion by the steps taken in issuing religious tolerance full blown?

By the same mixture, but seen from a slightly different angle my own standard of reconciliation by whatever means, does involve our religions leastways our monotheistic ones for now. To demonstrate there equal malcontent with the necessity of a real concerted political drive in its own colours if momentarily driven from an American political perspective. By the same token that my first and second dimensional religions had been asked to assert their own impetus but in religious relationship, is now offered to be forthcoming from the difficulty of my third dimensional religion of Islam. Not in separation from the other two or at leastways not classed as so, but only accepted because in open reality Islam does not have one voice if through the Pope or Archbishops or Rabbis. It has millions in the same contact, but in the vision of concept expressed through the faith proper and such other circumstances to have been laid upon us all over the fourteen hundred years of its history, which is not problematic, save the self value of millions of opinions?

## CHAPTER 101

By being a very strait forward religion, Islam presents us all with a great many changes of attitude, which need addressing first religiously and secondly politically. Therefore it will have as an important say for us all by using in grace this opportunity by taking three prospects and one active in overlay of self, to be expressed in political reasoning, but by our religious attitude first. Even that there is a vast difference in the number of the faithful called to the same tasks, with approximately fifteen million Jews along with over a billion and a half Christians and the number of Muslims slightly less at a billion and a quarter? There is to be no comparison from that value, save by my own representation that the three religions are irreversibly intertwined by other connections than just coming from the same area of geography as the Middle East even if at different times?

By the very nature of Islam through its form of structure, makes this journey different and more difficult for the task in hand that I offer. Yet for some reasons and in some cases the very cause of this plan will have a more open response from the religion of Islam than we might expect from Judaism and or Christianity. By structure Islam has a better style of recorded history than the previous religions of area, which means give or take our own ability to naturally differ by the factors of personality, being laid back or dashing or forward looking. In Islam we do interpret the same standards of written text differently without malice of intention or disruption or for other ends, but seem able to come together in better organisation than other religions might, which is never to be a bad thing?

Without direction I had previously stated of the Islamic concept being of two main parts Sunni or Shiah from very early days of the actual instigation of the religion its self? Although the terms are representative to the people involved, that separation it is not to be viewed in the same way that the catholic reformation was to keep other subsequent religions Christian. The Sunni and Shiah relationship is a feature to the Islamic religion only seen by the factors of there early definition to events at about the time of Mohammed or shortly after his death? Having both set their own course, there relationship unfortunately can have violent interactive phases, which have no bearing at this time and are not under test by any of us now, which is not to side step the issue or ignore it for matters of taste of a political or religious opinion! In testing the theory and reason for such a split we would need to take the time to dissect every reason of why all religion have likewise expanded or contracted there own opinions, yet remained to the same central core. Like the picture from Christianity, Islam simply focuses to its core when religious matters come in debate either from what is considered to be in the negative from other religions or political mediums?

Our view from this exercise to be conducted by the same format offered to the Jews and Christians finds difficulty? Whereby for those two religions I had set the plan in charge of the clerical heads in the form of priesthoods by historical connection? Islam dose not express the same method of control to conduct, nor the same view in reference to a structured priesthood to define behaviour. Its controls without direction are from the written word, taken in actual explanation from there time of delivery and set to be learnt in study by the individual adherents to the faith. Although directed by teachers and imams in the rate of learning, assumptions run to the effect that each of us individually can only draw the same meaning and therefore react in the same way. Being no slight to pure religious concept, especially when enough was to have been considered to be done to limit the changes of directed faith than when Jews had first come to God! Later to be challenged from within by the codes of new Judaism if to be cemented into Christianity by the removal from source of its seat of reference in the Middle East to Rome of Europe!

All intentions for and to the Islamic faith when set by Mohammed, was to fix its views under control of the pen, which were vitally set in text to be of first relevance to the people of Mohammed's familiarity being us all? Not to have the religion usurped by what may have happened to other religions, writings were set in the primary language of the Arab tongue, being no wrong thing. Unfortunately that display was of its day, holding that we were still tribal communities, therefore local language was naturally to be the choice of the written word. By such encouragement and in not being able to express a level of thinking to encompass the consideration of politics by inclusion save as a vassal. Our first Islamic view set in our first splendour of personalised equality was unable to translate overall intent to everybody, because everywhere we had thoughts of tribal definition before the standard of humanity, expressed by later understanding if as late as these past sixty years plus!

Islam was a huge success in its believed true form and from source, I can say with open conviction that because of the original format and layout, it on its own account transcended such barriers of normal tribalism, which might have been complicated later with the coming of nationalism? All of which can be brought into focus by todays examples where by some political reasoning, many different sects have been joined in the funnel of modernity and poured from the spout of reason in confusion? If not led in the confusion of our different religious attitudes from the past, then the outpouring has fully brought us up to date in this last modern era since the close of hostilities at the end of World War two?

By representation of some events in the Middle East, I had previously set the observation of the Arab/Israeli conflict in the mid nineteen sixties and from the same conflict by reference, had indicated to Palestinian reaction with a direct note to the formation of that reaction, but had not named its style. My point given is to the possible outcome of any events which might yet affect us all in the future and not to consider how some histories have been recorded from whatever point of view. I have not avoided the issue, which can be seen how I simply suggest how the Fifth Dimension is to be used in exactly the same way by our three Middle East religions. Even that I have surreptitiously moved an emphasis of how I had directed attention from my cited Israeli/Arab conflict and mentioned Palestinian reaction then dropped that inclusion?

Beyond the limit of a half measured proposal, which would automatically lead to the steps taken syndrome outcome. My direction has been set to avoid one adaptation and the ignoring of several others not in complexity, but because histories have already been recorded in any case? Because of so, reactions are sometimes set from the imagined outcome and meaning of what was recorded in the reflective form by relation to what was deemed to have been seen, even if the left hand was raised and the mirror showed the right hand in response. By the same justification, un-judgemental, in referral to my own use of the Palestinian image of a reactive force to the Israeli victory in the mentioned conflict of the mid nineteen sixties? Although by trend all of the misfortunes of the Arab/Israeli conflict fell upon the remainder of the Palestinian people, their reactions by an uneven review were taken from the political emphasis to have been laid upon Islam!

From the above and by previous reference there is to be no difference to the consideration of how Islam may approach our assumed political standards bearing the three taken and one active prospects of the Fifth Dimension? Although I have set my own agenda in format, for us to consider how and why our three Middle East religions might use the same action over the same matter, can only be done by how I personally have indicated there main and only real difference? In order for needed support I have used this last modern era from nineteen forty five and through the United Nations politically as another point of reference. I had also used some past examples of our previous conduct to highlight certain points, not least by expression in religious terms of how the first or my first two dimensional religions had a common relationship with gods own commandment of "thou shall not kill"? To that concept I had allowed for the reasoning of the dated reaction we at first applied and for reasons of how Mohammed had reviewed the matter from a different perspective, were by my interpretation only?

Allowing that the concept of human death was a viable possibility and could be now used as a force to defend or promote in prospect the new set desire of direction from Islam, even if laid by the same medium that had previously forbade the incident twice before. I had intoned that the act under acceptance was to have been now permitted to compensate for the reality of our personal conduct expressed by the values to have readily developed in many societies at that specific time. Not least to compensate for the fact that our political renditions had never been able to leave that feature of human death by self by law, which was assumed to have been a requisite of many of our legal systems?

From what Mohammed had heard in record and transmitted to the pen, was done by or at leastways to how we did actually live. By further references to the matter of martyrdom, by this reasoning and from the descriptive and colourful means the idea was received, the image if not directly intended, could be taken that this was a desired obtainment. Not withstanding that in general terms it was perhaps only a compensatory feature likened to the time of history in relationship where conversely the early Christians were taught to turn the other cheek? From which they went for in expectation only? In core and by the natural steps of progression, the basics of the religion of Islam, were the most intricate update to all religious concepts, but more to the former and existing religions of the area then. In the real form of taking religion back to the top of our agenda by human conduct, the feature of ultimate control was to self judge in the matter of human death attributed to the direct need of the religion and restored an aspect of human dignity?!

By that reasoning I have set our course to the acceptance of that very act in all cases, but have set the divide by interpretation, which has readily been shown in these recent times from an Islamic perspective, whereby that ability of self determination without a direct priestly lead is still open to any interpretation from any individual? By those criteria it could lead to the possibility of prompting all religions to reinvent the standard of human sacrifice by any means, unless now questioned in all cases by our taken prospect of 'Natural Body Death'. Prior to our standing of now, earlier attitudes are not to be taken as a slur on one religion to another, leastways by the common thinking of our three monotheistic religions which have all created dark self histories?

Engrossed in the same mixture of religious ferment expressed by all three of our Middle East religions religiously, changes of an unreadable character over the difference of some of our political attitudes need comment. From within this modern era, but pre our Fifth Dimension in study, notice is to be given of how the Islamic faith was able to compromise its self study in good faith? When by purely political means the countries of Iraq and Iran were found in military conflict, which now by reflection I examine within the rules of hindsight. To say or suggest that the matter of conflict was to its own significance belies some of the cruel events to have taken place in the same standard of what might be linked to all the victims of the political wars of Korea and Vietnam?

By the style of the conflict between Iraq and Iran to be only political, the fact is always to remain they were both Islam by religious standards. Unlike in Korea and the later conflict in Vietnam which by historical recording on all sides was an epic struggle for political ends only, because in both examples the theory of conflict was to become politically communistic or accept democratic support for commercial reasons? In the Middle East although instigated by a servant of politics through open commercialism, other factors led that the war was more of an Islamic struggle between fundamentalism and if suggested, modernism, if including commercial intentions? It is not to say that any war between two nationalities of the same religion has a particular significance. Leastways if noticed in record that in far too many cases since World War Two we have had wars of similar dexterity, if on smaller scales to have remained political by definition?

Our case in question and by my interpretation only, is worthy of separate study if for no other reason than the sad fact that it is a recent local conflict reminding us of the two gigantic struggles of nineteen fourteen eighteen and nineteen thirty nine forty five in Europe. Those two conflicts almost in repeat with the same combatants with the same basic religion of Christianity, gained no knowledge except to compromise the original concept of the one religion expressed nationally. With both sets of antagonists in open declaration that God was on there side and in carrying out the most atrocious of bestial acts of man upon man, their religious acceptances were conducted without fear or favour to the stated commandment of "thou shall not kill". In turn, in these modern times, our religious suppositions are to be contained by another form, the one which we have set in the measure of equality, which from the barrage of religious tolerance, has proved to be a false premise. From the imagined intention even as new as the theory is by representation in these past sixty plus years only, results of actual conduct have in cases overstated our original intentions, not least by how many Christian and the one Jewish country have steered there own course religiously?

For the same reason but, by another choice I am to put the Iran/Iraq conflict under scrutiny because it can be used to show us some of the methods in which we have erred by the structure of society we have attempted to create in this last modern era. I add that at the time of their conflict, we did not hold in prospect the ideas of the Fifth Dimension, from which we could rebound in echo some of our religious acceptances. Nevertheless from the very mixture of events in the Middle East since the mid nineteen sixties in world terms, the voice of Islam was again first heard in some judgements through the voice of Palestine though not literally. What was transmitted by them was a will with a religious foundation, that sufferance be allowed for there actions or reactions by whatever means!

In the face of the results of war which did not go their way for political or military reasons, in answer, all fields of resistance now counted to try? I make no direct study to the level or measure of any form of considered reaction, even if to include religious perceptions. By much the same code that any people in the throws of battle might find justification of what to do to gain their perceived intentions by now working inside the range of religious settlement? If that was the case for Palestinians being first of Islam, then by outcome to die in the execution of their perceived objectives, could have been done for the defence or protection of that held faith? My observation has not been made in a fanciful manner, but to make a less than obvious judgement of the style of one of our Middle East religions and in this case of some of the group of people to operate it! My own suggestion covered by their particular attitude laid down by ancient circumstances of geography, climate, diet and other reasons is to show our separation of attitude only?

By any methods in the use to balance, such situations that did occur attributed to Palestinians, if their style of self sacrifice could have been covered to be pronounced as a method of martyrdom? Religious answers were put to political solutions, but not in conclusion to all religions, even that all were prepared to kill under there own acceptances, but some not to the Islamic format if so! When Iraq and Iran entered another phase of our brutal ability to war amongst ourselves some years after a particular style of Palestinian reaction was affected, a different picture was to be painted if using the same colours? By being mixed on the pallet with a base of Islam in standard to each side, what was to emerge from the created cloud and mists of reasoning was the eventual death of over one million people who just happened to be of Islam!

In previously stating that our histories are always recorded by whomsoever has the true meaning of being done by a personal standpoint in all cases. It is not factual that the victor is always the one who accurately chronicles the events when in this example the war was reviewed by the world in deference to the level of technology we have gained in these years since nineteen forty five. Opinions formed by some international reasoning from there own standpoint, drew most conclusions to be tied by political perception, the rest is not yet history, because from some quarters the Iran/Iraq war has been marked down in status. But still allowing or being able to have cleared internationalism to the standard of being observers only just in case? Even if many separate political entities were involved by its tentacles of material supply and demand, marked by commercialism, then in a final sweep after the conflict, adding moral judgement to victor or vanquished who were all of Islam, could only be offered stating religious tolerance?

My own look in religious terms is done to make a specific point so that we can create a standard to some aspects of futurism from that perspective, but in open relationship to us all. My first set of circumstances are drawn from hindsight, because that is the same marker we all have without the adage of reflection in relating things of the same intensity, but now with the reasoning of the Fifth Dimension and its three taken prospects and one active, which are not to be used in changing the past, only our perception to what the future might be. I have to walk in the cinders of recent avid emotions and ignore what the feelings might have been to the families in grief to the loss of sons, husbands and fathers, not to protect my own emotions, but in attempt for us to rationalise our differences, which I do not subscribe to in any case. We are a collection of peoples who have been sculptured by the winds of change through the circumstances of geography, climate, diet and other reasons, who are now coming at last to recognise our similarity by the open features of connection when seen at Shanidar or our cave paintings or by our joint inclination to ponder in the mystery of death?

Leading that we were from the one base in being 'Out of Africa' has often been used by my self to point our direction coupled with our other collective pasts. If only three and a half million years later we are to stand at my reference point to how I might have interpreted some events to have occurred, tied by my Palestinian association of being to a specific religion and the same from the Iraq/Iran war of a similar time date mark. My objectives are to be measured as ours, not by inclusion of my own reasoning, but because of the way I have been able to assemble the flotsam of our joint knowledge and make the connection with the jetsam of our political and religious reasoning in general? If my point has been to state that some Palestinian activists or re-activists took to the end stand of martyrdom to endorse there perceived objectives. It was done from a religious flush in good faith, with no other defined intention?!

By forms of connection, but unrelated, my association of the Iran/Iraq war to the Palestinian struggles has been done to link them by some forms of attitude, not in condemnation, but again by reflective means that could affect us all by fallout? Although I had put it so that some Palestinians within there struggle closed in, like with many other peoples throughout the world who were ready to die for there cause no matter what religion or politics they followed. The connection of how things have been further clouded by cause, reason or intention from others standards imposed by several different moral, political or religious outlooks. Now takes its own complicated forms of why I use the Iraq/Iran war as a modern standard, although erroneously misread by vast political misunderstanding and also by myself, but taken to view our conduct!

I had mentioned in reference of how some reasoning was attributed to the means of both countries opinions over religious matters, from one people taking to be more fundamental to realign to truer and past acceptances. With the other still tied to a very robust form of Islamic doctrine and was prepared to settle by some processes of change even if politically motivated? What ensued was similar in conduct like so many other struggles throughout our histories had done, and could not be told apart, save that by some political observers it was noticed that a different interpretation could be put upon religious events by political reasoning from perhaps the wrong perspective? Here from our first modern example, peoples were again prepared to take to battle full face and die in open combat, but now to be determined as an act of religious conduct in purpose and style for religious ends, but alas only seen politically? For both sides were of Islam and not divided by that standing to be Sunni or Shiah, when in the execution of the act of suicide or more precisely an act of self sacrifice, whereby a man may have no greater love than to lay down his life for his brother! No real connection was made that these deeds had generally been preformed in most wars since when; amongst Jews, Christians and other religious groups, allowing that in most of their philosophies they were not to kill of self anyhow? Yet in recent cases the connection was made that in some modern wars when religious groups of the same standing fought, in being allowed to kill in the defence or protection of their religion by written concern, by the natural link with associated text. Then martyrdom was a standard if called or desired by those of that one religion only, which was done without a finite edict!

## CHAPTER 102

By direction the only one, two or three statements I am making through these pages are for this modern era and the future. All said has deep connections with the past but in cases most of our pasts have been duplicated and duplicated one upon the other and again. My own reference point has been through our three Middle East religions so stated, with also all religions counted of the same value except if self lacking. My focus in notice has been conducted full balance, with all the standards of geniality by measure of how we are to self appreciate each other by the fixture of standards I have been able to lay down through the Fifth Dimension and its associated prospects. Of which the one active is our single voice over all matters of concern through query, question and suggestion from/to/through 'One for the whole' in being active and fluid.

By my final first connection to religious matters and in the name of modernity in using the three Middle East religions I have made no comparisons against each other from one to one to one except by perception. I do however relay how I have connected myself to all of them in representation by side stepping our political attitudes. In calling by request that the leaders of all religions are better equipped to settle the matter of our differences, separate to politics, has not been done to form confrontational standards. It has been done in our name and for them to emphasise the importance of our individual lifecycles now expressed by the vulnerability of a fifty plus or seventy plus year span.

In order that they/we in our representation might suffer good cause, they/we are to operate from our taken religious perspectives in our relationships as they are from religious understanding only. In order that this matter can be viewed in its best light it is for them/we through each of us to address our religious expectations in the above terms of our lifecycles in relation to any expectation religiously? This by intention is not to confront our other order of control in politics, even that in cases those of that sphere are connected to some or other of our religions or not? Our overall intention is to show understanding of our differences if promoted by the aspect of our religions and politics in the same control of guidance, even if from differently developed perspectives generated by the circumstances of our climate, geography, diet or other reasons?

Although I take that we are the same by virtue of our final outcome, this is not fully true by some other means of our natural circumstances of how they have been compromised, by what can only be determined the result in the expression of some of our religious attitudes? In taking this opinion through the political medium only, whereby some of us might follow in the pursuit of self gratification by whatever means, leading to the matter of crime to achieve our own ends extending to the pinnacle of murder! Then by that expression our political aims are to be more compromised than those of our religions, which are to be yet made clearer by us through and from future religious leads. Even that by suggestion, I have laid it so that the Fifth Dimension will act as our cornerstones to the foundation of change, our full examination is to be now, expressed in looking at how our politics has set some standards of acceptance without background?

From some of my own challenges to any measure of our situations I have perceived, although connected by both religious and political tendons, my direction has been loaded in favour of our best solutions being from our religious perspectives. But in following a coded preference to operate outside of political reasoning first, than into the realm of religious observations again, not of their standards, but through the three taken and one active prospects of the Fifth Dimension? In overlaying both influences with the stated prospects, the most open conclusion I have set to be drawn from or by concern to our religious standing, is from our reflections through our taken prospect of 'Past Gods Words'. Meaning to all religions, our standard is purely of human extraction and from that demeanour, we are to be in for the final preparation of conduct of how we might present ourselves to God or Allah by our determination 'Past Gods Words'?

In making that move, although our political outlook is the real master of our reactive head by interactive means expressed through all of our systems, notice is to be given by example of how we have used them in the wrong order, leastways since the ending of World War two. My own measure in recognition of which is considerable by how I have been able to focus on political matters, falls to be heavily critical of some, because in addressing the route of our many world problems exposed from that time, our energies have been misused by over statement, from which I would deem to address in the abstract!?

Counting these past sixty plus years gives us a new era point of reference, along with the other marker I would use for our best considerations, from our ability of all being involved with our ritual burial at Shanidar or else where? Between those two marks or periods of time from sixty to one hundred thousand years ago and sixty odd years ago, is encompassed the full period of time where the measure of all of our progress had taken place throughout the whole world. So much so that it might be considered that anything we have done for local or global control since the end of hostilities in nineteen forty five, could only have been done in or with complete global understanding, which sadly is not quite the case?

My own relation to these past sixty years has come about by the same circumstances to have affected us all in all cases, being those delivered by the circumstances of geography, climate, diet and other reasons. If I have suffered any advantage it has been in the way I have been able to recognise the obvious for what it is and offer the medium of change without alteration? I have queried in thought and doubt as we naturally do, and in most cases like the vast majority of us have done so in the quiet belief of our future. My main fault of personality in open confession has been of my personal doubt in our individual ability to make a difference. This feeling has been expanded by the most recent of times by how our religions and politics have been forced to interact and react by our own shortcomings!

In placing this work at our disposal, I have set it to be a vocal point of our connection far and above our separate qualities, not least by its very formation from our shared wisdom and our ability to be positive in all open exchanges from now on. My hesitant lead at this time relates to the standard of our general political outlook, which is solid by all comparisons except in structure, simply because it has learnt to doubt its own morality and in haste has made do with rough repairs of any nature. I take my personal stand to be formulated from our conduct these past sixty plus years only, but counting all other inclusions. In so, the most violent notice I can give to thwart political reasoning in balance to our religious standing is by the core question of "are we truly civilised"?

In asking to place that question again has been to point the difference between our religions and politics on a massive scale, for by one we are, and from the other we are not? By previous mention along with other features and acceptances our standing is what it is, but to consider that we might have come to a fairly recent conclusion that we are truly of the political opinion whereby we can classify ourselves into being of a first, second or third world status. Is enough to question all of our abilities of judgement, when at the same time we assert that we are all equal by religious tones given as shown by the use in application of our political classification to be religiously tolerant?

From where I question all of these devices is done by the time span of these last sixty years, which by coincidence is in rough alignment to our individual lifecycles where we each react by our own past judgements, which propagate our differences unseen? By taking the idea of the Fifth Dimension in overlay to our two defined guiding bodies, we will more readily see how much of what I have said has a credible thread without being judgemental. I offer no competition save that which is generated by how we might apply the same option over our two governing/controlling bodies even if as individuals we are only attached to one? As I had stated that the outcome would be different, because of how we had promoted our attitudes to our politics and religions as separate. It is from that study we may see the level of how our attitudes have changed even within this sixty year period!

By any and all methods of how we have conducted world affairs, if by some extent through the United Nations symbolically, our methods have only carried the suggestion of remedy without substance. I take this as a failing, but not by deliberate intent when by the construction of the rules of conduct we have clouded all issues. Like by example, how from the adoption of our human rights standard, which unfortunately has not been an asset, because from its very motivation our approach was misdirected?

By the ending of World War two as a social marker, our main aim was to create the situation whereby our future judgements of each other was to maintain the process of equality through the heading of human rights or even basic human rights? In all of our noble intentions at the outset, by our excellence of mind but error or thought, our original plan was compromised by not being able to openly address our human relations in other forms? My own described opinions of the nature of our differences has been continually mentioned in part throughout these pages, which in our representation is valued, because they will make us see matters from a different perspective. My use of mention in relation to some individuals or of some nationals or of religious groups, has been of the same intention ever done by many noble and honest and worthy people from amongst all of our histories. Above intent, what I do expect, and am prepared to drive for, is for us to draw the one open conclusion that what is said here by the written word is totally of our times by this very last modern era, posted since the ending of World War Two. Leading in a matter of fact manner, which was probably the only time when all nations were of an opinion by act, deed or sufferance leading to change by future reference in all ways or in one way or another to suit a particular group, politically or religiously?

Allowing for the haste that was in promotion of meeting all needs to the shortest of routes, even if at first bestowing basic human rights upon all without levy and without the real understanding of the concept or how it was balanced?

Human rights will always be a massive promotion to the good, but will fall in its own shadow unless we strive to understand it beyond conception. Of its biggest weakness seen in the hidden recesses of its actuality, and from where the concept finds doubt has been by offering the same plan to the same level politically, religiously and in neutral. Allowing in genuine claim that each of us holding to a particular source might invoke its principles, which compromise others and others of the same standing as we, but with there own outlook! If by adoption over these past sixty years, we have been able to exercise our own perceived right to practice the art of genocide on every living continent and at any given time, then needs must in an order of change. If Europe by its own manner of sophisticated and civilised behaviour from which many supported modern principles have sprouted by our own representations and input? How were we able to practice the updated version of religious genocide under our own political heading of ethnic cleansing, not only do we question our standard of moral control, but further weaken our first fine modern concept! By reflection it is important to note that the done deeds of ethnic cleansing with the implication of more to follow in Europe, was covered by pseudo political objectives! From the religious perspective, the matter in those terms was seen to be between two religions that had originated in the Middle East one from about two thousand years ago and the other from about fourteen hundred years ago. Victims of either brush were counted as the dead!

Having said that our religions had always carried their own sense of equality if to be human rights, does not question the balance of our religions and politics by interaction, even by the cases that I have just mentioned! Where my judgements fall are in our existing time lapse so that we can relate our acceptances as they had matured without proper consideration. Having named other countries and America in particular by influence to our modernity, the matter was to bring us to recognise open thought about our progress, if that is how we like to believe we have moved? By some associations that I made in reference to how America was formed, including notice of its brutal inauguration. By due process I had stated that for us all to be in the same achieved state as America now, is a worthy goal in respect of how she had sprung from European progression, which itself had sprung from our progress first made in Mesopotamia? I had also indicated of a crucial time in Americas own development being of huge significance to us all by reference to the changing world we are in, this phenomenon also occurred in America who had its own trial in the mid nineteen sixties?

Having stated in request that the feature of human rights and basic human rights were aimed to be installed for the individual no matter what there circumstances of politics or religion were. If to have been set to taunt the side holding to be communistic and in theory ungodly by inclusion to all individuals, even if some had there own religious standard, then by licence of the United Nations we had a manner to check the theory? Except when the situation in America at roughly the time stated within this last modern era, exposed itself to be nationally internal but with international ramifications? In keeping within my boundaries set by recent times, but not to undermine previous desperate situations in America itself, the real question in meaning of human rights was again raised by Americas own black skinned population?

Being a vast group of people who until they raised there own voice for the general good of basic equal human rights, not only to oppose communism, but were now to created as a bi product the first proof of modernisms fragility. America even while on the moral high ground were prompting for self determination to former colonial or tied nations to other forms of imperialism, seemed to forget here own internal obligations. Black Americans, a pseudonym for some Americans, were held to a different account than others even with the vast majority adhering to Christian religious principles, had equal rights, but were not able to exercise them? By sheer coincidence this picture was repeated throughout Europe! What emerged was a style of correction to a bad deal, but alas by the wrong formula?

I have taken the following view aided by my ability to use the abstract on certain levels. By no exact science nor adhering to any formula, methods of correction taken from the American and slightly different European perspectives, was to legislate the need of correction into the stated different coloured nationals, which by some terms was a new, if not fully recognised condition of fact? Both factions in laying internal laws to suit various situations, but aligned to the cover of internationalism, failed to notice the obvious obstruction we were placing to better and future forms of reconciliation? By having black Americans, black British, French, Italians or Chinese or Japanese, we were only being divisive and at the same time compromising the best intentions of our standard of human rights or basic human rights for all. By the same feature of legislating one way or the other and in adding the fixture of being politically correct as another token of our understanding, we were once again setting standards of our differences. Which have since that cut of time in the mid nineteen sixties, shown how some of us can and do react exercising our, our, our human rights to ethnically cleanse our own area of importance for what ever reason locked to be within our religious codes or from our political standing?

By the introduction of any word in law to describe a person other than being 'Out of Africa' has allowed the fact of our singularity be seriously, cruelly even brutally questioned by reason that other than our own claimed national standards. We are not Black, Brown, Yellow or White, we are who we are, and who we wish to be by being Russian, new Zealanders, Kenyan, Chinese, Canadian, Iranian or others and others. This is to bear upon us all so that no more can we enter a world war by the ethos of being chosen, to be superior or naturally selected in being superior. Like the situation in Europe through the German Nazi party of super humans, or from Asia with the Japanese militarists of the same inclination through there own imperialistic outlook.

To the matter of our standing by introducing some anti racist laws pertaining to colour, more than any other feature, we have elongated the process of time when we will simply accept each other for what we are measured by our same fragile lifecycle. All of which is not, not to say we must abandon our efforts to better enlightenment, what is asked is that we start counting tomorrow by a fifty plus or seventy plus year lifecycle? For us to think on those terms in particular is nothing more than we have been doing at hard labour since the ending of that final last great conflict of world proportions. From the political perspective by due compensation in how we have managed to keep our footing over this last two hundred years has been nothing short of being religiously miraculous. It has not been easy to notice our faults because of the effort put into correction, but by first notice in this last modern era is perhaps the best chance we can take to channel our civil means of control to the expectation of our new religious objectives.

However we might tackle the management of change required to implement better direction is to be set not on the horizon, but what may yet be seen from sailing into the sunset or by reflection in sailing from the sunrise! Each giving a different time of when a new image is to be recognised, but in all cases from now, our load can be lightened in not carrying the encasement of unnecessary laws or legislation at first politically?

## CHAPTER 103

Politics is never bound to lie down, nor should we attempt to lay it so without deep and fair thought, however, when placing a different emphasis to one or many of its tentacles we can challenge effect if shown from another perspective such as from our religions. I had already indicated that the difference between our two main governing bodies would be highlighted by the overlay of both with the idea of the Fifth Dimension and our three taken prospects with one further being active. By mention of what may be considered our first taken prospect being 'Out of Africa', because it alludes to the beginning, we can readily see the stated difference by some of our attitudes? With religious opinions that prospect is already in acceptance by style if not by reference, because of the expected outcome by divine decree if suited to any religion? By other standards as we are all by our mode of connection to the same religion, when by most philosophies we are to be equal in death, therefore our equality in life does not come into question, save by our own means?

Where politics has been able to question that aspect of promises made on the religious sphere that our best rewards are to occur after life from death, by the matter of that reward system, some religious attitudes are sometimes compromised? Not least by show that it was religions that were our driving forces for all change throughout our histories, but it was our civil order and later political institutions that covered the credit. By the example that I use, like how a foreign body of reason like the Catholic Church, was able to overpower such a mighty beast as the Roman Empire. Only to be turned into several national bodies of the same core religion, when from the political decline of empire, and in watching the final replacement of such political order mimicked on several different national levels particularly in Europe. Making the twist, from a one only imperial religion to become several national ones, falling to the line of dictate from several political opinions nationally! By many processes of blending, this situation has been further manifest by these recent sixty years giving an overall mix, whereby we do lead off the same foot, be it American/European led, with an in ground or from an imbedded Christian outlook, but now more politically?

My mixture above is made in relation to these very modern times in judgement, but without sentence, I raise the point in the obvious of where our leads are from, politically and religiously, also to bring into focus of how some of our religions are reacting to each other politically! Our Middle East picture cannot be overlooked religiously first, because it is/was from their that our new last era is of its greatest concern?

By whatever political decisions we had or have or do use to address any situations in the whole region politically, is only the latest, but we have not yet realised that feature? From the Iraq/Iran war which was politically inclined, although fought of the same religion, if that was the case leaving aside motive? From the Israeli/Arab wars and attachments before then, there was deep religious sentiments applied or rekindled at the end of the first world war in the name of Jewish and Palestinian self desire, but in the prospect of justifiable realisations in the near future one sided or both ways?

Later when at the end of the Second World War in forty five, Israeli nationality was established in the modest form of the worlds only Jewish state, then a direction was set without being realised? From previous pages I had made the mark alluding to Jewish terrorism in claiming their sighted objectives against the same standard of Palestinian nationalism. By many further processes of Arab to be Islamic involvement in open opposition to such a situation of such a situation. I had also indicated that early Israeli support was given by American and European influences with political seasoning, because at a specific time when there was the/a bloc culture motivation to the emerging global situation commercially and economically. But of paramount importance to that total area of the Middle East in general and as such features of human rights and religious tolerance were also proffered, mists began to cloud most issues particularly of the religious vein. So much so that little consideration was given, when for a dangerous length of time, some Palestinian action groups came to be termed fully terrorist before the realisation of being nationalistic was given by any sort of comparison, which could have be related to the former Israeli/Jewish struggle?

For the same open consideration when through the Iran/Iraq war in the slaughter of over a million Muslims by self upon self, no generosity was offered to another of our human tragedies. Perhaps because that particular conflict did not have enough substance to worry the nerve or conscience of our own United Nations, being our own voice of this our global community, or perhaps it was perceived to be a conflict of religious tolerance, permitted by all decrees! From which if I draw the comparison that by another realisation from when in conflict, without specified merit, some individuals in heartfelt self harmony, defied life by destroying their own conscience, when destroying enemy factions in the proclaimed name of martyrdom?

Alas without the sentiment of remorse, fed from the regions of abstract thought in mind of my own ability of reasoning and sadly supported by other methods of study. A huge outfall from the Iraq/Iran war in terms of our own outlook had been then or now since, related to expose the factor of a specific trait owed to the religion of Islam only, or! By the acceptance of equations submitted from any quarter, but in the said case that I now personally submit from our borrowed ability to self reason by different counts. In taking that the specific war I mention led to the supposition that Islamic martyrdom had any difference in substance than similar acts recently described by poetry. Whereby "a greater love has no man than he might lay down his life for his brother", in kind only! But by the twist of the pen and in the blinking of an eye, deeply wrapt in supposition, possibilities, probabilities or imagination or even here say, to die by/in martyrdom could become suicide, could become murder?

Could become suicide, to take away a credible form of religious remorse, and be replaced with a political interpretation of being motivated only politically in the terms of murder by the means of practiced terrorism fully in the political medium seen half measured! Inducing our renowned steps taken syndrome where all new responses are from that medium, even carried by a collective focal point for reconciliation through the United Nations in support of the same standard of human rights and basic human rights, endorsing all religious and political acts by quarter to be civilised, and in following the reasons of conduct to be politically correct and religiously tolerant!

Once we begin to move or think on certain plains intermixing our influences of politics and our religions unbalanced, because we question either from different values, we err in automatic drive by the assumption of what we now do is correct? From our past and because of what we are ultimately trying to achieve, the above is an example of how we have progressed, but also a warning shot across our bows. In desperately trying for success by all honest means, our haste has driven us to react in the immediate to maintain what we have earnestly considered to have been considered progress, which is commendable. But now can be seen as fragile by examples such as this work and many others over these past sixty years with others from before, setting perfect direction, though perhaps in being too early for there day or time of publication. By best example I can use as a side dish, Darwin's The Origin of Specie publication, had a positive message, but how was that read?

His legacy much interpreted by myself, although a simple statement in explanation of concerns to evolution involving time spans far in excess of those needed to the accepted aspect of creation, was obvious and simple. It was we who have complicated the matter beyond all reasonable doubt by taking his thoughts and studies and for some of us to then manipulate them to what he actually meant, instead of what he might have meant? Reactions were heartfelt by both camps, but even until today have split some of our reasoning by what can only be described as our own half measured approach inducing our steps taken syndrome? Whereby through the human mind we sometimes follow a chosen path from what amounts to habit by previous conditioning. From where I had set the two features of evolution and creation in tandem was done to give us all another perspective and perhaps show that we do always have alternatives of reasoning?

I had previously mentioned in agreement of those features to be fully complimentary in that evolution was of god's time scale and creation was how we could think in terms of that mater in relation to ourselves having been directly created? Being simply a matter of our own human indifference then ago by our natural habit of study, we were pushed to give a structure to our very means from all that we began to remember in the proximity of Mesopotamia and the Middle East in general. Like for the finding of god, being chosen by god, and so mixing the matter to be of our understanding these four thousand years and more, save by the recent intervention of politics which has progressed our excellence of mind into further error of thought on many levels, even now unnoticed!

By a cruel mixture of events and somewhat backed by the results of fourth dimension study, our standing of today in some cases has been factionalised into the two same camps that had showed open contempt to each other from the time of Darwin's publication. By that mark religions and creationism has fallen to be the weaker of the two edifice of our judgement, because from the other view being evolutionary and political, that feature has grown to be more real and acceptable even unfound. But by the economy of reasoning, not least that when we speak in judgement by religious terms in relation to most serious world religions, they are answerable through a feature that the fourth dimension has proved to be unviable, like anything being able to be done or recorded as done from any of us deceased!?

Leading that we have been able to assemble our methods of conduct as best they should be, but in concern of our religious concerns to be religiously tolerant of each other and attend to those matters religiously! All of which without force has enabled the dangerous situation we are in today continually develop. Whereby on political terms if we can address what might be seen as a matter of religious penance through the act of martyrdom, be translated into terrorism? Then the face of religious tolerance has been compromised, not against intermixed religions, but in subservient abuse by some political reasoning unread! I add in haste for anybodies consideration, that the above situation does not exist, does not exist, because from perception with any care that that situation might raise is covered by how we have allowed erroneously allowed evolution and creation become separated!

## CHAPTER 104

Without any charge in cost, I would offer that one of the best and least complicated ways we might sanction our timing features of evolution and or creation is reflect upon our own single lifecycle by whatever measure. From which we will best be able and realise that it is only from these few short years that we have open session to meet our best realisations? All of which is not to deny the effort and understanding of all of our ancestors even from the period when we were first 'Out of Africa' some three and a half million years ago. By all the greatest of any realisations the best is yet to come, which has always been our stated obligation of study by all means, but only religiously? That is why I have attached more importance to the understanding of us individually, with our interpreting the difference in overlay on our political and or religious standards with the three taken and one active prospect of my Fifth Dimension!

By attaching my own title as a first reference to the Fifth Dimension now, is done to be a better means of protection, because of the suggestion that ultimately we must cut through our stated aspect of religious tolerance. With most religions having been formed within the time band of these last four thousand years, for me to have suggested that instigation was first contemplated some sixty to one hundred thousand years ago by signs at Shanidar. Has not been done to side either with evolution or creation, what is offered has been the notice of our better judgements already to the fore before any other consideration. In full recognition of how we must have had a prevalent means of civil order then ago, is as good an indicator of our direction to be separate from other specie. All of which shows that as that/this single specie we are truly unique by all other comparisons.

Working to those assumptions from our time span of these last sixty years we have lost the notice of how our political controls have overtaken those that were to have been our religious ones. So much so that in mixed comparison we have and do allow our political standards reclassify some of our religious acceptances under cover of having a lead if by our United Nations? Not to have remembered some religious acts of terrorism, like ethnic cleansing wherever throughout the world, but allowing some of us to classify what could have been misguided martyrdom into former and new acts of terrorism religiously, but by political determinations! All of which crosses all barriers of our own codes of conduct when we allow religious tolerance by the same political reasoning that we use to make political judgements of what to some could be the proper practice under the definitions of religious doctrine if?

Of the questions I have offered in reflection to any matters of the difference between our religions and politics, I have seen the worry of our concerns spread from recent times by how we have let politics rule the day for expediency. Because we have shown reluctance in the belief of our own religions fed in part of how we are supposed to manage the aspect of religious tolerance permitted by the range of political understanding fed by our half measured reasoning and our following steps taken syndrome.

By issue, that we can tackle some cause of our differences having been consolidated by our mixed outlook into the avenues of understanding that we believe we have generated by these past sixty years. My own take about the Fifth Dimension will help best to our religious determinations, but not until an open view is taken of all of them, leastways our three Middle East religions of Judaism, Christianity and Islam. All are closely linked by agreed terms of reference but are capable of misleading us not from their cause, but our own interpretations. Of which the biggest example is between Christianity and Islam politically, which stimulates my own desire in our contemplations that are to be expected when using the Fifth Dimension in even pattern to all religions.

Without obvious definition, it would appear that by there own good claim and religious judgement, the Jewish religion by dint of the political instigation of the state of Israel in the area of my note over much of this work. Can and do practice all the connected forms of permitted conduct through our political standing of religious tolerance and have been ably supported by many Christian mediums through America and Europe politically. While from the same geographic area of the world with the closest of links to our first civilised stirring, even before religions were in open prospect, our political standards grew by relationship between us in the land between two rivers? To be now focused by political perceptions if to ignore other aspects of our political religious tolerance in owing to the religion of Islam seen to be of separate conduct?

What has been achieved albeit unseen and even un-noticed is that by modern political assessment, we have found it possible to categorise almost the same religions of Judaism and Islam under different outlooks, and have been able to asses their perceptions differently by expedient political acceptances! Without the critic, but in open observation, this misdeed pivots on or from where we have set to draw the line, which in cases has offered the political argument whereby some religions and in this case Islam, can be falsely misjudged and or vilified or not! By that score, but not of single extraction, I have set to open our realisation of what we have done even unguided but in judgement if unseen.

A danger to the aspect of our standing by religious tolerance has always been that we have left most religious choices to have been self regulated until such a time set, whereby if a religion by way of its adherents was to offer its own solutions, we were in the balanced position to question all aspects politically? I cannot dress our situations personally as so, for to do that is to deny our natural progress which is ever to be shared. What I have set out to do throughout this our work is to lay the same standards in all cases by all readings, but with/in/from the good fortune to have recognised the three taken prospects and one active of the Fifth Dimension, to be able and have equal influences upon us all. Most likely, with that as a base, it is fairly possible for each of us to question each others religious perceptions, without forfeit or condemnation, but only in the aim of query!

From that kind of reasoning, unblessed questions to be asked, can broach all of our religions for how they had and do question there own motives and have done so for a four, two and one thousand four hundred year period. To now question any religion by and with use of my Fifth Dimensional prospects is done of its adherents only, and not the base or standard of the particular faith. This has to be so, because from whatever standing a religion has, it is to address our relationship with God in Allah or the Great Spirit by our own self conduct bounded by our own lifecycles and nothing else! We as individuals are to be our own judges of self conduct only, of self conduct only, of self conduct only! Not in question but acceptance and then not necessarily in compliance of doctrine or our own doctrine. Because when entered, all of the Middle East Religions answer to the theory of life after death, and therefore cannot legislate what our political conduct should be in relation to our religions while we are only living, necessarily!?

Making it in first notice that it is only now possible to make the differences between our religions and politics with the same overlay of the Fifth Dimension, because although the three taken prospect are the same for both entities. In one case they refer to aspects of living only and in the other the same three refer to life and death by the course of our beliefs supported by the ultimate in hope and expectation! Because of the addition through our religions that the three taken prospects can also refer to activity by our death in expectation, is why and how the Fifth Dimension will empower our religions more so than our politics, redressing any imbalance that we have seemed to accept without objection.

## CHAPTER 105

To be able and stand apart from our political controls at such a time as this is of the utmost importance, because by that deed we are setting future operational needs? Our religions without fanfare are to be rightly elevated above how we should review our political expectations, not in competition, but simply as a means to set direction? By reflection of how I had pointed to the fact that our politics were to have been our best providers, because it was only them who actually delivered all sorts of materialism, therefore we owed them most heed? My concern to this situation has been highlighted by how from the management of our conduct in order to cope with the ability of our religious acceptances they had been dealt second hand because of there promise to deliver after we die!

My case has been supported by reason of how by addressing our religious standards, we had supposedly met upon them all the same limitation of conduct. But in reality had reacted to some more than others not for reasons of our religious wellbeing, but in order to make political observations, because at particular times recently we engaged in bloc culture stand offs. Sidelining some religious attitudes because at certain times they had not had an obvious influence on the outcome of what we had tried to arrange politically first? From such recent times as these influences were felt, when different religious behaviour was seen to be a normal method of reactive conduct, then in misreading the outcome, our reactions were not standard, but bias in favour of now, instead of all tomorrows!

Instead of allowing solutions be manifest by supposed standards of religious tolerance in expectation, our own political reactions took overall direction, because we had not considered the possibility in depth of some religious feelings or motivations? Because by all political standards, religions were deemed to be of the imagination only, so much so because of the heart of Moses, Jesus and Mohammed and others were not in compliance to the fourth dimension or scientific community! Equal acceptance in the form of religious tolerance was offered even if not really supported by factors of acknowledgement to the range of our different religions. By giving our religions that equal status we effectively reduced there overall base in respect of there real influence to modernity. Because of the veil in which they all moved they were not to be recognised to be the greater part of valour by there open moves to establish full credibility of our uniqueness in relation to other specie in as short a time span beyond all previous imagination!

Our recent sixty years of endeavour to rationalise our civilised behaviour has been well spent to make all acceptances real, but by political standards only, most ably shown in how by other means, reasons and from our tainted ability we had began to mix our religions and set them to political acceptances. All of which can work by treating them so, providing that in the act we are not selective in choice to support one against the other, because we have had no other means to react differently until this time of the Fifth Dimension? In as far as we have come along any roads in the future, before the advent of our three taken and one active prospect, we had come to our biggest crossroads by how we showed political reactions to religious problems. Allowing for the inclusion of the Fifth Dimension proper, is set to redress any so acts marked in triplicate and all events from the mid nineteen sixties?

By the very real factor of not paying direct attention to the voice of Islam, at first expressed to most general nationalities at times when the world was in effort to accommodate all needs one way or the other? If from our United Nations moving in slow forward, we allowed ourselves slip into neutral when exposed to some Palestinian reactions later endorsed to be of particular reference to the thinking in or of or from the faith of Islam! By turning our focuses to the over importance of our political judgements in faith, and by reducing the standing of real faith by the fixed offer of religious tolerance, we began to set the fixture of real confrontation?

By any means when at any particular time we began to call some of the Islamic political reactions being other acts of terrorism, in as much as justification by that standard could have been argued over, we destroyed that avenue of communication? Not unlike when el conquistador was permitted to challenge and change the practice of human sacrifice, had we been set query by such misguided reasoning, because el conquistador set a political solution to manage a religious determination or problem if accepted as so!

By the abstract and in tender care of our well won religious concepts, in suggesting by the plethora of recent events over these past twenty to forty to sixty years ago, we had settled upon politics over religious choices first. Then in some semblance we had the choices to think and react for now, over our ability to consider the future from the insights of our own death? In choosing politics we had no other recourse than to diminish the full sway of religious interpretations, which had seemed to have worked reasonably from an American/European based Christian philosophical outlook? But was found to be wanting in recognising the Jewish or the Islamic considerations by their own interpretations? Ably expressed politically when from the outset Christian based religious philosophies supported a Jewish based philosophy only by appearance, which was never apparent even now by some Islamic suggestions if by reaction?

From the stated times when Islam reacted religiously, which has to be seen in the same level of reaction that the Jews of Israel had and did and do. Their condemnation was of the political vein to be cast in political light only, as operating beyond the fringe as freedom fighters or country makers, but invited to destruction by stature from political determinations only. Allowing that even now if Islam voices any religious opinion in relation to events of this exact time they are read politically. Making the full circle in half measured judgements by alluding that all sides are treated the same in the measure of religious tolerance, but that for one to be devout as the Jews of Israel have fairly professed to be. Yet the other is fundamentally revolutionary by disruption and in cause to permit human sacrifice by the reading of suicide murder into martyrdom. So might work some of our political steps taken syndromes in solution?

From similar connections of how we can read some of our political ideals into the wrong method of tackling solutions, has been the main reason why I have made it so for us to take first notice from the overlay of our Fifth Dimensional standard upon our religions. This is to be done by the leadership of each set or sect of all divisions of all religions, which is not daunting, because each hold to the same head in God and Allah and with belief in the Great Spirit even without specified leaders! Nothing really changes except by our own ability to rationalise life with death in any case, revealed in understanding to our own form of religious acceptance. In looking more closely to the three Middle East religions is done in the same representation to all and all other religions without special favour.

With only the minimum of direction, but taking to the collective standard that all procrastinations are to be set in representation of what achievements we might gain while we each are still alive, brings us all to the same task. All important issues to be raised by those that lead in our religious standards and standing are to be fitted to our span of living by its own set limitation bound to the clause of 'Natural body Death'. From the Jews who have a collection of rabbi councils in representation in many countries, even if at first now fitted to the western style of thinking. Then the message by all findings through the acceptance of the Fifth Dimension in overlay will not challenge what might have become like their original perceptions? From the caution of time with the maturity of age, I have not implied that the Israelis/Jews are now of a western outlook by choice, connected to anything different than other natural changes. There condition is long winded and similar to us all levied by the circumstances of climate, geography, diet and other reasons. My first caution to their leaders is for us not to ignore our pasts nor live by them, when it will be seen that the Fifth Dimensional overlay is of open relevance to our singular lifecycles?

Jewish realisations when done, and by us representatively, can expect to have a better collective understanding of what is meant by the idea of the Fifth Dimension conceptually. Although recognised to have only one world country as a full Jewish state, there are many strong Jewish communities throughout the world. Of the total world population of between some fifteen and twenty million Jews world wide, and fitting by the general inward looking religion Judaism has become. Consensus is more likely, because in the best of positive attitudes, in this case their priesthood falls to be representative of the total congregation in the most part. Which could mean that the limit of acceptance of my proposal in heart, could be better seen not to be challenging, nor, or against the essence of any fair religion asking us to forebear with God in mind and think in human terms only, 'Past Gods Words'?

Moving in title to my second dimensional religion of Christianity the picture broadens but is held by the same frame of my/our Fifth Dimensional overlay. By open recognition to the vastness of the many sects in the proclaimed service of Christ, even when taking Roman Catholic to be the biggest and headed by the Pope. From where one man was to represent the many, some voices of derision would automatically be raised in question and query of what might be said or suggested. Seen from how the catholic religion has already fractured into many thousands of Christian religions, the picture is not lost, but shows our ability to self reason. Although the Pope in our service would have all the means at his disposal such as a feast of bishops and cardinals to reflect upon and or with, all tasks are easier when seen of the three taken and one active prospect we are to consider. These would not and does not compromise his standing of religious infallibility if so, because they are self explanatory by code. His brief like to the Jewish rabbi is of our use by human terms only, but to be able and support religious perceptions over political ones!

Without secondary consideration the religion of Judaism from is stated history of two thousand years before being changed into the new Judaism of Christianity some two thousand years ago. Hold in direct relationship by the connection of many religious habits or from or by the supposition of religious text, albeit in recognition of intention by other parties. But have allowed their original concepts be swept by political ideals, which has broadened my haste to the instigation of religious activity placed before political overtures as standard. By my own means of connection I have linked Judaism and its own following of new Judaism in Christianity, by one major original topic expressed amongst the Ten Commandments in basic form to be of base standing to both separate religions?

However expressed, "thou shall not kill" is of direct meaning and in direct relationship to us in our human form, whereby we do not kill of self no matter what! Although I accept that the meaning is of personal recollection it is bound by Gods covenant in all cases and in direct relationship to those and all religions? From where I myself have set the standard for us to heed of another taken prospect in 'Past Gods Words' does not nullify the understanding of my own and our intentions? Even if expressed as civil law, the original commandments delivered unto Moses by the hand of God, were delivered to us all, no matter how or by or from concept, they were 'Past Gods Words'. All choices were always to be our own if only expressed in part by our ability to exercise our own free will, relayed by the circumstances of geography, climate, diet or other factors, to have broadened our base of acceptances and understanding!

Our own priesthoods are now of the ability to relay the management of our religious affairs, but in the collective fashion that each had been shaped by their own examples of time and from the studies within those specific time bands. Their other main objective in relation to each other is to quell any competitive urges that have been recently introduced by some of our political practices. Like from how in this very modern era we have made political decisions to have had serious religious consequences if not always determined to have been of any or a particular religion?

Without implication I have relayed the hidden open connection between our first two Middle East religions that I had referred to as my first dimension being Jewish and of about four thousand years of age. The other of Christian characteristics and of about two thousand years of age, has likewise been referred to as my second dimension in the same method of connection that I have used for all of my dimensional references not in comparison, but in line by numbers. My last naming of specific religions as the third dimension has been related to the religion of Islam, with one particular difference of opinion, I had alluded to at various intervals throughout these pages, but in the best aims of discovery? Not to set any negative standards, only those that will show our true connected intentions whereby we all stand for our own welfare and those of each other known or not? My inclination is that of unity above separation, which will become more apparent over future pages, not least by our connective mediums above those that have forced us apart?

## CHAPTER 106

On a very simple level, the main difference I had forced to opinion between the three Middle East religions, stands not from the structure of doctrine or belief or faith, but in the form of there existing leadership? In prompting that the Jewish rabbi and the Christian priesthood structure were to take first lead of direction with the incorporation in overlay of my three taken and one active prospect of the Fifth Dimension. Leads to the question without the obvious line of an answer, who can be called upon to deliver the same study for the faithful of Islam? When from that faith and even to its credit at instigation, we were each expected by self lead to make our own standards, given at the time of instigation this was to be the best way forward when introducing a new religion with the same God in Allah that had always reigned in the same locality?

In having stated that Islam holds to a priestly set up, is not misleading when noted that their role is to check doctrine, to keep a similar line of direction for the adherents of the religion, but with most assistance to our young. Imams by some misinterpretation from holding to the role of teacher and leader, can mix in confusion for some of there own faithful the supposed final outlook of the intention of their own religion, not unlike other stated religions throughout the world. I cite the difference in the case of Islam to have been over complicated by the actions of others in separate, though linked religions, originally of the same area, but now led by our political determinations!

Because of the complicated means in which our outlooks from whatever stand have developed over these past sixty plus years since the end of World War Two. No means except speculations has been available for us to rationalise or standardise either political thinking or religious intentions, save to interact in the clumsy way that we have so far done. By approaching the form of decision making with our usual half measured standard politically, then allowing the outcome form by our steps taken syndrome. By working in assumption that our religions can take care of there own complexity through the inclination of being religiously tolerant of each other, has made its own dark hole of reasoning? By mixing that somehow we had now got it right, has only compounded the measure of how we had looked into the future with reactive eyes only, which has led us to form in structure this method of political management only. Any picture to be drawn from the above, are to be matched how by reacting only to political events in like manner, we accounted for all our religions to be self managed by the aura of religious tolerance? All of which has not been done in or through the intention of deceit, but are unarmed in the event of knowledge or understanding?

By reference I had given much time to the importance of these past sixty plus years of this last modern era. In and from all of our faults we have tried to understand the accent of purpose in relation to living and dying, which has been no departure from any time ever, save for the first time in human history we became aware of our own frailty through the open knowledge of our universality. Meaning if one standard was to be set for the human condition all were to be open to the intended benefit of it. That is why in our enlightened ways of expectation we introduced the balance of human rights with the method of being politically correct in conduct by it. In addition to the fixture of our religions in thinking that they were controlled by there own doctrine. We introduced civil laws within our separate national communities supposing to cover or be encompassed by the range of the aforementioned international controls, but we misaligned all of our best intentions?

From my own measure of study within the realm of true ignorance, I have been able and separate the vexing means of our differences to what might develop more seriously and what is of the moment? With all world religions included, but by standards, most in the highlight of my own observation has been done between our three dimensional religions of the Middle East, but more pointedly there relationship over these past sixty plus years.

From my lead that all political systems and or idealisms, along with all of our religions are to be overlaid with the prospects of the Fifth Dimension to stand in examination by separate code, bears reason. The three taken and one active prospect are there to fill the boots of all political values, because from all cuts and sizes they are to be the collective systems mentor. My intended religious mix bears no alteration when the very same three taken prospect are added, because for all time and in all cases most heartfelt religions have carried them from instigation, not least our three Middle East religions, even 'Past Gods Words'!?

By referring to my one active prospect so far at various times, gives no licence that I or in the same representation others as self might like to offer a better definition to how we can address change if needed. One active prospect in 'One for the Whole' is all we truly need at this time to promote any aspect of change, because it is our forum, of which by referral we can only speak in terms of our individual lifecycles religiously, far and above how we might speak politically! That in part is how and why the Fifth Dimensional overlay to our religions is of far more importance to be yet realised, because it will enable us to converse in open view to each other on our own religious grounds without forfeit? That is how I can make the offer of compliance to all our religious leaders in heed of our own expectations for the future in our representation. Because life as it is now is fifty plus or seventy plus years long for us individually within those spans, and not from a four, two or one thousand four hundred year history. But we are forever to be able and maintain the measure of our histories by meaning of then until now, by the query of circumstances then until now in all good faith!

By thinking through Moses, Jesus and Mohammed, there is very little difference between their motives. Where differences can be found amongst them has been misinterpreted, not by the process of religious awareness, but from the prevailing political situations. In tracing their three paths backward in Gods own time mark of evolution to the events that occurred at Shanidar. The picture seen then, would have been of three men looking into the prospect or possibility or inevitability of a future expectation managed by the prospect of Gods own creation. All arising from where we ourselves by all counts had thought in terms beyond our own management of laying a dead clan or tribesman to the comfort of death in expectation? Although done between some sixty to one hundred thousand years ago, the meaning in prospect was not rediscovered until some four thousand years ago. This time though we only focused on creation in partnership, which is of noble distinction, but when followed later in concept gave politics a wedge through the fourth dimension to question our endeavour allowing us only to be religiously tolerant?

Without the elongation or forced compression of time, from this modern era having worked to the belief that all was to be revealed in the comfort of death in relation to most religions and in particular our three Middle East ones. We have come to stutter in the expectation of our realisations. In keeping that I had regularly stated that our politics and religions have often worked or generated action from different perspectives, the same applies in how by suggestion, I had put the case for our religious leaders to even consider the Fifth Dimension? I again say that they all already employ the three taken prospects in principle without forfeit, alas to my one active in 'One for the Whole' there are divisive clauses, which are not to be of a serious nature in open review. Where I have been hesitant to mention in attitude, that from the Islamic stand, there has been a more pronounced attitude to the claim of martyrdom, stems not from how I might interpret the fixation of death in glory by religious determination. But in how it has been fixed or reacted to politically in these recent years, making the case I have taken, whereby in order to show effect we are to lay the Fifth Dimension over both issues separately. From what I deem to be unbalanced if not by political perceptions then religious ones also, even if uncounted and unfound for which in study the conclusions might always remain clouded unless we employ our active prospect in 'One for the Whole'?

From making the case that of the three Middle East religions, it is only Islam that employs the readily accepted standing of human death, other than by civil law. Is only representative in the respect that the other two in Judaism and Christianity had at instigation used the clause expressed in one of the Ten Commandments "thou shall not kill'? Even that both of these religions, one for two thousand six hundred years and the other for six hundred years before the rise of Islam, claimed to obey those ten fixed rules of God. Each had conducted there own affairs by civil and religious law in tandem, which on that very level compromised at least that one particular commandment but uncounted?

My interpretation that through Islam, it is possible even necessary for us to kill in the defence, promotion or protection of the last of three united religions from the Middle East, hangs by the threads of reasoning bore by these years since nineteen forty five and later. When first told by the writings of Mohammed, the proposal was to strengthen the whole dignity of man, mankind or by terms relative to the most prominent people of area, Arabs! From the very structure of the religion and in bearing to the other two that had gone before, his lead was on many fronts, but by considered calculations. Continuing in the vein of Moses and Jesus, the very formation of the new religion had almost all of the others collective traits, save 'my' interpretation of our ability to kill of self if necessary?

Without determination that has not been my feature in relation to the actual formation of our third dimensional religion, my concerns have only spread from these recent times, but bear deeper examination of how I might have viewed such matters. We must look again to our collective histories without bias, and agree to any collective consideration to what might have occurred? By connection I had alluded to how the Jewish religion had taken a lesser standing to its own newer form of Christianity. While the latter was expansive in the most generous fashion, but had allowed its ruling body to be moved in geographical terms to another continent even if unseen then? All propositions were to have reduced the main thrust of the religion by the factor of less direct contact, no matter that the religion was actually in practice to its own standard to be universal or catholic by concept? Therefore its point of focus of reference was to be impartial, when all were equal in all places?

If from the outcome of the actions of the Catholic religion, given in view by its own unfamiliarity to the real area of its instigation by connection through Abraham and Moses and others following. The matter was settled by taking the physical contact with Jesus through the first pope in Peter allowing infallibility on holy matters by all Popes, but in the foreign lands of Europe, coincidently? Allowing only recognition of the holy land where prophets once walked and ate of the very products of the land in question, now to be places of pilgrimage if directed through the eyes of Rome? Although in shrine, much great concern was given to where Christ Jesus had walked and worked and preached to the real product of all mankind, we still held to the missing link of Rome being the director of religious affairs through the Catholic Church which had been born in the Middle East?

If as it might have been or was seen that the management of such shrines could not be held in proper accord by a ruling force sited a thousand miles and weeks of travel from where events occurred some six hundred years earlier. Then in the proposition of suggestion that it was to be a further aim of Allah in God to so promote the previous situation, had we now/then come to a new bigger, brighter and more defined beginning or the beginning of an ending to old religions, with Islam rising anew in the old lands of the Middle East?

Un-sequent, but through Mohammed and in finite direction, new teachings were received and by interpretation in the force of command, were claimed to be from God to man through his own agents. Allowing that we might renew in covenant all of Gods own wishes, but now endorsed by the extra six hundred years of our own trial and error, which could only equate for us to be in the position of better reasoning and understanding. From all the failings we had shown by whatever management the one truth in several sections, was to be from now, carried as our pillars of wisdom into the future.

By methods and means of clarity, Mohammed in his own broad view knew well of the impact of the faiths of Judaism and again Christianity. By his own better understanding he had to relay these matters to those that he had felt directed to lead by much the same forces that had prevailed in the future from Abraham, Moses and Jesus in order of time but not necessarily by degrees of direction? Coupled within his own study and from what he was to learn in conduct, he was able to form by opinion for the instigation of a new religion from the remnants of the two main previous from the same area of geography that had always been, but from now to remain in high representation to us all for evermore, but how? Unless to employ the very tactics of the day by the mediums of power as force, education in learning or guile to fill the open needs expressed by our own feeling of separation to the overall aspect of the real future and what that might bring or was expected? Other religions gave to the imagination the great wonder of death incarnate, but burden free, what else can be expected?

Mohammed's whole demeanour was in our interest by the best of intentions, but vastly driven by his own personality, so much so that he felt comfortable in liking his own efforts to those of Jesus by status. By personal direction and in pattern to what had happened to the Church of Christ it must have seemed inconceivable to Mohammed, that Jesus, although in claim to have been the Son of God, could have been so! Not being judgemental, but by the same fixation and now after the six hundred extra years of interaction. He could see in picture of how Jesus might have got things a little wrong by claiming that he was representative to be the Son of God with new terms of enlightenment! Because now, Mohammed had the same real secondary contact that Jesus must have felt, but in the wisdom gained over these past six hundred years and from the way Jesus intentions had been interpreted, the field was open for renewal or replacement by direction from the last prophet directed through Gods own agent!

For the best of motives if not his own, Mohammed in line of conduct allowed the same fix to Abraham or Moses and Jesus the balanced status to have been prophets. Not least because in open declaration, if Christ Jesus was the Son of God, his force of power had been diminished by our conduct in all cases, because we had not been able to associate ourselves with, supposedly Gods own direction? Hence here stood Mohammed set to present new and better means whereby we could all reach the intentions of Abraham, Moses and not least Jesus as the last prophet before himself?

By one connection, no blame was ever attributed to the standing of Christ as the prophet, save of our adoption in error of his original intentions? It was from us as early Jews and gentiles who acclaimed Jesus in need or not. When hearing by word of mouth that the story of completion was in the making to the fulfilment of the Jewish scriptures bringing all men together as one in the face of God. Then it was almost obvious that from the main source of such findings, the speaker had to be himself of God! By the case that this man was indeed the cause of fulfilment in all of its best features, the certainty of the fact did not lead to the exact recording of what was said as it was said. In direct line to what had been in the understanding of the Jewish religion, whereby God was to the imminent release of all the burdens of humanity by that tribal connection but now to encompass all? Euphoria led in the direction of understanding when to be included in this fine and finite new form of Judaism all that was required to be done was offer self consent and open commitment. Even if the lines of endurance were somewhat blurred by sheer emotion?

We do know of how the early Catholic religion grew from its own quarters even in the light of sometimes brutal oppression, with the rest of the picture being informally committed to history in version. From the events by later reference and of slightly different perspectives and because of other better records kept in parallel, six hundred years later the now outcome by result was open to question with better reasoning having been added to the structure of that six hundred year history.

Mohammed had not set his own task to fill any gaps in the thinking of man or of our thinking collectively. We know of the direction he was inspired by from result, which has carried these past fourteen hundred years in its own strength and of our inspiration. Even from the un-clarity left to interpretation by people such as we in being at first mostly Arab or from the continuing populations that had first emerged into the quest of knowledge from the lands between two rivers. In statement that his new religion split into the factions of Sunni or Shiah by early connection is not to condemn the essence of the faith by any comparative form. It is said from my own interpretation on all of our behalf, because we are all in the position to overlay all religions or religion in particular with a product of these past fourteen hundred years of united human development?

From where Mohammed tackled the devices of query left over by other attitudes in the past six hundred years of stable religious history, he worked in union to relate the reality of life to the force of the individual by their conduct alone? From where and by how this was done, because of the methods of record, for the first real time in all of our histories the question of stability was fully tackled but short of finality. Because of the next six hundred years and the next and the next two hundred years bringing us to just about now?

Not unlike the Jews with their expectation if imminent unity with god virtually on the morrow? Not unlike the Catholics who from the same reasoning were prepared to accept the cruellest of persecution of their own lives because in the reality of death that expected unity with God would be fulfilled. In the new religion of Islam all of these fractions were to be made into the whole. But in order to gain those assurances other factors had to be brought to bear, not in measure of the future but in relation to the past or so we in our own acceptances had taken it to be? Effectively, whatever Mohammed was to say or record was at the peril of our interpretation and has remained so because of whom we are, not being he or Moses or Jesus or others?

Much of his fortitude gained from his own feelings and in reflection of our needs expressed by the actual times in which he lived were laid down in conjunction to accommodate by replacement all of our requirements!? With the general if real acceptance to the creed of monotheism from the whole area in which he then lived, only spoiled by our own behaviour if delivered from another perspective? Amongst other features he allowed was that by the first reason of maturity into this new religion, was to be done in the preparation of forcing the hand of doubters one way or the other. Did he see a better clarity, that for many of us in keeping to past offerings and religions, we were in mortal peril of our own future unless drastic steps were to be taken to move to the new religion?

By all channels if this was so, then it was by the order of the day and not much different how the other two main earlier religions had acted, even if not being able to equate to their own religious doctrine, because political/civil decisions had nullified their bequest of not to kill of self. By the order of the day and fitting to the rational, for Mohammed to express the feature of martyrdom in relation to the promotion, protection or defence of this new religion, no covenants were to have been broken then ago, then ago! What was to occur beyond all appeal was the matter of our own interpretation taking heed away from some aspects of what he had tried to fully achieve. Not least expressed by how from within this new religion soon after instigation and directly relative after the mortal death of Mohammed, we came to hold of it by two main opinions.

To be Shiah or Sunni with their own ramifications, some known to have been tragic on all of our accounts, at the specific time of any split to the core religion of Islam there was virtually to be no difference between either of them in terms of settlement by how the structure of the mean faith was interpreted even until today. Form within that structure both keep to the general methods of conduct laid down by Mohammed all those years ago which bears good testament to the value of the written word recorded at the specific time of described events or for conduct. By a cryptic note the religion even got more involved with, if from outside influences during the span of these past two hundred years, to bring our politics and religious of different standards? Coupled with the acceptance, that in principle, I have set to promote the tangible reason of our differences generated by the features of geography, climate, diet and other influences, allows for natural different attitudes! For some of us if excluded from one medium of what now has to be of our concern in politics then in some semblance we could be inclined to generate a deeper attachment to our own religion when maintained by a forceful written text that could be seen to be semi definitive?

Coupled with the brevity of necessity in so mentioning the feature of religious martyrdom, expressed by myself and by my own terms of reference, I do not speak for the faithful of Islam, save that the religion is open to my interpretation inclusive of how I might wish to defend, promote of protect its image? In turning full circle by several loops over these past fourteen hundred years, nothing from the core of intention of what Mohammed had said or was written on his behalf falters! Save our, interpretation from then to have been Sunni or Shiah or from now to be Jewish or Christian or of any other religion, except changes in the modern details of our existence, from which some of us have tried to repair the past with actions of the now?

In the finer detail of my thrust in relation of how we are to use our religions as the vanguard of change by employment of the many separate religious leaders we have in representation. For them to voice our opinions as a means of preparation only, needs also to be done by a headed lead from Islam, other than the fine writings of Mohammed, because now it is we who speak and interact in real time to the image of the future that Mohammed was to be hold! In harmony of how our religious duties had been set from the instigation of the religion, a fine point of that duty was emphasised by the fact that we were to police our own actions with the minimum of guidance from a structured priesthood.

All of which can lead to mighty misinterpretation of the two differences I have been alluding to, by trying to relate for all of our benefit, our own differences in attitude generated these past fourteen hundred years. But in highlight these past two hundred years and now in further reflection these past sixty plus years of this last modern era? From the cut off without qualification, it can be again said that the two basic differences between the three Middle East religions are both holding to Islam. They have not been gained from the effort or thinking or writings of Mohammed, but from our own conduct necessitated by the medium of interactive behaviour generated most in the two shorter time spans just mentioned. By first association in considering the level of our own conduct to the standard of the religion being our own responsibility, secures the basic standing of this religion in particular and any religion in general. It also helps to maintain the same general standards of the religion when the check in progress of our conduct is carried out by a mainly advisory body tending first to the matters of conduct in our places of primary worship, in our Mosques. Second by list, but first of consequence, is the standing of our own ability owing to that same individual status as we might see it. To defend, promote or protect the faith unto personal death and encompassed by the living feature of martyrdom. All of which had great bearing at a time of natural blood letting by character of the time we lived in fourteen hundred years ago, but has lost all meaning of content by this modern era and how we are trying to live together even if at times unbalanced. Our actions of today are to reflect our expectations of the day religiously first!

## CHAPTER 107

From all measurements of how I might have tried to interpret the past up to now, bears good human standing of how we have not yet completely destroyed our own efforts in the unity I as we are now involved in. At no time from any of our pasts had we been able to overlay either our political or religious emphasis with the feature of my Fifth Dimensional swing, generated the past five to six years, but in owing to these past sixty plus years only! Slightly before now, but within these past two hundred years by shear coincidence, using Charles Darwin as one of many corner stones will be seen how that even today we dispute over something he never even mentioned in his great work The Origin of Specie. He intoned to evolution in concept without direct reference, but not as to the main intention of his work. Yet we had readily used his references to or by explanation that the strong might be the best survivors under terms of natural selection over elongated periods of time for our own definitions and directed them to our own human racial superiority. A feature of which we are still trying to unravel these past sixty plus years by political means only if through such agents as the United Nations of which I have taxed in unfair bias?

In our close working and from our own desperate haste we have erroneously painted ourselves into a corner of discontent and labelled our religions in blame for most of our ills? Before we have even considered letting the paint dry, our drive has forced the pace whereby we are set to continue by the means of the fourth dimension in automatic, because it alone seemed to have delivered political bounty whereas our religions have not. Because science through the fourth dimension can talk to us in the terms of discovery and our religions not, it has aided our politics limit the show of our expectation from our religious desires and hopes, to languish in the realm of accepting tolerance of each other. Of which any objections if, have to be carried to the political forum with the force of meaning and to be interpreted of that particular religion as no other means of expression seem to be available. All of which has only compromised the political aspect of our directed means of religious tolerance keeping in check some of our non existing religious extremities? Unless some of our religious practices even if now and forever had been misguided, were to show by the very means of acting by religious tolerance again?

Politics in drawing a line in the sand has only been able to operate to the feature of scientific discoveries and expectations and not cope with our true religious belief from which all religions have been working since instigation. In order to equate the two aspects of our greatest human concern, by overlaying both with the three taken prospects and one active of the Fifth Dimension will be seen the measure of our divide but more importantly the best method of solution to many of our doubts. From the simple equation of setting our religions in the primary position to prepare for the advent of Heaven or Paradise of Valhalla or to Nirvana or fulfilment, but only from the human perspective we will reactivate new political initiatives. Just as if we were laying one of our clan or tribe to there final earthly rest in expectation, like at our burial at Shanidar!

At that time we only had the one fixation of what we were doing, but without any measure of explanation, then we were asking all of the questions that science has only ever been able to consider and from where they sought answers, we knew far better of the results? By referring to Shanidar I make no attempt to set any time reference from the event, save by its own time mark to have been far before the anointed time of humanity through the told stories of the bible. Although the bible is representative in terms of our first recording of what Gods intention was to be, was done in the error of our own importance making no other case but that of creation. By the dint of the same time, but spread over our own generational longevity and coupled with other religions of the area by there time spans of four, two or one thousand four hundred years, our religious errors are the reflection of our main political or scientific ones?

Whereas all religions operate by the conduct of the individual within there own lifespan even the eastern ones, all findings are related to discovery in death from which we cannot directly measure or so it is so by scientific objections? Science in turning to politics with is own bounty of answers has allowed the latter form its own opinions and so direct our conduct by religious terms with regard to longevity. But usurps that image by attending to our daily needs, yet is planning time beyond the limit of all single lifecycles in taking the political mix to be the most important by almost siding with evolution instead of creation. Allowing us disregard our human element in such cases that a political system turns us in habit to care more for King and country or the motherland or the fatherland or even mother earth.

If our needs are to become clearer instead of being our medicine only, then it is to be done by the stated heads or leaders of any and all of our religious sects to maintain better standards than our politics has shown over these past sixty plus years. For those ends, as an individual I have been generous to our own attempts of all in laying understanding open for us all. By serious considerations so as not to feed or nurture the misplaced, I have hung my own ideals upon our humanity, through our best religious acceptances, that is why at times I have been cautious when tripping around our three Middle East religions firsthand?

By reasonable and simple study, there is or never can be any denial, that all three being Jewish, Christian or of Islam are irreversibly linked by much more than their original area of advancement. Without study, because it has been unnecessary even ongoing, I have made one of their main differences between them, that two by line of assent had taken Gods own commandment not to kill of self, and I had allowed that the other does? Which is to be stated that the idea if used or accepted from all of our histories or not, has only come into relevance these past sixty years!?

Much of my above reasoning has been sited by the act of martyrdom in the religion of Islam primarily, care is to be taken and not use my words as enticement, for they are empty when only used to forward a point of view. For me to have intimated that of the three Middle East religions in keeping Gods covenant, with Islam as the only one to have openly and piously mentioned of its intention to protect, promote or defend its own value unto death by self or of self. Then the square was made better than from the two previous religions until perhaps these most recent of times in this last modern era, but at least now with the further reckoning of the Fifth Dimension and its worthy attachments we are allowed review!?

By referring directly to these modern times, we are not set to embarrass or admonish any religion, save by our own conduct from within, but now to be measured by our political judgements also. If through my own connection at first to Palestinian and of a similar time slot to the Iraq/Iran war, by suggestion to aim that martyrdom was first raised then by either cause, belies my own intention. My views in relief are not to the politics of the East West bloc antagonisms nor even to religious antagonisms, they stem from how such acts if claimed as martyrdom are to be honestly policed? Again I state my explanation of some events captured only with advancing modern thought in relation to these very past and recent times, not in condemnation, for that is offered by self delivery in our own human terms as ever. From where I might offer my own delivery stems how by doing one thing if wrong, it leads to worse by reaction and in many cases greater wrong of which we are not really yet equipped to cope with, unless we learn to apply our Fifth Dimensional prospects in waiting!

Many times I had mentioned of how the Islamic religion had been covered by the maxim of defending or protecting or promoting its own doctrine even unto the offer of death by self upon self. This feature I had stressed at first, was a necessary proactive means by which the religion was able to grow considering the nature of and the types of society we were in at those times some fourteen hundred years ago. By carrying the idea into this modern time was not done in relation to our then living but in mirrored reflection of how things began to pan out by our political leads only since the ending of global hostilities in nineteen forty five? From that picture I do not feel obligated to give a refined account of why we still continue so, except to explain the means of possible cause.

Although Mohammed was deeply involved with the initiation of the faith in all of our names and by his own visionary lead gave us insights into the future with uncanny accuracy. Some of his foresight was clouded by uncountable and unimagined events not least contained by or limited by our obtained standards in all the fields of learning since his time. Enough to be aware that we had grown in knowledge since the time of Abraham, Moses and Jesus but never at the time to be aware of how our politics would eventually take full control of our expectations in the following centuries. After all as a mortal man like his own interpretation of the other prophets before, how could they ever be aware that as of then we were not aware of new worlds on our own planet yet to be discovered bringing us to be able and form global opinion even if still only learning?

While still learning by our own generational sway in relation to our particular adopted religion the most difficult thing for any of us to do is be aware of the relationship of those religions to our ancestors. With some of them four thousand years old and carrying the same message in supposition. How then was an accurate measurement to be taken in conduct by the same standards, but counting by different and more developed attitudes induced by our natural progression over any time span? In mentioning that Mohammed could not have been made aware of how the future was to eventually work, is no slight to any form of imagination. His very fine work was in the stabilising of at first his and our final religion from an area of our familiarity, what was to become an off shoot of his first set of standards only manifest itself in these times of now.

## CHAPTER 108

Without counting the direct relation of how our politics had gained the overall lead to all matters of our real concern even to bestow upon us all the allowance of religious tolerance as a barrier to some of our extremities, when first done it was delivered half measured? By that means each of us in our own field was able to interpret our own actions and reactions expecting the others to always turn the other cheek, but in not understanding what was the most important of our influences. From which, I might have indicated what our line of thought should be from the formulation of events these past sixty plus years. It is more than essential that we consider the events of the past two hundred years in relief. So that even if we are from different camps we can understand why by adding more of my suggestions in supposition related to our real differences when we are in fact the same?

By sheer coincidence, from the nature of time in our own human terms and from our own frailty always similar, we had developed under the circumstances of geography, climate, diet and other reasons to be who we are now, but still do not realise that fact. In as much as I have given note in reference to these past two hundred years of our united histories, only two small factors really bear upon us all today. One from where as Europeans in first take again, we expanded our full global knowledge of this our earth by extending in representation of humanity to the new world by expansive and progressive means to form the new countries on the southern and northern continents of the Americas. From the second small factor in bearing to us all today, while at the time of our connected global expansion elsewhere, in the lands of the Middle East, the Islamic religious codes that had been adopted at and since the time of Mohammed, remained in the same situation, which is no bad thing?

Comparisons to be drawn are not to be taken of a competitive nature, but to my own interpretation of some events of which there was no difference whatsoever between us then and now. For what ever we did and by what ever rule, all events are only to have been measured by the code of humanity of which we all hold equal status except those that might be affected in other ways by carrying personal ailments? Again in open review by my earlier statement whereby I had offered that the North American goal was to have been the measurement of our united obtainment, was in abeyance of the rising situation? Because of other factors in common with the normal progress of this our single specie were brought in highlight by my own judgement, but contained by how things actually did develop again from European and the new European/American attitude in North America.

By statement that Europe had become progressive even if only politically, but tied in hand to religious thinking as it had developed since the time of Martin Luther in Germany over five hundred years ago. With credit or condemnation laid upon him for starting the Roman Catholic Christian reformation, our situation in Europe had indeed given us the drive to think more in political terms nationally also. From my own reading of how I have viewed other matters, my holding that peoples of other religions and because of our own special relationship through the three Middle East religions. By the circumstances of geography, climate, diet or other reasons we had formed to have developed different attitudes even to be laid back or less forward if only in political matters?

In pacing past the ramifications of el conquistador of over four hundred years ago, and in looking to the events of the past two hundred years in brief relief, in saying that most forward or progressive moves were done by Europeans, particularly sited through events in the North American continent holds by the rational. It is not a statement referring to betterment or superiority because at all times and now, any progressive act if, is ever done in representation of us all and by us all because of our singularity in being 'Out of Africa' and that we were collectively involved at Shanidar.

However Americans and Europeans began to see themselves/ourselves, the method of human ingenuity kicked in on being challenged by Darwin's great work The Origin of Specie. Without following a furrowed lead, by my own inclination I have seen the situation from amongst other reasons to have developed how it did and why. In referring to one or a mixture of political tentacles in trade, commerce and commodities, our modern thinking emphasis shifted from the morrow by religions to the now by politics, leastways by connected association to natural development now supported by natural selection commercialism?

Much of what I am alluding to is of a comparative nature between us, but not to be judged by a points system. My first aim has always swung by our own reasoning and the full acceptance of how we stand by our own standards in relation to each other and why. My insights of these past two hundred years are not laid in an order of merit, that is why I keep referring to our time at Shanidar which as the case shown was and is always to be part of our united histories. With the blessed advantage that in recording it by any measure save to its actual age of being an event carried out any time between sixty to one hundred thousand years ago, none of us can claim a special awareness of it from a clan, tribal, nationalistic or religious attitude by separate determining than any other tribe or group! We were no different then ago, although still in first fracture from our united origin 'Out of Africa', when now we traversed all the continents we carried in unity the prospect of God unrefined?

All of which has brought us to these recent sixty plus years from nineteen forty five onwards and continuing in the same way of looking to the future in answer to our first posed questions of all those years ago. By most of the questions we might have to yet ask of our progress, half measured responses and our induced steps taken syndrome are not going to be able and resolve issues if indeed any exist, which will always be challenged by our own stubborn or laid back or progressive approach to the future. Unless we are able to recognise the pivot of where we most differ and oil it from the bed of our own human rational and reasoning.

Without submission and if only by example, I have been able and view what I alone have considered to be a base of many of our historical settlements and air our opinions by them from the one perspective of us always having been this single specie. I claim this honour by self so as not to excite others in anticipation or to lay antagonism to others who might at first agree with my sentiments, for this statement is our own collective management, first to be expressed by the true status of our religious leaders. No one from any hand is to be disposed to kill of self by self, for no religious reasons have ever been ready to accept that standard, save through the spoiled imagery or adornment to the secular in terms of the sun, mother country, fatherland or objects other than in relationship to ourselves in living.

If I offer that the balance of society in all terms today, stands between the fixation of Americanism by representation of our/a political idiom and a single aspect of one of our Middle East religions, even wrongly described? In using how I have interpreted methods to defend, protect or promote the religion of Islam set in full time by modernism even from a laid back posture. Then the message to be carried is done in honest and unbiased appraisal owing to us all and in representation to us all, which is to be set in balance by the uneven perspectives of which both of these features have naturally developed!

America in representation holds by a western or democratic approach to all matters of political expediency, so much so to have been able and divorce herself from the human aspect of humanity. This is not a vane offering, but shows the full extent of human progression from how we have moved our natural drive which had previously been from our religious fortitude these past four, two and one thousand four hundred years of involvement. If ever we were embarrassed by the style of human death carried by the first and second world wars through political means, supported in most cases by the same religious standing from Christian outlooks, then other factors unseen were unavoidably brought to bear in these recent sixty plus years of now and still counting.

I have no desire to account for what might be the correct way I have seen how the European or American attitudes have come to override most global decisions in the political field. My first guide is of our human relationship world wide which is heavily reliant on many of our religious beliefs. By our own good fortune I am aided by the vast majority of people on this earth who hold in good faith and in finite acceptance to any religion, any religion! Save those that are self debasing in refuting equal liberty and standing to us all in all guises, but only in reference of the same specie connection. Religions can only be expressed by the terms of similarity in the mixed relationship between us living and in being what we now accept as being dead? I do not intend to qualify that statement for it is self explanatory by all views. In essence it is the explanation of what science needs when otherwise the fourth dimension uses the consideration of up to eleven different, new or invented dimensions to nearly explain the same issue?

From any time I have challenged science, is done in honest pursuit for our own understanding that is how I now set this open experiment to our judgement, not for cause or effect but for solution. By mixing and shaking the contents is one way to hold an experiment which is what we must do now with the two attitudes that we need to analyse? In now making all our judgements from the political perspective, we only further feed its tentacles of supply and demand. By pretending that religious tolerance is a sanctified form of care to what should be the main frame of our intentions and the base of all standards, does not give sanction of what is right or correct.

My case, or better intended, my area for a point of reference in making the experiment I hope would clarify small matters of our discontent, by choosing the Middle East is almost perfectly representative for us all, but not from ancient history, only these past sixty plus years? By the same query we might always come to, in such situations asking an open question of "who came first the chicken or the egg". Answers come from the level of thought already held and not the depth of study in owing to the question. In other words even if presented in an unbiased way, any contentious questions almost automatically draw standard replies or at least replies attuned to the previous imprinted thinking of the individual, who is us?

I had referred to the matter of the Jewish, Arab/Palestinian interactions before the state of Israel had been formed, to have been savage and violent not least because of the murder and mayhem meted out by both sides. Just post world war two I had also referred to the Jewish method and style in forcing the hand of decision to come down in favour of their claim even when using the open violence of terrorism then described. By pointed reference I had made much of the relationship of how the new state of Israel was more than ably supported by the West. But in the real belief of supporting a western style political outlook to be amongst other views that might side with the other bloc culture developing in an area of growing political importance, religions aside!

Without any change to the basics of any situation save by religious determinations, Israel on her own account conducted her affairs with political expediency but allowed the feel of her own peoples religious experience find new ground by expression. All of which was much a matter of her own concern in those terms, but to be spoilt when other opinions if generated by our representation in the United Nations had labelled us all to be religiously tolerant of each other?

Being a means whereby we had actually taken religious concerns out of politics in lead to be able to deal in our bloc culture standoff positioning by political means only. Which was set to work or be allowed continue while the strings of control to the international situation were of the one mind in being Western, democratic, European or American who just happened to run to mainly Christian ideals religiously?

So much so, that when from the mid sixties, and unfortunately now continuing by all manner and means, when at first the Palestinians directed the method of their dispute with Israel to be against the West politically. Any acts of violence were charged to have been of the nature to have been terrorist only, terrorist led, and unmotivated by any other means other than those to destroy or disrupt or spoil? No issue by any comparison was given in status to the people who conducted any of a number of vile and murderous acts other than in being terrorist or belonging to politically motivated groups by definition who were deemed to be politically motivated!

This, or at least then, that situation was managed by all political charges when by profile the Palestinians when being moved in unsettlement by different comparisons, were cast to be a disruptive force in the area by other means which bears no judgement by religious ties, unless later found in them being generally Muslim. Because by other related conditions but not necessarily connected to some political actions, other Muslims even if to have been from Iraq and Iran, found the same possibility to behave as Palestinians had and might still if to oppose Western, European, American or Israeli positions not even religiously but politically!? How these measures were to have found a base was by reconnection to the Islamic ability to defend, promote or protect the one and last religion of area as foretold even if to have been misread by several factors?

## CHAPTER 109

Suicide, murder or martyrdom or both or all three, stand by definition to who writes the history or who has written the fable. I had made the observation on several occasions in referral to how the first two dimensional religions had taken in offer the commandment of God delivered by man in Moses not to kill of self. I had made the observation by my own reasoning that Mohammed in more than suggestion had allowed for the fact that we might have to kill of self for various reasons. Also by generous definition for those that did or were forced to, in the best name of the religion only, our cover was the immediacy of recognition by God in Allah to our act of pure martyrdom? For that was the most open and obvious way of describing the act of dying in the event of trying to defend, promote or protect the faith that was reasonable, explainable or offered purpose, in excuse or of cause?

All the time I have made no comparison in merit to the feature of killing by self or not between the three, my described first, second and third dimensional religions of the Middle East. By squaring our thoughts again to be amongst our religions only, directed by the medium of religious tolerance, we are not to compare notes but acknowledge each others religious sovereignty, which was meant to stick from political directives since nineteen forty five. Although we understood our own hopes and aspirations since there separate times of instigation even as recently as then ago, we still and now could not equate our differences on religious terms let alone to our political determinations?

Religious tolerance was not or is not the answer, leastways when from it or by it we give and take to our own prescribed political actions in any case. My own direction in making the case of our different attitudes to killing of self has had to be done in and from the veil of abstract reasoning again, so as not to stimulate the wrong responses for each of our own good. So as not to call into action or reaction the very thing we know is morally, religiously and politically wrong, unless persuaded by ourselves through the medium of the Fifth Dimension and its one active prospect in 'One for the Whole'. By religious terms, the one and only divisive difference between all three and all other religions is from how some of us might see the acts of religious murder to have been religiously blessed, only now stimulated by the ineptitude of our political reasoning?

From where I have indicated that one side of our scales in balance is political in the name of progressive Americanism supported by the reality of market forces and other political attachments for her own good, but in that self same recognition. Stands on the precept that her status is at this present time the top of any tier compared to any other system that we had tried, because of its actual success. Because by all standards we all flock in political terms to be covered in the same image representatively, with the fact of her main religious attendances to be Christian has been seconded for the expediency of other political stimuli. Her real success on those political terms is truly to the best example of what we have achieved even if tainted that from her growth on all of our behalf by the means of progression, we have been unable to see where we now go wrong, because we are in a state to assume we are always right?

From where I continue to question our own morality stands by the open failings of our political drive run in forward nonstop. In using our three Middle East religions and in cases tied by some of our political actions there, I have laid the real matter of many of our concerns to review on the same level. I openly cede that the only contentious remark I have made so far has been in reference that form the one beginning and one hundred thousand years of active and interactive history under the same terms of belief in the substance of death? We had changed not to be able and be aware of such matters until these past sixty plus years when for the first time again we began to see our similarity, but had lost the vision of it to political notoriety instead of religious forbearance.

Almost in the same manner that politics has led us to our successes in the progressive feature of Americanism, the hole has to be filled for the delivery of our religious excellence which might appear somewhat disjointed considering the range of our religions? By reference to the mid nineteen sixties from at first the Middle East, whereby martyrdom was inclined to be reactivated by some Palestinian political groups and further extended through the religion of Islam in general terms. Is to be reacted upon by no other means than from the religions its self, not so to lighten the burden of political reactions how so ever, but to admonish such deeds in place by both reference and reason. In calling upon all religious leaders in the name of Popes, Cardinals, Bishops, Rabbis and other reverend people to be Monks or Guru, by extension, on reaching Islam, in calling to the Imams, differences amongst all and others would be no greater than from amongst the imams as well!

By reason that I personally had made the classification whereby it is only from the religion of Islam it had been sanctioned by religious order that we can kill of self by self, has been done because of our relationship with each other these past sixty plus years. By addition, with my own judgement, that from the political image, that is now prevalent, our leads are best to be taken from American influences representatively. Leads us to the corner of where I see on our behalf the features of most of our discontent driven by terms and conditions in the Middle East now, but not necessarily in blame of some of the conditions in how they have run. My overall view is of the misbalance of our political and religious standing only, further complicated by us not being able to show that we truly understand the route cause of our differences to have been our geography, climate, diet and other features, other than our religions and politics?

By fair reason I have given notice that by our different attitudes always allowed by who we are in any case. Our mixture of recent times whereby we can antagonise each other on the two fronts of being political or religious, are to be now addressed by our religious leaders again, but under the new terms of reference laid down by myself with the overlay of the Fifth Dimension upon both mediums. In all cases judgements in offer have to relate to our own living time span in all normality to be regarded as our own lifecycle in all reference. Nothing is to be deferred unless only politically, because as of yet we are unable to act by the assumed codes or imagined standard of civilised behaviour, when seen in replacement by suggestion through 'One for the Whole' offerings will further show our differences, unless we are prepared to try change?

What is to be of immediate understanding by our own reasoning bound with the events of these past sixty plus years, is of our ability to overplay our connection with history, contained within this same time slot. From which our measurements and memories are to be vivid and heartfelt and in far too many cases tragic and sorrowful. From that situation, although I speak of the immediacy of now, our first concerns of tackling our problems of misunderstanding are so that we lay firm and fixed foundations for the prospect of our young to be met in there own lifecycles. But with less troublesome baggage than we have had to contend with by our own corrective mismanagement so far, but now from the starting point at the end of total world hostilities in nineteen forty five! Even that I have marked a new starting point, although our intentions are to be set to future uses, it is essential we on our own account address the matters of our overriding discontent here and now and by result for our own use. We are to measure our own importance in relation to one another by the way we can adopt our religions to the same holding they have always had, guided by the three taken and one active prospect of the Fifth Dimension. But from now, even with an eye on the future, our intentions are to be of this time, not by our view in what God wants us to do, but by our view in how we might present our case in how we might behave to attract Gods own interest in us for his intention!

From my own observation in notice of how America and Americanism affect us politically in relation to the reactive element of martyrdom expressed in voice politically, but from deep conviction within the religion of Islam. Will forever hold us all to ransom unless we begin to apply the overlay of our Fifth Dimensional standards in representation by three taken and one active prospect, given to show that there is the means of change without alteration or altering our histories to accommodate the same in equality. My own take on the situation to arise is for us to lessen our own burden of shame or blame attributed to some past deeds and or the consequences of some past deeds or actions, but never only by amnesty!

I add in haste that I do not refer to specifics unless by this one cord, which is unanimously tied to all of our histories and corresponds to the time of referral in mention of these sixty years plus of our changing? Of the main and final acts to close the Second World War, was done heralding the new age of atomic energy, measured as a milestone to the efforts of our scientific community. Aerial bombing to the destruction of the civil population had been used in every theatre of war since the mid thirties. Being a condition whereby in many cases antagonists would send hundreds of heavy bombers to destroy cities and demoralise the populations in an attempt to force a result. Unfortunately those actions almost came to be an accepted condition of modern warfare with each side counting the cost in dead by how many bodies there were or in how many enemy planes had been shot down. By other reasoning the only real way to destroy an enemy at that time, was on the field of battle, which also bore a dreadful cost in human life for no real worth.

I have not attempted to make light of our driven attitudes of those times by the general indication of our intransigence, but more to show some measure of some of our attitudes, which might better be seen in the case of Japan and her eventual surrender by the very modern approach we finally applied against her cities. Japan knew she was a defeated nation by no more than the economics of the situation, with all of her natural resources used up, which could not be replaced by plundering other conquered nations. Yet she resolved to fight to the bitter end without any regard of her own population or of the enemy at the gates, who in this case was in the most part the armed forces of America, who were in the final preparation to invade the Japanese main land?

America in closing the war in Asia opened the new era of atomic warfare. Of that incident by using one heavy bomber armed with one bomb only, she attacked one Japanese city and wrought more death and destruction on the population than if two thousand bombers had been used. Japan did not respond immediately to this devastating blow, and three days later America repeated the dose on another city reeking as much death and destruction as before at such very little cost because in both cases the single bombers used for both attacks had returned to base safely. Making the score American plane loses out of two, zero, Japanese civilian losses from two bombs, two hundred thousand and counting. By association, those two hundred thousand lost souls who died in the instant of atomic fusion only counted to be part of the total war losses of fifty five million. But by other cases their atomic deaths did bear for us to think morally, in the name of human morality!

In this atomic age and not unconnected, our sciences have truly taken there seat to be an influence upon us all. What has happened without first sight to the aspect of our religions is that they were side stepped by the following course of events. Our bloc culture stand off concentrated our efforts to be more fed from political perspectives, because amongst other things science was of more concern to political motivation and readily fitted better to its objectives. In myself counting these recent years to be our last modern era even to also be our atomic age, makes no charge on how others see the same situation providing all objections and arguments are equally balanced? From nothing more than my own observations of how some events have occurred I have set standards by indication of where our main concerns should be and continued so because of who we are?

So that we can ponder in the same field of reference it is still open by the arguments of morality to the supposed justification of America using the two atomic bombs that ended the Second World War in political confusion, yet resulted in the confused restoration of religious tolerance in almost all cases? By the simple fact of wanting to know, any arguments to the bombing deeds are encountered by political determinations associated with, to, for or against scientific excellence or political notoriety and not by or through our religions to be eastern or western? Save until our religions raised their own head in the Middle East not in response to the bombings then, but to the now in the mid sixties. When all used old and trusted measurement from old readings from times before the atomic age, which could not have been pre supposed by future events for pre atomic, Palestine had not been settled!

## CHAPTER 110

From my own uncharted view of events in the Middle East over these past sixty plus years, beyond the stated success of many countries there by their own determination. The pointed direction by this atomic age and last modern era is of our ability not to see the written word or worse not to use the sanctity of our own free will? By all connections my own view in relation to the representative deeds and acts by self to self, only fit under political determinations not least through its own arm of religious tolerance. Having pointed out the way the ground rules had been interpreted by lead of the westernised outlook of America, Europe, with Israel, likewise to be western, but in able form to promote her own religious obligations as seen even by the distortion of historical events recent and far past! Against which stood a communistic outlook, politically supporting peoples holding to be Islamic of faith, who coincidently, at the time held a far more vivid religious connection to the area and had done in continuous form over the past fourteen hundred years!

Even that I had indicated the westernised conglomerate, was only an opinion of basic ideas we could better interact from, policed by the United Nations as a forum for all, which was so new it was unable to interpret all of our differing attitudes with some lost in the haste of its construction? With the Second World War just closing, yet allowing political antagonisms continue, it is easy to see how we could call to the idea of religious tolerance as a means for us to self manage our religious thinking against one another, without the inclusion of political intervention. Because from many political determinations and thinking, our religious fortitude had been reduced or in cases ignored or persecuted. From my own viewpoint, I have been able to see our changing attitude to the balance of science against religion in this modern atomic era. Not yet unilateral by acceptance but we are being drawn to the immensity of science in terms of its volume against the simplistic view of all revelations to be forthcoming in or by the death of one of earths frailest specie? Unless we do show it is we who operate science, it is we alone who began to control the elements like when using fire for warmth and cooking and light as needed. It is we who better used the caves of nature as places to ponder the exploits of the morrow by deliberate plan. And it is only we who have released the whim of knowledge to turn the ability of wanting to know into the answers of knowing all that we know and all that we need to know to our religious acceptances!

In leaving the last question open, has in no way diminished the force in pending of all of our religions working to the answers we ourselves have directed on our own account? Politics has blotted its own copybook in these recent times by not giving due credit to all of our religious aspirations, no more seen by how some religions had reacted religiously or not, but had been classed to have acted politically? Even if acting by their own standards even if misinterpreted even if against political judgements over religious ones wrongly misguided or misdirected, we continued!

When asking of our religious leaders to overlay their very own religions with the three taken and one active prospect of the Fifth Dimension no charge is to be made in cost or expectation. I do not really expect the Pope to give this work his personal attention nor so Cardinals or Bishops or Rabbis. Differences if generated by Eastern religious thinking can just as well take on board the idea of the Fifth Dimension, but will find it hard to follow its suggestive pattern through various changes. But can agree that in point of fact they already embrace the three taken prospect and are open to use the one active in 'One for the Whole' as a means to check some seen excesses, if seen?

By my own previous reference to the Imams of Islam, the same quarter is to be given to the Pope and religious leaders in general, their intervention is not required to examine the fixed writings of their faith or to question motivation if driven from different historical attitudes. What I expect them to suppose with all of the other religious leaders I have mentioned, is of our, our religious expectations in relation to the fixed lifecycle that we all suffer in being either fifty plus or seventy plus years long and not Eastern or Western politically. I know that we all hold the honour of our faiths by their specific historical adoptions by time; after all in terms of the three Middle East religions they have the same direct connection to God through Abraham and Moses. They are further connected through the prophets of Jesus and Mohammed, who from both we have only learnt additions to the same means of study. But to account for the changes in attitude that might have occurred in the two thousand years between Moses and Jesus and the six hundred years between Jesus and Mohammed. We had been given updates in reverence to all of the others who had used the extent of there own lifecycles but by standards and conditions that were then different and changing from generation to generation without a mark of improvement for our children?

Finding now, that over the past fourteen hundred years we have again drifted apart, but had consolidated the structure of our original concepts, has raised these current issues? Which could not have been tackled before, because we had never gone through these past sixty plus years together, nor did we have the Fifth Dimension as a means of reflection gathered over these past five to six years? As much as I talk in different terms about some of our religions, all of them including Eastern philosophies, are subject to review or by the very least to the reintroduction in format of the Fifth Dimension under my own presentation of now with three taken and our one active prospect. Although covered by many instances from the past they are now, done by accomplishment of Moses, Jesus and Mohammed as used to update us all together of this time by these conditions and for no other reason. The overlay is for now so that we can relate to our histories without the necessity of change when seen that in many cases even if the deeds done were unbalanced they had been set in general terms by the specifics of the time when done.

Having laid my own observation of Gods commandment not to kill of self and how the theory was accepted by my indicated first and second dimensional religions of Judaism and Christianity. From how I had further cited that both were able to redirect the intention, even the meaning through the parallel and working medium of civil order before it gained the accolade of political management. In setting the standards even if to have formed over long periods of time and have many means and methods and styles of execution, both religions in some ways had compromised the original concept that had been delivered by the hand of God through Moses?

If that was the case or how things had actually developed, then the relationship of the situation as it might have been was in first order of thing as they had been generated by the circumstances of the day and the expanse of time involved, fuelled by prevailing attitudes? Even if a direct edict had been set by God himself in full lead, we on our own account, through our own medium religions of Judaism and Christianity saw ways to redirect Gods will, and were prepared to continue and alter the written rules? This done, was not seen by either group to be odd, for even from Gods direction in religious matters, we had the staid controls of politics to contend with, which did allow for the general attitude of killing of self by self, but under direction in us being different nationally or tribally or in lawbreaking? If Judaism and the new Judaism of Catholicism could have allowed the compromise of Gods own commandments leading to what could be seen as a general deterioration in the area of there first influence. Then for a new religion to form by what ever means had two thousand six hundred years of religious history to make better judgements for new religious history?

Islam through Mohammed or by the reflective image that we have used for broader understanding, Mohammed through Islam, had set to rationalise modern thinking of the time to account for what did go wrong by direction from or by earlier religions. His new religion was not going to be tempted by the whim of any political over rid? By allowing acceptance for setting standards to kill of self if necessary to protect, promote or defend the faith, when in the arms of the faithful he had set obligation to its best meaning? He had not laid personal plans of conduct, but in connection had recorded the structure of how society through us alone should behave in conduct to the aims he had submitted for us on that level?

If it had been seen of the way we had followed the unrecorded words of Moses or Jesus which had led to our present situation, where by stages we had only created confusion by attempting to follow religion, but had yielded to our civil obligations in taking our political leads. Then by examination if unseen, when at such a time the feeling was of a divine lead, how was that to be relayed to the same people who had already mastered the concept of religion? But had shown our open failure, by having altered the same meaning to suit different understanding if even some two thousand years later?

Without defined instruction or direction, how by relaying a message were we to behave differently than we might have before, unless we could use the lessons of the past in relationship to how we had behaved and under what circumstances? By the same balance we could only have been aware of how we had behaved in the past and the reasons thereof in using the lessons of the past six hundred years, to have noticed the difference?

By the full reasoning of what Moses had brought to us. By the full means of the changes implemented upon the same religion to include all men as equal by the intervention of Jesus as the Son of God incarnate. Common ground of the two separate religions as they had since become, was divided and kept so by the circumstances of the day and in holding by their own specific priesthoods? In the use of those six hundred years of their united histories the same people at a time had worked the same religion by two different codes. Yet once again in the same area to the geography of the Middle East and allowing for the time gap mentioned which covered the amount of further development needed to allow new methods of reasoning at our own quickening pace. Mohammed's own light was to account for the best way that we were to meet this definitive version of conduct to Gods own vision in the name of Allah!

Of the three factors with some relevance that I have alluded too, like how Islam had changed slightly compared to other religions in the same area, is carried in my view by other accumulative features. But to be described in such a way not to be a bar to our better understanding fourteen hundred years later, carried by the Fifth Dimension and its three taken and one active prospect in 'One for the Whole'. Without note, what Mohammed had recorded was in direct relationship to what was told, save that as mentioned in the recording of such text, certain meaning only was given in the way that the written word was to be translated from its birth language of Arabic!? Another feature of great importance and in conjunction of how the structure of the religion was originally intended was in keeping for the base structure to be relative to the Middle East and radiate to the then known world from that fixation and by style!

Perhaps from our ability to now balance how some things might have been and of how they were intended to be, in looking to the factor of maintaining the universality of this new religion. One ideal is to remove the appeared failings of those previous religions even if used in context, style or habit and with objectives supposed? Even when taking any other reasoning that I have alluded to, the one fixation I hold without charge is in deference to the actual structure of the written word as recorded, even by or for group, tribal or of clan connection. Which when offered to be in lead by personal conduct as in other cases anyhow, but by a style to be better informed than how others had already shown themselves to be. If any failing from the past was to have been from how the priesthood had led the structure of conduct in keeping to what had now become there own interpretation anyhow? Then with a better formed written structure in open study, the need of a managing priesthood was diminished if not already seen as unnecessary and less than connected to our given objectives of self management by the individual!.

From such a broad yet channelled structure by conception and in the best traditions of how we then thought of ourselves. Another factor that I have given much time and concern too, is of how we might have been left to make bigger judgements on our own account without really understanding how or why or of later ramifications? If and so by the concept of real modernity then ago, and in conjunction that we now had a text that was in common use by all, the only lead given in relation to our temples of worship, now to be called mosques, derived from the Arabic language? Was that personal management was in guide only, and by Imams to lead how we were to conduct our own learning, but for us to be devout and learnt by the intention of the now standard! Being able under structure to adjust with the future in mind of change, but at the present time unseen of its complexity, because projections had been led from the ills of the past and not the possibility of how change would come or be managed by anyone in the future!?

Perhaps the only lapse owing to the serious, devout and conscious thinking to have been set by Mohammed, was of our conduct and not by his intention? I had stated that in the name of Christ it was said that one of his observations to account for all of our behaviour of expression, was in the humility of turning the other cheek if being provoked? To turn the other cheek, and without contempt, shows of our ability of understanding; and in turning the other cheek the questions raised are not of submissiveness, but of notification that we are aware of our fate and others not! Being aware in promise by the spiritual over any suggestion that we had failed secular acceptances, our demonstration in turning the other cheek was to refute the argument of brutality. Not to be the final means of communication, showing we are now privy to the understanding of knowledge forthcoming by and even through the only thing that ignorance or fear could rely on, the threat of our murder!

Turn the other cheek had been done in chore for the first three hundred years of its suggestion by Jesus Christ as a means of gratitude, perhaps for what Jesus as mortal man had done for us in humanity. We as the first Christians suffered at the hands of the Romans until our turning the other cheek had turned their heads by other means apart from preaching old Judaism? Which for a time slightly disjointed any perceived intentions of Christ, because although to be more expansive than Moses, who was to have been appointed in relief of being chosen, Christ's monotheistic belief, was for all men over all areas?

If this representation was to have shown its own flaws, then they were of there time, although we were becoming more aware of the different groups of people throughout the known and growing world. We were unable at any time to project into the future of how our interactions would continue while working to Gods own plans in progress, because of other missed readings and because we lived under the developing conditions that we were in, in any case? By the time of Jesus we had become too sophisticated, so that we could class ourselves to be highly civilised. But the missed note of the exchange between Jesus and us, not unlike with the Jews at first error? Was that we settled for the fixation that the future in explanation was imminent and all would be explained in conjunction of our own living to our own conduct by style. Without the explanation of how we were to account for our time in waiting for two thousand years or for six hundred extra years in a world that we could only assume of?

If God had said turn the other cheek, then by effect the sounding of that verse must carry its own weight. By all religions the expression does indeed carry through the core of most philosophies attuned to the mixture in the deeds of man and the desires of God. The only line of objection we can hold to any such expression falls to how we actually lived at the time and by what conditions we lived under and by what mix of self expressed peoples lived or coexisted together one way or another, even through religions or of secular interaction? In making any case and I include us all from my own year mark at the ending of hostilities in nineteen forty five. We can only review what might have been the situation when presented with open events even if misread at the first time of commitment, four or two or one thousand four hundred years later?

My own balance in part concluded by how I see the picture and pick up the words as they fall to ear, is done without turning the other cheek, because I have been able to better see of our conduct these extra fourteen hundred years of enlightenment owing to us all. By that very mark and in due reckoning of our conduct since and not least by cause of our Second World War, my acceptance of self abuse is to be challenged by all of our best traditions and from all quarters. That is why I have set it so for us to overlay our religions and politics with the same task of explanation carried by the Fifth Dimension proper, but best defined to our religious understanding? Not least by attending to what has become one of our greatest religious anomalies last/first carried by Islam if ever!?

From core, Islam was introduced to be best managed by the individual in terms of their own conduct in relation to what message had been relayed to Mohammed and further passed to the new converts standing to this faith. From how I have indicated by my own study in relief of previous intentions, all said has to be relayed on the same terms for a practitioner of the religion by all intentions, and of all deeds, is also of we using our own free will. Not unlike how or from what cause the religion of Islam took to be read from the two studies of Shiah or Sunni in as close proximity to Mohammed's death, as to any other prophet, but by a different style? When in each case with such a defined split to the aspect of the faith, but by inclination only, any reactive forces to the base standing of the religion was done from the same base, but without sited approval of Mohammed, only claimed!

Islam by my choice is always to be my third dimensional religion from the lands of the Middle East and therefore will always carry its own vitality being of vital importance to us all, but is in need of its own admonishment also carried by us all for the clarity of future reference in time to be reflective of a single lifecycle being fifty plus or seventy plus years long? If this is to be so, then without examination and only by perception it is for us to fit all of our squares into there marked circles of alignment, if for no other reason than to show our religious awareness above the mediocrity of our political or scientific objectives. By that lead and from the mist of our own personality I have written in the abstract to broaden our own base of understanding whereby from such study the meaning of an event is not always related to the cause or how it should have been interpreted?

If under that same heading I have made the comparison whereby Judaism and Christianity have taken to Gods commandment not to kill of self religiously, but do and by civil law and by political overtures. And by the same reasoning, I have alluded that Islam by connection or indication can kill of self specifically for the protection, defence or promotion of the faith, then the die I have cast has to be rationalised to cover these very times of now governed by how we have tried to coexist these past sixty plus years in this last modern, even atomic era?

Our religious structures set from instigation in the Middle East as described by my own interpretation have been from some of our own learning or more openly joined to the factor that we are all supposed to behave so. The three religions by various and previous descriptions in all writings from the past, have been treated in the same manner from this work, if not least by their own text, but always to have been carried forward one from the other and one from the other?

My own interpretation of events has been spread throughout these pages, but of the deliberate intention to show how we do not differ, but have done so by misinterpretation? Not least in part pointed to some of the imams and others who might seek to influence others in that faith, our faith, to be separated from the full intentions and guide from Mohammed himself? Simply because they have not been aware of the extra experiences of these fourteen hundred years of our united histories to be able and sympathise with the course humanity has taken; and not to have been directed, because in the time lapse of generational differences, we had been unable to record acceptable change by or from? All recollections were to have been or are now to be made and take into full consideration our own longevity by the actual expanse of our individual lifecycles. Then in the course of future reference our preparation of or for that equality has to have the same bearing to us all individually, both by example and of choice no matter what religion we suffer, but not, not by tolerant means save on religious terms only and not political ones.

All of which can only be done by ourselves and through each other by standing with an open front to show Allah, God the Great Spirit or the wisdom of Nirvana, that we are now prepared to be considered being the one specie to have broken into the realm of mans time by creation from creations time of evolution!? But only in the form of our unity of specie expressed by first choice of religious connections, which themselves are not of evolution but only in expression by the terms of creation! All of which is to be carried and in highlight by the last means so far in our attempt to account for our difference in life that we hold to the millions of other specie, and why?

My own show in connection from what we already know and have learnt is to be done by disection of our religious orders and of their relationship to us and each other. My best method is taken to be from this our latest modern era, because it is only now our reaction to each other have gone full circle twice to complicate the matter of our concerns if only generated by several factors of the past sixty years? My focus has been to the differences of what can be seen between the three linked religions of the Middle East for no other reason than by their own pointed reference of their own pointed connection!

By pointing to what I have given to be the one main difference of our attitude to each other by our own religious sway, and further indicating it is a concept of error, stands on the same grounds of when we first ever made our mistakes to the structure of any religion by including Gods contact with us direct, even from the time of Abrahams thoughts! Unopposed to that time, because when the last prophet in Mohammed gave of his own experiences, all was done to the same people, like from any other time of our histories, but this time again, being chosen as Arab, as the recording or his revelations were of that tongue? Without intention, but allowing for our different attitudes, which having been added to over the past six hundred years of our histories since Jesus in relation to the Middle East! Even through God's, own lead by way of Abraham, Moses and Jesus which brought two separate religions from that same lead. Mohammed being directed using the similar ideas, brought the third which ever since has been ably documented in good and honest voice, but not in notice of conduct, only of intention, bringing the created vortex of movement even if started by the correct intentions, but now running free of its own institution unchecked or uncounted!

# CHAPTER 111

By opening the pages of proposal in the Fifth Dimension, we will be able and see how less complicated our life's conduct needs to be related to our specifics in actual time by our own lifecycles to be fifty plus or seventy plus years long! Which is the extent of our living in the same comparison to a tortoise that might live to be two hundred plus years of age or a tree to be six thousand years of continuous living or an insect that for no other instant than the driven need to procreate, if lucky, will then die in a single time span of less than twenty four hours?

From our human quest for knowledge in at least wanting to know, we can readily accept that by style all studied specie, weather tied in concept to evolution or creation have settled to their own habitat by the made and now suiting conditions of existence? By connection, not unlike Darwin's finches, who had generated their own existence, guided by need? From other examples we can see the extent of any developments which ably match the life form with its own habitat, not least controlled by climate, geography and diet, sometimes limiting further immediate movement by any particular specie. Like for example with the crocodile, cold blooded creatures living mostly in warm climates and not the Arctic or Antarctica, conversely, Polar bears living in the frozen north and not the tropics? Of where any creature finally ended up by the processes of evolution, they all had to have a fixed starting point from where the conditions best suited there ability to further develop?

'Out of Africa' best suits our own conditions, for in the same representation to other specie by my time reference, it is the first place that we appeared in the connected form to be bipeds? Even without romantic subtleties, our impressions first made by walking in soft volcanic ash indicates that we were in the care of our young, that the record shows footprints by comparison to be an adult with a younger, smaller member of the same group in travel? When next we were at liberty to examine any other links displayed through the fossil record after those finds, nothing was obvious enough to have changed my opinion of our single beginning. Simply because over the past three and a half million years since our direct ancestors inadvertently left a clue to a beginning? We have only been able and compliment that revelation by what was to be our normal conduct of living by the same terms as all other specie, but from us, in all places of the world, contrary to the habits of any other single specie? Endorsement of which was to have been carried by us alone expressed from wherever we found to live, we jointly carried the separate study in query to the eventual happening to ourselves after death? Apart to any other creature, we queried the wonder of life's ending in the terms of there being a purpose beyond our own ability to understand? We looked into the future past the reason of utilities?

In direction, how ever we began to develop and change described by our colouring, shape, size or smaller differences generated by diet, we continued in the same wonder to the eventuality of our purpose? Perhaps revealed some sixty to one hundred thousand years ago by the result of our burial rite at Shanidar giving indication of our sign to consider life after death? With grave goods left by any habit, even if scented to disguise the smell of what was essentially a rotting corpse, our reverence to the situation showed impatience not common to any other specie without the same forward looking motivation! From which I have submitted that haste of intent was to be the dynamo of further thinking in the same view, but totally unmanaged at the time, because methods were carried out by the same people, but from self associated motivation, unseen or unconnected?

Further indication to our state and of conduct stems from how I had previously referred to the funnel effect, not least from the case when at places like in the Land between two Rivers, we began to show the open acceptance of our differences before our similarity. Allowing that as we poured general knowledge into the funnel at its open end, the extraction at the spout from the same input was treated differently by some and then others? Perhaps culminating in our first real acceptance of our differences by the first inclination of/too the reality of creation as our source, without any consideration to what evolution was or ever had been? Excellence of mind but error of thought is only to be a stamp to the good upon our human integrity if viewed in context to all of our liberty, because it was attempted in all of our names but not fully known at the time? By that expression I had given by my own time mark, unjustified, to our first monotheistic religion of Judaism which itself is wholly owing to us all in humanity.

By that first drive we were given insights to how we must develop the idea of the future, but not known at the time, because already then ago, we began to construct the idea of the future from the assumption of errors that had been made in the past? By that very rational we then had to account for the mistakes that had been and to why and what remedies were needed, even if only in understanding? From my claim that much was to be unseen bears the understanding of how we saw the need to construct the past to the fitting of what it must have been, above what it actually was, but in keeping to be of related context to those that were to construct it being tribal or clannish by status? By the same delivery, looking back we could enhance any relevance owing to the future by our own status of being chosen to interpret the means of the unknown?

By direct association to Shanidar, even if by my own assumptions, expressed by cave paintings, rock drawing or in some cases ornamental jewellery or objects of artistic desire, each society everywhere throughout the world. From such signs of recorded history, had shown the instance of a cultural direction aimed to a time not yet to be fully understood, but set in expectation? Mesopotamia and the Middle East were in lead on all of our behalf to find the first speck of rational thought for the means of enlightenment! By our legacy from one of the oldest countries in the world, that of Egypt, we can see the measures we had gone to in the sanctifying of our unity through the building of the pyramids. Great edifice to the glory of expectation, but ignobly expressed through the/our leader of men the pharaoh our king in representation to those of us being of a lesser standing by Gods own choice!

First contained in area as always, by some of us being able to switch the lead of direction from the stated acceptance of our king or of Gods link by our king, who was to have been our lead line by paternal connection to our beginning by birth, and therefore as a route to our end for continuance! Moves made by that one standard only, were to have rationalised that original concept into our following one God belief. By relationship from one of our religious considerations in gaining hold by the principle of God to king to us the individual by defined and divine management? Then in turning to our direct conception from God by the weighted means of creation, for those of us to that relative and connected thinking, the step to take by reason was in case of Gods actual appointment to those of us who first became aware of that situation by reasoning? Is to accept the situation for the chosen and in being chosen by means of reward in coming to God by being able and decipher his real intentions even if from before Shanidar? But the reflection is of four thousand years ago!

By the same concept of choice the one and only need of confirmation to any such consideration was to have heard Gods own words of spoken verse to the meaning of now and to the concept of why, driven through the continued connection between God and mankind under the previous flow of time by our known histories? Written in context of what was said and done by unconnected reasoning even if by our direct ancestors, which in line were always to have been of the same chosen tribe? From the creation of our first monotheistic religion if through the thoughts of Abraham carried in connection with tribal members as seen, then the precedent created then ago, had since been carried in the same good faith unseen? By that whim for any connected belief to have formed later, if from dire need and from driven consequences, all further contacts have since been driven by the same connected reasoning. That we have to be told first hand of Gods real desire, because we had failed in previous meetings to acknowledge our instructions! Action from now and to avoid further confusion was by direct consultation with God if to Abraham and if through Moses until later!?

In the simple realisation of change by a connected society to one that practiced religion on the level of God to King to man to God again in death, like from some Egyptian examples, we moved with some clarity of exchange to make the realisation that in fact it was Gods lead direct that we must follow! Making the standard that we were to follow Gods ordinance direct, because it was enlightened to us so, to be ably managed by our priesthood, who had now become the go between God and congregation instead of King and birthright? So much so that in the confused air of excellence, our theory was devised that we must now forever follow Gods lead as told, then obviously by direct intervention through or by or with or from Abraham? Through or by or with or from Moses, through or by or with or from Jesus in the supposition that each had or actually carried Gods own words to us? Leaving the enigma locked in history, how are we to recognises Gods own words when told by several mortal man that they are, and how are we to be translated to understand our commitment or how might we tell the chaff from the wheat when blinded by the pure certainty of delivery!

'Past Gods Words' in prospect takes us back to the reality of how we ourselves can understand, of how we ourselves can understand, simply because it is a modern term to accommodate our previous errors when some of us were able to believe what some others had said not, leastways politically? Since nineteen forty five it has been taken that the political road is to be our leader of choice in matters of fulfilment, from that choice we have set the standard of politics to its own meaning of delivery as it is supposed. Our intention is to make it spread the boards for all in and on the same level, so that we can maintain the theory to always question it of delivery in check. That in part is why we have taken political decisions to meet our ends by design in this time. By overlaying all political systems with my ideals of the Fifth Dimension we will be able to keep them in check even 'Past Gods Words'.

From the religious picture the same image is not transmitted, because for example owing to the three Middle East religions spread by their particular standing, the images translated are ever the same but different? Not least demonstrated by how they had differed over or between the times of one being moved to replace the other or others, which when done was unable to allow that we had changed ourselves. Two thousand years of Judaism brought its own matter of change by those two thousand years, but more importantly not in notice of how we had changed and of some of our altered attitudes generated by our generational differences which occurred in that time set?

By that very comparison and in relation to what actually did happen in that period of time, was that instead of meeting all of Gods own requirements as told to be expected or at least read in that direction. In order for us to recapture meaning of Gods intention delivered through Abraham and Moses, a new voice was heard in the name of God to account for our own misunderstanding that we ourselves had generated over the past two thousand years? Gained to further alter our religious emphasis for the same reasons, but to account for how that change had not been seen by introducing new rules to the same religion? Following cases generated from the same area of vital historical importance to us all, the Middle East, had continued ever since the time of Jesus but no more important than from the time of Mohammed?

From the six hundred years since Jesus where the mood of events could have changed by the same difference as the first two thousand years of the Jewish religion and from the following fourteen hundred years since the time of Mohammed. All were set amongst the opinions, events and occurrences of vastly changing circumstances determined by our actual methods of conduct by ourselves and from the duel sway of our religions and our politics. My own focus today if on our behalf by open intention is set by and since how we have interacted to my understanding of events. My only caution to some people who might doubt my genuine view is that I have been fortunate enough to have learnt the meaning of application from the concept of the Fifth Dimension. Although I have on occasions made my own reference to it again and again as our product, this is true because as with anything in any form produced by the efforts of one man is the property of all, good, bad or indifferent?

On that same basis but entwined within some of our biggest stumbling blocks to the best direct method of how we should try and get along, I have had to strengthen our collective resolve by what has been said in this book and must continue to do so in the same shared manner. By the same connection in direct focus of any reference I have made to any religion, was in the show of how each has a political bearing and how they are similar in outlook. In the case of our three Middle East religions which are the same but set to be different, not from the time of instigation because all additions were simply to purify intent if needed. Our differences have been made by us as we who practice the religions by self determination, even if priest led no matter what, which has drawn my own serious concern to be of direct interest to these most modern of times not even clouded by this atomic era?

My own view and actions in the pursuit of study religiously, have been contained by the only means I actually operate by, which on a personal level is no different to any other person with the same fair minded attention. Also, by the same outcome, I react most to prevailing circumstances which allies myself and all others in the name of our very ancestors who each had the exact same limitation of personal lifecycles? My own additions again on our behalf related or in connection with the mid nineteen sixties is offered for the consideration of solution.

Without comparison, but having said that our first two dimensional religions from the Middle East, those of Judaism and Christianity had come in personal settlement to equate for self understanding the means whereby it was possible to alter at least one of Gods commandments! By using the concept of a mirrored reflective image I had made the case whereby from Islam this was not the case, because the understanding amongst the new faithful delivered by Mohammed if read so. Was that Islam was not to be open to the same matter of reactive challenge to what God had said or to what had been delivered by Gods own agent and or his appointee?

No open judgement can be made one way or the other in respect of my own delivery as to how I personally have interpreted some historical events, because my statement is that of an individual in the same consideration to us all. Although I have made serious reference to the three main named people to and of their respective religious countenance, my efforts have maintained the overall reflective image of taking to our standards only, in all belonging to our own religions. With that background it is easier to concern yourself to the way we as individuals see things especially if being fed by an able and active priesthood. By that concern even if not fully attuned to such a situation our conduct can be in compliance to how that priesthood might better see things? From Islam, and by the way it was originally set up from a base of having almost every consideration be reviewed by the written word, the shortfall by means of reflection under that construction, was in reverse of not having a positively defined priesthood? In stating that Judaism and Christianity with there well defined priesthoods did, bears only from construction and not the follow up which only becomes apparent in review.

# CHAPTER 112

Fourteen hundred years of our history in connection to and with the religion of Islam in representation of its own length by time has been taken by myself to have brought us all to this very modern juncture. From where the design of any religion was openly meant to be constructive and forward looking in relation to how we as individuals were to behave, nothing had changed because of Islam, it in form, simply reminded us of our obligation to continually update? My picture presentation of now is how I have seen fit to reconcile Islam adjoined to Judaism and Christianity at its time of instigation and to be considered of how we view it and the other associate religions seen today?

From where I have focused in relationship to our thinking of then ago against our thinking of now, has been directed to show how in some cases we differ in attitude by our own expectations. I have even used the cover of evolution to explain some reasons to our physical changes to have occurred, that still remain from the instances of the difference between where and how we live, encompassed by the elements of our geography, climate, diet and other reasons including induced attitude? For the steps I have been prepared to take, if it has been necessary for me to stand on some of our nurtured memories, then this will be our cost without chastisement. Most of my theory has been between then and now, but I have deliberately missed all cases to be reviewed, because in doing so I would only follow all the mistakes we have always made and still do.

My approach is best to be seen through our Fifth Dimensional use of only three taken prospects and one active so far at this juncture, because both my own style and method of thinking is illustrated by them. Although ultra modern in concept, I have ceded in fair honesty that the prospects have always been used by us all at one time or another. But have never been used together in this form of construction, because our circumstances of today have never been read how they actually are, until these recent five to six years of interaction? Under those conditions, always to be ambiguous, when different meanings are taken by different societies who are essentially the same, laced with abstract thought from other fields is how the prospects of the Fifth Dimension can now be our judge and jury without sentence of internment, but in balance? Again and from now in connecting the Fifth Dimension to our first line thoughts by the prescribed use I have indicated, we can hold either our religious excellence or our political fortitude in review, even together. But my caution on that acceptance is that we really need to be aware of our political intentions as a separate entity to those of our religious choices, because of the cross referral nature we already accept of our political lead without notice?

Following the line that we are all the same, is a nice modern humane approach to our equal standing, but considerations have to be made to assure and lead to those ends? From where our religions have covered the matter by deaths own knowledge, has been a wedge into what we believe and what we do, making for the status quo today! By my judgement that we have hurried in haste to settle matters of our political concern only, by substituting our religions to be concerned with there own religious tolerance, has in effect balanced the books by the standards we were prepared to accept. That is until matters of this last modern era had shown us the quality of our separation, not least that instead of offering universality by our own specie recognition, instead of turning the other cheek, our understanding had been different. Not least when many of us try and forestall that difference of growing apart by attempting to tackle the matters of concern with what we thought we knew?

In taking to the challenge on my own behalf if for personal reasons, then all moves I make are representative of our deeds, which is to be the broadest feature of how we are to achieve our success in this matter which are due. One of the key factors I have been able to recognise is of the factual difference between our political overtures and our religious acceptances. That is the main reason I have opted to overlay both influences with the four named prospects of our concern to better discipline our political objectives. Not least by comparison that they deliver more directly and are heeded better by our natural attendance to personal comfort?

From the same study it has been noticed that these past sixty plus years have been the most influential to all of our histories no matter who has separately recorded them in personal interest. Because for the first time in all of them we have come together globally politically, even if misdirecting our other means of unity by setting them to cope by the unconcerned means of religious tolerance? On my own account in delving into some of the effects of these very recent times and in due deference to our pasts, which by enlarge had been conducted in religious terms by the maxim that we were right by whomsoever. Therefore our sharing is on the one level of piety to bring all thoughts within the same range of reference to spite what other influences might have left for us, because of the failure of war and the Second World War in particular?

Without the means to formally separate our religions and politics because they are both of huge importance to us, my additional study of them has been to try and examine there points of ignition which cause the sparks or flames of our discontent? With my own focus on the Middle East generated by the time reference of the past sixty years plus, from where much of the matted reasons to our discontent have been to willing to have been awakened! By looking to what I deem to be the reasons of our main split by how we have reacted, it is essential to focus better on out religious perceptions by all studies in order that we can better judge our similarity?

It is a known feature of how I have referred to Moses, Jesus and Mohammed by direct connection with our three monotheistic religions of the same region but of different timescales. One feature I firmly connect to all, is of their basic makeup and the factor that their relevance is to the individual and generally by self conduct with all concerns to our own lifecycles. Of this there are many slightly different takes, therefore I have tried and widen our general comprehension as to the real meaning of that particular interpretation, not least by how I have continually referred to our individual life spans. The answer in understanding is waiting at the doors of our one active prospect in 'One for the Whole', nevertheless my view can be evenly taken without forfeit.

What many of us find very difficult to do is accept our own religious standard in context. By that reasoning we have led ourselves to assume that in the humility of practice, it has been from our own standards that others are to be judged. Not unlike from any study we may have made of the others and each others religious base. Not unlike how I have brought a single feature to the fore of how we can react religiously to some political decisions, but not be aware of the reaction being forfeit to our intentions originally or not, but misplaced by all standards?

In order for us to be aware of the prevailing situation as it is and not by my reasoning, is why we must get and let our religious leaders to speak up in our representation, but not in the face of normal preaching. Only in form to illuminate in first order of our ability to be ready for Gods own interest? Only in form to show of our awareness of what evolution and creation mean in respect of the totality of purpose? Only to show that we at last understand if it is to be from any connected specie in life today to be part of this acknowledgement. Then it is from this single group who first left footprints in volcanic ash some three and a half million years ago who are best ready to be noticed and not the reverse. Then it was from this single group in that short time span, who had the endurance to spread out to cover the whole earth. Against the possibility to other creatures having shown no inclination whatever to be involved in the interruption of Gods own time of evolution no matter if connected with two hundred and fifty million years of personal history, save we!

Our religious leaders being part of the same religion we all belong to, are to show our resolve on this matter and by so, change the emphasis of our political headings or at least some of its tributaries in the name of gain prompted by market forces and commercial interests. In leaving our political devices to there own settlement at this juncture we have created many of our problems by the rapid methods we have tried to bring about the image of unity, which is emphasised by our political standard of religious tolerance. Not least highlighted by events that have certainly come of age from the one area of our particular concern in the Middle East and of these recent times by practicing religious tolerance unopposed?

In as much that I have made of our bloc culture political standoff just passed, by notification that with the two main political antagonists of communism and democracy, there were three religious philosophies closely involved. If sided, Judaism and Christianity were mostly associated to the western/democratic camp and Islam to the failing of communism in its original ungodly concept. Even if the above has no connection with reality, then the outfall of the situation was of a dominant westernised political approach to management of the specific area if not least run commercially? In following, although it was not supposed to happen along with the prevailing political situation the religions most associated with that dominant voice also seemed to gain self advantage?

Although of the same comparative nature by tolerant religious terms, many of the religiously inclined countries of the area to be Islamic by nature, as the

Israelis were generally Jewish by nature. Had no real connection to the spirit if internationalism not least if some were to have been classed of a second world order predefined in acceptance and theoretically voiced through our United Nations? Yet all nations were equal by the standards of religious tolerance, which although told before by myself, has a different slant when I introduce what I have considered to be the Sword of Damocles hanging over all of our heads.

By the same token I had also mentioned the importance of the past two hundred years of world history by reference to the battle of the Nile, a river in Egypt, yet the conflict was between the old European antagonists of France and England. All of which by one form or the other allowed European conditions of change be manifest by newer circumstances within Europe herself, be transmitted throughout the world on European terms. Which was of no particular consequence to some countries, because internal standards were met by their own religious excellence? With European or other foreign influences filling the newer recognised means of a political system or ideology and introducing the gain factor with new commodities so far unknown in the very countries they were actually seeking to control and by military power, little opposition could have been offered then?

After some one hundred and fifty years of this kind of association between the first world countries and the developing nations classed to be second or third world, their actual differences came into more defined relief by these modern times on religious terms, although they were to have been the same in all cases! In the mid nineteen sixties by my own pointed route of affairs between the Arab countries, Israel and the Palestinians, I had indicate that by forms of contact to be mainly political, no grounds were set towards the concept of equality even later policed by the United Nations. Therefore any reactions were not viable on the political arena the only options were to have been religiously motivated even if misread?

Even in the supposition of being well voiced by consent or the appearance of it, when no heed was given and conditions had been eroded from the stated start point through the actions of political war. Then religious fortitude was the only shield and sword in the method of reactive sources. If and only, the base of the religious philosophy carried enough in tenet to accomplish the means if not to turn the other cheek, but remain in the practice of defending, protecting or to the promotion of the held religion. How was it to be expressed if the only need to decision was from the individual or those disposed to set the standards of reaction internationally or religiously! All of which by what ever reading makes for the importance of the religion or of religions having the grace to accomplish the same standards for all within the same framework of the one similar religion. That is why we must look to the facilities we have in some religions and spread the concept fairly through all without forfeit, even if it means changing all religions without alteration!

Although the last sentence just written expresses terms of equality, it is to be heeded by association. From the fact that I had called for our religious leaders to embrace the concept of the three taken and one active prospect of the Fifth Dimension, even in religions where most in conduct is to be carried out by self, the expression stands. My objective past the balance of suggestion is for our open commitment to the spirit of change being led by religious delights, even now offered to be only relative to the issue of us waiting to be of Gods own choice by the suggestion of our own preparation. Waiting on our own account by all reasoning, but making aware that we do recognise that we are different than all other specie in all ways and most obvious by reason we have found the route to remain active even after 'Natural Body Death'? But the experiment is still to be made which includes Gods involvement already to be determined far and away above any thought of activity in structure to be carried out in the name of formula and by the fourth dimension?

For any plan that we are to show, we are ready to be considered to meet God on his own terms, has been enacted these four thousand years by us all in relay through our appointed religions and through our learnt religions? What I would seek from the abstract, is the clearance to accept that we all are actually the same save those of us who have been miscast in the reasoning that some are less important than others and therefore are expendable? Therefore are expendable past all considerations of 'One for the Whole' by self judgement and in counting time beyond its own limit of our own personal lifecycles held by the natural act of 'Natural Body Death'!

For the height of each experiment I seek by overlaying our religions and politics with the same understanding of the Fifth Dimension prospects expecting different results, is done to acknowledge that by some terms we can and do kill of self politically? But can never do so in the name of or for any religion, because they are all connected by the self same dual features of being responsible for our personal relationship with God in Allah or the Great Spirit or to Nirvana and from the span of time owing to us all individually by our lifecycles? If the importance of our own lifecycle is to make us self important, that is best done now and can only be done from the changes due in the near future from the full realisations exposed by this recent past!

By the same method, but in much less time than by cross referring to our three Middle East religions, much of what I have so far written has already been taken up as standard from now on, but to our acceptances. By how I have set the first two dimensional religions of area in connection not to kill of self by acceptance, but do religiously, then that case will be nullified by the accepted separation of our religious conduct, but led now by new circumstances? From how I had indicated to the fact that Mohammed, even before the official appointment of the new religion of Islam, had endured to create its doctrine, by how I had written in undertone that part of the reasoning put forward by my own interpretation of the situation, was set because of the normal brutality of the times of which we then lived. Was done not in excuse, but to show by enlightenment of how our thinking had changed in the additional time span and to be better prepared in the positive, we should think proactive and not rely on the standing of being reactive which has ever been a slight to real advancement.

From or by the same terms I had alluded to, and now mention again, it must have been comparatively easy for Mohammed to translate the same words of God that had been offered to other prophets not least Moses and the son of God in Jesus, who he saw to have been of like status to himself, mortal man! And in so at a stroke diminish the evidence of the gainful flow of our civil ideologies becoming political and taking the potential lead of our conduct to diminish religious fortitude as it had already happened twice before!

In order for this to be determined, civil law was to generally remain in conjunction to the wishes of the ruling house if so, but a new emphasis in the control or management of our destiny was to be put in the capable hands of our new congregation? They as we by the means of self conduct were to be enhanced with the sole property of defending, protecting or promoting this our eternal new faith of Islam. By taking to what was previously only a civil law edict, whereby we could kill of self for other reasons of disobedience to our running methods of order in control socially. From within other societies working to, with and from older religious concepts our religion could now update all of our affairs by taking to understand things as thy were now operated some six hundred years after the last time we were perhaps told to turn the other cheek?

From our own fortitude being able to kill of self in defending, promoting or protecting these new intentions, raised the standing of the individual. Because the aspect gave us all the surety of our own judgement above others who might lead us religiously, but were no different in terms just because they were administrators as in the priesthood, as in our civil orders? Islam did grow rapidly and it is not for me or any of us to state that the measure of her history is the mark of her or any other religious standing today, save by interpretation?

My own observations in quote are my own, nevertheless because of the nature of them and my time reference, from standing, they are to have no impact on any of our histories, if to be used to change the past, then that is done without alteration of past times extended beyond the actual living of any individual. In other words although I and many people speak in similar terms about past issues, our local comments are only active of events we are in contact with either by a time set or involvement! I am in contact with the mid sixties, I am only in reference to four, two and one thousand four hundred years ago, but can be aware that each noted time was mixed with people of changing attitudes?

## CHAPTER 113

By effect of not appearing to want to change the past, it is easy for people such as my self to say anything about any aspect of it or by pointed reference to a specific item, and make my own comments in any way! Then by using such stealth indicate so by stating that "I understand the reasons to have been such and such, because the circumstances then were different". Then using what ever suggestion put forward different than before, to attack whatever suggestions were put forward to diminish the standing of the person or persons to have put such original ideas? By a strange anomaly in cross referral to the human condition and to account for our differences even now, that is the very methods of conduct we all should use in the future. Replacing the unwarranted aspect of being politically correct by allowing full access to our united historical references by all means and circumstances to us all in cross referral. It is also to be used in access of how we can enlighten the proposition of religious tolerance which is to be done in all haste anyhow, under a new heading and will be done soon by our own religious leaders whomsoever, not least by our method of interactive communication through our one active prospect of 'One for the Whole'.

If replacing what to some societies has become a byword in expression to bury our heads in the sand by not looking at serious issues in the face of real value, then in losing our adage of being P C or politically correct is an immediate plus. The Fifth Dimension of this time now, is our first best step in that realisation, not least because it is of this very age, but has encompassed all features of the past without pointed criticism. What already is open to emerge from its standing is to replace the mixture of our acceptance of being or acting politically correct with the fixation to the reference C D; common decency! Not even to be used as an idiom, but a natural outfall from thinking within the terms of our three taken prospects and us all being in voice by our one active prospect in 'One for the whole'.

By taking active and vivid steps into the past, the one rule we must abide by is not to take our understanding of the now with us, which in all cases, because of recent events and the immediate collective use we all have by the excellence of our communication systems. Leaves us the ability to pack the wrong references and overlay them upon the wrong times marked by the standing attitudes of the people then and where. In doing so from the simple error of mixing the experiment of the future with the formula of the past our recipe in too many cases produces bitter tastes. From how we can take some past written text in its own lively and elegant hand and give it the prose of satanic verse if, belies our modern understanding of the past as it should be?

For comparative balance viewed in mirrored reflection by my own tainted observation to what I have laid to be one or perhaps the main difference between my third dimensional religion of Islam and the first two by date reference to be Judaism and Christianity. Was to their respective attitudes to the means or matter of we killing of self religiously and how the religions had differed or had come to differ by attitude or interpretation of Gods law first or their own law second or P C or human rights!?

To have made the open observation that in my opinion Islam had additionally differed from the two previous religions of area, by the original set of leaving the tasks of religious study to the conduct of the individual only, with some pastoral guidance, as opposed to having text or the meaning of it issued by an established priesthood, holds! From the picture that I had built of the matter of human death in terms of killing self by self in faith, supported by the process of defence, protection or in the promotion of, holds! But only in the context of the day, early days, because of what we did not know, even that we had had the briefing of two thousand six hundred years of religions in the area. Even that we had the same connection with God although recorded differently than from how it was done by previous accepted prophets, this time we simply adjusted to what was better suited by the prevailing circumstances which was no bad thing!

Even beyond my acceptances, any view to the reasons for us of any religion to kill of self will always hold to different meaning, also within the same religion, which can be tested by recent actions from the Iraq/Iran war in the last quarter of the last century. I make the case that when the same actions were done in and through the catholic reformation in spreading the same words to have other meaning, but in the name of Christianity. Our actions in the unity of murdering the same people who held slightly different viewpoints to the same entity in any case, which was not to be proved until death in any case, all erred by the reasoning and understanding of the time and day historically? Even if some of the actions were very recent and had no history by measurement of time on there own account, then the view has to be taken, of under what premise were such decisions made and by what errors?

Within the matter of the catholic reformation dated in time some five hundred years ago, when Martin Luther as a cleric wanted to nationalise some aspects of the one faith officially. It was to be from the already formed priesthood, working by preconceived ideals to determine from religious law, how these reformers were to be treated, but most assuredly by the case of execution, even to be killed by self upon self? All of which was to be carried on by religious fortitude until by the running of time we again began to heed the exploits of the scientific community in the name of political reasoning, leastways in some westernised communities. Which in bringing us up to these recent five to six years when we have had no real way of looking at how we can offer treatment to all of our religions equally!

In my own observation right or wrong and in looking at some recent events within the broader time band of these sixty plus years, but in notice of some events in the Iran/Iraq war. My take has been from how some issues had been interpreted and to put them in order of justification for all of our attendances, but by our religious interpretations, because of how in specific cases other meanings by attitude have been added? In concern that we as a world community have tried to force that acknowledgement since the ending of the largest conflict we have all been involved in, the Second World War? We have driven the concept by political reasoning in the main, which is only part of the route to take. It will not be made clear until we lead by religious directives, but by political guidance even to be better enacted by the United Nations with that same newer background of considering the later thought objectives of our new religious drive.

Unfortunately we have had far too many little wars amongst ourselves and continue to do so for some of the reasons that I had mentioned before. In order that we might change this sad and unfortunate situation, instead of looking to the possible reasons and cause, we have diverted our attention to the result with the victor writing the events of history and keeping us all to old values. For me to have cited the Iran/Iraq war in serious reference, stands not to the outcome or the morality or justification or even to the fact that in both cases the standing religion of each country was that of Islam? My concern is to how one very small incident of equal standing first came again into the political arena without intention, but has been maligned by the commentary of ignorance even when not properly addressed by myself. I do proceed on the grounds that my view has been tainted by our three taken prospects of the Fifth Dimension, but claim no duty from its prospects, simply because they are representative of how we all might be able and view things by this very modern era. Not least with particular reference of how we can expect to receive better representation relative to us all and contained to be met within our individual lifecycles. Not so that we are entitled to put up our feet and rest, but from soon to take to the attitude that it is our obligation to show of how we accept that God might be interested in us by what we are prepared to do for self, but more importantly, for all representatively!

Even when we can question the delivery of instructions to guide where we might suffer the death of friend or foe by concept to protect, defend or promote our last delivered religious ideal taken in the lands of the Middle East? In taking the Iran/Iraq war as a point of reference there, I make no other political or religious observation from the point of view of comparing specific events with any other religion at this one juncture. Both sides in holding to the Islamic faith worked all of there concerns to how that standard was to be met by other interests be they commercial or directly political. Muslim to Muslim at war with each other was no different to the tragedy of other wars, not least the first and second world wars of Europe internally being made up of Christian armies against Christian armies?

I offer the next phrase without definition, because as an act it will have been recorded by one side or the other and copied in reflective form in all justification from sound reasoning. But from the justification to have applied, the condition of martyrdom by any individual in the throws of conflict fits well to Islamic philosophy without charge. If and when first done it was by a soldier in the fundamentalist view of nationality expressed through the heart of an Iranian. Then in the same succession martyrdom was again repeated in the reflective field by an Iraqi soldier, nothing was done to compromise the good standing of that faith in particular or any faith in general, save those gathered to human sacrifice?

Again without definition, by statement that martyrdom is akin to suicide can only be covered by the rational of the individual! With the act of suicide frowned upon by most Christian religions also, who also have a fixation on martyrdom being a passive form of death in the face of adversity? Not unlike where by other presentations when Jesus said to turn the other cheek it was at a time where it was accepted that judgement day was imminent, therefore the pain of martyrdom was accepted because rewards were at hand by that very delivery. The secret of which was to readily accept that deliverance was indeed at hand for the reasons of commitment and to thwart those that rejected the words of God when ably shown of his intentions to immortalise those that were to die in the name of this our faith in holding to the objects of purpose! Any configuration we might mix between a Christian understanding of what martyrdom and suicide might be in the same comparison of what suicide and martyrdom might mean to a Muslim. Has to be done in reference to how some political objectives were seen in need of stabilisation either nationally or internationally and at what particular time of our own review!

I had likened the same features to different religions from different time slots to have different rates of understanding, some generated by our own attitudes or particular traits, which themselves developed over time guided by the hands of our geography, climate, diet and of other reasons. In taking on the Iran/Iraq war by way of example is not to separate it in section from the tragedy of the Korean conflict or the Vietnam conflict or other less noticeable, but as said wars on all of our continents! It is to divert our political awareness as it should have automatically directed itself to do, if not least through the United Nations, but in the correction of our religious tolerance stanza! Because to not have done that has allowed part of the reason for some of our present situations arise from where we cannot define what is of political concern or of religious significance?

By brief mention to the concept of what martyrdom means or is to most Christian religions, I have only used the example of understanding translated into modern thinking by many Christians, but of old meaning and practices? Against that, to take the modern Islamic understanding of the feature, it has forever been carried close to the Islamic idiom which is at liberty to be so maintained since the idea was first aired and from first instance delivered to Mohammed? Not by comparison but in relation to what the understanding is or might be by the separate congregations in both having different methods of clerical leadership! Allowing for the later religious use of self sacrifice to be turned into the feature of martyrdom, irrespective of the consequences to others not included by the measure of our standards?

By setting standards, there are none to be set, save that we incorporate the meaning of what Moses, Jesus and Mohammed had delivered to us by means of wisdom that we didn't really understand, leastways given that we are to take in use our own free will by all examples? From the outer rings of some types of reasoning relative to all of our personal religious countenance, I have made a specific connection between Judaism and Christianity. But have carefully aligned their thinking in terms of how to operate the mandate, tenet, covenant or open instruction given or carried to the individual of each religion by their own priesthood in relationship of what to do in or over or from situations that might arise for the means of murder upon self by self! Without the bounty of discovery, I have taken my own stand that the questions raised have only been delivered in answer by political standards these past sixty plus years?

From the same vein of enquiry, again without any form of direct comparison, similar questions are not applicable to the religion of Islam. Because her standing in the name of Allah in God was of full self promotion to replace the vagaries of earlier religions associated with the same area geographically. Not least endorsed by the holy writings recorded by most notice for all in the Quran, from which we ourselves were to take religious sustenance guided in essence by our teachers, linked in association to be set from our place of worship our mosques. New places in the same fashion that synagogues and churches were used by earlier religions. But in this case our religious leaders were only to oversee our progress of study, because at instigation this new religious text was up to the minute, it covered every aspect of conduct and even gave reason and cause to that end. Although written in excellence of mind the error of thought to have occurred if, was unforeseeable because all revelations then only accounted to the world as it was then and from the thinking from then? Even allowing for our maturity of mind to have occurred since the first time of the Jews and the Christians it was impossible to have known how events would be changed from the times of our united global expansion that did occur, because we did not really know what the future might bring?

If much of my attention has been to show the paradox of how thinking can be changed by the simplest of means, then to have mentioned once again in context of our global community, there is no possibility that some future events should have been foretold fourteen hundred years ago? By the best of good fortune I see this situation as one of the few positives that we can all manufacture to take forward in a collective measure to help aid our changing resolve. But first again we must look to the negative in some respects to examine why we have in certain cases reacted as we did and or might have?

By the cautious way I had approached the matter of martyrdom and suicide and in taking measured care to speak in open appraisal of either, has been done on our behalf. My most obvious style of approach has been through the strength of our religions renewed, but now contained by these past sixty plus years by open review. By cutting to the bone I had made the assertion that a basic difference between Judaism and Christianity compared to Islam, was that from Islam we had better or more openly accepted to the condition of killing self upon self. But I had contained part of that belief to have been raised from the sorry conditions of human existence that had prevailed at the first introduction of this our latest monotheistic religion from the Middle East!

By first choice in normal review, the religion was planed in a slightly different form than to others as well indicated by the standing of the new priestly structure. It was also constructed from or entwined with a base philosophy to expand by any means within the accepted standards of the day, which in case meant open conflict between city states and national tribal groups. It was also constructed from the standard thinking within the area and related in writing the text to be of the Arabic hand for the Arab ear first, but not to have preceded the exploits of Martin Luther who was never connected with the faith. My comparisons are not in light of old mistakes but to show that by that first connection even within the Arab hand, some meanings of particular situations hold better to the Arab mind only, without the episode of the critic?

Martyrdom as an expression even fourteen hundred years ago had a huge connection to other monotheistic religions, which coincidently had been given good grace by all codes as a means of piety. Judaism and Christianity in the short six hundred years of there concurrent histories until Islam had been even further swayed by the prevailing circumstances? But all was ever covered by how from those first two religions, politics had used the conditions of our living in expressed harmony begin to take a firmer hold with its own secular rule, while allowing the religious priesthoods take command of the spiritual in counterbalance? By whatever system of who did what and why and when, the predominant collective religion of Roman Catholicism allowed the mixture of our civil and religious leads intermingle not least in the Middle East until Islam was to show?

From the growing standards of interactive nationalism contained by a single religious force, the cult of personalised religious interpretation made its own rule, which was further highlighted by the absence of the original Roman organised form of civil command. Even before open recognition at the time by fixed attitudes, the same faith amongst different nationals began to show uneven values even expressed through what martyrdom or murder was or meant? Even without any classified or fixed heading from any controlled historical evidence this/these or those examples just mentioned, left there own mark, not least to be genuinely highlighted fourteen hundred years ago by Mohammed. From the main reasons that our religious signals of excellence then ago had begun to show a confused barrage of intent and were now beginning to be offered nationally well above the concept of universality. Then such grounds that might have arisen had done so through the minds of men but in need to be interpreted by Gods own agent for our understanding, but this time again with the means to copy the correct methods of fulfilment? Even if again some of us were to mix intentions with desire and in reverse?

## CHAPTER 114

The crossover of our secular and spiritual reasoning has been sorely buffeted by the means of the fourth dimension in just the same way we have been from religion to religion from country to country and from age to age. Mohammed's intervention on our behalf at its own time was unaware of the fourth dimension in concept, although he personally was fully aware of our human laxities. Yet was ready to take us and enhance the methods whereby we could obtain in achievement all old promises by any previous religious objective. In order to do this it was obvious that at instigation of change, our attention was to be drawn to what was said, written or communicated in the best order of understanding. By our good fortune the language used was readable to the actual situations in which we lived and reacted and also coherent with events in living memory, but also this time again touchable? By all the best means of delivery the same text in fresh order was carried to all, even if at first take the learning of it was difficult, not on content, but of its relevance to all, even to the majority of us who had little formal education!

What was of best advantage to all, was to the fact that even in full realisation of that notice, we were accredited to be of self conduct in our learning, ably aided and directed by a new clerical set up? Although dissimilar to other religions so that our clerics were not to act as our mentors, but only our guides, not to be heavy handed, only to point the direction for our own lead in understanding. Working the theory that in operating this religion was mainly by our own efforts, and therefore in control of our own due rewards given in consent like from other religions, but more defined by our direct behaviour, and why not?

What is to be of huge significance to us all, and I personally say without notice, that as I had made this particular remark in association to very recent events marked in time by occurrences in the Middle East these past sixty plus years. In relation to fourteen hundred years ago, the significance of which is totally intertwined with the past by direct connection. Not in case where we look to the past and because of modern attitudes or the like, we try to change it without alteration so as not to expect to change the future. Not so where we guess what some conditions of the past might have been like by this thinking and therefore expect to be able and change the thinking of either then or now, but simply to acknowledge the past and accept some of its traits by what ever standard? Having moved to the future, we are supposed to have a different approach not for its sake. But to accommodate the factors of civilisation or other attachments to assume that now our status, even if to be half measured or politically correct or holds to human rights by religious tolerance and or to other means of limiting personal conduct, that is true secular progress?

From which by those standards we can implant the future as it is now with some attitudes from the past either reawakened or at best re-valued to fit in use, because of todays standards aligned with some of our attitudes? If looked at, even without acceptance that my own view of a major difference between the first two and our third dimensional religion of the Middle East was to the latter's acceptance to kill of self by self. Then the study of that remark without conclusion is to be made in and from the abstract by the same methods from which I have been able to make open the relevance of such a particular accusation or the suggestion of a similar attitude?

Allowing for what might have been well developed theories in the first place and allowing for our ability to rationalise the obvious to suit certain situations that had arisen or might be repeated. In order to avoid similar mistakes and relate the situations that had occurred relative to the situations that had arisen and might again? When and if Gods own agent had relayed the solution in text of how to meet the answer of these/those deeds then the language had to be direct enough to follow in justification over all times. Not least being able to meet up to the minute thinking in any case and when turned to the explicit ideal that it was to us individually who had the ultimate choice, but contained for religious purposes only, the table was laid, but was it then for now or now for the future?

Having set the place so that we were of choice, because of the situation historically was set to the one option? Our objective was classed in justification by religious judgement but to include political standards if necessary. By how the new faith was to grow, contained by the human conditions of conflict if misguided by materialism or gain or desire, our own lead was to be gained in the context to defend protect or promote this new faith! But always in personal judgement, because we were to hold our self responsible to Gods own desire of us, read, but even if not yet fully understood? Perhaps not until some fourteen hundred years after instigation of this our last monotheistic religion from the geography of the Middle East, had brought us up to date by this time now, yet from recent actions given to have been directed then ago, and therefore was in compliance now, or from some it might seem so? Without too much danger by our own means of comparison, we can compare notes from the same status, even if being first second or third world in stature, because even from within the flawed concept of religious tolerance we all have the same voice in any case?

Without the volumes of libraries from where we could find the evidence of our similarity between the three monotheistic religions of the Middle East, in comparison have I chosen one item to highlight or express their differences? Many times have I mentioned the reflective unity within my first two dimensional religions and at the same time seemed to hold Islam in task because by my interpretation, that she alone accepted the standard of killing of self by self! To further connect that because of her priestly structure, the choices of task in this matter were to the individual as opposed to similar choices from those of other religions of the same area, and then elsewhere. My lead has been more than obvious, because of some leading circumstances exposed by this our work, composed of this very time and able to review all of our histories in part or in full if necessary?

By that standard nothing yet has been comparable, because from nowhere shown have we been able to review the past without alteration in cover by the three taken and one active prospects of the Fifth Dimension! That is to be our tool of conversation from where one to one and one to all can exchange the same religious text 'Past Gods Words' without forfeit, condemnation or even criticism? When in our own right and from our own new judgement, we can all express the reflective means of conversation through the medium of common decency or C D in replacement of being politically correct or P C previously. By first indication the change is not to gain any advantage but openly diminish our acceptance of some political standards over our religious sentiments even if by name only!

By overlaying our political and religious structures with all the same headings of the Fifth Dimension, will be seen their relationship to each other and the importance of that relationship in connection to the future. Even if politics continues to support in the matter of materialism and the best that religions can ever offer is the possibility of the probable, our endeavours of the overlay are to purposefully realign our true sentiments, if no more than to explain by experiment of how and why we acted as we did at Shanidar!? How and why, when we did not even understand the experience of night or day, hot or cold, rain or shine, but we queried to the comfort of the rotting corpse of one of our own clan or tribe or group?!

# CHAPTER 115

Without taking a giant step for mankind, in setting the proposals of now and turning them into active realisations, we only need to take little steps of change that might yet be hardened into a causeway for the next three and a half million years. All of which can be opened to our equal understanding being aware of easy going and uncomplicated reasoning expressed on our behalf by our religious leaders? By connecting to the faith of Islam in relation to how the faithful fulfil their personal religious commitment in the habit of study and conduct. With only a guiding hand being laid upon the individuals own judgement by their imams, how then can the same judgements of an individual be questioned, queried or steered by other influences than of self, if exercising our free will? Meaning in translation, who is responsible in the name of any religion to the act of any individual claim from within the supposed or proposed structure of that religion?

From where might we have a Christian or Jewish religious leadership endorse all the obtained and practiced changes in judgement over the time spam of four thousand years concurrently? Yet operate to the desires of nationally chosen political directives, even if similar or not, but enjoined to political interests actively above the standards in the collective use of religious tolerance to manage others of a different political motivation religiously? Even to the fact that in the world today, if we have very strong religious motives we have lost the ability to express them, not least by our own political influences drawn from these past sixty plus years. From any suggestion that I make for our insistence in the near future now exposed by the means of the Fifth Dimension, the one always hanging in the air refers to the meaning or to the understanding of what is or who are our religious leaders?

By at first referring to the Pope, Cardinals, Bishops and Reverends of the Christian faith adjoined to the Chief Rabbi, and Rabbis of the Jewish faith, because of there connection by area, was done on its own account ably covered without distinction. Obvious connections were maintained with the Islamic religion, but separated for the exercise of this very interpretation by myself, through the questions, of who leads it and how and why. So that our Fifth Dimensional overlay has the exact same impact on all three monotheistic Middle East religions without any sort of bias or manufactured means to separate my own directed form of appeasement between them. I intend to be wilful, but only because I am someone belonging to the same query first raised at Shanidar and who is running close to my allotted time span of fifty plus or seventy plus years on this our planet in my present form?

Although Islam in text was covered in delivery from Gods own agent and delivered to Mohammed, Gods last prophet? In the same comparison to the other earlier and similar religions of area, all accumulated writings of all three religions were 'Past Gods Words'. Because in each case the conduct of, which either described of hinted at, were delivered in text or voiced by our own compatriots to have been teachers and or of our priesthoods. That being so, our object of obtainment was to have met with Gods own first requirement of our conduct being a means by which we showed our commitment. Other than being led like ants or bees to perform acts or deeds without individual purpose, other than in compliance to the will of evolution. When we alone among all specie had unlocked the secrets of the universe unaccompanied, we had turned the clock in time from Gods own span of evolution into the matter of creation also by Gods own meaning, but even now not yet fully seen!

From when earlier religious writings by the Egyptians, Jews and Christians were recorded can be seen the exuberance of our delight in making everything compatible to how we actually lived and in the manufacture of what the past really must have been like to get us to our present situations, even while changing in attitude. Ever reflected in the act of replacement or renewal by the supposed means of progress scientifically, morally or socially bound, even reflected by how some four thousand years ago we experimented with the will not to kill of self, but still acted 'Past Gods Words' in any case?

If to have used progress as any sort of mark, then from Mohammed being personally driven or selected to be able and associate himself, with as he saw them, the earlier prophets of Moses and Jesus! Direction, if taken by direction, was for us to safely and generously acknowledge our pasts from the connected means of better recording history in any case, then the writings of Islam were of their day only by first impression. That is why amidst some of the noblest current religious text, they stand out in the reality of content, because they were expressly written to accommodate our ability to change without alteration as projected into the future! Because the work was written in context to the standard, style and conduct of the individual by their own lifecycle, so much so that in order to defend, protect and or promote our faith as it was actually introduced as a medium of change in any case! Other than by first lead, the individual was endowed with the means and matter to kill of self by self, depending on circumstances, even then to have covered the means of political conduct by others if in the level of crime against family, community personally or religiously? Which now honestly drew its own rewards if the risk of action or reaction was martyrdom?

Any enlightened act from this new religion or even by interpretation, were structured to suit the conditions of life as it actually was and where. Centuries of hard and bloody history have been unable to alter our taken attitudes by whatever means save by some religious interpretations. From the nature of working half measured by mostly political means, we have allowed our steps taken syndrome become the engine of its own fuel in continually offering those terms of half measured reasoning produce the same outcome? Although it is not difficult to give example and example of the incidences of half measured working the concept has to be maintained in order for us to continue in any case. Leastways while we are learning to gain our own self respect if a little shaky since the ending of the Second World War in Europe, which in first place, leading to this last modern era, has ever shown our biggest noticeable example of running half measured?

By that example of when the main antagonists were of the standard to hold very closely linked forms of Christianity, which at that time of area, in theory was to be our most highly developed form of moral control administered by an able priesthood. No notice was given in respect that virtually the same priesthood by determination gave the same blessings to the cause of victory to separate sides in the name of the one God? No notice was given in respect that it was political determination that we operated by to the old form of some of its recognised tentacles, in the name of trade, commerce, and the commodities in the shape of raw materials to progress politically only. Alas no connection at such times ever seemed to enter into view of our better religious objectives not least because they were to be significant at a time past the management of political deals in all cases!

Political half measures ruling came with the misapprehension that the future was all important for the individual, therefore such actions to be taken or driven at, were to that end only, supposing that others wished to deny us of our noble intentions for our own national population? By the most modern of associations we had arrived at the situation whereby in order to set the best means of progress in all cases we were to allow our steps taken syndrome rule by consent to the origin of our first half measured reasoning. In moving in this fashion beyond notice we were creating our status to be different ethically, morally, socially, racially and more poignantly religiously!

Even by our best methods of reconstruction headed by how we had hastily constructed our standards of human rights and in the same measure our allowance of religious tolerance we have been unable to cement in fortitude our own wellbeing. Because instead of opening the conversation, we have successfully curtailed the movement of progress half measured by not listening to all that is said under cover by the interruption with and in following our steps taken syndrome. Being a matter to be altered by our religious leaders in all cases, but not to or by their existing plan, only now bordered by the Fifth Dimension three taken and one active prospect in 'One for the Whole'.

Without example, we will never be able and connect in the form of human unity either religiously or politically because we automatically hold our differences to account, giving us false readings at all times. Therefore our religious leaders again are to lead to our united point of reference although we speak with many tongues, but from now in the same language of the Fifth Dimension, not in clarion only in check? It in concept will be our last modern era tool to nudge and cajole our political reasoning to realignment and also speak in the same tongue? Not necessarily of democracy or communism, but in the voice of our specie unity by result in carrying our expectations by religious aims determined by ourselves even if under guidance, but exclusively by our own hand.

In the matter of any direct offer for us to alter on a global scale indicated through these pages, it is to be done without sweeping changes in order that we maintain the vital concept of our specie unity and more importantly endorse our specie singularity in all comparisons with other creatures. Even when operating to the accepted political fixture of our standing to be first second of third world in terms of our commercial adaptability, there is no designed difference between us? Because our political standards are only a byword to our actual specie celebrity and by paradox operate in the first cause of service to us in the matter of simple civil control. Being a means of community decision making from as far back as we might remember, even by being unaware when decisions were to be made of how we were to bury our own clansman and by what medium in places like Shanidar?

Politics in our name, although a judge, is to be set again as our advisors, which is to be obtained by overlaying it in the same manner that we are to use in overlaying our religions and with the same references from the Fifth Dimension. Why I have used this approach in the form of insistence on our behalf is because we have made great efforts to show our awareness to many things. Our half measured approach although the best option in some cases and at some times, has allowed the outfall of our steps taken syndrome support what ever results were expected? Not maliciously, but in the context of expecting a result from a particular action almost as if to have been laid down experimentally scientifically, therefore if the outcome was not quite as expected, to justify the original commitment, we will run with the outfall in any case? Half measured experiment, steps taken result, condition to remain, at least politically, because we have not yet had an alternative?

Religious matters are of a different complexion which has been shown by my own approach even if half measured, I have led by indication to the differences between our first three dimensional religions to have come from the Middle East. A difference I claim on all of our behalves by accepting that if this approach could be seen as half measured, will be carried by how we can address the change in attitude I seek to promote in open understanding and not given to be steps taken, because the remedy is not half measured!.

## CHAPTER 116

These past sixty plus years of this modern era we are now in are to be pivotal to our collective welfare, not least because on a day to day basis we show how we do not understand each other. From amongst the mighty armada of objections we bandy between ourselves by working only through the political forum, our steps taken syndrome results are moving to split us more than contain our real similarity. Supporting this or any specific standard we are moving to, only allows us compromise what was supposed to be a political cornerstone of morality by our human rights mandate. Further endorsed by acceptance to the statement of our religious tolerance in respect of our equality set against our first, second and third world standing politically, offers imbalance and a means to overturn our intended objective to highlight our real equality. Producing quicker half measured approaches, manifesting harsher and blinder steps taken syndromes of which we will still react to, because we have no alternatives or we have not really thought of alternatives?.

Unbalanced as it might be suggested to have been by others, my own take to the situation whereby from the Islamic faith there has ever been the open acceptance to the personal decision of the individual make their own judgement to kill of self by self, stands! Not only from the historical connection as alluded to in past pages, nor to the specifics of the written text, but in as much to the necessity of the deed or act to simply round the situation in context to the existing standards of the day some fourteen hundred years ago? Other than to the obvious of total inclusion of the individual to this new religion in the making, even if at first offered by intention that the making of Gods last religion was to be done at a mixture of time wrapt in uncertainty, then from the order to defend, protect or promote it, the normal means of adversity applied? Not only was this modern, but it allowed the converts be totally involved to the concept of the overall proposals of the intention behind the new faith. They as we in the promotion by example of personal conduct, had now been given the ability of that personal involvement to make things really count, unlike similar religions that did not have the true written text for guidance. Although others were created in a similar fashion before Islam in time, there concept to the individual was of a second hand lead, whereby control was to be led by an established priesthood, in some cases with there own style of delivery?

Islam, if to be a more rounded religion than others has always remained so, but only to beauty being in the eye of the beholder. From our own histories and in dimensional form, Islam and Catholicism have had a brutal history on religious grounds over religious grounds? Although resolved in the course of time, intermittingly both had managed there own affairs in conduct, generally in the situations of there best development if in time to have been the Middle East for Islam and Europe for redeveloped Christianity!

From my time mark of some two hundred years ago with present day relevance, came the re-emergence of both entities this time from the European side to have been more with political emphasis. As previously stated but not in context, when Charles Darwin published his great work the uptake by some Europeans in rapid connection was to ignore one premise that we had developed from apes, and accept the principal that we had advanced by natural selection? A means whereby the stronger and fittest were the cause in standing to the welfare of the whole group if at first to be taken by homeward settlement, allowing for later additions to the idea of natural selection make its own course by definition to the eye of the beholder and by there own judgements?

By leapfrogging some of the intervening years only to notice by worlds apart, that the European German's in the name of politics and the Asian Japanese in the name of raw materials, came to promote there own concepts of being naturally selected to be superior to others of the same specie that we are. Although transcending real religious concepts both of there conditional marks were in the name of political fortitude above all. Setting the case for the standards we have tried to unravel since the ending of hostilities in nineteen forty five. By any unfortunate twist to any matters of history; the recording will always tell different stories than the events. But from the Second World War ending, our general pattern has always been led by what we see to be of our best interests, our political determinations!

By the coincidences of time the relationship between Germany, Japan and the rest of the world operate on the more general scale of our political medium intertwined by the pursuit of material things only politically. Although both countries are better involved to the general good of more open political management than pre war, they are equally involved to our commitment to all things political simply by association. Not unconnected from the specific involvement of America contributing to there total defeat in the war and in the responsibility of her financial support to generate a different political standard after. Even allowing that in the case of European Germany, Russia had been immensely involved to her defeat, they were gazumped by developing events, because Russian communism was not associated to any religion, therefore denied the concept of religious tolerance?

To carry matters to all corners of the world, for Japan by defeat, America was totally responsible for the downfall of her despotic conquests by bringing us all into the new atomic era. From that position although supplying aid to Japan for her commercial recovery, America set the pattern of her future political standing to be paramount to all of our wellbeing. Unfortunately the shortfall to this situation was to endorse every thing that Europe had been doing to the rest of the world for the past two hundred years, setting her/there standards to be unquestioned by application!

Although just and honest consideration had been offered in the restart to political and or religious standards and standing, we ourselves were not ready or aware that we were sentimentally different. Because we immediately became judgemental of many of the new countries if not least to classify there efforts to be second or third world by definition from those of us already first world, yet by the same score for other reasons we marked all heads to the same standard of human rights but without practice? Although much of this reckoning was done in haste, we can not use that as an excuse because once done half measured we implemented the application by our steps taken syndrome. Making that we were unable to consider what to do if questioned on any grounds, not because we didn't have answers but simply because we really did not understand the questions if offered by a third world country to a method of control used by first world political principles?

How ever we have poured our collective learning or our collective behaviour into the top of the funnel, what emerged from the spout was always led by thinking to be of our first world determining, that is why the records have been noted by assumption? By example, from very recent times to events in the Middle East, any clash of wills is always noticed, but is only ever seen politically? That is why I offer without objection to our religious leaders, to take the given prospects of the Fifth Dimension and overlay all of our religions to enhance our awareness of how we have differed by the slightest of moves, but to the biggest of results in challenging our recent tried concepts of genial equality generally? With the expression in use about religious leaders, our automatic response will be to connect to our own following in separation to each other. That is the nature of how we differ by acceptance without the study or concern to others or of their opinions, it is a human trait held by us all without objection, which in its self is no bad thing.

To refer of what or who our religious leaders are, is a defined move to show us all in recognition some of the ways we have all failed by religious concepts over these past sixty plus years. From which some of our problems of today have been given root to influence some of us who effectively are leaderless or are made to feel impotent by other features we are expected to take for granted, like our political determinations if fed only from one source on the assumption that other matters are cared for automatically?

Our religious leaders previously described by my self in connection to how they were to actually lead or guide us will always find controversy from the aspect of our adaptation of the faith of Islam. Leastways by this last modern era we live in, shown first hand by how we react to each other on purely political grounds religiously. Not disconnected to how the Jewish and Arab Palestinians interacted in the time of mandated partition in one side representing Judaism and nationality, with one side representing Islam and nationality. Where was the difference unless by political intervention if to have been unbalanced, even biased, but allowed the working and promotion of one religious viewpoint take a more prominent stand politically, over another religious standing left uncounted!?

Without further classification, for me to have intoned that it is only from the last of my three designated dimensional religions from the Middle East who morally could rely on the tenet for us to kill of self by self and fix to my reasons described in terms of defence, protection or promotion, stands! My own fixation is not with the death of any human, because sadly outside of our connected standards all too many people die by the hand of politics and or its fixation to support economies over nationality. All too many people die at the hand of the stunted or criminal mind in the pursuit of personal gain being a seen exchange for the measure of that gain only connected to the moment. For as of yet no criminal or those fixed with the desire of want have been able and extend their own longevity with any soiled profits they might be able and connect with as winning! Even those on the grand scale who had avidly chased there own concepts with or without religious fortitude in the name of the fatherland or the mother country or were to follow the adulation of the little red book in the name of the early communist Chinese, all failed to recognise the future because they never reached it!

By indication that the last half paragraph was to underline some of our political reasoning and to be covered or endorsed by our steps taken syndrome. Shows how from within our political structures we have lost the matter of control in not being able to see other opinions because we have marked the means of debate by us all to be religiously tolerant. From which has sprung the matter of our objecting to some religious responses being second rate or worse or second hand or worse, while at the same time supporting some religious standings historically even unfound!

If from how I have been able and determine by my own judgement what has poured from the spout of the funnel knowing what the mixture at the top was. Then I as you as we can value my own reasoning, value my own ability to rationalise the intentions and interpretations of other recent judgements even if connected to a fourteen hundred year history, recorded, but without alteration or the supposition of alteration!?

By consent that we were in the mid nineteen sixties at noted times of conflict between the Israelis, Arabs and Palestinians, extending to the issue of conflict between Iraq and Iran of some fifteen to twenty years later, although the wasting of human life was to be the only real victor in the practice of carnage! From the first instance if, and following, virtually all political decisions of area have been influenced one way or another by small religious antagonisms. From which it is more than our obligation to alter the basis for further and similar acts of human murder in the light of all of our religious determinations!

Having stated that support for Israel had been offered by America first and Europe second since nineteen forty five, in having stated that most of that opinion was because of the considered effect of our global political formation because of bloc antagonisms. Then by outward expression that aid and assistance to Israel was of political aims only and in light of that situation, less than measured consideration was given to the situation then in the Middle East, because of the Jewish and Arab antagonisms under mandate.

Having acted to there own best interest the American and European partnership standing together against communism to protect there own commercial interests with or through the alliance made with Israel, assumptions made were only of a westernised perspective. Not least because although the Jews and Palestinians there under mandate, who had entered in the exchange of brutal murder with the same view, most of the Jewish propositions to be entered for partition were put forward by Americanised or European orientated Jews? Who already had most of there standards to be read politically already being western! Therefore although there was a huge gap between the Jewish and Christian philosophies, in that case it was nullified, and to remain of a political deal primarily?

## CHAPTER 117

After all, with fifty five million deaths and counting in full recognition that they had been delivered to us by political means, by the further political means of expediency now generated by the clumsy way we had allowed politics take centre stage. In settling matters by those political requirements, in settling by the direction of community on political terms only, we were stretching the matter of our similarity in the haste of half measured thinking. Although now in this new and infant atomic era our thinking was set to be no more different than at first when we met others who had the use of flint and we who did not? Our inclination then as now was to be in a position to settle the matter even by horse trading?

By the smallest of unnoticed features but by obvious circumstances, our deals were forced to be only political by tone. Mainly because of the developed and developing bloc culture standoff situation, which is not yet history, but remains in influence because we allowed our communities concentrate on our political differences. It was also partly from those circumstances that in the name of placation driven half measured, we hastened to accommodate some political actions over the feature of our religious concerns. Unless genuinely unseen by how things might develop in the name of how our steps taken syndrome would drive the issue or some issues of a religious texture?

By open promotion of the first renewed awareness to religious matters, if only proposed to thwart those in the bloc camp of communism at early stages of this last modern era. From those that were deeply involved in our settlement theories, being American led, with the same western European religious standard of mainly Christian thinking behind them, allowed the misuse of religious tolerance as a political entry to highlight our political differences. Was one way to allow open support of countries such as Israel and others to follow the formation of their own standards to God! Which to the Middle East in particular we are still paying the price for, but why? When at much the same time seen as the biggest issue of religious settlement when looking to the sub continent of India the outfall was not treated with or from the same westernised viewpoint nor with the same tact or concern!

India is to be of special note to us all, because from her example can be seen the shortfall of any political horse trading that we as a collective body of people were ever involved with. Past her own deep history, which is to be just as valued as that belonging to Mesopotamia and to the Eastern reaches of the Middle East. She on her own account, even from her own style of hard won horse trading, having gained her independence from the colonial rule of the European country of Great Britain by promise after the ending of the Second World War. Found her self left with perhaps the greatest religious turmoil of any era in time, even eclipsing the head to head suffering brought about by the crusades of Christian Europe against the Jihad of Islam centred in the Middle East centuries before?

Prior to independence and unravelled by the talks to have been between Indian political representatives and the civil servants representing the British government, from deep set religious convictions. Soon to be former colonialists set the case for partition of the sub continent into separate self governing states determined by religious sentiments. Although Great Britain was more than aware of the deep set religious differences between both factions, when the case was made for partition the style of the split was made in haste. From the one sub continent of India came three separate countries. India named so with the state of West Pakistan on her western borders and East Pakistan on her eastern borders, with the Pakistan's holding to Islam as their standard religion to India's Hinduism, the stage was set?

Although the split was made in general agreement for the time of independence, from the matter of partition, the outfall of human tragedy was tremendous. Not to lessen the effect, but by the means of how the borders had been drawn up on the general basis of whether people were Muslim of Hindu the errors of accountability were never to be accepted by any of the three parties involved. Great Britain in her form took guidance but led in pace whereas the other parties split by religious determinations aimed to gather the best ground to there own advantage by accepting the principal of partition now religiously, and further determinations later by the means of referendum politically?

How ever the agreement was hastened, the result was the mass movement of up to some sixteen millions of people having to uproot everything and move to be amongst there own kind religiously in supposition? Sixteen million people who in the shortest space of time imaginable were disposed to abandon all that they had achieved through the rigors of a very hard lifestyle, to settle anew by the political desires for a religious solution? To say naturally might be an unfair expression, but it has to be said in context because of the human trauma involved by such deeds. Naturally, because of the extent of what we were expected to do and in keeping good faith with their/our political leaders, the open desire of the human condition in all cases of such complexity has been for us to find a medium of blame or a means to vent our blame capacity. This was done in fervour by about eight million Muslims moving in one direction against eight million Hindu moving in the other with the medium of blame transmitted to those of our mutual contact. So much so that for any number of reasons but not least religiously, vicious human frustration was vented on one religious group by another, resulting in almost the same number of murderous deaths to each religion. Nearly one million of the sixteen million in transit died from religious zeal by the same people as we, but by the counted reasons of indifference to the new world order we were attempting to reconstruct out of the political cause of other recent blood letting!

Connections that I make to the above tragedy of events is of a comparable way that the whole sad episode of the partition of the sub continent was done, was not by how it was done, because at the time we were operating to the standards of judgement we had then obtained. What we missed was the opportunity to notice continually how strong our religious feelings were to be ever. Without counting history, but in context, sixteen million people who were cajoled to move politically brought their own reactions to the fore under the guise of religious determinations even unseen. Yet by the early planning to the structure of human rights if to be implemented through the United Nations, we allowed the heavy hand of religious retribution take its own course. If seen the United Nations were not connected by direct association with the standing religions of those of us in first setting the standard of partition!

That is not to say there was any deliberate cause or desire from the matter of reconstruction that we were at liberty to ignore the religious standing and acceptances of others, for we had began to implement the study of religious tolerance as our means of equality of voice, religiously? Although submitted from an American/European Christian perspective the matter was supposed to be general, but as can be seen, when falling short, we had forgotten to implement a fair means by which we could query the very concept? Not unreasonably, because we did not take the time to consult with our religious leaders dealing in the spiritual ideal, save from those who were automatically committed to the political thinking of the day in general! With many views of a Christian texture by association, because much of the matter or our horse trading even in India/Pakistan, was centred to accommodate opposition to new communism already contracted by Russia, irreligious again after victory in Europe. Also in notice to the emerging Marxist communist regimes in China the most populous country in the world and contributing to the bloc culture standoff. If I had put note to this situation that the western philosophy was geared to thwart communism by any means, then having a religious standard of any sort like religious tolerance offered by the United Nations was as far as we needed to go morally, even if to forget Islam and Hinduism!

## CHAPTER 118

Islam from its own standing and to our equal benefit was the means to have exposed the anomalies that we had created and still react to at this time, because we behave by our steps taken syndrome from our original half measured action? What I have to say now is to be viewed in a reflective manner of how and why this work has been presented if not least through our first use in activity of our one and only active prospect in 'One for the Whole'. Even as presented the matter is open to immediate objection and correction as it should be, that is the main aim of our one active prospect, allowing us examine our own ideas first against the query of doubt applied by others and self for mutual acceptance!

By mentioning our religious leader in line to present a better case for our real unity, then the first steps taken beyond the muddle of our steps taken syndrome, is done on our behalf ensuring that all are included, but only by classification, if religious leaders?

If from first sight Mohammed was to be the religious leader in the name of Islam, then he himself was to dispel that note by the simple and open fact that he died having already fulfilled his own lifecycle to the chimes of 'Natural Body Death'! By then his and our new religion of Islam had been working and growing at pace, but not without query and questioning from others with there own vested interest for Islam to fail? As part of the means to prevent this, Islam was pushed forward at the determination of the faithful to accommodate opposition and further its own ends by self expression of ideals? Although to have been led by the existing means of civil control to have prevailed at instigation, Islam by structure carried a different message than had the first two monotheistic religions of area.

Even if induced by early opposition by some people close to him, Mohammed set the format of the new religion not from personal recollections but by the induced lead of Gods own agent in bringing his message once again to mankind. From the best signs of the political structure of society then ago the means to the discussion of change was generally force of one kind or another. From the religious perspective, in order to enable the drive of success prevail for further religious expansion, it was made abundantly clear that some lives of the faithful would be forfeit. From much the same thinking that had befell earlier supporters of Catholicism just after the time of Christ, the same mood gripped those engaged in the promotion or defence or protection of the new word in this new style?

Where I had indicated that many earlier converts to the teaching of Christ were prepared to suffer death at the hands of whomsoever in the certainty of reaching haven as Jesus himself had. Then from the addition of the next six hundred years of modernism and interaction in progressive stages the offer was changed to be more palatable to more discerning people if so claimed? Not now to the acceptance of death in cover of being religiously connected, from the very way that the individual in Islam was to address his benefit of the new faith by self conduct. And with the drive to openly promote the religion in the name of humanity by way of correction if necessary, we were at liberty to actively promote, or defend our beginning. Because of the nature of interaction in the present time the sword was to be the voice of many in the battle for converts. By the same token and in proper counterbalance, if to those ends one was to fall in sacrifice to that struggle, then paradise described by all the splendour of our imagination was offered in consolation.

By open connection, but unseen then ago, charter ahead of anything thought of before, was given to the individual connected to this new religion. Although the many types of civil or political controls were to still patrol our consciences, here in an instant, we in the best name of this our new faith, were at liberty to make utter and final decisions, which in time were to be of no significance? By the advent of our generational coming and going, as the religion grew, we who were amongst the greater body of people to be together as Muslims had no need or consideration to the fact that we might at will, still take those steps to kill of self by self? Even when to remember that Mohammed had expressed in lament, that a single regret to his own emotions, was that he had not been given the chance to become a martyr in open view of his own acceptance that Allah in God might have made such a choice for him? No connection was taken by association that the act of martyrdom was a ready part in need to the proper conduct of, in or through the faith at that time, but never now?

Perception might have become only myth if all of our paths were not to have crossed in these recent times of our international awakening these sixty plus years? From where I have tried with the fortunate good relief that many and many better informed people than I had done much to guide us to certain joint awareness. Will soon be endorsed by how the same better informed people as I can give lead to a better general understanding by our taken or developed attitudes. Our religious leaders are to be those people because from now and with the one sentence in the name of this book we can all move in the idea of one direction because we are already the one specie!

Anything said, written of even hinted at is to be ever all of our property, from the fact that our politics now rules by default is to be a cautionary observation not least because in voicing much of its own opinion its natural tendency has been to proffer its own interest first. Not unlike from how it sets our conduct to promote some of its own aspects in the form of commerce, trade, commodities all serving monetarism, which has become a benchmark of our civic dealing over and above our permitted religious tolerance?

By assembly and from my own and therefore our suggestion that we separately overlay our political and religious mediums with the taken and active prospect of the Fifth Dimension, is done on the one side to be most effective by our religious leaders. People like you and I who in open contest have always had the option to choose our own God providing we are all correct? From politics the talk is only of systems, that is why she, although in lead at the moment, will find her powerbase to be curtailed in reasonable fashion when operating anew. She from within will be able and allow her own alteration, because our future setting will soon be done better by our religious overtures.

Religious leaders by standardisation are not to be so easily set free of some of our shortcomings. Although my suggestions are from this moment in time generated by these past sixty plus years, our histories in cases have been clouded by the different opinions our religious leaders had held previously. Now we all have the Fifth Dimension as our equal guiding hand to put forward all propositions but this time irreversibly connected through our three taken prospects and one active being 'One for the Whole', from where we can now balance our different judgements!

Religious leaders by standardisation will have to include how I have, and we must examine the standing of the Islamic imams if to be considered as a means of contact from which ordinary Muslims are to express the essence of self opinion along with the intimacy of the written text? In actuality the opinion of some imams has been to make the case, not for Islam, but against other religions and or political systems, which by view is no bad thing when considering that from some, they are operating like my self and or far more worthy authors with better informed presentations than this. Some imams are taking to the personality for the promotion of there own opinions drawn from the first days of Islam, when it could have been under threat? Perhaps that we might forget our fourteen hundred year history with its accumulated progress stored in all of our names, then the matter is not to be desperate only to appear so by some of our opinions, which might have been too literal in the attention of study to some present situations?

In expression to the obvious opinion that to the vast majority of Muslims the matter of death of self by self is of no consequence to any aspect of the religion today by any means, is an open acceptance that the religion itself over these past fourteen hundred years has outgrown its own fears and the need of previous safe guards! Its huge success is testament to that fact, it competes on no level with any other religion, it does not have to? Our own confusion has been generated more by the clumsy way we have addressed some of our political deviations and to the reasons of why? From those decisions our religious leaders in the case of the Pope, cardinals and the reverend of Christian religions along with the Chief Rabbi and rabbis will have to rationalise how judgements had been made in the past and still, whereby we cannot kill self by self but do! If now to be bound in the liberty of our own Fifth Dimension acceptances, that picture is at last set to change religiously?

Imams by the same expression will have to show cause and reason, of how some in equally harmony, have seen it a fit and proper thing to encourage the ordinary of us induced by the splendour of Islam, throw away its bounty by being kept misinformed! Those of us in ordinary array to the faith have never, never considered the feature of murder or even to inflict death from self to self in any capacity or for any reason? Yet at the same time many of us guided by the same emotions would die for the good offices of our chosen religion. Our points of reference never to have been looked at in this way or for some stated reasons have been brought into being questioned by my version of abstract thought and reasoning. In making much of the past to have been serious to the style of our assumed progress, has brought us to where we are under somewhat different circumstances but still with the same objectives!

In order that we can change the past without alteration we must look at how the future is now and lay our thinking upon it as we might have assumed it would turn out in plan? Not least by how we were committed to our codes of behaviour as they could have been expressed in guidance by our priesthoods religiously? Or even how our conduct was to be measured by our own conduct generation upon generation, until we were past the notice of how the future actually was in comparison to what it should have been under our own religious expectations. Without direction, what could have happened and might have been the cause in any case to most of our troubles if indeed we have any? Is through the simplest method of reasoning to how we touched on the matter of religious tolerance to be a measure of control to our difference of religious attitudes, yet bullied our way in progress under the illusion that politics was the first medium that we should always use?

From one serious pointed avenue, I have set our thoughts to the matter of who our religious leaders are or should be, I have been deliberate to make the separation between our first two dimensional religions and their priesthoods, balanced to the imams of Islam? To show the manner of how we have differed while expecting the same outcome under all of our claimed separate religions and more so in this same period of time, because of our political behaviour, which has been set to be the same?

From the open turmoil of these past sixty years it has been painfully clear that we have not been able and adjust to the fact that we actually are this our single specie, nor that it has been from our religious endeavours that has made the connection be possible. By moving forward under our own recommendation and in using the Fifth Dimension to be our watermarked guide, the picture will clear after overlay, simply because too much has happened in these past sixty years. That I have connected to the conduct of Islamic imams and some of their conduct is to be of great significance to us all, but not in comparison to our other acclaimed priesthoods.

In taking it so that in places I had covered the connection between Europeanism, Americanism, Christianity and Judaism mixing and interacting through the intrigue of all the changes in politics over these past two hundred years, makes no case for any other comparison? In suggesting that there was a change in attitude by some people to come out of the Middle East on the part of the Palestinians/Arabs or Iraqi or Iranian opinions, is not done by condemnation, nor is it to ignore in the negative what had been done by European, American or Israeli ideals religiously or politically?

From the division I have already made, no balance can be set against Islam on religious grounds from the matter of the comparisons I have used? My own personal theory in the conduct of some imams, is based not on actions or deeds, but in the face of reactions, of which by some of those they have inspired to now become part of history, yet in the most modern of times, not because people might be encouraged to become martyrs, but because some of the suggestions are made erroneously by mere men as we! If made by our religions then that is on a different account, but if made by any part of the core in guidance to doctrine than the balance is unfounded? But in cases understandable by the very people we are or have become, at first determined by our geography, climate, diet or other features giving us some of our differences in attitude, followed by other un-listening, politically motivated views!

**CHAPTER 119**

Whatever my take is or has been to the matter of how I had described the consent which Muslims could kill of self by self, has deep rooted historical reference. My view on the situation today has been brought to the fore to show some of the problems we face in the face of trying to accommodate our united wellbeing. I have given fair chase to the shortcomings of many of our political attitudes which when expressed or worked to or from, have always created there own result not least by enacting to our steps taken syndrome? Unfortunately this is compounded into the further error of total acceptance, because amongst ourselves, we already hold different plans to the same political initiatives whatever. From which I can say our overlay of those political attitudes with our Fifth Dimensional prospects will be ready for the possibility of change if required.

Religions with much the same track record even with most following political leads in any case, have a better chance to manage change because with our Fifth Dimensional overlay, they will automatically be empowered to guide politics by renewed judgements? In order that we are to assist our own efforts of what has been erroneously taken to be a standing error of our own religious judgement, I have set the scene that we should look at some of our Islamic perspectives separately. Not in the divisive manner of which some might claim my intention, but for our own enlightenment and to be set in relation to each one of us bound by our declaration of the absolute relativity of our own lifecycle in all comparisons to self. My notice is to be given that by religious terms first, and in all cases throughout history, they have been our one uniting philosophy of intent!

All that we were to do then ago was to serve God by whatever religious tones we felt required to adopt, and from whatever lead we felt comfortable with, or through whatever circumstances they were delivered by. Nothing has changed save the needed interference of our politics which aligned itself better to our scientific discoveries exposed through our fourth dimensional outlets. Nothing has changed save that we found it easier to heed the art of science and politics over our religions, because the former delivered in the name of commodity, and nothing else did that above our emotions? By the factor that politics was able to lead us on a route that didn't really suit our original search has brought us in full sweep to why I on our account have moved that we apply some small suggestions to turn the tide of many of our histories that now lead in the name of mammon?

There is no doubt that we can all be swayed by the open delivery of ease, comfort or excess, leastways those of us fortunate enough to be of the first world by political terms and allowed maintain our religious sentiments in equality with all others? All of which would seem to cover all exits whereby we are well off politically and religiously in equal respect. Unless our concerns are interrupted politically, which is what has happened throughout our mixed histories until these very recent times when our assumptions took over again to assume that our political path was indeed the correct one? Until challenged by some of our religions, not least all of them who wish to take the emphasis of time to the significance of our own lifecycles, but how ever seen, even if not to be quite understood just yet?

I have always maintained it was our civil community that was the key drive to our progress, but viewed with the events at Shanidar the case was then ready made for the matter of our religious concepts and political management to go hand in hand for our best hopes. Having now arrived at another pivotal point in our progress we have been able and question some of our intent on those terms to be recognised of political or religious importance. Leading to how I at first, or we collectively, have brought one group of people from one religion into headlight focus. I have turned the supposed actions or the supposed suggestions that some imams in the religion of Islam might have offered solutions to redress some actions taken by other religious or political factions in these most recent of times in this last modern era and set the possibility of accusation?

By connection to the Middle East, my intention in all cases and examples has been to show that we as the same people have always been able to make different judgements over the same matters, none more than from our religious perceptions. In addressing the three main religions of area in terms of my dimensional heading, has been done to refocus and refocus our attention to the same matter. But now to be recognised for their differences from the same people who had been treated by the ravages of time and the influences of climate, geography, diet, attitude and other matters giving us different perspectives, even when discussing those differences? By my own connection to the ultimate concern of the matter that we of the same specie do kill self by self, I have been cautionary in the depth of what to say? Not by guide, but declaration, so as not to direct my own hand against myself, because in holding that all religions are to the prospect of 'Natural Body Death' is an end statement! From some of our own heated passion in concern to our own religion if seen to be under scrutiny, we might see that the answer to some problems is to kill of self! If that means to kill at random, to kill wantonly or just to kill, but making the same connection that in all cases the acts were in compliance to the issue of 'Natural Body Death', is the wrong pattern?

That very source is why I can see caution between our three Middle East religions, but not any others or our collection of Eastern styled religions, because 'Natural Body Death' by referral is to how we die. Not by intention, but in the form that our own lifespan has been won from evolution in the first place, and in the scope of creation we have changed its length of span by the conditions we have extracted in the awareness of God and the reality of time as it might be? In the fall of time to those ends we have irreversibly arrived where we are by status, yet still conflict over religious matters before political ones, which is yet to be how we can alter the past without change!

In the partial separation I have conducted between our first two monotheistic religions and the third in being Islam. I have focused on one very small matter which for no other reason represents how our difference of attitude has manifested itself. Before I commit further to the cross examination of Islam I had started many, many pages ago. One factor of balance to be noted is how the two other religions of area have to account for their standard by proclaiming not to kill of self but do, which in like form will be iterated by the stated leaders of those religions with the overlay of the Fifth Dimension prospects!

Islam to some ways of thinking by other people who might be more politically inclined, has no direct religious leaders or a specific leader by example in comparison to the Pope of Roman Catholicism. In parallel, because of the structure of the religion, the influence directed to be upon its adherents by the over sears hailed to be in the mosques and generally accepted to be Imams by the common eye of the outsider. Lead in the matter of interpretation to how the text should be read or listened to depending on the circumstances of what other situations might prevail and where? Generally being no different to others, because the religion was first laid down in a volatile place and in violent times when actions had to be the voice of argument against adversity and indifference. Being an exercise to us all in another example of how we had changed to suit the circumstances that then and now exist?

Because I had read past records to state that the religion in new form and for eternity was to be capable and hold the sword to protect, defend or promote cause, stands! Because I can recognise the age in which it was introduced to need any and all methods in the ability to cross lethargy, indifference or violent rejection of anything new no matter from what source? Tied with the general human perception that if you did not belong to this or any other exclusive society, your standing of worth in all terms by not being counted within, was of little or no consequence? With the first and final concept of Islam being in the Arab tongue, the tenets, covenants, laws and guidance were laid in that pattern and for all time, because this religion, different to the other two of area had the perception of time built in. Not unlike the existing Roman Catholic religion the intention was to come to the whole world or all of mankind, but through Islam taking a longer period of time than to the first consideration of the individual, already covered by the matter of self conduct!

Success for the religion of Islam is assured, if it has brought itself to the matter of conflict in these modern times, not from the style of doctrine or from the written word, any drawback is how we as men have failed its outlook? By keeping to the idea of martyrdom as a shield to those that might fall in the defence, protection or the promotion of Islam, in context some of us have failed it. Some of us have confused the issue in the matter of faith alone by the adapted inclination that along with the duty of personal conduct to the religion, we individually can now make our own judgements to kill of self by self? Allowing that in any incident to that end, others that might die by our deliberate intention are of lesser value to Gods own judgement, because we individually are set to manage the act in Gods name? But have still been able and address the matter, that if in the pursuit our own suicide to our own judgement of God's desire, we inadvertently kill those already to be of Islam. Then they also carry the shield of martyrdom to the open joy of paradise which alas lacks all substance!

From any time, by any imagination between sixty and one hundred thousand years ago, when by collective consent we buried one of kind in Northern Iraq at Shanidar in the Middle East, a beacon to human liberty was lit. In leaving that corpse in loneliness and by the small task of adding tribute to the scene, a voice of mighty proportions was raised to the heavens. If ever that act has to have a bearing upon us at any time, it stands that we of now were awakened to transform evolution into creation, from which the rest is history in the making. Ably shown in the progressive way we have always been inclined to make forward movement in the fields of moral and social advancement. Geared by that ability we have had to turn the wheels of our own religious perceptions leading in all cases to the image of God, the Great Spirit or understanding in the mind of Nirvana. We have constantly refined our own images to suit the relevance of time to any given situation, so much so to fit the arrangement of our own understanding to be gained from our own deaths. Effectively we have set the biggest scientific experiment of all time without the need of up to eleven invented dimensions to issue what is suggested might be the correct result?

Our own human rational has brought us to this situation, no matter that we ourselves have left questions to be answered, because by our ability to recover we can find it easier to understand that we are all connected to the same expected outcome. The only prod we might need is by the means of our religious determinations yet to be delivered by the effectiveness of our religious leaders under what are new terms or at least by my interpretation of what is needed and how? Unless we have or reach a defined status in permitting the full voice of Islam use the same feature of the Fifth Dimension in context. Such minor problems that I had raised through the connection of martyrdom and suicide linked to be an arm for the defence promotion or protection of Islam will continually fall to be misread and misused!

Continue to be misread is not a final statement, but indicative from where the reading should be done, which is in the direction of an open invitation to all of us, but now to be gauged after using the prospects of the Fifth Dimension? All main works in this example are to be carried out by inclination of our religious leaders from wherever. Any problems that I have supposed of from Islam are quite small in context, but made bigger by submissions of or from some of her leaders in my example to be imams, right or wrong? In the specific area where I have made issue in relation to us all in respect of how we manage the feature of murder on self by self, extends to all religions and political situations. My example is to be met in the same way that I have set the situation for all religious and political leaders handled in separate fashion, but to acquire the same outcome if now to be aided in concept by the lines of C D or the principles of Common Decency!

Imams in name carried by the ignorant such as myself, have not, not been selected other than to the very personality some are. Looking for Islamic religious leaders in terms that have been proposed of other religions is not a test of one religion against the other, even from my three often mentioned religions being monotheistic? Other religions in the same light also using the three taken and one active prospects of the fifth Dimension are set the same standards to fulfil. Not in self acclimation, but to show the reality of our unspoken similarity and singularity. For me to extend in reach to one group is in relation to the difference of behaviour exposed by the past sixty plus or from forty plus years ago, not from the critical. Because such incidences if, are the property of our own consciences which highlighted by some religions from others, have allowed us all bigger personal involvement.

Recent time is the key to my drive for a different study to how we have so far reacted and interacted by/from recent events particularly in the Middle East. From which I see the open danger of our own entrenchment which is a means to keep us apart when using the existing features that we already apply in control from our assumed base of being civilised. By setting my own examples to this last modern era is done from its age anyhow, but most importantly because by the fortune of circumstances the time span of most of our world conduct has been focused into this era! Coinciding with the fact that by personal relationship in conjunction with all others, we can relate with the most important era of global change by being here at the start, and watching the process grow, warts and all? We, I, can relate to the vital importance of the individual lifecycle in relation to anything and all things, not least our religious convictions, far and above our political thinking!

Therefore by intention it is our aim to highlight our similarity religiously first and now with our three taken and one active prospect in 'One for the Whole' as a unifying element, even when the same overlay on our political bearing is to produce results of lesser value? Having put much importance to our religious influences in offering our religious leaders a tentative means to lead by choice, it is for them also to react in relation to our conduct in the specific time band of an individual lifespan. Bringing the whole sphere into the same context that the really important expectation is to our religious elements, not because of the past, but now because of what might be? When for another first time, we can declare that our objectives are to show that apart to the conduct to all other specie, we alone are in preparation to show a standard to be like what God in Allah would expect from us by our own inclination without previous direction?

From much of our conduct having turned and turned since Shanidar all those centuries ago. Who amongst us from any time since then and including these past four thousand years of our united religious histories worldwide, could have imagined the importance of these past sixty plus years? Who amongst us might have been able and see that one of the better ways we can direct our own opinion worldwide is to focus on these years since the ending of World War Two. Not only to examine every thing that happened before in due humility, but check on what might be further divisive measures in this time, even if born from the prospect of the drive to establish our united wellbeing, but being compromised by the standing level of our natural and in places political ignorance? Leading to the factor that by having a two tiered system whereby we can recognise our first, second or third world standing politically, commercially, yet endorse our equality by the suggestion only of us all being religiously tolerant? Allowing on that basis that objections to that situation are only from those who would be disruptive when our avenue of exchange the United Nations, was not used in proper exchanges even if those wishing to object were not recognised by implication?

## CHAPTER 120

From the many roads we had travelled in this enlightened last modern era, even allowing for the horrendous act of war set on international levels, through to the equal horror of religious divides creating internal civil strife leading in the extremes of genocide or ethnic cleansing localised! We have addressed them with the air or compromise only, unfortunately dressed from at first having taken half measured steps to any particular situation and then naturally following with the steps taken syndrome we had produced. By the same code to events in the Middle East from forms of connection between the Israeli/Palestinian/Arab interaction and the Iraq/Iran situation, our political conduct has been impotent on more than the obvious level.

Our political conduct, supposedly neutral if policed by the United Nations, had unfortunately taken on the air of being western led by the properties of that branch of our political reasoning in the forefront of world opinion. Therefore if any opposing propositions were to be forwarded they had to comply with the standard western political outlook. Making politics to be top of our agenda by first considerations, therefore in policing any objections, again by the United Nations, who had expressed judgement or sentence accordingly. Was done from the political, because of accepted understanding, which in turn began to leave our religious objectives impotent, because the matter of their concern had been considered to have a different standing than politics?

If our westernised outlook had allowed us maintain political standards over religious ones from the wrong assumption that religions were nice, but only secondary to the importance of our real drive now seriously run by the tentacles of commerce? If objections of a religious nature felt by one side if not seen by the other were forthcoming in any way, then the resulting attempted solutions are needed to alter the error of direction we had taken politically. Not haphazard and in haste from the recent past, but by our duty to it from where we have learnt all that we know in any case. Also by consideration of how we have acted recently by the means of localised war with local genocide or local ethnic cleansing; and allowed ourselves endorse those disgraceful features to have been either of a political or religious motivation? Allowing ourselves look each other in the eye of agreement when holding the same political objectives in the form of control, yet by that standard accepting our religious differences be the same by religious tolerance!?

For as long as we are fortunate enough to continue on this our planet as the defining specie, the mater of the future by no other criteria than the length of a single lifecycle has to be attended to by the first principle of changing no more than our attitudes. That we can overlay our political values by examples of the morality expressed through the three taken prospects of the Fifth Dimension. Also that we can overlay our religions with the same standards even when using our one active prospect of 'One for the Whole' will allow us notice the difference in standing of our religions and politics. From which we will be able and set better judgements to realign some of our future objectives cast from honest awareness of ourselves, but totally reliant to the measure of a single lifecycle!

Having made the case that the most influence is to come in direction from our religious leaders being our best means of representation, is done so that they are to take the lead in promoting political progress from the measure of our religious expectations. For this to happen under its best circumstances we will need terms of reference with a linked or cohesive approach first led from our three monotheistic Middle East Religious bases? For this to happen under its best circumstances I have put the case that we need a more defined lead from the religion of Islam above the matter of guidance delivered by some imams? Because for the first time in fourteen hundred years, even when confronted with what could have been serious setbacks by events in our human relationships these past two hundred years. We have created the conditions of forum by our best reasoning even if a little wanting!

That the three taken and one active prospect of the Fifth Dimension are to feature as a point of reference to us all in any case, I have set the picture of us heeding our religious leaders to endorse that point. Not for us to heed all they say, but by our representation which is automatically adjudged by our one active prospect 'One for the Whole'. By token my judgement upon Islam is not of a religious nature but through the behaviour of some clerics who have been very serious about there own duty. But in cases have reacted to the impetus of some foreign political reactions and turned them to have been religiously condemning, which can be seen as honest assessment if endowed to defend, protect or promote the full image of Islam as promised. It is only when the actions acclaimed in the name of reaction, forfeit the basis of the new understanding that we are trying to acclimatise too, that generate the measure of query that I personally want to introduce!

In selecting the individual character of imams in this as my own core collection of observations, is not to ignore the separate influences that some individual members of the catholic or Jewish priesthood has tarnished or enhanced our/their own religion. In not testing them against imams by balance is done because my case is made in direct relationship of what has happened over these past forty plus years in reality expressed by the beginning of deeds in the Middle East.

If Israel as the only Jewish state in the area has shown her own success to have been carried because of her associations with the western political system of government, if she has been able and give her own religious praise for cause in the prevailing situation, then conclusions drawn by others can follow in pattern. If in the same area communism working from the position to thwart the efforts of capitalism even by supporting those connected to the religion of Islam. Lost in the confrontational bloc culture standoff against the western political outlook, what might be the result in attitude between the differing religious entities?

Quite simply nothing was to be due for change, because the original situation had been addressed under terms befitting to how the world was evolving to the truth in concept, like for instance might being always right, determined by commercialism only! Acting from that principal and in compliance that the best changes were due to political endeavours with religions held in restraint by the maintained concept of religious tolerance, all systems were go. Any further mishaps along the way were to be curtailed by the concept of carrying a big stick and using it to supposedly change or direct the situation, but under who's terms?

Without direction, but in an inflamed area of political and religious intrigue such as the Middle East, no matter what recent events to have occurred by political signs, there can be winners and losers, because in most cases politics is fluid. Unfortunately it is unemotional to the human spirit or condition, that is why from it we can kill self by self to the tune of twenty four million dead in the First World War centred in Europe. Then to come back within the span of most lifecycles and start the whole process again in the same place, though this time our tally exceeds fifty five million dead. Although both conflicts were just outside this last modern era they were within range to influence our thinking when making change now?

Without any signs of connection, but in reference to how politics managed the matter of allowing, even encouraging our attitude to the aspect in killing self by self of in cases like from our bloc culture stand off position, with each side preparing for the same battle? The unnoticed grew of its own volition from an aspect of some of our religious determinations even if only assumed to have happened by myself. Islam being a more rounded religion and expressing that fact almost by a deep seated resentment of direct political commitment, might have found some of the results of recent interactions unpalatable through the devises it was obliged to operate under, which is not to be used as an excuse? If any description could be levelled at what might have been and still is experienced by some holding to Islam, then full accord has to be given in credence to those feelings, but not necessarily by Islamic interpretations only!

If a picture of doubt or uncertainty was raised because of the feeling that the religion in its home land of the Middle East was threatened by the political wheeling and dealing going on all around. Then who was the name for listening too as a united voice of concern when raised in connection with how the source of disgruntlement was to be dealt with? Past the obvious and in keeping with what was originally laid upon individual Muslims; the matter was to be met by the existing code of doctrine. No particular voice was to be heard in unison, because for many people wrapt within cases, the confines of the religion as their own first point of reference did not see the necessity to object? Islam, religiously, was able to keep its own distance from others because all in line was actually going to plan in style. It was only when the new order of confinement by the aspect of freedom through nationality that new world religious problems began to arise!

Without notice even over the past two hundred years the wind of change was upon the Middle East as well as many other places in the world, it was only by compressing that short span of time into this last modern era that new problems arose. In so far as we were to gain our political freedoms by political means, even to the extent that our religions were not to be directly challenged by the matter of our religious tolerance mandate. When the dust began to settle the picture to emerge was nothing like the architects original drawings. From the voice of argument in any direction, if it was not met by the threatening fist of political sanctions or more severe measures in support of the original half measured outlay? They were met in conduct by one side following our steps taken syndrome in supporting the uncounted, because it was half measured, but run politically, economically! Leaving little scope for rational argument when argument had always been religiously driven!

Without setting direction half measured, but in stating that by my notice I had seen a difference between us under several layers of circumstances laid down over time by geography, climate, diet and other influences. Attitude stands more prominent because that can stimulate reactions to almost anything from any rational, but no deeper than from religious convictions, so much so to actually break with some religious traditions? I had made much about the fact that Islam differs with the first two dimensional religions of area over one stated observation. By referring to the commandment held where we all had been directed "Thou shall not kill" I have felt committed to go over the details again and again to distance us all from half measured thinking and the subsequent steps taken syndrome?

Much of my plan has been to state the obvious in monotonous array to direct all of our reasoning into the same pattern for our own brevity of effort. In direct and obvious evidence when Gods commandment "thou shall not kill" was delivered, it was done to be a positive mark to the sanctity of religion and at first instance to the Jewish religion delivered through Moses? Even though at the very time of delivery we openly killed of self by self through crime or in the form of justice, which was to remain in conduct for two thousand years religiously? New

Judaism expressed no different a sentiment, the extension was of the same delivery with the entwined concept to turn the other cheek, where from amongst its many meanings, was to reaffirm the commandment of "thou shall not kill", be of direct relevance to us as men upon man?

From the above and with little other evidence, I had asserted that from the events of conduct and or by other judgements the first six hundred years in new Judaism with the last three hundred in Roman Catholicism had set no standards of conduct religiously if turned that Gods desire had not been met? Without doubting Gods own ability and in the certain knowledge that by his own intervention we had twice failed in reasoning, had the last time now come to enlighten mans own reasoning again, which has not been counted by myself from result, only attitude!?

I had stated on the main cause of difference between the last of my three monotheistic religions from the Middle East and the ideals of the first two centred on the set up of our personal obligations to the matter of death of self by self! In giving sway that Islam adopted a proactive stand for the determination of the very times of which we now lived with the addition of further human experiences adding to the well of knowledge. In coming to the opinion that each new member of the faith through their own religious obligations were also tied into the aspects of defence, protection of what was and the promotion of what was to be. Were obliged to act in the best interest of the new religion come what may, because in concept all were now the same and had the same means in the air of decision making to maintain the positive forward drive of the faith by all means.

In line with the fact that the religion was envisaged to eventually convert the world as then known, in direct conformity with the style of most interactions of the day where human life was to be forfeit for many reasons in society. Why not address the matter of religious excellence to that consideration for those who would compromise the future, not least if they were to oppose the new verse written in all of our names in any case? In taking to that mind set, allowing for the means to kill of self for the overall good of self, then little can be said against the theory at the time of instigation? Especially if the tone of the reasoning was to divert us from the cause of the failure of other religions if indeed they had? All pointed to this last form of enlightenment issued through the name of Mohammed, who still endorsed other religions in concept by association, but had been set to make tangible improvement even directed or not so that Islam would not fail as other religions had, if indeed they did!.

Certainly by intention this religion was aimed to appeal for all men, but like all others had to fit into the ready structure of civil community and political acceptances even though not set by today's standards, therefore then ago greater responsibility was bestowed upon the individual religiously. Most of which resulted in a changed perspective in the relationship between Islam and the other orders of existing civil and political control. Just like elsewhere we were all subject to the whims of our civil leaders by whatever description, for one master was much the same as another? Now even if by terms we were led to the necessity of battle, all concepts had been ably tied to the cause of intention including through our religious doctrine. Not only were we in duty to our own civil leadership, but were replied to by our just deeds religiously, even through battle, even if it meant we were to kill of self religiously!?

In keeping all matters to the conduct of the individual by the set standards of religious dogma within a civil structure, in keeping the awareness open to the individual, that by his own conduct, even if initially by regal direction to settle forms of dispute? We were to fall while in the execution of our defined duty, then the obvious blessing of martyrdom was to balance the issue! Although societies of the day were indeed violent, this new religion did not advocate it as a means to an end, only as an alternative to what was considered the irrational acceptances of others who operated in the realm of indifference. Working to designed ends while being directed by a forlorn priesthood or a dedicated lord and master to drive their own ambition in conjunction with each other, separate to the full inclusion of all? Islam did not compare itself to other religions, but enveloped them either by habit and now doctrine, which fitted more to the very time of its instigation even directed or told by inclination!

## CHAPTER 121

Without comparison, but formed by my own reasoning based on what the differences between us might have or could have been in attitude or by inclination having formed over four, two or one thousand four hundred years. By as much as we came to our last dimensional religion, by as much as we were ready to adopt a new approach, when done or from the doing of, the accomplishment drew its own benefits. Not lest by forming in our personal makeup a standard whereby we could look to be laid back in a disguised form of attitude, stimulated by this new religion of Islam!

In so far as the means of success was determined by our own conduct, then the outcome could be assured through the prospect of 'Natural Body Death', when all of our aspirations were to be met? Not least that we might be called upon to expand in the form of promoting our religion to the benefit of all, not least that we could kill of self by self to do so and when the untimely came upon us by circumstances, we were automatically recognised as martyrs? Why not feel comfortable when we were content to hold the same status as all in death, even that we knew and could see that others were in many ways better off than us by materialism or civil social standing while alive?

From what ever any actual opinion was, my submission that in general terms those of us who had taken to the new religion of Islam by what ever means and under what ever circumstances, automatically began to feel comfortable with there own situation. We still had an accepted means of leadership for the layout to continue in normal conduct by normal practice. Our civil laws were administered in normal fashion allowing for the odd tyrannical leader submit there own forms of retribution as they saw fit, but not enough to break our mould of Islam, which was self determined.

What ever the case was by the hand of any secular leader we had at any time, by my own loose interpretation, Islam was to prove to be its own judge in concern to what actually went on because for the first twelve hundred years of its growth? It remained unchanged no matter what the political edge was? Until from about two hundred years ago when fresh European influences came back to the area of my own noted interest in the Middle East? All that really happened by first contact again was the meeting or two generally opposed philosophies? One European and Christian by religious attitude, presenting the range of their own political desires to the outcome of achieving their own political objectives? Sidelining peoples who were already in area, even to the extent of overriding the standing fused religious attitudes of the people there!

My point of reference being that as the English and French fought for the aim of setting there own influences as part of there greater plan, they were unaware of whom they were amongst? If as I would indicate that the peoples of the Middle East at that time, were to present a laid back feature of character socially, then the balanced uptake by these Europeans was not in recognition to that, only to their inability to stop the inclined national European drive. Although from the course of recorded histories each of us knew of past conditions and interactions if somewhat tainted by our own records, our future recollections had not yet been entered, because until now we were unsure of the new conditions or how to record them?

For me to have suggested that Islam had introduced a laid back attitude to its adherents, converts and followers stands, even in the face that some people of area and by there own tribal connection were socially volatile by nature. Laid back in context does not mean being lethargic or uninterested, it does not mean for the necessity to accepting tyranny as a form of government, of brutality as a form of normality. Nor does it mean not taking immediate advantage of all the technological discoveries in the relish of business management. Being laid back in the cloak of religious sufferance means the one thing only, that all decisions made in relation to the religion of choice are to have been the correct ones? No charge can be laid against any person of a laid back religious standing, when by literal choice they/we can form in mind by our own determinations to kill of self by self on the fixed terms of our personal religious standing for personalised religious bounty recognised! From that frame and for the past twelve hundred years, Islam carried that feeling within most individual members totally unchallenged? Which by reflection had allowed us accept some of the harsh, mean and cruel judgements made against us by the avenue of our civil/political leaders. Because if there conduct was outside of our joined religious perception Gods own retribution would equal the matter of cause by his judgement as foretold!

Why not feel good about yourself even if in volatile argument with a friend or ally over some small family matter or the matter of claim to an object over business deals, which might mean civil intervention to settle, when the final picture is complete by the style of our own conduct religiously? Why not feel good about yourself when even if by foreign intervention nothing can be done to replace our own feeling of wellbeing intoned by our original and now generational religious beliefs. Why not feel good about yourself when even in context but controlled by conditions if bound by technology, others even to be foreign become our overseers, it is known that Gods final judgement will be upon them by natural justice in time in any case. Feel good by being good, even if tied to our own ability to be able and recognise when we might need to take the final step in solution to what we perceive individually, individually, individually as a course of action like killing of self by self and when?

Twelve hundred years of the one setting in some cases has carried Islam apart to other religions, which on its own account stands by all considerations in all examples, because there would not have been any comparison made by any individual in Islam as their/our religion is ultimately self policing! To ask for a change of direction if by overlaying all religions with the three taken and one active prospect of the Fifth Dimension is not the question. In asking by suggestion who are the leaders of Islam so that one first voice could direct all of us in faith to further doctrine, is not done by misdirection to compromise either individuals or the religion, which as known like the first two from area is unshakable? My aim is to equalise all religions, but not comparatively, because in the field of discussion and from open exchanges by that measure any points to be made, can be done collectively and if taken, acted upon in the same manner.

Without defined leaders as an accepted standard, like with a well structured priesthood, Islam had suffered no form of disadvantage, but from my own observations tied with or from the way we have interacted over these past sixty years has brought an urgency to the matter of how we are all to better communicate through the intricate means of religious intercourse? Not that there is to be any form of coercion in the matter save that examples are to be drawn from how we had interacted without being force fed. On my own part and included in my own scheme, when suggesting that we overlay all of our religions with the same mixture of our three taken prospects and one active in 'One for the whole' is done in one aim but stimulated by other recent events.

Examples used in the following form by means of experiment, are to the essence of my intentions, therefore what I say can be tempered with genuine fair judgement. All religions are the same in the eyes of the beholder, anything said to for or against them usually finds response from the good offices of the deemed administrators of each one, except perhaps Islam, who by my own interpretation being un-led, or uninspired from an active focal point! In placing one of our taken prospects in the form that we are all to accept like 'Natural Body Death' is not done to weigh our opinions one against the other. What is to be seen, is something that has always belonged to us all in any case, that the outcome of living, the results of life, the advised picture of reality, are to be exposed to us in equal measure from the event of 'Natural Body Death', no matter what religion we belong to!

Taken to interpretation by its most direct route 'Natural Body Death' is our offering so that we might address Gods own intentions in a way befitting to our specie integrity, drawn by self example apart to how other creatures had displayed there own readiness in a similar fashion? We unlike other creatures are the only one to have included God into the aspect of normal life, while at the same time through the scientific arm of politics created self doubt for how we should tackle the developed situation as it affects us? My charge in part has been to suggest that we overlay both politics and religion with the prospects of the leading Fifth Dimension! Not least because the former contains political outcry against religions expressed through the fourth dimension, for which I want to place its referral tactic of experiment in the correct context. Also it is for us to embarrass political reasoning by the introduction in equal overlay as upon our religions, with the same prospect of 'Natural Body Death', so ends war!

In like form with our religions, but charged with a different emphasis from where we have the matter of rules run by political systems or ideals, none have been able to equate the principle of maintaining human life in the same value for all, only half measured? Our religions by my statement are to insist to the matter of preserving life, because in plan the process fits to all of any religious intentions delivered, even if only read half measured? That is why it is becoming more vital that we can each address the matter of our political conduct from our religious perspectives. That is why it is becoming more important for our religious leaders to be able and communicate on all of our behalves at least now set in the one direction through our Fifth Dimension. Because of how politics is beginning to fail a higher percentage of people daily, expressed in no more detail by how we have conducted matters and still do in the Middle East!

Taking the offer that I propose on our behalf and in looking for a more defined structure of a guiding influence to those of us already to have chosen Islam as our defining religion, no compromise has been set. Nor pre conditions, not even expectations, because the structure of conduct already taken is not under question, it in general terms is insurmountable. My only idea and leaning has been for us to gather in line for collective religious reasoning, not competitively, but in line to show our ability to maintain religious tolerance in its true form from religious perspectives not political ones, but through religious forbearance within reason!

## CHAPTER 122

By any act of reason even unbalanced, even reflectively from a mirrored image, even if by mixture and poured into a funnel, the consequences of what is produced at the spout will eventually have to be addressed one way or the other? My overall method, unquestioned as is due, is to include the task to be done by conduct and query from our religious leaders, not to be self examined or investigated but for them through us to show full awareness to all future prospects! My first difficulty is to be self addressed, because by my own line of thought even if presented collectively is offered to be compromised by our last monotheistic religion from the Middle East?

Again without comparison, Islam has become my focal point because of recent events generally in the Middle East over these recent sixty years. Without comparison, for me to site any aspect of the religion or any faith comparatively is done from my own recollections, is done from my own reasoning, is done always from my own meagre understanding, and is vitally always done 'Past Gods Words'. With virtually all major and honest religions falling in category to the settlement of there being a supreme reason outside of any form of real understanding or our ability to really understand, sets the picture for further study. Taken not to say in denial that some religions have indeed had Godly contact or have been in direct contact with God, but at one specific time only, perhaps like for our three monotheistic religions originally from the Middle East!

Without close examination and not to compromise the ethic of any religion in concept, for us to take the standard of acceptance to our chosen religion 'Past Gods Words', expresses our own devout reasoning positively. That is why by inference and to compress self doubt our religious leaders in having cleared there own consciences can cover for us all by the means of expression of how we are to continue in doctrine 'Past Gods Words'? That is why I have focused much on our last religion of area, Islam, again not in comparison, but because she inadvertently has too many religious leaders? Expressed almost one per mosque for that is from where the main seat of teaching emerges in the Islamic calendar, but uncounted by the fact that in those mosques our attitudes are sometimes managed by who we are? Having developed over the rigors of time from the influences of geography, climate, diet and other reasons sometimes unseen, because even if we are supposed to operate by one fixed set of standards, our own preset influences take over?

For me to have considered in fixation that one main difference between Islam and earlier religions from area is represented by our ability to justifiably kill of self by self, rooted in concept because of how those earlier religions failed to manage the precept of "thou shall not kill". Has been used only to express the different standing between some of our political and or religious determinations by mixing the spiritual and secular in some forms of decision making? By switching the emphasis to that one activity of religious forbearance and in relation to how it might be operated today is my intention, so that our open effort in the field of the spiritual is not forever compromised because we begin to think and act how we suppose how God might want us to? Instead of preparing the ground from which we can show our own awareness to the situations of contact but in the human form 'Past Gods Words', because any other reading in vanity, compromises Gods own timescale of evolution!

By the full circumstances created by ourselves in this last modern era encompassing these past sixty years expressed by all the means in the field of creation over evolution, we have come not to a turning point if seen so. But to an area of consideration whereby we can question our religious judgements because they have been made under our own power in terms that in most cases although guided, we are responsible for our own conduct?

By that case alone with other consideration already dancing on many lips,

I have focused my concern to the matter of the Islamic faith being somewhat leaderless if so, which from such a well written text, does not need leaders unless now to make the point of being able to make points internationally, politically! Without admission, when that happens and if it has recently been done by any members of the faith of Islam, the actions are from the one point of view only, being no bad thing? What I seek to determine is when we overlay politics and religions with the same standard of the Fifth Dimension prospects the desired outcome will be soon seen of the different outlook between our religions and politics. Which from the latter change will naturally come, because of the better alternative produced by the three taken and one active prospects of the Fifth Dimension. Seen as an aid to enhance such bodies as the United Nations with a clearer definition of how to manage our political dealings by what ever means and from whatever prospective, except religiously!

Without sign, in wanting to overlay our religions in the same way we will upon our politics, differences will follow by attitude more than outcome, which is to be my first intention to show the real importance of our future drive. Religious leaders in this instant play a more vital role than our political aspirations if claimed in falsehood by any individual, because for the first time ever internationally, we will be exploring our relationship with both entities?

By my own examination, when looking to the religion of Islam for the line of leadership, will be seen without comparison to other religions, the face of contradiction? Being a mater to have been spread in very early days of its instigation and to have carried the distance of its total time, judgement is to be made against the outcome so far. From the internal doors of our last religion from the Middle East, looking by sight to the two main factions split to be Shiah or Sunni the complexity is of its own character, which in case is not for dissection but can be used in evidence! No real difference between them has been displayed other than the similarity between the Roman Catholic reformation of some five hundred years ago when Catholicism became Christianity, but much later than Islam's duplicity!

Sunni and Shiah have remained to uphold the written text in general terms as it was delivered those fourteen hundred years ago. There own conflict of idealism on or by religious terms is carried almost in complete replication from the time of there specific differences. By the same token in those general terms the foundation of the split was made by men tied to the same advanced thinking that had superseded previous religions of area. Again by the same dissimilar terms, from the out side, little difference would be seen between those two religious factions. Both operated to the written text almost verbatim; both held in principle to defending, promoting or protecting the religion by the process of killing self by self if necessary, but now what are their respective political aspirations and expressed by whom!

By more recent advances of time owing to all religions and all people always, that I have set a standard of difference between us spends no favour from one to one to one or reversed. My drive in all case has been for us to realise that we are the same people in the full pride of our deemed, seen or claimed differences, where I have put a measure to them is in our realisation that we are different, but think, hope and aspire the same? We are different if only measured by our possible attitudes stirred by the events of climate, geography, diet and other influences, but think, hope and aspire the same? All of which from full historical connections can only be determined by our religions, that we are the same people who left footsteps in volcanic ash that reminded us of their presence when they turned to stone from any time up to three and a half million years ago. Later to remind us again of that singular step some sixty to one hundred thousand years ago when in reality we aspired in full representation for all at Shanidar?

## CHAPTER 123

Religions we have in abundance, religious understanding we lack in the same quantity, but only by reflection in terms of how we view each other, which from when, has always been from the wrong perspective? Although we all profess to heed in the order of command in most cases by the final order delivered to us by all majesty, it has only been by our own interpreting of events that we have clouded all issues. By the most intricate of relationships we in our haste in coming to certain realisations by our own good strength of mind had discovered the means to the future in meaning. Yet under the same collection of concerns had almost brought the whole ideology crumbling to the smoke of dust by our own exuberance in accepting that such contacts that might have stimulated us, were for us!?

By four, two and one thousand four hundred years in the aspect of all religions has not been enough time for us to establish the correct modus operandi, not least because at no time ever during those periods of time had we become aware that we are the same specie. Religions by there own best relations had hinted by effort many times that we were special, but in reality had been unable to allow us make that connection as told by Judaism into Christianity. Then by further exchanges of Judaism into Christianity into Islam of our western monotheistic texture and in expression by several eastern styled religions like Hinduism leading into extended forms of Buddhism of there own multiplexed collective texture. Pleased and delighted that we might have been by our achievements, our progress had been so stable and static, that other events in the name of science began to cause inner doubts amongst many of us, which has been unchallenged until now?

All through the extended time period of four thousand years we had indeed expanded our knowledge but by the several setups of how we maintained our separate religious fortitude. To the good we did keep active with religious momentum, but had not really realised how much the advent of time was eroding concept? If our most serious doubts were to be raised, it was beyond intention by Charles Darwin when he published his celebrated work The Origen of Specie. Without comparison, criticism or objection, holding that his concept of natural selection overlaid to the human condition meant that there was to be differences between us, not least by reference that the strong were best able to survive, at a stroke our maxim of creation was reduced in formation to be held in abeyance until further specie development, perhaps until now, because now we can become more open by choice?

After Darwin, within one hundred years of publication until the start of this last modern era, we toyed with our own concept in translation of what he might have meant by those terms even to be aided by more of our own misunderstanding of the concept of evolution? The worst of what we did has been given in example many times, but at no time was Darwin ever responsible for any of our mistakes or outright blunders in consequence to the human condition. From the strong surviving best, to natural selection through to evolution on each subject we erred by our own interpretation and in cases, half measured reasoning!

By first choice with some of us accepting the time span of probability owing to evolution, we directly challenged creationists, which although the theories were thought of on an intellectual level, unfair doubt was cast on those of us religiously inclined. If fraught with inner anguish that some of us might have felt under attack, then how were those emotions to be measured and by what criteria? Apart to initial outcry, and I take most of the expressed examples to have suited more the developing European mind set. Impetus was turned not to one side or the other over the terms of evolution or creation, but to the matter of the strong being in the best position for further decision making by whatever choice?

From the result of some serious bad choices made on the stage of the European and latter day Europeans by that time to be now American, with their own developed similar attitude? Arguments that were essentially of a religious nature were of less consequence when an arm of our political development in the name of science, unarguably sided with the terms of evolution! From our defined fourth dimension we took to the wrong settlement, which by some very small instances acted in terms of being self promotional for the avid in study sometimes wrapt in technology? Not least from its use by deduction, it alone was capable to lead to the discovery of our first forms of bacterial life on this planet, almost as a final push against our own made theorised concept of creation!

By the same measurement on my own account, I can accept evolution to be a factor of Gods timescale in relation to all life, even if from as far back as eight hundred million years ago? Our well defined concept of creation entered the arena some three and a half million years ago, was strengthened some sixty to one hundred thousand years ago at Shanidar. Then tidied up in concept about four thousand years ago, accepting that all things had a relationship with God the creator! But through that understanding we were prepared to ignore many aspects of God's intention to all life. When it became safe to assume that by our originality we had found Gods ear then ago and in good faith maintained that ignorance until perhaps now or soon. When to realise in reality it is still our aim to excite Gods interest in us?!

Without matching comparisons between what we think evolution or creation is, one of our first writings in the Bible given holy accent by us, created its own aura from which we moved to endorse what was relayed to us by it, but tied the events to the matter of God and creation above the unknown. From which virtually all later writings took the standard of creation for its total value to the aspect of humanity in our various forms, which again in terms of the length of our own lifecycles suited us all in concept, being no bad thing?

Placing Darwin, and how ever we guessed at his meaning to one side, for our own review of the modern situation as it might still be undergoing examination. For all the uproar that his publication at the time produced, by the same reasoning that we were able to attribute the concept of creation only to the bible and following religious writings. Evolution has since been able to forward its own case at first supported by science as an arm of politics, for it to have accounted for our human delivery in status and in argument to our own first and better reasoning to the matter of creation! Not least formed from some genuine interpreting that as we had progressed over these four thousand years if by the terms of creation, we proved to be lacking in our own show of what Gods full intentions were for us, save by personal acceptances, hence our multitude of religions, hence our main belief for creation!

Evolution fitted better to account for the longevity of all life forms in cases like the crocodile of some two hundred and fifty million years of existence by form. Even if to have propagated with slight difference as Darwin's finches had. Crocodiles, since and through their total time span had shown no signs to want to change, nor the ability to look for change. They were living by evolution without question and had done, long enough to supersede many interesting specie like the full range of dinosaurs, which only represents our interpretation, because the dinosaurs superseded many insect kinds, all without any display to aspire?

By mention that we came to certain conclusions if four thousand years ago and then drew the picture of creation, was done by a unique drive exceeding anything else done throughout the universe, which in any case above science holds its own mark. We alone put a sign, even if the first of many at Shanidar in present day northern Iraq, giving indication past the ability of rational or evolutionary thought as it never was? Giving indication of our acceptance that in terms, evolution had at last ridden all the waves and was coming to settle on the sands of content, even that we did not know why we acted as we did, except that it was outside of any inborn drive to only propagate? Evolution was Gods own time scale and still is counted in the un-measurable until perhaps the crocodiles in turn show the will to aspire beyond survival, having had a head start upon us of two hundred and fifty million years?

For us to have looked into the future with parallel concerns of living and dying along with what ever original query was raised by whom; when our concerns included what the future might be in acceptance by us even when dead? We shifted the mantle of time into the realm of measurement by seasonal persuasion different in kind to other creatures. There conduct was determined by such factors in the normal manner of supply and demand submitted by the seasons in some parts of the world. Then, even from other cases where the seasons had no real difference, like in the tropics, all groups did enough to maintain there own particular lifestyle but to the one end of organisation only for propagation?

Naturally that was our primary objective, but in our drive under the same terms as all other creatures, while expanding in territory from an area in Africa that had enabled us to survive. We inadvertently changed our status not to be recognised in the blink of several thousand extra generations of the crocodile. Where they had developed to there own particular situation by changing certain physical features, we had developed to challenge them? From the small time slot of only three million years, our travels took us to be able and accommodate all seasons and all weathers in all places, along with the ability of tool use and using the comfort of fire?

There is little hard evidence to my particular theme of reasoning, save by the fact that we are here now, save by the fact that without compromise we all have a defined ancestry. Save that when all of now in human terms is put into the top of a funnel, the result pouring from the spout will be the same for each entrant? For me to have connected to Shanidar as an equal point of reference for us all has been done to focus us into this very modern era. By turning the switch from evolution to creation we can now account for the need to get everything right instead of being half measured and conducting affairs to the preconceived steps taken syndrome outcome?

By taking the step of recognising the fixture of creation on my own account, I can see the case better for us all to be able and accept our specie singularity in the form of us all being equal, even if selected or not but to be inclined by those of us in representation for us all religiously! Because on those terms we counter any fused concept to the meaning of evolution by representation of being selective one above the other than by developed circumstances. On those terms the natural link is formed how at first we were of evolution like all of life and with all of life were connected in fact to Gods own time span of evolution! We were of the same status to the crocodiles, to the insects, even to Darwin's finches? But in one instant without duration we made a connection, if nothing more than by thinking of life after death, which in all confidence has been the result of all experiments in proving our singularity? In proving that from a small beginning if like at Shanidar, human unity was born from an example of death making us unique. But also by connection to the fact that if we were not all there at the one event of some sixty to one hundred thousand years ago it has affected us all as though we all had been there!

By inventing purpose, if that was what was done then ago, by further extending the process into hope, we had an immediate break from other specie. By connecting the matter of our own short lifespan to the prospect of continuance we irreversibly set the clock to the real matter of creation in any case, only to err by the making of our own human frailty. From which we assumed the matter of creation was directed at us because of who we were and not as an extension to Gods own time span of evolution for all life, until at least one life form was near ready to make a better connection?

By the modern idiom of gambling or betting, in setting odds to or against the prospect of who was to discover the route to the future in reality, even to include the excellence of character displayed by those in group who excelled to be the strongest and fittest encoded by natural selection? We as that single group to have created the principle of creation, did so against each and every other life form that had ever lived and still does, from over a time span of some eight hundred million years more or less. Because in evolutionary terms we vied with millions upon millions of other creatures who by Gods own terms of evolution had the exact same chance that we had, under the exact same conditions, proving the odds of my theory in relation to evolution? But in acceptance that our sanctity was not to be endorsed merely by us, even if individually centred upon any relationship we found with God or Gods to lay homage upon?

Our vital connection was not through the habit of procreation on beastly terms or from swarm behaviour like bees, whereby some were only to be part of the whole, without any connection for the individual being able to conclude on there own account? When we came to our first Gods from any time between Shanidar and four thousand years ago, creation was already in theme in all cases and by all means. Four thousand years ago by our own excellence of mind but error in thought we settled the matter askew, not least when by distraction we compressed evolution and creation as the same?

## CHAPTER 124

Although some things are broken and need fixing and some things are not but get fixed, in general terms that is how we might have managed much of our actions to each other by our religious determinations. Not least to have arrived at this very time in this modern era conducting religious matters under or by or through the concept, of religious tolerance. Failing us all by essentialities, because it is a two sided concept whereby we all are supposed to have regard for each other religiously backed up by the political determinations of human rights, of behaving politically correct? Shored up by a raft of other essentials by law, determining that instead of us being the same people first collectively 'Out of Africa', our very differences are guarded by the imposition, half measured, of innumerable race laws? Even if designed to ally us, they are ever divisive in character, because they fail to have noticed our equality in any case through our different, different religions.

Without cause, but through our steps taken syndrome, when acting within range of what a law or edict laid down half measured is supposed to mean, any actions that we take can be led from the perspective of the initial law makers in supposition, not least because they are supposed to be the most civilised politically? Not least because in judgement, if through the United Nations, when proffered by suggestion that we are to be religiously tolerant of each other, we bandy to the support of some religions and not others. Sometimes with political support supposedly only because of religious tolerance, but again by our steps taken considerations we can make political judgements over religious matters outside of religious tolerance, which if weighed, tip the scale at a dangerous angle?

What I can only suggest from the above stands on grounds that because of half measured actions we can induce the steps taken syndrome for more than one quarter. Meaning that although if a political decision was made half measured, it in case would always find support because of the steps taken syndrome? Then again because of the nature of some political religious decisions there is uncounted scope for reflective action or reaction if from one to one to one religiously or even now politically. What we have not yet been able to see from all the good that we try, is that we are indeed the same people politically and religiously. Therefore there can be no query to why any of us might suggest in the good name of our religions, that we ask of our religious leaders to represent us in the first place politically. Meaning no more than to react to the simple realisation that some/most political sentiments have now to have a core fixation by the same manner we have to out religions!

From where I had set a minor problem between some of our instinctive religious attitudes by stating that it is mainly from the religion of Islam that we do not have a defined leadership code. Has been done in relation to us and the religion itself, because even when stating that we are responsible for our own conduct within the confines of the religion. At times of personal uncertainty, if brought about by the newness of the situation we might find ourselves in, who is to decide action or reaction even to the extent of killing of self by self. When it is only by these very modern events, circumstances have created the need to react, respond or counteract if thought so?

Islam when first set amongst our plans had the vision to make a better start than other religions in close proximity of area, although connected by the avenue of some earlier prophet thinking and delivery? In making new vows, as well as considering some aspects of the past, secular and religiously, by also being aware of the new differences of now from then. Any visionary concept to have been laid before the flock could only have been done by the projections of what the future might bring given the supposed circumstances of what the future was to become?

By the reasonable assumption that the world was already round, with the better connection that at last we had met with the one true religion able in all cases to defend, protect or promote its own image under likely circumstances. Our case for future recollections could only be proved with or from what we already knew geographically about our planet. By the same matter in terms of science and the future, plans could have be made in the respect of how things might improve or develop in relationship of one to the other from the example of Archimedes. Who when expressing his personal delight of open invention by discovery, cried "eureka", "I've got it, I understand", leading from that time of new discovery to following progressive ideas continuing to open the realms of formula.

What is to be noted by my example is not in cost, because all I have done is try to rationalise how some messages of the future were to be translated only in expectation to some physical matters. Science apart, for we were beginning to know even at the time of Mohammed that we were on the path of progression, after all what step was it to introduce Islam if not to be definitive and expected and progressive? Even if from within the structure of the religion, included in the text were specific references to some matters of future learning not yet fully understood and in rare cases not even in the human perspective as of yet! Ideas were recorded by the listening to Mohammed's words for reference, not least to show by indication, that when the understanding of such matters is of common knowledge then any references to now would gain better ground by reaffirming the standard of the faith in all cases. Mohammed in bound duty had recorded what was heard but in due credit to his own thinking and that of his scribes, a fixation was put to the written word that has lasted in earnest these twelve to fourteen hundred years under joint circumstances!

No uncertain legacy was ever intended to waken at any time in the future with regard to how Islam was to react religiously or politically to unseen events or the possibility of unseen events. Which by every dawn from now or over these past forty years of this modern era has given us all the matter of concern, because instead of adopting an approach of unity religiously or politically, we have chosen the wrong middle ground in both cases? Which with the aid of chipping and scratching at the structure of our religious and or political standard we now have the ability to meet all expectations if given our religious head? If listening to the guidance we are to receive from our religious leaders delivered from us through our one active prospect in 'One for the Whole', our paths are not set to cross, because we can only move in the one direction?

For this exercise to begin to work we need to look in necessity to find the voice of Islam with regard to all that was left to our own judgement in any case? For by the ability owing to some of us, unfortunately my self not included thus far, no matter what background or interests in open display to be shown? There is a huge silent majority who forever have conducted there own habits in life far and above many of us who subscribe to our different religious or political entities. There is a huge silent majority who for generations and generations have aimlessly conducted there own habits above all religious standards and political drives, who have lived and died without once raising there voice in any form of discontent but enabled the less talented amongst us make the loudest noise on all subjects.

Although religions effectively have been the source of our continual drive throughout our histories it is from that picture, even to include the silent majority. Our religious leaders are still needed to guide our own suggestions in the best cause for the result of change that will be needed for us to rekindle our first recognised endeavour outside the picture of evolution. Their lead which shown, is from our thoughts, is to insist on our behalf to tackle the first main cause of our differences explained politically and to show in expression the means of our religious unity above politics, and by intention in all cases also to represent all silent majorities.

Religious leaders are needed in full bloom as our representatives because in reality we all practice to the same standards governed along with our own free will. Islam is not complicated by this view, not because I have been looking for a leadership style similar to the priesthoods of other religions, but to draw her expansively forward and in doing so give all the best leads. I make no comparisons but in order for Islam to meet our goals we have to play the game using the same rules, those shown, even if by political determinations these recent sixty plus years.

In calling upon Islam, our words are only of encouragement so that when we speak our tone is to already have altered from the comparative references I had previously used in relation to the imams and their duties by connection with mosques. By leading the faithful who are proposed self led by the tone of their own learning from self conduct, imams seem to make up the shortfall by definition when outside influences came upon the religion? From past issues this style of conduct seemed to have held its own ground owing to whatever other influences came upon it until perhaps these recent years.

By not using the religious example of the Iraq/Iran war barley a generation ago I have not compromised the standing of any religious leader if so called or claimed, because in owing to us all, Islam on both sides of the conflict worked the same rules religiously. Not unlike the massive Christian conflicts earlier in the same century in which who's case, both conflicting armies were blessed by the same religious priesthood and leadership!

My indication of the recent wholly Islamic confrontation centred on how some Iranian soldiers soon to be matched by equally committed Iraqi soldiers, turned the matter of self sacrifice into the means of justifiable martyrdom? Later having the events of that self sacrifice directly counted to be acts of martyrdom and from all standard reasoning justifiably so seen in the eyes of the beholder and or endorsed by a third party if or from whom? Even if I classify that third party to be associated as personal religious leaders, generally known as imams by the same code of difference between mosque and mosque separated by land borders politically defined. Further comparisons with how earlier Christian European conflicts were to have worked out, cannot be given because as the Christian image was becoming deeply politicised, the Islamic picture was becoming religiously entrenched! Not so for reasons to be different, but in line to show that the new ideals professed by the new world order since nineteen forty five and now, was more than frayed at the edges because in enough cases we had chosen different options right or wrong?

## CHAPTER 125

Swung from that measure of our separate views with one side religiously entrenched and the other more politicised in arriving at the junction of now, we have to pick the bones of argument to our acclaimed differences as I personally put them. From using my own grace of connection with the abstract, my views are to the mixture of being my own and ours covering the same topics from the same angles, but with different tools? From where we now can and use the three taken and one active prospect of the Fifth Dimension and little else, we will better see our likeness, but by continual mention of our differences?

However abstract considerations are brought to bear they will always be from abstract abstraction? That we can look out from the reflection cast upon a mirror and when the right hand is raised against us it is always our left we raise in answer, which in part is how we have operated some of our religions differently at different times throughout the ages. In all times up until now we have always taken the reflective image of what we have seen to be the answer to the questions we have asked, but made the same mistakes by our own misunderstanding in the last place?

Not unlike from the three monotheistic religions of the Middle East who, all profess to be in Gods own conduct, yet have managed to allow the one unifying feature we have through them, lead us to misread some reflections from conception. For many of my own examples, I have directed later consideration to our religion of Islam, not least because it is the last of three from area, but with many of the same habits intertwined with the two earlier versions of the same faith? But I still allow for their particular styles to be in command of personal situations to the adherents of any of the three religions apart.

One of the main pointers I have used, but I remind us all, it is still derived from my own inclination to review it from my own perspective to how we might consider the situation of us killing of self by self? On one side I had laid the matter in council to the Jews and Christians by their own accepted code not to kill self by self at least religiously. The case is not rested there, but raised by a different tune when through the same mirror I had held it to Islam, that at times it was profitable to kill of self by self reflectively, which is to remain un-compared to either the Christian or Jewish ethic? We are not in a competition, but all that we gain from the study of our own conduct is to a winning formula, not least because we have never given up on ourselves shown by personal endeavours marked by four, two and one thousand four hundred years of religious histories. All of which has brought us to these very recent times by countless generations of the same different religious thoughts, but to be effective by the last generation in all cases as us, being you and I!

How ever we might like to address the history of we killing of self by self, has been attempted time and time again even if meaning that in order to stop the bedlam some would seek to destroy nations at a glance for future consideration. Some of us had openly sought to destroy those of a different religious setting, those of a different political, economic or social setting. Some have sought to destroy those of a different tribal, racial or colour setting and in too many cases attributed the actions or idea of, to Gods own desire? From all of our cures to the case, none have been forthcoming, because we had not yet arrived at the situation we find ourselves in now, being the age of our first attempt at real reconciliation. Alas flawed in concept by attempting to rectify all matters from our political perspectives first, if partly under the assumption that religious matters were self catering through the admission of religious tolerance! Bringing us in full circle again, but now being able and use the three taken and one active prospects of the Fifth Dimension with rational or from abstract reasoning to make the same points about the same reflections but now from different angles?

Although we profess to be civilised, the measure of the standard cannot be used in our present situation, which is to be seen in change from the use of our delivered prospects. Politics apart, the matter might best be decided from how we are prepared to acknowledge the matter of our civilised differences through religious determinations and delivered by our religious leaders! In the example from the Jewish or Christian religions the matter is of simple extraction for our leaders, being our Pope or cardinals or bishops or the reverend or the Chief Rabbi or Rabbi's in general. By representation of both religions they have already carried the letter of the law "Thou shall not kill" and have done so for four and two thousand years continually!

By my own observations of the Islamic religion encased in these very recent times, what I have seen or interpreted to have been near fact, is to the matter of self preservation on religious terms of how we can and do kill of self by self in justifiable homage if so addressed. But wildly misaligned by either those that would be martyrs or those that sanction the act in relation to the code of conduct we have professed to obtain by being civilised, in maintaining human rights in part, or in being religiously tolerant in full? Allowing by those very features that in being over civilised, licence has been given in all terms to the individual collectively but not sustained individually!? Creating the gulf between the easiest of our recollections in our religions, but as of yet unseen until we fully apply the concept of the Fifth Dimension morally! Which simply means we regress in attitude whereby we maintain fewer personal controls by lessening the amount of political laws or edicts to prove our similarity, but have only gone to maintain our racial or social differences by at the same time being cast as first, second or third world people commercially?

By the cruellest of irony, the picture of martyrdom has had the loudest voice over these recent years, but in dissection and I add not by the surgeon's knife, the remains have caused more to show our difference of attitude rather than enhance our unity. Because again for the first time, religion is being used as the form of our conscience from a closed perspective and not in open exaltation as it should. From this last modern era in or from my open observation of some acts of killing of self by self, the split between murder and martyrdom is of such a fine divide that it is impossible to sanction the act religiously!

Looking for my own answers, which are ours by consent, having given an example of how modern martyrdom had its roots through either the Arab/Israeli/Palestinian conflicts or the Iran/Iraq war, is not of any issue what so ever! Israel has to be dropped from the equation in terms of martyrdom and take judgement in the field of killing of self by self by her own balance to be religious acts of murder not unlike most Christian behaviour through the same connections. From the ethos of the westernised political philosophy in normal presentation of the maxim of "In God We Trust" that feature in supposition has to carry the words of God as delivered by Moses to all of mankind, but the Jews and Christians first, through "Thou shall not kill"?

Islam by her code of killing of self by self has much the same structure of the former religions of area, but had carried the huge advantage of being able and set individual moral standards from the additional use of our progression over the last two thousand six hundred years at least in the same area. In putting a modern fixation because of prevailing circumstances to the first consideration that each member of this new religion was responsible to self by self conduct. Was heavily endorsed by the factor that under certain circumstances, even if uncontrolled, the individual was obliged to maintain a core tenet of faith to protect, defend or promote the same religion, even if at unseen times it was necessary to kill of self by self for those ends?

One of the key factors allowing for compliance of such deeds were sited to the core of the religion itself expressed by conduct of the individual taken in the main from the holy Quran. Being the mainstay of the religion these fourteen hundred years and holding all to the structure of its verse, which has not changed at any time since, nor will? Although carried in the best of good faith, the whole time the questions that are open to be raised about its intensity, can only be directed at some members working through the core belief to the faith. Perhaps not being able and see how the holy work has not changed but the reading of it has by the addition of fourteen hundred years of living and generational attitudes changing just as peoples attitudes had changed in the preceding two thousand six hundred years?

No mark can be made against Islam in this instant save to the keepers of text if through our imams, because as the religion was structured to the conduct of the individual in any case, it was generally to the keepers of conduct in the mosques to determine some of the finer points of doctrine if that is the case? By my own determinations only, I have cited imams to be a class of religious leader in Islam which is not a true picture, for in place there is a structured set up for the upkeep of determining doctrine if needed. There use is separated from other priestly setups in respect that for various reasons their intervention is not to promote a particular change in attitude brought about by whatever circumstances. All messages are moved through imams because they are closest to the point of community worship which in cases is not to alter conduct, only guide us in line of future personal behaviour if by the theory of maintaining the set standard of the religion in all cases, save by some misinterpretations?

For Islam to have held its base of religious doctrine virtually unaltered over such a long period of time is commendable not least measured by the Quran and other writings of similar intensity. For me to have suggested that nothing has changed except by interpretation over at least twelve hundred years lends to the opinion of how and why, and can lead to the less generous of some of us to criticize the religion unfairly. Because we might suggest that although committed to the acknowledgement of some future events in discovery or the prophecy of some future happenings if expected, we had failed in mission to be able and declare what might be the biggest threat to Islam by unforeseen forthcoming events. Not through the act of failure but, in acceptance of other truths which were more understandable even that we did not quite comprehend intention fully. So how might we have reacted if told in concept that perhaps the biggest threat to come upon Islam ever if, was to be from peoples at the other side of the world who were not to be European or Indian or Chinese or African from the continents that we knew of, but from peoples as our selves from lands yet unknown? Without surprise, with the aid of hindsight from all that we were told by indication to some of the wonders we might reveal in due course. When in the matter of recent times we were to come into full internationalism by the haste of events since nineteen forty five. To come into the realisation that the new rulers of the world were matters of commerce and politics in the first order of delivery from continents not to have been revealed to Mohammed, how might some of us react to the story even when rebuilding after fifty five million murders!?

## CHAPTER 126

Without connection or comparison, if and when events were to have occurred to be a threat to our last introduced religion, then the process of defence, promotion or protection was set to automatically prevail, or were they? Nothing by reaction was particularly set down in the order of conduct that is why we need a more defined leadership core from the religion of Islam, in order that we can all refocus to the line of events to have occurred inducing reaction of the above field?

From my own studies of the changing picture I attribute to the religion of Islam, is born not from text or intention, but the frailties of man, meaning from our own failings some of us have not reacted to the benefit of all, only to the destruction of some! In its worst form, what I have seen has not been acts to purify the religion of Islam under any heading. In its real worst form what I have seen is an open effrontery to the religion itself from within, and by the nature of some acts headed to be in the line of martyrdom, I can only see uncertainty felt by some of us about the integrity of the faith in all cases? In its worst form, what I have seen also from such acts of martyrdom, separate and un-associated reaction has been introduced by others even if outside of the religion, to unintentionally drive a wedge towards the very heart of the religion? Even if the form of that wedge has been the standing imposition of religious tolerance accepting the honesty in core of all religions, but ignoring the essence of some! Which is why I have found the need to personally challenge in honesty some of the people in the form of our imams who might or would or could or have directed others if, to be enlightened and follow acts of martyrdom in glorification!

That I aim my attention to imams being inclined religious leaders, has been set by two small factors, one concerns their own rich study of all text in its pious form, and the other of how they by personality can have a direct influence on some individuals of the faith? If we are all obliged to follow our own conduct by connection with the level of our own studies in the matter of our faith, no matter from which age it had been delivered to us, which applies to the three religions from the Middle East. From the first two in particular, reliance of control from established priesthoods has always been how the line of command worked good or bad, although some small measure of our personality has been taken under control having at first been applied by that established structure which is no bad thing.

Unless we had already given our own judgement to others by acceptance for other matters, whereby the means of our religious control have been entwined to other aspects, like for us to follow political or autocratic leads of civil conduct? Which eclipsed our religious fortitude, unless badly led!

Working to my own format in general belief that Islam works as it does, then from the core of modernity there is no direct way that we can enter into the faith structure of any religion, save by effrontery? Although all religious aims are to set the future in finality by the exposition of all that we have believed in and practiced for, we still have to contend with these very recent times or at least be able and recognise our own intentions if only politically? By closing on this last modern era my outlook has been to focus on how things actually are rather than they are supposed to be. For that exercise, it is not to be forgotten how or why or from whom any of our paths had been set in the past? Some of my own caution has been aimed to neutralise our ability to react by best intentions from the smallest of sparks without knowing why the tinder box had been used?

That is why I have set the task for all religions to be representative to our needs enabling us carry out normal daily activities and to better show what those daily chores are to be. Without complication Islam along with some of our Eastern styled religions like Hinduism and Buddhism set for us the widest gap for general community, mostly because they are almost complete by chosen concept! Our eastern styled religions make the aspect of realisation to what all consider fulfilment more difficult to judge by comparison, but never of less value. Those fine religions as with all in belief, falter to a range of thinking from our monotheistic ones in current study, because it is impossible to rationalise one to the other type, except in our final expectation. Again without comparison, but if one was to enquire to a Buddhist or Hindu for settlement, although led in concept, the answers forth coming have to be unconnected, because of the nature of the regenerative concept owing to us all?

In many instances by general terms, both of the main forms of our Eastern type religions starting in line at about the same time as the Jewish religion some four thousand years ago. Automatically married in concept the processes of evolution and creation, also unknowingly, but better organised than from the Western styled religions. By taking to the perpetuity of time and the scope of all life around us linked by natural interactive processes, our Eastern stand by including all life forms together had no need to accommodate the progress of man alone with any significance. By the same token to include the interactive connection of mutual dependency between many other specie and we humans, some sects were to determine that all life is sacred and therefore we do not kill of others or of self ever! Making the challenge, that but for the advent of time, no answers will be forthcoming unless awarded in using the full concept of time by God or collective Gods grace, evolutionary or creatively?

Although I do not feel any range of conflict in concept between the two religious ideals, my standing still remains that our Eastern branches of faith in hope and purpose took the same spark that stirred a mind to react as we all did at

Shanidar? Even in taking sixty thousand plus years before both religious concepts were not yet finalised four thousand years ago and still? Leading to now where there has not been such reaction to political laws or edicts from the nature of the Eastern religious concepts, except to claim religious tolerance because in some eastern cases that feature is an open necessity from most angles!? Eastern religions in a sense do not have defined leaders as so, because by appointment or not the task for a leader would also have to include the understanding and nature in cause for all creatures by their own specie codes and of its natural relationship with mankind?

Imams and Islam from a different jigsaw of faith, have a need to be self defined by my inclusion of imams only for closer examination of their relationship with the latest of our Western styled monotheistic religions? From no act of disruption or from any intention to discredit any individual my direction in terms of imams is representative to us all, not least because some of them are inclined not to consider the all, but nobody instead? Also by connected influences, Islam if of its own council foretold by its holy writings, but already of an age to have advanced past its own expectations by various self committed degrees! And no longer needs to be protective or defensive to the faith, because the matter of self promotion at a time recently was almost in full drive? Any back peddling done has not been so, because of the religions growth and success, even if some from within, have not been able to keep pace with the natural, if not always agreeable processes of change, even politically?

All notice if driven by uneasy methods of implementation have been reacted to by standardised posturing, if not being met by some methods of political reflective posturing from the defunct bloc culture standoff situation, has from now in this last modern era been taken on through religious determinations sided? But not necessarily facing in the right direction, because even if some of the root cause of our religious differences, have now taken on political influences, there is no case for religious reaction only when the field is mostly political? Not that we are to justify or protect our obtained political standard or necessarily thwart them, but from now we have the means to change the relationship between our religions and politics using the same element of the Fifth Dimension?

By many examples politics on level ground is self policing, because even if over the spread of many of our lifecycles through two or three concurrent generations it can and has changed its spots, religions do not. Religions either have to use the same text or rules or guides by different interpretations displayed by us not in the means of change or alteration, but from how some of us see things differently because of natural advancement over time! By some cases more so through the catholic religion as an example, when change was seen to be valued from the original if now distorted, change took place in the form of the reformation still active over these past five hundred years. Although in place on that scale, we now have a broader based collection of Christian religions, who in general terms have maintained the priestly structure owing to the even earlier form of Christianity in Judaism?

Islam on several other hands by not going through a similar reformation simply because the written text was binding enough to withhold whimsical changes that might have been done in some cases and else where. Had what I have been able and feel with the emotion of great concern for us all, cast her seed to the wind from a very real and sound base in comfort, if by no more than listening to the acclimations of some holy but errant imams!? In other words Islam has sought to reverse any method of natural advancing we might have collectively done politically by determinations to stand in the name of fundamentalism religiously without fair consideration of the emotions of the faithful? But never politically only by the religious flag, which through this very modern era has been done without a defined leadership and at worst without direction and at worst to soil the very religion in concept because the first principles of the faith have been seriously challenged?

Hopefully not on a broad base but only by the disenchanted, who have not been able to gain the value of the religion from its untold and complete history, who have not been able and further support the essence of faith in context to the individual. Who have not become aware that the religions foundations are so beset with history and purpose and intent to have remained unchanged these fourteen hundred years endorsing its scope, because it was designed to be futuristic? Apart to the first religions of choice in area who's general expectation was of imminent discovery of God's intention by now? Islam's own consent was to bring the whole world to the one true word of God already knowing this deed would take many generations and generations of specie/tribal living. Also being aware that we live in a changing world in any case, with new discoveries bound to have influences upon us, some ideas at least having been tantalisingly revealed to us in normal pursuit just showing how we are still to advance though act and deed in any case! Covering the view that indeed this religion was for the long haul, leastways until Gods own determination whenever, but still holding the religion in fixed concept even through expected changes? From which the picture is yet to fully emerge!

Whatever the changing world was to bring, most changes were brought in the abstract, because some reactive responses to what might have occurred were met with the wrong reply even if carrying the right emotions. All of which can almost be described in relation to events of now induced by the example of forced change, but under the best of intentions to have occurred throughout the world these past sixty plus years. It is only when we consider how individuals might or have reacted to some events that the scene is set for further reaction in the same wrong course. Ably presented by my own example in using what might have been expressions of attitude displayed by some imams undirected to deal with matters they personally did not understand or over understood!?

On re-entering the story of passion induced by religious attitudes in the Middle East, our aim is not of condemnation or support or understanding. By looking into the mirror of history what is seen is not how things are, but how they were intended to be, which determined by area is of crucial importance to us all religiously. In looking into the mirror of recent history will be seen how the guise of politics has taken over some religious intentions and left others in abeyance, or worse, ignored some through religious tolerance! In response to either and or the other feature by political or religious reactions, cannot be done in fair and open judgement. Because in all cases when looking into any mirror, when the right hand is raised the left hand responds, even if a friendly wave is transmitted, what comes back is not of the same substance. All of which means that even if some good intentions were used to drive change, the outcome was not what was expected, and brought excess in response?

Taking us in full loop of the situation to have emanated in recent times from at first the Middle East by religious or political tones, has compounded some of our thinking to be channelled into the wrong mould? Not unlike general situations when we add the ingredients of thought at the top of the funnel half measured then simply take the result pouring from the spout by our steps taken syndrome? In the case of our religions which can be charged with super emotion, because of there connection, we have to honour their principle and include them all for all and not by reflection!

## CHAPTER 127

In compliance to the bland expression that we are here now, I have fixed motivation to the will of some individual imams who by small similarities are of the priesthood to the religion of Islam. Without historical connection I have also focused my own observations to determine how some have failed in mission without any inclination of comparison with any of the two former religions and of their priesthoods, who have the same difference to account for, but not of this particular time!

By fair judgement I had indicated my own take of how I had attributed the sole connection to the religion of Islam whereby we can kill of self by self separate to the method of style operated by the former and ongoing religions of area. In all cases without objection this particular trait is only of a modern reflection best illustrated not by the Israeli/Arab/Palestinian sad and sorrowful disturbances of recent times. But in double focus by and from the Iran/Iraq conflict in the midst of other local ongoing political dissatisfaction!

I can make that judgement on my own account, because of the effect of such masterful tragedies to have occurred at the time of that specific conflict and because of the reflective mismanagement of the same events by political determining, even if wrongly sighted. Already stated has been my own comparisons to the Catholic Christian reformation, where change was seriously instigated by determined people even under the threat of death for the breakaway of protestation, but in keeping to the core of Jesus Christ's assumed teachings!?

Here in ultra modern times by any standards and in connection with the new world order committed to the processes of change by assumed realignment, two separate nation countries aligned to the same religion of Islam, conducted either religiously inclined or politically inclined war? Alas in conjunction on a considerably small basis, both sides took to and acclaimed the process of killing of self by self to hold renewed understanding if only connected at the time to or through the self same serving religion to both camps. In telling of the heroics owed to both sides whereby individual soldiers deliberately gave their own lives to the good of cause, what balance could have been used to accommodate either personal or directed thinking if that was the case? For what we must now do as a first, is not look at what the conflict was about but how and why or for what reasons would a fighting man give up his own life not necessarily to save his comrades but simply to destroy self in the face of the enemy and why again?

Politically, all soldiers are prepared to die, but hope to live even if not to achieve their personal religious goals or other distractions for the welfare of family or mother country or fatherland! Then to be part of the flotsam or jetsam of eternity undefined, but still carrying our native inborn ideal to the future and settlement! My association with the Iraq/Iran war of discontent has to include the aspect of imams in conjunction to personal judgements made by individuals. Although at the time the war was publically noted by all the measures of modern reporting through our acclaimed technological ability to maintain up to the minute news. If a particular heroic act was recorded in detail the word suicide was sometimes implemented in use rather than other definitions.

Suicide in the modern idiom then, became the forerunner for all expressions in relation to us in pursuit of other objectives untold. Suicide as a by word from some quarters came to mean an act of deliberate violence without due reverence to the open fact that in most of all cases in the Iran/Iraq war. Death of self was offered under the terms to defend or protect or promote the same religion maintained by fixed text, but at the time unrelated with other events! Suicide seen in those cases to be acts of commitment in the name of martyrdom was given to the acclaimed ending of life in glory to Gods own name! Where war induced deliberate self conflict to personal standards had been proved throughout history, in this case there was no difference. The same people without our taken prospects of the Fifth Dimension, though then undefined in there new fashion, carried out the means of brutal war, but in similar fashion failed to see our similarities more than they felt our differences!

From and on a personal level, when an individual is committed to an act of any type or description, but especially of a religious texture and when from within that same religion there is open support to what is intended or done. Any actions self warranted have to have approval from within that same religion if the religion itself had been structured so. Islam offered the case whereby we were all responsible for our own actions and conduct. In addition we all had the alert ability to self judge, but in the extreme of circumstances from unused ability in the quandary of doubt we had the option to seek better advice if from our imams? Keepers of unified conduct in our best places of worship in compliance to what we already knew, but might not have needed to use for centuries or at least centuries until other events came into focus over these last two hundred years or these last forty years or so?

When, and it is not for open judgement, but when individual Iraqi or Iranian soldiers by committal sought to die by other means than from the expectation of

'Natural Body Death' in the throws of battle! Using religious codes as a means of expression, the task in overt display being self sacrifice or of killing of self by self, was assumed to carry the end product in the name of the same religion of Islam, which it did! Without pointed assertion for us to look upon any individual who pre advertised the action decided upon, the only question wanting, is to know that if either an Iranian or an Iraqi soldier decided to die by their own hand, even against each other, where had the assurances come from that the act was blessed?

By my own lead I have committed the case to be hailed in the distortion of a political act to endorse the aspect of might is right, in this familiar case of peoples going to war by political leads and using religious tones to back political assumptions no matter from where they were directed, Although used by the broader means of political objectives the standard of self sacrifice gained its own liberty from within the thinking and style of the individual religiously? If their own thought by the means of media interest and financial blessing given to their families in retreat could have been of small consideration, the crux of the decision needed a final over comment by religious terms!

If knowing that the morrow was to be life's last own exercise in preparation for entry into paradise carried by all the majesty of expectation, we as individuals although also knowing what we contemplated was indeed right. By our own human element of query we would still need to be reassured that our actions carried all that we expected from devout study? Apart from the certainty of the war we were waging, if Islam against Islam religiously and politics dressed to be westernised by support and fundamental by belief on either side, what now without that understanding?

Although to the devout, even if decisions were in the making and had become of some political interest of how both political systems or styles had used the feature of martyrdom by connection? Greater reassurance in all cases would have been sought from the closest means of religious countenance by us wanting to include our imams advice! With more than the probability of this having been done on both sides of the conflict by individuals for reassurance from the offices of the same religion the new take on the situation was to set new reasonable standards, but of a different complexion than what had been happening through the Arab/Israeli/Palestinian interaction of the same years and earlier. Almost as if we had attempted to alter the misreading of history tied to other incidences but now gained from different circumstances that have been accepted by some to be real and relevant!

If and how and why some imams were able and reverse the service of history in calling again for reassurance of holy war at a time when the desire was to expand our social and political and religious awareness, if a little misguided? There guidance when given was to turn the matter in reverse again, by alluding to something that was no longer necessary to promote the tide of expectation? Islam by all its own majesty from within had already passed its own need to promote its own standing by eliminating those that might object to her code. Islam is forever here and here to stay forever. By any connection that some of her imams have failed her and decried the text as delivered by Mohammed, can be seen now without judgement through our one active prospect in 'One for the Whole'. So that we all might wash our hands in public and not to have washed our hands of other deeds we might have encouraged others to take by their best judgement and our flawed understanding!?

By my own reflection upon imams I have not sought to apportion blame, in this case I have married them to all other priests by connection that the same are to the same standing in all monotheistic religions generally? The faithful in all religions have the same point of first contact through individual priests/imams. My assertion to their different standing is that to a large extent the imams are far more representative for the pious than are other priests, because of the original structure of the religions concerned!

Where some would seek to confess all to a priest to standards of conduct preordained, the range of consent to such a situation is still in abeyance to that conduct which might or could be changed from higher levels of leadership in some monotheistic religions? That picture is not quite so within Islam, simply because the onus of conduct is to the individual preordained by the matter and delivery of the text carried in its entirety by fourteen hundred years of history unchanged. By the same reason of that conduct and to compliment its standing, Islam has offered to its followers a means to remain unified in practice by the orchestration of behaviour within the mosques. Delivered in council by imams, who in type have remained in full status to be able and fill in the fuller or clearer extracts in understanding for the devout. Like for anyone who might have trouble, not in translation but understanding, not least because Islam is for all people. Therefore in appealing to those of us who have been marked by character traits delivered by the influences of climate, geography, diet or other features giving us a standard to be laid back or volatile. When we approach our imams who we may well regard as confidents, some of us might take to their own influences gained within text but tainted by their own experiences or the reading of their own experiences without higher leadership guidance even upon the aspect of personality!

Without any complication religious conduct is religious conduct, only very few people would not need to seek further guidance in faith throughout our lives, which in general terms would come from higher authority, which in general terms means better informed people! That is why I look to our religious leaders to carry the Fifth Dimension and overlay it upon there respective religions or not, either way the consequences will not go away. What they will effectively be doing is giving fair warning of the shortfall of how we have controlled our political management on a global basis. Further to that, we will be setting the standard of future political settlements to have a religious flavour far and away above the mediocre of being religiously tolerant, once we have examined the damage recent political settlements have caused?

By further reference to the structure of Islam and its own code of conduct held by the holy writings, what has emerged from principal political dealings in the Middle East has been a serious block to the ideals of community, either politically or religiously? Although in terms the Iran/Iraq war was purely of personal concern to the politics of the warring nations. Because of how we had formed to act internationally throughout the world, too much notice was given to the content of behaviour, rather than cause, especially because it was conducted in the Middle East and therefore involved world opinion?

Without a specific mark, with the bloc culture standoff feature now almost curtailed in area, the standard of change was shifting to the shape of a cultural separation or one which was so, but not yet seen until perhaps soon? The shift if it is to be defined, was to the same standoff pattern one against the other, but this time the selection uncontrolled, seemed to be one side of westernised democratic principles if, with the other side determinedly Islamic by nature! Not for the first time mixing religion and politics, but now for some to be carried to the final matter of confrontation by the process of impasse. Which under the abstract flag offers more than a glimmer of hope to all, because it shows how much progress we have actually made over these past sixty plus years since the ending of other national hostilities that had covered the whole world!

Sixty plus years to begin and organise the fete so that it runs smoother than could be known is no mean feat. When in reality little was achieved over the last four thousand years of our best defined methods covered by our own multitude of religions already owing to the strength of our political divisions. Sixty plus years has now given us the grace to act better reflectively, but rearmed with the three taken and one active prospect of the Fifth Dimension. With which to overlay our religions and politics in the right fashion and to gain the right priority from now on but not determined by what is supposed to be, only what might?

**CHAPTER 128**

For any of our future intentions to begin to work, my own serious take has to fall upon how we address consideration to religious matters over political ones. That is why I have laboured our opinion to be more certain about our religious leaders concerns in representation? My view is that they must grasp the nettle in open fashion not for their personal concerns, but ours, which is best illustrated from some failings from how Islam has presented herself to us all, and all? By connection, as the other major religion of Christianity had survived its own split without disintegration the message to Islam by uncompetitive features is for Islam to offer a main channel of communication between us all and politics and not rely on how things have developed by the opinions of some worthy and unworthy imams if that is or was ever the case! But having been said if only directed by my own thoughts, it is from Islam answers are to be given us unbiased, but from whom?

When I use the word Islam, the word just as in Jew or Christian is we! Not for effect but in relationship to how we have been brought up in all cases and under all circumstances, even including the broader scope of all religions with those of us who might transmigrate, the picture is ultimately the same. We are all here encompassing God's time span of evolution and our reflective time span of creation as a means to turn two hundred and fifty million years of existence to the importance of having to survive for fifty plus to seventy plus years in space?

Having done the maths and explored the experiments owing to the fourth dimension by touching science and her tributaries. That it is our religions that have been the means of our solar power, it is from their judgement that we must make better assessment of them than assuming, because they are religion, they are automatically right? Which has brought to us the need of being religiously tolerant politically, automatically devaluing each religion, because we are forced to think of them being the same when we know we are different on all sides? Not to be a contradiction of terms, our religions in bulk are the same, it has been we who make them different because of our own experiences and or the time factor involved between us learning of different religions other than our own in competition when we are generally on the same side connected through our own lifecycles!

Without connection again, but not in the field of brevity, when I had personally offered cause to doubt the integrity of some Islamic imams apart to Christian priests or Jewish Rabbis, was not done for effect but from cause? I have well made the case for how Christianity and Judaism had led us in the wrong direction over the aspect of us to kill of self by self or not! My connection had been made through unquestionable detail by association to Moses and Jesus respectively. What followed without direction, is still tied up in the annuls of our history and might never be released without consent of both separate religions leaders at this present time!

My own reflective case in a form of opposition to some imams stands on the cornerstone of fair and open assessment owing to us all no matter what religion we might support. It is based on the one common denominator to all in religion, that of open faith, open faith, not blind following, but in the openness of query? Because that is the way we have been happy to design our outlook from and since Shanidar come what may. In following the total mystery of what happens after 'Natural Body Death' we had unearthed the only question outside of God's own time set of evolution?

If driven by such examples without concern to how Christian priests or Jewish rabbis might review any similar situation, I have put onus on imams to the religion of Islam separate, but not apart, because our core beliefs are the same. But in this case and by recent association again, some of us who would seek to become martyrs might have become murderers instead by no other case than following our religious teachers, who would aid us by discovery of our own doubt? But only wrapt in their own understanding of what should be, based on other events of recent times because those judgements had been made from the thinking stimulated from now and recently, instead of what had already worked over much longer time spans? In other words, some imams might advocate older teachings of retribution in Gods own name but by promise of recovery by Gods own means against the reality of the situation when far more imams do not offer that course, because they can see honest intention before the grief of self pity?

Islam being one of the most complete monotheistic or western styled religions can lose its own charm by being too competitive if only self led. Not from the written text but from the acclaimed text of others, which might not have the same history, but has the same intention? Although our first, second and third dimensional religions have all emanated from the Middle East unconfined, their tone with exceptions has always been international. Except in full modernity when Islam before others has relented and become fixated to ancient or to the time scale of instigation regressing to age some fourteen hundred years ago attempting to ignore the same time span of progress?

Meaning by intent, but without direction, that in her own guise she was complete as has been said, but in representation she only reacted complete for those that decidedly followed her, just like the chosen? From that perspective by normal understanding we had been led to believe that we had been selected more than others and by inclination were now to follow all signs to the point of confirmation of those facts. Leading to the obvious statement that all designs from now on were of our own significance come what may? Until we came to the facts of nineteen forty five and the following sixty plus years in able pursuit to create the real world from the debris of the one we had assumed was real!

From where I had led in the belief that justifiable claim had been put to the matter of world unity through the lead of America at the ending of world war two, no comparative exercise was used in politics and or from religious perspectives. What I had done and stand by was to try and relate our similarity above our differences without the chance of comparison. Not least because from where I had seen cause, I also recognised where others might not, and further noticed that from the difference of where that notice was taken, different conclusions might be drawn which indeed they were. Without definition the result of standoff now, has only been derived from now, no forward plan could have been worked out, because nobody knew the components that would make up the world situation now, from whatever went on in the past?

Therefore when we are confronted with what might have been our religious intentions against our political ones, no balance can be set, because for however much we like to think we have things under control, the only case for doubt is brought about by total modernity, that which we cannot allow for? Because at this very time we still do not know how to treat politics or religions in the same matter of concern for different futures, which come upon us from different perspectives depending on how our pasts had been cast outside of comprehension or at least our own understanding?

## CHAPTER 129

By talking today and thinking of tomorrow in any sense covers the smoke of our past, which is of great importance, because in virtually all cases of it, ground work had been done to broaden how we should think for the future but in forms of confinement. Much of what needs altering is already known, but we must now intensify our study of how to settle the future better than it is already expected by some of our religious determinations! One of our first steps has already been taken but at no little cost to our own dignity or how some of us began to think in terms of the case for self promotion as a means to show our understanding of all matters?

Science by name is a collective clause for the study of query, leastways those that can be answered through the organisation of experiment and formula. In most of all its forms it also is a beacon to show how we alone are separate in style and inclination than any other specie on this earth and in the universe. By that fixture it is reasonable to classify science and all of its bearings to be the fourth dimension in recognition to the standard created by this work as a living memento to our understanding. Even that I had referred to the fourth dimension in terms of science, was representative to the aspect of time and space in reality, that attachment alludes to the spirit of human advancement unconfined. By notice I had turned the issue to be of only part concern by human terms when time is of the only concern to the individual by our own lifecycle in whatever time span, to be fifty or seventy plus years long?

Bringing us again to this very time of now whenever, but started by new visions created from the ending of World War Two, but in the haste of retribution, more than for the open necessity to really change in attitude and outlook by us all? By statement that the best or most vital means of accomplishment for such deeds is in our own hands by collective use of the Fifth Dimension, is purely reflective, but definite. Because it more than any other offering so far has far reaching influences into the core of our religions, all religion and all of our political systems? Without compromise we can lay the three taken and one active prospect over any and all political systems and curtail there excesses. Parallel in time we can overlay all of our religions with the same diet and revel in the changes they invoke because although attached to different sects, all reasoning is the same. From which we can then compare the implications of meaning without comparison, because no matter what our previous settlement had been by either camp of religion or politics, the self same issue is there for consultation relative to human life and human death!

No matter what religious or political camp we belong too or what out personalities are like being open, closed, forward looking, fundamental, godly or ungodly. We will have the same stepping stones so that we can all move in a similar direction but maintain our different standards now realised by our similarity. Naturally in using stepping stones great care must be taken so that we arrive at journeys end having learnt safer means to get there without affecting from where we just came, but with a different outlook towards what might be? By projection and in preparation, when arrived at, even expecting to have used only three taken prospects and still with the ability to induce change by our one active in 'One for the Whole'. Our first new guide without addition can be for us to behave with common decency C D to all, not by class or favour, but in standing! Even if it means nipping some of our considered old values of behaving politically correct, or complying by religious tolerance, which in cases has more contrived to put stress upon our relationships, than ease them?

When first offered the Fifth Dimension was a means for us to form better understanding of each other, but in a way for us all to remain to our own chosen paths if led or not, if from deep feeling or not, if appointed or not, and not much else! Because of the restraints of time held by us all in the form of a single lifecycle, we have to rationalise all of our histories into these meagre years. Knowing now that although we are generationally different than those who were before us, our and future thinking makes the case that we always will be, because we can add the additions of learning from the future and not the past! That in part is why the Fifth Dimension is to be so important to us being a product or these very times we are now in, but only looking to the future, because of now, and not the past!?

By further additions we are never to forget the past or challenge its total integrity, because everything done without comparison has been done in our name and charged to our account, which exemplifies our relationship with it. Which is not to say we can ignore the value of only some things to have occurred being un-tasteful to some of our opinions especially when in some sincere instances we carried out our duties in the faithful belief of doing the right thing! Which only stands by religious determinations and not political manifestations which are natural and instinctive from evolutionary beginnings, whereas the former had been inclined form the medium of creation?

## CHAPTER 130

Nineteen forty five and 'Out of Africa' are to be forever joined in realisation, because it was after the close of the Second World War this last modern era began, which is to be thought of as our first new beginning by every standard. Unfortunately somewhat damaged by political influences, but fortunately held in place by our religious countenance best used to curtail excesses leading for us to devalue the very core of purpose in the name of science? Which by playing second fiddle as an element of politics, but leading politics to some of her judgements to be made against the essence of purpose. Has been able to mix the wrong solution to our previous ailments and dished them up in the order to be religiously tolerant, taking full control of that situation in the assumption of equality. Yet on the same scale tarnished our understanding of humanity by encasing first, second or third world standards to the same single specie we are. Even by best intentions to have considered that we were to move forward in this last modern era by forcing the issue, we have only smoke screened it. But remained in acceptance that we had made the right decisions not least in having to make separate national and international race laws to endorse our singularity which remains fragmented and misunderstood?

'Out of Africa' in a single sweep takes us to the zenith of modernity by acceptance without qualification as we have now done. It is the sign in root of all of our family trees from where the first signs in time of an erect biped as we, was recorded to be later counted. Even that I know of the tenuous connection it still is our best reason of acceptance to forestall the query of experiment as to the exact moment in time we emerged in our present form and by what connection. What was actually done to leave those celebrated footsteps in volcanic ash, tells a new story even to contradict later concepts of evolution?

Of course progression is involved even natural progression, if we think of the incident as it was and why it might have occurred from the direct evidence shown, which viewed from my perspective allows us to relate the event to our ancestors directly and not the even more tenuous link through the ape line? One small feature of the recording makes the case other than probable, simply because in the record there are two sets of foot steps side by side. Already taken that the probability in connection between them was that of an adult and child walking side by side bears a better relationship to them being left by a separate group than indicated by the ape record of similar times found in parallel? From as much as we know of the status and standing of apes at about the same time brought forward by projection today, whereby in comparisons made now under scrutiny. No connection can be made between adult apes accounting for the difference in size of the footprints left, to be those of a male and female walking side by side?

By the rules of natural progression, discounting the process of natural selection where indication has been taken that it is the strong to be the best survivors. Standing erect taken either way to have been an act in continuance of an ape group or troop, over the actions of a completely different specie, swings in pivot to what the expectation of such actions taken could lead too? First signs taken are of the obvious that we were in transit, in walking over a volcanic ash field shows that the local situation was precarious in terms of gathering food because of the blanket put down by recent volcanic activity?

Of course such a situation is owing to all creatures faced with the necessity to feed by being hunter gatherers in the same situation, when the call was to move or die? Our biped effect can be viewed in terms of being separate to others if considering that walking on two legs allowed us to carry what we had accumulated to our needs while having to move to new areas? Against the signs shown to most of the ape and monkey classes having the innate desire to carry young at times of insecurity in keeping to the feel of instinct, which also cemented the standard of using there knuckles as forward feet while moving distances in a quadruped style of motion?

Without claim, whatever the circumstances of connected likeness between us and any ape group was/is, the style had since been dislodged by the one factor that although we have many and varied groups and troops of apes throughout the world today in comparative unity to Darwin's finches. We as specie are separate, even that our geography, climate, diet and other features had conspired to give us a different appearance by looks and attitude, we are we, because above the ethos of natural selection, we are the only specie to aspire!

'Out of Africa' has not been nailed to the mast as a factual means of appointment to any group, where I have used it in relation to us all is to take the idea one justifiable step further. Even that the accepted standard already maintained by most working in the field of hereditary study is that we did emerged from Africa first, is becoming unquestionable, which is indeed no bad thing? Where I take the issue further is that not only did we first emerge from the Dark Continent, but that our first ancestors were indeed dark of skin, suiting the then situations to be inclined and aid the best specie adaptations by what ever means? Even if meaning that it was necessary to migrate elsewhere and be changed by the new prevailing circumstances into the need of requirement whenever, but in carrying the image of inquiry that allowed us profit by the very changes we ourselves introduced in accordance with our needs, even by religious determinations which were to follow in style?

By keeping faith with that taken prospect of the Fifth Dimension what can be achieved is nothing more that we have been struggling for these past four thousand years of our religious education and expectation by that course. Without coincidence, over vast expanses of time when we had either colonised most of the world and were still in the process of migrating. At the top right corner of Africa by connection to the vast continent of Asia at there junction near the land between two rivers, our similarity was again endorsed by our first committed religious act by the burial at Shanidar! By which through our human chain we then carried hope and purpose throughout the whole world but now only by human terms!

Although some of our religions will or have assumed that in or by the conquest of reasoning and tied in story to the fact of man and or the causes of the obvious, already shown before the bench mark of monotheism also from the Middle East. From the noted area of Egypt came first signs of our God to man through our king philosophy, leading to the creative and creation element of most of our cultures following. So much so that by influence the great pyramids were built as a stamp to the authority of death and more so for what was to follow connecting us with the ideas imagined at Shanidar, but not sufficient to be held up as the core philosophy! In as much as the Egyptian story referred to the ideas or creation and the substance of how all was derived by grace of the king in appliance to the will of the heavens. Our story was to change by using some of the Egyptian plans and overlaying them to other ideas in consideration one way or another, but following in time!

Throughout the world in every corner from every time since any recording of history by any method, we have always been connected to the form of belief in the unknown and or the mysteries by some means? I have put the Middle East to the fore without comparison, but simply because so much in the name of beginnings came from the area. For any religious connections I have made, in cases to my own level of understanding and from the same comprehension. I have used blocks of our united histories as my own stepping stones in the form of understanding of how some things have worked. Although by my own suggestion in many cases, I have been profound enough to recognise our general but most decidedly powerful intellect. Many of my suggestions are ours, but I have allowed myself traverse our memories and have been able to note what has been of importance to us, which in all cases have continually come down to the aspect of our religions first. My signal for all is to also use the abstract as a tool, but not forgetting our own sensitivities nor so forgetting that in all cases we all have opinions!

Although connected by footsteps in volcanic ash some three and a half million years ago, tied by imagination of a single connection to events at Shanidar, endorsed by our cave paintings and rock drawing in the same lead to our unity, honed by events first laid down in writing anywhere in Mesopotamia. The one core through us all unquestioned, is that we emerged to form a drive unknown to other specie from the one collective start 'Out of Africa'.

## CHAPTER 131

'Past Gods Words' is our alarm call, and even better is to be the final open acceptance of our specie singularity. 'Past Gods Words' is the means to tame all of our excesses, but is to be best delivered by our own religious leaders from whatever religion we adhere too? Making further reference to our imams and the role they must now play, although mentioned often, my aim is not to praise or challenge any particular religious leader, even if not religiously defined or if so, not to be used in comparison to others, even that the comparison is there? My separate clause to imams is for all considerations, because their particular status is more based on assumption than with any other religion that operates a leadership structure?

So that we might not pick and mix in to random a fashion of what the phrase actually means, it is to be surrounded in the abstract, which by any delivery allows for us to look at this taken prospect by others understanding of it? Because in our religious fields, Gods words and the sign of acceptance to any religion, have varying degrees of intensity, not least to be of first note of delivery through the religion of Islam! Why and because relates particularly of how after several attempts by preceding religions, this was the first time records were made from contact even if delivered 'Past Gods Words'. But in all acceptances the recording was attributed to what was actually said by terms within the memory of sound, recorded or so it would seem, instantly?

Therefore does the expression as one of our taken prospect mean we are to think and operate outside of the limitations of what we know and accept to have been God's words when delivered? Does that mean in the case of the Quran which is probably the best kept record in time of holy revelations, are we now to consider not to take what was written or alluded too, because we have to think in terms 'Past Gods Words' to comply with a dubious case of establishing our new unity? If not delivered at any time until this first real global attempt since nineteen forty five and endorsed along with other prospects delivered in these recent five to six years through the Fifth Dimension what meaning is otherwise held!?

What is meant by the prospect is not to what has already been done, because we all already operate 'Past Gods Words' in any case. None of us living today have experienced the same contact that was done and understood to have been done by those we accept as representing our own thoughts in the name of Moses or Jesus or Mohammed! What we have in the three cases represented dimensionally, is conduct of what God actually did say, but four, two or one thousand four hundred years ago. Still allowing in reverence that we actually do operate past the written record or spoken words of God, only this time discussed in there particular form of intensity as delivered, but with modern opinion without modern thinking?

With some better open realisations of what we are trying to achieve in moral pursuit these recent sixty plus years, delivery from this source is for one aim only without any hidden agenda. For any individual to state in cross reference that by all terms inclusive of all our mediums of interaction, none may profess to offer enlightenment or guidance by terms of direction from or of God's spoken desire! In first notice this standard is by religious terms mainly, but allows for us to include the same limitations of acceptance if used in a subservient manner to false claims. Like in cases whereby even through the arm of dominant political regimes the cry of "in god we trust" has been used universally to imply direct religious connection for purely political aims. Illustrated by how at the same time we allocate the dome of religious tolerance to partially indicate political tolerance by the same means, but to a different understanding?

'Past Gods Words' by delivery is to be best seen in context whereby up until at least five to six years ago anything could have been offered to answer questions and solve enigmas? Now our real terms of reference can only be offered by political standards or better religious understanding to be equally overlaid with new phrases in offerings, because all of our original written text is done and has been for some considerable time. Meaning that even if any more representations on the same scale by Gods own instigation are to be made to us through any person, all can now only be done 'Past Gods Words' by the two meanings of full interpretation. One, that at this time and I state hence forth, no person or individual on their own account or in representation for us all or a single human life, can claim new or later uttering's to be from the mouth or idea of Allah or God equally! No person as an individual in like commitment to us all, can utter in compliance to have ever heard Gods own word even if delivered by a golden glove or emerald voice or any sign to be other than by total imagination. Because of the second feature, being all to have already been said by God without my qualification or acceptance has been honestly delivered to us in all cases at times of our best understanding in any case, when we could relate to God without the complexities of now, owing to these times!

Four, two or one thousand four hundred years ago are three equal times of our ability to partially understand what were given as direction or named revelations to us then. Although in reality we had always progressed by the very advent of time, but even up to about those fourteen hundred years ago, we had not become sophisticated or civilised enough to be able and deny the reality of God, Allah or the Great Spirit or the picture of Nirvana or the real future. Change in our attitudes and learning being a feature to have been expressed by how through Islam we recorded the words of God for posterity. Not half hearted or misplaced or misaligned but to right the wrongs from earlier times when our recording of the same features had been done in most cases from the memory of personality or in other cases from the memory of how things might or could have been? Now/ then fourteen hundred years ago we set the reality to be finite even to the extent for cover in the future, given that this was our last heavenly delivery as foretold in any case by our last prophet!

Then ago we were fully aware or our surroundings whereby the only possibility of delivery to some of what we saw had to be maintained as mysteries, still allowing us hold in full reverence to the holder of the mysteries in being God; simple, we were as we, God was in our acceptance by reason? Even through many reformations in time and in conjunction with our advancing knowledge generally, we were able to hold our religious convictions, but now always under threat from the exposition of some mysteries? Eventually when our pace of development got into second gear about some two hundred years ago, mystery expositions in becoming the norm, decried some of our religious sentiments? Arriving to allow those sentiments languish in the realm or religious tolerance, but now fully subservient to other determinations as seen.

What is to be achieved by us now, taking the taken prospect of 'Past Gods Words' through the first means of our religious overlay, is that we are giving notice in challenge to political expediency. We now accept where the future is, even if through our demise, but always in terms by 'Natural Body Death', therefore our expectation above further challenge is that our religious leaders will expectantly explain in support of our real options by directing political expediency?

'Past Gods Words' is to be taken now by its own intention, but forever remembering it as an adage is of this very modern era stamped in time to be recent by these five to six years. Covered by the brand of the Fifth Dimension as one of its taken prospects, we will from now need reminding of that fact; mainly because as an expression, it straddles the two greatest influences to be upon us in our religions and politics. If from some ideas through our political standing, they would ignore the prospect having no bearing to be political that is a case for the individual conscience? However, allowing for the haste of judgement assumed by first impressions, I am at liberty to push the issue to be political but only by tolerant acceptances equally with religions!

Accepting that if the balance in some matters seems to be against us, either personally or collectively it could be our own doing by remaining with judgements we had already made to fixtures that had been presented many years prior when our general thinking was not the same as now? With our level of understanding not the same as now, when our means of interaction were not as open as now, but from where we were better attuned to the final reality than now? All that we need to be able and consider onward is that it is possible for us to take control of the running of our lives to the possibility of the future in total relevance to us all, even if it means considering our single and separate lifecycles paramount! By another definition that we must be able and consider ourselves only; but by open consideration to each other, because of how some of us are in the position to force change by being misdirected or misinformed or misaligned to what ever degree?

For us to openly profess to operate 'Past Gods Words' in context, is not to sign up to the alternative, but give open assurance that we accept in provision we must act to a standard of how we should now re-ignite Gods own interest to any single specie that might show promise? To be in the position to proclaim that is our quest, it has been from us only that the image of creation has been presented to be understood of what the future might hold? Without competition against the crocodile or tiger or ant or finch, but in promise of our self definition, we alone have conquered evolution and expectantly wait in turn for what creation might deliver! Although guided in the past to where our direction was to be, we have at last self asserted by our own discovery delivered in these years since Shanidar?

'Out of Africa' is a statement to be used in force of recognition that we had a single beginning and for good measure were pointedly black of skin at outset, because of prevailing circumstances. The only division which can be expressed by that statement and constantly is, stems from the question of were we appointed or did we just evolve? Did a female ape in laying eggs produce one from which a human was hatched? Without levity but in good spirits and why not, the last question was not to forestall the probable discovery of the missing link but to dispel in part the necessity of the search other than to know? Because in using the study of science we can pretend that nothing was ordained and therefore we do not even have an imagination. Scientifically proving that in fact we do not exist because up until now we have not found the missing link, therefore where did we come from, but never why!?

'Out of Africa' is a statement of fact and fully covers the nature of our yellow, brown, white or any number of intermediary skin colours that we share delivered over time by the circumstances of geography, climate, diet or other influences! Along with personalised traits in the form of character definition, although the mixture is mighty and complex it is pulled together in all realities of how we interact religiously and more so determined by these past sixty plus years. Without compromise, that image is to be cemented in charter by our United Nations in best representation of our political awareness to the effect of instigated change over these past sixty years, even if in another vein by working with our taken prospects?

Not least because in our other field of religion we have taken the bull by the horns and grappled with the nature of how best to heed God, then declared that from now in open awareness we are to operate in active judgement 'Past Gods Words' on our own account for presentation? Simply to fully equalise all of our standing not to be of a first, second or third world classification on any grounds when factually we are all of the same standard religiously! From how we might think in terms of the future by any means, we have to set the parameters to include us all, wanted or not? Leastways to protect the work we had started by breaking free from the shell that first contained our single image?

Meaning in its entirety, that as this single specie, our representations to it and our interactions are not to be dismissed by personal choice, which in no small part is why our religious overlay is to hold greater force than political ones. By taking note of this feature we effectively are stamping our endeavour to all for the same outcome by expectation. But not yet fully realised by how we have recently come to interact by the same violent means to each other but undefined by noted intention, because until now we had lost direction?

Following the taken prospect 'Out of Africa', with another by reference to our core belief cut with brave heart to understanding, when no other creature by specie was able to do the same, is proof of our individuality. By working the prospect 'Past Gods Words' properly, is another example that we are indeed different to all other specie, standing on the grounds that none have been able to make similar signs. Born of one beginning, 'Out of Africa', blessed with one objective in the general name of wanting to know and then expecting to know, 'Past Gods Words' by unimagined means, makes that we are unique. In the same habit, but from a different perspective the third of our taken prospect also endorses the above, but to another style, because we are involved through it in a different fashion!

'Natural Body Death' only refers to us humans of this same specie, not least because of the level of inquiry we have shown, which places us far and above what any other creature has obtained or are likely to ever obtain by thought? With them linked to natural instinct in cover with the one driven role to procreate, leaves them all on a different level to us, but not lacking by some determinations even if considered only by us. Other creatures by whatever denomination will just as assuredly fall by death in the same manner as we, but there deaths will be by natural means in all cases because they live and die only to procreate? Although our intention is the same by given example through the finding of purpose, we have altered the reason for death and how to handle it?

"Though shall not kill" is not of the same meaning, of defending, protecting or promoting any religion by the ideals of commitment to kill of self by self! Through civil law in the supposed defence of murder by any means and from whatever angle, cannot be covered by the same shield for committers of, apart to victims of, which evokes much greater strain? 'Natural Body Death' in statement is purely a measurement of time, but wholly on human terms and only relative to the situation of the termination of our individual lifecycles being fifty plus and or seventy plus years long!

Nothing can be simpler than the acceptance to the fact that we live and will die in due time. What has become a huge issue in meaning from such an obvious statement, but not in human terms, has been manufactured by time from our own misunderstanding to the assumption that we are civilised? Although we seriously show greater and greater awareness to our surroundings than any other given specie in group, like the insects, birds, land mammals, reptiles, fish or water mammals. In cases we show ourselves to be less than aware of what our surroundings actually mean to us because we can live and survive anywhere allowing our sense of purpose show we are different by type. But have the same drive of concern to matters outside of the general picture of natural survival, which in seeming has generated our claim to be civilised first among beasts!

To hold in concept to our taken prospect of 'Natural Body Death' on its own account gives no licence to any by acceptance that we are civilised and therefore right. From where I would challenge that claim has only the remotest of historical connections but is worthy to be revisited to redefine and check if a step taken was a step in the wrong direction? More concern to the matter of our civilisation has been generated for my personal notice over these past sixty plus years, which is a road we must traverse to understand the need for our Fifth Dimension overlay to our politics and religions!

## CHAPTER 132

Civilised or uncivilised is that the question? Although I have created a small picture for us all to peruse, my intention is not to lay query on our sophistication as it actually is or the level it has formed to. What is to be done is in the reverse, whereby none of us in any guise can use any part of the feature of being civilised as we assume, to block or change what is bound to be forms of natural selection, but only religiously inclined! Where politics is a means to its self and operates so, is to be of considerable concern to us, but likewise no political verse is acceptable without up to the minute religious considerations. Which is to be taken in honest array from now on, because we can be involved by conduct befitting to the lifespan we unerringly have? Where all new political considerations are to focus to those ends of improvement which are wanting and wanted, shall be conducted by the same people as we, but defined by the three taken and one active prospect of the Fifth Dimension.

Without the means of judgement but by reflective intones, my charge that we are not really civilised has been overcomplicated by the full means of our general sophistication even if supported by most of the revelations supplied by the fourth dimension, helping cloud the issue by some of our character traits? On my own account I had indicated that much of our early modern forms of interaction were cast in Mesopotamia from about eight to ten thousand years ago. One of the first and oldest of our world countries was formed in the general area of the Middle East and was to give us a huge first lead in being civilised by social terms of reference. Egypt had openly displayed huge advances in self representation to all of our ends, but in drawback by religious determinations, because of the method used in style and how it was to naturally pan out?

By brief reflection for us to have considered the reasons or the style in which we interpreted the flow of life by creative terms, although unable to interpret purpose by the means of any religion then ago, we automatically endorsed civilisation? By doing so we knowingly took a step further away from other creatures into the realm of knowledge pertaining to the mystery of creation? In that form by the ideals of being civilised we aimed our considerations to the study of the why and how, from which we were able to establish the criteria for being civilised? But never aware how things would change, because even in direct ancestry almost five thousand years later if only through the main religion of Islam, we would not have been regarded as civilised at the time of the Pharaohs because of our spiritual/religious views!

Once again dressed in the unknowing value of speculation and supposition, earlier than any other group or tribe or clan from area, the Egyptians in setting standards, began big with their godly concepts in almost being monotheistic. By maintaining the direct connection in style of the individual to the one big God through the regal offices of the king through his ideals pursued and ably managed by a dedicated priesthood! Standards had been set for others of us in surround to compare, enabling us the ability to interact by general and normal means with the ability to add to the open standard shown us by the sophistication of the Egyptians then ago.

For some of us the rest being history, has been recorded on our own account, but for others, if in case, part of the regal house entwined with their own very able priesthood, found the necessity to be moved on, or move on because of severe circumstances locally if by forms of opposition or persecution? Other circumstances to come into play by whatever conditions of time might take an assumed very civilised culture, and destroy the value of being civilised, by coming from a more civilised community in the first place, but only by attitude!?

Again, to walk in mind along the road of if, down the avenue of maybe, onto the roundabout of probably? By looking at situations as they were and introducing the speculation of change, we might be able and satisfy our own queries on other solutions which is not always easy, but does not have to be that difficult if we use abstract reasoning? Simply to give us the fragrance of solution without the need of experiment to explain in formula what we had already guessed at in theory? By the simplest of connections in explaining the two general types of people found inhabiting the new world of the American continents some four hundred years ago, any person can speculate through the image of stepped pyramids?!

My own view being not entirely new has offered a type of speculation proving that we were to become more civilised only to show its fickleness by others who had become civilised from the same source of the same time, but were all changed by circumstances? My examples explained only by supposition relative to what might have happened to the royal and religious house of Egypt, even before the first beginning of true monotheism had made its mark and being controversial, is of its own consequence by other means? Beyond conclusion, from my own viewpoint I have been able to the see the Egyptian and stepped pyramid influence having spread to the new world by the means of imagination I had already found reasonable? Even if from abstract thought, I had not needed scientific proof to count the odds of the similarity between some pyramids of Egypt and in the new world thousands of years before Christopher Columbus?

My point was for other reasons to explain the two basic religious cultures of the new world and the reasons to follow by being connected to the Great Spirit one way and human sacrifice as part of an alternative belief? By most accounts the source of occupation of both continents was by migration, with human tribes using the same northern land bridge from Asia and carrying the connected wisdom of Shanidar? To have already suggested that the later influence of human sacrifice was added to the Great Spirit belief was not done to manipulate the facts, for they are always open? What I was indicating was in relation to the theory of isolation whereby those in that state or of that status, because of little continuous general interaction like it could have been in Mesopotamia several thousands of years ago, kept to what was accepted. Without the lack of compressed interaction, we compounded our reasoning by means that seemed to work by producing very sophisticated societies who all maintained the similar Great Spirit philosophy. Until perhaps some of us became excited if told of a different even more plausible means to finality through the amazing building of stepped pyramids and the inclusion of human sacrifice for further improvement!

If we may yet find that it was a somewhat disenfranchised Egyptian sect that made such a fantastic journey across the Atlantic Ocean always towards the setting sun in adulation. Driven or hounded out, who kept on with such determination of belief to imagine that whatever they encountered they could overcome, because of the guidance they carried in the two chapters of being in connection with the giver of life through the king in line of divinity and from the activity of the living sun in the heavens. So much so that by terms of nature it had become necessary to offer token to the cycle of the heavens, which so obviously ran the essentials of life in general. By comparison when seen that if some blossoming fruits were cut from the branch in early growth, those left were more vital in production and delivered more, to maintain that vitality such sacrifices made, proved more bountiful?

By the very size of the Americas when first inhabited via the land ice bridge, remaining sparsely populated was the norm? Allowing some groups wander with seasonal trends to determine habit with a general comfort of a belief in care by the Great Spirit. Outside contact with such people might have had no bearing as it was rare and of less benefit than the maintaining of establishment! However, from my own suggestion that there was/might have been first contact near the isthmus of the two continents, any influences from then has to have following considerations? By any length of time for each or any community to have settled or further developed in self study, the two lines of conduct also went hand in hand by civil order and religious belief. But if the Great Spirit answered man desire on the American Continents, how was a different picture seen by more recent travellers?

In as far as I might wish to go, carried by speculation of how the found situation in the new world was at the time of the conquistador. I use my own ideas of breakdown for understanding so that we might move at a pace of reasonable speculation to be able and arrive at suggestions of how we can continue from todays altered appearance, but to guess how things might have been? In accepting that some Egyptians under very odd/strange or difficult circumstances of travel that might have occurred while crossing the Atlantic Ocean several thousand years before Christopher Columbus? Then being responsible for establishing the human sacrificial culture in the old new world as a means to placate something wholly un-human, like the sun in the heavens? Takes credence in comparison to what could have been considered of isolated cultures in other parts of the world up until much later times even as late as the last two hundred years. Or in extreme cases even relating to the Great Spirit belief of small groups of people in the remotest parts of the world like the great Amazonian rain forest of south America or the lost worlds of the great Asian islands separate to the vast island of Australia

In such places the Great Spirit culture seems to be predominant because from origins, little contact had been made with other cultures apart to similar local tribes. Therefore there had been no natural interactive exchanges that had occurred in other parts of the world between similar tribes who themselves were forever interacting due to the prevailing circumstances of geography, climate, diet and other reasons. By my own consent I view the continental island of Australia in almost direct comparison to the North American continent in terms of religious settlement by the similarity of belief in or to the Great Spirit concept. With the one exception in Australia at any time they had no other inclined travellers to bring in suggestion a belief to the sufferance of God by the means of human sacrifice!

**CHAPTER 133**

'Natural Body Death' is to be a vital component for our full understanding of the human condition from today, even if we are not fully aware of intention by it joined with the other taken and one active prospects delivered by the Fifth Dimension? One of the cornerstones to reason will futuristically stand on the measure of how we apply ourselves to that particular prospect outside the supposed standards of civilisation that some of us profess to understand? Along with other equally uncomplicated measures that we can adapt to style, our three taken prospects are almost all we will ever need from the very near future to accommodate many of our hidden desires. All they are intended to do is show our similarity over our differences beyond comparison?

In order for this to be done and achieved I have put the task of us all being involved to overlay our political mediums and our religious aspirations with the three taken and one active prospect. But also for us to be aware of the fact that although we have life spans of different duration, all attention of focus is to be set to our individual lifecycles to maintain our similarity of purpose. Meaning at this particular time, some of us even if attached to the same political ideology or connected by our religious belief, have a shorter or longer period of time to come to the same terms? Meaning even in light of that division our first considerations are for us to address our sentiments to what is assuredly the total objective of what best suits our condition and how we have made it so? And how we have made it so? But not determined by our own political, religious, national, racial, ethnic or inclined belief or even the lack of it, but what might be the indicator of our difference to all other living creatures upon this earth past or present. Even extending to the rest of the universe, where by deed, it might be us who are to be Gods own herald upon those other vast expanses?

Without the need for debate unless to listen, some of our first steps to be taken are to alter the future without changing the past or its valid representations, when some are seen to be on religious terms more devoutly than to other political holdings. Lack of alteration to the measure of history is an honest charge, because by token we have to be able and trace our path not to see from where we have come, but to look at the direction we began to travel and were able to record the due changes that were made on the way! That way we can accept how our attitudes had been able to change by the advent of time if recorded to some measures of official reasoning four, two or one thousand four hundred years ago. But in forgetting that more change had been automatically added to our living in the past fourteen hundred years than ever before even highlighted by these past two hundred years semi naturally. Not allowing that in these past sixty plus years in the haste of further progress we have regressed?

How we have fallen back without recognition of some ideals, is not that we have failed by consent, but that we have always tried too hard to rationalise past events by means of correction into future conduct. I will submit by further reflective terms that the means to our ends are far more reasonable than might have been supposed these past sixty plus years, not least from where we have set our parameters. In trying to please all of the people some of the time or and some of the people most of the time we have alienated most of the people some of the time because of our style of how we have tackled some of our finer details to or from our first world standing. Being the only way we could have started but not by style, because from that position some of us have always assumed our own righteousness on all matters?

One of our greatest errors, although just alluded too, has always been how we can each talk at the same time but only hear what we think, which is of small account now that we are to be forced to listen in any case, even that we are told the wrong answers? Although I still lead from the shadow of the abstract, what I have always maintained is our own expression of impartiality now borrowed from the assumption that we are deeply civilised. Although in the same manner I refute that measure of our achievement, it is beyond the ability of any single person or group to question that we are truly civilised or not. Where the query is to be set is in how we can speak to act for others while being civilised in life and at the same time have no plan for the eventuality of death? All of which only gains substance through the doors of 'Natural Body Death'!

Civilisation by any reasoning stands only on one leg, which in effect means that as we live we are all civilised? My own rules of concern to the whole matter has been spread by how some of us from whatever level of society are able to use the concept in excuse of deeds by whatever means. Not so to challenge any particular group by modern definition but so that we can refer to the past for the purpose of alteration without changing emphasis. Simply to show how by mixture our own human experiment on all levels is to have been purely human led by us and or our ancestors. Meaning that if and or because we were stated to be civilised or more civilised than others we used its fair office, to rule over any other society or group or clan that we came across, particularly from the past; and passed civilised judgements upon them. Not accepting that the same measure of judgement was to be upon us all possibly by the same core of 'Natural Body Death'!

Although our leaning at this time might seem to question how some half measured rules run by taking assumption that some of us are more civilised than others, allowing that statement in belief carry our steps taken syndrome! My plan of design through this work is to set general standards of our future behaviour but un-tainted by our pasts, though not without the open expectation of fair retribution from uncivilised means only. To be taken not in the literal sense, but by measured judgement uncivilised, when using our current means of conduct by our current religious and political settlement. But never to ignore the fact that we only have fifty plus to seventy plus years of individual lifecycles to settle all matters, even if only using the sixty to one hundred thousand years since Shanidar as the same marker point to us all. But to remember that if all are not civilised then none are civilised!?

In using some of the above by example the rules or methods of change open to us are enumerable but not inexhaustible. By keeping to the core lines that we have each already made for ourselves, based on our religious belief and or feelings or our political doctrinaire, we will be able and focus to that specific time span that is of the most vital importance to us all, but not yet fully realised. By first record it might be seen that our religions are the most considerate while our politics are the noisiest, because essentially politics operates through far more mediums than our religions. Politics in carrying its own tentacles of commerce, trade, industry, law, conduct, even expectation through the money belt of desire in using the apparition of the fourth dimension, has come to overshadow our true religious commitments! From which we have been set to doubt religious aims which in reality have always been our united driving force, but not even realised today by using yesterdays thinking by what ever form? But now to the matter of correction with the Fifth Dimension!

How ever future intentions are to be set, our first lead is to be taken by the depth of our religious convictions, not least because they have been our biggest driving force to spite not having given us any weighable dividend except that of purpose. Our religions by blessed desire and from now in spoken direction from all of our religious leaders ongoing, are to carry there own weight but not in competition. That is not to say they are to be all the same, where I make some comparisons is drawn from the depth of all of our histories in time, not least by the first sight to the expectation of 'Natural Body Death' at Shanidar. In carrying that image as it was into modernity, the picture fits well with all religions thus far or at least any introduced these past four thousand years. Save perhaps one kind of religion or similar to have taken to the killing fields!

## CHAPTER 134

'Natural body death' has always been a big feature to all living things, because in all cases to us, even if achieved honourably, it speaks of fate without definition? Although by inclination I had referred to those of us who were somehow disconnected from normal behaviour. Not to put too fine a point on this group, even if it includes us all at all times, my direction is for us to only include this one specie being human. In factual terms all other creatures have always been held and encased by the natural conditions of 'Natural Body Death'. Without the pinnacle of God or afterlife before them, their terms of achievement are only to procreate in the spectacle of time by monotony and monotony? There role is a product of evolution met in all quarters by the sole monotony of procreation and procreation, which by deed is no bad thing, unless being set as a condition for humanity!

My considerations over such matters are for us, we are not at liberty to down cast any other specie against our own achievement. Their evolutionary trait will always be our anchor, for without it there could never have been the spirit of creation. Just as we in our specie tribal forms have borrowed parts of each others intellect, the same remains from how we have used other specie, even if not to follow their example! By the whole mix of all creatures in any case, by the numbers and time periods involved, we all have that interwoven character of life and living to contend with. From where their format developed to procreate first, such serious involvement was sufficient to maintain that specific heading only, so much so that in fair reality the incidence of 'Natural Body Death' was a natural outcome because of naturally contained instinct.

In our short three and a half million years not even explained by how we were designed to travel the globe in a singular way, part of our own trend was interrupted without explanation until Shanidar? Although that sign is extremely positive by all readings for the future, its best signs were to be interrupted by what was to follow shown in depiction on our painted cave walls. First noticed from anytime between fifty to thirty thousand years after, but better defined to have first been done thirty thousand years ago. Showing unreasonable items of inclusion yet to be determined, not because we lack the ability to open the past, but in some cases because we have only been able to judge it by some of today's standards?

Without the standard conflict of interests my own reasoning of Shanidar and our cave paintings carry the same fair burden to how we might begin to really understand the past. First without alteration and second with out the need to alter any thing, save how we can now judge it by any number of ideals created from the decision of ideas? From how and where I have seen Shanidar, I have put the notion that most of our cave paintings were done in tombs, cannot alter the past and therefore the future because of events? What I can do is place us all in a better position to realise that we are better connected than we ever thought but by purely modern reasoning not from the old past but the new future?

By reason that I might give credence to some of our cave paintings having been done in dark places allowing for comfort in death, stands without alteration? Even when I had alluded to most paintings being done in mark to the dead and by connection were done in places to be tombs? My definition is to be indicative of how we seemed to have been more driven by the prospect of death over the excellence of life, but now to be combined in collective union. Even in following the positive trail left by many assumed different tribes emerging to accept that as a standard at first in the land between two rivers. Our real and well recorded symbolism in or on religious terms was done by the same means of collective reasoning for me to be able and refer to the Egyptians of much the same area having cast the roots of monotheism within complicated avenues?

Continuing in the same, vein although after hundreds of generations since cave paintings were superseded, by following some of the messages left in Egyptian tombs, what can be seen is the full organisation of new religion? But more inclined then than now to have been man made and not yet appointed? Meaning in brevity, that instead of our previous standard of mysterious expectation by the event of 'Natural Body Death', our new reading if at first through the Egyptians, was by inclination to the prospect of creation!? Following that society was ordained by forces locked into the benevolence of the mystery of real creation? Being a feature when turned by reasoning from the role in society to be of the priesthood, who were able and lead with some measure of direction from what they knew by already being involved being the keepers of the mysteries was a mystery itself?

All that I have depicted above is nothing more than our first signs of working half measured and therefore following our assumed findings to be worked through our steps taken syndrome? I make the case not against one of our oldest defined nationalities being Egyptian, but in self representation of their own point in definable history! Not to labour any point, save that although the Egyptians led us in many ventures to the beginning or real understanding religiously, there case has been rested by our take up if of the following three dimensional religions of area and the supposed effect they are yet to have upon us all?

'Natural Body Death' by terms of reference is of its own meaning but only through the avenues of creation in line of reference to this our single specie? Although not to change in the future by steps taken, as an expression it now covers a different standard than first proposed in earlier pages in this work. My overall view is for us by majority to accept the standing, because we are here by fact. One of the only distinctions I make is of today, that I have given us all the same reasoning in quality, but there are still those of us who are unable to use the rational in the same normality. Therefore if any of us by personal choice or in being led by direction would seek to kill of self by self without any consideration to what any God might have delivered. We/they belong to an edge, not yet fully defined, because until now we have not had the wit to be able and decide on their/our other classification?

Our biggest historical challenge to the fixation of 'Natural Body death' has already been met, but not by layered recognition. In making the suggestion that somehow there was a direct association between ancient Egyptians and or the style of there beliefs, to have been implanted to the new world of the Americas even before Moses had heard truer words of direction in the lands of the Middle East? Has not been done to question theory or rustle into our own areas of doubt, my points made have only been offered so that we can consider later facts to have had a point of reference in this time. So that the important energy of time is not wasted looking for the holy grail of understanding in the allotted time of our lifecycle achievement under any wrong heading?

Breaking back into time some four hundred years ago by being aware of our European discovery and exploitation over peoples of the central Americas and the eventual use of both continents North and South, our judgements are to be readjusted without alteration? Meaning that although we know much of the past we still must try and relate to it by its day, meaning that my own stories of how I have been able and reconstruct the possibility of ideas can be applied by prospect of the now and future? From not knowing anything about the people we first encountered in the new world our only actions and reactions were based on and or set by our used civilised behaviour.

From the European picture of having discovered new lands excluded from the previous structure of history, no fixed comparison was set against these people who seemed not to understand our own obvious intentions. Apart to the open aims of gain from the business of commerce that might have been originally expected through the tentative travels of Christopher Columbus, other helpful and disturbing discoveries were made. By coming across peoples who readily amazed us by there standards of national and social organisation, we were relieved to recognise their ultimate weakness by religious standards, of which the rest is never to be history?

Having discovered new societies who seemed to practice the art of human sacrifice by religious tones, any merit in owing to best conduct by the means of organisation, were to be ignored or sidestepped for other reasons? Even by finding the obvious, European judgements could only have been drawn from the level of their own standards whatever, even to be claimed only by our general level of development up to that time in any case. Hence that we can travel back in history, but must do it by working to the developed standards of the time! At the particular time of European intervention, no comparisons could ever have been drawn of the peoples of the new world having formed any part of their religious philosophy pre delivered by the same peoples of the Middle East?

In looking north and south by means of further discovery, over time the whole continental land mass of the new world was always to be judged religiously first, that all two of the core standards were lacking and wanting to the full Catholic/Christian intrusion. With the found standards being to the Great Spirit and of sun worship culminating in human sacrifice, by intrusion was to be a mark of time! Not now used in condemnation as the findings might have been some four hundred years ago, my connection without balance is one of reflection? I have not laid the proven case for how a society might have come to the open skill of human sacrifice by the assumed romantic connection of travellers from the Middle East to gain empathy. My flow is designed to cover the possibilities of false assumptions that a desired concept like human sacrifice from sun worship, could ever have originated from the concept of the Great Spirit!

How ever I had indicated how the Catholic/Christian conquerors of the new world were to behave to the situations they had deemed to have found in these hitherto unknown continents. My inclined observations, that we were then and now, proved to be the same specie 'Out of Africa', stands by all comparisons, even that some of us took to human sacrifice as a means to promote the benevolence of our own existence. From the meeting of minds which did not take place, led by other standards we Europeans acted by and to the standards we then thought to have been guided by! Leaving no other alternative to what we did in shame other than the actual outcome?

I have made that real case by example to highlight how from four hundred years ago we carried the exact same reasoning to have already led us to two world wars before we began to realise that change by the expanded reasoning from the standards of the fourth dimension could be used to take us in opinion of the Fifth Dimension, even before its form created intention. From the very same time if only four hundred years ago we brutally challenged the practice of human sacrifice religiously. From the aspect of religious standards, it was almost to take us the whole of those four hundred years until the ending of the Second World War, before we set our opinions to change. Alas some of the reasoning that we brought into the arena of change had been made by our encouraged approach to view all matters half measured because that is the way most reasoning has so far worked.

Standing as we do now on the eve of all tomorrows, once again our task without cost is to look at how we have laid our path yet to be by natural development owed to the past? When using abstract reasoning from where we can lay aside some of the events to have occurred since the first known time at the meeting of minds in the discovery of the Americas? For me to have put any connected link between how we might have come to the art of human sacrifice in some places of the new world then ago, stands?

If we consider as we must, that by sign some forms of development of the human condition only occurred by the means of tribal or clannish interaction through the effect of general contact, for example like in places such as Mesopotamia. Although from a period of time first noted to be about ten thousand years ago, in the land between two rivers there had already been notice to the fact of our tribal differences. In comparison, but to different circumstances, not least because of the time difference from where we can find evidence that the migration to the Americas occurred much earlier. I had used the pointer that the Americas were first inhabited by humans through the land/ice bridge between the continents of Asia and North America but from any time well before the interactive phase between tribes or clans in the Middle East, even in the land between two rivers!

By the same token because of circumstances, not least to the vast size of the continents such peoples as we to have migrated there, naturally kept to the primary need of provision for the necessities of actual living. By any time mark when we did travel over the land/ice bridge we also carried the experience of Shanidar with us, buried inside almost as a source of energy, allowing us continually move aimlessly for purposes yet unfulfilled? I have made that direct observation because of the outcome, without using my own felt important value of the fact that we individually are here now? By connecting to the viewpoint that there was indeed a general acceptance for the peoples in the Northern part, where the original land/ice bridge was. Peoples from the same time in general, adopted the same core belief by extension to an acceptance of the Great Spirit for religious contentment?

If because the different tribes and clans migrated under the same circumstances by our growing need to follow a food trail and or simply that then ago we were natural hunter gatherers. At the time such moves were made we did carry the full experience of Shanidar even if unknown. From that dormant spark of our growing spiritual awareness our limited interaction with others was driven in time by mutual interaction and cross referral. My mention to the size of the Americas when used in comparison with the size of the Land between Two Rivers was set in excuse to give credence to the acceptance of the amount of influence tribes would have on each other in close proximity, when others might not!

My own connection with now and then has been to use the one example of the normal interaction between people who have similar outlooks, which I have subscribed to mean the same for most of the people inhabiting the Americas at first. Having settled for the belief in after care to be by influence or control of or from the Great Spirit, by the normal course of events that style of religion had settled on all? By that general acceptance belonging to all peoples, even that we were aware of our differences, there was no defined need for a managing priesthood to associate us more precisely with the inevitable! There was no need, because all were readily suited in concept to what meaning and use the Great Spirit was of when we encountered death!

By encountering two religious standards in the new world after Columbus, we were unable to equate what was met by any form of understanding because at that precise time Europeans were in self religious conflict? Although forms of progress were to be the overall outcome of this newer colonisation of the new world, much by way of importance to our own developing attitudes was to be clouded by what we actually did or thought we had done by cleansing one religion and absorbing another?

For me to have created an imaginary situation from which we might ponder in the abstract, the whys and wherefores of how we found only two basic religious formats with little relationship to what was known in Europe or the Middle East. Has been done to show in relief how other half measured thinking need addressing so that we might support all new endeavours of these last sixty plus years since the end of the Second World War. Although four hundred years by time lapse has no real bearing on the progress we have made since the coming together of our newer and more open acceptances of each other. I look to compound our start less half measured than we had operated in the past, even if by accusation we are to find ourselves less civilised or uncivilised?

## CHAPTER 135

Egyptians yes or no are to figure only from the abstract, because without defined links by fourth dimension reasoning through the proof of experiment, at this time there is no way we can accommodate what was found in the new world from about the time of Christopher Columbus. Unless to accommodate formula we have to make our own association to explain the two basic and main religious holdings at that time and in that place, one being to the memory of the Great Spirit with human association. With the other belonging to a static belief in one of the elements to fulfil our needs, even if it meant that we were to abuse our own form by the wanton and needed use of murder by self upon self without pointed direction, because death could not stall or change night sky predictions!

Points I wish to make, but in being warned not to judge history for alteration, are that in the case of the peoples of the new world then. Europeans were able to make judgements of whether any of these peoples met, were substantially civilised or not, wholly cast by there own religious standards which at the very time were undergoing serious change. By using the one single classification to the fact that some of these peoples used the aspect of human sacrifice as a religious tool to the full assumed desires of an innate object like the Sun by day and the stars by night. Gave us the line of a further split we could drive between us until this present time when the damage had already been done!

In following the trace by any desired route we might wish to make for our studies into the past, my reasoning above that we are all here now, stands to the fact of our own lifecycles from which we have assumed by proxy that several features to our wellbeing are already in place one way or the other?

In as far as we have dealt with those cruel habits of human sacrifice found in the Americas, by being able to quash them in principle, was done in our own shadow to ever improve our situation ongoing. That was or had always been an unsighted objective of the human condition since Shanidar from where we looked to join the mysteries of what was to be the answer of what is to our own acceptances. All that has happened in the past four thousand years is our looking for the same answers, but by following different leads, which has been no bad thing unless we remember to keep changing some of our views in the best balanced way that we might?

One of the several unfortunate outcomes to the result of the banishment of our human sacrifice as a theme when found to be done on mass, was that to its destruction we were also able and destroy the concept of civilisation in comparison that had already been reached in the new world under its own drive! One for one the standards of civil organisation expressed by community conduct in the new world, outpaced the standard that the flood of different Europeans were to bring. Unfortunately because the many different European nationals had had vast periods of time to intermix through their interwoven histories by their own close proximity. Each were able to copy the range of standards shared, therefore with each able to classify themselves of a similar standard of supposed civilised excellence. Followed in no small way by how generally at the time of Columbus all European religions were interlinked through the similar religious core of Roman Catholicism. Even when being broadened to the full concepts of Christianity, Europeanism on competitive levels was generally of the same civilised standard throughout or so it was thought?

From that very fixture although we addressed all matters as best we could, even as best we thought was necessary by religious codes. Our standards of general civilisation were so low and lacking that from the positions we were in, we naturally assumed that we could not be wrong? Therefore by connection when we came across a society who by general outlook were very well organised, even to our own standards, in order that we might pick an avenue to diminish their flow against our own. We were able to connect whatever they did to be of a primitive nature, never mind that in cases they organised the civil structure of vast cities better than had been done in Europe up to that time.

With using the factor of human sacrifice religiously, we were able to undermine other concepts leastways by implementation, when we likened their attraction for gold to be only of aesthetic value. When we could compare our use of the wheel in profusion, against there's only for mathematical determinations in demonstration of the structure and movement of the heavens? When we had developed base metal weapons and had found an open and general use for the same metal used as cooking pots and the like. We used any differences to undermine there potential, but always first religiously? Not least because we could use that lever to enhance the general requirement to change all of their concepts fitting neatly with all the properties of guidance we Europeans were beginning to excel at. Not least by the way and frequency we could change by our own familiarity!

## CHAPTER 136

Having already stated from the example of how the colonisation of the new world actually took place, we could do worse than try and emulate the standards now obtained there, particularly on the North American continent? By asking why is not to question my observations, but only offers self doubt? From where I had stated that the North American example directed to be the national American style of life in eventual obtainment, are to be used as the end example of what most of us have already strived for politically and endlessly.

By the first instance in some of us travelling from Europe after the time of Columbus and over the following four hundred years by example, what was shown to be envied was a certain style of life that could be obtained by following the American dream? Even not ending then, because pictures to have been created by the political and materialist model, were not to all without effort, which to some of us that might have felt a personal lacking or an unfulfilled objective, remained still empty and in cases to be also envious!

As we were to become more aware of the possibilities of what could be achieved delivered by the ever expanding means of communication to have opened up to us all these sixty plus years. Our own measurements by some political reasoning could be set in balance of what might be? Any checked or imagined progress that might have been put upon some of us, were automatically offered by our consciences or stirred by aspects of our felt religious obtainments as seen? From which by others, those reactions although heartfelt were nothing less than regression and sour and worse, even to compromise the means of our religious excellence by standards to them or more poignantly, to each other?

Making reaction across the board to be forced by some of us, all of which was done by hearing without listening, by seeing without looking and by actions without thinking? All was done most vividly by these recent times because we have all used the means of reflection wrongly? We have alluded to matters by thinking that when the right hand was offered in future congress it was the left hand making the signs? When political solutions were offered, it was taken that religious sentiments we expressed in relay leastways in some places in the world, because at such times we knew no difference and all was to be well? When the American dream and European plans of similar construction were pronounced to be upon us all in all places, even if only because of recent events such as the Second World War? Our thinking from those in the organisation of some of the future plan was unaware of our real religious fortitude which was only set to be accepted by degrees of tolerance, which only posed questions before answers? Although no particular drive or route had been set save come what may, our reactions upon each other come what may was to belie our original objectives, come what may?

All of the time up until the ending of World War two our political emphasis was continually growing, it is only since then we have had to think about our full relationships religiously. But not yet to be defined by fundamental beliefs determined by our note to religious tolerance even if not by belief to the concept? That is why in part we have to take to any suggestions whereby individuals have to stand back a little. While our connected religious leaders give us our due acknowledgement to what we are doing in being right and forthright, delivered beyond the edge of competition!

From the one acceptable line whereby any act or deed done by an individual is a product of and for the whole, from which we are all to be culpable in our own names. From such a connected sentiment I would more than expect our equal involvement in some matters to take a positive form. Not by terms to always obey blindly our own particular group if defined racially or socially or religiously, but in accordance to the matter of our first connection by being of this single unique specie?

Without compromise, the abstract has placed us all of the same importance by position, but that factor has not been noticed from any of our religious or political perspectives yet. Our standing is to be readily underlined by the maxim that we are here, and here and now, allowing that however we might be able and interpret any relevant acts or occurrences from the pages of our histories. All are only relative to how we alone of this one specie are prepared to alter things from how they are now into what they should have been if wanting. Because we had already been able to work out what the best route was, but in some cases we had used the wrong fuel leading us to the second alternative?

How such matters might change can be developed in theory from the importance of our religious acceptances above our political actions. Not so that we are to create conflict because of how some of us already think, but simply to accept that we can think differently but within certain bounds that although long used, had become hidden by other developments. Without direction or even without full consideration, some of our commitments being on common ground can relay to some of us the wrong messages. Allowing that by the feelings we receive especially of religious inclination, what ever we do is right. What ever we do by our own definition is not only right but given to be righteous through religious judgements?

From how we had previously set most of our political standards through the tentacles of trade, commerce and supporting the obesity of profit margins, we have further clouded the importance of we the individual. By committing further political additions in various degrees of assumed success and by not using the example of the meeting of two of our religious philosophies some four hundred years ago, that we cannot alter history as it might have occurred? From that experience or similar joined experiences of area at the time, what was done was our surrender of the meaning of religious doctrine and open objections to the same by what was to follow unguided. Whereby our political aspirations were to overcome or overtake the essence of previous religious instructions. We effectively personalised our politics without fair judgement, because we let our religious messages lead in the destruction of those of us who would involve human sacrifice as a religious offering?

Allowing in the light of European standards in the first place, we had come to accept religious reformation politically, not seeing in context that by allowing our regally appointed civil means of control! Automatically by our own error of thought, but politically led, took impetus to the matter of gain as a standard above what was previously meant as our entitlement through our religions? Four hundred years ago and counting was in general terms the mark of time when our politics really took its own course. In as much as the catholic reformation was going on with the standard of our mysteries under guard. By opening the question book onto our religions, we unavoidably gave better credence to our politics. Leastways for a time through the European and or new world colonisation that was to take place. While at the same time in the Middle East mostly held by Islam religiously, attitudes contrived in the same general way by being done the same over the past one thousand years, by peoples that I had cast to be laid back. Allowing that no changes even without example would ever take place for the same four hundred years in the change of attitude by the merging of Europe into the new world!

Apart to other influences like the peeling of many of our mysteries under the umbrella of the fourth dimension, some general thinking influenced by the medium of forced change, if induced by the discovery of the new world and the way it could be exploited. Almost set the very conditions for the present day situation to account for the way we manage our affairs internationally. With the divided structure of society more centred by attitude of where our biggest split has shown its colours back in the Middle East, but not always counted so!? From the above examples even that I had made pointed observations, many of my considerations in turn have been led to refer to events in North America or more so to the country of America on the northern continent and the Middle East connected or unconnected?

Without balance by the simple fact that we are here and now gives all credence to what I might say representatively or what any other person might. Any difference claimed or suggested by myself or from any other, can be balanced with the same intensity, but only from this very moment, because this mark is the beginning of our new tomorrows without the inclination of change. All that we might hope to realise is beyond any claims so far laid on our behalf referring to our personal longevity. My own take on that vivid proposition is relative by that particular timescale being fifty plus or seventy plus years long, stands to be a binding historical fact?

Although all of our personal histories are intricately linked and have always been if only spoiled by our own deaths, for which by the best standards of recognition we have made it so to allow for how we have not been able and understand God or our sense of purpose? Although we have been the only specie ever and for ever to have reckoned so, all done was for a greater understanding which is further proof of our unique style? From which one of the best ways to connect with our standing has been for us dare and allude that any so clamed reasoning is due to be in our own domain in any case! Unfortunately by the largest of coincidences we had been able and find full justification in purpose, but had only been able and work to our own standards for Gods own ends?

By promotion, the name of this book as the Fifth Dimension will be seen one of our best ways to reconcile our different tenures in relation to God, along with a better placement of our political drive and or excellence if assured so. Not least to curb the latter from its own induced zeal created as a product of how we have ably mismanaged some of our corrective challenges over these past sixty plus years. Without further challenge I do ask that we are to be fully self serving by intent, but not fixed to how some of us have acted in the past or might yet? For all of our personal sophistication which is bound or locked by the level at the very top scale of our religions and politics, our new and tentative Fifth Dimension creed undefined, is on serious good ground we all must traverse. I make that claim without embarrassment, for although I promote its content, in due course it will always be a product of the whole in conjunction to all of our other acts and deeds good or bad?

Once said now is ever held in the yolk of time, therefore the Fifth Dimension as a product might not exist, but its plan and route will forever remain by constantly reminding us of our limitations and at the same time giving light to the solid reasoning that at last we have conquered some of our ignorance. Not through fear or from the whimsical, because from now on, in offer through conversation, we can each express all the elements of the future by our acceptances of the past having always been under the same intention in any case? We are assuredly this very fine and noble people of the one and same inclination, except that for some of us, we have tried to be separate these past four thousand years for the best of intentions assuming we were right, alas implying others were wrong!

Where we will be able and address some of those greedy tasks fermented by our own arm of politics, will again be exposed by the same shields of understanding from the Fifth Dimension presented in overlay upon our politics and religions? Having mentioned this task again, I still make the distinction that our reasoning will take the same balance by all communities. That for each group it will be found priorities will automatically be given to how our overlay is interpreted religiously above our religious standards. Even to those communities that lean in domain to the un-Godly status by connection to forms of civil control, those not in any supposed acceptance or inclination to any religion?

Without condemnation I labour the point from some political perspectives highlighted by communism and some of its practiced versions, because many dissociate religious acceptances. How ever Marks and Engels first promoted the ideals of communism being un-Godly, was done not from the medium of balance. But simply from some personal whim if primarily held by Marks as one of the pair, by whatever key he was an irreligious man, which is no bad thing, save by conscience if improperly employed. His own take was by a political drive for fair exchange for human accomplishment expressed by the physical output of the individual in terms of working for his/her/our daily bread. Although he shone in the theory of concern his comparisons were not for the human sector but the innate?

So that we could better promote his objectives, we were to be striped of our religious duties good or bad. We were to be isolated from the concept of purpose and given to the aspect of a drone, whereby our will in use of concern to human dignity was to be abolished to be given over to the one element in the supposition of self service? Unfortunately how ever the coffee evenings between Marks and Engels were spent together in our name. By the time when the first flushes of there plans were introduced some short years after there deaths. Application had moved from the implementation of an ideal to the brutal imposition of a regime by others, who were more self serving than we were supposed to be? So mush so that from the given plan of a political system in rule of the people by the people for the people ex of religious inclination. We came to believe that the essence of reference to the people meant the state, the state meant the political leaders with similar standing to some priesthoods', in turn replacing the element of human purpose to the standing of state alignment or service? Meaning that purpose now was to be directed to the mother country or the fatherland, meaning that all the elements of human sacrifice were promoted to those ends! Allowing in all future observations the individual was not so, but only a small ever insignificant part of the whole leaning again to our committed service to the innate!

Unfortunately the timing of the new ideals to communism as it was to be, even if called Marxism, coincided with Darwin's great work, giving off the aurora of racial concepts without the need for the assent of purpose? Whereby natural selection was by appointment to an end for the best, mixing that evolution itself was cause? Therefore our forms of religious concept by doctrine were misdirected further allowing the ideology of politics as a better means of appointment even where religions were fully maintained. Bringing us to the position we were in at the mid nineteen fifties with all the features of required preparation already done, meaning that our political systems were almost at the maximum extent of there development, while our religions had already been folded up neatly in religious tolerance!

Such positions that we found ourselves in, were unavoidably self motivated by all the leads we have to the theory of assumption, but not in a harmful way? Not least when by our own motivation we have shown how we differ, not to be different, only that we look at the same problems from many different angles, some acute? Being a method of conduct we do unknowingly, but with our full acceptance even if we carry out our talks from the starting point of being half measured? Leading that by or to or from that trajectory we conduct our follow up by the steps taken syndrome, meaning that we support our original suggested or indicated action to suit the original outset or outlay?

Not unlike when at the ending of the Second World War, both of our bloc culture setups followed their own steps taken syndrome to their plan that had been set by their different types of government? Both professed to act in the same best interest for and on behalf of their people or those associated with their trade obligations or similar alliances. But fell short on the two counts of not properly allowing for our previous religious settlements, making them to be either a political tentacle, or have no sway politically, unless by association through tolerance? Secondly, by ignoring the advice on offer by classifying religious without fixed weight, but automatically applying the rating to all groups, giving off the wrong message from one to the other? Who although working in support of communities deeply involved to their own religious concepts accepted their commitment of faith, while denying the same to there own people, unless to the advantage of accepting the ideals of religious tolerance to other agendas or for other times?

Although such matters were conducted in the heat of true political antagonisms by direct association, influences to be had or made from and in owing to the Middle East are in part responsible for our situation today, even if not fully seen? Communism by association to many Islamic countries dealt its own hand, but remained unable to connect with the style of religion encountered as Islam, by any comparison to older self held orthodox Christian views! When the bloc culture divide ended nearly twenty years ago being replaced with the present day standing of our Middle East syndrome, the off shoot of the whole matter remains today. Unfortunately by the western conditions held then, including a religious feature, for us to have displaced some of the ideals of communism in view. One of the outcomes was to assume that religious tolerance was just as sufficient for Islamic ideologies in its place of settlement by tolerance, and therefore accepted so politically?

That I am prepared to pick at the bones of our recent histories centred on or to events in the Middle East has been done in the most strait forward of ways, even if covered by abstract reasoning or thoughts? Allowing for sudden changes to our ideals, by exposing some of our own anomalies along the way, I have also allowed for the ability of each of us to make points of our personal interest, but turned the focus to the future for personal satisfaction only. That is why the abstract is able to come to the fore because by its self reasoning to the cases of presentation, it also has a case to offer without having to justify the standard from historical events. Also by abstract reasoning we can change the effect of starting half measured, therefore changing the fixed standard of steps to be taken, creating misaligned elements to any original intentions? Even with said or suggested theory the main and unending thrust of my intentions are for us to relate to the one fixed medium between us all, that in terms of our own lifecycles!

Although on many occasions I had given the ranges of our life spans relatively, the picture to be taken in relief is that by the times involved, we each have only one attempt to reconcile the meaning of purpose which can only be translated religiously. Making for us all the heaviest of pressures, when by applying that philosophy we have a different depth between our own understanding of having full relationships with all of our pasts? But still only vitally relevant to those fifty plus or seventy plus years owing to us all of one group or the other all ended by 'Natural Body Death?

Although we have and do blindly accept this feature even if unknowingly at times, my expectation is an offer not a command? But is timed to show that many of our past efforts have already shown there weakness of tenure. How the picture is to be better clarified is due to how our religious leaders from now on take to the needed elements of change. If by no more than using the three taken and one active prospects of the Fifth Dimension, they can offer across the board standardisation without direction! Because from now the direct emphasis by conduct is confirmed within the bounds of our individual lifecycles only, allowing that from all, all conditions we/they/them have no quarter to determine otherwise. Even if in cases the obvious has fallen to the best of our imaginations only?

## CHAPTER 137

If by my own fixation I had come to make the proposal offered through these pages, in order for us to accept the general meaning of my intentions and the open need for such a work, has been born from our histories, therefore under no circumstances could I make light of any past events! What I have suggested is that we leave most of the events to there particular time of relevance and influences bound by the circumstances to their making, even if only understood by the limits of our general knowledge at the time? By working in that vein I have tried and allow for our differences then ago and now, even if to have changed by other means or been reflected upon by other deals read from different angles? That proposal is to be our benchmark for change without alteration allowing us to change future expectations without changing the past, unless necessary!

By using only some of what I personally have found to be of specific interest, my course is to include us all because of the actual prevailing circumstances. Even that our political influences are huge, without balance, I have given that it is to be how we challenge our religious acceptances first, having been generated from influences from the Middle East over the past ten thousand years via Mesopotamia! Then directly over the past four thousand years by direction of our first, second and third dimension religions and how influences also came from our almost earlier monotheistic religion practiced by the Egyptians?

To have stated the importance of the last ten thousand years and then jump to the exceedingly important period of time in global terms of the past sixty plus years, is not to put a block on how we individually act or react? What my objectives are is for us to place our own importance upon each other, but by reflective tones. Meaning that from now when we look at how we think things should be, we are to fully consider the effects upon others because of how the reflective medium works. Also that by our actions even if born from necessity from the thinking of today, we have to meaningfully show general consideration of how others think, but not by or through the false medium of tolerance by any degree? Then again the exact same applies from those that might be first to react, assuming direction one way or the other was planted against a particular group or another type of person as we?

Our intentions are to show that from the background of all histories, our first steps were taken some three and a half million years ago. When we allow for some of the pointed references I have made, the only way is forward, but by my own warning, we are to move so in a wide line as before, but from narrower perspectives. That in part is why I have pointed to this as our best modern era and by the time reference to be about sixty plus years long. With the further overlay by other leads, that very time span of sixty plus years, accounts for the amount of our differences accumulated over these past four thousand years by civil leads from religious leads and finally from political leads?

By using the Middle East is to take visual advantage of how we still generally differ by our political and religious standing. Although I use the area with many of its own historical references, I do not take full advantage to criticise any specific group. Simply because of how by its own introduction of our three dimensional religions was by the same people to be you and I, even through others as we have not made that connection! Perhaps because we are laced with different natural attitudes, although driven the same considering our true link from Shanidar, we seem to want to be different in achieving the same ends, even that we roam the whole world, all of us reflect to the Middle East?

To turn heads I have not been sensational, all I do is take some recent events and personally reflect on there effect in cost to this single force of humanity that we are. From much of what has been written, even if only fleeting, I have harshly set out to clear our path for further travel. My one case to use the modern idiom is to show more easily that we have not checked our instrument panel often enough on route. Although many have reset our course, what has been done has not failed, but allowed us reach where we are or might be by our stated acceptances. Meaning in translation, because at specific times when we did not have full steerage we inevitably made our move, but half measured, being no bad thing? Unfortunately with that style of our half measured approach working by degrees, by naturally inducing our steps taken syndrome, we have been content to proceed even if appearing to be laid back by character?

Our failing and what could ultimately be our downfall in relation to conduct, stands to the full measure of what steps taken syndromes are to bring forth under other similar circumstances and from other perspectives, like those created actively now in the Middle East! Especially when we still use half measured reasoning and our steps taken to ably promote the pastime of international war, and have done so, virtually every one of these sixty plus years of this last modern era?

From all that we may yet do and have done and are prepared to do, on my own account under the veil of our abstract reasoning, I have been lucky enough to identify two comparatively small reasons for the open adherence to the full belief for some, that we are still more different than we are alike? By that self proclaimed observation I give notice now that the vital truth is that we are one! Therefore it is to be determined on our behalf that the case of proof is that we are the same, but vitally different only because of the fact that we alone are the only creatures of existence to have a future above and beyond any inclination of all other creatures past and present. We were never appointed, never directed, of which we only realised it so, and it is for us to show self awareness, that by all combinations it is for us to spread our willingness to prove this one group worthy of Gods notice!?

I accept that any observation made by any of us is open to various interpretations, but will not argue the notion that we as specie are different other that by the whims of geography, climate, diet and other factors. Black, white, yellow and brown of skin are simply conditions of the above, being progressive or laid back religiously and or politically are of the same stamp. Not to put too fine a point on it by being repetitive, our open likenesses are generally too similar religiously to doubt our difference in comparison to other creatures? All call for the perspective or a different condition for the aid of our own ability to eventually be in the realm of understanding, far and above the best theories of the fourth dimension? Even that I think of us as one and always will, there are those amongst us who will be forever not be able and grasp the concept of our singularity, perhaps those who are ill mentally or those who might only see the advantage of personal gain as a driving force. Nevertheless we are one of kind and can deal with our own split ends as necessary without disrupting our own good grace, perhaps by readjustment at this precise time of now?

## CHAPTER 138

From the two small differences I have brought to our notice between all, without direct association, because of the range of our wonder and query, these observations will help us notice some of the actions and or reactions we take. Not to be final or the only two values by my interpretation we should study, but from how I have been able to look at our united histories unconnected? Along with our natural developments un-paced or un-compared, has allowed me continually see more of our likeness than our differences, balanced by the circumstances of geography, climate, diet and other similar features to have automatically influenced us one way or the other, but all exactly the same!

Not now looking by those effects, but to the fact we are black of skin mostly in Africa, that we are brown of skin mostly in central Asia, that we are yellow of skin more from the Eastern region of Asia also. Taken that we are white of skin in Scandinavia then percolating southward to the rest of Europe bears no separate value to any of us. My connection is simply that over longish period of times our standing and status had developed differently but from the same core. That particular connection from the white perspective I have personally made, is to illuminate one of the processes of change stimulated by human query as said? Illustrated the past four hundred years by our own migration and linked colonisation of the new world continents of North and South America where all people mix 'Out of Africa'?

From the period of time being only four hundred years young, not enough time had elapsed for us to naturally develop the best required texture of colour or correct attitude supplied by much longer periods of time elsewhere. No new specie was created in the Americas to be predominantly white of skin or progressive by nature, who were mainly Christian by religion? All that was seen was self expansion owing to us all of this single black, brown, yellow and white skinned specie moving by degrees in what we have naturally perceived to be forward in a progressive loop?

Unfortunately four hundred years ago, ideas and ideals were not open to the mature and normal reasoning that we are now happy to use, therefore by the additional coincidence that much of our scientific reasoning was given to expose many of the mysteries also coming from America by connection with Europe, later endorsed by some new areas of thinking from Darwin's publication? American and Europeans were connected with the latest form of social and or scientific reasoning, placing both to be in lead when at the closing of hostilities in nineteen forty five, they, but America first, set certain standards even if misaligned, but for our own best and equal welfare? Translated in some cases by the introduction of our United Nations who by association took an ambiguous lead from appointment in representation of us all even that some of us were defined to be second or third world citizens? Although by population most people of Europe and America were white of skin and generally Christian of religion and progressive in the business come scientific sense? Little adjustments that were needed to be made for our collective welfare were left in abeyance until a new collective was brought to the fore like the Fifth Dimension!

In the mean time, without complimentary criticism, when I had precariously stated that we should all accept or even strive to connect with the American perspective is exactly what has been happening in any case? All over the world people love to hate the American prospect of her own dream, but do all they can to copy and in cases copy her out of blind envy? For me to indicate that America had achieved our ultimate goal politically by how she has translated want into the desire of delivery representatively. Is not a claim of or any form of condemnation of others to a different political persuasion, nor others of a different religious persuasion? My only case is for us to express the acceptance that as a single human unit by natural development, our best achievements are to the American example, in proof? Because of the way some of us have reacted to that situation, we have only managed to regress the general standard of human development by promoting further half measured solutions, even if religiously delivered by the basics of fundamentalism and from where and to what end except envy!?

Without excessive desire, from where I stand, all I continually see are footsteps in volcanic ash from three and a half million years ago. My next memory cementing that image is a picture of one of our number being laid to rest at the unnamed place of Shanidar some sixty to one hundred thousand years ago. Even with those two images my next view is, and I make this statement without the fear of successful contradiction, because the range of some communal events to have occurred to us all, certainly marked by the closing of hostilities world wide at the end of the Second World War! Is through our lifecycle now known to be relevant to the one aspect of purpose, illuminating the full feature of creation above all else! Even when viewed reflectively the factor of that timescale cannot be balanced by or through any indicators from the sciences. Not least when most features to that name need uncounted years before they even come into the picture of study, like for example the decay or half life period involved for some of our radio active elements. Making the case to the obvious difference between evolution and creation once again?

Although I have continually referred to our individual lifecycles, by pointing in the direction that we have the inbuilt ability to naturally change by attitude over time. Effectively extends our time here on this planet to be of universal significance, not least that unlike the abilities of the sciences and other creatures, we alone are unique? Where I have drawn lines of contention has been to firstly maintain the aspect of our difference to others. Then extend by experiment if necessary the full and total range of how we developed our separate attitudes if that was included to set us upon the path we have representatively taken for and to our separate religions.

When viewed from or through that perspective, the picture shown is complete, but has to include the unbelievable to form that complete image? From where I had stated that science in all cases by formula is a matter of a given number of circumstances under a given number of circumstances reacting to give a set number of solutions, remains finite. Although we are still in the processes of study to the final solution of that or those configurations knowing only some of the answers, our expectations are based on credible assumption? Whereby to facilitate the picture to know all, we have created the possible means to solutions by beginning to imagine again, that if we invent further dimensions we will know all?

Not having realised under religious interpretations, by coming to use the dimensional medium of death, more vivid answers will be put upon us all. From where science has readily grown these past four hundred years forming and finding better means of understanding and enlightenment for many of us if required. From that impetus, for some of us, our impetus; we have missed the total connection with human death first placed some sixty to one hundred thousand years ago! Along with misplacing the standards of how we should reason the unreasonable, yet through science we have now created the unbelievable by turning those assumptions of wanting to know into the belief of being able to know by our own study? Almost putting most of us generally to the expectation of waiting to be told by the priesthood of the fourth dimension, preparing to lay the whispered laws of study before us all in due course, but having to rely on the aspect of evolution only so that we would understand all!?

From the above I do not propose to enter in argument on whether the principles of extra dimensions should be weighed against the addition to knowledge from human death, because we are all here and now living within the bounds of our single lifecycles, and for what? Although linked in style to the span of life with all the creatures of evolution held by the rules of pure instinct, our lifecycle has more to contend with beyond their single link to the ideology of procreation. Within our meagre period of time, we have to contend at all levels with either political or religious influences one way or the other, that is part of being human, to be involved with aspects or life outside of the element of pure need. Even if that drive alone draws much of our concerns and living energy, we are each held to the feature of control or guidance from our political or religious interests. It is only from huge international events such as the Second World War that we have been able and focus upon our singularity, because while it was going on in violent disarray. The best outfall in all cases was to our full and linked advantage by looking for the first real time to our singularity, but not yet realised, because then we had winners and losers?

From the habit of reconnection to some of our best or worst past events, will prove to be the most open way we can all progress, even if seen not to be required by some of our thinking from some unless to be misaligned of mind by circumstances? When we are able to level our standards by inclusion, even if only in overlay by our Fifth Dimensional prospects upon our religions and politics, the clicking of the light switch will bring some of us to eureka? Simply by endorsing what many of us had already thought about, even if we included the full history of our religion being more relevant than the image of our own lifecycles. It is only from now and the near future we will come to a better understanding of ourselves, by being aware that in link to evolutionary forces, we are no different to other creatures save by expectation?

Although I had previously broached the subject again and again with my own particular representative interest to the mighty crocodile. My comparisons to them allows that over their connected two hundred and fifty million years of existence, even while little changing like us through the same conditions of geography and climate. Their standing was not to change, because from amongst how alteration ensued, they still carried common ground by their singular drive to procreate only to the requirement of instinct. We on the other hand, from who we are, have managed to alter the value of our three and a half million years of continual changing to hold reference to the importance of purpose above the future. Understood from within these short years of our own lifecycles?

# CHAPTER 139

Two small points I have made, and will make again, are to be our pivot into the future, because even if from different perspectives, they have emerged because of these recent sixty plus years of this last modern era. By whatever reasoning I use, the following suggestions are our product, which is not to place any significance upon them until they are questioned? Deep into this our work I had given point to the idea that we in various ways, even of desire, have been constantly on the path to emulate Americanism, leastways politically. Because on that level only we have been able and imagine the aspect of the American dream, representing their general national goal in the pursuit of materialism and for good measure, notoriety! That or those aimed for goals, being first or a winner, as opposed to only taking part, which without the critique, allows for developed character traits being always forward as opposed to be laid back, which is not a judgement?

Without to much dissection of that or my view of a particular theory, my indication is purely to show in limit how we have progressed politically, not to betray our other emotionally felt and used connection from our different religions? Without comparison, when at first and then through the merging of time, Roman Catholic to be catholic Christian colonists went to the new world. We carried the same progressive loop that had started in the Middle East any time before and to include how we began to progress from our interactions in Mesopotamia! Even when our first progress was recorded to the adulation of our many Gods, becoming to the eventual realisation of one God, our path was still not quite clear?

By the means of succession to the fable of Egyptian travellers, which has been given a ready significance by my own thinking, became lost by some naturally progressive steps we were set to take? If what I had suggested by my own delivery of thought about regal Egyptian migration having occurred for any reason, like to travel in a party to the unknown new world pre Columbus? Then the balance of the event is or may be set from some unfinished conclusions that I had drawn and given notice of.

By comparison I had stated that if and when an Egyptian message was carried to the new world, reactions by any time scale could only have been made between the peoples then involved, and at or from the level of the thinking then ago? So if the general standard of settlement was to the edifice of harmony by acceptance to the inclusion of purpose attributed to the Great Spirit. Other ideas were to remain stagnant if there alignment was of a totally different complexion. Even if new advice was to find its way to the central Americas first, any mixing of ideas would be incompatible without the catalyst of further opinions that might have occurred in the Middle East at about the same time?

Conversely by using the Middle East model composed of many cultures using there own separate Gods of everything, because that was the multiplex way things had evolved. Even if at the time of the Roman conquest when they brought their own Gods to be amongst similar cultures and now including one to be monotheistic by choice, when and if by connection a second monotheistic creed was pronounced, over suitable periods of time and from time lapse occurrences. These two monotheistic religions even if carried by the latter form first, when bringing influences upon the Romans, spread the effect across Europe expansively. Being robust enough to absorbed change by intermixing with other religious Gods, but remaining to be monotheistic and at the same time progressive, but soon to be tied by tradition not to focus on the life span of the individual being important because imminent was late!

Overall the success of the Roman pattern was not even challenged, when the last connected monotheistic religion of Islam came into view in the same general area where Abraham, Moses and Jesus had felt Gods desire? On the one comparative level with what could have been happening in the Americas with the inclination of a no need to change standard. From the Middle East the growing idea of change without alteration was now/then delivered by Mohammed with directed and connected fortitude within at first his/our area of vital religious influences. Although if to be viewed that this new religion was the final step that we all should take or at least by presentation we were all to come to. If it was to miss one vital connection it was to the realisation that we were required to change for acceptance. Being no bad choice when the open intention was that we were to improve, but also making that we should step in two directions at once? Making the step, even though we could openly agree to what must be or should be by intention, was to be a rejection of our own pasts without direction. Making that although Islam was to be our final offering from God to man, it was done in case, that we would and must come to the religion in any case, again being no bad thing when counted it was for all?

Without connection by some of the descriptions laid down by myself through these pages, what could not have been addressed at the time of Mohammed was of our actual conduct as opposed to our expected behaviour. Having already come to and reacted with two similar religions of the same area and intensity, Islam by offer was a decent and honourable task, far in excess to some of the other means of thinking at the time. What was to take the wind from its sails to eventually check its rapid growth, was the unrealised opinion that the earlier religions had inevitably come to be involved with the total aspect of living, also before 'Natural Body Death'? Not least delivered by the political management of many different peoples of the new scale through at first the Roman Empire and now national groups, even that many held to basic Catholic beliefs set down by the Holy Roman Empire. On the decay of that first form of collective political rule, with self determination and nationhood calling there was the progressive lead for peoples to follow there own religions and developing political systems. Leaving it so that to some of us by our own set standards this new religion of Islam was to be opposed, because at first theme it would upset the corners of our/my developing views politically along with our held religious views in any case?

For new ears in the Middle East, those who had not yet set choice or inclination of what was happening in Europe, choices were far more singular. Therefore when taken, we could hold better to the new faith, because it presented itself to be the final choice and in being final, accepting the processes of politics already having induced a laid back outlook towards the political medium? Because the feeling of fulfilment sat well through the new faith, its original form rang well above the limitations of any political system then proclaimed. By reflective tones Islam having set our methods of conduct by edict, which was a better defined means of delivery than had been used before, although well connected with the past, made a less obvious case for mistaken identity than had the first two religions of area? But could not then or now account for what would actually happen if and when the mediums of religion and politics were ever to meet head on; which is now?

By taking it we are the one progressive specie, no matter what has been endorsed by our political systems culminating in the word democracy, I make no claim to the level or standard of meaning to that single word, save that it can be representative to us all. I make no claim to the standard I have used to express now, when by collective advancement we do strive to meet the American dream if only for material gain right or wrong? I make no claim to the merits of such a suggestion that I had referred to previously. Save that my direct inclusion of the theory of how America grew, is one side of the scales that we can hope to balance at this juncture of this last modern era?

Almost joined in fair union and close unity, when I put it so that America and Europe had similar outlooks further endorsed by their Christian religions and continuing progressive nature politically. My case was to express one of our conditions today, but only politically, because by the range of that progressive path it has been our standard to follow by connection progression, expressed scientifically only! By exposing any of our mysteries through experiment, our fourth dimensional lead in part has induced us to call a check on open religious acceptance by opening the door of religious tolerance? By using such a medium as a mwans to express our religious ideals and not listen when coated with any sort of political comment one way or the other. Has not been an open method which we can communicate by even politically, because effectively we are opposing our own wishes and desires by using our own wishes and desires in one medium and openly rejecting them in another! Making my case and our desire that we need to review our own conduct, but without criticising intention, only reasoning!

## CHAPTER 140

If religious tolerance from being an imaginative forward steps taken syndrome idea; began because of such matters of the Catholic/Christian first referendum some four hundred years ago. It was done in effort to allow differing religious opinions interact without the direct prospect of conflict between basic opinions of the same stamp, quite separate to wider religious differences, like what was met in the new world roughly the same four hundred years ago! Later limitations of the idea and the strength of our religious fortitude were compromised without notice by the political overlay we had allowed envelope our general thinking? Because it as an entity began to offer more and more to the contribution of self wealth even if only in the hands of the few. Making the matter so, that in cases some of our politicians came to act as our mentors, like others of the priesthood by connection with some of our religions? From the same feature almost without notice the final switch was to take place when about two hundred years ago European influences came/went back to the Middle East, notably Egypt as a start?

Two hundred years ago the standard of conflict between the European Christian churches and Islam was never rekindled like in the days of crusade or jihad many centuries before. Without consensus, Europeans forced there own political agenda expecting compliance as was the norm from the European perspective then ago. By the way I see things, that we can mix our histories without alteration can carry the fact, that opinion then and now can change by us, but we must not make the mistake of likening our views of now to our opinions of then? Quite simply from just before Darwin's publication, our world standing by most instances was of political domination one to the other. Afterwards, although much by interpretation was given to our spiritual awareness or the making of it, because some of the theories associated to evolution. Even while putting the fight forward for creation and God, we allowed ourselves think in terms of Gods appointment to have been meant for us, for us both politically and religiously connected by the means of natural selection!

Alas by our own good offices of imagination, some of us positioned so, connected with the idea of natural selection being selective not least by those of us first to draw such conclusions. That done the die was cast, so that in turning to other ideas there was no turning back, because of the general disarray exhibited by some religious standards being left to there own argument, bigger cases went unseen? In quick succession some of us also attributed the clause of natural selection by direct association to the leading elements of society, whereas Europe over the past two hundred years was still fast developing in the name of political ideologies. Our steps taken were to assume politics was more inclined to be a measure of our natural selection by evolution than religious hopes of creation! Again stacking the odds against our best forms of reconciliation even in nineteen forty five when we tried if a little jaded to make amends on all accounts?

Although I had made much or the Arab/Israeli conflicts reaching a point of impasse through the one sided approach to Israeli independence. My view to be expressed without lead has been to show the case of how the spirit of conflict was nullified half measured, for we were not of the ability to manage all of our interactions then or at that specific time? From where I had honestly stated that my own valued recollection was true to the facts that the Israeli case was supported purely politically by Europe and America. No realisation could have been given to the Arab element in being first of Islam, because the signs of claimed nationality shown then were expressed to or from a different perspective than the assumed western plan with European/American input half measured?

Allowing then, that the country of Israel could be fully supported on religious grounds politically, when the only signs given off by the Jewish people in most, was being inclined to western and or American/European thinking, whereby there alignment was therefore much the same? By that scope and to the imagination of some western thinking, it was to be right and correct for the religiously inclined Jewish people, set up a Jewish state from ancient historical reasoning, but only to be sanctioned politically? Although the above in short delivery indicates how our steps taken syndrome can set false prophesies, even if only allowing that religious desires can override political sentiment one sided? Our conclusions from such matters by reaction are not entitled to be handled or dealt with from the reflective image of looking into the mirror of the past, from where the picture is given off, though real, it is not a true reflection? Leastways if for one side religious settlement is encouraged by religious tolerance and the other is only to be considered to be of political significance?

Because of such happenings to have occurred even by reflection were to draw reaction to half measured decisions, what form could we expect those to take in the modern field even invented? For my self to have referred to outcome under different names was to set the standard in no uncertain terms of the open difference between the methods of politics and our styles of religion. Unconnected, I had been drawn to the course of events emanating first from the Middle East covering these past sixty plus years by the happenings there in this last modern era? I had well documented the representative position of America by political involvement if so in balance to the religion of Islam? My connection to the pairing is purely sympathetic to our own delivery, because both features are deeply connected to us all one way or the other religiously or politically!?

Having mentioned time and again by this exact time of now, in our last modern era, the two vital confrontational aspects of world society extend from events in the Middle East, which represents the total expanse of our differences elsewhere throughout the world. Because from there it represents the disconnection we have made between our political standing on one side our religious standings on the other and one against the other by specifics. By expressing that void in terms of how social American development wholly relies on political aspects of representation with our religious representation, having been adopted to control others becoming fundamental again if mainly expressed through the religion of Islam? Presents us with our major challenge of this last modern era covering the whole world, because within, we either operate by politics or religions or both or sometimes by science only, but not in the full understanding of their relationship to each other yet?

If on one side our representative America has taken only four hundred years of existence by our new world image then any improvements made, represent us all by some circumstances. Because once an act or deed done, once an achievement made, all causes are ours by connection. Not least because when the America new world was created, all of our histories were taken there, having reacted together, already through all the mixing that had continually occurred from any time since Mesopotamia at least. America as a nation now was us then, and will remains so in conjunction ongoing, simply as proof to the fact of our mixing. By just the same coincidence in happening that her/our selection to political control leans to democracy, republicanism and monetarism is of the same value! Even that she is closely connected to Christian religions in the most part and in our representation, the picture changes by means of true definition, because in counter she/we have opted for the aspect of religious tolerance to allow the separation of religions and politics when cross dealing?

By direct balance if only reflectively talking, from my personal choice of pairing American/Europeanism to or against Islamic fundamentalism, can only emerge from my own standard of how I take and view matters through the abstract, but in all of our cause. By first mention to Palestinian reaction in or from or at about the time of the mid nineteen sixties, I had pushed my claim that their demeanour was close to some of the end philosophy of Islam, to die in good and open cause for cause. I had suggested that if any person might so feel committed to such an endeavour, that self in terms of sacrifice, if felt that was the only option to balance the emotions of discontent, was only apply to apply to the individual? From that I make no case to the theory that a man has no greater love than he will lay down his life for his political or religious beliefs!

What was conducted by some Palestinian movements, was to bluster there own opinions for the same sense of justice administered to others if by political arrangement only! By the open fact that virtually all Palestinians in our representation were to the Islamic faith was to bear heavy on how we all were to proceed in future and further conduct? By unfortunate unavoidable coincidences that Islam covered vast expanses of the Middle East with the sheer volume of its adherents. Islam under the so acclaimed concept of tolerant behaviour had no single voice to express its will other than by using some or one or any of its defined means of defence or protection to the faith if threatened! Palestinian's who sought such actions in reverse to other political decisions made around there own desires, might have fell to the sword in consequence of there own responsibility, which on the broad scale is to be a loss to us all. Not least because even at the time of measure and countermeasure we have allowed our political decisions rule by consent ahead of religious tolerance!

Without papering over the cruel and sad measure of events in the Middle East in this last modern era, Islam was and is and has been a voice directed or taken as a source of expression or commitment by driven circumstances? Our heading and balance have been put under review by my own expressions of study based in the abstract but openly aware of real features. Having used the idea that forms of self sacrifice in comparison to acts of defiance, had been done by some Palestinian politically motivated groups? Over close and short periods of time the standard of those actions if charged by intensity and in name, were directed against the same enemy, or in direct assault on Americanism and or westernisation? Most moves in that field were valued but uncounted and unconnected with the human cost on both sides being an open result of our steps taken syndrome?

Unfortunately by no more than the emotions of fear, jealousy, greed and the open spirit of half measured thinking, we had allowed ourselves react on the same levels reflectively, assuming to allude to the same outcome? With one side by my classification in representation to be Americanism, prepared to wantonly kill of self by self to save opinion only. With the other side now to be defined Islamic fundamentalists, additionally wantonly killing self by self through the first act of suicide, from which some thinking was that a single life given as self, nullifies the slaughter of the innocence! Keeping the picture that both opposing sides leave in aftermath the odour of death, just to balance opinions on political assumption and religious settlement without conclusion!

Without connection of deed, from the above the only lesson we can learn and have been prepared to learn is to perform better and finer and more devious means of the same conduct? My own voice on this matter caries our value that all deeds done are a matter of all of our involvement. I know that the vast majority of people feel desperate by such situations I have pointedly described. Not least millions upon millions of American/European Christian political thinkers, coupled with millions and millions of multi national believers in Islam. Who in holding the open power to defend, protect and promote there faith by forceful and violent means if necessary, do not, nor ever would feel the need to be drawn to that end under any circumstances, not least because of the very depth of the religions code!

From the above and through these pages, the conclusions I have drawn are made and tempered with my own emotions. Having watched some of the shocking events to have taken place over these last six decades plus, any conclusions I have made still fall within our scope of understanding. Whereby from our ability to be only spectators to many a sad and cruel occurrence by any scale, allows that the deeds done however bad and from whatever direction include us all by connection. Not least because of the actual extent of our own lifecycles in common for common achievement. Our lives have been automatically drawn to be of the same worth in theory, which is the total plus of our best efforts over this dark period and which should not be forgotten!

Unfortunately without the spirit of personal or international agreement we have seen and acted upon our ability to disagree. Not least perhaps how by general conduct unseen, our style quite naturally has been to act from our own positions of comfort. Making the exact case that although we are the same, our differences sculptured by time and effort aided by the circumstances of climate, geography, diet and other features have made us who we are? Sculptured by time and effort might be a good point of reference we can use in self study and examination, because realisation of that maxim is a bridge from the past until now. From where it cannot be given that as it is now recognised we are all the same as I have laboured the point in title to our specie singularity. Although we have been made different enough physically to approve that we can be different emotionally expressed by our personal traits in being laid back or prominent?

It is only when we look in earnest realisation by reflective means, that we will be able and see there but for fortune is you or I. Accepting that the mirror of circumstances to have given us our separate tasks of which by this last modern era we resolve to make the same assumptions from those different perspectives.

Even though in taking to the single standard that Americanism and Islamic fundamentalism are the main features of our differences, which are not to remain eternal. My case for change is rooted in the very recent past without any deliberate changes being levied on the structure of our separate histories, because on day one three and a half million years ago, we first stood on our own two feet?

## CHAPTER 141

What to do, what to do, what to do, is not the name of a new song it is an open invitation for us all to be better involved in the future? From where I had seemed to simplify any style of our varied means of conflict and dispute any where in the world, by naming only two features in confrontation representing of us all, stands on all counts? Not least because of their structure and our own relationship with being political, counted as Americanism and religious to Islamic fundamentalism? I have said enough for us to be assured that both terms are representative, even knowing that we share a multitude of political and religious opinions!

By means of natural selection in all cases and for coming to some agreement to my own definition of a defined reactive split between our opinions? It will be necessary for us to consider many of the attributes we have designed in order to bolster our own opinion which inexorably overlap and intertwine ever? Americanism as a concept, or even a reason, has been directed by my own determining, so that the points I make are relative to us all. With the religion of Islam, my standard is somewhat different, because from where it touches our lives by the aspect of natural living, it like all religions has a more powerful force yet to be revealed. And until when, can only be put to our own imagination to the best means of what our conduct should be? On that subject of religions and from the depth of our studies into all aspects, our crossover is better defined than is any of our political strategies.

Therefore with that principle, any person such as I, might bear our own opinions to the best effect we can put on any of our religious hopes, therefore by my own reasoning on that plane alone, by using all of our desires or hope in purpose to the aspect of God and the sphere of creation. We can question any and all concepts by self and community connection to the open acceptance of how we alone as specie might redress the theory of time by evolution and creation? In making ready to be worthy to receive God's full intention using the concept of politics in aiding us to the better choice of religion and creation, separate to the attitude of religious tolerance by our own human terms through politics?

I have always made the case for our religions while giving fair and just recognition to the full wealth of our political advantages. How I had come to generally view both naturally, I have taken first sight to our religions as have we all? Leastways in comparison of how we have always been prepared to kill of and die by our own hand only in gods name, which is not a judgemental statement. How such a feature came about has the same roots in all histories, not least by imagination when looking at the timescale of human development between footsteps cast in stone from volcanic ash and a tribal burial at Shanidar.

Between the two we have found evidence of tool use even if only to have been flint knaps, nevertheless worked to a deliberate form beyond the realm of nature. So begins the food of politics from the creation of material want, even to the time when we began to distinguish the difference between instinct and natural progression by natural selection, civil order had come to be a basis of the way we could advance? Even, and my own suggestion about our burial at Shanidar being our first conceptual religious step, stands that the actual event took a deal of civil organisation, allowing notice that we had come to better organisation than any other specie!

Making the case again for politics and religion, by giving in review some if our recollections, I have set the standard of how politics had come to grow out of our shear need of better self control? So much so that by today it wears its own coat flooded in memories unconnected to reaction. If I had indicated in part that a primary result of our political desires has led us by uncounted circumstances to hail Americanism as the zenith of our political achievement so far! Our connections to that feature stem from Africa, Asia and China, Europe and the Middle East, bringing our own lead forward right or wrong. By the same count we began to make religious progress by answer to the search of finding our mysteries, progress being written all over by us, was still unable to explain purpose? From where was made our first split or parting, between the concept of civil control turned to politics and the way we approached religion through God because we could now see the difference?

Although I had given the idea of Americanism as one of our main objectives defining the limit of progress in human obtainment, the idea on our behalf and by our own and equal input has fallen short by allowing its main force, lessen our religious concepts? Allowing our religions the same status, when clearly we should not, culminating in our perception of some religious beliefs being different to others. Leaving in case that if any said religion was to hold in reflection to the American political concept without using the considered religious grounds of tolerance, then they as an entity were to be kept out of the forum of change by referral and through our steps taken syndrome?

Taken in concept, if because of some open political situations, for instance in the Middle East, were to be offered to counter a reflection seen, then by paradox the second form of reflection inducing reaction becomes the norm. But only by classifying the opposing element to a different title or name for convenience through assumption, which is not to say directly that Islamic fundamentalism is a product of American indifference! When America on all of our behalf claims that some matters of religious concern fall to be of the political significance of terrorism! Not uncommon to many other events throughout the world from when ever, but from this time, stimulated by the haste and pace of purely political change, even if the overall intention in all cases was for our general improvement claimed, but only for expediency even if to overshadow religious emotion?

By again referring to the Middle East and what might be the route of many of our world problems? Without acceptance but by way of recognition, when dealing with any form of discontent or opposition, we have only been disposed to operate from a weakened political basis. Even from when I personally gave reference to what or how the distorted principle of religious suicide came to the fore first hand highlighted by actions in the Iran/Iraq war in the second period of these past sixty plus years. Has to be taken as an exercise of self discovery so that we might balance some of our personal views?

Personal actions of suicide were conducted in answer to the benevolence of the same religion, but erringly viewed from different political objectives, not least enforced by opinions shared and spread by wider political interests from whomsoever! When any final act of self sacrifice was laid in the religious name of martyrdom, both antagonists supported the same standards politically first, but also by tolerant religious means of justification, self perpetuating the need of change if in the course of reformation! Because both were from the same religious base actively, but preferred to act in the unclear cause driven by unclear political motives taking much of the supposed virtue away from the commitment of self sacrifice, because now the acts belied the same religion. Although not unlike earlier episodes from other religions of area, but now/then they were conducted in this last modern era held under the rapidly changing ideas of conduct offered half measured!

By open translation without comparison, under changing situations that were and are rife throughout the world today, from using any religious standard to make a political point will always fail. Not least because when a religion looks into a mirror the reflection is always the same but religiously different seen by the same followers, but from different time perspectives? When a political system looks into a mirror the reflection is always the same, but ever changing when the tentacles of political means always give off a multitude of different inclined responses reflectively! Leaving it so that in part the two issues are not at present compatible, because from the same mirror, politics will never reflect religion or its concepts, nor will religion ever reflect politics, making no case for religious tolerance politically! Making the next case for the avenue of change which cannot be entered, until we first acclimatise ourselves with the reasons for change and the best means to carry them out for the future, without altering the past!

In counting on the future for reference we have to view how we would like it to be before we understand what it is going to be like? By our already accepted standards multiplied by changing opinions in these recent times, from being aware of what we want, we now must look to define our aims more clearly. While at the same time being fully aware that the conditions of our successes are determined by the criteria with which we began. At the same time in looking into the intention of further advancing by way of following our natural trait to progress, by taking to the acceptance that for all deeds our politics is in lead, has to change, but only by delivery? How first signs are to be given have fortuitously always been with us since the times we first controlled fire. From that act beyond notice would have been set some of our first levels of civil order leading to the necessity of control. Of which the stories to be told in support to that possibility can only be colourful and convincing. Like when done to the cause of clan or tribal organisation with the next memory belonging to Shanidar when we were really on our way?

Having already indicated almost by decree that some forms of political order expressed as civil order, means our civil control did exist from any of our earlier memories or before? It is no wonder when we realised that driven by our own progressive character some of what we did not understand, some of the void we encountered in the name of human vanity, was attributed to being held by the keeper of our mysteries? From which was made the first mark to the partial acceptance of society having two forms of control and or mutual leadership, now, determined to be what we hold through our civil control as politics and the prospect of mysteries as our religions? Even though by that time I have given a deal of importance to the second of our developed traits, that of our Gods and our religions, delivered first some four thousand years ago? Although by inclination they have become subservient by having formed priesthoods to cope with controlling religion within a structured political society, already that format has remained unchanged in general? Even if understanding how politics was developed and religions were appointed, to be a feature which even up until today we have not been able to fully quantify, being a feature from today, that we can begin to quantify that standing by accepting how, what is where and why?

Looking to understand our first means of open control, working from the assumption that in name we are a civilised society held against all other creatures, who because of their natural instinct, have not the ability to self determine. Hence, if we can determine for other creatures, we can make judgements for ourselves, even if to set, our own civilised standing, coded to be first, second or third world? Yet further endorsed by bigger gaps in nullifying that accepted concept through our secondary standing of proclaiming human rights as a right to all people as people? Then by rubber stamping that shallow concept, in treating all by the same standard of being politically correct in our dealings with each other, along with us all being religiously tolerant on a different scale? Endorsed by a plethora of social, racial, ethnic and general international laws endorsing our singularity by the exactitude of our differences, allowing even beyond intention that in order that we might voice opinion. Our acceptance of any of the above approves that we can kill of self by self through the avenues of our collective political reasoning. Or that we might kill self by self under the badge of holding in shield our blessed religious arms, with the difference of only one kind of killing having been anointed by the one and same God!?

# CHAPTER 142

By using any word in an effort of definition to cover the range of our obtainments, like civilisation, we are fully obliged to make its description an objective within the span of an individual lifecycle covering the period of fifty plus or seventy plus years. That is essential so that we might claim its actual value by description when we obtain it to our own worth, especially if the word I would like to raise in question is as heavy in claim to us being civilised to any standard? No voice is to be obliged and classify any group or principle to be of that specific force without setting standards, even if in full and fair representation for us all through such bodies as the United Nations.

I make the case not from my own generalisations but in conjunction with our collective thoughts, even if at this time un-shared? In doing so, there can never be a claim by such as myself or others of similar sentiments to be anything other than one of group. At the very least taken under all circumstances one of group is openly translated to one of specie, even if only connected by times outside of our normal lifecycles through footsteps left in volcanic ash three and a half million years ago. Having made that connection in cases throughout these pages, by turning in reference that we might not yet be civilised is done as a means for us to better reflect, not on the total times involved, only these past sixty plus years of this last modern era?

My own case for review of deep concerns about being civilised might begin and end in the Middle East, starting through how we acted at Shanidar, but is carried on at first from the land between two rivers when by comparison we could make such early judgement, but in one case only? For that is how the title works, simply if there is civilisation in claim in a case or in cases, then equally there is not? Therefore when dealing with people from other tribes; and by the fact that we did primarily under political influences, each group bore the burden of being civilised without consent, which is how things remain today. Although we had tried for the first time since anytime to make proper generalisations over these past sixty plus years of this last modern era. Until we can decipher the standard of reflective intercourse, our own conduct will automatically mar our view clouding our overview. Which has allowed the situations develop as they have over this same period of time, whereby on every continent subdivided within every separate country the same civilised groups tore and scared and murdered self by self. Even that in cases much of the motivation was religiously directed or irreligiously directed, but all deeds done were by the same people all civilised? While the rest of us who coincidentally might have felt more civilised, watched or governed events in the name of market forces even religiously?

Civilised we are not, but must be, although we are aware of our own standards, we have allowed these prevailing situation arise because by our assumed and dedicated existing behaviour, when spoken, we each know what we mean by our own inclination? Even that the picture is fully clear in looking at it reflectively, better judgements have to be made to determine what those reflections really mean.

Knowing now that when the right hand is held up the reflection is of the left hand being held up, is not sufficient to assist us to change course. With now being the time for the acceptance of the blind force of the Fifth Dimension, we can all have a better and equal base to view new horizons. Bearing in mind that by its delivery of three taken prospects, we can all reflect to become of the same standard, even covering our accepted differences? Taking the result of the overlay of secular thoughts and our spiritual thoughts with our own after mix from how we are to use the active prospect 'One for the Whole' in all forms of query?

Bearing in mind that our continuing monitor of any change will be done in the matter of change without alteration, so that we can realise what the basics are and for what reason we are involved in this self study? Not least by how the effect of the overlay will automatically spur in us the ability to understand the basis of specie singularity on one count by human natural progression! Supported in the utmost of our ability to understand the possible concept of human continuance beyond animals, who to the one product of procreation, are driven by the wheel of pure instinct. We by accepting that we are one, but having been dispersed by the conditions that geography, climate, diet have put upon us, will remain the one specie we are!

Knowing that some changes within our own attitudes are vital as a reasonable point for us to make a start, we will have to emphasise that there is no standard to how we are required to meet the aspect of driven change, only that it is necessary. Change for change sake will not work nor would I expect it to. Where I lay my hat has not been to the open necessity of overt change, but well expressed in the need of it without designed alteration, it has been from that perspective that I have been able and refer to our separate realisations of history not needing to be changed. Only that we might broaden our personal interpretation of what we were doing then and the why?

From the secular aspect, which in partial result is an array of open planning, even if designed somewhat askew. Our objective in recognising our political shortcomings is necessary in order that we do not regard other features in the wrong light, like through our feelings to religious tolerance? Because from our political standards we had tried to introduce the essence of equality through features as human rights, yet carried that banner to first, second and third world classified peoples of the same standing! Allowing that in cases where there was no supposed religious standard, driven by some forms of communism. All would seem to be correct by interpreting the same influence to human rights to the same excellence of civil rights, making that inclusion unnoticed.

Because we have used those expressions in the means of communication for internationalism, then by other indirect inclusions of old maxim such as operating politically correct even religiously? We took the standard of our half measured delivery to follow by reacting through our steps taken syndrome, which although active and until now, had never before been defined in such a relationship? But by its own format has been able and generate more half measured reactions perpetuating the cycle until we find the correct method of realignment? For which now I have been able and see some of our needs if not least first brought to our joint attention these past sixty plus years by some events of the same importance to others even if not of the same significance?

Before making the case by the need of positive intention untimed, having talked about religion and politics being of secular and spiritual significance is done not to make them separate, because even without acknowledgement of one to the other by personal desire. They are the two most vital of influences upon us all without choice, without our classified dignity of human rights or civil rights or being politically correct? Without the guidance of how we are the same viewed from different perspectives, not least of our characters being progressive or laid back, judged by the same two features of our secular side or spiritual side? Having prepared and armed ourselves with a new thinking range, what we might better do to begin and solve some of our continuing and our perceived problems. Even if the cost is to lose some of our political reasoning, then the change is minimal when viewed in respect of the benefits we are due.

Although in turn some of my suggestions of changing or replacing what had become fixed assets to our reasoning, like viewing our separate opinions to the examination of being politically correct. All we are choosing to do is place the concept in its correct order of reference by lighting its shadow. Because in being politically correct we have set the standards from which we have falsified our account in treating all the same? No matter how we had tried to place the matter of our equality on the one plane, and in doing so we have had to falsely legalise the concept! Thereby bringing the standard of our differences to the fore? Not least when viewed from the angle highlighted by example of how America by full representation for us all, had to make the force of law carry the image of our singularity. But in all cases made the same mistakes we had always made by noticing our differences more than our likeness!

For example in having to make laws leaning to our equality, but in reference of our differences, supposedly concerning our racial orientation, even if and also borrowed from colonial Europe who by much the same haste in allowing self determination to new nationalities still gave of the wrong impressions? Our classification has remained in terms of being racially inclined through the colours of being Black or White, which by direct connection has allowed us all promote the concept of further differences by tribal or clan or group notification. So that instead of holding to the image of those separate white, black, brown and yellow groups to the extent of how we physically differ. We have inadvertently encouraged political inter colour factionalism while at the same time each group pretends this is not so. Even that we might have attempted to level the situation with the ideals of religious doctrine through its referral to unity, we have muddied the same water?

By the same standard, but only in example, by not being aware of the obvious in relation to our lifecycles to the fact that we are all 'Out of Africa' in all ancestral cases. America on our behalf openly accepts the principle of her/our own defined make up to be politically correct, intoning that even from the fact of having been formed internationally and intentionally over these past two hundred years.

With the inclusion of peoples from all over the world making her general mix of population, many of us still hold old national habits and traits with mixed racial and religious differences, but in type are now classed to be American? Equally held by the new order of human rights and civil rights but still being different by self, and some of the classifications above?

Although in all cases except the obvious, within or without the prefix of attending to be politically correct and built by the concepts of civil rights, human rights and religious tolerance. By that very modern presentation of representative Americanism, we show our blind spots attending to our half measured reasoning by reference to self, either being American, Native American or Afro American. Although complying with all the activities and attributes of being politically correct or alluding to hold human rights and or being religiously tolerant, are we? From the false premise just described, whereby instead of addressing some of the factors that keep us apart, we assume we are not? Unless now we can and might use the new/old principle of common decency C D in conveying what we mean instead of showing what we are supposed to mean. By the one principle of that maxim C D, we will soon be able and look into the mirror and see the one reflection to the same picture!?

Americanism is a worthy goal for us all, not least because by my representative expression of how I understand some things relative to the world situation. Her standing is the follow up to what we have all been doing all along in any case, to improve the past in understanding, also to create a wide and hopeful outlook to the meaning of purpose, which has to be carried on religiously, yet with political input. What I have been able and achieve on our behalf by linkage is to show whatever the individual does, it is the property of the whole. For which although having praised some aspects of Americanism I have given fair notice that in general she has acted no differently to we or others save by interpretation.

Even so, to the edge of caution I have levelled the point of how she internally classifies her own people to be American, Native American or Afro American in our name, followed by the more recent rift to address those of us received in religious tolerance to be acting under violent religious means politically. Even if beyond coincidence most definitions of that name by activity are centred to us of the Middle East or in being of Islam, laid back, second world and always concerned of being wrong when we are right? Without to much effort the switch is there to be turned for us to bring a better light to many of our present situations without increasing the heat?

We have little to do other than overlay all of our political determinations and most of our religious hopes with the same three taken prospects of the Fifth Dimension, allowing that from the fourth active one being 'One for the Whole' we can still have discourse. Because it allows us have the same accelerator and break pedal to fast forward or hold still and reflect. But in all cases with due deference to the past because that holds the product of the last four thousand years at least, we must aim and turn all deeds for the benefit and understanding relevant to our single lifecycles. Not for our self determination because we are each owed our own time, but to set the standard within the one common framework, of what intention is by any description. Even that we might use the standing of those in the greatest of assumed success, against those on the lowest rung? Our open intention is to ultimately come to the picture of purpose to determine our faith in the divides we use of the secular and the spiritual!

## CHAPTER 143

From where I had given vocal reference to Americanism in the name of our political drive by degrees and in balance, had set some deeds of opposition to that concept, even if not only politically motivated? By my own personal reference that the counter weight of such in the name of our secular representation was to be the force of Islamic fundamentalism gained from the same motivation, but from the spiritual? For and because of those two differing perspectives we can make our case for change even by not suggesting we have to alter anything. Because those features in my case by suggestion are the points of highlight to have emerged from our close contact these past sixty plus years. Also from where I have been deeply saddened by many of the disgusting and brutally cruel events to have taken place throughout the world elsewhere over the same period of time? But not necessarily classed of the same intensity, because some of our influences in other places have not been of the same thinking? By not necessarily involving the core of political management balanced against the core of religious fervour, but turned to the greed of men to ethnically cleanse for gain only!

Although many such cruel acts were of the balance in competition of being religiously or politically inclined, some acts of pointlessness were political on political and others religion upon religion, our loss is not to be dismissed by anybody with comment. Even with my self alluding to the political aspect of religious ethnic cleansing mostly in Yugoslavia of Europe, within our recent last modern era? Islamic fundamentalism was counted for some of the reactive cause which was always false, yet the United Nations tried to make amends through Americanism! One case of redress is to be seen from what is yet to come through these pages not least stimulated by you, that is you the reader; that is us the listener, because any case for general and public ownership of open thought to the reason of purpose is ours in class. It belongs to us all no matter what our input is even if only by listening!

Although I use my own two opinions to be the height of difference even if generated by actions these past sixty plus years, my, to be our best solution of resolve is only futuristic. We are always to be permitted and lick the wounds, which in cases of mere observation, have always touched the heart in sorrow from many of the monstrous atrocities we have seen any time on this our single world? Best intentions in relief to solve the vile and inhuman acts we have cross committed, have indeed worked in the short term, but from any standard of our best intended standing, we have only been able and operate them half measured. Resulting in the inevitable steps taken syndrome, which by inclination has left the opening for further reflective reaction, because the picture has never seemed clear in not being looked at by how we can do so now? Not least because in many cases we have been given the offer of solution through the concept of debate stimulated in part from past standards unconfirmed but from now with new impetus!

For the first step into this last modern era, if using the good fortitude of our United Nations in adopting a fair policing policy for interaction politically, religiously or morally, we can count too many rules and laws in abeyance for our upkeep. Meaning that from the simple standard of allowing the concept of religious tolerance work, we could not listen to any religious reply. Because by means of being politically correct or suffering human rights or civil rights, we had opened the door of discourse and closed all others of equality? Without realising the picture of when the right hand of political suitability was raised, it was from the left hand of religious tolerance responding, there was a mismatch? Meaning in all likelihood by the new means by which we have now come to make world decisions politically, our religions have been legislated by overkill, sadly in more ways than described!

What to do, what to do, what to do is to be done by reason of abstract thinking, is a feature of all denominations and or persuasions in all cases. Because it is so loose fitting that the style can be altered to almost any shape in us all, therefore we can give the idea credence. Not least because from its first principle we can each communicate by processes of thought before the fist, we can each communicate to the means of personal acceptance before jihad or crusade. Then we can each communicate by the processes of needed change, before the bomb delivered by the hand of martyrdom or suicide drawing its own reactive process of obliteration delivered by the forces of civilised justifiable war?

If the abstract was to indicate a lead, then the ultimate steps taken to be taken beyond the reaction syndrome have to be moved by you and me which start now. In walking the same path that we all have done this past sixty plus years, much good has been done to spite the little bad we continually do? That remark is not to lessen any vile deeds, but to give open praise to the vast silent majority! Those of us in virtually all cases who only wish and hope for our welfare and the same for all people, without seeing self made obstructions leading to where they may. Those of us who are content to be led in political disarray if that is the case, because we have ultimately found spiritual solace in an understanding of self, unmatched to any other medium? Also broad enough to carry us shoulder high amongst the turmoil that befalls some, because we have had the nerve and resolve to stand by what we believe of tomorrow and the future in general. Above and apart from what we must do today by some standards of our own political direction given to be taken one way or the other in necessity?

Perhaps the most vital of unnoticed influences to be upon us all is how in cases some of our political impetus has been connected to the self perpetuation of that political impetus. Meaning if all causes are to be challenged then the voice with the most noise is the first heard, but in order to support the best hearing, that voice must carry the best delivery of necessity, to the sublime! Supplying all that we must have, but do not really need, which as a fair philosophy was delivered for the best reason to how we might have judged ourselves to be civilised. Perpetuating that masterful loop of us servicing the server better than the quality of service now received, accepting more dross instead of essentialities, which exemplifies the idea of success. Being no bad thing, unless our own impatience has turned, that by our accepted political reasoning, we find it easier to War, War than jaw, jaw!?

Becoming political from the meagre means of civil order by deciding who was best to operate our cooking fires in the depth of our united histories has helped us no end to rationalise our existence. But when taken, and I truly made the point that we started on a different road when we buried one of self in kind memory of the future, beyond the thinking in limit of what politics might be? Even if the act of burial included a means of organisation delivered by civil order amongst those of us who laid a comrade at rest some sixty to one hundred thousand years ago, the case for politics was too early. In taking further steps to be separate, this singular specie, by what we thought more in doing than the why, has delivered us to this third modern millennium, coded by political judgements made some one thousand seven hundred years ago by our Roman masters!

Then ago by the means of our political achievements so far, at that time we were the first set of people on a continental scale via the Middle East and Europe, to rationalising our political thinking to include different peoples under the same means of control. Who were also the first group of people who took an associated monotheistic religion of area under conquest at the time, and rationalised the concept of church and state in a vital way that had not been done before, yet were to suffer the consequences on our behalf? Some of which was to have stimulated great debate, great activity and great sorrow throughout the full extent of empire, but by the luck of choice from what must be held as Roman determination, generated the concept of political equality of sorts? Being in part a benchmark carried by us, whereby we hold in region to our controls and aim to be either politically or religiously generated. Switching from time to time even up until now, when we can automatically think in terms of one or the other unconnected, even that the dual concept was first set by Roman intervention. If some of our actions as close as today give no indication of that fact, but might yet, when we remember what politics and religion are supposed to do!?

By indication and for better clarity the one political lesson to be learnt from the time of the Romans carried today, is in the lesson of personal reference by no other means. When from the time of acceptance of the Christ religion to become the state religion of Catholicism, they in we by no more than the rules of expediency, first set our international year date standard. By linked coincidence and even from more intriguing political motivation, in taking and making a faith universal, in deference of its delivered hand to man by the hand of God through man? They in court paid tribute to the very man they had crucified some three hundred years earlier by anointing us all to the existing year calendar we now carry, which was not a slight to any religion from before or long, long before and or after or long, long after. What was done was to address us all to the importance of our personal notoriety, both religiously and politically. Simply to comply with the run of thinking even on a scientific basis, for instance, whereby our references of when we were born stood throughout the then, and I use this word cautiously, civilised world!?

Having taken some of the romance of how we began to use the birth of Christ as an international year time marker? My case has not been to promote Christianity apart to other religions, but simply to indicate the properties of some political decisions made to indicate our singularity unknown? In making the case by suggestion of American Christianity is not to indicate a religious measure of American superiority if supposed in anger or jealousy by some. It is to simplify the essence of the past to the fact that our world standing of now, stemming in change from about the year nineteen forty five is of collective reference to us all and in cover of how we interact.

Proof in kind of the above references that I had reiterated, will bear out my own view when seen how it is best for us to precede and overlay the condition of all of our political and religious systems with the first and all examples of my Fifth Dimensional levy! Which without design will gather our own open opinions however held now to be collective by natural selection, because choices made will be seen to have been in need by collective desire. So that we might begin on this new journey even timed of this last modern era, we are the same people to be in transit all at once, but given to move in our own time by style drawn from the experiences in example? Those which have been bestowed upon us by the conditions of diet, climate, geography and other features, even being laid back or progressive or only politically inclined or from the newer religion in the study of our mysteries through the avenues of the sciences?

## CHAPTER 144

In political theory and in religious fortitude we abound, which are vital in us being unique as specie upon this earth. In all cases we know the how, but have never been sure of the why, which again is owing to our uniqueness whereby we move in realms of thought mixed with the necessity of actions? Without comparison our successes have been outstanding, but on course we have also been blinded, not by success, but because of the ease of it? Our one limitation making us again separate to all other creatures is from how we fixated to the open concept of purpose, before we came anywhere near to understanding the means of purpose? Not a bad start when we consider other creatures in fine example, have been on the same path for two hundred and fifty million years or longer. When held against our standing to the beginning of understanding from about one hundred thousand years ago we are indeed proved quite separate to other specie leastways by motivation!

From the whole picture all that we can take in unity from that length of time since our burial at Shanidar, is the extent of any time scale being of singular relevance to our personal lifecycles. By turning the key in open review and admitting, which we must, that the length of those personal times are spread amongst us to be fifty plus or seventy plus years long. Has been and was always a product of our secular living and not attuned to our spiritual inclination with the two views never having been aware of this at any time over any recorded history. Hence in style we have operated both entities to different standards or for different objectives, surprisingly one to the here and now supposedly, and the other to the hereafter continually?

What we can do in respect of our obtainments and in belief at this precise time, is overlay all concepts with the spirit of the Fifth Dimension collectively, even separately, but most definitely. What will be seen without direction but in order, is that although the same three taken and one active prospect are to be applied in all cases and to all venues. Some of us will and can determine the propositions differently, but because of the structure of the said prospects we will find it easier to reconcile our politics with our religious determinations. For they are better informed in leaning to the incidence of our religious philosophies having found the first sense of purpose to spite any and all forms of opposition even religiously? Even that I had predicted a greater degree in the split between the same overlay on two different subjects. My case for allowing the above form is not to be read across the board, nor is it to be confused with the factor of how some political and religious relationships are today. My standing that our personal interpretations might draw different conclusion, will operate by present standards of entrenchment but are subject to alter when the openness of our intentions are fully realised.

Our Fifth Dimension being constructed with three taken prospects and only one to the service of debate or query or study, is of total modernity in style, which could only have been formed or developed this past decade at least. For no other reason than to change our emphasis on how we think we must precede on all levels to any goals. Although I have seen the need to limit the number of corner points we need to develop further into the future. My agreement is to set the one standard for all people who we know we are, but have not yet seen the picture. From my own personal study, I have shown a true human weakness of being opinionated on other peoples' behalf, not unlike from our political systems or our religious achievements. From that, but still in experimental styles I have put it that in overlay of our political and religious standards with the same limited number of taken prospects, different pictures will emerge which is to be first taken as a possibility come what may?

By making rules which are not included and using a limited number of positives, I have been able and see that such proposals that I make are set to set new standards even without the need for drastic alteration. Our first deliberation as always is to the future, but by strict notice to the present. Although on numerous occasions I have set the standing of how our politics works on one scale to be democratic and or styled to lean to Americanism only, is done in representation? In balance or counterbalance, by implying that the one and main opposing ideology to any standing is by the religious element of Islamic fundamentalism is to take us to a point of reference from where we must look to take the next correct step?

By further duty I do say that my comparisons of above are purely representative of all political or religious systems or how any two given entities, one from either school might react, because? By example I have also not used our histories in a malicious manner, covering at the moment that over some deeds already to have been done, need counting? By the same token nothing has been dismissed or been kept from our sight, my tone has been open and remains so for the duration, but always remembering we are of the same human stock!

For those of us holding serious political convictions along with the measure of more serious religious convictions the picture we are to paint will draw the same result for all including those just to politics, just to religion or those not committed either way? Even if only having to follow the winds of change by what ever circumstances or from whatever source. No matter who we are, some one some where will have counted our head and recorded on our behalf what we think, because of the way they allow our study or because that is now the way we have developed if only half measured? Without the critic, by referring to that term again just at a time we are to address what might be new codes of conduct full measured, is to show how entrenched we have become simply because in operating change by first suggestion we had no rules save our steps taken syndrome!

Whether the Fifth Dimension by being implanted to our leadership styles of mandated care instigated from our own sources these past sixty years, will have the desired effect is not in question? Because in cases if designed to present our open intentions, it is our steps taken syndrome! At last revealed by no more than the course of events throughout world histories, but encapsulated these recent six decades plus a few years since the closing of one style of hostilities on two continents in nineteen forty five. Although in quick succession, because of the rapidly changing international situations we hastened the restructuring of the United Nations, fitting in mostly with the primary ideals of the victors from world conflict. Working by some hopeful assumptions that the force of law could be managed by the theory of open debate chaired by lead from its own constitution but not being aware at the time of what might yet happen between peoples?

In order to take better control to the means of delivering equality at least politically, then to best thwart any so nation who would seek and declare there own atheist state on behalf of her people, so as not to delay the point and the move in what was at the time considered steps taken! By proclaiming the principles of religious tolerance, one serious possible issue of discontent, even violent argument was nullified for a time but without consequence? From some of the localised wars even to have included the United Nations in aspect through the Korean conflict of the nineteen fifties. Without any religious theme confrontationally, our thinking again if fluid was to fall to half measured standards. From where if only pointed at, the United Nations was to get involved in some of the principles of human rights, civil rights and or the syndrome of behaving politically to rationalise concept, but only from political perspectives if to carry there cost? Creating in debit the balance between some of our religions if to be against some of our political blessings?

By putting our tolerant religions on the back burner, we sometimes began to forget the first trigger of all thought looking to the intention and meaning of purpose? By so and in open representation, even today we have been able and set targets, aims and objectives and the elusive connection with hope, purely politically, which is fair? Unless the case could have been made from a different emphasis by some religious standards, for which we can now look for in the comparative comfort of the Fifth Dimension!

## CHAPTER 145

In using our three taken prospect in direct overlay to the two main features of our drive and or influence to be upon us all within our own lifecycles is totally unavoidable. Being a very fine thing, because the Fifth Dimension in concept is a means whereby we may at least lay down our arms and by style only enter virtual war by means of normal and general discourse in relation to each other, because now we are the same?

Having previously said my stated intention in overlaying our politics and religions with the taken prospects would bring different results. Although a personalised statement, it was said to propose how we should think of any supposed results, but not comparatively. From where I did mention that the results were to have a difference to the good by our already prominent religious thinking, was done to redress lost ground to our politics. Not least to balance our account of how it has been more through our religions in accepting the principle of God that has provided us in content of our wonder and query of all mysteries? Religions have been the heart of humanity even if in cases we got it very wrong at times. This view is not just by my own reasoning, because religions point to the total prospect of understanding? They have created veins, which the heart could pump the responsibility of knowledge into for us all! Even if some of us remain unacquainted with what that might be, religion remains our crutch when needed, and if not, will remain at hand when it is, which is inevitable!

Politics being an ambassador for us all had blotted its own copy book even as early when we were at Shanidar? In its infant form being a means of civil order amongst our small family/clan/tribal groups, when allowing that we could apply all the necessary requirements to bury one of group. From the act of delegating which would have been a certainty, was created a structured class in our group? Of course no parliaments were sitting then, not even committees, but decisions made as to who would scrape a shallow hole in the ground for a grave. Who would arrange tribute to be left with the dead, created the system of delegation which in short led to a leadership syndrome, with the rest not being history even then. But once a system had been set up with additions added over vast periods of time in the thousands of years range, growth of clan or tribe automatically meant growth of delegation. Leastways into the spectre of organisation and if adding to that theme the more expansive a group became the created tentacles of management grew to be the arms of political management proper.

With religions not following the same case, because we needed better civil control to assist our growth as a group, even that some of us might have been assigned to bury dead or at least dispose of the bodies how so ever? Even if that was the case, by the widest of imaginations, those of us who were to be undertakers found that from the additional role of being close to the medium of death. By association were able and organise the deed by instruction if necessary, but with our own personalised input. Although not supported by any record, but through sheer imagination as I might have implied when suggesting that we could assume that some of the caves we have found from thirty thousand years ago and closing. Were in fact tombs, in full respect of the dead by depicting scenes for the living, which we assumed then was to be inevitable again, otherwise why the monumental efforts of producing such memories, if only from the civil perspective!?

Because from the cases in example and what followed if studied, might show how we did emerge from our first concerted tribal interactions in the land between two rivers with what could only be described as priesthood concerns? Those to have become accustomed to the organisation of the properties of death to the individual, creating the aspect of the spirit if unknown, but recognised from the memory of distance? Just like with the same of us now in trade among our own community, who were not fire lighters or hunters or gatherers, but as we by the expansive growth of our needs? Like when we became carpenters, or stone masons or farmers or tailors and sandal makers or potters or weapon makers. All eventually in need of care or aftercare by our own tribal undertakers who in dealing in the aspect of death grew in thought to the prospect of continuing life and from where the means of such a happening were to be found if told?

Without comparison, without direction, but under the most vivid of personal experiences, when we know, we know by a different understanding, perhaps like when through the open expose of a meagre spot in a huge scientific field, a key was found in honest exploration? Like when Archimedes by the small chance of natural interest unlocked an opening in route to solve an unabated mystery. Such an air of excitement created its own form in the word eureka, "I've got it" or "I know" or simply the answer has been put before me! The case represents the time removed, but connected when seen that any step into the dark that brought light, flooded the whole area. Whereby science had its own reins of limit by experiment, other routes that had been in the dark and are now drawing reflected light allowed that we could expand our dealings, in the expectation, because we had broken into the light we now had the means of full illumination? But in following the trail to its ending, that picture after several studies can only be lit by one ending opening another beginning by the medium of death, which in all realities had never been certain! Leastways since Shanidar when we came to imagine that there was something extra in need? Without the conclusion of that extra value belonging to the order of science, but more to the emotion of purpose, we continue in the means of natural growth, and now natural light?

In lacking to the source of any open answers, my suggestion of a link that we might be able and draw some answers by the means of the events that resulted in a burial rite some sixty to one hundred thousand years ago, will come like eureka to some of us? Which is a noble occurrence when using an event from so long in the past to rationalise the now but by connection from other leads also? Meaning that from the events at Shanidar under time delay, but by the very good references that had been planted in the depth of hidden memories, just like the way we had conquered fire and never knew the how, the same example can be cast over how the mystery of our religions developed? Fire use as a natural connected factor, enlivened every day spread amongst our many differing societies still contained the basics of originality, but can mainly draw its originality belonging to the secular, which was civil?

Almost from the same example of using one thing, item, task or objective we can quite simply rationalise our full religious perceptions, covered in good grace by extending the reality of our differences drawn from the circumstances of geography, climate, diet and other features. By the same standard that fire use never really changed, our religious perceptions kept the same mix. Whereby although we formed them from different viewpoints noted to the spiritual only, like to the standard of Eastern styled religions to the continuing cycle of life, balanced to the Western style of direct crossover by way of 'Natural Body Death' more so? Both styles could draw on the heat of comfort that the final show in all cases was lit by the events of our deaths? But inclusive to us all has been the order of our impatience, by not addressing true possibilities, in wanting to eat the cake before we grew the wheat, by wanting tomorrows light before the sun would rise, our haste sometimes clouded the compassion of our singularity.

Fire tending is not a comparative example of how we did come to God, but by the reasoning I have applied, the thought even now, will generate comfort for many of us, from which using the store of how we always progress even unknown by natural selection owed us all? In as far as we look to science today because that is where one branch of our query has led us. Much the same happened some four thousand years ago when by historical liberty the core of our Eastern and Westerns religions found their beginnings?

As stated in terms of the Eastern style of religion the opening answer was for us to be kept in the dark because at the time we were incapable to fully comprehend what we must eventually? Therefore because of our real limitations of mind, by perpetuating the rhythm of thought it was seen by acceptance that other mediums would also be required, in the same good order that we might have thought of ourselves to be special and created, like we might have thought of all other creatures? We could quite naturally include other specie as we; to be part of the whole, whereby they as we along with us, were included in the full cycle of lives to regenerate, reincarnate, to eventually come into the appointed realm of full understanding? By to or from any one of the linked/joined specie who were always as we in all cases past singular death into collective understanding?

Almost at the same time about four thousand years ago, though not necessarily connected by any deals from our interactions in or from the land between two rivers, if the code for our monotheistic religions of our Western persuasion were first read. Then the style was settled by us holding a different temperament if that situation was to apply? In this direct area to be the Middle East, perhaps owing to a growing tendency to be impatient even if laid back or to have become laid back by other standards. Our first connection was to be connected with what we might have come to consider the inevitable, even if by previous example from area?

Having already had the open practice of worship to several Gods for our assistance, through the standard we had began to apply, paying tribute to many in their separate fields and from our own way in which we could measure tribute. Without concern for some of us, the inevitable drew its own course, from which one of group, any group or tribe, first made the obvious observation? That as we could and pay homage to many gods, even that some were adopted with creature features, that we had appointed them all by separate connections in the same manner as the many, then there must be one to be the God of Gods? So might have been heard the cry eureka from earlier days, but with just as much conviction and this time from the spiritual perspective? All similar associations having deduced of the one God concept to be the one and only God, claimed for the makers of discovery the finding of our first key, spiritually? Now to turn the lock to unravel the open mystery of creation, whereby there was indeed a tangible connection with all life created and God along with the one benefit owing to the discovery of that fact? Making the one conclusion that Gods appointment was indeed associated to the first tribe who could see thus far so far. Creating for some of us who felt openly involved to such situations, that the finding of the key to our mysteries was endorsed by Gods own humanly words to our own agent in our ancestors. Aligning us to the full aspect of creation and from where and by our own hand as from when, but if recorded by how, we simply thought or thought simply, in comparison by todays near understanding to the theory of everything?

Having explained what could have been some of the picture in setting a small group of people in tribal fashion to the ultimate realisation of what was gleaned as duty on abeyance. By our own excellence of mind but error of thought we had created in balance a religious style that was to lead to be now classed as being Western by description? At the same time by the understood nature of understanding, we were forced to compromise the concept of human life in relation to the concept of human understanding of intention? Being only now challenged because of different orientated events that began again about two hundred years ago, coincidently in the Middle East, but to have led to closer events counted these sixty plus years on scale to be world wide but centred in the Middle East!?

## CHAPTER 146

'Out of Africa' by connection to be a starting point for any deed, has been used by myself for all of our measuring in relation to where we as specie first set foot upon this earth? In consideration that we are here and now, where ever throughout this earth, my taken prospect has particular reference to us all having been black of skin at our outset. I make no appeal to sentiment nor do I stand for opposition to any objection that might be put forward by claim in due of selected natural selection, of which there is no example amongst any of us in this single specie we are. My standing is our history, upon which a clearer picture will emerge when this one taken prospect of three, is overlaid on our politics and religions?

In the first case of our civil controls, because that was the first case of our influences by taking us to be separate from other creatures. 'Out of Africa' has only been questioned about many of the avenues of how we had developed but only measured from the political perspective. In casting doubt to the prospects viability, our political enquiry by its own assumptive style has weakened its own case. 'Out of Africa' is a natural standard with the small interruptions delivered by the circumstances of climate, geography, diet and other features. From where I had made the strong irrefutable case for our singularity from other measurements in our character if only realised half measured, in treating some of the gains and benefits adjudged to have been morally earnest. We become automatically inclined to continue in the manner of our steps taken syndrome, which done in the case of our skin colour has drawn disastrous consequences. Of which this work is a small part of all the means we have had to employ to reverse an open error or an open wrong.

Politics alone is almost responsible for our misguided attitude about skin colour as our main noticed different character trait, only politics. By further reference to the American situation of the mid nineteen sixties when civil rights became an issue over that primary concern amongst joint communities. How the subject was to be dealt with in good cause was unfortunately half measured by our best intentions but the worst of our solutions from the results of our steps taken syndrome! When both factions, black and white, even at a time overseen by the spirit of the United Nations, were quick to accept the judgements of new and hastily drafted race laws to accommodate questions asked? Although intertwined with the spirit of human rights the concept was set to be always further questioned? Because our natural progressive spirit took us to better general acceptances, but were to be contained by the mixed interpretation of race laws or racial laws to supposedly suppress the difference between us, but once again they only highlighted them because they were race laws? Then ago we had to acknowledge new laws made in haste to help squash inequality and in doing so helped complicate our emerging human rights movement? Because by the political example, trying to operate on the one international level, compromise was forced to lessen our own intentions when we expected from other whole countries to do so, but could barely operate internally, perhaps because of some character traits!?

From concerns of which I have tried to throw light upon, 'Out of Africa' has been accepted partially politically, but channelled to be only evolutionary? If we strip my bland taken prospect of its depth in meaning, it will still hold to be a primary indicator of our beginning. By connection, but separate to the ideals of form from imagination without thought, many of our political regimes bowing to the arm of scientific query have taken to accept the expression fitted to the charge of science and formula? Allowing that the concept has its beginning by association in direct link to one or some of the separate specie of apes, therefore, because of fossil finds from relative time scales to when we left biped footsteps in volcanic ash some distance of time ago. Politically we can imagine that the new undiscovered missing link remains, will prove the loose inevitability of several different starts allowing that we are indeed 'Out of Africa' but of first, second or third order? Turning a fixed pattern of thought carried in belief through the certainty of progression expressed by the certainty of influences under natural selection, that if we had developed from different types. How now of our similarity having continued all that time to the zenith of human development, if represented by my indicative choice of how we behave, after the discovery and the changes we did in the new world continents?

If the taken prospect 'Out of Africa' is to be tarnished by or owing to the fact that we have indeed progressed and therefore must change? Then our political determinations have to be shown the most direct route of connection with scientific reality. Although we might overlay our politics with any form of judgement, we must at this time begin to value the way we decide differently. Not least to organise what our real intentions should have been over these past sixty plus years. Instead of what we have actually got to show by vane assumption that some might voice, even on our behalf, of course we accept the principle behind 'Out of Africa' as a nice maxim? After all we have supported the assumption with our standard of human rights, of civil liberty even to be politically correct. All balanced by how we have driven into each nationalist community the standards of racial and social equality under the force of laws and laws. At the same time allowing the said same designed communities interpret any so law in guidance too or from there own particular standard, not necessarily using the example of 'Out of Africa'. Because by most political standards the saying is applicable in all cases providing it does not mean we are the same? Which without direction indicates how we might never understand the game if only overlaid to our accepted political standards which are administered by us?

'Out of Africa' of the same standing, when overlaid upon our religions, automatically carries better weight, but our shortfall may be seen first by the means of how we acknowledge the ideal? From religious matters including the spiritual, it is our taken aim in all cases to listen to the inclination of our religious leaders. Not least because by best use, even from some of our priesthoods, the job of our care is in their hands, sometimes only in the interest of expediency, even if a little unrealised!

Once again, so not to personalise some deeds and actions, in the beginning if we attend to any religious and or any political events from the past, in order for best inclusion even if we are somewhat biased. We are to openly mix our thoughts from events over these past sixty plus years. By that direction we can put all opinions into the same arena, not for the means of comparison like being first, second or third world. More so that we can realise there are over two hundred national viewpoints politically, but generally only two religious aspects if we look at there classifications. One being Eastern and perpetual until the cycle of completion, the other Western to be monotheistic by style and expected to gain full understanding by the cessation of a single lifecycle through 'Natural Body Death'! By acceptance of those definitions even if only by view, it will be easier to understand why in the overlay of our religions with the Fifth Dimension in full. I have put the case that first reaction or judgements although not binding, can be made in representation from and by our religious leaders!?

They in turn are not a medium of God, but our human representatives, therefore as the leaders of Catholics or Jews or any defined Christian faith are to translate the deed of overlay. Any answers are ours, but from our perspective when taken in this instance to our open connection of being 'Out of Africa' referring to the single fact that we all had the one and united beginning 'Out of Africa'. Proving our line of specie singularity, therefore endorsing any relationship we had made in the error of our own judgements with or by directed association to God. Has to cover in referral that if one comparison was through a direct link, then all were involved, or as one comparison was made that we have yet to forge by that link, all are to be included!

Saying we must expect and demand in retrospect from all of our religious concerns parity, even directed by our own ambiguity, because for many, their standards had been set in guide from rules laid down at instigation of their religion when ever? Being already tied to fixed traditions that might have been spun from previous direct relationships with the sound or image or representation of God, but by the understanding of past days only shown in relief of how our religion did change? Not for the sake of it, simply to accommodate newer thinking to have become the norm by natural progression. Being no more than in the case of how we were to receive the wisdom of Islam to one side of our split religiously by my classification to be Western by style, but monotheistic by form?

From overlay with our religious leaders giving court to the taken prospect of 'Out of Africa' in reflection to our singularity. How might Islam represent us in the same way if the faith itself maintains there are no defined leaders, save we and the duty of compassion and understanding in relation to the will and wishes of Mohammed? For me to paint the picture of what might happen is done without comparison to the two previous religions of area, even if delivered to the condition of earlier human realisations? Although I had made reference to the open fact that much of the structure and set up of the faith drew on previous examples, with the biggest share taken from the Jewish styles of habit. By no more than the given temperament of the mixture of peoples and ideas in the same area of the Middle East, who had behaved much the same religiously for long periods of time! Even through prolonged conquest by peoples who would eventually form there own styles of belief being European and of a different outlook if generated by the circumstances of geography, climate, diet of including other reasons, how to proceed without open opposition?

By expectation that we use the society of our religious leaders, even defined to be established priesthoods or gurus even mystics, all of our intentions are to be the same. From what I see to be of considerable difficulty with some of our future dealings stem from the religion of Islam without cost. By natural progression in any case, Islam through doctrine automatically shows no conceivable problem with the prospect 'Out of Africa', aided form inception when all considerations were set to and through the individual. Because without direction the religion is laid to the expected capabilities of man from which the thinking is automatically attuned to the global concept of mankind without exception! In such a way that at no time by any degree of study has any text been recorded to dispel that certainty, all verse has been recorded in exactitude of delivery, which allow no other conclusion to be drawn?

Islam was openly for all men without distinction that in part is why once first acceptances had been made, all were treated the same. So much so that even if my particular assumption was not met to my derivatives, then balance would be achieved by the extent of our laid out obligations through the written word? With minimum guidance from our imams who generally oversaw personal conduct as prescribed, but more so for our young, at least in our mosques, but always set to rule by the rules themselves seen, heard or directed? Although taken to the highest standards from the thinking of the day only, based on the knowledge of the past, being no bad thing unless we err from overlay in using modern thought to alter how we thought then ago!

So that we might balance things in the correct order by mixing the past with the now, there are some imams who involve themselves in the task of direct guidance to some of flock. They in turn are not to be yet classed as religious leaders, which in turn makes my proposition ring the wrong peel with the right intentions. 'Out of Africa' is not an issue of concern to any person involved with the religion of Islam by any terms. My reference to religious leaders might pass on some occasions in the order of importance among all religions. But will remain to be of more pointed importance than the overlay on our political regimes. Our test without trial begins past the obvious, into one of the next two taken prospects of 'Natural body Death' and 'Past Gods Words'. With the former becoming a deposit into the future, indicating how so many religions think of continuance only from reaching that goal. Quite naturally those concerns are more important than any political view of that situation, when in turn our religious leaders come to express views in our representation about that particular prospect, there is only one voice?

'Natural Body Death' has three fields one is my own and applies that when we die the causes are to be what they are without personal input. By my same code, suicide is never an item by that definition, not least for the other reason that if my intentions of prospect are to be met, then the act is proved only to be an act of denial. If I have supposed that intentional purpose is for us all to make ready to be attractive enough to draw Gods interest against whatever interest might fall to any other specie. Because no others have shown any kind of wonder to the possibility of existence that we do, therefore from the other side of their legacy to us, their example of living to 'Natural Body Death' better supports our singularity. By our deliberate testing of the implication we openly display our own linking to the spirit of God continuing, but yet unseen. We still drive in that direction knowing that our end is eventually and the same as all other specie as all other creatures, but for us even taking many wrong roads we still try to improve!

In overlay to our religious views and or political determinations 'Natural Body Death' carries no weight by incline to any set, type or political style. Because all can and do interrupt its flow by causing types and the means to promote serious death by unnatural means even if based to be fully civilised. Without praise the best example which we all offer is about the matter of how we still War by means of invention but only by or from full political determinations.

For the one example that I had used from the Iran/Iraq war of barely twenty five years ago, in statement that the ideals of suicide were renewed in case by the felt duty of some opposing soldiers, even that they were in the same general faith of Islam. Were purely political acts of which I had considered other means also, but because of the confrontational situation involving volatile emotions, my take has been to view the results in a political light only, but with the cost of serious misinterpretation? In passing judgement by a half to one view and a half to the other that Islam itself was in the throws of reformation belies the extent of the acts carried out in war. My only joining statement has sprung from now, but only by being able and overlay religions or in that case a specific religion, with all the prospects of the Fifth Dimension.

Young soldiers did not have the base of query not to commit suicide, any act carried in judgement by the times and the day fortunately cannot now be given grace. For one of the best medicines that we have had to take since when, are these common three taken prospects of self reasoning and judgement to be delivered from now, now in time that has been in first review these past five to six years,

'Natural Body Death' is total to all political aims, it is only we dressed in the vanity of self civilised propositions that allow it so that we can take the life of self upon self. But only from the top end of being civilised, so much so, that in display of being fully civilised we have reached the zenith of human development to only recognise the study of nation as the last resort. Creating the myth that humanity is only part of the whole, but in many cases that we individually are to be servants of the medium of statehood through our politics in virtually all cases, making the case we have no other role? By mixing the results of how and why we saw the need to create the cover of religious tolerance so that we might only operate by political standards. Covering the aspect of the distance between the full extent of some of our irreligious political systems and other politically inclined systems following some religious obligations, was not done to the measure of balance even by very modern times. But more in the case to confront different aspects of political standoff created by our bloc culture system formed early in this last modern era.

What our study of 'Natural Body Death' will bring is needed modernity, because we have not ever looked to the event in the openness of real situations. By adopting a position of overlay and using the two camps of the secular and the spiritual to determine for the first time we are ever moving to a position of our proven singularity? By meeting our true obligations head on which are only relative to this our specie, but only by declaration, not any claimed example? Comparisons to be made between the political and religious means of influence are to be likened in there correct order from now and how we should better proceed to equalise our standards of humanity by reconnecting to the importance of our religions first. Simply to highlight how we have rediscovered the danger of serving them by tolerance?

In using the separate case for our religious leaders, different than our leaders in politics to be involved by acceptance of the principle of 'Natural Body Death', we are stamping a claim to the obvious, which has always been an open distinction of our religions even if unread. What we are to further show without confrontation is that although valued, our political perspectives are not qualified to take the lead in our service when all there objectives are misguided and encased beyond the scope of understanding the principle of purpose! From the same intermixed viewpoint, but always in open review, my set direction of religions has not been made to compromise our religion of Islam? Nor any of our monotheistic religions nor is there a case made against our Eastern religions, from whom there is a likened format of service to Islam without a heavy priesthood representation that there might be in some other of our Western religions?

'Natural Body Death' has a different bearing to some Eastern religions, because in representation there might have to be several cycles of the same to include other specie before the religious standard of the prospect would be realised? Eastern religions do not compromise the factor, but illustrate in an enlivened form of how we are to better organise our political and religious thinking? When through there principles we can live to political regimes even stamped by these modern practices of the past sixty plus years, but not feel as challenged in quite the same way that some of us detect in following Americanism, to others being of an Islamic fundamentalist outlook? For some of those small reasons my focus has been to our western styled religion and politics and the specifics of there relationship, especially over these past sixty plus years because something is not quite right?

'Natural Body Death' will not go away, but in our set task to try and understand ourselves better than we show to each other, we have to bring into the order of control of where some of our religions lead us by the most serene set of deadly examples that we could imagine? Holding three taken prospects is as much as we will need for community so far, leastways when we have our one active prospect for backup when our reasoning hits to be upon our own ignorance, which I would conclude that there is a germ of that gene in us all!

'Past Gods Words' is an actual example to us all and from the western religious perspective has been proved time and time again over many centuries, not least to be an example to have held these past fourteen hundred years. When given direction for the last time in God's name, even that the exercise was conducted to our best standard up to that time of our united histories! Mohammed's role in delivering the scope of reason, set in text the finest religious writings to date and for all inclinations. What was done was on our behalf, but in like form to Judaism some two thousand six hundred years before and Roman Catholicism some six hundred years before. Mohammed offered what Abraham and Moses and Jesus had done, but because of seen laxities in the actual developed religions, his path by direction was to offer us the Quran, maintained as the last, but never final word to the method and style of our future conduct, although heard to be recorded, all written was 'Past Gods Words'!

Without the means of contradiction, in using the term last and final, my direction when considered in relation to what was recorded in the Quran, being similar to other text of any religious persuasion 'Past Gods Words'. Stands by the very time to have elapsed since any words were first recorded. Last but not final, allows for the fact that we have always progressed in style, even if at times wanting. Advancement has been to the good in comparison to how we have now dared consider the religious feelings of all peoples in concern of our own? Irrespective of politics and how by reflective means we have let them question some of our choices in comparison to others. 'Past Gods Words', although not aimed in direct relationship to our politics, has been taken in force as a maxim to minimise the full prospects of our true religious standing, displayed by delivery of our religious tolerance manual, from politics, not to be 'Past Gods Words'!

In overlay to our politics, no wind will blow nor will a head turn, because that has always been the norm by how much of European growth was first achieved at a time of our united histories. In the order of our own natural flow by some progressive means, 'Past Gods Words' grew without reference of how we managed political attitudes? Therefore my own directed pressure to be applied upon our religious leaders, is first to set that fact, but by religious tone, 'Past Gods Words'! Leading from now and earlier, we can add the process to our conduct of these past sixty plus years of general advancement in the same balance religiously and politically 'Past Gods Words'. But also from now in the same way that we must not or should not or cannot alter our best words of direction if delivered some fourteen hundred years ago, because we as individuals have never had a brief to alter the past and its implications, nor can politics, which has no call 'Past Gods Words' save by future opinion!

What we must wear, even when carrying the open delivery of Gods given words, is to accept the process of advancement by natural human progression unseen, which was to have generated our will of acceptances 'Past Gods Words'. Not to have been called or taken by my own delivery of the Fifth Dimension, but from the relationship of our politics and religions having been made over the last fourteen centuries or four since the new world or from changing direction over the past two centuries. When our political drive was to drive change in the same area from where a large part of our religious sociology also had originally been generated over thousands of years?

'Past Gods Words' is an open declaration of our own free will working by consent with our religious aspirations alone, although I had set it in line as a taken prospect to overlay every aspect of our lives. From the political perspective, not to divorce the framework of holding differentials than the open and supposed concern of our welfare, even if coming from one direction only? My direct message to our religious leaders in waiting for there own judgement of the same case, is to expect that when any declaration is made it is to set the decree nisi so that in settlement. Politics can only now operate 'Past Gods Words' by not referring to the status or standing of any religion. That politics in no case can lay judgement to be upon any group by operating 'Past Gods Words', that any type of liberty is to be denied to the essence of our free will, from which politics was made! Then without final acclaim politics in one voice cannot ban or bar any religion from any type of regime when standing in corner to the overlaid principles of religious tolerance, civil rights, human rights or even in being politically correct. Unless by fully adopting the lead from the fifth Dimension and by type our religious leaders who are set to discover new ground that we might walk?

## CHAPTER 147

I have no personal intention to establish any kind of set ups that will cut into the case of decay we have already achieved? Change is not a necessity to openly follow when we already know all the answers! It is quite obvious that our religions and politics although interactive are separate entities. But because of the structure of the human mind apart to all other creatures, we have been able and visualise the future while operating in the past. Hence much of the cause to the situations we operate under today even at this late stage of our development, have induced an opposed lack of self?

Expressed in the first by our politics operating by presenting the future in the expectation of now, but without claim, that all cases are made for the now, leastways by those of us who feel connected with the primary drive of politics itself? Then more so by how some political systems have made part of there objectives to be only of secular demands which not only regress humankind. But cut a wedge from where that tool may be driven to split an already forward movement of human destruction and make us tread water in obscurity! Simply by suggesting in force if necessary, our commitments are only owing to the innate. Hence 'Past Gods Words' in maxim support that view from our business and scientific perspectives which known are new tentacles of our first means of civil control!

With much of my own and our suggestion that a greater charge is to be set to our religious leaders, without placing any in a dangerous position, the label of standing up to be counted, has to be worn, but only by example? Allowing that by following to a new direction without that labelled charge, some might feel appointed by what has been now said or indicated! But now, even if from different perspectives, all religious leaders can make judgements or representations for or against the similarity of our political systems having all been coated with the overlay of the Fifth Dimension, but in harmony?

I place no doubt that politics and its means will always be the foremost of our leads, certainly until we die. Much of all I would point our direction too, can be gained by very small movement and the poetry of that situation abounds from one of the smallest elements in our change, us! Fortunately from very close to the time mark of now, by little more than acceptance, our switch does not need to be severe or drastic. Although said many times we have always held the ingredient of change through formula, but not knowing how to set the experiment religiously or politically until now?

Sixty plus years of serious effort to the means of improvement amongst ourselves had unfortunately been looked at from the one perspective of politics giving off its own aura in display of what should have been answers, but were only half measured ideas? Leading to the fact that because choices were made by our half measured attempts, they were obliged support by our steps taken syndrome, which was in automatic backup every time. To the one extent at least, where by tone it was to equalise social accounts with the disguise of human rights, civil rights, if partly policed by behaving politically correct? We could call all religious issues to be of the same tolerance, but alas not in the realisation that most in operating beyond political standards or political input. Unfairly disqualified their intentions of concern to secular decisions over those of a spiritual vein, endorsing the political levy of setting the standards of religious tolerance, now to be vigorously opposed religiously?

That I have asked of our religious leaders to send us in a different direction than might have been asked before, is done so that we will have heard a voice we can reply to, instead of from politics that always leads by set directions, generally half measured, but now to be steps taken? How can we expect our religious leaders to set new standards 'Past Gods Words' when and if most religions had been built through having a direct connection with what was actually said by God first hand or through his own appointed agents first hand? Setting direction which is not to be questioned, yet at the same time are to be directions we must follow under our own free will questioning, employing our wrought free will, even if only appointed first used in Gods own garden, the last time we were all together!?

From those first and last times we were all together, be it in the Garden of Eden, no tale of evolution is compromised. For then we were directed to have the ability of free will, which in recorded time has been marked to have been religiously counted some five thousand seven hundred years ago. I personally know that particular length of time has no real bearing to any standard of evolution excepting that it is taken to be inclined to the aspect of Gods creation! Even without us having to consider the maxim, I had previously delivered, that evolution is the measure of time by Gods own hand in relation to time and the spread of change. Whereas creation is still of the same mixture being from Gods own hand, but by the specifics of relativity to mortal man! Marking for us all, the opening to fully understand to the meaning of time, when all measurements are directed from within ourselves, but only in relation to our own fifty plus or seventy plus years of full time, counted to be our own lifecycles!

From where I do claim the fact of Shanidar being of vital importance to us all from any mixture of people. Although in essence it compromises in appearance some of the taken guides of one of our main branches of monotheistic belief. By having occurred some sixty to one hundred thousand years ago, Shanidar endorses the picture of evolution and creation in the mixture of unity we can all consider. Because it proves our ability to have been able and rationalise far in excess of all other creatures tied to the whim of evolution only, which is not a scientific statement, but makes our age over five thousand seven hundred years?

Making our own judgements to have been created and by the one true God, has not done anything to damage or challenge the mysteries of science. Which being one branch of politics had made that claim against us on several occasions, because from science finding self answers, religions found self doubt amongst themselves right up to now? That is why I call for another challenge from our religious leaders to reconcile politics and religion and set the screen on our behalf. Not to any competitive edge for we are to be mature enough to recognise the mystery of death by all colours and for all reasons. Again religions do not lead by evidence, because there is none, not even to the ascension of Christ. Evidence is not a question or answer in relation to the timescale of creation, which is another chore for our religious leaders to deliver understanding of if necessary! Because working in the light of open study now, they as we will have lost all of there competitive edges by our natural progression into the full mystery of death, but only by 'Natural Body Death'.

## CHAPTER 150

Having already displayed the army of change needed through the three taken prospects of the Fifth Dimension, accepting that the overlay is of joint significance to our politics and religions, but of a different emphasis, not from content but application. My charge has always been upon our religious leaders and will continue so for the sake of our own experiment. If I have been able and identify a crying need for all by application it has been for us to rationalise the spiritual and the secular by all modern times.

In using my own fixation of these past sixty plus years as a core means to look at an even better means of change, including reasons, all that has been done is the mild statement that we are the same. By link to that association is to be found by the small agreement that without the scientific answer of proof my conclusions are to affect us all. So that irrespective of any personal political holdings or religious settlement, our one goal by unity is to set the record strait. I acknowledge not for the first time, but by my own vision of self creation this task if a chore, is to be done without the need or inclination to change any part of our history and histories. Which might be held in deep consultation with features from the past having been set to change matters without the grace of real humility by the extending of irrational reasoning?

My tenet bears on religious extraction, so that for the one end experiment I have suggested, we make to turn the head of politics and feel the heart of religions. Making ourselves ready to be in the, a probable, possible position to attract Gods attention for and in human life if rather than owing to the means of what we expect from death? This simple standard although tried before could never work until now, because we have never had a time set such as these past sixty years plus to consider reconciliation. Because in the leaf of that time our drive had been wholly political to such an extent to make all matters half measured and bring all conclusions to follow our steps taken syndrome, both ways!

I have made no choices for us to alter how we are, what has been one of my aims are for us to fully recognise who we are and of our total connection with anything? It just so happens that the two giants of our character are measured by how we think and act politically or how we think and act religiously or how we interact between the two. By that quiet statement we have to come and realise much of what I have pointed to is done in conviction of our complete uniqueness. Now to be seen and challenged by the one task we have yet to follow and equate our consciousness to how we had behaved in the past to each other. Followed by how we can better interact in the future under the directed overlay of the Fifth Dimension upon both sides of the mirror?

Making ready for Gods direct interest is an odd standard when on one side we have already made that case through the structure of several different religions Eastern or Western orientated. My view in terms of further definition is for us to turn the tables on our politics even without charge if necessary. By first choice now that is to be done by gentile persuasion again beyond the critic, avoiding the usual means of dispute from where we have always had the same objective but from slightly different opinions. By taking to the standard I make with the Fifth Dimension change can be taken as and if necessary?

Although I have presented much of my case directly and on opposing fronts, no orders have been given, save we view anything said full measured, even half measured, but under no circumstances do we follow the standard steps taken syndrome. Which have been the steps to have caused the biggest means of keeping us on very different paths, while aiming for the same objectives? Any ways or means under proposal to make change, when done with or on or through the three taken prospects of the Fifth Dimension are guarded by the one active prospect in 'One for the Whole' to be our first real forum, which in office can be employed through the ideals of the United Nations in all cases. It is an unsophisticated means for us to reach the point of some of our objectives in the shortest of possible times. But will allow that we might repent in haste and reflect in leisure to what some of our proposals might have been, had become, were intended, but now viewed from both sides of the mirror! Making the picture seen in full relief that when the right hand is held up the reflection shows the left hand in motion, alluding to make the case in all cases against some the standard beliefs carried under the wing of the United Nations even now?

Through 'One for the whole', we will be better equipped to make the case against any means in the cause of genocide, ethnic cleansing, criminal murder, social murder, political murder or religious atonement of us killing self by self in our one Gods name Eastern or Western! 'One for the Whole' is the full question of not doing upon others what is done in retribution? Not only in answer to the acclaimed injustices of life from what ever cause, not only to respond in self doubt by any means including reflection. But the query of questions posed by what must be first done upon self by self as the only means to justify cause! It is a phrase in clarion to intimidate us all by forcing the answers before the question. So that at its height no person might use the sweep of ethnic cleansing in excuse, save the deed is not first sampled from reflection upon self by self? So that none can point in aim that another should act so on our behalf, unless following my example from the very suggestion I make! So that none can incite harm to others claiming protection by self immunity induced by the self assurance of being self led by other forces? Expressed as an entity in personal judgement of what might have been the intentions of the first holder! So that none can call in substance by using that first holder to have been Moses, Jesus or Mohammed or others and others to incite actions for others to do instead of self, so that all claims come into meaning by the obvious in 'Natural Body Death', without the pin being bent to fit the cradle? 'One for the Whole' is only an answer on receiving the right question, even if guided by the three taken prospect of the Fifth Dimension?

However we might try and set the standards of 'One for the Whole' in use, the one memory we are obliged to take from its intention is by its order of suggestion. Meaning in part, although it is to be recognised as a piece of our conscience working, our main use is to be full measured, whereby no leads are to encourage steps to be taken, unless by full enlightenment. Enabling us to openly recognise the intention of its worth as a string to the protection of our consciousness in relation to how we use the two mediums of politics and religions, 'One for the Whole' is to be our hand break on life and thought!

From the many different angles that we can look upon the results of our politics and religions through the mist of overlay by our three taken prospects, all belong in root from how we ourselves feel from the dressed emotions of several pasts historically. Although ultra modern in approach the Fifth Dimension as a catalyst for change, is only to operate from the results of our interactions over these past sixty plus years. That way we will be better able and focus on what might be intended or was intended in all cases from all the cases of history. Not least when seen of the open difference of how our politics and religions do work or have worked or are intended to work?

These past sixty plus years and our living memories of Shanidar are as much as we need to address the future by understanding? From both times we had the influences of religion and civil controls to be our guide. Any difference between then and now was how we developed over the last sixty to one hundred thousand years since the early experience of that tribal burial up to about the time of the Egyptians, nearly losing another missing link of our historical recollections? Without review but in range to much of what I had already indicated before our developments in the new world up to about four hundred years ago, by having the obvious connection of our human elements living and dying to and through the order of civil and religious control. Even measured by different standards than we would now use having the God feature expressed by acknowledgements to the Great Spirit and or a more vivid connection to the innate planetary or solar object like the sun, our aims are the same!

Indications of where, how, why and because, I had again referred to some of the old new world religions having originally come initially from the Middle East, stand by how my delivery was made not to include speculation, only the probable and or possible. But contained by how and when contacts were made or not. Limitations automatically applied by the nature of developments in place or by the level of held belief if to introduce new intentions. Allowing that my suggestion is beyond the standards of guesswork and in the name of experiment, because in my sight we had made formula without previously knowing the nature and setup of the experiment? Making for some proofs to how we think or are inclined to think without full use of our ability to make first contact by though?

By not then having the open use of our one active prospect in 'One for the Whole', some judgements made and deeds done were to continue the myth of our differences until these past sixty plus years. Making once again some of my reasoning in the assumptions I had made on all of our behalf unguided?

My double connection with the spirit of the old world in new days is to give another time mark for our studies, so that we might relate to how things have now developed? From the time when the Conquistador went to the American isthmus spreading North, but mainly South onto the continents proper. Although most European forms of expansionism and colonisation were generally stimulated by civil demand for the means of material gain. Our religions in those instances being mostly Roman Catholic by definition were closely linked to the civil and growing political influences of becoming more nationally inclined even religiously?

Outcome is not always the results of events, and for me to have suggested that much of the drive to alter events found in the new world, were prompted by little more than the emotion of greed. It is not to be inflammatory, when the open distraction of a culture that practiced human sacrifice, gave route for at least the opening for outside interference. Which although not intended directly was to draw unmerciful recompense for the two colonising standards of politics and religion with politics the eventual winner? When from study without looking, conclusion drawn about the peoples in the new world, we, allowed that the new arm of invasion in the name of God and state could further fragment the issue?

What was done was through the un-dramatic, by using Gods own will to have led to the destruction of the barbarous practice of human sacrifice, through the offering of human slaughter! In order to promote those acts suited to the situation, an additional split was put upon the same cultures beyond count. Whereby we European invaders could now cast our opinions to be the saviour of souls religiously and the bringer of wisdom politically? Not least because the standards we used having been set even while changing, were done in a civilised manner. Always now to improve from the European perspective while we had the open charge by religious direction to continue our political view first? All of which is not to be excused or accepted as progressive, but only considered from whatever the motivation our thinking belonged to then? Being a case in proof until the end of our World War Two, when for the first time we have become embarrassed about our pasts but have not found how we might look to the future except in using old tools until now!?

Although much has been earnestly done to paint new habits, all has not been smooth these past sixty plus years, except that we try and try, but by papering over the holes that have occurred naturally or unexpectedly. Much of the repair tape itself has been made from the ingredients of human rights, civil rights, laced liberally with the concept of being politically correct, topped of with the icing of religious tolerance in honest attempts to make the taste better from the product of living by our wits over any time. With the final picture having the form of we being fully civilised, but in the standard of being first or second or third world components politically, then in following all leads to be steps taken, because of how the list was set half measured? By first entering the arena half measured the concert has not been played with the intended score because the orchestra has not been positioned in place to give off the correct sounds in time with the correct conducting!

Giving weight to my endorsement of us having to try new methods so that we are all tuned to the same rhythm, even if at times I have to kick start the style of thinking I believe we need, but tied to the abstract? My motives are ours without any form of denial to any person, group or gathering, excepting those who automatically disqualify themselves?

## CHAPTER 149

If all that I would lead too, is the statement of our representative differences shown by the standard of Americanism and Islamic fundamentalism? My personal case has been built only from these past sixty plus years. Any certainties that we might hope and wish to draw accuracy from, have to be related to some personal experiences? Even then we err unerringly, because we have not been capable to see by understanding, opposing points of view if offered from different quarters at the same time? Hence part of the wonderful mix of the people we are, carrying the same mixtures of hope and expectation, although at times we may not see the same picture? Our standards have always been looking to the future in positive ways, even for those of us who when we die, expect only that the light goes out permanently. Are included in our full singularity because those signs are of human extraction also; and separate to all other specie?

Americanism and Islamic fundamentalism are to be my personal points of reference in cover for what is said in hand over these next pages. I take that standing on my own account as many already do, so that we might not exclude those lacking in our open ability to understand the concept of humanity spiritually. By placing the two above opponents in the situation of each looking into the same mirror faced off, whatever image is returned can now be challenged in question by the three taken prospects of the Fifth Dimension. But only working to my personal interpretation of what Americanism means on its own account balanced by the impartiality of what Islamic fundamentalism means on its own account. Once again, only by how I have seen events by looking back into the same mirror they are to use from now, so that we all might look to the future or what we should hope for!?

Both phrases above at either end of the scale, have no meaning unless we add the image of people, not to do would be a gross error full measured. Working without people continues our steps taken syndrome, because by first stand accepting Americanism and Islamic fundamentalism by name only is half measured? Being the unfortunate way we have come to operate by several standards over these past sixty plus years. By only taking a single aspect of any situation, generally politically first and working from that emphasis to draw final conclusions, half measured approaches equate to our steps taken syndrome?

Putting a human face on Americanism is complicated by several features some of which I have mentioned and some I shall now gloss over for expediency. Amid all the complexities of any national standing, America was created at a time of some of our most monumental changes by any historical standard. From her growth, all past images and records were side stepped in the very name of what was perceived to be progress. Done in all of our names, changes were carried out for us all in representation, but at first from old habits, which was the norm. Nevertheless America in her later turn brought the idea of some different conditions to be amongst us at the close of hostilities from the Great War or First World War ending in nineteen eighteen.

Without notice, by some standards although the Great War was the ending of many forms of autocratic or monarch led forms of government that had been standard in Europe at the time. Our American form of democracy was not fit or able then ago to force change, even if tried through the League of Nations. America by our own drive and flair was fast becoming the most powerful nation in the world economically. This tool was to draw the interest of us all by inclination, not least because how it worked was to give in return the idea of improvement materialistically, drawing many of us to the product rather than the deed, which was no bad thing. For in fact the system did not compromise our natural drive which was always in a forward motion.

If in some forms of counter measure, what followed in Europe in times to become know as the inter war years. Most forms of government were challenged to change, sometimes violently through revolution or through the fist and the ballot box or through national civil wars. Following by whatever code, European conflict was raised again to involve all nations in the new Second World War, with much the same reasoning as before, like expansionism, materialism even under the frosted cloud of idealism! All systems of government through the eyes of communism, fascism, imperialism, democracy and in some cases democratic dictatorships took sides to the general lean of self advantage.

After the Second World War even that I cite the wrong order of events, through personal involvement on the two major fronts of world conflict, America once again was our lead to the political reformation that was to follow. By the many anecdotes I might feed for our consideration, because of circumstances aplenty, we were more prone to listen to further American proposals, even if through the renewed League of Nations into the United Nations.

By whatever means we were to promote our own standards post the Second World War, we were automatically drawn to the outcome from the circumstances of consumerism and commercialism being the main means of our expression? So that in seeming we were best involved with the global community politically, which in some terms of my presentation leads in balance of my earlier statement by reference that America is representative to the concept of Americanism? Which has been only able and work by our desire of ownership, being part and parcel of how our natural progression had an inbuilt aim for improvement in all fields, after all what we only want is better, or more or more?

A small example to one meaning of what we do is illustrated by how we naturally treat our young by general consent, wanting to provide better for them than it was for us. Unfortunately this in cases means we inadvertently drive the process of materialism first following the skills of the secular? Yet at the same time allowing many of our religious instruction lay fallow in afterthought, although a personal observation, the above allows the format of what might follow or has already followed? When considered that the same forces of political materialism when it works, lets our civil means of control allow us set the empty charge of religious tolerance? Further allowing that we might operate in full operation of how democracy and Americanism works following those ends, but steps taken!

Not to highlight the obvious lacklustre of how some of our westernised political systems have inadvertently diminished the value of some religious determinations. Not to emphasise what Islamic fundamentalism means or is supposed to mean or has been taken to mean even by myself. When calling for our Fifth Dimensional overlay religiously to set open reaction by our religious leaders. The concept of fundamentalism cannot be equated in modern times unless Islam finds a means of expression to compliment the Quran along with other just as serious and devout meaningful writings!

I make that claim in the order of control, because since inception some fourteen hundred years ago. Islam, although prepared, had not been able and notice the volume of change in us as the one people were going to undertake by natural progression. No more complicated by how from the new world discoveries of some four hundred years ago those new lands in our lead, were to produce Americanism. Being deciphered by me to be an expression in cover of how all peoples in general were designed to reach the same desire of Americanism if natural progression continued only by civil demand? From the other side of study America who had the majority of earlier settlers hailing to a Christian background, yet in more recent times has a growing number of generational American Muslims for our mixture?

Although Christianity can justifiably assemble to its many leaders tied in aspect to innumerable Christian religions, having sprung from the one heart of Christ over the past four hundred years of her total history, even if driven by Americas own interpretation. Much of the reasoning behind the Christian reformation was because of the events in Europe leading to the push for increasing nationalism then ago. There was nothing in text aimed to hold or limit how politics might develop because the core Christian religion of Roman Catholicism alluded to the universal aspect of the faith being catholic. Plus allowing the consideration because of different circumstances that being Roman the faith was both national in part and political by type? But overall all, was set in the one direction laid by the heads of the church, perpetuated by its own leadership structure, which was to remain universal amidst the concept of nationality!

I do not make the same comparison about Islam because when first brought to us the study was to reassert Gods original plan, proved by the direct connection with earlier religions of area by mention of Abraham, Moses, Jesus and others and others. Mohammed, even if not to our plan gave us a worthy goal by returning us to the real concept of the spiritual being of primary concern over the secular? I had stated many times that a core reasoning of the new faith was for the individual to be responsible for their own standard of conduct in the pursuit of moral fortitude as seen. A mark to enhance this approach was that in proof of the standard holding its own weight, there was no seen need for similar priestly structures held by other faiths, who had used the like good offices of Abraham, Moses, Jesus and others? The faithful of Islam in general were to be the masters of there own conduct ex of a defined religious leadership structure.

Without comparison but in looking at Islam today and or any century by century of its existence so far, nothing has changed save the general growth of materialism or the gaining proof of consumerism and the influential field of political controls one way or the other. If the first or middle and now the last generations of peoples to have come to the faith were inclined to be laid back, even if only from the circumstances of climate, geography, diet and other reasons. As a general connection, what might have forced our hand in the making of other opinions outside of the normal which might have been anything, could have been nothing different than self conditioning, induced by the natural flow of change generated by the past fourteen hundred years in any case? Only now to have been highlighted by our collective actions and reactions over these past sixty plus years when change is the master of all, right or wrong?

With the aspect of change so prevalent by my own observations of late, what has occurred to me, is what I have relayed to us. Although change has worked unforced, although change was inevitable, although time at a time was temporary set to stand still by 'Natural Body Death'. Up until now we have been able and do nothing that was to show the cause of our differences, other than be different? Unfortunately in far to many cases of creating the only form of reaction to those differences is because we had not yet found the way to control them. Our first step was of denial with the second of destruction preserving the misplaced mixture of defending protecting or expanding our personal concepts having being delivered with no other alternative? Being only a defined term of reference about us all in relation to how we actually are, as opposed to how or who we think we are?

Since the realisation, that for those of us here now is representative to all peoples from all times and to us all, all past knowledge is our own, meaning that our link is complete by that and those historical connections yet to be balanced by our realising the fact. My story belonging to us all sets the open display of our differences in these modern times to have been rooted in some of our historical associations, which is no bad thing. Because in any case we can recognise the essence or cause of them, we can suffer the means of change to accommodate the necessity of the need for change. By having laid the case for what I consider to be a pivotal point of reference between how Americanism and Islamic fundamentalism are prone to interact or react. All representations although our own are culpable in name to some people directly involved in the cause and means of actions and reactions form either side of the equation?

By noting several aspects of how politics will be brought into check. On the same scale, but from a different court, we must equate what Islamic fundamentalism means in terms of how some people act by its lead. Islamic fundamentalism is part of the pure case of religious tolerance, but in my eyes through the lens of the Fifth Dimension, actions from that core are to be changed. But not too or against Islam, only those of us who would use its good office for devious ends in disguise of being religiously found?

My representations of our natural drive to be ever progressive, has been put forward for us to strive to reach the American dream. Meaning nothing more than the open connection of what our lives are about without religion. From the times when we were hunter gatherers the same objective was always on hand for us to be released from what we were doing, to be in the position to have things come to us on different terms. One of our first proofs of this situation can be collected from our efforts of depiction in many of our cave paintings, by sheer presentation they tell us of the difficulty involved in creating those images. We are automatically told of the complexity of simple societies even by their standards, but the obvious is clear, by divide such societies were naturally progressive. Add several millennium of time to our cave painting days and the internal split between the same tribes and clans had become almost inborn?

Society at about five thousand years ago carried this split to have developed as a normal medium of conduct. Having now structured our styles of life in the form of what it had become in line by the falsity that we could not visualise? Our assumptions had led us to abide by all the means of civil order structured by how individual tribes set there own standards. With the close interaction between those of us to have gathered in the area of land between two rivers, our first inclination was to follow those civil commands or the making of them by whom so ever in terms of how our tribes were socially structured. By our own close proximity extending to the Middle East full, we have had the constant use by our interactive ability the ability, to change and follow the best leads of conduct.

In using what we had in any case, our social structure was best expanded by first following civil/political leads if and through how the tribes of Egypt first spread the religious connection by a meaningful link between king and subjects and cause! Even to be aided by an established priesthood to be cast as our first set of religious leaders because of the role they had acquired apart to the rest of us being the faithful? Without comparison, even if, and I raise the point again because of the obvious by two standards I would like us consider. Being the very standards that had developed, but so far without explanation, which is required but not necessary if they are held dormant. If any Egyptian migration was to carry some of us to the new world pre history by my own suggestion, any points I have made are to show how a particular method of conduct might ensue from the one start!

However any contact between my Egyptian travellers pre history might have been made, what they brought to the new world, could have ended when passing the knowledge and reason for building stepped pyramids and nothing more? At the same time allowing that any form of religion to have developed there was done from the time of Shanidar, but inclined to have evolved in there own shape from that hidden wonder about the mysteries? Accepting that the one core religion was to the Great Spirit fixation, so much so that by all proofs the concept was always so. Until as was found by later visitors that the tribes of the Americas, although having a vast continent to stamp the excellence of all opinions in a progressive manner to their own interest, later had come to change by others influences?

Making the case if this was so, what was the means of change and if it had been introduced by others, what was the lead? By some means if the Great Spirit concept alluded to a force of final control, an end being, controlling the mysteries for peoples who were one with nature in any case? How would it be if an introduction was made not to change this motive, but highlight it to an acceptable and reasonable form? Like for example, if others, even visitors, settled to stamp there own impressions on this new world in building pyramids and showing a more defined religious order controlled by a dedicated priesthood. What could be more like the power force assumed to belong to the holder of the mysteries than the image of the sun by day and the sky by night? Then to have our tribal leaders more directly included to the welfare of us all by decree if supported by our readymade link between our civil control and religious lead from our dedicated priesthood as had happened elsewhere before!?

Without the suggestion of result, even if concluded from abstract reasoning to some of the above, our alternatives to what happened in the new world from any time, are for us to ponder without consideration of how peoples in several tribes managed to invent the extraordinary or a new concept unaided! My case is not a case of accusation to be negative. What I would hope to illustrate is how a set of circumstances might have occurred not from the means of operating half measured, but to the consequences of any so provision later carried out as steps taken!

If the fact of human sacrifice was to come about only from the existing standards that we have or can assume to have been the case in the Americas at times pre Columbus. Our assumptions have to include and allude to the realities of certain other situations and circumstances that might have been. In as far as we are honestly prepared to accept some of our ancestors in first settling the Americas post Shanidar possibly forty thousand years ago. In myself using an ancient burial rite as a means of connection by all creeds, accepts that in the Americas we did indeed translate our thoughts of what mysteries might be into what they fully are by accepting the spirit of the Great Spirit as a means? So much so that by first sight and relaying on the immensity of the geography of the continents, allowed for the fact that we did not need to adopt to live in city states as the geography of Mesopotamia demanded.

Our American dream was stamped in Braille in that we understood by touch that being there was all that was required. So much so that to the vast parts of the northern continent by our own influences we found in answer that the Great Spirit was enough in requirement religiously. So much so that our political structure our business acumen our materialism was not to the fore. We had what we needed; to live in the full compliment of life and while doing so interacting by the means and habits that we had acquired in due course. Our situations in many cases allowed that some of us were semi nomadic by seasonal traits because of the size of the land we lived in. Knowing that all things eventually were in hand by the Great Spirit allowed that we had found all levels of understanding, so much so that our individual standards did not demand the necessity of a driven demeanour, because our first recognition of self was to the realisation of death and the afterlife.

By briefly looking to the tribes described above connected with how I had interpreted their situations if only by example. My point of observation is for us to see by example of how things might have developed across the world from the Middle East. But with no knowledge of one to the other until starting afresh some four, then two hundred years ago and finally now in us having to make judgements based on interaction from now, determined by events over these past sixty plus years? Although I have made the choice of some of our studies to be what might have been the situation in America before it had any name in a global sense. By statement in effect that one philosophy stood on firm ground to the open acceptance of the Great Spirit and the other did not? Opens for me a study in why we have come to be slightly different by very modern reasoning!?

## CHAPTER 150

Accepting that the first human migration into what was to become the Americas; was done through the northern land/ice bridge between the extreme North East of the Asian land mass and the North West of the North American continent. Then from the general migration expanding south ward through the isthmus into the Southern continental land mass proper, also pre history. By how the system works, bearing in mind these peoples as we carried sufficient in knowledge to naturally progress and definitely survive. Without the influence I had described that we were able to gain from each other in such conditions of interaction in the land between two rivers. In the new world first time around, I can speculate that the original line of contact from first entry over the land/ice bridge was maintained by all peoples to have made that journey and at the same time grew from those settling to there own choice of area.

Without tangible proof it is also safe for me to speculate in terms that we of those in group by clan or tribe occupying the new world, had come to the Great Spirit philosophy earlier than shown elsewhere. Hence no tangible historical records like cave paintings or grand city states like in other parts of the world. Because in having a final outlook we began to see the importance of life and how it should be done, until the full knowledge of the Great Spirit through 'Natural Body Death' became revealed to all, but in kindness to our own ability of acceptance! In being an end philosophical outlook that standard in no way compromised others as self from wherever. Unless to awaken our natural drive in another direction from other sources, perhaps some four or five thousand years ago came travellers to the new world arriving at the isthmus? Arriving from the direction of where the sun always rose and by carrying different styles by several counts?

My suggestions without the lead or intention of argument are not bland nor are we to assume I have other motives. All I would wish to relay on all of our behalf is how we are factually the same. By making note of the situation in the Americas pre history, against our behaviour in the Middle East of roughly the same period post the change of emphasis if gained in Mesopotamia. By the outrageous connection I had made of Egyptian travellers to the new world in proof to have changed something of the standards already there? My own and personal conclusions are that although we are indeed the same people by all considerations. It is crucial that we note in context what I have alluded too. In order of not being able and accept our similarity, I have given good reason of who we are in any case, but under the caution that change even if unwilling will be necessary so that we can progress in the correct order. Allowing at last that we might all draw similar conclusions even if through our one active prospect in 'One for the Whole' providing change without alteration!

If Egyptian travellers moved to the Americas in success, then by following the sun in its own ever quest to always set in the west, we have a mystery worth beholding? So much so that the mystery of what the sun was and did, had now/then become our history? Leastways for some of us who already thought on those terms and were now ready to suggest our own thinking on others? So much so that others like us found it easy to follow a branch of thinking because such a style of presentation suited our natural inquisitiveness?

Not least when by presentation in using the accepted habit of the sun to always set in the West and rise in the East, a form of signalled meaning could be presented to apply of an overall purpose to this phenomenon. Especially if given to have a relevance to the meaning of life through purpose delivered by the contained motives of the sun? Managed in case by an appointed collection of people set in task to relay style of presentation by the cause to be in understanding of such matters in the name of their appointment? Even if strangers, not as conquers to destroy the comfort in habit of feeling to the unknown concept of the Great Spirit, but to be enlighteners, to overlay an already accepted standard with a new medium. Whereby the civil leadership was enhanced, allowing that they were to be involved with the structure and fate of tribe and clan, absolute! But within the confines of how the motives of the object of worship through the sun by day and the heavens by night were to be translated by an able priesthood by whatever conduct advised?

My own observed reading of a cultural priesthood in the old new world, has been done to bring to the edge of balance the mere suggestion of my own ability to have made judgements of the past by events that occurred over these past sixty plus years only! By calling that we had arrived at the situation whereby any individual might pick and choose any number of failings, that we have achieved, over the same period and stated in claim, has been made for us all to face the mirror? Even that my choices are direct to the political aspect of Americanism and the fundamentalism now claimed by some in the field of religion through the religion of Islam, is not divisive!

From my own open concerns of what the future is worth and of what we might expect is done in the newest form of honesty that we must all employ by our best reasoning even at this time unfound, but discovered? We are a mighty force to be upon the influence of all universes, not least carried by our obtainments up to the ending of World War Two and since. But carried in the new face of reflective management, only now revealed by the concept and hasty application of our Fifth Dimension overlay upon all the aspects of the two matters of our previous guiding leads?

Accepting the two are owing to or from our political and religious systems of use so far, that I might have refined them in representation to the two items of my own interest? By claim that I use Americanism and Islamic fundamentalism in balance and counterbalance is to be a matter of all our concerns, when I have only asked for the reflection to be judged by what our religious leaders offer above any political leaders influence, is made from the sanctity of the abstract! That I would ask for that ideal to be our main concern enhances our joint position pre 'Natural Body Death', but to the caution of us all not to fail in the understanding of where I lead with abstract reasoning!

Although I have forced the connection of how the possibility of human sacrifice in the old new world had come about from the dying ember of belief to an original Middle East philosophy that had originated before my defined first, second and third dimensional religions from the same area had developed. Is to be considered of vital value, and was done by reflection and in cause of how we are to use the good office of the Fifth Dimension? With the concept of human sacrifice, no value can be given to or for or from any religion with such an aim! No value can be given that any human being in this age of time, being this last modern era, can make those judgements whereby we might kill of self by self religiously, which is always a personal choice?

If and so I have cast the gauntlet before our religious leaders in all cases, how might it be returned to me on all counts by the good faith of Islam, of who have no defined religious leaders, except only the faithful? If and so I have not laid the same gauntlet before those in what I and others have defined as being of an Eastern religions texture by style, the reasons are before all of us in any real case by us being who we are? My exact formula works in all religious cases because our connections are beyond our own real understanding and always have been since Shanidar. Which has allowed us formulate all personal concerns to how they have so far developed?

Buddhism and Hinduism and as stated all other eastern and or western styled religions have not been excluded from any offers, nor in the abstract has Islam been singled out in opposition to the two previous religions of area. Save that it followed both and took on the legacy of not having to have defined religious leaders excepting we alone, even if to be guided only to the extent of our own judgement in all cases, in all cases!? Therefore Islam has singled out herself for at least my own particular notice, but only in futuristic terms relative to the individual!

Although I will never be able and express my own religious passion if I carry such a tool, my views if expressed by the power of the pen are to have somehow fallen on fallow ground, if the gauntlet is never picked up because of ignorance or fear. However, even if offered by objection my standards cannot be moved by the negative. Islam by not having a defined collection of national or regional or tribal set leadership, in terms of the religion only, is not empty of responsibility? Her case by all examples in certain terms is of our last representation by determination to have been felt from the fine intentions of Abraham, Moses, Jesus and others right up to the last prophet in Mohammed, which is not to be seen as the final step but only the last place before all tomorrows!?

Mohammed always in the dignity of human emotion continually expressed the views relayed to him on our behalf. From which we were guided to and through the method and styles of personal conduct in self control within the framework of the new religion itself. By definition from having all that was relayed set to the written word in the exact time span it was done. Was enough to delay the installation in format of what had gone on before the standard of having a defined priesthood to determine what we might self read or learn from actual written text. Here was born the final method of contact and the means of contact we might now use in all cases of further adulation and understanding to what might be the last stand as it should?

Although mentioned before, from when I had previously indicated that this new religion of Islam was to be complete in terms of guidance for the faithful. No plan could ever have been full in human terms with regard to the future if the idea and the ideals were for peoples not yet born? Which is an open challenge I have continually levelled against all religious leaders from the abstract, or the use of reasoning so led? How then might I deliver in fair command what my own meaning is and in relation to events of these past sixty plus years to all religions within or without connections to the Middle East?

My case in our representation turns on little more than what had just been written by the expression "not yet born" in reference to us in membership of these past sixty plus years? Although all religions pivot to the objective of 'Natural Body Death' in full depth, some of us had allowed our own intellect and self importance draw for us separate paths to the final answer when obtained. Not least from how we at a serious time of our united histories had drawn conclusions, that we might in need, offer self by death through human sacrifice!

By re-encountering the structure of that image even if to have travelled back some four hundred years. My standard of recollection is by reflection to now of this instant, remembering that human sacrifice to have come to our notice in a different form than previous or accepted standards of the day or any day unseen. Had been worked at or on an almost industrial scale, so much so as to undermine the civilised conduct of those in the possession of that practice, even having been compared to a different style in any case right or wrong. By the clash of titans in religious terms through the meeting of a devout religious come civil order that revolved around the habits of the sun by day and the heavens at night to a living belief of the hereafter, sacrifice had to be sacrificed!

By the same standard, but from a different outlook, in dealing with the cases of human sacrifice as our human travesty. We displayed all the venom of real civilisation called by all from Europe, leastways those who were to lead as religious conquerors in our own self belief? In giving the same space again to the same solution of events of some four hundred years ago and since, my personal claim, which to me has to be successfully challenged by others, is to balance what civilisation really means? Is also that there is no case of proof that any faction of human society can be civilised by those events then and the same representation that I perceive to be owing to us all today, by my own historical and or ancestral link with our pasts? We must from now, always be ready and challenge what we mean to be civilised, if not first from the joint overlay of the Fifth Dimension upon our religions and politics from then and to now but of how we behave in any case?

From that very challenge in setting the standard in representation of what Americanism and Islamic fundamentalism mean. My conclusions are ours, but by and only through use of our one active prospect in 'One for the Whole'. By the same token, that I have set the task for our religious leaders to review all new standards is done on one basis only, that we all live under the same new expectation even that our personal time spans are to be fifty plus or seventy plus years long. From within those particular time spans for all of us, we will find our true unity. Although in constant display we have not yet seen the picture because we have only been looking at the reflection?

In as much as that image showed us the left hand being raised, we only responded to what was meant by raising the left hand, forgetting that the original form, was us lifting the right hand either in exultation or condemnation, but by the sign of the right hand! Any reflective response was assuredly how unnoticed we had now and then automatically taken the un-obvious and reacted half measured. Making its own product of our steps taken syndrome to be the norm, because in all cases one wrong has never been able and make the representation of ever being right?

By connection but unforced, that we might respond better or quicker or from the limit of our understanding to the reflection of events, before the original devalues the concept that we can be truly civilised? But not in finality only by my own means of being or thinking in cross referral or naturally from the abstract! I do not question the value of civilisation because it in faith is the same substance as the very skin that covers all of our forms being black, white, brown or yellow? Even the many other sub titles some of us would like to be known by, like for example the red skins of the new world as first taken at a meaningful time of our united histories. Although created in error from the time of connection the standard can be held if to be in the desire of those in who the error if, was made about!

Without a clause in proposition my mention for intention about us being civilised has to be brought into question as a means of unity. Although we might or at least some of us might think of ourselves being civilised, the question is only raised in the positive if we acclaim we are not? Not to compromise any situation but better enable us open the doors of human unity. By example of at least the United Nations, when next we speak without the compromise of having fixed standards being first, second or third world economically, who is civilised!

From my own intention showing that we are to accept the case that we are not fully civilised, is to be supported by the fact in limit of our personal lifecycles!? They in turn to be of a fifty plus or a seventy plus year span are our judges. Not in how we behave but how we think in terms of the future in so far as that is to be the judge of our conduct by the said time differences but not on a sliding scale. From which in meaning without the study of the difference of twenty years I had previously continually quoted, sets no value between any of us because of prevailing circumstances!

For me to temper the value of us being civilised is done in the positive, but alas some four hundred years late. If I have led us in what to think in terms of the human sacrifice syndrome found in the new world then ago and by whom? In connection to what I had laid in balance by Americanism and the other opposing faction of Islamic fundamentalism today. My overt comparison is to show the vast difference by similarity between us then and now, but in case of how we automatically accept different standards of civilisation between then and now. Especially when now through the good offices of human rights, civil rights and in operating politically correct we have inadvertently moved the goal posts. To show that only some of us were civilised then ago but we are all now civilised excepting those who would set to compromise the concept in meaning of what religious tolerance is? Falsifying our account of accountability expressed through how we now behave against or through the full limit of our personal lifecycles?

Our religious leaders are now best qualified to give us religious instruction, but always now in full modernity. Because at this time for the first time ever futuristically enclosed by these past sixty plus years of this last modern era, we can now successfully cede to the ability of national understanding in a difference than ever before, because we can relate to the individual over the means? Meaning that we can all relate through and by our overlay of the Fifth Dimension upon our political and religious standards with the same intention, but allowing for different readings of each element? When from politics we have individuals to interpret intent, yet at the same time through our religions we can cross refer through our religious leaders the meaning of intention balanced by the object of purpose? But in both or all cases now, to be refined to the importance of the individual?

Being ultra modern has no value across our human divide because of the one standard that we shall all hope to meet with the good offices of 'Natural Body Death' by my own preferred terms! I personally cannot change or challenge what that might mean to us individually, save that in my own case I am now even closer to the situation which I see as being no bad or troublesome thing. If, and I make this statement without claim, the only objective of life is death, then that is a fine reward? Because lets face it, for the vast majority of us all, our standards are utterly magnificent religiously!

Even that some of us which might number in the hundreds of millions have to and do suffer almost intolerable indignities, laid upon us by our normal circumstances, broil in our unseen misery? Although endowed with the wonderful concept of hope, laid upon us by our religious determinations, those of us in such misery are also to be brought forward using that image in guide? Now to be set in balance by our religious leaders, of who we are inadvertently involved with, but this time from now through the Fifth Dimension! Being our own tool from which we can set the temperature of required change, when we may begin to look into the mirror and now see what is real even from the double reflection?

## CHAPTER 151

From our four billion year old planet to our three and a half million year old specie, makes all of any arguments to the time scale of evolution turning into creation, but running in parallel. By further consideration to our own lifecycles being fifty plus or seventy plus years long, the total case of creation going back to evolution is balanced?

That I acclaim we are indeed Gods own clan, has been done in full reverence to what that expression means, has been with due deference by human terms only. For it is we in all of our loving complexities that has cast the colours of the rainbow in the sky, it is we who have painted the graphics of the land and the sea and the sky at night. It is we who have dated in time the lost dinosaurs the lost ice ages with there land and ice bridges. It is we who had seen and recorded how some other specie lost to us had more easily changed to become or remain crocodiles and or remain crocodiles and become birds from the lost? We are the only counters of time in the whole universe, how ever made, our total concept of now in this instant, which in comparison to some, is like the blink of an eye. But oh what a blink, for it is only we who have looked into the face of God and seen the right reflection of ourselves to be able and consider all of life's beauty, while living in its ugliness!?

Only now after three and a half million years and the sixty to one hundred thousand years since Shanidar can we wish to see the light of our intentions. Especially when for the first time ever we had come to accept in concept that our views were similar leastways politically expressed through the United Nations. But unfortunately compromised somewhat in assumption of our deemed acceptances to the standing of natural religious tolerance?

From where I in our representation had taken issue to such matters by formulation, my claim to our differences instigated by the circumstances of mostly climate, geography, diet and other reasons. Was done to show our changes without alteration, were made from the abstract without interference, had been done to express how we had indeed changed, but were the same in all cases, was for the matter in representation of common decency, C D, simply because we are who we are. Even that in cases far better people than I readily allow our colour or creed to have had self expression on our behalf, instead of listening to there own separate requirement? From now we are required to only listen on our own behalf, but here for all, here for all no matter what our perceived differences are, leastways now with the overlay of the Fifth Dimension!

In paying the piper and calling the tune, no tricks have been laid in any case, save we are all of one face? By collection, any standards I had previously laid for our attention are still ours, any ideas of unity I had proclaimed or suggested, are still ours, everything written from every pen that had been used in all cases, are still ours. But from this example of now, because of now, and not fourteen hundred or two or four thousand years ago, what had touched the paper in the form of ink or paint is not what we must do. But only what we must consider, even if to question what the same hand in we had written four or two or one thousand four hundred years ago unchallenged. But by indication to the thinking of four, two or one thousand four hundred years ago, against us alive now, having gained that extra time in what should always have been progression its self or natural human progression always!

By example that I had connected to the importance of our past four hundred years of our united histories, whereby for the first time we truly became global in our own human concept by the discovery of two new already inhabited continents in the north and south Americas. By the same errors of time owing to humanity, from then was our first path to now, but under the importance of our direct connections with and through and or connected to Shanidar? Making now my own case for the utter importance of humanity above the distraction of our clan or tribal or social definitions, but further and finally expressed by any division we had allowed grow between us by our own religious divide. Making the case, because of how things had developed, instead of what might have been predicted? We each can look to the future without altering the past by any means, which means we do not have to accept all of our past values. Because they had not had the ability to consider or predict how the future could be without the full knowledge of the two continents of the new world or for the new born in us!

From that connection which I had laboured, how I would direct any further conclusions from what we now hold in knowledge and understanding, even supported by our best efforts over these past sixty plus years. With the additional factor of the three taken and one active prospect of the Fifth Dimensional overlay, our reflective image will naturally grow in definition. Not least by the range of understanding we can now give in comparison to what is meant by Americanism and in counter or balance, Islamic fundamentalism?

By representation, what I have delivered by the word Americanism stands for complete modernity without deviation. My concept of it has only the hint of religious influence even though I had offered the picture of the mainstay of national culture to have carried Christian perspectives! Americanism by count stands for our success in most of our political fields, so much so for me to hold the final standard of our achievement to follow that plan in title of Americanism? Done for the reflective image of what we all have achieved but by reference to from where our modern standards had automatically come from or have been projected from historically?

Political or commercial success will always have a lead, but in this case we must take final note that although I personally had attended to these past four hundred and latterly these past sixty plus years in reference. My conclusions in all ways about Americanism are ours, by name it is representative to us all no matter from where or how our standards of today have come from, even by other circumstances? When in reality we had actually migrated to the two continents from both sides of the world through the land ice bridge in the eastern end of the land mass of Asia and across the ocean from the East. In both cases having taken our own but different perspectives of how the simple event at Shanidar had affected our thinking by different primary circumstances. Nevertheless even that we had drawn many different conclusions about our religious heritage now unknown, we now act politically?

Americanism has now come to represent the idea of humanity by secular activity first? Even that I set the standard to suit my own reasoning, my view is tempered by our Fifth Dimension overlay on the subject of the political, separate to the religious? In pressing the point, my standard is of the same value we all must use, even when in balance we can view or review what is meant by my own other view of Islamic fundamentalism?

In so far that I had made on our behalf, a lead to the feature of Americanism, one of my aims is not to be divisive but show our real cohesion, whereby the term is to represent us all even if some of us might be laid back? With the progressive edge of what Americanism means owing to us all by our own direct methods of natural progression in the secular. Whereby we have naturally looked to always progress by the physical means of materialism, which is no bad thing. Unless it is offered in the failure of what commercialism means without all aspects of our religions, but not only by tolerant means?

# CHAPTER 152

In using only hand held scales to balance what I mean, by having one cup loaded with Americanism and the other cup filled with Islamic fundamentalism, is done from the past sixty plus years only. In so far that the two are offered in one being representative of the secular and the other stands for our latest religion is to bring into focus the gaps in our very modern thinking. Which I promote to be of vast serious concern when within my own lifecycle incomplete, I have been able to see intention when alas some of us as we avoid the issue?

In placing Americanism to the secular all I have done is to set that side of the scale to be of our final political position of standards so far achieved, right or wrong. We all know much the same thinking has been set upon our own religions, but also from that political example we must take note that the main religions on the American continents seems to have been Christian by texture. From where I have judged them to be from the Middle East first hand by association, which has to be our conclusion with other considerations counting?

If Americanism carries religions form the Middle East in being mainly Christian, then in link without balance religiously, our standards have to include the aspect of Judaism because of Abraham, Moses and then Jesus Christ. Also to that connection from and by various Christian doctrine some of group or groups, will use the essential part of the Old Testament of the Bible, quite the same as some Jewish users, but from a different perspective?

With the further addition that other Christian groups will only use the part of the bible know as the New Testament, to disassociate from Judaism so they appear to be naturally progressive by the nature of events to have occurred in the Middle East and when!

Balanced on scale by the exact same margin is the factor owing to the religion of Islam, which also percolates from the Middle East, her adherents carry as many political objectives owing to anybody elsewhere in the world. If I had given for many in Islam to be laid back, was done for our enlightenment, not least to indicate again all might not be what it seems? My standing has been by observation in respect to the open display of many people as you or I who are there? What I have not been led to say, is I make my assertions not to how many individuals act, who could seem very active and volatile in the process of normal daily life. Look at any market place there, when from the colour of sound it might be given that these/those people in we, are indeed very excited while bartering daily business? Which might sign they are not to be classed as laid back, when from virtually every aspect of normal daily life heated and volatile exchanges seem to rule the norm, save when religious obligations are raised, because in out final belief, we can relax, laid back!?

It is from such a mix I have set my own standards to many groups being laid back. Of course we are entitled to be excited and vocal about many of life's smaller dealings, like when we go to war for instance? But my take upon Muslims being laid back from the aspect of their religious doctrine, stands in case, because that is the picture many of us present through the satisfaction in the finality of our chosen religion?

Mohammed had placed the concept of Islam very well and in the full area of modernity that then prevailed. No wonder that on its own account it was able to grow in form and by connection to all people. By previous mention that this concept had similarities to both Christianity, but more so Judaism is a statement beyond the requirement of comparison and is not to say it was directed from there? Quite naturally the peoples of the Middle East had obtained most of there physical attributes in mask of the normal circumstances of geography, climate, diet and other features. Which is also explanation for some personal habits of hygiene or indicators to thought patterns, alluded to by the prevailing circumstances that were and in place? Therefore by reference only, to the standard of circumcision in conduct by Jewish men being adopted by Muslims, associated by my own speculation relates to area! When for Christians or Roman Catholics at the time of Mohammed, did not see the requirement of such an act. Nor did they see the same dietary considerations associated to the slaughter of some meat stocks by blood letting or some particular animals seen as unclean, which in idea could easily have been borrowed from some parts of the Old Testament, which did not necessarily apply to those who used the New Testament?

Making the case that with a ready made priesthood led religion, now coming to an Emperor led empire, from the massive influx of converts to be spread throughout Europe, no set plan as before was laid down whereby males should or would need to be circumcised? In addition, for all Roman Catholics on a broader base, normal local dietary habits were run by different tribes, because of the factors of local climate throughout empire, there was no area link with unclean foods as there might have been in the Middle East proper?

From the first time when the Islamic religion was to be set up, although much of its primary doctrine was quite naturally aligned to some aspects of the past, it was a wholly forward looking religion? Perhaps to take all others one step forward, not least from with the natural feeling of exuberance of being involved with the intentions of the future, new converts took in good heart to the task? Although praise for the religion is worthy, we must call in question some other features of its benevolence? I have called into question some of what might be shortcomings, if measured by my own reasoning or what will be collective reasoning. When we all adjust to what is on offer by the three taken and one active prospect in the overlay of our religions and politics with my Fifth Dimension! When the first open realisation of that feature stands to support all religions in balanced power to the denial cast upon us from the fourth dimension, which is our levelled critic, who criticises from the arrogance of rejection, but now our prospects will trump that issue!

Although Moses, Jesus and Mohammed left our associated respective religions in good order and in the natural link that occurred by progression. From Islam being the youngest religion by date, shows in concept it was the most modern? Therefore of its time, was able to apply any newer learnt benefits separate of the other two progressively over two thousand plus the additional six hundred years of Catholicism and Roman Catholicism from Judaism. Although in general terms all of the faithful to each religion was to be responsible for there own conduct of behaviour. Contention might have been drawn to the thinking of Mohammed, not least about how the order of conduct had previously been passed on by a dedicated priesthood?

Even being fully directed, any leads set by Mohammed are always questionable in context by those of us who question the mystery of the rising sun daily? My own connection to the religion of Islam proper, is that no direct high priesthood structure was ever instigated, not least because direct emphasis was more detailed with regard to self conduct of the individual not only by suggestion? If modern thinking was the case, then the idea of reducing the need of a dedicated priesthood would have been prominent. Again when their role was to continually remind us to the nature and reason and method of self conduct? Instead, why not use the spirit of progression by example and rather having to continually remind the faithful of how to behave. We could strengthen their hand to the faith by better self inclusion in the means of conduct supported by self manifest pillars of wisdom! Whereby using the aid of support offered, we naturally acclimatised to what was meant by being self involved. Rather than being led by a feature that could have held or showed political influences also, if the habit was to suit political direction!?

Again to the same image by reflection, no man is an island, although capable by self conduct we do need some formal leadership to decide over what might seem small matters? Like for instance how we should conduct ourselves while gathered for collective worship in our mosques, which is of particular concern to Muslims? Although ready and able to manage by self conduct, we still need guidance for communal conduct, even by a dedicated group to be known reflectively as Imams in general appointment to our mosques? Which in one role were of a priestly definition, but generally attuned to how collective conduct should follow in our mosques from another role of being ushers, teachers and keepers!

What ever started in good sound was sure to change by the advent of time, which in a way had been compromised by or from the very roots of Islam. It had been mainly through the Quran that conduct over virtually all deeds and imaginations had been presented to us all by the works of Mohammed. Seen to be a more defined form of reasoning than the hearsay collected from what Moses or Jesus had imparted to the faithful. This new progressive style of presentation was to be final, therefore while not presenting a dedicated priesthood as the directors of wisdom. We had been left with a situation capable of working continually without change, being of great importance and value to the thinking of the day and the time then ago. But not being fully able to consider what might happen a thousand years later like after the rediscovery of the Americas or the events fourteen hundred years later in these vital times over these past sixty plus years!

We cannot judge how the course of events in the name of progression affects any individual, but must look at the range of possibilities from which some actions or reactions are drawn. Islam by its own structure of finality, by its station to represent us to the end game religiously, has no charge to be met in the same way that some factions of other religions also are not to be challenged. Where a gap has been made for all considerations, is balanced to how our religions and politics had come to react together from the time of Mohammed until now in these past sixty plus years?

By implication and by reflection of Americanism, the same finger has to be viewed to accommodate what I see Islamic fundamentalism to be? In the case of how the faithful are to conduct there obligations to the last Middle East religion, I had stated of the link between imams with involvement to general conduct in the mosques and with other duties to oversee the learning patterns of the young. Leads that in role, they are at liberty to pass on their judgements from within the limits of there own understanding or more!

Without pointing direction, but not to have a higher or a more defined form of direct leadership than had been submitted some fourteen hundred years before, belies how some of our reasoning of then and now can be coloured due to the changing circumstances of then ago and more recent times? Without compromise and in the position to act liberally on our behalf by some events to have occurred in recent times again. What I have viewed in balance is our relationship generated by events in close quarters. Following my judgements that new world decisions generated from the concept of the United Nations have left the order of equality wanting, not least on religious terms!

For us to have covered the seriousness of how we take our religious commitments within the envelope of religious tolerance, does not give credence to the nature of most of them in spirit? Not least when some, or by my own acknowledgement to the religion of Islam, have shown objection, stands by the very circumstances we have ignored it by the false manufacture of our political determinations! Seen reflectively over these vital sixty plus years of this last modern era showing our raised standard of attention to the aspect of politics alone, allowing that we give false praise to this new prophet!

If our religious leaders are to be involved to the matter of change generated by the aspect of the Fifth Dimension in overlay, how can Islam self represent her own standing in deference to the past fourteen hundred years of changing attitude without change? Even that we might take its standing pre the ending of World War Two to have been settled enough to meet the impending political changes professed. All we are now required to do to separate opinion from deed, is look to the mirror of self, and study our own reflections?

## CHAPTER 153

Even by a cursory standard of looking at the basics of the religion of Islam, how might my charge in recognition of Islamic fundamentalism be set, upon the most settled of our three monotheistic religions to have emerged from the Middle East be met? That I had used the expression as a means to balance our political standard has or had been adjudged to mean by my own reasoning. By now looking from a different perspective to the religious element of that balance or what I have perceived to be of essential importance! To take either case from different standards is also weighed by fact that the same approach of understanding is laid the same way upon both elements? Because now by my definition of being political and religious will have the overlay of the Fifth Dimension to be upon them. Allowing by its own means to be able and draw different conclusions from the different standards again, because of there different primary outlooks?

Islamic fundamentalism up until now had been a phenomenon to have raised its own form most prominently from the time of the Iraq/Iran war, also again by connection to the earlier events to have been given the misnomer of being politically driven between the Islamic people of Palestine and the Israelis in the mid nineteen sixties!? Then by my own description to have derived in the new format, we can now see from events just after World War Two ending, when by political terms the religious state of Israel was first created?

Such connections that I make are not to be cast in stone but in part are to show how conclusions drawn can come from the reflective edge of our own observations. In claim that my case having been described in balance to be of our political emphasis through Americanism. With the other cup holding to my own definition of what Islamic fundamentalism means, is given air in the utmost of caution, because of who we are. Although I consider ourselves to be of the highest order and worthy to work in claim of Gods attention to us by the value of our own actions and deeds at this present time, even to focus on these past sixty plus years? We have arrived at the point where on pivot we will either progress as intended into new avenues or regress to the thinking of how we were prepared to act from any time in the past, only to remain there?

Islamic fundamentalism might prove to be the counterbalance whereby either of the above prevail, not least that it represents us all from our religious perspectives? Even from earlier when I had indicated the difference by comparison where Islam in religion was established without a directly defined priesthood of the same scale as earlier religions from area. Allowing that certain patterns of duty in conduct from individuals, mostly our young were overseen in there religious obligations in part by imams. Those of us who from the benefit of our choices to have studied in time all the events of what our religion was to mean, who could now impart what was or had been meant originally from text. But in style to have laced the study, our study by the human element of changing attitudes gathered over the past fourteen hundred years, then latterly from the political influences of the past sixty years fail, if we allow the transfer of religious reaction respond to political claims or assumptions?

What I have taken religious fundamentalism to mean by reference has to include the basic instinct of returning to what was easier to understand, or more comfortably to follow? If in case, I have associated Islam the same way, then it is possible or has been available to mix the standard, like from or like with no other religion? By taking the steps of imagination to compare some supposed actions of today with what was permitted in the early times of first contact. When the standards of protection or promotion to the matter of defending the faith, Islam's own adaptation, whereby honour to become a martyr to ones belief, stood, but from first instigation only!?

In the time of Mohammed when life and peoples exchanges could be very hard and violent in the fixed standard, with the value of human life of little consequence, some of our reflective responses were mirrored to how we though of each other by any set of standards even religiously? Then in direct consequence as it might have seemed, when the new order of conduct was first promoted by Mohammed, in following the style of persuasion by the means of force in due? Even when the same methods of that persuasion resulted in the physical deaths of some on either side to the conflict of ideals? If from the death of one person to be of no consequence while opposing the first ideals to the promotion or protection or defence of another's new ideals, how would a death on the other side balance? Unless now to die while in the defence, or protection or promotion of this newly appointed faith, active martyrdom counted?

Applying a newly contrived form of conversion, whereby in some cases after battle, if against city states or those in the unity of statehood, was to result in victory for Mohammed and followers. Those to have survived the rigors of open conflict, were offered different standards, but in or from the general line of thinking of the day, whereby the vanquished were offered conversion to these new ideals or obliteration without consequence? But allowing for the same consequence for any deaths to have occurred on part of the victor, acclimation to them in being martyrs to and for the good standing of what this new faith has to offer. So much so, to carry a greater accolade than ever before displayed by other religions under the hand of Moses or Jesus!?

With reference to the style of Christian martyrs without comparison, but in line with the thinking of or at the same when Mohammed first operated, already throughout Europe the term of dying for ones beliefs, to stand in meaning to die in cause of faith or empire. Covering several historical anomalies from past records even pre Roman by adoption, when those mostly persecuted on religious grounds were the Catholics who were later to conquer there conquers in any case. But until so, any deaths to the followers of Jesus were called the same martyrs as later, even that Christ's martyrs were to turn the other cheek!

With the political overlay of Rome to follow, standards of conduct could be and were devised between a dedicated Catholic/Roman Catholic priesthood and the civil administrators of imperial management, who were already there for the new thinking now needed? That I had referred to the new religious standards of empire having remained mostly in Rome by name, even after the fall of the city. But continued to stretch and reach throughout Europe proper, held together by a dedicated priesthood using the ability to mix and match the ideals of a single religion to the name of many different tribal groups. Who in general terms still had some tribal members martyred for their faith even when allowing political management grow in Europe to a different style than in the Middle East! From where new ideas and ideals rose as ever, but not yet to have been viewed on my terms on our behalf? Because it is only from these very recent sixty plus years of natural achievement we can look back in balance?

## CHAPTER 154

Our own modernity is to be the cause of our collective reflection, by which we will each have to be accountable for what is seen? But from now more importantly to be doubly accountable for how we are disposed to act over certain observations or how to react to other reflective reactions? Meaning most simply that it is we here and now who count, it is we as individuals who count, but from now by being actively involved in this last modern era, it is the pure standard of our ideals that are to matter. From which by translation they are to be formed from the dedicated efforts of all our religions above, but in conjunction with the meagre presentation of our political standards. Both influences are to run together, but from now in the correct order to be better seen and understood from having been charged with our Fifth Dimensional overlay!

In all cases we do readily recognise the full importance of our political management, not least now to be seen for the first time by many people, only because of the way in which we have become global these past sixty plus years. Only because in the same time period we have become aware of each other for the very first time through one of our political tentacles in the art of global communication, generated from the standard of commercialism. Even that this commodity had its main roots from the wars just gone in bowing to the requirement of needing to know of events quickly. Driven by the flurry of our own progression, stimulated by my feature of our fourth dimension, that of science and its own tentacles? On purely political terms we are here where we are now in the time of instant access to information by processes of natural development, but at a speed where sometimes reactions are triggered before we can evaluate all points politically!

What we have not yet been able to recognise from the measure of our scientific developments so far, is that they were going to happen in all cases from the same source as we anyhow. But on a different timescale had we not undergone the devastation of two world wars and or the mix of unsettled political grievances those conflicts left. Because at closure, even when led from our collective American/European perspectives, our standards of proposed settlement even if through the United Nations were politically wanting? Simply because at the time our thinking was not to the nature of our unity, but to the fact of our differences, most ably expressed by how we still standardise classifications of us being first, second or third world politically, economically! Yet to be the same by other political classifications through the promotion of the standard that we were all to be religiously tolerant?

By taking my own standard that we are here on this earth now stands, therefore, change if necessary starting from now, is not to be called! The error of our ways, have been ably demonstrated these past sixty plus years from which we can look into every corner of our world and see the mayhem without the idea of fair solution. Therefore without design we have to develop plan B, at first to be politically led which is what we have been doing, but from now to be religiously generated? Once again in care I draw our attention to the fact that we are all here now, all six billion plus living souls?

Without slight of hand we have a wealth of people politically ready and willing to force gentle change. But alas they have been contained by the structure of the political engines we have made, most notably by the silencer fitted form the fallout of exhaust gasses curtailing what should be said even if not heard. By standard what most of our open politicians have been doing for us generally, has been to try and improve matters. But again in like error even with excellence of mind, they in we still use the old ingredient, which only add to restrict the true menu some would like to serve.

In note that I set to overlay all political regimes with the three taken and one active prospect of the Fifth Dimension, is done from this last modern era, simply because it could not have been done at any time before. I do accept with open gratitude that my own opportunities have been won from the fine and able and noble efforts of others as we, also from the past? But in the same context, I must offer the critic in measure for some ideas we had instigated pre to the prospects of the proposed Fifth Dimensional overlay?

Having already stated that on many occasions in our haste to rationalise modernity into or to represent our specie equality, by forgetting what old standards or intentions were, in order to put matters on the same keel, we worked by assumption, fortunately without malice. Even though our leads were given by or from our westernised, American or European perspective perhaps unknowingly some new matters were tainted. Until we began to see in full notice our purely political similarities when we all expected the same bounty from our wholly political exploits, measure for measure. After all is not politics the main supplier of all of our material needs as sold from the reaches of its tentacles of business, commercialism and materialism? Our dues were considered the same even that we are brown, yellow or white or black of skin by those naturally evolved differences, or were they?

Through the political spectrum early in these past sixty plus years, we were not thought to be the same value as the one people? Until actions like the American civil rights movement focused us upon that standing bias. Whereby peoples of the same nation, the most powerful nation the most industrialised nation in the world, the richest nation in the world, were penny poor to openly recognise our true similarity. Until awakened in course by the means to the demand of civil liberties by mostly peaceful means but also through violence past the wishes of hopes in mind of having a dream! Perhaps leading to the rest of the world in the general awakening to other matters of gross misunderstanding and or open brutality of our behaviour like ethnic cleansing or the practice of genocide without realisation!?

By political determinations, if in hand of discussion at such times, the United Nations endorsed on mass, many of our national stipulations through the obvious from our civilised standard, by that same code under the need to change. Change was to be set by the civilised standard of government. Taking the query of dispute in hand and setting new legal precedents, disallowing the hitherto unnoticed standards of what was loosely termed racism. By such method to legalise against racial abuse if ongoing, but in term only from the time set of how the world was then in the mid nineteen sixties.

We should not look to the failings of any such exercise when the intentions were noble and honourable. But in consequence what most racial laws of political tone tended to deliver is to express the full nature of our differences unwarranted, leastways to be a false premise of thinking. From which in term of result by political definition most openly, we have black Americans, which are also ethnic Americans! Who generally also originated from Africa to be afro Americans, all now coded by recent times in attempt to exercise the principles of being politically correct. But in the confused fixture, not to see we are the same single specie? When by those examples through purely political reasoning of these recent times, any measures taken were in the haste of commitment to appear to be doing right. Unfortunately at the same time all approaches were half measured, inducing our steps taken syndrome seen from above.

Without cause of blame, I have centred on America and Americanism for our own measurement, whereby because of the circumstances of time in world history. Her influences upon us were generated over these past four, two hundred and latterly these past sixty plus years politically. Simply because the time set of discovery of the two continents in purely political terms, although European led from the aspect of greed, was complicated in part by the timing of the Roman Catholic/Christian European reformation? In its simplest format for the case of North America who in term was to carry more reformed religious influences than had occurred in the Southern Continent, or even Europe by reflection!

America began to settle first to a religious style of government more associated to the changing attitude in becoming Christian instead of only Roman Catholic. Allowing for the birth of a nation arise from its own conscience within, without any direct representation from Rome of other kingdoms. America on her own account for us, because of set circumstances of the time and from the progressed human thinking to that time set. Was in our presentation to be in the position of having a new canvas in the aim of nationhood, but alas like us all her new colours of paint by the theory of chance, had been first mixed on an old stained pallet. Of which by past reference, we are brought up to these very modern times where with the influx of a new colour chart in the form of the Fifth Dimension politically, we can touch up errors?

**CHAPTER 155**

When we convene to tackle the biggest gaps in our political thinking so far, without charge one of the features of our concern has to pivot on what is meant by 'Natural Body Death', not least because religiously or politically that is to be our fate. How to tackle that mystery is of the utmost importance on all readings on all levels of understanding to it, so that at last we might be able and express what is meant by specie singularity?

From the very fine measure of our total achievements by exposing in magnificent use many of the mysteries of the sciences, so that we can benefit to a very high standards of comfort to our welfare. Unfortunately some have been a smoke screen to how we should be thinking of ourselves in terms of what evolution would have deliverd at least after the reflection of another four billion years of Earths history? What ever the results of that doubling in time of the planets existence, we as our single specie will still count as being first 'out of Africa'. Unfortunately having reached that future time, our total achievement would be by the same division where one group of us will still have a general lifecycle to be of only fifty plus years with the other of a seventy plus year life span? So much for political reasoning if some would claim in due, that they had heeded by exercise to be purely political and operate 'Past Gods Words' but to what result?

Having stated on many occasions that I would set it so that in all cases, all of us should operate 'Past Gods Words', has been meant by me alone to be set in standard for all of our religions as a means of equality, stands! Not so to challenge the name of Moses or Jesus or Mohammed, not so to blank other religion that are not purely monotheistic like the above form the Middle East. 'Past Gods Words' being a taken prospect, is directed to us as individuals. It is an expression of human excellence without error of mind. Furthermore it enhances the liberty of personal choice bestowed upon us all through the medium of discovery our medium of personal free will. Being the master of all expressions which in all of its finest habits shows how we on our own account, just like Moses, Jesus and Mohammed have already expressed our own free will by standing in line abreast with Moses or, Jesus or Mohammed 'Past Gods Words' in countenance of our own consciousness and from the experience of our own conscience!?

Another aspect of the importance that we must take our personal actions from a position 'Past Gods Words', stands from the times we are now in. Stands from all the examples of how we have got things wrong politically, continually? Because in acting by far too many cases to the steps taken syndrome from half measured political decisions, we have compromised all rational thinking because many of these half measured decisions would claim to already operate 'Past Gods Words'! Fortunately in the wisdom of our time, we have been able and reflectively compromise any political claim, because of instigated religious tolerance! Instead of ignoring religiously motivated acts of political aggression, because of the standard of religious tolerance? From that same perspective religiously, if we do take that standard of all religions then our methods of conduct are purely for the overall intention of what religions are for? Not least to bring in circle and up to the minute reporting by the standard of these past sixty plus years, by getting ready for the end game again!

From my own understanding by altering the past without change, I feel free to sidestep many serious issues we have amongst ourselves, not least from our own held religious perspectives? I do not and have never dismissed the past by any count, my good fortune is from where I have been able and look into the mirror and determine what the reflection really is. Without any pretence I do not look to the mirror for answers in change, I only observe and see the volume of how we still manage to get things wrong. Because we have smothered our own personalities with the complications of accepting our standards of display to the meaning of being civilised as taken? Only to be divisive, in the same vein we have added race laws to join our similarities, but instead they keep us separate?

In the same way we govern ourselves to be politically correct, confines the intention of what living in a diverse world should mean, but under no bigger banner of how we pretend to manage the core of our religions. In setting any standard of religious tolerance we throw progress in the form of natural progression into the mire by the core in understanding of not to understand!

Religious tolerance in concept belies the dignity of humanity, because in all cases it purports to the standards of separate groups, when in case all of those separate groups beyond the theory of expectations, have and do propose the full concept of human unity. What our political device with the standard of religious tolerance induces for some is the excuse to force supposed unity by what ever means, by what ever means! Unfortunately from now, this done under the umbrella of unconnected political action is not to be of any religious persuasion morally, because we are all supposed to be of religious tolerance!?

In foot note, I have set the aspect of our standard of religious tolerance to be of this modern era, speaking for these past sixty plus years stands in link with all the other modern connected thinking I have proposed. My standards do openly recognise the factor that the concept had been met earlier but under different terms and were always changeable. So much so that from those earlier times by the force of some progressive moves, we in connecting to that advantage, done so for political expediency and continued to accept the undeclared ruling of religious tolerance. Like from the eventual recognition, leastways in Europe, that it was possible or it had become essential to allow the standards of Roman Catholicism and the new ideals of the reformed church co-exist by tolerant means?

Allowing in supposed concept that religious tolerance was a political feature by its own means and therefore could be reconciled by modern thinking post world war two, but not by being able to see from a reflective image how this feature would or could be taken or accepted by all other religions? Even for example to the matter of conduct with the defined codes of many Eastern religions? Also in giving creed to the matter of tolerance concerning the religion of Islam without first being able and set new standards because there was no ability to discus intention with a defined leadership? When no recognition was taken in any case, because Islam in core was already the master of religious tolerance and had been in idea from long before the catholic reformation leastways politically?

Although I have referred to modern religious tolerance being an ideal under lead through the political force of the United Nations since wars end, my case is to illustrate how some forms of modern thinking in the term to be always progressive on that particular scale, the image was only illusionary? By trying to set the same standards in all regions to be politically and religiously aware, without balance by use in concept from us all to be religiously inclined was not to look in study of what the real reflection was?

By the very way in which most of our political reasoning works, any standards to cover any particular interest of our own making, is a feature of our steps taken syndrome, unfortunately by fact the original pointer was set to be half measured. Half measured from most political viewpoints has always been allowed work, because in general terms such moves were in the form of commitment, but generated by the need of action? Allowing that some systems have worked, is a sign that when instigated, ideas at time were of there day, like from where and when the issue of religious tolerance was raised again in this modern era. It was done to the specifics of taking the matter of religious consent to or to oppose communism. Not professed atheism, but the aspect of repressive communism towards the original religious beliefs of existing nationals where change was instigated or forced. From where the giants of that political philosophy in Russia and China were set this latest standard of religious tolerance as a means to check there excesses, pictured by the open brutality they allowed one of our political systems act by? With religious tolerance introduced, even if only to be a cosmetic feature, allowed that from all of our political perspectives, heed in type was given to religious matters at least in show?

If the root of the matter was similar to what has just been described to cover some aspects of our modernity. Then by result we now have to equate what the total measurement of religious tolerance might have led us to? Not least by any connection we might think of or that I have made with regard to our religion of Islam in its present form. By previously stating, that in effect Islam had already allowed a fair amount of religious tolerance to other religions from an earlier time than that of Roman Catholicism, stands! My notice in the form of how I had viewed the balance between then and now is cut by the vast difference in the state of thinking and the standards of attitude between then and now?

I do contend that because of the measure of those main differences unmeasured, while Islam in core was carrying on to her own standards of normality. Various political influences cast in these very recent sixty plus years began to override other normal acceptances, which in terms of reference we needed to be able and answer? Political theories to our religious standings have now left us with the measure of how we differ religiously by association, by balance we do not, we are the same! Which is to be explained and played by our separate religious understanding through the concept of religious tolerance, which although now named politically, aids the matter of our differences? Creating further assumption that our religions are to be subvervient to the full standard of our assumed political excellence one way or the other, or all ways?

In the case of politics, having its own field generated in its new style these past sixty plus years, taken so by my own judgement? That it has allowed our collective agreement falter our plans, not least covered by the assumption that because we need civil control in any case, our current adaptations were sufficient to be voiced over any other precedent, including our religions, which were to be sidestepped in the name of progress and modernity. Because although the outfall of our general plan was for everyone to come to the one code of political understanding being democratic by example. Assumptions were the only result of where our religions were, even if closely connected to the theory of democracy mainly having a Christian perspective?

## CHAPTER 156

Without any open opinions, my assertions with regard to the concept of religious tolerance stands only in full context through these sixty plus years. From what ever path we have taken, from what ever our defined motives might be, our case in point to the fact that we are not religiously tolerant. Is held up to question the one religion that understood the concept long before others, even long before political compromise? Islam by name to be the religion of peace, by causing the very root of being laid back added to our character, has now unfortunately been compromised from within?

Her own core of creed cannot in face be challenged except from within, first done when in view early days, when it became separate to be of two main parts by the one basic belief? With one to be Sunni and the other to be Shiah, both truly of the one faith but by that same feature both expressing the modernity of now then, at the same time displaying the frailty of man? But also by the biggest of conclusions, with both serving a distinctive positive attribute of man in expressing our own free will! Allowing that we are all entitled to project our own thoughts to what God's desire was? Also in event to ever question the theory by the additions of thought generated these fourteen hundred years since Mohammed instigated the last world faith in the Middle East, but to spred!

Without further pointed insights, yet on the value of Sunni and Shiah or Catholic and Christian by our forms of interaction, our first matters of concern religiously have to remain in that vein. Not so that we might compare one standard to another or in reverse by looking into a mirror, for in the left hand being raised will show the right hand in response? My quarter of now is given in relief of how I personally have taken by my own vision in what and from where and why Islam on our behalf has caused the action of some deeds that are to be separated from our range of further thought!

From my own rational, but in the same context of thinking owed us all under the umbrella of our own free will. Is taken from what has to be up to the minute events progressively. Not on my account alone, for all that I have, is from all that I have been given in the same style that applies to each of us including the providers and the doubters of the fourth dimension? We are all still connected by the open hand of humanity, allowing that by the tentacles of our many religions and with the tentacles of our many political systems our hopes and aspirations are even. Unless we can account for some of the reasons we have come to differ in particular focus in this last modern era? Shown by my own expressions to the balance of Americanism and Islamic fundamentalism, held in reference to both by looking into different mirrors for the result of our collective dealings these past sixty plus years?

If Americanism while looking into her own mirror was to see the image of Islamic Fundamentalism, what sight in reaction might be taken, except the unvalued aspect of religious tolerance!? At the same time using the other mirror, what balance when Islamic Fundamentalism looks and sees Americanism, what is the measure of that reflection? Save that we see a political head in belief to rule in control a religious heart?

From where we might agree to what Americanism is by being representation of a standard to what might be the zenith of our achievements politically in terms to obtain the objects of materialism. My own feature of Islamic Fundamentalism taken from this modern era sets to be the image of our collective conscience, but not for this era! From where and when Islam was put into our domain some fourteen hundred years ago, without debit, but by open fixation, the core of our conduct was in our individual hands with tentative care of conduct overseen by imams? By those means the religion survived and thrived, if only until now at the edge of change, even unseen?

Islam in core is the same now than fourteen hundred years ago, perhaps, and I say again, perhaps its only weakness has been exposed by some of the actions allowed then ago in choice, against what might be done now by suggestion? Carried in definition by the differences between our social thinking of then and now, bore in kind by thousands of generation of people each with a varied perspective of the time in which we lived against what had been by reflection and what was to come unknown? Which in case up until now, we do know what we know, but do not know of the future as always, yet for those of us who had followed the defined offices of our religion in Islam. Some of us applied habits of old thinking about the structure of the religion in attempt to solve problems of today with the allowed remedies from the past!?

Although having a very fine code of conduct captured by the pen to its original and unchangeable definition, leastways from the structure of the accurate recording? By other overt happenings throughout the world right up until now, a body of our concern by standing in point to full support of the basics of faith, had evolved? What was to occur and become of concern to the national standing of most countries gripped in faith to Islam by first choice? Happened unseen, but under view of the standards we had allowed develop over the full sixty plus years of this last modern era. Not least because in allowing or creating religious tolerance and with Islam having no set defined leadership structure. Many assumptions to have come from the early bloc culture standoff prevailed until the questions in the asking began to be parried, not least from the general area of the Middle East!

Without self choice, but in grace of how we must each look in reflection on what had been done is by consent of our young. Being a case I have made in the objective, which is set to ally the image of politics to religion and Americanism to

Islamic Fundamentalism. Grace upon our young is the best link in opportunity we will have to reconcile the past with the future, but more importantly through the events of these past sixty plus years. From the closest examples but in full consideration to some of the horrors some of our young are subjected to in the name of obtaining their daily bread, only heightens my intention. Because in drawing the differences between them of three parts is the most direct way to proffer needed change even at this time to be undirected?

Of the three examples I might choose at this time, by first count, how are we trying to evaluate the flat and basic thinking owed to far too many of our young? Whose only consideration is to be at there begging site as early as possible each day for best advantage, with others whose only consideration is make hast to work in the despicable conditions in sweat shops. Working long hard hours for a pittance of pay, in many cases providing the only family income at below subsistence level to barley survive? Whose only consideration is to be first in line following rubbish trucks while they spill there collected waste from many of our cities into landfill sites. All tied to the nothingness of hope the bareness of desperation and the fullness of despair!

Held in case by reflection to the other class of our young who are overfed with the aspirations of what Americanism and Islamic Fundamentalism really mean, one supplying the false economy of materialism as a duty. With the other fulfilling all of our supposed proper objectives of spiritual desire expectantly? Not to be divisive but in core to show that if our young can draw such conclusions of necessity said by me in the paragraph above! Then as much of any failing to be accountable is owing to us the older generation, who by taking what we have and translating it into what we believe, over time we have lost the plan? But at this time until now, we have not been able and see our full goal, only the reflection?

Taking full care of our young by fact rather than conversation can only come from now, not because of how far we have come. More precisely because from the full depth of our emotions we are still trying to accommodate the changes necessary to balance what should be from what is. Of the many good things we are prepared to try, our attempts are not to be confused by applying old remedies to new problems, even with the right objectives. In all cases we must insure that the end is achieved before the justification of the means. Unfortunately I see the picture in full, whereby in my determinations of what Islamic fundamentalism is and means. Neither or either the end or the means are to be given credence by any consideration, leastways because of events to have revealed themselves to us all over these past four hundred years? When from the first time in all histories seen or unseen, we became global in terms by the discovery of our American continents, which truly set the world to be round, even to wait for Australia to make us complete!

Solutions from situations are not always possible not least because even though our direct aim has always been to learn the correct methods from specific codes. Like in the form of Americanism, if that is or was the recorded intention or balance or from my point of view an offer of what the progressive edge of our politics has become? What I may use from that political example is by illustration for how I had set the results of these past sixty plus years to the excellence of our politics and religions in balance, but not in harmony? From the very exact supposed condition of what Americanism is, our total failing upon our young is wholly through all of our political avenues and for no other reason. That in part is why by overlay of all political steps taken with the three taken prospects of the Fifth Dimension we will have better intercourse with our religions. All of our religions in concept, belongings to the one field of contact, which will be led in further contact by the image of our religious leaders for reaction to the full standing of all religious also carrying our Fifth Dimension overlay. Our comparison can only draw the best forms of progressive reaction simply because the full strength of the reflective emphasis of the three taken prospects apply to the spiritual and secular in parity, but to cover the change of there own perspectives? Which by no less than open coincidence opens to the door for all time in concept of our future determinations for the benefit of all, mostly to the new guided hopes of our young?

Islamic Fundamentalism in corner for some of our religious representations might hold in full issue what I have considered to be of vital importance to all of our religions. In order of claiming back the full balance of what is to be our first line of approach to the real aspect of modernity and how it truly stands in our case? Already well established has been the nature of how we have come to interact, but in case, as I have always stated? From both of our perspectives we have only been able and further interact from recalling in error the conditions of the past, but in first reference to our religious attitudes?

## CHAPTER 157

On its own account, Islamic Fundamentalism is a reactive skin of what is considered to hold without charge or change, which is what Islam in faith is meant to be!? Although, and I make these assertions on my own account, but in company of the one active prospect of our determined and now defined Fifth Dimension. 'One for the Whole' is our authority to self question any line of proposed conduct that might only be planed or has already been done on our behalf. It is our conscience to the value of historical connection, but in first line to accommodate how we must think only in the real terms of the future by delivery of what we are to expect from our own political and religious institutions!

By making the Middle East its own focus point by modern times, is done to represent the importance of what our fifty plus or seventy plus years in terms of our lifecycle really means? Not least when by religious connection of four thousand years we are all connected directly in time with Moses, Jesus or Mohammed and 'Natural Body Death' to all? Bringing the spectre of the Middle East into modern focus by the way I have offered this scheme. But not to forget that in the same representation I have been making in all cases on our behalf to the good? I have also made the unique connection in fact, that the role we played at Shanidar to be undertaken, was the first spark in any age that we could connect too the aspect of purpose, but at what time historically?

From then without the knowledge or awareness of the situation by deed, until rediscovered within these sixty years plus, we have had no real connection as specie to purpose? Although again in reflection to our first considered insights to the realm of our monotheistic belief first installed some four thousand years ago, we had been content to accept that principle in core for all other leads. Even that we were squeezed by vanity to make the direct connection with God and understanding long before we were ready. Because from no age in any attempt, could we have made the open leap to realise that we alone are in type and kind the most grand and gracious specie yet to have felt the sun in comfort to warm our physical bodies and spiritual souls. Because until these sixty plus years marked to be this last modern era, we were not sure of how to proceed to the end of understanding that we now deserve!

I have made no case against Islamic Fundamentalism, but have made the assertion that in concept the holding is fundamentally wrong? Because in time, although now regarded to have been a modern medium created under the umbrella of our reflective standing in good cause from need stimulated in this last modern era? We have given it so that all of our suggestions and ideas are final in terms of how we review the ending of today in relation to the separate paths for the hopes and expectations of tomorrow!

Although by the standard of three very fine religions from the Middle East respectfully, each setting it so we are to interact definitively, we have failed on three counts by the same definition in setting all standards to the one end. Whereby in all three cases and by all three, religions, conduct will be measured later by what we as individuals have done within our own lifecycles. Each person would come in name of consequence before God by the range and scope of personal conduct coded by our religions in the same difference of their similarity!? All making the one mistake by excellence of thought but in error of mind, by the same fixture on purely religious terms, that when we die the matter of our ending would have been solved by how we had lived, by how we had lived!?

By the true sophistication of any age relative to any lifecycle, even that religions, and in case, our monotheistic religions throughout the world being Eastern or Western, no factor is realised until we die. Which on all counts by time, is to be by the grace of God, not included, but from our true perspective owing to 'Natural Body Death'? Therefore it is vital more than necessary our religious leaders of now make ready so that we know we are worthy to have attracted Gods interest in who and what we are? That I expect our religious leaders to show us the human and humane people we are, we as individuals have very little to do other than what we already do in abundance. By first showing from them, that the only errors we have made and still make, are in our own hands and for individual rectification?

They in all cases by the best will allowed in becoming our religious leaders, must now apply all aids at our disposal to determine our political decisions. By fact that we have come to the state of open knowledge to a small act of burial some sixty to one hundred thousand years ago is the route we can all take in harmony. Not least in review of how we have behaved on all accounts in the Middle East these past sixty plus years and still do, but by not being aware of Shanidar, only these past four thousand years!? From where we put the wrong case, that from or with or within or without the structure of the three religions to have grown there, we are not different, but in all cases to be the same religiously and have accepted that we always have been, but have taken the option to keep our separate religions!

By not now taking the direction we should in allowing the status quo continue as recent, our divide will ever grow, but led not religiously only from the political machine of expediency. Whose wheels are only self serving because in all cases we have allowed our civil means of control take issue in its own interest. Having endorsed it with all of its own tentacles in its many forms like the ones of commerce, business, commodities, market forces, profit margins even science and its own tentacles. But mainly in how we have let it set the pattern of social conduct rule over the aspects of spirituality by the half measured standard of setting us all to be religiously tolerant!

Fortunately from now for the first time ever aided by the standards of the Fifth Dimension being overlaid upon all political aspects and all religious determinations. We will have set the same fixation for all people, not least because our three taken prospects with the open role of our one active, now gives us all the same voice. For the first time as a product of these past sixty plus years to the matter of fair and just change, we have created in us a tool. So that politics through the moral offices of the United Nations can speak for all the same, so that all of our religions do likewise but in hand of our religious leaders, not least because in picture, politics is, whereas religion might be?

Not to be that we can use past images from any individual, setting for us their own interpretation of what our conduct should be. Our religious leaders are obliged to make our case, but done full measured to all past standards, except making the open mistake of looking into the mirror of what God in agent if, had actually said, and from there to notice when the right hand was raised we move our left?

Without charge our religious leaders in all cases are no more than our political leaders, but better connected by station, because by the same number of cases, we have all got our own free will? Being an entity from which we can self evaluate personal choice by not having to accept what we are sold politically now revealed by the Fifth Dimension. But in the same way what we are to heed religiously because of emphasis and not in the direction of behaviour, but from our choice of direction. By the correct form of steps taken in being offered full measured appraisal by our religious leaders, who in case are to acknowledge the same standards for us all with the same overlay of the Fifth Dimension. Even if it disallows some of what some of us have been led to believe, by not having in term what I can only determine, not having the same leads, if proposed by our religious leaders?

Because in sight, because of the following and overall structure of the religion of Islam in the first and last place. Our steps taken were complete, except not to have fully recognised the spirit of our free will, because up until then our conduct had not been displayed so. That we were fully led by the accepted or forced standards of religion and politics if told by the conditions that up to the time of Mohammed had already prevailed? New innovations that were carried to us were for all time but restricted to the thinking of the day, and the natural circumstances that had developed in any case? Another feature of their concern by rationalising the inclinations of the morrow, although introduced in their own total modernity. Perhaps in order to make the almost right moves by creating the final religion to be self serving by our personal conduct. Our personal past connections were made first part by connection to Abraham and all the same associations with man by connection?

Taking in first point the past connection, we in line had a better alignment with Judaism before Christianity. Not for a divisive intent simply because such associations were easier to make in context with another religion of area who had all the same ideals, except our addition of the added two thousand six hundred years of further development? Form that association and it does not have to be taken, but Judaism and Islam had a natural connection. Not least by the similarity at instigation to the same general diet by excluding some foods or meats and in case much the same way of preparing food in general?

Also Moses in time brought by hand the reconnected edict direct from Gods own agent to be our lead in the Ten Commandments. Mohammed in kind by the pen, reconnected edict from Gods own agent to be our lead in the Quran! Whereby Judaism could not have a forward effect on Islam, Islam in reflection although readopting/adapting much of how Judaism had been and was. Again by the most strait forward example of diet alone with the hygienic habit of circumcision as another key, a prime example of how at first current and new conditions were localised? Even if the pointed connection that I have just put forward had any particular leaning, the one primary condition new to Islam, put it apart to other religions of area by the extreme concept of placing our conduct apart and separate to the guidance of an established priesthood?

Islam by reflective intercourse with Roman Catholicism could not see that when Jesus spoke, his words were not of Gods voice in the son, but only those of a mortal man? Although in fair warrant Mohammed acclaimed Jesus in the same class to be a prophet in similar status as he. Further connections were limited by standard of approach between what Mohammed had felt and what Jesus had deemed to have said?

To turn the other cheek in terms to show no form of open hostility, might have been a viable standard of approach to indicate to ones conquering masters of the Roman Empire, no directed form of aggression will be used while bringing the new word of God to all men. Although we have it that Jesus was indeed of such piety by character, his display was not of the day or suited the natural violence of the time.

Six hundred years later without being impressed with how the Church of Christ had moved in centre to be the Church of Rome? Six hundred years later the standards of violence in normal interaction had not abated. Although the

Church of Rome had adopted the catholic image, no claim was put in turning the other cheek. With Mohammed finding the state of affairs to be as they were in the general area of the Middle East having sectionalised back to smaller tribal groups and city states as pre the roman conquest. With small Jewish groups and small Christ church groups following Moses and Jesus through the uncontrolled order of failing priesthoods if?

Feeling to be drawn to rectify the open needs of man in God, might not it be seen that the cause of such religious decay was by and through a disenchanted leadership of previous religions? If from the comparative message of Mohammed's own thoughts holding the greatest respect for Moses and Jesus, but for the latter, allowing for the fact that Jesus himself had become over exuberant by his own good feelings from within? So much so that to catch the attention of total strangers, but of all creeds, his style was to promote himself to be more than man by connection to the general and broad religious thinking of the day with multi God aspects? Not least to now highlight the funnelled view of thinking monotheistic as the one faith of Judaism did and of which he was familiar with their codes? Of which was expected the final triumphant link with god imminent! Jesus to his personal responsibility by the expanded emotions of the full two thousand years of Judaism saw now to include all men to be of the one God?

But in order to maintain that force in terms of communication to both Jews and all other peoples invited to this proposed new Judaism. His seen method to extend Judaism, was not for him self to be a normal man, but to be as Gods direct family representative, expressing the full range of our open similarities through story telling from the outside? In style his stories were an attempt to open the closed thinking in the vaults of Judaism, because of there continued expectation of unity with God on their tribal terms alone? He in offer to broaden their base of acceptance was not to be an affront, but show the open ability of expansion to readily include all men as all men had been at fault in God's eyes originally? Jesus in the form of seen man in local normality by existing standards, was done to give insights to what God his fathers' real intentions were? Then it was from Jesus to be the expected messiah as he might have seen it, that the cause of change was to be introduced in such a way not to antagonise any situation already there, but at the same time include all of us equally without separation, without violence, to turn the other cheek!

His seen victories were of his own doing and eventually for the convenience of our political structure that had always held there own ground against the proposals of all religions. So much so that with that political influence to have always overridden the concept of the spiritual by the event of materialism, Jesus was to settle to continue with the God belief for all, but maintained the political structure of a body as all priesthoods, to manage the spiritual end of our expectation? Leastways for the past two thousand years right up until today late in these past sixty plus years. Which has been double marked, because again we had not noticed the value of Islam over these past fourteen hundred years until now, late in these past sixty plus years?

From where many will see the aspect of fundamentalism carry its own cost by the cloud of disassociation it earns because of the wrong approach or from taking the wrong leads. Is set to create a situation that might have been first encountered some four hundred years ago, has yet to be realised, but only now through our styles of reflection can we begin to see the issues or see issue? Understanding is at no cost, but must remain open because we are not challenging Islam, only our own misinterpretations of some aspects to the original plan for promotion or protection and in defence originally!

To cope with the noted conduct, whereby in some general terms, might was right by whatever means, above the proposition to turn the other cheek. Islam being forthright and in facing daily standards of general conduct, took to the obvious and natural condition of being prepared to die if necessary in support of those considered normal standards? From the very condition of the promotion of the new faith expressed through the means of self sacrifice. We took to those habits because in plan it was indeed a more defined route to Allah's or God's own paradise than seen by either the Jews of Moses or Christ's Church of Rome, with the added element to suit human emotions of the day. If and or, one was to die in battle to promote or defend or protect this new standard, martyrdom was the name?

Instead of being cast to have followed Gods will to turn the other cheek even to the point of death as Christ might have thought. Both in knowing that he must die for his/our fathers new faith, but also being aware that in so he would be now next to God in heaven because of that act, which appearing to have been done in placid acceptance. Set the first standard to follow the law from Moses "Thou shall not kill" but with my and our terms of reference to be of self by self in specie recognition which is to become a modern era standard!

From the time of Christ, although Jesus had to die in self for all men, his first message through his new Judaism was to fully modernise mans collective role by the new thinking, again of the day, but covering the change in attitude that had occurred over the first two thousand years of Judaism? Which originally had been set to meet the expectation guided by the standards of those times in the world historically, which was to reward the chosen for their prime discovery, of the image of Gods own revelations?

In following the same expectation, Jesus had now turned the expected Jewish immanent hope, to the last reality of discovery by ending the Jewish story in plan even though there was no such objective set? He also took that from now, our future was to be secured in heaven as expected and in duty already foretold. If a case was lost or a picture unseen then the difference between Moses and Jesus in the two thousand year gap was owing to how the politics of empire had a better outlook of control, than only the possibility of tomorrow?

Without direct comparisons, but in style with what might have been, using any examples I might from the base of theory. Is done from the abstract, where I am at liberty to use the context of possibility in our representation. I consider the only way to clarify what might have been our styles of thinking from any time in the past, was bound by how we reacted to the event even before we had the real ability of forethought? From my own case in turning to what was the style of our thinking four or two or one thousand four hundred years ago spins on the scientific edge of the rational. Not least because of how I have thought to look into the mirror of the past and have been able and examine what the reflection meant by our collective reasoning when for instance if Moses or Jesus or Mohammed had raised there right hands for acknowledgement. I had been fortunate enough to have seen intent using abstract reasoning in turning the four, two or one thousand four hundred years of our religious histories into the truth of living for fifty plus or seventy plus years!?

From Moses, and in following Jesus, had given the direct impression that the end, our end of normal life was indeed imminent. Although both perspectives were slightly different the standards were set or had become set in style by both able priesthoods. Who perhaps tied to the naturally formed thinking of their day could only see from the standard at instigation of both religions in the one form of monotheism? For that relay to be handed on, could not have been done by God, but in medium through Moses and Jesus allowing for our own charge of free will to cover all the anomalies? Priesthoods were to organise the order of conduct in able religious management allowing for our connection between Gods own agents as prophets, and we! Because of how we must think today, they might be our best link by representation as our religious leaders best in leading us by their own realisations with other religious leaders and from us through 'One for the Whole'?!

## CHAPTER 158

Mohammed on our behalf took on a mighty challenge in bringing Islam to the fore in replacement for what we had used in the past as our religious determinations? Even by the links that I had put forward in the name of its similarity more with Judaism than the Church of Rome, are mere observations from this very time we live in now, bracketed by these past sixty plus years, not because of anything Mohammed might have seen or foreseen if suggested? But in the most open of statements I make on the same terms, that any individual is responsible for their own conduct!?

My view is ably represented from this last modern era, because its unique standing, is that we have with us the full picture of learning over these past fourteen hundred, two thousand and the four thousand years previously. Our age is of now, therefore we must rate our examples to be largest by the full care of our religions rather than from our politics. Here we are in serious need of having a dedicated core in the form of our religious leadership, representatively to promote all the standards of intention set by Moses, Jesus, Mohammed and others and others, including from the base of our Eastern religions. Not so to always question in debate each other religiously, but by that aspect of or to the intending to/for the aims of finding the light of purpose. Not least first attempted when we were all the same and sought to extend the wonder of life even through the experience of death in sign by our/the burial at Shanidar of which we know nothing only growth!

By translation with many things considered right up until the edge of today, with direct connection to events of some sixty to one hundred thousand years ago. By the same level of review, for us to have considered separate initial events whereby we were to have felt led some four, two or one thousand four hundred years ago. By the open connection we have made or imagined about what our future is to be or holds? I have set the edge of our achievements by standing our opinions face to face of our political endeavours, in reference by my own reading to reach the pinnacle of obtainment through Americanism!?

By the same reasoning again, owing to my own rational, I have pointed that in counter balance to no more than a political standard of achievement, by coming to the opinion of Americanism, is a state of mind and not a personality! Being created inadvertently because we had come to use half measured judgements, particularly of how those judgements were made in world terms if through the United Nations from within these past sixty plus years since wars end, but also rooted in the ideals of the future!

From the reflective edge of that image with the full picture showing the extent of how we had come to be ruled politically with the abeyance of religious tolerance. I submit on my own account, because of how that assumption if through settling some religious differences politically, particularly in or around the area of the Middle East. From where we got the centre point of balance to our open standards for everyone being individual or to be individual collectively, but from the root in having choices from our free will charter, earned? We had come to use only political references in that time period again from half measured reasoning, but had increased the error by working the results steps taken in order to balance our first half measured attempts? Because we are still not aware, that when reading the reflection, the standard is to be set from the position owing to the eye of the beholder!

If that eye was carried full by religious connection as it might have appeared to be from Islamic intentions, then the point of reference for action or reaction in term by the result of how the essence of religion was told being done by any side at any time. Leads to what is vital to say or comment upon, but in style to be futuristic beyond what might have been a slight observation but for the serious nature of reactions only after the deed? From which our intentions have been to condense at least four thousand years of our united histories into the two different but same considerations, of how we have come to manage all our affairs most recent between politics one way and religious tolerance the other?

By the first recognition that we can and have only now operated by the recent events to have happened and are still occurring within this recent sixty plus year period? Coupled with how we have come to review our same political desires proposed to be democratic in choice, but on our account, along with the full tolerance we have supposed owing by the same political assurances we are not yet sure of, yet again have come to ignore? We have had double vision in not noticing the importance of our religious hopes realised soon, as always from now to be maintained in thought to the realisation of our fifty plus to seventy year plus lifecycle? From where we can review how individuals are permitted to interact or how we might impose our own will upon those whose choices are wanting, if they are to be from any religious standard of fundamentalism?

## CHAPTER 159

Islamic fundamentalism if now transfixed to support in operation the standards of suicide blessed to the desired ending in martyrdom? Has now to be looked at from the abstract, where it will be seen there is no balance to the situation or any part of it whatsoever. Unless we propose to make the standard again with another episode of our histories hitherto unconnected, but to remember martyrdom is not to cleanse human sacrifice? Because until we come to remember the wastefulness of suicide in the cause of martyrdom, most decidedly relative to the individual, any individual, we cannot view the intention!

Unconnected that events had been and to have remained unnoticed about one thousand years after the beginning of Islam as our last religion from the Middle East proper. When at first Europeans had come to the Americas and found the open practice of human sacrifice to the liberty of the people there. Our collective reaction even to be covered by the advantage of greed, was to end the pointlessness if such deeds, because of the open awareness of its failings. Even that the church of Europe at the same time was undergoing its reform which included the wilful murder of peoples with differing interpretations of the same religion.

In the Americas the political drive of commerce gave way to the standing that things could be done if blessed by the holy mother church. Therefore some proportion of the method to eliminate that cruel practice seen to be human sacrifice, was by the same standard of murder. But passed to be allowed, because the value of life in one scale was by mans own judgement direct from God, if directed by able priesthoods? No let was to be upon any people who by show clearly adjoined all credibility to the motion of the sun by day and the sky by night. Not even to the feature of sacrifice was inclined to the idea of regeneration of self for future purpose or offered as a levy to continue in normality!

If the physical image of human sacrifice can be offered to deter the deed, then we must look further without examination at the task of killing in sacrifice our own young. For which there is no defence under any example or in any cause that stands to human dignity. Not least by Gods own council, when on three occasions our intentions were to be made clear that we as individuals were to be accountable for our own deeds. Being a matter whereby on the first two of those occasions our conduct was shadowed by the order in organisation of an ordained priesthood. Who by representation had taken in duty to manage the forms of sacrifice needed, but to other creatures of God in the aim of placation to the image of our God or Gods at the time and not to be upon self by self by any connection?

From the third and last religion to emerge from the Middle East no form of human sacrifice was ever offered save the value put on dying for the new religion politically until the precedent was truly established. Although the style was new, the advantage of thinking had the extra accumulated six hundred years of our own general knowledge. Probably more acute than what we had gained over the preceding two thousand years, but linked in any case because in the same two thousand six hundred year period we had enhanced and endorsed our monotheistic beliefs!

I had given notice that in good and true style, not unlike from Moses or Jesus, Mohammed in talking of the future of the individual in terms of unity with God also spoke imminent? By the same reasoning perhaps because of how the other religions were or had become. He took the challenge of death owing to us all over the next few years as a time of consolidation allowing its own set of rules apply tied to the effect or outcome of circumstances? Without direction, his standard if led, allowed that because of such circumstances come what may. We were to be given charge that the final acts of we, in the defence, promotion or protection of the new faith, were duly to be rewarded in the name of martyrdom! An adopted means to take honest care in duty to those who might die soon outside the un-thought of feature of 'Natural Body Death', those of us who would do all we could in name to bring the one God in Allah to all now, were to be protected?

Martyrdom was never an option, nor was it ever included in creed, it was merely a consequence if called upon. As an act in term, it was not sought as a means to an end, but in the consequence of, it was not settled to the standard of being a reward by inducement. It was settled to the same status for catholic martyrs having been granted that name as the result of being persecuted. Unseen by comparison, although the catholic form of martyrdom was offered in reward to be in heaven instantly from result, because persecution was not a choice only an outcome.

By new more modern thinking about the matter of death and cause related to previous martyrs. With consequence under review the connection between dying before ones time, but in result to have done so in the promotion or protection or defence of ones faith, carried new perspectives? Reward was to be set to the same standard from premature death, with others in life eventually going to God in glory. Therefore those who were denied that expectation of living by dying naturally, because of the nature and cause of our death, were to be carried in memory to the same future. Even if to be enhanced on reception into paradise because of the open standard for being willing to die without self concern for the full faith of all men foreseen!?

The biggest example of consequence was relayed by the message of the days by time in which we lived, where in consequence the loudest voice was by the sword and shield or from the arrow and the bow? In many cases commerce was conducted under the shadow of intimidation from purely political leads. If religion was ever to gain the high ground which in comparison it might have seemed to be loosing. By reference to the local Jews along with the concept of Roman Catholicism now led from lands unfamiliar to the original prophets, who from area had created both arms of monotheism, the third view was now introduced?

If Moses and Jesus were now to be under question about the office of power in what they had left for our guidance from the establishment of our own priesthoods. Beyond the critic, in certain cases of the same approach from where Mohammed was to have cut into defined codes that had lasted in running for two thousand six hundred years. His only choice was from the political advantage of using the same arms of expression by using the arms of battle politically for the best ends in result!

Fortunately beyond the range of open comprehension owing to some of his opponents, even if guided by the priest end of there community who would try and incite morale when faced with what was essentially new religious war? Determination was to be the victor in the real sense that as other religions were or had shown signs of decline for whatever reason. This new vital robust religion was making its presentation in full modernity on several counts. Here was a religion to be, that offered the best chances of reality because from within it was prepared to allow that human life could be valued against the extent of religious commitment by the individual, providing that person in we, was fully aware that the measure of responsibility was personally conducted?

Our new religion of Islam allowed that we could take on board all modern attributes to suit the actual conditions of life as they were, without connection to the supposed finery of turning the other cheek? Now at last, but again, we as self were to be in control of the future for how it was intended to be, with great expectations that was our mark to be made. One of the problems in making that great leap for mankind, was by time, because of its own time set of introduction, Mohammed could only lead us to take small steps for men. His full tone was by perspective, in first drawing our attention which was ably done because in the exact same class from Moses and Jesus, Mohammed joined our religious fortitude but now to be apart, if above political principals?

Far bigger than to render unto Caesar what was Caesar's, Mohammed in the concept of unity recorded for us in script to cover correct conduct now anew and for the future, but with an eye on all further futures also. If then, cover was given in eye whereby if was necessary to die in the promotion, protection or defence of this our new faith to be, then in reward martyrdom was to be marked in worthy praise done to the exact time of then, but with the same eye to the future? From where in time with the world or the image of the world then known, would have eventually come to Islam in any case by the best example of conversion. Therefore martyrdom in case was more to be an incentive then ago to the expedience of need above the effect of religious duty!

This time around different to some views expressed by earlier religions in commandment not to kill self by self or to turn the other cheek also, we were accommodated to the violence of the day which had been continually generated by the politics of the day? Any reactive response to political violence, was carried by religious charter but only in balance, only in balance? Islam through Mohammed gave us the latest update to the vision of Gods own desire ably expressed in the Quran, in which there is no charter to deny Gods name to any other religion, nor can it ever be so read, even in translation in reaching all peoples as we had been told it would and can.

Fourteen hundred years after the death of Mohammed being marked by the two thousand and connected four thousand years of direct religious views expounded from first choice in area open to how we as specie had developed by many traits has found us to be religiously adept, but politically uncivilised!? The measure of which has called us all to the funnel of choice, but from now to be used with and by the thinking off the day, today, created in mixture these recent past sixty plus years of this last modern era. Our own new steps to be taken are with the established overlay of our political and religious perceptions with the three taken and one active prospects of the Fifth Dimension. Although we each can be involved or not politically by personal choice, my standing until we find our own feet, is that in relief we would be better of to take heed of our religious leaders to represent us. Not to any finality, only for how we are all to direct the better head of politics from now on!?

## CHAPTER 160

For myself to have put forward our political connection with the term Americanism stands by inclination, but is fully relevant to us all, leastways by how in all cases we are subject to some influences from the concept?

Islamic Fundamentalism is to be looked at in a more positive light and by heed to the reflection of events and by looking into the double mirror situation, through the reflection of the reflection? Although I had pointed the two concepts in relation to how we have come to react over these past sixty plus years. I have left the full importance of all previous histories to show the value of opinion in force over the apparent past for better study without the drive for fact, which like many of our histories carry there own seeds in the eye of the beholder!

By my terms, to focus on this last modern era, we can each try to or limit the necessity of the past because in type what was and did happen to the reactive specifics of the moment, I have never claimed that we are to ignore anything from the past, nor do I accept that our thinking about some events should be modified, nor that any person should have to be first to make standards change? What I do expect and have always done so by this work, has been to make us more aware of the future in the possible, to be more important than the past in the probable!

Therefore with the hope of confidence in us as this specie, my honest connection and direction to the political example in focus to Americanism is of relevance to how we can stabilise all of our thinking politically, which is not to forget the North American association to Christianity and Judaism more than other religions. Even so, by reflective means, for me to put the case for Americanism is to put the comparison on purely political terms that I could use the balance in expression for or against Communism as a means of our pointed political desire. I would have to allow for the fact that communism although used in part to change the standards of life for hundreds of millions of people, had not delivered in the same abundance of materialism that had been done in Europe and America through democracy, which in part had released commerce to work towards Americanism?

At any time before and even now, for me to justify my own connections between the balancing of Americanism with Islamic Fundamentalism if the only opposition voice on the religious side, has been used to establish what has come to be the ends of the spectrum between the two major influences of politics and religion in the area of the Middle East? So much so that rather than be concerned with any individual lifecycle, we have heightened our own emotions to mix the metaphors of religion and politics cross referred, but from the wrong perspectives? Instead of taking now, for the mean route of communication, we have taken the features of our pasts, being four or two or one thousand four hundred years long and attached the wrong relevance to those dates politically?

All of which has again stirred in us the emotions of justice and fair play and morality, which are elements of religion and not politics. When on the other hand our political aims are only to the spread of materialism, and in consequence our vivid emotions above are left in the unheard, unspoken corner of religious tolerance? From which, by what ever means, we have sought to match one to the other, from our deep pasts until now, only to be left with the view of confrontation?

If all political channels are charged to this modern era by my idea of achievement in the form of Americanism to be our fixed objective materially, even that I have called the standard as an opinion without further or deeper examination the concept is valid in world terms. Because of the number of our political judgements to have come from the same base since wars end connected with America in the abstract! Without separate praise on any concept except our own, after the Second World War, order was led by political opinions before any other means. How ever dirty many had become, it was our own endeavour that has carried us thus far but by the glint of Americanism, materialism and commercial politics?

Our balance by time from within the same sixty plus years, set on the religious front, has not yet been realised until now, because we had not been wise or quick enough to give full credence to the value of our religious sentiments. From the first aspects of Europeanism to Americanism over the past century from which was delivered two savage world wars, because of the concept of national politics commercially, overbalanced the fixture that virtually all of the main antagonists carried to the full Christian retinue of religion!

After the Second World War which was histories worst short term tragedy, in counting the loss of human life for nothing more than clouded political ideals. By the extended good fortune of trial and error we as a world community came to this last modern era, covered by these very recent sixty plus years. When at last we began to consider the real aspect of humanity in process, but unfortunately curtailed by staid and old fashioned views held by habit. New days did not always carry the best of responses, but from the new winds of victory even if having been worked from the times of the League of Nations translated to the United Nations. Our open concerns were for what should be, instead of what had been done, which was the right approach on the only level of suggested improvement. However, if the core was set half measured and the core was only political, then being led by our steps taken syndrome, we set the sail to the wrong wind with the only means of correction to come from the last Middle East religion but untimed!?

From any of the signs we saw or imagined that had been, we missed the obvious one to be now classed in name by me as Islamic Fundamentalism? Although I had intertwined in mixture some of the relationship between Judaism and Islam, most pointed through the spectre of the Israeli Palestine round of very sad and painful interactions that were only acknowledged to be political? Although I had made the same issue of the Iran/Iraq war with both sides of the same religion, in core it was also political? I had never ignored by half measured means the full religious significance of our interaction between each other on those occasions or by inclination to any similar events over this last modern era.

My first open objective is to give an account of how we have developed and the reasons in directing our thoughts to the two main avenues of our concern and now turning to the most important of the two, that of our religions. By focus on my own terms to what Islamic Fundamentalism is, and or has become to mean even by representation, will go better than half way for us to really visualise what had been attempted over these past sixty plus years. Even that I do not set the questions to be answered I take the asking to represent us all, because on our own account that is to be more than the value of purpose!

Fundamentalism in my own eye has led us to the acts of wanton murder for no reason what so ever. Martyrdom through the body of suicide as a means to promote unity is our last abomination. Where by time we might sit down and end such acts over the drinking of cups of tea by swinging the whole concept to the open reality of living above the carnage of dying. How we might first come to be aware of ourselves is to look into the mirror of 'One for the Whole'. Giving in kind all plausible reasons for martyrdom or suicide to be relayed to self before any one else? How we are to change willing into activity is to maintain the sanctity of living by all religious creeds!

This is now to be best done in choice by our religious leaders, whose open brief is also aided in shadow of 'One for the Whole'. In having stated that Islam by direction of choice had not the same structure of religious command, has not lessened her position, unless only by our ignorant use of standing to religious tolerance. In choice I ask no question to or of the integrity of Moses or Jesus or Mohammed, from who each had the same ideals save to the opinions of the day and the instigation of three similar ideologies? By pointed direction of this study now falling upon Islam alone, is no check in comparison with other religions, because of the very different thinking at the time in which she was created.

Although I had indicated on many occasions, that by those standards of the day, plans and decisions were also of the times, but by my own standards of recollection and reasoning, even now tainted by abstract thought? For me to have suggested that because of the natural violence of the day, licence had been given in text, that we as we lived then we could also behave the same for the means of creating this new faith? If physical violence was used in any consequence it was not a religious edict and never to be tenet. All we did by best measure in not turning the other cheek was stand full square to what was forming to be one standard. Mohammed in person was to be our best judgement of intention not least because reality was endorsed from where we were to stand against all attempts of denial to the prospect of this new venture. Any and all replies to the promotion, protection or defence of doctrine if not done by word alone was open to other means in like for like?

If the force of battle was necessary, then by balance to kill of self by self or not, was standard to the need of intention, by killing those who might never come to the written text was to be a form of relief for them, that henceforth their fate was now in Gods own judgement, which cannot be measured by any man! To have been killed while in attempt to defend, protect or promote the new ideals was duly covered by the object of religion it's self. Whereby our glory was aligned to the realm of paradise, to be attuned as never before to God in Allah's own intention with the additional praise in having died for God, the rewards were multiplied! Not to remain hap hazard, but by standard, special rewards if applied, were to those of us who could now be called martyrs?

Although there can never be any constructive argument against the above by description, even if I had left my theory wanting. All related events were of there day and fit for the thinking of that day. Cold blooded times demanded cold blooded thinking, which when done in the heat of the moments carried there own references. Back amongst the cooler atmosphere of reflection when looked for, it is not difficult to make the open comparison between the past and now. Because of how we would naturally look at both situations simultaneously, but in relation to now being the manager of how we thought, all rules in referral to the past can only carry the meaning and interpretation of past settings. In case where Islam has made a burden, cannot be seen for when, because the concept was timeless although projected into the future. But not being able and rationalise our thinking or actions or reactions, because again from the time of Mohammed until about four hundred years ago the world had not yet become global until the Americas were exposed!

From that expression, "the Americas were exposed", was set a limit upon the extent of how we could ever look into the future with old values or old thinking of the day or any day? In so far as Mohammed might have hoped to lay our path including futuristic thought, was limited by the extent of what we did not known before we knew? In so far that Islam by religion had no early influence to the standards of conduct found in some places of the Americas, bears that the spirit of her religious martyrdom was no longer workable on old terms or that from now, it never could be again!?

In the same way we cannot alter the past, only by reason, far and away more important than mere speculation we can change it without the need to alter anything, except opinion? In as far as Islam came to deal with all manner of political ideals and groups with other spiritual priorities, allowing in concept that the means of battle was general in the need to face and force opinion for most conclusions. Then if in trial, seemingly done by Mohammed for our best guide, to have set the one religious code for now and project it for the future was a two step forward move but creating a one step backward reaction.

Even that we were given other futuristic insights some parts of the mirror of reflection if influenced by the other revelations through the fourth dimension from its tentacle of science. No sign could have been given, if and of when and of how, we in the name of Islam might have behaved, had it been we who with our faith had made ancient journeys done by some disillusioned priests and or a persecuted society within there own beliefs created in the Middle East?

Not to be taken as a representation of what might have been, but the association has to be made in speculation. Not least from my own drive because it was I who showed part of the reason of how the main body of beliefs in the new world had been turned from the peace of mind associated with the effigy of the Great Spirit, into the reasoning of human sacrifice? Although by times if to be examined even with no direct connection ever being made between Egyptian travellers locked to religious authority by a vivid priesthood, who could have had associations with the feature of stepped pyramids? My case stands because of other connected branches in comparison, if not recognised, are well seen?

## CHAPTER 161

By examination of events in the greater Middle East of now or at any time ever, in relation to how we interacted when our familiarity became entwined by necessity at first in the land between two rivers by the unseen force of circumstances. When several assumed different tribal groups began to form the habit of fixed community. That done, once a measurable base of competition had been established by natural behaviour, we as specie have never had the time to look in wonder over our shoulders and see where we had been? Save by modern thinking which unfortunately is from the effects of today, which does not have the valued opinion to alter the past!

From the same stable of thought, we have compounded the core of our errors by first assuming in far too many cases that religion or the theory of religion had been achieved, even if we allow that in the theory of religious tolerance we can be decidedly different when in fact we are not. Except how we have made ourselves politically attuned to spite the supposed unity of religious tolerance, to be able and classify ourselves to three different groups of the same people from the political perspective of economics? Created through the political standards having been buoyed up by the feature of my fourth dimension classification relating to the sciences, showing us the best way to increasingly supply commodities, aiding the misbalance of that assumption? That the level of won commodities to each group being first, second or third world was proof of our differences!

Not that our politics is ever to be cast in the negative, but the case is to be made that politics has allowed us use different values. From which we have planned the future on a separate path, so it doesn't have to cater for people, only its own tentacle values! First shown in reflection those often referred to, approximate four hundred years ago when we politically exploited the Americas, but were guided by the necessity of religious sanctification. Losing for all time the extended ability for us to act on the two fronts in the correct order relevant to mankind first through and with our religious needs to fill the overall hunger of expected purpose!

If what was found in some places of the Americas to the open acceptance of human sacrifice, then all that was revealed to us was a giant failing in our own perspectives. But sadly endorsed by how we were to tackle by removal the considered abomination of what it was? Let us remove this clouded failing of these strange people by destroying them and or offering the open choice of conversion to Catholicism in the realm of unity with God, but to the more open acceptance of the real political determinations required, then unseen, but set by the times and thinking of the day then ago? The rest is not history only ever speculation because the past cannot be altered by the thinking of modern times reflected?

From my own reasoning in first making the suggestion that the shadow of human sacrifice was carried to the Americas by very early Egyptian travellers by choice or émigrés forced to flee? Was done not least to show our similarity expressed before the written word was necessary to record the means of our differences. When by comparison we can look into the mirror of Egypt pre history, and see even today that in the deep past her religious acceptances were locked into the considerations of the after life? Although structured by political reasoning through first connection with the King our Pharaoh and through his mortal death to be accompanied in burial with grave goods not unlike at Shanidar! But in those cases which were managed by the civil authority of an able priesthood, at this time to be given full authority over society because of the death of a king?

Priests who on authority might have had the additional role to prosecute for the sake of society, the need to be able and order the sacrifice of some builders or pressed men, aiding the final entombment of the king? Allowing for the secret and peaceful transition of the Kings' soul make its final journey unseen, but in belief to the desire of the sun God in the sky by day and of the heavenly sky by night! Rounding in perfect harmony the cycle of time relative to the individual through the symbolism of the King our own representative of God, told in full honour by our priesthood!

For each of us to continue in view of thought to what has just been written is of personal desire but has been forever challenged through and with the character of our one active prospect in 'One for the Whole'. All reference from now can only be forward looking and only in compliance to the standard of 'Natural Body Death' which is a result and not a condition, but remains to show how it is to be for us along with the Pharaoh!

By further first reference, any person as we in this case me, will always be always able to speculate on the past or not, even if only for the turn of imagination to make an indication, providing 'One for the Whole' is not carried before the edge of any other historical event now spent? 'One for the Whole' is of this last modern era and is not yet vintage even over these past sixty or so months of its first incorporation by myself as a cornerstone of reference for us all. In turning to the reflected image of history our rules of engagement are not to alter or ever challenge each others version of events. But only accept that when done all were by the misconception that we differ of kind of type of thought and even of appointment, held in the double error that our manifest differences do account falsely that we are indeed different, until we realise our true fate!

'Out of Africa' was not a consideration at any specific point to any of our histories not, even at wars end, some sixty plus years ago. By now, living in proof with that reality in all of our tomorrows, we can look not to change the past only the results of it by the first consideration that we are indeed the same single specie. Without expressing the condition of our differences now known to have been caused by the circumstances of climate, geography, diet and other reasons even some unseen, we can better challenge the results. That some of our young grovel in the means of abject poverty previously described, just to be family providers. While others of our young without blame or condemnation had not had to do the same and will be fortunate to inherit the efforts and bounty supplied by loving parents who had worked extremely hard through all the mediums of politics to be providers. To gain vast fortunes not least for there own young, but by a system of politics set without balanced control to look after all of her/its people, beyond the realms of impartial sharing to the fated concept of purpose being achieved!

Leaving in all cases the outlook of the future to be of now, in which we are all involved and in which we can all be involved at no cost, only to show being involved. From where I had said many times it is for our religious leaders to stand up and be counted, stands in full part by our overlay to be upon all religions past any of the intentions of religious tolerance. Fifth Dimension prospects are of human interest only, in overlay to our politics there is to be no separate standard than what might be claimed by some of those I trust to be our religious leaders. My definition by citing the difference of our political and religious league, stands in corner that by recognition politics has no interest in man! It in form is the machinery of conduct only. Whereas our religions are deeply involved with the human spirit, unfortunately at this time more than ever to the aspect of death more than to life, helping create the measure of our rift to everything of real human contact?

One of many places where our religions fail in there true character, stems from the same political link I had just made between wealth and poverty. Although made in soft tones, I do not lessen the seriousness of the gap between our selves created politically, but naturally agreed by religious settlement. For me to have made the remark that some very, very rich people in making their fortunes by extremely hardworking endeavours is only done by and from political endeavours, stands. Although I had expressed the view in the same honest pursuit that the results of their efforts being passed on their own young, carry the rewards of that hard work. Is to cast no slight on either them or their gainful young religiously or politically! What has to emerge without balance, is that under force, the world we have created, is wanting on both the political and religious fronts? When considered that between the efforts of an extremely hard working rich person who is generally inclined to bring benefit to many. And the incessant efforts of some of our young to purposefully beg or work in sweat shops of rummage amongst city rubbish dumps, bears no difference in commitment or determination, but only by the circumstances of opportunity!

It is from the same spirit of circumstances owing to the same edge of opportunity that our religious leaders, without direct concern to such specifics, I expect that now they might turn the wind of change so that we can begin and reset our sail. I have set their task without the final say going to the last religion from the Middle East, of which I have coded to be the third dimension, being a choice of title to raise the standard of all religions, but by need to be able and refer to them collectively rather than individually.

Having already made the case that in general terms we have set the two types of belief in the hereafter stemming from Eastern by one style relative to some aspects of perpetuity, symbolised by the wheel or the circle or the hoops of life? With the Western religious style being monotheistic, stemming from the Judaism of some four thousand years on to the new Judaism of Christianity with a two thousand year past. On to our second largest world and monotheistic religion of Islam, to be over fourteen hundred years old, creating by reference my own defined first three dimensions. Cast by age and number as the first three elements before the fourth dimension of science with its own political connotations leading to the corrective force of the Fifth Dimension, naturally and progressively? In presentation for the purposes of open reform covered in need of recognition that in term we each are held only by a fifty plus or a seventy year plus lifecycle!?

By indication to have stated that our task in hand is for our religious leaders to make the same pointed direction that we must take. Is questioned in the name of Islam by who is our voice other than the basics of fundamentalism!? From where I had made much for the case that Islamic Fundamentalism is a filled cup of balance to the other cup of political Americanism sets a heavy burden to the religion in order that we may discharge the words of man 'Past Gods Words' into the living face of humanity!?

If, and we must bear in mind that the face of our Jewish and Christian religious leaders fall to be upon a ready established vital priesthood. How might we consider in link, what might yet be told by the first voice of Islam to the personal desire of imams, in cases who are of no more religious commitment other than to the envy of some opposite political influences? All that is exposed by the above is nothing more than the open need for Islam to be able and utter intent through collective involvement to all, come what may, but in the full expression of modernity accepting the global concept of mankind!? Automatically to include even all the wrongs and rights of the world being global, but still using the ability of change to ferment by its own personal and human desire! In the same manner of how Mohammed had foreseen events accepting his own failings as a mortal man. Not least when without the benefit of his own hindsight his new religion almost at source came to have two hearts without blame or condemnation to be Sunni and Shiah!?

Mohammed as the last prophet in the form of Moses or Jesus, through the conversation of the Quran, always had set the last standard? But without the offer of full insight to the possible final condition of man, later to develop from or after the time we became global exposing in the broader sense of how we had come to conduct ourselves with other unconsidered religious habits? From which no provision could be made to an unbalanced future at first unknown. Therefore was unable to read the feelings and consequent reactions of some religious lieutenants having been left in guardian to the written text to certain set standards, but not being able to set reaction not fully knowing of our future global state that was to be! No provision could have been made for how some of us might behave as imams in the name of Islam but through the eyes of men in the broad new world!?

From my own rational, working through the mix of how our politics and religions have been drawn together, is easier to follow from the abstract. Allowing that we can address any feature even with emotional intent, but always on line to be in the realms of full reality. For me to have come to the picture of what Islamic Fundamentalism is and intends to mean, is so explained by my own terms in the reason of fact, but not against Americanism? Even that I have used both elements illustrate what I consider the most startling examples of condition by influence we have come to on world terms. I make no direct comparison of each to each. Save in the necessity of what is and to where we are now by time marked in effort these past sixty plus years of this last modern era. Making the claim that it is to be from now we are to always look forward without forgetting the importance of the past or without altering it in any way to try and affect this future?

## CHAPTER 162

Islamic Fundamentalism and imams are cast with no judgement by my self or from any religious quarter, save the judgement of time producing the flower of our young to the expectation we are all adjudged to herald! This code is to be our next open and full realisation, but is only to be presented in the possibility of evoking Gods interest in us as the single specie we are, individually or collectively?

By what ever rules, tenets, guides or covenants we suffer upon our own religions, all are for all people, which is to be better realised when our religious leaders can now talk from amidst the three taken prospects of the Fifth Dimension. Additionally allowing our own voice along with their own views be expressed through the one active prospect in 'One for the Whole' which in reality will be little used save to quash our excesses!

Fundamentalism in reply will not work with or through the office of the Fifth Dimension, because the past in all of its wonderful finery will never be able and replace the excesses of the future once revealed. Of which has been done here and now without condemnation and of which will follow everywhere with the same lack of condemnation. Ideas and ideals from the past have always been the building blocks of knowledge. Unfortunately because of the nature of man, even from our own vanity, we have always sought the desired glory of our judgement except in a few examples like from Moses, Jesus or Mohammed!

I do acknowledge the three mentioned above are pivotal to my own case in question, but maintain that the example I use in cover of reference to the Middle East is a modern juncture. I have used the connection to show how errors in line can create monstrous misconceptions, which can percolate from one source to have global connotations? Our Middle East connection in terms of religious aptitude, is as solid as the rock on the mount, our codes even if not seen are of the same issue and of the same meaning by intention. What our own wit must gather in use of our own free will is the full extent in meaning of religion, not just Judaism, Christianity or Islam or others and others. What we must determine in open contact or conversation through the standards of our religious leaders and imams? Is the realisation without cost, we all have the liberty of thought through our one taken prospect 'One for the Whole' as a finishing tool?

Without example I give the effect of that tool to be our strongest means of proof to our specie singularity. Not least when offered to our crocodilian or any type of our finches on the wing or to the splendour of the big cats expressed through the majestic tiger, all in reply is their hush of silence. Other creatures have never, never shown any signs of the spiritual in longing, none have been able and challenge the spectre of doubt by the same image of wonder created in we humans? Other creatures as we have only been content to follow the codes of instinct to the one aim of procreation, unknowingly to continue the same line to the same ends, and ends, and ends?

Our own break to the mould of life, although a long time in coming after three and a half million years, once began some sixty to one hundred thousand years ago at Shanidar, had left a mark on every single tribe that we were to become. Purpose beyond the raw edge of instinct was our burden and in case it is only over these past four thousand years we have been experimenting with cause? By the same modern measure of those years when we came to the last fourteen hundred years, on entering the era of Islam we had not lost the means to err. Unfortunately taken in the first order of being the final of all monotheistic religions but by an emphasis I had already questioned? Not in balance of what we thought of the faith but in the way we have practiced its clear message of unity expressed in a more forth right manner than had been done by Jesus?

Islam's main unseen error if, stems from the fact that in all of time its basic plan had been misread by some of us with very unfortunate consequences, even that I would offer now that we can sidestep our errors of judgement in becoming Sunni or Shiah before we were either!? My connection in comparison without balance stands in corner that the design of Islam although taking on the face of many Jewish practices, which were not of religious copy, but an example of our own tribes and connotations of familiarity to area. Is little else than the use of that familiarity generated in mixture of what had developed with and since our interactions living in the land between two rivers many centuries before?

I had stated with some authority that Islam although set in area by call had the addition of planning for a bigger screen than at first allowed for in our ability to understand? If by contest I had expressed the view, my view that Judaism was cast to be imminent by outcome from its offset. That Christianity was/had also given that impression whereby death was imminent if only by the normal standards of living then ago. In so far as we all lived under political systems of emperors or kings who automatically held sway over us. Therefore why fight the inevitable why not turn the other cheek to get on with what is important in life, expressed by this new Jewish faith? If death is to be a consequence of our good and pious behaviour from the bullying and overwrought hand of oppression and conquest, then in just reward we will attend all the goods exposed by our release from original sin first committed when last we were in the paradise of heaven?

Imminent and imminent in being the expectation of at first Judaism and then the new Judaism of Christianity, was a term of reference but not then able and mix the meaning generationally? Although expressed in similar kind by two able priesthoods always to the conduct of the individual even in lifecycle terms. No mix was ever made between the forces of government be it of emperors or kings and those of us to be believers other than with or through those already established priesthoods.

By any form from such formal relationships our paths on the Jewish line of some four thousand years ago, along with the Christ path of Roman Catholicism starting two thousand years ago in line. Both made the unavoidable error of attending to our political expectations in terms of longevity. Without making the most important of cases of how we as individuals stood by personal conduct of having lived and behaved in covenant for our own life spans? And or to what was that relationship to all else, and again most importantly how were we individually to pass the religious message of our religion to our own family young, which in style is to be expressed by the double heading of our physical family young. Through that very real connection and or including all of our young of faith, the same faith in each case, who were led in love by all the elders of that same religion only? Allowing in outcome that in all cases, we encourage our young in the fields of use of their own free will, in attendance to the political matters of life including religious direction? Allowing any alterations required of the same be made, but to attend to our first standards of religion first!?

Four and two thousand years ago, from the impatience of ignorance had led us to take giant steps, but unfortunately without the direct connection of events at Shanidar. Being what I have determined to have been a trigger of mind to the event of thought if measured to be a case using our own free will in comparison to all other creatures of this earth. Free will as any thought or action outside the requisite of living by instinct attuned to the bells of procreation, free will seen, but unheard by not looking backwards only to tomorrow?

How ever we might wish to argue the issue, we are all here and now by the one means only? Holding in faith all of our different religions and or political assumptions no matter what. We are all influenced by politics and religion or both or none. Even that some of us will argue in denial to either and or, we are all living and co-existing here and now?

Of which I have set the highlights to have been latterly attempted over these past sixty plus years since the ending of World War Two. I have further been able and connect this last modern era to the past four hundred years of our united histories when we became global. From which now stems the measure of how we actually interact, but not counting the individuals choices, unless through 'One for the Whole' to become a personal voice!

Of which to be able and understand one of our main attempts in the case of individuality, attempted again some fourteen hundred years ago through Islam. Not to be a trial or a test, from which for the first to any of our major monotheistic religions, consideration was given to the aspect of the future beyond our own lifecycle? Because it was seen to hold the aspect of care to the outcome of learning for our young and the same prospect of family into the future by our own generational codes? Quite openly the tenets of the new faith in the supportive role to be the pillars of wisdom for all generations yet to come, and for the conversion of all the reasons of personal conduct. Allowing in open display the broad concept now realised then ago of the faith being carried to all, even including through conquest if necessary to enrapture the whole world in time!?

From that standing and standard, was set true futurism, when even if unseen by those of us close to Mohammed, the plan of text was in display of conduct for all in all times to come. With the past if necessary being swept into the corner of what had been, with due deference to the efforts of Moses and Jesus, but in open display of how their efforts had fallen short by not giving direct command to individuals? But had allowed the civil or political leadership continue in form with our religious link run in secondary unity by our priesthoods in also following state or national commitments first?

Islam by concept at instigation had failed on one small issue to that end. In not knowing of our true character as men, because all that we had to go on was past conduct. By working on and with and through the open assumption that all told would immediately be open to the changes demanded in compliance to the essence of this new and final faith. Mohammed, if in error, was too generous of his assumptions. Not least by who he thought we were in being men, hungry to have the prevailing doubt removed and replaced with what should be, by the means of direct action to spread this new faith, which although violent were by our normal standards of the day? Allowing that we could indeed operate in character but for a better future, making the un-assumed errors of judgement set by those times, although all considerations of the day were now met. Including to the picture of the faith having to stand the test of time by applying to future generations in the same way as it now applied to us then, because the written word of then was the standard of now. But forgetting or not finding it necessary to include that we would gather more and more understanding to all things just like we had made forward steps since Judaism and Christianity!?

Although Islam in structure was not to change because of applied text, it was we who changed by sets of unforeseen circumstances even that it was always possible to forecast what might be from what had already been? Much of which was done in the directed writings of Mohammed for our own good benefit. But almost as if in test, avoiding the feature of us not then belonging to a global entity that we knew or did not of, being no fault? Save that the feature become disarming when we did become global and have so far taken over four hundred years of half measured attempts to rationalise that fact either politically or religiously only to fail, because now we are more two than one!!?

If I had set the first target of the new rational to state in the uncertain clarity that politics now is a modern deed described by the end product to aim for, to the American dream being Americanism for all. That case stands on the open grounds that I personally use it so to present the single argument representatively to the standard of desired material gain by name from politics? Now to have outgrown its full use to our real benefit, Americanism is a standard not a requirement, but for our purposes now and in the future discussions I use the reference in balance to the religious side of our necessity? Accepting that both features can be included collectively, unfortunately causing the biggest flaw to our understanding of self to have been exposed by our half measured functioning through these recent sixty plus years!

For the open study of any new or repeated errors we have made in this last modern era, for less concern to the direction we might have made, I have laid the case for the difference of our two functions of law. One political, one religious, both fully to this modern era and both to be treated with the same medicine in the structure of the Fifth Dimension, to enable us recall the measure of error we are prone to continually make. Because until now we have still not found the route of how to change all of our habits issued by half measured delivery supported by our steps taken syndrome?

Without separation from other religions my own point of reference to Islamic Fundamentalism and what it really means, can only be viewed from now through newer formed perspectives of having all of our religions overlaid with the prospects of the Fifth Dimension! Fundamentalism by style is to revert in concept to original standards first laid down, which is no bad thing. But only holds if those very real standards are adjusted to total modernity in link before duty, because now we are the same people reading the same words by living generationally different lifestyles? What we are not entitled to do by any means or reason for regression, is set or reemploy those original standards of when, even if recorded in the original final definition, expecting the same consideration after an additional one thousand, then another four hundred years of actual living!? Not least when in all cases our continuing generations had naturally progressed in other fields of knowledge. Until finally some four hundred years ago we became global adding external opinions to our broader outlook?

Without the defined structure of leadership to Islam in previous monotheistic religions originally from the Middle East, but not to be a comparison, because that leadership feature is not always necessary for all interactions? Although any new or different code of conduct is to be employed or any defined original method or reasoning is considered? How are we to take leadership on our own account!

## CHAPTER 163

Having made no comparisons between imams and or the established priesthoods of other earlier religions from the cauldron of area I have referred to as the Middle East. Without that connection I now state that some imams of the faith, without wisdom, have contrived to curtail the faith from original intention? Although I speak words of serious criticism, my tone has to be unbiased, because on the one hand to the case of politics, my concerns have not been shown to be as deep about politics, than with our religions, but why?

Islam and politics have run the same course over these past fourteen hundred years, each contributing of there best by our own interpretations? It is from the past sixty plus years of this last modern era we can and must refocus all intention. But now rather than before our two cases of control in guidance to be our politics and religion are both overlaid with the Fifth Dimension, giving both the impetus to change without alteration. From our political standards this will be very easy, simply to adopt the three taken prospects of the Fifth Dimension will be enough to stimulate the correct order of events at all levels and to include all of its tentacles even the probing fourth dimensional one of the sciences?

Our religious changes, are to be just as simple, but with deeper thought, because the new addition by expression to be our one active prospect in 'One for the Whole' is the unalterable link between all religions. Which in any case has always been applied by all religions, but until we were to reach this exact time of now, remained unseen to each other of different faiths? Again by making my own observations clearer, how I have formulated my opinions will lead to a more open approach for us all to our own religion along with others. At last without the restricting knot of religious tolerance, which has been used in the wrong context virtually in all cases over these recent sixty plus years of continual change?

Having lived and conducted too many of our offices from the political perspective only, our time of reason is dawning, providing we can come to the one case that is to be of permanent memory to us all even by reflection!? I speak of moments on how I have interpreted the considered events to have occurred at Shanidar some sixty to one hundred thousand years ago? I speak through this pen as is all of our right to do, providing the mark is of a naturally progressive nature, which only applies to the human specie and on living terms only politically, but from now always to include religious perspectives first!

Shanidar brings to us factual evidence of how we broke the confinement of pure instinct and its tributary of procreation. By whatever good fortune we were able and reconnect with the incident again is only due to our natural progressive nature in any case. Not least in our open query of the past by all avenues in not only wanting to know of then, but also the answers to when and the possibility of why? By quick succession my reading already told of that burial in which grave goods were first left and so found, stated in clarion of who we are in terms to be different than all other creatures. Not least by reference that our thought could not have faltered to any other time in relation to self than the future through 'Natural Body Death'.

In leaving the incident at Shanidar pre history and returning in line to the time of Mohammed some fourteen hundred years ago, our direct link with the future is still maintained. Any differences then were by small influences of other similar religions, except to suit the natural violence of community then ago? In allowing for that fact and in order to establish what was and is by many considered to be our last choice by desire. Islam by feature, if on course had always allowed the fixture of defending or protecting or promoting our third dimensional religion by any means necessary, even by our own determination to die in martyrdom for those proofs. Not unlike what Christians or Jews had done before, but this time the comparisons made with death, for or against, had different connotations either way, allowed the opening whereby choices made to that end were led to the matter of an individuals own standard and choice. But to know much of direction was fed by the open exuberance of the faith in its first flush!

Making the reversed situation of two completely different outcomes having the same intention before operation, which I personally challenge on all of our accounts!? I have mentioned the features of suicide of martyrdom and of murder in the same context before, which cannot be challenged by any force of knowledge. By the same course I have laid command to some of the reasons for some such acts and some of the reasons thereof. In dating the concept of martyrdom by any style my point of direction in all cases has been to confine the incident if, to the realm of fantasy which by all counts that is all that it is!

Martyrdom in concept is unviable it is a true falsehood in any of the languages of any of our religions. All that the prospect of death before the incidence of 'Natural Body Death' ever was, stands in the mixture of what is permitted by political standards only, by political standards only? Martyrdom in link is of the same value to what had been formally adopted in the far off lands of the new world pre full globalisation. Human sacrifice is the full epitome of what the feature of martyrdom means by picture and by act.

With the other element of suicide in context to the modern approach of personal suicide used as a weapon against a supposed enemy, the concept belongs to the same shallow and hollow essence of murder when other lives are taken universally. Suicide by all measures is to be measured by political standards only, by political standards only, not least if taken on to the final act of committed death of the individual by choice. Of which in those circumstances the only shadow of doubt is cast upon the open necessity that we rule all living by the standard of 'Natural Body Death'. Unless we can unlock cause through and with our one active prospect in 'One for the Whole', being the ultimate feature whereby the aspect of necessary suicide can be argued with confinement of self only. If there was ever a desire in need without personal insult to self or others or the final settlement, whereby we hope to attract Gods own interest in us as individuals collectively. Whereby if we came to believe the aspect of Gods interest to be only stimulated by this single specie, might not we be charged in single contact by good heart to acknowledge Gods own light of consent to what we would or could wish without harm or hate to be upon any subject?

Suicide in the culture of our own hand is a free agent laid against the prospect of national suicide in the form of herald? Nothing is seen to be achieved from the grave in death to any form of it by any political impetus. That is why through the grief of war we might lay fifty five million souls to death without knowing the why except to our own faulted concept of greed, any greed in general. In so doing such monumental deeds in the name of politics through the eyes of civilisation, our remorse stands, even that by wars end we had not ever found the reasons for what we had done, save in counting the physical gains and losses of profit. Unfortunately being the tool which was to take us further into the uncertainty of now by processes over these past sixty plus years.

When next we looked at the reflection of the reflection in the mirrors of life for the object of reaction, we had gone full circle in understanding of how the world had been fourteen hundred years ago. Although change was counted we could not see the full necessity that our religions and politics were to be one of each leg in our progress. Both working at the same pace for us to move forward as we assumed we had always done under the political cloak of civilisation and the religious cloak of taken or suggested tolerance?

By the same connection long mentioned in these pages, our religions in remaining passive under the bondage of tolerance have denied us our full voice. Save through Islam if measured by what is meant to be the edge of now, but only kept in flight by being carried on the wind of earlier monotheistic faiths from the homeland of what some of us consider the land of God in the Middle East? By cross referral and in taking in the effect of reflection upon reflection, when seen or deciphered by the same difference of opinion, even if by some to be our priesthoods or others to be imams? Any picture produced in what might be the reactive pose has left the standard of our priesthoods to be tolerant and political? Yet by reflection the standard of some of our imams is to the only voice in what some see to be fundamentalism!

Being a corner we must round in order to understand our own perspectives, which are to be obligations for all or to the benefit of all. Covering in blanket fashion what any attempts of progress had always intended to be part of the continuing growth in understanding along with our ability to understand! First seen even in this last modern era through our politics if taken for granted and some of our religions in activity, only if to counter the lame posture of politics? Updating virtually all of the reflective actions and reactions by how we are now and from who we are now, most exemplified by my own coding in referral to the political aspect of Americanism, now rebutted by the personal reading of some imams on the religious front. Who in the regressive development of fundamental reasoning, have used the wrong tools of we or by worse means, some younger members of we, to attempt and correct misalignment, if that was the only case!?

From what had been done, was nothing less than challenge the roots of all religions, but most pointedly the very heart of Islam. Not I add by open consent but from the darkest reaches of envy, spite, ignorance and in the modern context fear! Not in order of presentation above, but in the applied use of encouragement to others in action on our behalf, taken even to deeper depths of depravity to have encouraged our young act and do the vile deed of murder on our behalf. All from the impression only of posting opposition to what is supposed to be the political persuasion of other religious standards. Which in case, are part and parcel of Islam linked through Jesus and Moses in line from Abraham and the picture of what should have been!?

By token, that I had presented facts that Islam was created from amongst the violence of normal conduct some fourteen hundred years ago, denies the prospect of the future laid down by the faith even from the lips of Mohammed. When by self mention he rued the fact that he had been denied the experience of martyrdom whereby he would also stand with others in the expression of good will but by Gods will? By code the expression was on many levels not least to connect the matter of life and death in relationship to the service of God in Allah? For the man Mohammed to have mentioned even in passing, that he lamented not having been chosen or not having naturally succumb to martyrdom by means to confirm his/our commitment to the new word of God, was a judgement beyond our understanding even by the day!

His message unread, carried the heaviest of burdens for us all, because in good and honest commitment to the ability of his standards in how we viewed things generally? Our additional step in conduct maintained that lament far past the use of any measure to the function of when and where and how we were/are permitted to kill self by self and for what!? Unknown then ago our religious progress had far outstripped the measure of how we were to progress politically. Which is shown in the full reflection without connection to the wishes of Jesus or Mohammed in being men of our times by attitude and opinion by wishing to represent our future by what ever vision? But also from where all things were to man by our differing attitudes stimulated by our generational extra acquirement of at first knowledge and then the best means to apply it socially and politically, eventually forcing the need of or for religious change. Which from mid ground was most vividly done through the Roman Catholic come Christian reformation of approximately four hundred plus years ago mixing all the emotions through reforming. But allowing the unseen progress of politics on its own behalf until challenged in the mid nineteen sixties to my own terms of reference?

# CHAPTER 164

Although I have called one side of our balance cup to hold Islamic fundamentalism against the other holding Americanism, my intention is to bring into full focus the standing of our politics and religion at this precise time in this last modern era. By the term Americanism shown as intended, I speak it to represent the top level of our political desire. Although Its' main positive side is that it allows us our freedom of choice, which to small extents are right or might prove to be more correct if we explore its true possibilities now that it is to be overlaid with the Fifth Dimension?

Islamic fundamentalism reaching to fundamentalism has been used by my self to stand in support of the intention to our religious excellence, but is to suffer in great change led by the accumulation of our growing knowledge set to the spirit of intentions of mankind. In making the case by suggestion that for Mohammed to have sought concern over the meaning of what true martyrdom really meant was done in style to draw our intention to the actual time in which Mohammed then lived. All considerations were set to the style, habit and acceptances of then ago. For us to wish and nurture those standards is not a viable reason to the cause of murder now seen in places throughout the world today, but to note they are only done on political footings, never, never to be resurrected from the reflection of the reflection in how we now live religiously?

By the same token of now and to the time while Mohammed still lived, imams had the similar duties then as now notwithstanding the fourteen hundred years of our changing if by political reasoning against or balanced to a religion that did not or was not supposed to change, but might by some of all the wrong reasons? From cause that to some of the role belonging to mere men as you and I, have imams in cases been influenced by all the wrong reasons of input to their own specific tasks in life religiously? As long as they might try and encourage the young of Islam, our young of Islam to carry out their religious duties and obligations none can offer offence. However some have erred beyond the limits of duty or care or consideration or civilisation, allowing that our young might become martyrs in the name of Islam but not to defend or protect or promote the faith at any time. Only to employ the policy of death by suicide by placing in replacement the vindictive thoughts of envy, spite and jealousy owing to some individual imams having lost all the reasons of creed?

Imams who can close there eyes to the open failure of our and there own political standard, by advising and encouraging the untried minds of our young in the realm of murder. Papering over the deed of personal suicide to be at best an act of martyrdom, have never looked into the eyes of Mohammed or heard the words of Jesus. They have been stolen of thought to be able and apply the reasoning and working of people to have failed in there own duty and our duty of humanity! Some of our imams have arrived where they stand without knowledge of wisdom, some of our imams have only been able and look to what Judaism is what Christianity is and what politics is over and above what Islam is to be. Some of our imams have been able and only see the small picture of Islam contained by the times and of area where it first raised new hope to the concept of purpose!?

In directing our young of now, although to apply the teachings of then ago as the same religiously to now has no difficulty in the honest terms of transition, all the hopes of Mohammed have already been realised. Islam is already a major world faith carried on doctrine, not now in need by the bomb or the bullet or the sword and shield. All that has been missed in what can only be described to the blindness of some imams is for the scales to be removed at first by our collective use in application of the overlay in the same terms to our politics with the prospects of the Fifth Dimension!

Without direct comparison of one to one with our political tentacles, our taken prospects in overlay are to be used directly to each of our divided ruling entities even with the decisions made for all, but by the taken prospects separately! In previously stating to the agenda of fact that our politics were a self generating force having little to do with the true picture of what humanity is. Is to be used as a point on which we must pivot, not least when we can look to see in the same breath some of our young foraging in rubbish tips for sustenance. Balanced to the full political success of the parents of some others of our young able and inherit billions or hundreds of millions of money which has no moral weight by the concept of religion or God! Used only by example to show the gulf between what politics is and what religions are. That from our young to be desperate and hold nothing, can in many cases equal the same piety as our expectant rich young with charge of material expectations, has its reference in death only, which is 'Natural Body Death' only, but never in excuse?

Showing the importance of where our religious leaders must stand from now, to apply the concept of our taken prospects forward on all levels. Our religious leaders have no need by request to alter or deviate from their own standard to our understanding of what religion is or what our expectations of religion are to be? Save in applying the overlay of the Fifth Dimension to be upon all religions for there own sake and not to be compared to how the political experiment will be self evident? Without any personal levy put upon any religion I have made the case for the importance of what our religious leaders must now do. Even that I had set the direction whereby we can look to our separate religions separately is done not that we might compare, because all religions are for the future. Therefore the importance of now is more paramount, because all the intentions of our functions are to the future even if unknown, and cannot now be compromised by what is done or is suggested to be done religiously under protection of our one active prospect in 'One for the Whole'?

By best example, again without charge, but in open error to what some of our religious standards had been by the silence of voice in the two major situations of world wars of the last century. In general terms the religious silence had not been drowned out by the noises of conflict through the artillery shell or the machine gun bullet. But by the political objectives of desire of some Christian cultures wanting only the material advantage of now, before the religious objectives of the future.

Silence in the voice of our religious leaders in both conflicts, unfortunately was the first step in moving to place our political requirements in charge of our religious expectation. Even carried and honed to first hand political objectives over these past sixty plus years, alas half measured, inducing what has always been our steps taken syndrome!

Steps taken, having mostly been conducted from half measured political decisions have been so successful without balance, that we have become so used to the situation in not being able and tackle the necessity of change except by the extreme if that is the case ongoing? In core and from the application of the modern understanding of religious tolerance now applied in blanket fashion to all religions? When first done in the last modern era marked by my own view to have started at wars end from about nineteen forty five. It was done from the European/American perspective to allow our political decisions hold sway over the supposed self care and consideration of now above the weakened prospect of our religious expectations. Not least generated by the absence of our religious leader corps!? Which beyond the critic was done in the prospect of moving in the right direction if the true situation of then ago was to be viewed in the reality of what was really going on?

Up to this minuet and in direct connection to some events of the Middle East now, what was done half measured was done without any realisation, but without the malice of hatred and fear? By previously stating to the existence of our post war cold war standing of our political bloc set up throughout the world. By dealing first hand to the existence and difference of two political cultures being democratic or communistic bears only some of the blame or cause, reviewed in area of the Middle East, where the three, my first, second and third dimensions still interact? But now in aftermath to the complication in effect of the previous cover and support given to the Arab/Islamic side since and from the result of mandated partition although from the Russian side of communistic ideals balanced to the one side of politics being democratic!? The concept of religious tolerance internationally was branded as a useful medium whereby differing political structures, were to have common ground religiously? Keeping those opposing philosophies under wraps and out of the direct line of oppositional conflict in theory?

All of which might have made ground to our equal advantage, but for the fact of the two inbuilt errors unseen at the time and since, the relationship by some aspects to the suggestion of religious tolerance was unbalanced. Taken at first by instigation upon Judaism and Christianity, it was done in part because of able priesthoods in supposition to be representative of our religions previously shown. For which the cut made was divisive, but based on bigger errors than expediency, because in all religious structures the future is cast to be the object! But further unseen by political reasoning the future through all religions was for all people in those religions, in those separate religions to be of the same dividend!

In case that Judaism and Christianity had many heads contained in theory by the leadership structure of all of them, religious tolerance was a genuine feature in attempt to promote certain levels of harmony? Its overall weakness was in direction to have been done to placate political desire then unseen, but also to cover the standard of our political differences in terms of wealth, at first measured between democracy and communism as a benchmark to success? But also to allow that in all cases through our religions we were the same and better in any case, if only another way to undermine communism?

By the same balance, because communism was not supposed to carry any concern to any religious matters, the concept of religious tolerance through the same political vein, was prostituted in order to accommodate sanction to the standards of communism in the cold war, and through the post war days of all excesses! Religious tolerance on edge was turned to the standard to cover by intention the proposition of human rights, which was wrought in the melting pot of change generated throughout these past sixty plus years, but always led by political overheads! By that referral human rights in apology for all the wrongs of any and all political systems was a token offering to the standing of the future?

## CHAPTER 165

Although I had cast reference to the serious nature of religious turmoil owing to this last modern era by the political partition of the sub continent of India, at about the same time many other countries throughout the world also achieved nationality? On the sub continent, with India being split into three main component parts, India, of mainly Hindu religious persuasion, West Pakistan to be mainly Muslim with East Pakistan also of Islam. All of that turmoil then created, and it was bloody and deep apart to the single cry to God or Allah, was political, that is the measure of our political depth by all standards. Unfortunately there is still a great deal of conflict in area between West Pakistan which is now Pakistan and India over political boundaries in disputed provinces bordering both countries!

Between East Pakistan and India, although social and religious conflicts also had the same brutal connotations, the conflict outcome was in mixture to be different than what was to eventually settle over the whole sub continent. Mostly because of internal matters between the two Pakistan's and because of the overall war between India and Pakistan in general? From the mixture of all factors even hidden ones, East Pakistan of then ago was to become Bangladesh being a separate nation to the other sub continental entities. Three nations at a time in total war which in turn was first stimulated on religious grounds, allowed us to be reminded, that even in Gods comfort to style religiously, we were able to act in our own interest politically. Which must be taken in shadow that post World War Two, we had lost ground to the reason of change?

Leastways by show that from the original demands of promised fulfilment, when these new countries were set up on purely religious grounds, to be underpinned by political motives from other quarters? Making the double case that we are able and hold religious standards to the depravity of murder and murder collectively, even to the lost advantage of collective change when turned to politics!

Unfortunately no real lessons to the full benefit of humanity, could have been learnt from the full range of interactions between the different peoples in the sub continent? War when raged was just as tragic there as anywhere else in the world. If politics was to be the controlling influence or really held the upper hand to the decisions of outcome after the basics of partition, which as we know was designed around religious division more than political outcome? Then politics was the only winner in the divide from the warring amongst India and herself, until East and West Pakistan were created!?

For East Pakistan to have become Bangladesh by her own code, however painful, was a matter attended to from area which by effect did not have the same reverberation throughout the world. In knowing that the duty of partition and the later wars of independence in claiming her own name for instance, by realisation although the population remained mostly devoted to Islam the factions of religion and politics seemed to have been kept separate like in other places also?

Our sub-continental tragedy stands on the equal grounds that although a single specie, we still hold serious political animosity between our separate countries involved? No let up to the standard of our religious beliefs have followed save by good reason. We have come to notice that the endeavour of our point of real conflict is not on wholly religious grounds, but over ground in the form of acreage? That which was fought over and still, but in places only accepted by the adage of possession being nine tenths of the law, remaining a high grade precedent, which alas has been misused throughout the world in this last modern era!

India, Pakistan and Bangladesh have been an open door to the matter of concern by and for all world standards. But unfortunately, by our stubbornness over our political determinations, we have found the matter of some of our religious concerns usurped by political thinking? When from the offset at patrician we behaved in terror of each other, not religiously only politically, without defining the nature of specific events which in equal measure were from the realms of human degradation? We have now been content in latent anger allow the dust settle. I say this to be representative without direct blame apportioned, save from we collectively. I say it unqualified, but all settlement when done was of a political nature. Almost allowing that we accept the standard of religious tolerance hold its own in relation to Hinduism with Islam counted between the peoples of India, Pakistan and Bangladesh as we, to the same style as we of the Middle East, but of different emotions!?

Meaning that although partition by the most outrageous act of half measured development set from religious desire, had been through the full political loop to induce war before any settlement, when all things were decided politically by our steps taken syndrome. In all cases without claim, for me to indicate by suggestion that neither group in religious terms had held to a specific leadership set up or priesthood standard. Stands by reflection from one to the other and back again, but to eventually mean by what standard were they measured? How was Islam to be realised against Hinduism or in reverse, too or against Islam, when neither held to any direct form of spiritual leadership?

Islam and or Hinduism were the same but incomparable because both had not acclaimed the full order of any direct form of leadership. If therefore any conflicts in this last modern era had occurred between the two cited religions, our measure of reflection has to account for what might be before what was or was intended to be? Not least because before any political acceptances were demanded, it was our full obligation especially after the length of our actual histories, to be able and give judgements beyond what was thought?

When we remember the cause of partition to have been done badly by the first standard of political decision making, we must be reminded that all desires although half measured, were political. Because no official body on religious terms were contactable unless only from the edge of political desires? Any designs to how the jewel of India was to be cut was done in heed to the desires of all of us who quite simply did not know or were not sure, except to political advantages if that was supposed!

With all the events just mentioned above, from the nature of our histories, which are complex and creative in the same fashion all others are, cannot be changed or altered by different desires. We are left in the same state of concern and confusion applying to the nature of our religions and the way they had developed and or the way we have set there course. Knowing of some of the outcome to the event of partition on the sub continent and some of the following events, our charge is to remember that the timing of those events were also to this last modern era.

Without comparison, by having laid or mentioned some of the facts, no final decision can yet be made politically because as stated there is still unreasonable animosity between the same peoples as we. What I can suggest even to be supported by the unreasonable authority of my own abstract reasoning, is to the standard and style of religious leadership displayed in time, at any time any where in this last modern era. Which must be done now, because in other cases of our standing, political animosity is carried forward by our uncensored religious conduct which itself is carried by self judgement from some individuals even if led!

By effect, any reference to the events of interaction between Hindu and Muslim in this last modern era is done, because from my own notice for our concern by either core religion, in not having a dedicated leadership structure, all interactions between, can only be political? Because in or from the incidences of personal contacts however made, there could not have been a directed intention unless led by outside influences, which is not to confuse the different religions being outside each other!?

Our point of interest to the above is to be held to account for anybody in we, who would use our own religious acceptances to be a burden on others. Not because we do not have a representative leadership structure, but because our own religious duty is personal. To mean in full realisation that what ever our chosen religion, even directed, none are self sufficient. None have been measured by what we have done, only by what we are supposed to have done? Although in cases throughout history we can present argument in any direction because of other events local and historical, we are beginning to see the necessity of showing our understanding of our singularity of specie. Even if at times we are still more ably to show by our actions and reactions that we do not yet understand the necessity of how and why for the first time ever we are to show we understand what specie singularity means and is?

Without praise of conduct from any of the events to have occurred in the sub continent of India these past sixty plus years. Without condemnation to either faction politically or religiously, any ruling of judgements made can only be done from the time and events without alteration of the past or of motives or ideals. Where we now stand by all events is to simply state in fact that from within all the human classifications of and shown by some of our differences of how we thought then ago? Our good fortune in cases has been spread by the result of some of our political determinations in working, but to the heavy cost of our religious sufferance in forcing us to make choices un-led?

Almost being the cause of our divide to be managed from political perspectives, which only operate and more so in the sub continent, because in such cases without defined religious leadership, apart to the direction of creed and doctrine, we can only follow our instinctive emotions being regressive in purpose? How this phenomenon forms, is generated to our obtained standards, has in far too many cases shown us of our ability not to understand what we should? If so, this can be improved from now in what we might do from or encouraged by the reflection of the reflection or only in part, allows what we do by action or reaction. Undoing the timeless form of our differences without us being able and see our similarity through the medium if be it political or religious!

From the one last common feature between Islam and Hinduism, our last remain is that even from both outlooks by religious terms, in a sense we were leaderless by the absence of an established priesthood, has not lessened our desire or determination. But perhaps that case compared to other similar examples has shown us the most of how we have allowed politics take hold to be able and impose in sanction our levels of religious tolerance? Only to be drawn from the core of where religious tolerance was first delivered if through the United Nations from only political leads. Who had tried and intervene in the India, Pakistan and Bangladesh conflicts of old and the embers of now, to support religious tolerance while at the same time surrendering to political whims if decided by those in conflict?

Allowing that all factions were to be classed as the same religiously through the concept of religious tolerance, unfortunately, has only endorsed the blind side of our openness in attempt to make better judgements all round. By one of the most unusual circumstances never to have been evaluated by reason at any time through the events and since, that the initial cause to the split up of the sub continent, was for religious determinations, but by political desire? Which was to lead to another episode of our own human brutality of murder by self upon self, holding the appearance of that singular feature belonging only to the sub continent, while at the same time much was the same throughout the world!?

# CHAPTER 166

By mention of the fact to some or all of the way we have conducted our affairs world wide and particularly in this last modern era, with my own main concern inclined to the wonder of our religions first hand? My care has been for us all to remain calm and fully use our religious perception in line to what was revealed to us at Shanidar or to what we might read from the event of Shanidar?

By what ever means our religions had developed over these past four thousand years when we came to evaluate any aspect of them through the mirror of these past sixty plus years. Without comparison by looking at individual religions all that is seen is the true and genuine approach we have all tried to make to suit by association the purpose of humanity! Except in the sad and open case when for a time our main objective in faith was to the belief alone, whereby without the object of comparisons owed to other groups throughout the world some religious practices were different and unworthy of humanity?

For too long a time we gave to the main objective of our belief, the practice of human sacrifice, not in cost to placate the image of God to any like life form. But to the assumed objective of the sun by day and the heavens by night, which is a standard, but not to be of value when the system was managed by mere mortals as we in our priesthood!? Although we at last have stepped beyond such an image we remain in danger by all of our religions who still operate past the prospect of 'Natural Body Death'?

When looked at with an unbiased eye if that is possible for any of us, our serious disruption of partition of the sub continent has offered us intrigue. Which in hope will lay a better force of reasoning to be upon all religions when viewed that such actions were wholly political with the off shoot of violent religious interactions through the veins and arteries of retribution? Our case is further confused by human frailty in respect of how we as self in the name of religion. Did not ever consider in purpose why we did what we did because we were content to let politics be the engine of our emotions above the hand of politics, but acted in cause through the eye of our own religions? Our only plan without guidance was to kill for kill no matter the consequences politically or religiously!

What had emerged without any form of reasoning has largely been ignored, because of how we are supposed to be only accountable to our God by name, leastways if carried in desperation by the affected? However, in part, without consent, although the manner of murder in the first flush of partition was horrendous by both sides, just as in war other stories can emerge? There were some vital heroic acts by many individuals of both religions in respect of how some risked life and limb in helping family and friends to safety as intended. But were to suffer the loss of self through the total tragedy that was to overcome us all? Again not unlike in the specific act of war where some combatants might suffer no greater valour than to lay down there life for a friend! Bringing a range of consideration to situations which in being tragic and horrendous, we were given a minute amount of calm for reflection. Drawing from the character of man honour to be monumental and putting into context what martyrdom is and always was!?

In the violent crossover of possibly a million deaths in equal to be half Hindu and half Muslim, countless little acts of self sacrifice were done amidst the mayhem giving us all a miniscule of human hope. Although we had behaved in abomination, the other sign of light that I had seen and see, beholds the tragedy but in no grace, except that from that one specific short duration in initial resettlement, a key was offered to the future which can help make a dark deed hold a brighter light by which we might better balance our judgements!?

From the first instance to the last by general consensus, having being forced to move and in doing so, what ever was done in its worst form was through the driven motive of political reasoning. Harnessed by normal human emotions if compared to any situation when a person or family were to lose a vital part of their history being forced to leave home in the name of religion by the cause of politics. Although I have not been specific to the type and acts of murder done, in saying they were political does not reduce the open horror nor does it explain the level of brutality. Nor I add does it lessen the full involvement of our religions in being of an Eastern or Western philosophy when both the main types were less than controlled by dedicated priesthoods?

Which brings to the point, although most of the first wave of murder done was by Indian upon Indian in not being of the same religion, when the count was done without choice, no clear examples were seen to the acts of heroism that occurred only in cases to be counted as full martyrdom and not suicide? Being a highly pertinent point as far as what we might have been led to believe by some of the held views of Islam. Leastways those expressed by some imams, who in case by other examples have openly promoted the specific act of suicide as a means to defend, protect or promote the religion of Islam, but in a parallel time to now, then ago? Therefore to gain from death of self the inclusion of being a martyr for the faith in claim to all the accolades allowed did not fit then of now?

Having no real evidence, although none is needed from my own abstract reasoning, that the deed of suicide to have been done in the religious conflict that partition had created, stands only in part to the view, even if only my own. That the standard of action, reaction or retribution done from either side, did not then carry the required doctrine of martyrdom, because the single balance of personal survival was considered first against the standing of religion, politically?

Meaning by translation, but without being directed at any specific party, our personal acts of heroism in the first flush of partition from where some gave there own lives in the protection of family and friends from either side. Were pure acts of humanity in display of humanity and in open display of our specie singularity by any comparison with at least other creatures? Ignoring self held beliefs in politics or of religion. Even to surmount the total error to the judgement that partition was necessary in any case, save by warped political motives drawn from some of the avenues of greed and envy. Scandalised by the mere fact that we were wrongly excited to the political prospect that our religions were to be compromised, which they were later, but under the umbrella of religious tolerance to be a common standard.

Although as a standard, if to be a method to police our reaction to the prospect of some religious animosity that might occur in the future from the events of now or from the distant past. By my own statement of now, but relevant to all dated thinking, is done in open reflection of those events that occurred at partition on the sub continent, which are ongoing now in the Middle East, but here the difference is from another perspective, which it is not supposed to be. Yet done by peoples of a slightly different outlook if to be progressive politically, overshadowing others to be laid back in that sphere, but are volatile over religious matters? I had suggested that in part plan, it was through the United Nations that the similar picture of partition in the Middle East was to translate into a worse sworn tragedy than the hurt in India was to generate? By the unfortunate feature of our forced open acceptance of political domination to have been laid upon the hope and aspiration to purpose through all our religions!

Making for the renewed case of now, whereby in full and proper use of our religious leaders in all cases, we can set the real parameters in which we are to operate without any fixation to the considerations of partition anywhere in the world? Our religious leaders!

By tone for me to have continually used that expression has not been a measure to set the balance of our belief in faith and our faith in politics to the wrong. All I do is set the standard that we are individual, but held in class to be of this single specie of finches! We are all 'Out of Africa' only and originally? We are all to be set and only die by the wings of 'Natural Body Death'. And in respect of all of our religions we are all to conduct the chore of life and living, 'Past Gods Words', which in context has only ever meant of our and all conduct in relevance to self only, self only! 'Past Gods Words' is for the faithful to be faithful in our own conduct, our personal conduct. In using Gods Words 'Past Gods Words' will show the extent of meaning for religion being purpose, and not the shadow of politics! Being of purpose will bring the reality of what 'Natural Body Death' is, simply being the test of time? Allowing in show of what was intended in any case without the need to hasten the death of others except if the poisoned mind cannot rationalise on our own and on behalf of all. From which our one active prospect in 'One for the Whole' is to be our mirror of reflection and double reflection whereby we might set the result as our target?

Before our religious leaders are called, in due deference to us, we are held to choice in who they are from our own past, which does not have to change only be seen with clear eyes? Without compromise I have laced many of these pages to the conduct of some imams of the religion of Islam for close attention. Because in exact parallel to the time when I first put pen to paper their role by some, had been conducive to challenge the authority of the written word from within, to be in the negative of what 'Past Gods Words' means by any name or by any reason or by any religion?

Like for like, I have not made the case for how our priesthood has similarly failed in respect of how many by open consent had empowered the aspect of religious tolerance, because it allows apathy reign? By extension, attitude can endorse any aspect of politics even the irreligious forms of communism linked to the standard of how we have proceeded by our half measured reasoning allowing us all comply by result of our steps taken syndrome!

My claim in balance to or for our religious leaders is not to what they have done up to now, but to what they must do from now. My claim in balance not to our religious leaders if some are imams, is not to what they have misread and by so doing have been able and encourage the matter of murder by suggestion? But for them in we to take the one step in always looking to what we saw as self purpose, before we were blessed with Gods words and look to our beginning 'Past Gods Words' when we were at Shanidar!?

Without connection, without test or query to be put upon Islam, my voice of opinion expressed about some imams is in query by my own standards demanded of any religion. But supposed to be different in those that are monotheistic, who by setting standards have done so by reflection of self and or by what had been shown by the reflection of the reflection. Which from now in cure, having always viewed old images, having been set from old prospects, we now have better insights through the prospects of our Fifth Dimension?

## CHAPTER 167

In the last case of time that we need review, imams and what some have done in the cause of all, has to be seen by how the split between their view, was made in this last modern era? What was to occur without notice, happened in the most part because of our political judgements at the time unseen, but to be given account that in the face of no other opinions or options the need for decision making howsoever was paramount because of our recent past? Quite simply we needed to make a world standard as quick as possible so that we would no longer be in spiral to world war in our next generation, like when had just happened in cause of the first and second world wars?

Using our standard of operating half measured our case of concern had to be directed by the one move that was truly international, our political standing! Even if only reflected, because of type we did not really notice that the past world wars had to be political, because the main antagonists always were of the same religion? From that core and time to be set from the actual end of World War Two, by citing the opinion or in reconnecting to the opinion of religious tolerance. What was done was we closed the ideals of purpose religiously to use the aspect of intention by political codes instead.

Knowing that the state of the world needed closer accommodation on all levels, but more so by our national political demands if indicated? To forestall any concern of our religious opinions if expressed, we opted for the standard of religious tolerance, which allowed that we all had a voice, but politics did not have to listen. Unconnected in real terms, because no body had listened at the start of the two previous world wars, nobody was expected to listen to religious opinion now, because in that present cold war situation, one side in operation had no moral religious connections unless to accept the principle of religious tolerance?

By never being exposed to the situation we found ourselves in or that we really did not understand, our conduct since wars end, had always been of political ideals. Considered right, but often wrong when worked half measured in the first case of introduction? Although I had mentioned communism several times, from where I had stated that many of our supposed moral judgements like the suggestion of human rights and civil liberties worked to be politically correct. Fall short because they were designed only to compromise the barren effect of how communism had actually worked. Therefore by steps taken, if communism was lacking or bad, then democracy and religious tolerance must be good, which of course is only an opinion but done to include communism into modernity!

Unfortunately from working the viewpoint that had been set by wholly political perspectives, our standards of humanity being civilised, were exposed to be of less value than what we had imagined. When considered how the world operated through the early days of our bloc culture set up to the one standard of political opposition, beyond notice, yet through settlement by our steps taken syndrome on that political edge. By being able and ignore the true depth and value of what our religions were supposed to be for, because of our religious standard of religious tolerance? When aspects of communistic support to some Islamic countries of the Middle East began to wane, what voice had Islam in self competition to the broad side of westernised democracy!

What voice and from what heading could Islam answer to the aspect of religious tolerance, when she in time had invented the phrase by religious forbearance in any case? Allowing other religions their case providing that case was of self, and not to override the teachings of Mohammed? Although similar, the picture in reflection is different because with forbearance we can still question the measure and style of other religions if only against our own, but to others also? Unless the misunderstanding hand of political judgement by winning a bout against communism in the Middle East, set that small victory to be the standard for the future thinking with regard to our political override of democracy and its style of political management to be the first winners?

Although the picture will always remain hazy, what was candidly missed, but only with political judgement, unlike the sad problems of partition in the sub continent of India, in the Middle East, enough was different to draw different values even unbalanced? Not least from how the western political standard generated by Europe and America were openly seen to support the religious objectives of the westernised country of Israel. Being no comment to the feelings and emotions of ordinary individual Jews who just wanted there own nationhood. But in allowing the die to be broken to what was supposed to be religious tolerance, how was the reflection to induce reaction on the reflection of the reflection be seen or accepted and by who else, associated or not!

Unfortunately with such an open gap to what was supposed to be our made standard of human rights managed to be politically correct, with the outcry of all religions contained to the corner by religiously tolerant understanding. But only from the political perspective, because that is how things worked for a time, given the stature of always being the intention and therefore to have been ordained by old and ancient political habits, but not counted across the board. Which by itself shows in example the deceit in value of our steps taken syndrome, which without voice or opinion has guided us to make decisions we do not understand, but have to support because we had held opposing economic outlooks spread by some of our political acceptances?

Making the one case that if politics is right or a style of politics is right, then all else is wrong. Making the repeat case that if a style of politics is right then whatever is endorsed by that style to have been set, is also right? Which although unseen is a fact of operation to most modern standards carried only by tunnel vision in accepting self acclimation over the broad band of the intended care for humanity. Being accepted as standard until our light at the end of the tunnel is hastened by the shine from our Fifth Dimension prospects? Not that any force is to be used, but the standard of experiment in setting all the criteria is to be known and recognised by intention. So that when the machines of experiment are turned on, all of our true expectation will be met, allowing that in case we are no longer to be wary of the fourth dimension, because we have surpassed its old claims in us being here and now!?

Because the Fifth Dimension has nothing more to say than what should already be, leastways by the intention of these past sixty plus years if claimed. In already knowing how we must operate by approach, shown by the gilded success of one branch of politics. When the pitcher is full, how might we look to extend our progress if not to look back in part only to check how it was filled and with what!

Although I had stated that most standards under the wing of our United Nations had been set from our European/American perspectives promoting capitalism and democracy. Allowing that by reflective means we in those forms of use, had set to curtail communism in part by addressing to the fact that communism must also heed to religious tolerance? Without prior insight we buried the good intentions of most of our religions at outset by the standard of religious tolerance. Except where and when we allowed the means of religious sufferance raise its own head by its own command, but shaded by the conflict of our bloc culture first hand, or in early times of our bloc culture standoff situation?

Without reference when we allowed the state of Israel raise her own flag, amongst other considerations this episode was international, complying with the better new order that we supposed. Unfortunately without measure, the case of her flag along with those on the sub continent of India was supposedly political, but in fact the three separate flags raised, in two to be of Islam and one Hindu on the sub continent, were decidedly from religious motivation. But our standards of definition were locked to the thinking that we were first, second or third world peoples by every division, by every division? With the double misfortune that we all played the one role of being who we thought we are, determined by the cutting edge of political decree issued through the United Nations in the colours of Europeanism and Americanism. Which at the time and we must remember this, that was our first best chance of maintaining progress we needed for the ultimate change for self recognition. Missed by Charles Darwin but seen in notice by the Fifth Dimension!

No claim is ever to be made against Darwin nor do I make a claim on behalf of the Fifth Dimension, save its outfall is to be guarded for all future times to the matter of being able and think now as we might have sixty to one hundred thousand years ago? Had it been available some short sixty plus years ago, what voice to be heard 'Past Gods Words', when at instigation, the countries of Israel, West Pakistan and East Pakistan were called to be of religious necessity but by political determinations. Making for the true standard of nationalism but measured in those cases religiously to different measurements?

By more than coincidence my choice to highlight the political aspect of the countries just mentioned is set to what standard of our religious determinations? By the same token but not directly connected, when the mist clears, we have two of our three main monotheistic religions in view. Judaism and Islam in the whole period of time since political determination in the name of nationality have stood face to face in the mirror of concern with both factions raising the right hand and seeing only the left move in response. Unfortunately in the case of action and reaction, we have overlooked the involvement of our third monotheistic religion of Christianity, held in parallel through the factor of religious tolerance. But linked to the reaction of Judaism first, because both are of a first world standing politically and therefore miss-associated morally!?

Conversely in the sad case of Indian partition, we all suffered loss in the face of what we planned humanity to be? From the partition of Palestine at about the same time, in open display our world reactions to what happened were somewhat different if generated from political considerations before religious ones? Which in plan was supposed to be acceptable by assumption, because the standards had been set to accommodate an emerging political situation before any religious divide was taken into account on its own behalf?

What we did not notice or failed to accept was how the scale of balance had been set religiously to accommodate our assumed political objectives! What was ignored to our total cost was allowing the fate of a majority of Palestinians of area count there cost of not being supported on separate account politically, as the new Israelis had been! What was to further count to our cost was even though being undirected we allowed the assumption by classification that the Palestinians were indeed second world economically. Because there support was only delivered from the ancient and misread rights to the proposition to defend, protect or promote their own religious standards first, not politics!?

In as far as rights were allowed to any side, the picture was to be decided in the view that might was right, carried better by the Israelis under massive support from a western political incline, but to be denied? If it was to be the case that the

Israeli temperament had indeed changed from its original standard, being the same to the people of area in the Middle East. Then through centuries of intermixing throughout the rest of Europe by forming separate nationally inclined Jewish groups generally following the one theme of Judaism pre Christ. If in root and by association in any case, the small Jewish communities managed to remain to there own religious determination. Then by the actual processes of interaction, that situation of small communities in other countries was to lead to the fixation to include the one standard of political determination whatever that meant, but from within any other country of Europe and carried by expansion to the new world of the Americas. Then returned unnoticed, covered by the accolade that all religions now were to be seen under the one standing of tolerance. Even if some religions to be fully monotheistic also, were not directly mentioned by case, but were included in the same manner, it was assumed that all religious leaders automatically accepted the fixation to religious tolerance!?

In point of fact, as hinted at, up to the exact time that the new partition of Palestine occurred, Jew and Arab or Jew and Muslim played havoc to each other in the preparation of counted independence. Both in preparing there own grounds had lost sight to many of the similar features of actual requirement, so much so that when the act was made, in almost the same fashion than occurred religiously in the sub continent of India. Here in the Middle East the offset was wholly political and had to be, because in a sense the mover in task was through the theory of the United Nations? Our body of law to set the basics of all new standards in supposition, which had been created on the two fronts of active political reasoning and the dormant focus of our religious determinations?

Instead to the fact of reality in Palestine, by effect the Jews wishing to be Israelis were better organised to there own intention than the style of opposition offered against from the Arabs of Palestine proper, or the Islamic countries in close quarter? Not least that the Israeli organisation ran too close to the order of settlement, supplemented by our western European and American cultural standing. Which I had only referred too to make the point, the said drive was simply from attitude delivered by the circumstances of climate, geography, diet and other things. Be it from religious inclination fermented by national desire, yet attuned to the one image of Christianity only, that assisted us all to be more different than we should have been to one example over others?

Even though in partial comparison at about the time when our second dimensional religion was undergoing the trauma of reformation in Europe , while at the same time dealing with the trauma of how to behave with an honest heart while heavy fisted in the new world. Having found a style of so imagined religious practices including human sacrifice or the cases of murder of self by self in the name of sacrifice or offerings to the image of the God through king and Sun and the oracle of the sky at night?

Not relying on the distant past which I seriously advocate, but had always stated that we can review it at any level without the need of change, is done only that we might be able and evaluate our own standards of now and the reasons of. In taking judgement of what was met by the expansion of Christianity to and through the new world, in having obtained the attitude that the religious standards there were wanting at the time. Our overall attitude unseen but through the medium of circumstances carried by the additional centuries of time along the avenues of change to have occurred on the new continent, unknown then and now were to affect us all later?

Allowing for the full mix of all the circumstances involved in the four hundred years from the time of el conquistador in the new world to what emerged to be my own case of Americanism. That is the state of condition to cause our present standing in the Middle East at about now? Whereby the standards, were to have been primarily set from the now western political perspective? Including with the conditions to have evolved in the belief that although the second monotheistic, the held religion of Christianity, was indeed also to be first world politically? Any so legislation if carried by the international factotum of the United Nations, normally were to express the view of the beholder!

Which might explain some of our indifference by some of us when India was partitioned to one standard, yet mandated Palestine was partitioned to another at about the same time? Viewed under the core of those different circumstances to have accumulated over the past four hundred years when we bludgeoned what was accredited to be a religion in the new world. Even that we should never wish to have learnt of it by understanding, instead of offering change by example unless because of reformation, we had not been able and see the one standard of Christianity? Except in using most of that time to note the difference for the acclaimed ideals of politics and religion, which were separate, yet we allowed one to become western and democratic, with the other being western and monotheistic. Without clear definition of who to include unless all religions also to the standing of being united through the new realisation of religious tolerance were to behave so?

Although a defined perspective clouded by outcome, what we had created at wars end from nineteen forty five was to the new beginning of hope, but built on the straw foundation of added progress only by outcome? Making the case of futurism politically, by not realising our future had always been planned and known from our religious perspectives only? Without malice of forethought by those circumstances we had all allowed ferment with little attention or consideration. Without any resounding echo of notice to what any standards were to be, by the silence in one hand of our religious leaders, the green light of go was not objected to? Making the case if through the growing influence of our fourth dimension in the body of science, was to step in to replace the vacuum that silence had encouraged, what answer could have been given from a dead tongue?

Even if we could check our own role at the time of involvement, when less than one hundred years before, Charles Darwin had published his mighty work creating in us self doubt religiously more than scientifically or politically? Allowing by Christian standards that from the religious perspective because his expose by suggestion was first to bear on European outlooks? With our perception toned by those outlooks having been formed from religious thinking whereby God in actual fact had set all standards in the first place? Through our interpretation of normal conduct, that Darwin inadvertently introduced by making the case that our beginning was other than by spiritual appointment! We all ran to our own corners of acceptance and or query, making our own case, culminating whereby we had to agree to differ in so far as our religions and politics were formed? Especially when part of his suggestion by revelation showed or did suggest that a condition of our existence applied to natural selection, being an easy pill to swallow if meant to connect with European religions and political thinking of the time, of that time ago?

By not suggesting anything Darwin wrote had a serious bearing on our own interpretation, is all that some of us require so that we might agree with what was really meant by him? Not least that the perceived missing link between mankind and evolution was determined to be natural selection! Applying in full measure to us of Europe first hand even extended to the reality of European Americanism and or American Europeanism? Who from that position in first notice, had arrived at station without Gods hand, only our own? Therefore although God in the name of religion was real and vital, in order that we could show our acceptance to the standards that had been historically set by God and his agents any time from about four thousand years ago and earlier? Even if to include what involvement the Egyptians had to the spirit of God pre/post monotheistic?

Remembering that Darwin did not have the insights we now have, helps us recognise his fine contribution in the name of science, he gave us all a view beyond the scope of pure religious mysteries so far taken? My caution in relation to what he exposed is that it was only a leaf in the tree of life shown in magnitude of how I have used his studies to make no other point than his view was only a part of the whole! But it is us who have argued his case under the false premise, that his work somehow exposed in falsehood the efforts of religion or the efforts of the effects of religion? Which were taken up mostly in Europe where we began to act and react making the core of his work to be something it really was not?

For me to have linked in time by cross reference, that Marks and Engels also worked in Europe from about the same time that monumental changes in the consideration of change were due? Is done to set the picture of how and why our main religious attitudes there to be Christian, began to lose its own lustre in modern terms by not realising that there was such a thing as actual progress?

Even that I had previously given credit to the open fact that it had been through our religions that real progress was fermented, will always stand coupled with the style of the thinking by many individuals within those separate religions? By fact that I had given status to the ideas of some peoples being laid back or progressive, holds in character to who we are being the one specie at all times! Without blame, for me to have used Darwin or Marks or Engels in the role of conspirator to what might be a standard in basics of our present condition today. Has been done to be an example of how we might use our skill in reading what the reflection intends before the image is presented?

Although Darwin never really used any example of religion for or against any theory his work might have produced, his single direction was to the matter of time only, which coincidently hit a cord with the fast developing community of self by first concern to social and or scientific matters? By show to expose that the process of evolution in any case was a protracted event, endorsed what our own creative image of evolution was? What the theory actually did, particularly in Europe was place religion secondary to the matter of progress. In view of how other matters were offered to be decided by the long arm of politics. From where the likes of Marks and Engels were offering the exclusion of religion by part from suggestion they were to be completely ignored by reflective reaction; better be a force of influence even if theoretically dismissed, than to be no force at all? Therefore religion in its newly addressed form easily fell to the suggestion of religious tolerance, just in case, because by doing so, we could pretend to have a voice!

All the matters above were in continual ebb and flow over the first one hundred years after Darwin published? What was to turn a head, all of our heads in the most recent of times, perhaps from the mid nineteen sixties just gone, more than from any other time? Came from a new dawning, particularly from a religion which had been around at the time and subsequent times since Darwin published his acclaimed work without consideration to anything other than European thinking, was first raised again for new reasoning by the religion of Islam!?

## CHAPTER 168

I had cited one main difference between the core of our monotheistic religions extending to cover all religions of consequence Eastern or Western. In time of presentation from my own designated first religion of Judaism. One of our early rules in religious law delivered by the hand of Moses if directed by the voice of God, along with other leads was earnestly received! Thou shall not kill in line of the commandments delivered, is all that I would seek relevance from on or through my own personal experiences of that law by result, more than effect? Because from the safe conduct of working from thought in the abstract, I submit that the reflection can better be interpreted than the subject?

Likewise in looking at all the proposals from the new form of Judaism in what was to become Christianity, in using all steps to have been taken in reflection to what had been delivered by the prophet Moses. Jesus as the incarnate carried the same agreement, not least because the tenet in covenant had been from the voice of his, our fathers, as might have been thought? Although our developed and progressive second dimension religion becoming Roman Catholic, which by slight of hand was almost inevitable to confirm the lead that politics was to gain by the reconnection of King in the form of Emperor holding first status over county and religions? Even if religion, like in the case of Catholicism, was given first orders of conduct through a divine lead expressed by an appointed priesthood, all decisions from now and forever were to be cast in lead from the government of politics!

Without recollection, by the time in relation to when Mohammed was born to me in a supposed less stable world than we had always lived in, even if undirected by any known circumstances, human life was cheap. So much so, that almost the normal for general interactions even religiously, the matter of life and death of self by self was barley a considered view depending on any particular political standing at the time! From which if and how this unseen acceptance was translated to be part of the thinking of Mohammed if ever? Bearing in mind the supposition of our general interaction at the time, even that the hidden ideas of a new to be formed religion were to be formulated? Certain features, but only now seen with hindsight, have created for us some serious problems for this particular time of now in this last modern era, but again never to be imagined from then ago?

My personal pointer to the open fact of the similarity between Judaism and our new religion of Islam stands by consideration to the area where they were first introduced and of the people therein, such as you and I representatively. Although we are in the continuing state of change ex to the normal/general standard of evolution, much of our attitudes are generally localised to area further joined by the influence of geography, climate, diet and other reasons. My connection that in the Middle East, most peoples were of a laid back character stands by reflection? Not least that in general terms we had all moved at the same pace with minor ups and downs since the time of our first probable religious global interaction by course in the land between two rivers, but always to progress?

Knowing the full facts of what our emotions are in terms of how we might react to each other at any time, my proposition that in due time we came to be laid back by style more so in the Middle East stands in open view unseen, but from religious influences? However volatile our actual conduct might appear to be in terms of how we do react to certain matters only, counts to the situation and driving circumstances. From where I can make the statement that indeed we are laid back by character in the Middle East more than in other places is because of the study we have done in the name of religious excellence? From that balance and by connection that in part is the reason of how we were to approach the matter of politics particularly in the Middle East!

By all processes of time it was there in first order that we had come to our theory of monotheism, allowing that there was a direct connection between us and the image of God ever more!? Although past the full wisdom of understanding, by making that first real connection of there being one God only, gave us the mightiest sense of power to be able and truly understand, to be able and understand everything? Because in forming the association of mankind and God was the unseen key to understand all of our mysteries! That they were naturally in the knowing of the beholder in our own study to be the creator of all and all around us, how bad was that to know we would know all by Gods will!?

By classification of how I might now see events of some four to five thousand years ago particularly in the Middle East. Has allowed me make the full observation that part of our character to have developed from where we might be considered to be laid back comes directly from our discovery of monotheism as a religious concept?

In parallel with many other societies throughout the world, but to be taken or considered by the directions I have laid, from the times of the Pharaohs in one of the oldest lands to be continually inhabited as a separate nation anywhere in the world? Our conduct by the abstract in understanding of our very existence, formulated the most amazing blue prints creating standards in asking the same questions we all continually do, by first putting set solutions to be the answers in any case!?

Even if what was to begin in ancient Egypt geographically of the Middle East, even if to consider the importance of what occurred at Shanidar in present day northern Iraq also geographically of the Middle East. No connection whatsoever could have been made between the events of some sixty to one hundred thousand years and or the events to have occurred in Egypt, probably less than five thousand years ago, but not later than about four thousand years ago marking the time of true monotheism and in knowing, to be still or first laid back!

By claim, I make no standard to the meaning of politics while on this theme, save in a sense it was also part of the key to understand from not knowing the answers either? By any stage of examination community has always been the first signs we had found to indicate that we have a modicum of civilisation, as our first call to be civilised? But only marked against the conduct of each other of what we now know to be the same specie. Even that I have put on record to charge us all in not being civilised, is made to be the best case for the need of general improvement. Not least from the best efforts of direction to be generated in these recent sixty plus years of this last modern era, but on all accounts to bring in the right attitude and standing of the relationship between our religions and politics at the same time, which is a new principle from now!?

Community begs organisation which drives for success. Although much in example can be gained from deep studies of what occurred in the Middle East at outset and by time from about ten thousand years ago. From the birth of modernity we have also carried the scars of ignorance. Not least from then how we were first able to recognise our differences before our naturally progressive selected likeness?

At all times since the point of reference standing to our conduct in Mesopotamia, we have progressed in competition on the same basis as other creatures had, but in our cases with unwarranted ferocity of each other. Our cave paintings and our rock carving or drawing carry this message from long before we interacted in the land between two rivers. Not least if in note that although we carried weapons for the hunt, they could be turned to any other use if seen to be required. Even to leave those times in the mystery of intrigue as we might, what is also seen from the cause of such events as cave painting is that in order to have done such work, in cases under extreme circumstances. Organisation to lead as civil order was required on high levels, far in excess of the open need to hunt in an organised manner to survive. By the example of some of our cave painting, by the volume of effort required to do them, so stands a well organised society on most levels. But if as I had suggested, that some of the work done, depicting very elaborate scenes, might not they indeed have been done to adorn tombs for the representative sanctity of the dead, even if never to be proved, does not compromise how we should think about our burial at Shanidar?

From being organised enough to contribute the pure form of art to our communities if carried to the standing of civil order in any case. From where and when can we see our first open display of what religion is was or might be, in order to explain how even if one of our number was to die by whatever means, how were we to account for our own memories of them? Like from which we could still see them in minds eye, hear them to the same and be aware of their conduct while they are not? By putting any number of ideas of how some past events might have occurred, forces the outcome of result in needing to understand how our religions formed?

In knowing that they did, fortunately is all we need be aware of, because how ever many years of our existence, we are the one specie above all else to have come to the sense of purpose? Reaching our pivotal time of these past sixty plus years of explaining! Because we are all here now and are still prepared to break with all the other conditions of evolution by calling to the understanding of creation at first through our religions? Which more than any other cause in the name of humanity have progressed the full ideals of what purpose really is? By our joint explanation on scientific terms, which of the same flavour have only reacted in the name of confusion, putting all cases to the result of what is revealed in the reflection? Not least from how we had set cause by looking to the mirror and in seeing the false image we have reacted to that, and not the proposition?

By time, to have reached the end of World War Two, has only clouded the issue of the importance of our religions, when one of our first acts of reconciliation was by connection to the avenue of proposed religious tolerance past and present! From any open plan to the observer and from any direction, what was done was from unseen motives but by inclination? What was done was from a defeated Christian perspective which was to allow the very roots of discord to the theory of religious tolerance run amok, resulting in the establishment to the theory of religious tolerance! With all other expressions opened to be determined by political judgements, which has been our full situation since nineteen forty five from where in attempting to stabilise politics expressed through nationality, we have become politically unfit to make further religious judgements!?

From any image of what are genuine religious judgements, in virtually all cases we have only been able and react from the image of reflection. Even if we have seen the original difference of one idea being placed and another in reflection, our pickup or means of connection to that image and reflection has returned us to what we know or think we know, which is no bad thing, save the cost of our reactions? Unfortunately in open view, but without the sophistication of direct intention, when we inclined to the full principles of religious tolerance it was in all cases to sanction the criticism of the extreme owed to some of our political systems.

What was gained to be wrong before we were right, was how by placing our religions, we also had reduced the tempo of our religious leaders, which always has been a regressive step counted most now. Not least when from this call I would demand of them in us, that henceforth there voice is to be ours and not the reverse, but counting that any command from, is only issued through our one active prospect in 'One for the Whole'. To always remain our gauge in this case to counter some of our religious excesses while expecting our religions take the lead to the aim of purpose before commodity!

By any clause what we have to bear in mind with affection towards the ideals of many of those mortal men who would be and are in class to be our religious leaders. By the name of priest or pastor or imam or guru, their task without question is in our service, but not in our charge? They cannot act or suggest that we act on there behalf to for or through the meaning of our religion to us, we do have our own free will appointed or not!? That is why we all must ponder to the result of how the standing of standards of our third and my third dimension monotheistic religion, is ready to accommodate our desires in meeting the three taken and one active prospects of the Fifth Dimension!

Without comparison, but in use of how events could have been at discovery of the new world post Columbus. By bringing only our religions to protect our exploration there, in finding human sacrifice practiced liberally as a ready means of communication from the necessity to address the meaning of life from cause to service. In our unseen effort to accommodate our own greed seen or not, because we had not had the scope then, also allowing that the act of human sacrifice to any cause should have been repugnant. How we did destroy the practice for any reason, no case then could have been made to offer solace or understanding that our main principles of conduct were not from the fact that we had superior weapons or ideals. But that we could work from the principles of ideals added to the sanctity of our held beliefs? Not least to a prospect that set the principle in what should be Gods, any Gods view, in the acceptance to the living belief in the standard of 'Natural Body Death'. Joined to be discussed through another leading prospect we did not have then, our one active prospect in 'One for the Whole'. 'One for the Whole', now our one active prospect whereby if a body, even a priest was to discuss the principle and or need for human sacrifice. Then they also were to partake in the act in the same manner as all to be selected for human sacrifice, as one for the whole!?

## CHAPTER 169

Without any statement of choice, for me to have made the separate note of difference between our first and second monotheistic religions, with there established priesthoods against Islam who in theory does not have the same working. What I have called in all three cases is their similarity and nothing less, by direct challenge to the source of what imams are and of there role. Has been taken on by myself for the one aim only to empower all religions to be of the first order over politics and or any of the political systems we have used for our ultimate guidance?

With politics from the roots of evolution and religion from the roots of creation the two are set not always to agree by our individual interpretation. Therefore because of how we have allowed ourselves fall into the unset trap of normal conduct by civil command or in being laid back or of a progressive incline. We have in point of fact compromised both issues in terms of how one system or style is to ultimately react to the other. Not least how one system has taken command of the other by the complete acknowledgement to the standard of      religious tolerance? Translating that evolution even unconnected by real understanding, head the result of creation by all modern standards, to mean our politics rule in such a way to deny any of the intention of religious purpose! Because we have allowed the standard of now by materialism, compromise the concept of hope through those ideals of purpose?

In being able and recognise what is a direct block to the principle of progression by the terms of progress which are yet to be centred upon. Finally we have to look in fair criticism of how we are wrong in the why from what we already do, which by term is what has been done throughout and through this our work. Even that we are involved in picking over the past to examine where and when we went wrong bears no evidence only history? Although the best means of progress are to be of a more subtle approach than by living from the effects of our histories most realised in this last modern era. By focus to what might be the main or most important matters of concern, we have less to do than imagined, especially in use of our one active prospect in 'One for the Whole'?

Having once again separated for notice, a principle difference between on the one hand Judaism and Christianity having defined priesthoods, to on the other hand Islam as not. My personal focus to what some, only some imams might suggest of conduct by some of the faithful, especially our young, requires urgent review on its own account even if my own judgements are wanting or misguided or simply wrong?

I have made the obvious case that both of our religions of Judaism and Christianity have continually compromised Gods covenant of service and reward by how they have looked at or to the ideals of the commandment not to kill of self by self! Which is never to be ignored, for in short shrift our religious leaders in us, will soon be reviewing how we actually act and or react to the means and matter when to kill of self by self, when we do kill of self by self, our reasons have to be accounted for past politics, even past religion!?

In separation, but reviewed by the same standards that we are all entitled to express under any religion, under any religion is the aspect of our understanding to events by our personal reasoning through our own free will!? Meaning in class that although we belong in choice of whatever form of political regimes and more importantly to whatever religious system, Eastern or Western. Our predominant choice has always been that of our free will? Our particular good fortune in all cases of life now, is that we, we in every individual alive today, have the choice to leave the outcome of evolution and enter the open field of creation! In the preparation of how we should ready ourselves to the standing of how God might want any one of the millions of evolved specie just like us. To show an inkling of how we understand purpose translated above the means of scientific experiment, but to account in the reality of who we are, enough to draw Gods ear to make that link of creation!

Not at this time to God, because just like in the annuls of creation spread to be over all creatures, their time has remained to the same drive of instinct in construction to apply to the one aspect of procreation only, so it must be Gods will if that is the case? So it must be Gods will in terms of the real time allocated by Gods own exclusive standing with the ability to wait and see if the purpose of procreation only has served what might be Gods own other purpose? Whereby any creature of any class shows in any aspect from any time span to be a two hundred and fifty million year old reptile by continuous line or from three and a half million years in walking on two feet!? That then in reality is purpose from creation, by no time mark, save in being allowed and voice opinion above the grunt or growl made any time during the past eight hundred million years long before Eden!

Without any fixture of type, even that we might curse or praise the style of our reactions to what Darwin revealed, it was only we of this single specie who were to query what was said, meant, revealed, implied or discussed. Although we in our own way were not ready for what the implication of what was said by revelation or how we reacted by camp to be evolutionists or creationists, our credentials then ago were not full enough in general understanding in any case? Once again allowing us without notice that any reactions by any group were to be done half measured and inevitably brought us to the parallel of operating by our steps taken syndrome. Because we did not have the means to improve or endorse our own standing at the time or from that time since, until perhaps we begin to see the value of the prospects from the Fifth Dimension!?

From our final fixation some four thousand years ago to become monotheistic, our rule of thumb had been to the thought that God had come to us, not the true fact that before all other specie, we had come to God! By crossover of definition through looking into the mirror of discovery our image in return was of the same type in all reflections even the reflection of the reflection. We saw what was acceptable even though reality was in front of us, our euphoria won the moment more intense than the feelings of eureka could deliver!

In finding that it was we who were indeed the chosen above all else was no error of thought, only a mistake of time. If the case is so, then our first objective is made not through Moses who was involved or not, but in us who were to feel involved. Because when first heard or heard again from Moses and the section of our particular tribe, with having to manage the complexity of our previous God beliefs? Those of us who were already locked into the management of our religious concerns towards our many graven images, were to successfully emerge as our dedicated priests. Culminating from when until now to be locked to the first and subsequent times through changes if in the full management of our religion by becoming our priesthood!

Even new Judaism if by review was counted the same by also having a priesthood structure. There case as ours differed in that the core of the religion of Christ had sprung in continuance of the words of Moses and others. Not least through the commandments from God, one being not to kill of self by self. Any decisions to where or when we might now kill were to be of religious selection throughout both of our first and second dimensional monotheistic religions. Leaving in test now, a review of how our religious leaders might reflect upon us for future conduct, so as we might show our open willingness to set the standard of how God in return shows an open interest in this minor but separate specie to all others. Who alone without individual merit has been able and call to voice the matter of how we fit to Gods own time plan of holding to evolution, yet also ready to expose the matter of creation if or when found. But not yet shown!?

Whatever further challenges we might have politically, which we require to keep in check. Our first intentions will have to be to concentrate our religions beyond the scope of accepted or standardised religious tolerance! Our religious leaders henceforth are our arm of deliverance to the important matter of our future intention by whatever means, even not from our own choosing? Which brings to the mirror again for further and further review in what we are to see by reflection from the last monotheistic religion of Islam and or other religions in between if connected?

Islam without a defined leader or religious leader structure, is put to task by my self because of that missing feature, not for what has not been done, but because of its self it has compromised our own base by the lack of two realisations. Mohammed in name to be linked with Jesus and Moses, all to have been prophets in the eyes if Islam. Put to us all the written text so that we might gain access to finality in a more direct way than was previously left by other similar religions. But I add for our caution only, all of his works although intended by structure to be used for all time. In order to reach our required thinking the works were set directly within cover to how we then thought and were required to act to reach our goals even if directed? Marked by the standard that they were also for the future, but in not being able and determine how our future attitudes might change if driven by the growing influence of progression, even guided by the civil command of politics. We had been left to our own devices to set what should be reaction to what the new reflective images from the mirrors of now had returned?

If there was a general failing then it was we, either as a collective body or individuals because of how we heard the word or saw the text and used them by our own interpretation, which has always been our will? One of the most prominent instances I offer is through what I myself have seen, what had been left for all, but by my own circumstances. Which had been from how we have come to interact these past sixty plus years, a time without ever having been imagined or thought of, to how we might turn out, ex to religious fulfilment from the time of Moses, Jesus or Mohammed? Save by text in what our religious leaders had said was said or how the read text that was said was to be read and understood by our interpretations from then until now?

These past sixty plus years of this last modern era by concern, represent what would be the average time we each might expect to live, marked by our own lifecycles. What has become different now from any other time of our histories, particularly in relation to our political standards? Has been set from the uneven balance between how our religions have not been able and operate change, because of there structure, being no mean feature, because of what they intend? But in parallel our politics being able and change generationally if required, sometimes mixed to accommodate some scientific revelations if by the fourth dimension, seen to expose our vaulted mysteries, but in time not, by religious codes!

What really had been revealed on our behalf was to the fact that we were capable of change by all of its virtues, but again as with politics, we have not learnt to manage some changes unless categorised in like form to religious tolerance? Our free political acceptances in tentacle form by division to be monetarism, commercialism, market forces have now drawn to exhaustion the principle ideals of government for the people by the people and or by the people for the people or from either port? Which had invisibly formed a structure more intense than ever before from the end of World War Two, when we tried to apply the concept or morality in book form or what politics should really be by this new opportunity!? Because in class we saw to the effect that religions were in theory to or of or for the future, whereas politics was supposedly of the now, but planed into the future supposedly!

Making the blind mistake that in qualifying both concepts, our selection by the political field of now laid claim that all matters were primary of political concern first. With our religions being locked to the one standard of being tolerant of each other, but by that clause to be also tolerant of political decisions delivered!

All of which might have worked one way or the other but for how our political reasoning began to lose reasoning on the main front of religious interaction. Not least at the time by the example of how our religious division on the sub continent of India was handled or controlled politically. Allowing the core of our religious divide be only set in reflection and in case to be less than reasonable, while our political aspirations set there own camps? Of which we had no way to understand how future or further meeting of minds might take place or occur from the same perspectives but to a slight difference of character! Not least when the naturally inclined temperament of both Hindu and Muslim was of a laid back style, but ultimately having to deal with world standards set politically, supposedly from a progressive attitude?

By the worst of all coincidences by the sad events to have occurred, even through what might have been severe provocation one to the other or reversed, by the ready human condition of highly developed emotions. It is more than possible that the same type of people we all are, 'Out of Africa', although following the separate standards owing to the Hindu religious rites and in the same way those of Islam. Without examination, both religions by structure create in all followers the open standing to be laid back religiously? Which is past the standard of written defence, because by religious terms and on there own behalf, both religions are as complete as we might ever need to follow our own desire!

If politics was to alter that supposition by my own classification, any change to our character however generated, has no critical bearing upon our personage? Even that in many cases, if we were to momentarily desert the true and proper intention of all religions by performing devious, mean, murderous or violently cruel acts of self upon self, the case is to be settled beyond the limits of cause and without mitigation? The cases are to be settled to our original civil law standards of control, locating those involved to the class of being criminals, laid back or progressive or not, Eastern or Western politically or religiously or not. But of criminal tendencies that are part of the effort by effect to deny the credit our own specie separation and flight from the containment of evolution had already achieved?

By normal or general cases of expansion to the supposed effect of progress generated by our modern political intentions, in every time band allowed to community over different ages, we have had to contend with measures of control at first delivered under civil contract? Although unseen from within those separate time bands then and since until our first opportunity at wars end in nineteen forty five. No realisation could ever have been presented whereby any person ever, even as my self on our behalf. Could lay the claim in reality that criminology is only a product of our civil means of control and or controlling and or!?

Meaning by extent, that criminals are only under classification by political terms, which in form is one accepted standard not yet realised, because of how we have designated all religions to belong by being religiously tolerant? Being a trait born from not being aware that the making of rules, laws or standards of conduct have several arteries from the one heart, which are never revealed until they become inflamed. Meaning in context that instead of being aware to civilisation and what we thought it might mean we have misjudged the issue, obviously not by intent! But once done in birth, the new specie of conduct and behaviour without the ability of self correction, heads to the philosophy of self betrayal by all standards, but religiously more so?

Without example, but by escape in using old statements in relation to how we behaved and interacted to each other when one group as we, met with another group as we, in the new world some four hundred years ago? Although politically motivated for other deeds, we were before our time to get the concept of religious tolerance wrong. Instead we used the consideration of being civilised to destroy the degrading, disgusting and criminal practice of religious human sacrifice. Done by terms of death of self upon self by people who displayed a far greater degree of social to be political organisation than ourselves, but never to be acknowledged? Unable to see how the reflection of the reflection created its own actions in reaction, all that occurred was held to the reasoning and thinking of the day, and of course greed. Specifically European greed!?

Specifically European greed, in accord by the posturing displayed to enhance our own forms of separation through our standing and emerging nationalism then. Europe by growing into more defined national groups in having and holding internal conflicts in any case. Became excited to be able and expand by the means of invasion to other lands because it was easier than continuing local conflict, plus with more to gain by the adventure. Although some of us travelled under our separate flags in the theory of holding a collective religious base. What was found in the new world, along with the benefit of wealth, was the terms of settlement by the people already there being better socially organised than we, but in the position to be exploited to our advantage, because of their religious terms?

Of the two groups of people there, naturally of the same direct specie as we, differed culturally on the one level only in the same fashion as many of us, through the aspect of our religious settlement? By similar reasoning, that condition contributed to their moral standing to be laid back by character, which in view, can only be gathered by hindsight and not from first contact? Which gives light to how politics began to claim the sceptre for further thinking of leading in the role of command to self nationally, but from the European point of view and to be extended if the opportunity arose?

Human sacrifice by religious decree, the Great Spirit in concept to be the overseer of all events and to the complete picture of purpose through the general picture of human death? No wonder we stood to be laid back because for good or bad both philosophies held in concept to the future, therefore the now could take care of itself?

From foreign intrusion little needed to be done against the spiritual aspect of belief in the Great Spirit and those who so believed in the concept. By slight of hand that image could be overlaid by the picture of Jesus to reverse the concept and force Christ's own principles, having done the rounds of interpretation? Even at the same time with the Catholic reformation being brought from Europe, Jesus was still at the core of all Christian beliefs and as so was to replace the image of the Great Spirit by being the Great Spirit also. But under the sufferance to those who originally thought so!

Impossible to praise, but without retrospective sanction, when human sacrifice was found on the scale it was practiced. Because it breeched the supposed level or standard of what was called civilisation in Europe at the time. It was dealt with hastily, brutally, uncaring but greedily. We did not need to look at the core philosophy practiced by these people of the isthmus of the Americas extending to the northern and southern continents proper. By there own fixation to acknowledge the sun by day and the sky by night in icon, set them to be idolaters. Allowing that we could treat them in the same way that we treated heretics at the time of reformation, also carried to the new world from Europe in like fashion of how we carried the aspect of European superiority!

By the very real coincidence that both new world cultures were laid back under the terms I have put Islam and many Eastern styled religions of the old world to be in. Allowed that each culture so described accommodated all the newness from the revelations of the sciences in any case, but at there own pace. Whereas if in Europe because of the squabbles between different nationals generally under the one religious core? Played push and pull me politically in a confined but vibrant area? In order for one group to take advantage over another first generated by who controlled trade or the management of some commodities in trade? An impetus was generated to hasten progress leastways politically, thereby dragging the culture of the one religion to better represent the success of national objectives to be over the aspect of that religion! Creating in open display that the winner in conflict of the future or now, was to be politics, because by its own tentacles, it was the medium of supply in demand to win the day of now, including the outcome of satisfying greed!

## CHAPTER 170

By connection but not really intended, the course the wind was to take us on, had a beginning about four hundred years ago and has not yet ended. From European migration to the Americas the outcome has been to set to fixation that the final concept if having been set only by myself for us to aspire to the American dream has been done representatively, but to the name of politics only?

Our case by fact is that world politics now is dominated by or from the roots of European expansionism to the new world, and in outcome by the success of that adventure, how ever understood? We have a world standard of politics, of pure but rusting politics, dominating all of our decisions, but to remember that sometimes in a normal family of two parents and two children of teen years there is rarely accord, allows us open our personal opinions? Which is no bad thing because even on that level is displayed our ability to differ sometimes stimulated by the fact that in all cases we have our free will appointed in all cases!

So much so for politics being able to please all of the people all of the time, nevertheless we do have it warts and all of which there are many? From where I have attempted to clarify the phenomenon that in order to make politics truly work, I have managed only to address the aspect of mammon at such a cost, to offer the query of whom it is supposed to serve? Its success can be balanced in terms of outcome or to the final level of its total achievement like when we buried the Pharaohs. Mighty kings that they were, linked to the heavens and by claim and definition, to be father of their people being us? But to return in cycle to their beginning, who we prepared from the time of mortal death to be incarcerated in there own tomb of mighty construction with the greatest collection of grave goods they accumulated in there own actual lifetime. To be reborn in the afterlife where presumably there goods were to be of the same value in use in the new rebirth!?

Knowing that our Pharaohs did live and exist in a time when there beliefs and ours for the same reason were honourable at the time? How politics of today might equate the same story when by its own wisdom it has delivered itself from the means of active religious involvement by classifying religion to be only tolerant of each others choices! When a mortal man of today, who by all of the best examples of endeavour, hard work, commitment and genuine seriousness has managed to accumulate a vast fortune of billions of money, how to sanctify that occurrence when only done in the name of politics? Being the same master of standards that allows our children beg, borrow or steal from rubbish tips or toil unremittingly in sweat shops, but are allowed the solace to be contented by the scope and range of our religious tolerance!

What I might only suggest by the above is not in the critic but the standard, to mean that although we have to live under political regimes, we have to! Because no matter how bad or severe our systems have turned out to be, each in turn set to improve the lot of the many against the few. Following old tribal habits which only worked when the effort and reward were of similar duration and spread across the whole group, even like when we were at Shanidar? Which in bringing the new unseen dimension to our consideration completed the full circle to the means of our break from evolution into creation, which yet we are to still unravel?

Creating for us the matter of such concerns that we were long to understand and even longer to put all learning on an even keel. So much so that in the interim over any periods of time since Shanidar and our cave painting days and the first conceived King God religion from ancient times. Leading to our first monotheistic religions some four thousand years ago, spreading to others as late as fourteen hundred years ago, when the image of religion was to become cemented into our standard. Whereby we were to take them forward in concept that religion or religious objectives were primary to our ends, being now laid down by the written word? Although cast to be the great leveller for all considerations, by laying down our methods of conduct, to behave to obtain religious fulfilment. Much was not seen that would allow our own predesigned character trait compromise our religious intentions?

From the time set and in the way Islam was first introduced through the intensity of purpose in need. By account, but not counted in religious terms, in order to promote the urgency of the situation, Mohammed was to lead us all to the last means in promoting the ideals of this new faith to be? Although taken to us in a forceful manner this was done not by the standard of creed, but to override all types of political and or civil means of control then administered? His placement was not to alter or change the structure of any particular style of government wherever it was imposed. But to bring our new religion to the fore by the open means of including everybody to the same standard religiously by the practice of our personal duty obligations, shown through the one unaltered creed!

So not to follow the error of our ways from other religious examples, his further intention in allowing that the leaders might lead of or to the civil management of their tribes, clans, groups or city states. His direction to us all in religious terms was that by being our duty to follow the now written word. There was no first need that we should be led by or even the need for a dedicated priesthood, when the text said all, but just in case, there was to be a line of contact through our and from our places of worship being our mosques? Doubled to be our places of learning for some of our young who might attend to be only directed in the habit of understanding by the station or our imams?

By feature, when our third monotheistic religion took root and expanded, all was done to induce a laid back manner even if unseen. Endorsing the exchange, that now, some thinking led to that fact we at last had arrived to the gates of purpose! Even that the same religion was operated under several different political systems, come what may? Not least because by how the best had been set from first entry, it by general definition had covered what might be the expectation of normal progress in life, always religiously, irrespective of how our civil orders of control worked.

From when the religion of Islam found the quality of its first adherents, by the very structured condition of the people, who like all of us had been determined by the circumstances of geography, climate, diet, and other reasons. In now having control of the situations of body and soul, having been allowed drop or not having to use the first violent means we could have used to defend, promote or protect our religion. All was in hand by us, even if in other places on this our planet there was serious types of conflict first judged to be politically driven. As there was to have been no political intent cast in how other political systems operated in war or for other reasons by us individually. Because our religious obligations were being met again individually as thought or done, with no need for further concern through an un-needed priesthood, all in the camp of Islam by standards was well? We in the name of all people were of that same condition when in picture our general feelings without open contradiction, was that the situation in being, was that the world was to take to the one final religion of Islam in the course of time?

Without any comparative religious contradiction some two hundred years ago when the political edge of Europe in the name of England and France had set there own political interests in the Middle East, Islam had no questions to answer? What occurred was part of the political switch then ago that was happening throughout the world. Even though both of those nations carried the banner of Christianity clouded by the view to accompany the matter of commercial invasion with missionaries of the Christian faith to do Gods work also? Being no bad thing if spreading the intentions of politics along with the religious incentive of genuinely helping people, this time a little more subtle than had occurred some two hundred years before in the Americas again by European influences and from there own particular standpoint?

What was never to be realised by our western approach, was relative to some of the peoples we encountered to be different than we, but not to be understood that the means of our difference was by the circumstances of geography, climate, diet and other reasons. Only the later study of what was argued over by translating Darwin's work and trying to understand what might have been meant by it? Showed for some of us in purely political terms, it was easy to use the concept of natural selection to endorse all that we did now or could do in the future, but by the false premise that natural selection through evolution means appointment!

Unfortunately from the best of intentions in delivering a scientific work to dispel some of our mysteries to better understanding, some of us plundered his excellence of effort and turned what had been written into an error of thought which for some is still the case? One provision or how Darwin's work was interpreted provoked political ideologies which were freakish in allowing some of us think that because of the feature of natural selection working in nature across the broad base of all specie. We in kind by the proven success of our own political system as thought, showed that we and our politics were to be appointed to the fore in determining further opinion? Taking us back through the poured funnel of life, back, back to the days when might was right without any consideration save to self by the means of being naturally selected! Meaning that to be so was not the scientific matter of specie success as Darwin supposed and suggested, but to the case that it was in the individual who charmed the idea of purpose, secondly his tribe, clan or group? But to be held in open competition within the same group unseen, until from the ending of world war in nineteen forty five we thought to reissue opinion!?

Taking part of the full legacy of what was left to us all at the world war ending and to the means of proposed change. In point that we could and do accept the standing of the United Nations, that we are politically classed to be first, second or third world, because of our economic standing, is a stain to the matter of our specie singularity. By token that we accept the principle of religious tolerance from the same office, is also a regressive step to the believed intentions of the progress we were to make since that time by at last becoming a global society! With the full reality of fact, we only ever achieved to be or have a global economy in open surrender to the principle concept of always putting politics first, even by any division, but not understood to be the case, needs us to look at how we have followed the wind of expectation only. By assuming that our first choices made after the war were our best?

From now, in charge is the matter of all of our political systems to overlay all of there principals additionally with the three taken prospects of the Fifth Dimension. Then to use the one active in 'One for the Whole' by reflection and or to act upon in theory to the reflection of the reflection to accommodate the need or style of required change! With the broadness of the Fifth Dimension beginning to become part of our consciousness in force, it will also be our conscience, because it never commands, but only reflects on choice from now on and for ever! 'One for the Whole' to be our one active channel of reasoning from that source is all that we need at close range to tackle some of our different approaches to how we manage our political systems.

By taking the same mixture and on the same terms, overlay our religions with the same from the Fifth Dimension, creates a new form of reaction far more important than from our political cover? When set to religious thinking in the first instance, our three taken prospects are already there and have been since the signs at Shanidar? Without connection what is to be stirred from our religious overlay is the matter of how our religious leaders indicate how we are to react and or on what level will they expect us now operate our religions? When at the same time they are to be fully connected and intertwined through our one active prospect in 'One for the Whole' evermore, but still able to offer different opinions if so?

No charge of any kind is to be set against any of our political systems by naturally selecting the three taken and one active prospect of the Fifth Dimension except perhaps the reflection of greed! From being a human failing allocated to the timing of now without regard to the future, although set from the standards of future markets! From that, charges to be laid against our religious leaders are in the one case of sloth, even if a monumental amount of energy has been spent by some individuals to the good, to our direct benefit, my personal charge is based on two counts but having been set on two levels.

From where it is to be easier to implement the fixture of the Fifth Dimension politically stands in the matter of how we introduce change at the end of contract. After all this is a new deal morally, we do have to clear our books before we set new standards which are to be minimal. Because from now, even unconnected, that some religions might appear to be minimal, our charge is to include concept, but not necessarily to hold it politically, but raise the standard above the image of religious tolerance? Our religions also to be overlaid with the prospects of the Fifth Dimension are to produce a better case for change or adaptation because they in core have been a better source to the true matter of how we as this single specie have managed to break the spell of evolution along the standards of pure instinct!

I have made the one choice that in overlay to our political systems and our religious attitudes, the one standard to always remain, is against any idea or consideration that the feature of the Fifth Dimension has ever been appointed! It is a simple refinement to the matter of concern about our conduct from how we managed the past until the present time. It is a refinement for how it would have been better to start this last modern era if we knew then what we know now? From that it is also a made case of how we can now tackle the future by collecting the refinements of our efforts since nineteen forty five. To use the expression of what we have found over these past five to six years to be able and overlay the two necessary principles of control in service to this our single specie!

I make no case to the failure of our politics or the success of our religions, all that has been expected is that we use the right ingredients for our progress in the right proportions to endorse the same standard owing to us all. For correction if needed, by inclination I have been able and judge that from now the best means to maintain our sought and continued progress politically. Is to set the one aim whereby all of the best efforts of any political regime is to be turned to set and keep the one standard in us all to be first world without expressing any particular style of lead taken from what had been our recent extremes through communism and democracy!

First world on the political calendar does not mean we have to sit down and share in equal measure all of our produced effort that might have been suggested by some forms of communism politically, fortunately to have now failed in context? What the open intention is for by hint, relates to a time even if only seen by reflection when that situation was a natural occurrence like when the event at Shanidar was jointly managed by group. Because from recollection that event was a mirrored occurrence frozen in time with the one effect to permanently line our memories of how things were. That all of group worked hand in hand for all activities, therefore all effort was shared as though it was the first? Although a scrape in the ground to lay a body was a small event in itself, it was somewhat magnified for us to have left grave goods, which are a further sign of our equal responsibility?

Because simple as the act actually was, the means of joint sharing and organisation was big and new!

## CHAPTER 171

With politics having been cornered, our next best aim is to set the deeds our religious leaders must follow? Amongst the other means to our joint order of control, we have already made better moves than ever thought of by the means of politics. Our only reason for apology stands from the aspect of our naturally selected human vanity. Not a trait to be confused with any aspect of the thriving imposition of greed we accept through the medium of our politics. But more inclined to be a pious acclimation from the reflective pause, because in cases we had discovered a better image of God if through monotheism, until then the obvious was that before time, we assumed that God was in haste to make his presence known to us and waited to see how we would use politics and or religion!

Without connection but since Shanidar our drive was carried in the false vision of trying to understand our growing number of mysteries, not least from the very event of laying in rest one of kind, lifeless, but active by our memory? Before ever we could have realised the full meaning of that event so far ago. When next we looked we had become sophisticated enough in group as country, to have formed the idea of an established priesthood, solely to attend to the matter of our newer religious appraisals, which numbered the management of control to all mysteries, unless by human vanity!?

To the fact that we had come to the point where we thought of our own king be he pharaoh or the common leader, his role was seen to be in the appointed form of our provider, directly connected with our gods or the ultimate God? This comfort of opinion was enough to massage our own growing self worth tribally, by bringing together concept that as our king was in such contact, this flow was passed on to us by our own and the kings appointed priesthood, to include all with other gods, but in the main to be of the one King God!

By the linked proximity of our clan or group or from our kinship, knowing that other similar groups as we belonging to there own city states or half countries so far? Even that similar social planning and behaviour was common and practiced by many others, from our own separatism as theirs, we each sought advantage over kind? Our pace to understand began to grow from the rational, so much so our cup was filled to the realisation that gods really did mean God. Not least if at first our priests were to feed our own inquisitiveness by the realisation that the connection between the God of gods and man was the link of discovery? Was in the clear decisions to have been made from the aspect of our excellence of mind but error of thought, by those to be our priests if, but without human blame!?

So began our first monotheistic religion even if to have picked on the bones of other religions of area, but assuredly from contacts in lands in the Middle East of now, but by happenings from four to five thousand years ago?

Although our travels along all roads since then have been rocky and unfortunately in time getting to be more bumpy by all events, because of how the wind has changed for the better since nineteen forty five. Although to be still serious enough to have induced unreal measures of reaction to the read first case intentions of our political systems? No progress can be made until the process of delivery to how we are best to further precede is realised? For me to have laid that task upon our religious leaders, is not that we can divide our standards because of the fact that we operate under several different religions?

My intentions are to assert the real power of what our religions are supposed to mean instead of more readily relying on the poorer and poorer delivery from the means of how we are controlled politically? I have set no standard of competition between the two systems if that is a balanced description, because of the fact of what they really are! My case is proved by setting the one standard by equal overlay of the Fifth Dimension and warning in fair honesty, that our assumption of each thereafter will be different? My point for expressing our first shortfall of either feature stands by my own recognition to the lacking in fact of there being no defined international Islamic leadership structure!

Also that under and since formation, many of our Eastern styled religions do not have a particular leadership form of control is because from base they are inclined to follow an intermixed pattern of reincarnation? With at times there being an unmarked sensation of the same counted individual transmigrating through our known specie mix in any order but without the means of reflection from one to the other which is no bad thing! Not least that from the absolute base mix of how the religion of Buddhism and Hinduism were set up again some thousands of years ago? From either of the collective Eastern group religions, there was never to be a conflict of interests between what we have tried to understand between evolution and creation!

Their basic philosophy is to the excellence of all life forms as self, but can be charged that the gap between the picture of God in our God and all life in all cases has been left to the ambiguous. Which again is no bad thing, because by sign it even overrides some of the monotheistic claimed errors of thought in classification to form the basis of needed change? Western religions by my standard have been set to challenge to the aspect of evolution and or creation if only from the original printing of Darwin's great work The Origin of Specie. Our Eastern religions without that distraction into the class of there philosophy, already hold that all life mixed through the classification of mere specie, are involved in the wheel of living naturally, and to be linked eventually?

If any waves or vibration was sent or even felt from European thinking of how Darwin had written and by what was essentially a Christian reaction? Then by choice without selection whatever might have occurred to any individuals from our main Eastern religions, was in fixture to what would have occurred to people to be sufficient enough to be naturally laid back by character? Why get excited by how others in far off lands thought when and if their religion was challenged by self doubt because of reactions to another published work?

I make no claim in relation of how people belonging to some of our Eastern classed religions may have though in relation to the doubt raised by some of us belonging to Western styled religions? Save that by modern times, in all cases we have allowed the matter of politics take on decisions that perhaps we all must query? In note that by character many people to our Eastern styled religions are or can be classed to be laid back, stands by my own view from the fact that there is no inner doubt from the concept of their particular religion. Because in all cases Eastern focus is to the actual conduct of the individual without the need of being directed by a specific priesthood or a religious leadership structure like in some Western religions? Being a case in common with the last major monotheistic religion of Islam, which as I see is the only Western styled religion not to have a fixed or appointed leadership structure similar to earlier religions of area?

By or from that connection in bringing Islam to that accepted focus by myself, I look to understand what that unconnected issue really means? But more to the moment of now I look to that situation as it may be interpreted in and from other events in this last modern era. But now counting much of our previous histories to be responsible for laying some of the attitudes from all of our thinking heeded or not, but without any impossible desire of altering any of our past!?

By the very aim of stating that Islam in comparison with other monotheistic come Western styled religious, has not got a defined religious leadership structure and then laying claim that this might be a disastrous oversight. Then to make the further suggestion that in duty it is for our religious leaders on our behalf, but to make all the inroads we need to enhance the desired progress we must make over the next generation, but now? Is not to be a measure of misunderstanding, but the sight we are to take in order that all purpose by the way of politics and religion are to be met?

That I had focused to the other dark days of our united histories, where events in the Middle East have been paramount to the lead in what is to be the ultimate of world standards in representation to us all! Because of our political motivations and how we have manipulated the texture of our three monotheistic religions all to have grown in station from the very area, but have now been addressed to be in the standard of religious tolerance to each other and other religions! Being an open betrayal from the medium of our politics, because in term while its only focus has been to the welfare of supply, with its sideline of profit and greed to have become its nemesis when soon realised, which in real terms now, means it has the better ability to destroy before it creates!?

Politics is not a sin, but it is a servant and all those in association to the class are to be our servants, just like from the standard of excellence from our religious leaders or servants are to be? By direction, once told and by structure of all religions, our only obligation is to self, for it is we on that personal level that all claims are made for. All that we need and understand is the total simplicity in meaning of that brief statement. Much of the rest in type of padding, we ourselves have continually applied to be the remedy of medicine we have required to understand any degree of the events at Shanidar or similar occurrences even if not as of yet found? Such a sign of that burial is the key to note the start of query leading to the start of what mysteries are, because we took the physical time out of requirement to do a task that was outside the element of pure instinct. Finding the means to pursue our open interest in an ending before we could imagine what the beginning might be? All of which has continually stimulated our mind in this single specie who are in open concern to the future?

From our conditioned state of now into the equal status we have assumed, little needs to be done to save course? By my own meagre input of introducing what are to be the three taken prospects of the Fifth Dimension. Although borrowed from time and by our combined efforts, the importance I attach to the overlay I have suggested. Is for the reasons that we are to adopt the attitude that less is more, which is not a challenge to any aspect of the events of materialism or politics in that view? Which is also not a challenge to our social standards so far achieved, but is there to openly question the thought in any case, we are right?

My own fixation of our specie singularity is to clear its own ground by setting the image of new standards without changing any aspect of the past, only to accept some of the reasons of some of the events, but not to balance one side of the cup in favour to the other in decline? My message of the past is that in too many cases, features of it had been driven by unclear motives always supported by the uncertainty of available reasoning? Like for example if one group or tribe, because of not being able and see the naturally selected phenomenon of how our geography, climate, diet or other influences were to take the one finch and make many alike from that original cast, where could our thoughts have been led?

In not seeing the effect of evolution and when we did if only as late as when Charles Darwin had indicated to it? My case is given credence when in turn it has taken us sixty to one hundred thousand years since Shanidar to begin and propose us all as the same if not in appraisal then from a burial? By some of the various reasons I have made the case of our total singularity although now, only now, supported by the connection between evolution and religion or birth and the religious concept! Has been done to circle the single feature of birth and human birth in breaking all the chains that had tied us to the endless, which at the time was the circle of evolution, which was the lot of all life created!?

Without clarity by more than one or a hundred connections our progress over these three and a half million years has been relentless. If we tip left or right by personal desire in who we are, no charge can alter the fact that it is we of this single specie to have awakened the wonders that surround us. Our own agent in science if guarded in keep by the honoured title I have bestowed upon all of its aspects to be the fourth dimension, is only a full servant of our desires and not the reverse. It or science by feature is no more than the understanding of how rain is made, is no more than how the wind blows? Science is proved so in my appraisal by what the mighty crocodile thinks of her? Two hundred and fifty million years of continued existence to the art of procreation and that great creature does not care where rain or the wind comes from, nor of night or day? Its full intensity has always only been to set in place its own shadow without knowing the why but mixed with the lack of care to the how also!

By trial, our mighty crocodile or from the beauty in the beast of the tiger, the same result is that they remain as children of evolution. Whereas before we had a chance to make our own choices other than by pure instinct we had turned the corner that the whole of evolution had been waiting for? If and so for us or our direct, direct ancestors stood up and walked on two limbs in fashion to be biped. By doing so a switch was set to hasten the process of evolution in an irregular way. Allowing that the turn done was in the new name of creation but not at first seen!?

Because from the very time in Africa when we did walk and leave impressions of our foot prints in volcanic ash? The deed was not recorded for posterity, but by the circumstances of chance, showed us that there was a similar biped in the midst of evolution when perhaps there should not have been? Therefore from that act being done then ago, evolution was simply in progress, if to compare Darwin's finches from what was to be considered the one type originally? How might we consider that we do not at this time have several plus several types of the same biped, who have adapted not to be the fastest or biggest or strongest type of the same specie, save to be the one single specie as we, 'Out of Africa'!

If standing upright and walking was a step in the right direction, then from other instances in note we can connect other deeds preformed by similar groups by type? Not least by the act of homesteading, in staying in the one place for time enough to leave remains to the effort of organised living. Like for instance in leaving the debris of fire hearths and the evidence of accumulated food waste in the form of discarded animal bones or shells that had been worked by deliberation to extract the full feast. At other similar dated times before the mark of history, in having found knapped flints to be used as tools, also the standing of walking on two feet!?

By the same connected reasoning for me to have associated Shanidar in line to be part of our united specie progress stands in abundance by the open variety of our religions! Not least that although varied, our one common trait as said, is to be realised from the conduct of the individual? What ever historical connection we have developed to by line of control to be Eastern or Western, we have some times used the office of the priesthood as a matter of personal choice until this very moment, from which I would only suggest that our priests are to become our religious leaders? With the one and only one intention, to alter the personal and collective welfare of every single person upon this earth, religiously first and politically second, but both timed together from this new mark of time and timing having been first set some sixty plus years ago now called this last modern era!

We have little to do to address our problems if formed by the way we have misguided the changes between our politics and religions? In most of our cases of concern we have been unable and look into the mirror full face and see the real reflection because some of us have allowed what we think we know cloud our reasoning for what we do know? Which in all cases is that without counting, we are truly the one specie and have always been, what has caught our breath and needs addressing is that we have not yet seen the ease of change?

## CHAPTER 172

How ever our care order of politics has taken path, our route to the supply of now, has been brought into question by an offshoot of one of the elements of care by our last monotheistic religion? Beyond the name of any individual, I have made pointed accusations that some of our imams, those who attend to the conduct of how our young in faith are to learn text. By the same reference I have alluded to the one pointed feature of martyrdom and by direct connection of how the act has been wrongly connected to the event of honourable suicide, which its self is an ideological mirage!

By my own judgement, I had put to note that from the time of Mohammed, held against the proposed flow of how the first two monotheistic religions of Judaism and Christianity had proclaimed not to kill of self by self. By my own reference that Mohammed was set the task against the now, then ago, to mix in open display against a background of difficult and violent times in which we lived and where? His natural inclination in preparation of what was leading to be the structure of a new beginning for us all religiously, had to hold in regard to the violence and measure of apathy and religious discontent to have churned in ferment there? From the prospect of the possibility to be directed and create the lost image of what our true conduct should be? He was bound to relate to our range of emotions and of the way we had come to look in bleak expectation of what the future might bring?

Without direct assent, but in order to be able and accommodate the needs of how his uncertainties were met. Without any level of spite of fear or terror, his readings in writing, allowed that in order to instigate the level of change needed and the matter of its capture. It would be necessary without the misplaced conscious value put by Judaism and Roman Catholicism to bypass the level of misinterpretation both Prophet guided religions had found in regard to the matter of whether to kill of self by self or not?

From this new standard in proposal and by extension, if by necessity it was possible to kill of self by self not least to defend, protect or promote this new appointed expected desire! No case was made in the name of murder, only in shadow of dying in cause to be a martyr to the faith! By this feature no let was to be put on the need to kill, save in full and open promotion at first contact. If the case was made in response to the natural violence of the day then the ready matter to defend, protect or promote our new concept was to stand to the extreme in human terms. If the need was there to open the city gates or the hearts and minds of local populations then the course would be or could be to the execution of that duty to the matter to kill of self by self as so appointed in the haste to stabilise the concept of the new faith to be, rather than not?

Born from the same matter of change for benefit in the name of the one true God in Allah, if by the appointment of battle, a body was to fall in the name of Allah, then in due regard to the broadness of this new concept, that loss was not to go uncounted. By instant recognition associated to that death, because the waste of life was in the set cause by following in lead the acclaimed new direction of promotion of this new concept. Status in the name of martyr bestowed upon the dead in the same duty it had been done in other times, but when the faithful had died for there faith albeit in the arena and for the sport of others?

Here and now beset by the said natural violence of the day in any case, our misguided take by some of us who did not really understand, was a matter of no, no concern to those we had to kill if to act in promotion to this our new concept? By the same balance for those of us who might be killed in the defence and protection of our faith now carried. The honour of martyrdom was readily bestowed upon us, with the much added reward that the event of martyrdom be marked in glorious fashion by entry into the house of God to be paradise. Taken by the thinking of then, that the reception was arranged to accommodate what was noble and honourable to a society who sanctified the male female husband wives relationship, if so. No less was to be set for those of us who were not unfortunate to have died in cause for our proposed faith. They as we were to realise the same edge of paradise but if a little less formally than our martyrs. Our case was in like form to the expectation of other earlier religions of area, but in there case the matter of conduct and reward was reminded by long established priesthoods?

That I had put in frame the picture of our first two monotheistic religions of Judaism and Christianity to the expected outcome that all living would see the image of God imminent, had cast those religions into the depths of self doubt. Not least by first accepting that the relevance of life as told was to be through death or earlier if by Gods own intervention. Judaism and Christianity had no need to cast any aspirations into the future because all of the signs so shown were to be of recent access imminent!

From which by separation, for Judaism to have lasted two thousand six hundred years and Christianity to have lasted in overlap for the last six hundred in parallel to when Moses first intervened by representation on our behalf. If by association I had suggested that both religions had lost direction. Without using the corporate narrative, when talking of Islam, even at instigation some fourteen hundred years ago, by my own rule of thumb and from my own imagination, in watching the reflection and the reflection of the reflection in accord. I can say that by opinion only, Mohammed in reference was more aware than had been indicated by his standard of what Moses and Jesus delivered as prophets. Not least because in plan, Islam was set to the open acceptance of what science might reveal insofar as some of our perceived mysteries would in due course be explained by that very medium.

In a sense by monotheistic standards of ready acceptance Mohammed's view was futuristic. No immediacy was set to the effect of what conduct would achieve in haste, because his overall plan was generationally placed. Apart to the one rather large concession on behalf of those of us who might become martyrs in the promotion of this our faith. We were led to believe in the future by deliberation and to include the existing and expected, unlike to the first impact of some other religions of area?

Another turn of page to that story was to the concept that the religion in natural and numerical terms needed time even generationally to advance and become global. Being a ready recognition, but of its time a very fast and vast concept. That there was no promise of immediate triumphant unity gave the honest opinion that reward was the simple necessity of time and conduct. If to create or further induce a laid back cultural approach leaning to fatalism amongst a violent world, then this particular pattern could have been justification of how we could be judged even now.

From the other outlet of martyrdom having been won in the extensive charge to continually promote the new religion, its force in use became more prominent from defending or protecting the same feature of faith, save by implication. By any theory of excellence, Islam as a world religion not only held her own course, but made inroads into all other cultured standards in places of direct influence and beyond. If less and less by the aim of battle then more and more from a calmed or laid back fixation by use in the consideration that in time all would come in sight of all that was expected?

Although Islam was no stranger to the necessity of change if only introduced by the revelation of what science could deliver. By standard, the pace of change had not become an issue on religious terms ever, but was given light to what some changes meant some two hundred years ago when England and France fought European political battles in Middle East waters and on land also? Both to be Christian nations in soul, if not in practicalities, because both had now moved in pace to allow politics rule Christian hearts. Without due recognition to the effect upon slowly led nationally inclined tribal groups politically locked into the full span of what their specific religion of Islam was to deliver!

What might have become the clash of titans over the next one hundred years in some cases was to be a non event by some calculations, but was held to be a plugged and closed boiling kettle? By the time of the First World War from nineteen fourteen to nineteen eighteen. Islam on religious terms still had no religious voice or the general need to express any views because although essentially a Christian war by European terms the conflict remained ex to the image of Islam?

By the end of the last world war in nineteen forty five, the pattern although similar to the last one hundred and fifty years of mainly European influence in the home area of the Middle East. All came to the totality of modernity and now; but dressed in old and sometimes ancient habits. By direct connection for us all in relation to how we have attempted to join all the pieces of this single specie jigsaw we are. Our best efforts are wanting, because we have sought to heal new wounds with old remedies. Even that we have now found the proper medicine albeit of a recent mix, but the positive to be like a welcomed stranger, is that from this batch there are no bad side effects, except the feeling of guilt!? Before we officially begin to apply our medicine in equal measure to the rashes or rashness of politics and the pain of heart we feel from the misuse of true religious sentiment. By allowing our personal perception overrule what until now was not known to be in open consideration of all by aspect of our one active prospect in 'One for the Whole', in line to offer general change. Without the hint of accusation we must look at the pointed edge of our real differences, which instead of the cure, will add to inflame the hurt, until we are ready as now for our waited treatment!?

Kill or cure is sometimes a lightly used expression in relation to small matters of personal concern, perhaps in discussion between several people or a group inclined to similar opinions. Even by negative tone it is actually a positive expression when used in the light hearted way it suggests. Like in relation to the completion of a menial task a person had undertaken if to assist a friend or complete a commitment on ones own behalf. Oh well here we go again, its kill or cure to get on and complete in time? By the same index and of the same tone it is a self confessed chance even that no risk is possible. Because the application only refers to self by the stated terms and is not a factual admission, therefore in essence it has no union with the concept of suicide even if told or encouraged by others!?

## CHAPTER 173

I hold no claim to be other than a minor cog in the same wheel of life of which we all belong. But as so have the same entitlement of opinion about any range of subjects in answer when the voice or pen are used for the terms of enlightenment right or wrong, but from the self judgement of inclusion for the greater good! For the greater good, but not to the advantage of, is cast to the same as we in ourselves. That is why our one active prospect in 'One for the Whole' is a store to hold the remedy of our ills? For any individual to have called that politics must now hold its breath so all hearing is done in the utmost clarity, has been put so that our religious voice is not distorted. Or should not be distorted by the course of some action or reactions some of us have been led to take, even if claimed as an act of the utmost morality? In the absence of the original link pre distortion, is to see that reactive deeds had not been prostituted to become acts of venomous murder!

All the virtues of Islam like a crafted work of art have been carried these fourteen hundred years without showing age. Save to the one pointed aspect of the future if now distorted by some of us to be classed as the keepers of conduct to the young of faith, otherwise to be some of us as imams in care of some of our young, but only one way? Islam has been able to carry her own coat these past fourteen hundred years, plus increasing in size of numbers by the sheer tenacity and clarity of doctrine. Set so to continue without any call to the wind of change that has blown in all corners these past recent sixty plus years, at times increasing in intensity to flick the sands of apprehension onto the eyes of some who in style had forgotten how to look! Although time has always been standard to the image of Islam, impatience is a rule belonging to us mortal beings.

If from the wound of being played by what appear to be the scales of ignorance to be ignored, if seen to have happened until my own time mark from nineteen forty five up to the mid sixties of the same last modern era? If for that ailment to have festered by the product of religious war conducted under political motives between the same coat of Islam through Iraq and Iran shortly after. If every lead expected to be taken, was designed from the culture to be politically Western, over the concept of what had been the normal means of interaction before such events actually occurred from about nineteen forty five? Then no case can be made against what was or what is, save by the opinion of any individual who must be prepared to clean and dress the wounds of time? Which involves all of us if to use the penicillin of the Fifth Dimension in its two strengths for the ills of politics and the ills of religion? Having already been explained, but to take newer meaning, that if we use other old remedies also, they are welcome providing they do not numb our true sense of liberty of mind that we require from any medicine?

From the delicate balance of what is justice or religious justice. Beyond change the case has been made that without the mirror of foresight, when looked at, we saw what was to be seen? Our differences of now are in full bloom to have been after the events of what occurred some four hundred years ago in the new world! Without comparison and adding no virtue to one side or the other, to waylay and destroy the feature of human sacrifice was an epic but deadly action? Although for and within the mire of mixed human interactions ongoing, but unknown and unfinished through the terms of the European reformation, when the practice of human sacrifice was ended by whatever means. The ideal of the concept of 'Natural Body Death' was created and is now owing to us all from then!

What was to continue after that one noted event, even if in protection to the prospect of the ideal of 'Natural Body Death', which was to be put on hold? While the Christian religions of Europe and now the Americas juggled the eggs of human desire until the feature of human concern to the individual generally was found smashed on the ground of memory! Because raising the right hand to catch the falling eggs in the same practice of juggling, from the mirror of life, it was the left hand that moved in the reflection!?

From the clouded resolution that we could not allow what was considered the savage deed of human sacrifice, our natural turn because of events, was to abolish that standard under a religious blessing but by the uncounted political force of greed. Led that the opening was made for an aspect of care to be given to a section of this one human community even if to be classed third or fourth world or primitive. Which as we know in all cases is unfair and unjust and an unreligious attitude, save by being political, which again has fallen to show what route we all must now take, especially the one of gain!

From the need to end the practice of religious human sacrifice it was easy to associate the same matter of beholding to the terms of how the Great Spirit was revered. Not least from how this form of worship in religious terms was laid to the open broad perception of all life, all life forms being part of the whole? Unfortunately without the display of cities or temples to be plundered of gold, our next interest for those of us to be called European adventurers was set to settle and exploit the very nature of this vast land in its pristine form when we first found it we carried to plunder nature herself!

Even with little concern given to the original inhabitants, the European contingent by the claim of natural progress began to lay the foundations of religious tolerance but only amongst themselves. So much so that when the first steps to be taken towards the declaration of independence of parts of the North American continent made in text, there was no open mention made of some past events in link from early settlement. No consideration was set to what was a long destroyed concept of human sacrifice as a religious deed or any mention about the standard belief of peoples who stood to the concept of the Great Spirit. Our only connection by way of reference was that before the terms of being second or third world people was a political standard later carried into acceptance. The existing inhabitants on the North American continent were classed to be mere savages, so recorded for posterity in the American declaration of independence. Part of which text was later used as a basic standard to put before the new United Nations as a blue print constitution of new conduct in charter for care to be given in the equality for all men being first, second or third world?

From any of the above situations or circumstances so put, we come to the end of an era whereby our real standards of everything were to be determined by American and European influences how ever devised or though of or instigated, that is natural progress! If by reference I had previously cited the case that there had been cultural and religious differences between the same peoples in Palestine of the Middle East at the same time and since. By the standard of modern outlooks or how we were given to be expected to think. Interaction between the two main factions of the Palestinian people, one of which was religiously set to be of Islam and holding a laid back character to be still emotionally volatile when called. To be balanced or set to oppose the other Palestinian group who although following an earlier religions from area to be Judaism. But with more than the open inclination to nationalise into the new state of Israel by the imagined, the imagined link of a defined directive that all they were doing was to return!?

Set the situation whereby the new beginning post nineteen forty five was to have all existing standards of expectation turned by political request into the image of settling those desires, all desires to be of political origin? Stimulating in haste to the configuration that at the same time we tried to settle some conditions of nationhood was when we were to operate our bloc culture standoff situation, to cause deeper confusion? Like from when the first step to Israeli nationality was taken, the cloak worn was to be Western by inclination politically, but the mask was always to hide Jewish theory! Carried so in effect when the standing opposition was to become communistic by nature if for no other reason but to take ground from a westernised political situation because of their own international drive for communistic expansionism at all costs, even if only carried out by whimsical desire of particular people already in the position to exploit the power of politics!?

So much so to exploit the delicate situation without direct input of how on behalf of the forsaken in Palestine, the Russian, if communist outlook, supported the ideals of Islam in answer to simply oppose? Further linked by support of the Arab cause who were generally all of Islam by religion before any particular interest in communism or the west, save to be involved in this new process of internationalism? Perhaps unseen by trend at the time, but for me to note that about then, without attempt to alter the reading of any national or personal histories. To now make the observation that even then the new Israelis were fully westernised and or Americanised by style with the unseen adopted progressive nature by or from the same connected circumstances! Standing in fierce and bloody confrontation to people who by another list of circumstances were laid back, how then to balance that concept past the image of might being right politically!?

If any suggestion is to be made to endorse belief that the case of American, European Christianity was aligned with Judaism on one side of the iron fence, to the other of communism, the Arab cause, and Islam? Then the idea is flawed when viewed to what was to be held through the United Nations on all counts. It decree was our referee point, but only made to the rules we were actively making as we went along, with some new mistakes we continue to make naturally. Because no path has yet been cleared until now, when the right level of the Fifth Dimension is found!?

Unconnected in the full theory of compassion, but if only because of the measure of self exultation, when the state of Israel was born politically, only politically, her status of formation, was made her own affair? Which is a law unto us all save by intrusion, exclusion, expansion, indifference and the contempt of not even listening! If and so from the status of intensity created in area to be our Middle East, one act or reaction was to induce over wrought stipulation. Like for instance if the Arab league or connected Arab countries or connected Islamic unions were to cast a dark spell to be upon Israel. Our conduct in plan was and had been excited to all extremes!

Not least political extremes, which are always to be managed, but amongst our religious extremes which have reflective images, save to the name of dogma being cast from one side to the other or one extreme to the other? Both forgetting that in all examples from being monotheistic there is no other case, there is no other case, there is no other case, but to be inclined to the same God on what has to be his terms by our terms! Like how we must each cast our lot to the one aim under all personal charge, whereby along with what our religious leaders will have to say soon? Our own judgements by definition are to be set from our individual plans to activate if for the first time. Gods own interest, Allah's own interest, in this single specie we are! Because of how we alone amongst all other creatures and life forms anywhere on this world have turned the longevity of evolution by Gods time, into the brevity of time to allow creation!?

In doing so and in self belief of that fateful connection, our case is to God and in time Allah and in time the full rational of existence coupled to the equal discovery of purpose? Even from the short time we are to spend on this mortal coil owed to our singularity through the one lifecycle? We can each make the full standard while in the normality of living, we can and try and do show a standard, specie wide, of what would be of interest to God if, if, if his design of person was inclined to any creature who so called his name!?

Making the title that we should expect to fall to the full ability to understand first time because of our expectation, and not as we have done so from the haste in desire, create God in the final form to be us! Instead of allowing that the reverse was the probable cause, leastways if evolution has been our first ally. That by the very grace of unmarked time from any theory, evolution was always there in source until the open discoverer of creation came, who in all realities was us!?

From any of our political perspectives, I do not ask or expect that we are to be disposed to change the matter of fact of our history. Even if almost like looking into the mirror of the past, fond memories might grow while sadness might look to seek the means of retribution. If those feelings had been stirred from the cause of injustice, being natural injustice. Not the offer in type we have readily developed whereby instead of the honesty of some kinds of retribution our objectives are compensation in replacement of natural justice? Turning the feature of concern to the matter of how we have come to collect and nurture the full sway of political dominance to be in the supply of now, instead of to the hope of tomorrow, if to pursue the icon of now instead of the future?

When previously stated this past sixty plus years of our last modern era in taking us to a new reasoning, even if using old tools. Without consent we have moved politically, not least because it is our first form of basic communication. Taken from the first times when by trade we did not have to know of another peoples language or religion, only that we needed there goods? Which in a reflective form does give insight to how we more readily accept the principles of supply and demand to elevate immediate need if that was a drive? By that example if needed, for us to have been attracted to have found an inclination towards the element of supply, then civil order into political drive was the made outcome of time by time. That is until we were able to fully value the intent of time through the image of our religions? To be of no more pointed relevance than from the mid period of time in these past sixty plus years of this last modern era, when for the first time again, our religious purpose was to shake our misaligned political urges!

Although our formed concept of religious tolerance had been made by various aspects in relation to how we were to permit balance the old of yesterday by the terms of all religious tenet, covenants or honest belief. Align to the matter of statehood in practice to our national trend by country, even if in that country there were many factionalised groups, but generally of the same religion? Which under the terms and concept of religious tolerance, they could in open belief follow both of our means of control of in or for or to life, to serve the theme, so that our conduct to each other has no adverse bearing by any degree? So that by show the obvious split from evolution was shown, so that the open sign from us all is that on both counts we accept our singularity while we know and feel we can be different, making us fit to be the same religiously?

Being our ultimate goal, but as yet unseen, because when we do look into the mirror the reflection has already been put on show in minds eye. Of which it will be under the best of motivation that in and through the theory of reflection, our religious leaders in us are to realign all the same standards that we already hold. Without the need to make them the same, but in remembrance that we all hold to the same God in Allah, and do, even those of us who have no concern of God in concept or the future, save to remain in instinct!?

From the twists of time and in relation to this last modern era even to be completely unconnected with very recent events. Some of us by the best of intentions but through the worst of all actions, have used these our events, to extract retribution in the full style of stepping back to the times of sorrow when we practiced human sacrifice by means to be a religious requirement? Hiding the complexity of the matter by clouding the pure religious event to be from political ramifications, further hiding the result of such forms of suicide in name of martyrdom to be no more than murder by human sacrifice, wrong, and God wrong!?

## CHAPTER 174

By our own good fortune and from our own good intentions, the one seed to bear fruit by effort these past sixty plus years from us all as a world community. Has not come from the crudeness of how we have tried and manage our political/civil innovations. But because of how we have tended to ignore our original and pertinent religious accumulations in all forms set to the displacement of humanity by any degree or method, or none!

By direction and without apology I have been able and focus my first concerns by whatever means through the incline of our monotheistic religions as they are, but not by any form of comparison to other religions, even if Eastern by style. In all cases of religious importance I maintain the connection of all religions by indication and have chosen to direct my inquiries from both perspectives? But have taken to the Western style primarily, because in being monotheistic, all relevancies through them are to the one and standard lifecycle in owing to us all of the same duration for the aim of fulfilment!?

If I had indicated to the one point of reference of how some of our imams have interpreted aspects of the third, by my own definition the third dimension religion of Islam. Which was to follow Judaism and now Christianity as my first and second dimensional monotheistic religions, no comparative claim is made of one against two or two against one. I have set them to differ on the one principle that the former two were directed not to kill of self by self, yet do. With Islam as the third and main last monotheistic religion never to have denied the aspect of death upon self by self if the standard was to promote the new concept! Or had become by revival and review the matter to defend and protect the faith always when and if necessary by the means of death by self upon self!

I make no charge to be upon any of our imams except in being odious, colourless even more un-appointed than thought through the devious mind of envy, spite and shaken fear in there own convictions ultimately set against the principle of personal ideals! That is if only and by claimed assent they impose their own, their own political will to be upon any of our young and young of heart. If they betray their position of trust to the welfare of all of us, but mainly our young, if they putrefy the openness of hope and expectation to aspire to the waste and costly service of human sacrifice in tribute to the featured element of physical politics like worship to the roundness of the sun by day and the sky at night. They serve nothing less than to be resolved to court the elements of evolution by denying the very concept of what creation is and of who it was that told us so!?

Imams in our best habit are no more or no less than we, each of who have the one and single lifecycle in limitation by range to be fifty plus or seventy plus years long. For all that imams in our name and what some of us in any faith might have done also, has focused all intention in more highlight than ever before especially to the Middle East representatively. I make that connection because of the reasonable grounds I had put forward to where the method and style of all our political systems and of where all our world based religions have come together in circle, to be able and stimulate one way or the other, future steps we might take and leave in trace for posterity!?

Without the image of reflection, save to political results, for me to have put the name imam forward has been deliberate. Which is to be obvious when shown it is not a personal choice but a collective observation in order to cut to the divide of what some imams without prior knowledge or understanding might be responsible for. Without implication of the type of deeds some encourage, my standard is that they have failed the concept of Islam and or of religion in general?

No balance can be held against the edge of reason or the other earlier monotheistic religions because Islam is a refinement of both Judaism and Roman Catholicism. If by exposing to us the full concept of the future instead of acting by the implications of now, which the other two did to the immediacy of the situation they understood to be imminent?

Islam's first choice was to grow amid the situation of normal life and interaction when the idea of introduction was relayed to us all from Mohammed. There was a new condition to include, that the religion was to be based in idea, and set up to cope for and to and with future generations. But alas what was not made clear or had not been taken in context was that once the seed of thought had been planted in the arable minds of us in line. Although never to be denied to the core concept of the new religion. Even when from the very violent times of initial introduction when there might have been the factual need to kill of self by self in situation? Until the very means of open connection even unseen, we had later come to a time in the rapid succession of a few centuries, where Islam had become rock solid in balance to all philosophies!

Outside of any infringement by internal forces only, if the matter of human life in balance to be taken or not, swayed between the ideals of Sunni and Shiah or were maintained to be of internal concern outlined by those differences!? Are not to be considered in like style of what happened in later centuries from the links of Roman Catholicism to the chain of reformation mostly in Europe? Islam although if to tear at her own inner unseen hurt, allowed that on one level of death by self upon self, was maintained by the meaning and understanding to the one and unchanging text, for ever? This was maintained against the assault of other influences from other quarters, who offered violent means to challenge a religion growing of its own accord?

From the very fact of how Islam has grown and expanded allows that we might pass comment to accept that the spread was maintained by sound ideals which were delivered by people naturally laid back! Aided in the overall outlook that Islam has followed all expectations and still does to the proposition that in the due of time by future reckoning all will come to Islam influenced by text alone? Save by the interruption of what might have already started from within the very religion that has always been open to all, even though at first some rejected it. But maintained an open view, because from the static terms of some verse, Islam was to be tolerant of peoples who already held other similar faiths!?

What has changed opinion, but has not, because of the nature of the written word in the Quran, has been by the measure of very small examples, some imams have contrived beyond requirement to take the forward looking ideals of our third monotheistic religion. To set it in standard of what had been the case by our first two monotheistic religions of Judaism and New Judaism particularly, whereby new converts were ready to die instantly in the face of persecution. Because the event had been marked by the very death of Jesus into the act of martyrdom, because in immediacy, that act brought about the unity of body and soul in death and to heaven, if later to be paradise also, but to be with God?

Such action in the other name to defend or protect or promote any new steps, was an outlet whereby the first service to and for the introduction of Islam was carried out. Martyrdom to for or by the deemed requirement of the faith was never to be an issue some fourteen hundred years ago. Unfortunately until perhaps what had occurred in these past sixty plus years of this last modern era and by whom?

No case is to be made on a comparative basis to what is meant by the term of our religious leaders, even if defined by myself and pointed to in classification? By the same distinction I do not operate on my own with a single opinion? Anything offered is part and parcel of our collective thoughts and learning and experiences by the one condition that all said and done is not, is not by self expression? In relation by example whereby some of our past political leaders were able and find their voice to the matter of all intentions personally by how they might have thought? Not in a collective vein that might have been claimed, they were to lead to what we can now consider to have been depraved ranting, where by placing there own standards, nothing else was considered! Far, far worse by the same lead of being in a position to lead, their view was final? So much so that in far too many cases of recent history some of our political decisions and or leaders, acting for the supposed benefit of all, were apt to wantonly destroy without the consideration of any benefit to understanding, peoples such as we, whose only opinion was to be part of the whole or taken to be, because we supported what we did not understand?

How some of our political leaders of recent history were to operate was on the basis of supremacy, running to be tribally supreme or nationally supreme or politically supreme? Connected with the same addition of others only relying to be politically supreme because the other aspect of our best means of controlled learning in our religions did not count? Therefore religions were to eventually become expendable but to the ideals of human sacrifice now to the innate standards of mother country or fatherland!

Politics at various reasons by timing, even if planned for the future by supposition, has always and does continually fall short in overall presentation, because of the open and general feature of supply and demand! No more shown than in concept by the reality of now, by setting all the features to settlement of now! Which from abstract connection has been how in some cases and in view, it has been possible to link early monotheistic thinking with the political prospect of now! Unless reviewed in painful delivery of what is open to be learnt from the opinions of some imams!?

Suicide of any personal desire is a sad, sad state of affairs in respect of any individual thoughts, because it refers to the system of failure which on any personal level is further saddened by the felt need of the deed? The act in its self and of self takes us out of the care factor of our taken prospect in 'Natural Body Death'. Martyrdom in link to refer to death has no kind of reference to the meaning of what suicide might mean outside of the case of its own sadness and misery. Martyrdom is an event of happening and not now ever to be made, which is not to deny the sincerity of any thought given to the event from the outside. Martyrdom if from first having come to be recognised first and foremost by religious terms, then in that case it holds a mixed meaning described in form by the event, the occasion and with what the reflection is, even the reflection of the reflection or what it really means? Martyrdom by first take to be a victim, is also to be covered in best taste By 'Natural Body Death'?

Human sacrifice in reality to be the sad, sad terms of being a victim, and because has no answer, except from the abstract, is an event by other ideals which fortunately will never now work when examined by our one active prospect in 'One for the Whole'. Not by choice, but because of acceptances that there can never now be an example when one person to another, to spite what there age difference might be. Could ever suggest by opinion only, what the act of human sacrifice in the name of suicide could ever mean or be for!?

Human sacrifice is never, never or never to be considered by any tribe, clan or group or nationality to bring the act to be upon one to the other by any means from when! Human sacrifice is not! It is also not to be a concept belonging to these past sixty plus years of now since nineteen forty five. But by the same time span we are entitled to look and see what was to occur domestically, politically or religiously by some other outside influences around, but not principled. Because by waging war with any different group it had not been realised that until the now accepted method, no call had been given to indicate the reality of a touch by hand deflected to deny what some consider to have been an act of compassion?

Without looking back to the misery caused beyond the open understanding that all of our applications were to the past and what it should be, we continued to act in past reference like to the religious beliefs we held and still do! But have made to bring the ideals forward into the past by overstepping the additional progress of thought we had accumulated generationally. In the case of Judaism, Christianity and Islam those spans are of four, two and one thousand four hundred years long. Time differences whereby to each generation newer aspects to the style and means of life were exposed, not least scientifically, but in adding more and more to our attitudes to our religions and politics we began to think of old times!?

From which without examination no conclusions could ever have been drawn about what any future conduct might be. If changes to occur or not were tempered by new opinions but in old hands of reference to former ideals. Like for example, but I stress without connection, of how the first Jewish faith from the one start of immediacy to the chimes of imminent? When holding from instigation the covenant not to kill of self by self and being able and diminish the word of God if over any period of time, but did! Balanced to what by the line of time might have started to be the new Jewish faith, who again if, was to reaffirm the same standard if, not to kill of self by self, but did!

Arriving in check some fourteen hundred years ago when the standard was not changed but renewed, not against other religions, not even that from the simple fact of having held to the faith of Judaism for two thousand six hundred years expectantly. Nor to have held similar ideals in being then Roman Catholic but of the same monotheism for the past six hundred years running concurrent with Judaism. With both former and still active religions being subject to the same standards of life that we now actually lived in by the first time of Islam. Even that they had accumulated knowledge of learning on the same general scale but now more tied to what had become seats of learning. Is that the gathering of such knowledge was managed through those appointments from a section of community doing the same vital job as builders or cobblers or fishermen or farmers or administrators. But noticed instead as our priesthood who now controlled the flow of discovery by first passing new theory through religious acceptance?

Leaving in void the situation whereby in actuality in any case, Mohammed was met with how things really were with us all belonging to a violent world. To spite the vital link through Jesus back to Moses and the ideals from Abraham representatively, to spite the added years of our generational plight with the situation whereby although told by the word of God in the past, even if not to kill of self by self. Including progressive steps made on the civil to political side of our development where the option to kill of self by self was carried in law to be laid against wrong doers. Then the overall picture did include the matter to kill of self by self in law, so spoke society without the contradiction of what might have been the supposed case religiously past?

Leaving the situation of now even from a new religion to be, if directed, the option that we could take human life, was not to cede to what was allowed in law politically. But to be able to promote the new standard by any number of generationally progressive steps, plus cover of the correct type that in the name of religion it might be necessary to kill of self by self in being told to do so!?

Whatever was intended or included from the time of Mohammed was nothing more or less than the same expectation that had been promised twice before, on which occasions the expression was to the immediacy of the situation, which I quote again without being repetitive. But to avoid the setting of or for misaligned agreement from what I personally have interpreted on my own account from the making of monumental events, all of which had and is done without any form of connection save human passion in our abilities!

For Islam to stand on the solid base of now, then, in also to proclaim, but not as doctrine, that we might kill of self and self by means and for needs, is not a comparative narrative with others or other faiths or styles of politics. Save only to matters of the day to emphasise the brutal violence of which we then lived under? Defending, protecting or promoting any new concept however received had to be passed to us all, to us individually in the known or accepted verse of the day, being set in a violent mood but not to be overtly violent. Only to countermeasure even if being pre-emptive if only to apply the possible remedy of pure misery if our new and fair and open advances were rejected out of hand in order to continue in old and wasteful and unholy ways!?

If Islam was to be the new brush to sweep away bad habits, then the vocabulary was of the day the time the people and existing lifestyles. By implication the first trend was also of now, like from other religions but not from the immediacy of the imminent, save to some reasons of early death by those in we who would seek to follow the word in the promotion of this our new ideal! Especially knowing that the element to turn the other cheek if only to find martyrdom was met in challenge of how we could apply the same violence that was prone to be used against us! But from our view to promote our new blessed religion to be counterbalanced by the factor that in or from the need of promotion, we were to die, then the case for martyrdom was ready made, but not from previous static standards. Were we to be forthright in approach not to die, because in all cases there was too much to happily live for even in those violent times, which in cause is to the very matter of life, to live! But if death was to come before our allotted time, martyrdom in faith was to seem the best of all choices given? Not for its own sake, but in the cause of living to have protected the means of life's salvation in purpose on all counts and to all of specie, not known then, but known now!?

To step forward in time beyond imaginings and into the moment of now, we must openly protect the image of all histories without bias. But only to remember that the facts are in the eye of the beholder and are to be of the most dramatic of relevance to now, but not always of now? Meaning in open translation that the importance of all our pasts if by indication, are for us in relation to our pasts, but only in time? Now is of now in relation to what has been added to the key mark of all pasts religiously, not least in view through our first, second and third dimensional religions, all three served by time in line, linked by accord but born from the age in which they were conceived. Having already indicated my own opinion to state that when the concept of Judaism first saw light, its outcry was of now, even if linked that in effect God was first discovered? Therefore by any reasoning coming at a ready time of our accepted God culture with our multitude of other gods, any group to so state in claim that this one God was the one God, could hold self confessed comfort!?

For any group or tribe or clan to come to the one conclusion even if in excellence of mind, but error of thought? To have found in discovery the route of purpose, at that very time of realisation, all doors were to be closed but were not. But were not, is not an answer only a condition, which is a shadow in overlay upon us all!

For any group, even of now, this very time, the suggestion done for us to be in a position and change direction in realisation religiously is made outside of awareness to the effect of the above overlay. Especially when we use the best advantage of our experiences to follow our own heart, head and desire in reason, but by connection to the spirit of religion only!

## CHAPTER 175

Four thousand years ago the same was the case but to the fact we were in a far more infant state to the open concept of the future? For a group and in this case again I set my route to that time ago, or before, when we came to monotheism? By the same standards I have maintained what might be the case today, actions and reactions are held to be the same as then but with the one subtle difference whereby attitude had failed to intervene?

In our tribal communities which had already expanded to be city states, when ever competition was offered our first and natural tendency was to gather to those original ideals of belonging, certainly not to be a dated concept. Therefore if and when our God trust was found to have expanded to the feature of there being only one head God, realised from our tribal excellence of mind. Then however that was to be interpreted was done from our ability of thought that at the first time, marked by the gravity of what was to be the beginning of real understanding, we began to make more progress in all fields!

To have found the one God was taken in the obvious reflection of meaning, what we saw was of particular relevance to how we alone were looking? By the simple meaning of connection to any sought reflection, it was less than easy to associate what was of particular relevance to us alone!? Matched by our own sophistication whereby because we had found a key, the key, all needed was a record of how we had been chosen and why and when to fit the lock, and turn?

Keyed by our experiences within the void of the past, all we now needed to do in Gods time of creation, was set the standard which was done through our writing the old testament of the Bible. Then also of how the future if attended, was to use its own conduct and did! Even to the one and most obvious feature never seen, but when we came into the first understanding of our single God, we were unable then to comprehend what we had done by reaching the doors of purpose? Again, in looking what was seen was desire, instead of us making the last connection done in the abstract, we made our view to what was to the immediacy of thought by the imminent nature of how we saw all events! Still not yet realising, that in our study to a beginning, was also found the end leastways by how we ourselves celebrated the issue!?

By finding God and from the actual study of creation, we had come to an ending, but to only belong to the tribes with Jewish connection however that was spelt? They in we had come to the end, not first seen to be the beginning? In finding God or the study of God, there is no future, that is on gods own terms, because the future is now in the open nature of Eden, which by all covenants is only made one way from man to God, from man to God! Our acceptances were monumental those four thousand years ago to the full extent that the opening we had found to the future in heaven was to be heralded by God in our name. By the imminent unity we might expect for all that we had done in first realising God to our tribes, later to follow by Gods desire, we made the circumstances fit our understanding!?

From the wonder and joy of the immediacy of our expectations some four thousand years ago by history in time, although weary, we in the name of Judaism were just as resolute as ever. Save to the measure of our failings measured by our successes, but timeless only. What had been done by us to the importance of our original expectation, was to blend into the mix of society that also had improved into the wisdom of time?

Even when Jesus as a Jewish rabbi gave us a newer perspective to lessen the burden of our wait, his realisation was refocused to the imminent nature of Eden again! Jesus also took to the same habit of expectation that had been gathered out of context two thousand years earlier, from when we were to expect the actual coming of God? Be it to ratify our triumph of study at that time or not, but in some of us casting our mind to the new words of Jesus, we also were drawn without trickery to the idea that our time was also imminent?

If and so our time with God was to set the final situation from where we were taken in hand to our reward, then the path was to be through the expected messiah? But thought of to at first be relevant only to and for the Jewish tribes? Which was a feeling by some to have become more entrenched over the past two thousand years with all expectations having been put on hold from what had been at first the excitement of the imminent nature of things to be, felt daily, until perhaps now, when without answer the question was to re-ask!?

Carrying no inclination to attempt and change the past even by personal opinion, my own reference by connection with the Jewish faith from the original time near instigation, relates in the same way to the time of Jesus and Mohammed. Although much later it was by the distinction of both of those men to have taught the same message from very different perspectives. Leading for us all to question the actual structure of things as they are today, marked by events these sixty plus years of this last modern era?

From my own fixation on religious matters, of our prime concern, I have never set to deny the involvement of our political doctrine in relation to how we have advanced thus far. Although much of my incline has been through events primed in the Middle East, any and all cases I have pointed to, can be blanketed over the whole world, but to local areas where we still have similar social, moral, political and religious disputes!

Our Middle East is to me the prime example of where and from where we can focus all of our major differences in highlight of how our first tendency is to be reactive before being proactive, proactive, not, not pre-emptive? What is done here is to work from the reflection then to carry on in minds eye to the flaws only of what we had seen, and introduce any means of reaction to what we thought we saw? For us to now make the standard to be of religious significance more than the limp and weakened hand of politics. My tone is to do what all religions should better do and look to the future before the past to bear in mind that all, all, all religions are only inclusive of we by measure of our individual lifecycles! Be it by fifty plus or seventy plus years long!

Making the case in full, that it will be up to our religious leaders in us all to accommodate the same measure of full understanding, but still to be maintained by our own different religious opinions. How that is best to be done stems from the mere attitude we can all generate from the pure open generosity I expect those who would stand and represent us in care, do! Not to be confused in how we are to be represented politically for the people and by the people or by the people and for the people. Religions carry a far greater responsibility to the full extent of what purpose is!

By it, no case is made that we might own one or two or three or four houses above need. Nor to be of a society that holds part of a country within a country as our own land. For which in the political battles of past times commercially, peoples had died for King or politics and country if to save another individuals claim, which has now become an item in the form of an engine to drive the other and to some the primary aspect of our controls in politics. My own term in respect of what I expect from our religions is not a weight for weight comparison to an effect that really is of less concern than we know. Because all religions have been guided to carry the same expectation?

What I do expect, without the need to raise my voice in the negative, is for our true religious leader make the full declaration for all people and in the name of all peoples. What is expected in and from cause relative to the one matter of fact that we are not chosen but self selected to be different than and from all other creatures upon this earth, not least because we think we are, not least that we think!?

There task in our name is to proceed in much the same fashion we have always thought of in Gods name or to Allah or to the Great Spirit or Buddha's vision or by Hindu acclaim. Our religious leaders in the best of good heart and of personal selection have always done as much as they in we thought was necessary and in cause to our linked benefit, but by religious selection! Now, by the call, armed as we all are with the three taken and one active prospect of the Fifth Dimension in overlay, we can be spoke of in the one language of unity, even that we may all travel our own paths but guided by our talent to the real ending by the full unity through 'Natural Body Death'!

My own standard to be ours has always been set to what we already know and to who we accept in being? All that is to be added or subtracted from style has in open contest been generated these past sixty plus years of this last modern era. From which if I have suggested we are to make the first cut then the incision is made forward looking which brings us to the important age of purpose as both the cause and reason to our drive however previously generated. But with Shanidar never to be left in stillness, like when found counted less than one hundred years ago just outside the last era I refer to. Bringing the pointed importance of this last modern era into the perspective, that until now, we had not had all the ingredients to complete the experiments we need know of for fulfilment!

No matter how much I suggest we look forward and hold the nature of our histories without change or the requirement of change. Can only be done if we set the plan how it should have developed. Bringing to sharp focus again what I personally call the route of what could be our biggest fall yet, by everybody applying the coat of ignorance to smother the need to understand all objectives. By all of the same coded messages I might have left not to stir animosity, the time of our full modernity has now to be met on some of the two counts I had referred to from time to time, but now left to the full study of abstract reasoning?

To look, even by reflection at some of our imams who operate to the side of our religions we know to be Islam, who in duty look to the care of some of our young to assist in study of this my defined third dimension religion. Will be shown how we are all entitled to care by the utmost sincerity of self by self over all matters of faith if they are due in cause to this or that being our faith! If not directly connected, but deemed so by other means, I have looked closely at the full intentions of our politics who in a blow for blow exchange will be seen all of our political shortcomings in relation to our religions? Because in the same style, I have also overlaid all political aspects with the same three taken and one active prospect of the Fifth Dimension!

Our religions are, and I do submit because of that very real occasion, it is now for our religious leaders to show full acceptance of that fact, without the need of a blow by blow account of our religious relationships. Of which there is no need because all have and do acclaim to the same outcome to our real beginning from the end? That I might reflect to that complexity in term through the event of 'Natural Body Death', refers to the mighty importance of that theory? Because it is automatically backed up by all religions throughout the world, because all show the prospect of fruition through the same element of our beliefs in the supremacy of God! If he is found to be one, two, three or four headed, all will be in likeness to be black, brown, yellow or white by character. Or also in the image of change to the only description to suit whatever specie in the time of evolution, might change that study for creation! God might have a crocodiles head, but in evolution, was he asked!?

For us to make any near judgements of the above, automatically instigate a series of changes completely uncontrolled to have happened in the course of full time and latterly in cause from creation! Although we might find various degrees of contentment in what we now know and in leading to what is to be known religiously. For all purpose required or not or seen to be or not, of the few steps we must take, have to relate into the realm of religious expectations, but from and by all of our most modern examples! If positive religion starts from about four thousand years to the first time we were able and show that we had truly broken from the restraints of evolution. Then we were carried to assume that from that find all else was to be revealed.

By fair return again, for me to have leapt past how Judaism and Roman Catholicism into Christianity, were first connected to the imminent, but were not! Has turned to be a matter of contention if first or secondly shown how Islam had not travelled that path, but was unfortunate enough to set similar errors by not at first taking that stand, which may have come to be a haunting over these past sixty plus years of our last modern era.

Additional to the main differences I had declared to be upon Islam other than previous monotheistic religions from area, that through matters of need it was possible to kill of self by self at first to promote the new religion. Seconded by the same division to also kill of self by self to defend and protect it under challenge is a viable rule. When added to the major difference between the first two in my first and second dimension religions and the third of Islam being my third dimension religion, by the situation that Mohammed was fully aware of the rolling future expressed generationally. Allows that thinking on two terms, one generated by the natural interactive violence of now, then ago, with the other situation with people now living under the effect of several and several extra generations into the future. Was accepted that as we had had our generational pasts, that we knew of in the most obvious of ways by our direct ancestors. It was obviously expected that much in the name of the future would continue in like form, but by what interpretation?

Even though Mohammed led us in direction by the sheer emotion in what he recorded for us, his fine example was to follow to the same formula that had affected Jesus and Moses before. All three of our great prophets along with many, many others, even if of less standing, some true some false, our story in all three cases after Moses, Jesus and Mohammed, was left in our own hands even if read from slightly different perspectives? As I had said without the blow by blow effect by different reasoning, by the same tone that we, we are all here now, bodes well for our three monotheistic and all religions, because it is we who have carried them all, Eastern and Western!

Making the best case ever for our generational support, save that by any tone we might regress to the call for deeds and acts belonging to past times! If by lead and I call it my lead so as not to point in blame, only study? In statement time and time again I had led to the one and extra means only, why Islam has a different modern outlook than other religions of like formation. Has been to illustrate the error of our way generationally and not religiously?

By deliberate take that Islam by my own referral allowed that we could kill of self by self while at the same time in balance that Judaism and Christianity said no, but did? Is not an accusation only a reference to what was actually happening or appeared to happen under clause while even now the, our first two dimensional religions are charged by conscience through the Fifth Dimension to reassess there own ideals, so has Islam been. But from a different angle to understand from, and at the open times of natural reaction to the violence at instigation, our generational additions have done enough to limit by expression that deed is no longer valid? In such a way to challenge in conscience our former and existing previous monotheistic religions in not having to kill of self by self bringing us all to the point of this last modern era?

Islam can claim no immunity to be above or better than by any standards of any other religion, even by age or its age! Because all and all religions are generational, all and all religions are generational, to make past efforts be able and improve standards. Not least, because all religions eventually come to the one importance of what fifty plus and or seventy plus years on this earth really means through 'Natural Body death?

Without comparison and by pivotal selection, for me to have offered the study of how some imams behave, has been done for our own merit even if a different case is shown by other imams? By type and without separation, without distinction, without the merit of setting the situation in balance, for me to point in direction of what some imams might be responsible for is warranted in the name of Islam. Because she in religion as our last major monotheistic standard, had been the only one to address the option of if it was proper to kill of self by self in overrule to the climbing importance of our political and civil motivation.

Beyond the aspect of choice when originally in the confined station of the Middle East as we know it today, when Islam first set foot, its mandate was to include all that was thought of! In also being generational by first sight to suspect what might occur was actually false and unpredictable for the two resounding reasons. One, we did not then know the full extent of the world in terms of area and population, nor could we accurately determine the extent of the differences our later generations would have to contend with by being in very, very different worlds? Most by the effect of what materialism brought to some people being far greater in what was due in heaven or paradise?

All of which in contradiction had set the marker whereby all events had been poured into the funnel at the top to gush out at the mouth of reaction measured religiously. I make the obvious distinction that the first importance must always be to our religions and of our religious perspectives. Because they only have had the inclination of the future by the real terms of creation and not the falsehood of what politics has always offered to the importance of procreation only fitting more to old evolutionary challenges in service. Making past errors that what we had was all that we got?

What I personally contend in the name if not to the false faith of some imams is to there conduct, but only measured religiously, never, never politically. Like of how, by ranting in name that Islam is over all else, and to protect or promote or defend the faith, she as we can take in name the status of martyrdom? By deep reflection and in better study by referral, all imams ever and without claim can only offer deed of action if posted through and by our one active prospect in 'One for the Whole', that is now!

Judaism and Christianity to me personally at this precise second in time when read now and more so before your very eyes have no comparison to Islam? Because in knowledge they at instigation were less informed as we are now? They in our name by all the best reasoning assumed the best for us. Alas without failure they have to purge their and our consciences, also without cost. Because the past was then and this is now! By the same token there is no charge against Islam except our own conscience on the same matter, but for different reasons. Not least the simplest in all cases, whereby we are all at the edge of the future and always have been, but now added by facing in the right direction!

By my own examination with good heart and good feelings, if I criticize how I have determined how some imams might behave or have behaved. Is done in equal clarity to all without bias, simply as I am not qualified as our religious leaders might be to balance the scales except only from my good fortune to have seen them!

I have personally given to discredit some imams for what is offered in pure speculation, but without fear or favour I can and will state that intention is culpable to the act even if encouraged to be done by others? Beyond the scope of how civil responsibility might have to view such matters I make my own case within all the aspects of all political law. Islam along with all other religions, are far and away above all civil and political laws, which stands in case if written in stone. Even without proof but tied to the time of Shanidar when the whole episode was fully religious but could only be done under the one name of civil control. Because that was under the only circumstance whereby such an act could have taken place, with time so vital to community? Whereby time had to be taken from the important acts of hunting or gathering foodstuff in order that we might ponder in minds eye the further fate of a dead comrade of who we now bury!?

Islam by the same relief had drawn the circle at point from the mid nineteen sixties by crucial times of religious conflict between Arab and Jew, democracy and communism. Mixed and spilled through the Iraq/Iran war of global importance, but all above happened in the Middle East from where the circle is not yet complete, but continues to spin? Like by the smallest of issues if and so a new division of death to self by self was first moved in distinction to the eclipse of what dying for a cause meant? Then in the first and final twist of or for what the design of tomorrow was supposed to mean these past sixty plus years, even if locked by the United Nations politically? When the call was to be made, nothing was there, because we had not realised until now what was needed to play the hand. Not least until now, when we have the three taken and one active prospect of the Fifth Dimension which imams did not have!

Therefore there only answer in discontent was for some to evoke old habits to the name of killing of self by self to other ends in making the ends justify the means. Unfortunately in tracing old ideals some fourteen hundred years of age, the matter of fact way to continue in the status quo denies any connection with fourteen hundred years of extra development socially. Which by code also includes in equal manner the matter of change predicted to the fact that there could be any number of extra generations to contend with over that extra period of time? Because Islam unlike Judaism and Christianity did not consider immediacy in the same way earlier religions had! Therefore in order to allow the effect of generations work, it was necessary to regress to defend, protect or promote this new religion even if required by bringing death of self upon self by returning to our generational pasts, but now unconnected!

Regression was always to mean that if it was found necessary to kill of self by self to promote this religion, then for that hand to fall itself in doing so, just rewards were already in waiting by the second feature of what martyrdom was to offer! What has not been fully realised by some imams is that our changing times already directed to be events by Mohammed did not compromise Islam in any way. By laying to the future his disposition was to tell us of it in preparation, but by using the tools of the day to insure there would be a future? He knew us well and of our fickle character of our fears held in doubt and of our strength of resolve if properly supported. Martyrdom now, Mohammed would shun it!

All of which in understanding became our charge, when by Gods own good grace Mohammed died in theory by 'Natural Body Death', if not set as a sign then the event can be and is taken by my self to show in warrant how it should be for all of us. Not least because of the future we were foretold of by inclination only so as not to turn us to the apathy of fate, come what may. If only to endorse that fact by offering those of us who might fall in battle while promoting our new standard by flag and tenet. Our reward in compensation to be martyrs was given because we would and had missed the full importance of what the future was ultimately meant to bring over any time span, even our standard longevity of perhaps fifty plus years ranging to seventy plus years?

Islam by sitting firm in the name of fixed obligations for all who would come to the new faith had missed one slight aspect realised, but not readily seen or accepted? What ever rule or obligation we fallow has to measure for whom the law is made and the difference if applied by other criteria? From which without contest I on my own account and in my own view offer relief, even if by reflection of the two codes of Sunni or Shiah! No balance in this referral is given to any other religion from before of since the time of Mohammed!

His legacy although better addressed than had been done by the prophets of Moses and Jesus, was done in the same heart to the same wonderful and singular specie we are, even that all three great prophets each tied to the one God concept, they at the time of instigation thrice over. Could show no sign of awareness to us all or the single connection that in place there might, might only be other means whereby God was of a particular importance to us? But not that we were made in his image and likeness, because once started the picture was to remain of its first painting!

Bringing us from what might have been our original excellence of mind but error of thought, from then until now to this last modern era. Where by in wholly religious terms, our open situation has remained to kill of self by self heeded in direction from some imams if from our last monotheistic religion! But not so claimed in Gods name by Allah, only in mans name in error of thought without excellence of mind! Because instead of opening our vision to the good of our young futuristically, some of our imams have sought the regressive steps to honour the faith and religion of Islam by the name of politics, using political murder dressed in the coat of martyrdom!

Without count, past the hope and expectation to die within the bounds of 'Natural Body Death', carried beyond the range of issues whereby we might die in the cause and determinations of political desires only. By observation that we do die by all political means only, is a fact of life saddened by the real means that we have disguised the matter in more crucial cases to belong to the vagaries of our religious determinations? Why we only die in the name of politics is an open book and is not supposed to change by the introduction in overlay of our three taken and one active prospect of the Fifth Dimension unless we become aware that we do not kill of self by self religiously from now!?

## CHAPTER 176

My first case for murder of self by self on religious terms is in operation by result but not cause. No religion of this day and age, offer to kill of self by self, save we individually and without any cause, because we are bound in tenet and covenant not to kill of self ever in any form of suicide? From so, that we do kill is also carried more from the base character of some in one faith more than others, but not to be counted in modernity. Simply because in time we had not seen the picture of disguise that had been laid down at or from or by the time of instigation of when Islam was brought to us all. Which unfortunately has been a fixture that I have been able and determine personally in representation of how some of us in number have also misinterpreted the same events from balanced observations and taken political leads to murder instead of religious ones only!?

My finger of charge although pointed in the direction of some imams, is representative of us all, but to the good fortune by the matter of change we can all interact, because of the overlay upon all religions with our Fifth Dimension fixtures? From which from now we all can draw common ground at least to see the importance of our religious perspective over and above the best of our political determinations or what ever that means?

I have seen out route and further stress without obligation it is set in the future only, because our past clouds many issues emotionally if not factually, which can be sidestepped because now we can better use the advantage of what we have gained generationally. Having grown in stature, because now the hurt of the past has been by reflection to show a brighter future for all of us by what had been the full intention politically since nineteen forty five by this last modern era?

For me to have turned all concerns on what appears to be the single issue of how some of our imams behave or have been given credit or condemnation for. Stands because of how they alone in some cases have taken judgements far above their station or for the religion they represent in us? To speak by religious determination only, there is never a case in this last modern era if and when we can ever offer to kill of self by self religiously. In this last modern era only to be marked by how when under extreme examples created by political ineptitude, we allowed the religious slaughter of Muslim to Hindu and reversed by the condition of partition in the sub continent of India. Although politically driven and of time, it was a sad, sad event, because it is to always remain unconnected politically, only to have been left emotionally and in blame religiously!

All because our change of gear into full modernity had not yet been counted by the codes we profess to use for it, if from the same matter of new international law? If coded by the United Nations by name, when a feature of indifference was shown not to be included under the wing of religious tolerance, but was clear to have been led by the religious intervention one way or the other. How was that or those incidences to be read other than by political management only so not to upset religious sensitivity!

To make the case that such an action from religious fervour, was not religious but political and as so was against the will or intention of the people by the people. It as an act from religious grounds was to be cast politically as an act of terrorism, to be a word meaning we do not understand? We do not understand because from all experiences we have tried to box the ideals of religious tolerance fixed and quiet. No discontent or reaction or discord or objection or complaint, with the only word to fit the query of question having been taken to be the last word of direct opposition. Which by the tone of modernity when classifying religions to be tolerant in all cases leaves the only path of question to be political?

Therefore when said that some of our imams give religious direction in the name of martyrdom to be suicide and murder, no political case is to be made against them if that is so! Unless by further definition they in kind suggest to the young of our society that it is their right by our fourteen hundred year old interpretation to kill of self by self? When all of modernity denies the reflective use of old, old habits that no longer have a bearing on any aspect of modern life, save to be political murder? Political murder, therefore such horrendous crimes committed in the name of martyrdom are subject to our civil laws. Having no defence to what some deranged imams might preach against all the full ideals of what Islam is about, might be about? Never to show awareness that also Islam is directed in open unity of faith following Jesus, Moses and Abraham and possibly the core rules in law from the early religions from the house of Egypt!

By pointed direction, but without looking and not wishing to give the comfort of personality to some imams who do not serve our valour, just like I have never given in name the leaders of our worst forms of fascism or from our communist world. Peoples who would allow the murder in death of millions upon millions of there own ranks politically just to make a point. I do not name individual imams who are the same if they wrongly code martyrdom to be an excuse for murder, perpetrated by the act of suicide over the act of self sacrifice which can never equate our short term of real life!

Imams have only one charge in our name, to cosset our young in the open task of learning holy and religious matters on the right path with the ultimate future in mind? All times are now past when and if any of our young might kill of self by self religiously. There is no tenet, law, covenant or control from which this act of murder might take place. All who would lay claim from the script to our faith of Islam have had scales upon their eyes if read so? Islam in its own law is the only progressive religion among our three monotheistic religions from the Middle East and anywhere in the world, which is a great bounty to us all. But with the heavy burden of being to all through our generational links, making us at times to look back in anger when our first direction is forward and has always been so. Although the feature of family and generation is of the utmost importance socially, morally and in conscience, we have to set in mind that when spoke of, we meant for all religions. From Moses, Jesus and Mohammed, our first point of connection was by self of self and through self in the order of personal conduct.

Our first three monotheistic religions although set in different times when our personable habits in connection to each other socially had developed only to the time we were in. All three gave the same standard of self by show, not to have realised the event of that difference socially, because in subtle stain we had developed slightly different by character not least by the advent of thousands or hundreds of years between all three religions?

By the other case of awareness we need to hold even without an historical connection, is not of the past by how we behaved under civil or political control but openly by now. Of which in the case of some imams without distinction who have not been able and show our awareness to what Moses, Jesus and if finally Mohammed had done in our name!

Which is to continually reaffirm our connection on the one plain religiously to the image of God, and to that image we are perceived to be as one. Because it is from us alone that all liberty has been given to all life by including the matter of time to have been set for all, but by God and not the haphazard experiment of what the unknown element of science was to reveal? Which in like form is our own creation from creation, especially with no sign of contradiction from anywhere in the whole universe!?

If and the case is only laid like in this now, that the future is paramount but only in terms of us individually. Then by all counts that is what we must look to by extension of these past sixty plus years laid down in time from wars end in nineteen forty five. My own standard in thought, has revolved around what some imams have done or said what they might encourage others to do, simply so that we can rationalise the best and most constructive steps to be taken for expected fruition. For me to criticise individuals is not the case, what I have done is criticise the station of a particular group in our last monotheistic religion, but for us all?

Apart to other matters, at the same time I have balanced my equation by what I have not said in a blow for blow account of other religions. Save to the open matter of how we might profess not to kill of self by self but do? In open show I have laid good ground to always question the integrity of some in our priesthoods, but never to be comparative. I have challenged no religion save any who would kill of self by self to the terms once used in the new world before we had gathered cause and reason relative to the connection to our human form?

My personal focus of how I see the behaviour of some imams falls to a question of doubt to there integrity. Because if and from any connection to support in suggestion or of promotion to the ideas and or the ideals that modern martyrdom has a place in our society, they have failed from core to justify the thought in any way!

Yet martyrdom is a permanent feature to hold in the utmost of sincerity, therefore we must save it from being hijacked in use or of misuse to promote only self importance, which is now unvalued if it contravenes the principle of 'Natural Body Death'. By case, although suicide has come to have been removed from some statutes to be no longer a crime to self, our review of that situation will have to be held in abeyance. Until we lay better foundations for where the United Nations can better represent us?

By the same sphere, if by misunderstood review, we have come to the situation by what some people class to be religious terms, where an individual, a group or a number of clerics in the name of priest or imam, were to stimulate actions to take steps and become martyrs in faith. I can only state that the term can only be influenced from the outside. It is not graced by being an open act of suicide in any name but murder! If used by exclusive command through the religion of Islam directed or not. No sanctity can be given by any other person outside of involvement to the act, like for instance if some imams were to encourage some of our young to so behave!

Because of the one open fact that Mohammed was not a martyr. His own fixation was that he was openly prepared to become a martyr in faith but by the choice of another's arrow or sword or spear when in battle. His open regret in not becoming a martyr in faith was a lament to those of us who were also prepared to die and did while helping to set up the last monotheistic faith! His foresight of direction was how he gave insight to the expected longevity of the faith to grow by time and from parent to child. Also, from his generational cover of how science was to bear on us alone in specie, was another message to the aspect of change. Allowing us in religion take that time factor in our stride, further allowing us hold to the less importance of our supposed civil and or political obligation to carry on as normal. Because in time, this new cleansing religion was for the future also, to be proved in time by the way it was expected to come to all people?

What was unfortunately missed by Mohammed, which had to be, even that in plan better considerations had been given by the burden of Islam, ratifying we were to self generate our religious acceptances by conduct, thus showing understanding and commitment, but as always by personality. Which then ago was from people whose daily life revolved in shadow of the violence of the time in terms to the overall value of life, which since then has been mistakenly brought forward by time and to be included in modern thinking but tied to older habits? Then further to the error of undervaluing life by the concept of not being of Islam for which there were several stories or considerations, then far worse if being of Islam and even if by the hand of another Muslim one was put to death by expediency, all was sanctified in being martyrs!

Fourteen hundred years ago much of our future had been set by the very good standards of most world religions, but all were constantly locked into the unknown. Although I say it, my words have been far better introduced beforehand by many, many other genuine people. Not great prophets or statesmen but people who have sat and pondered by time, gathered under all extraordinary of circumstances! Like when travelling to work, at work, while shopping, minding the children, preparing for battle, sitting on death row or maybe just having a quiet moment with a relaxing drink. Only to find all that we had thought of had been done a hundred times before, but then to realise that is how we progressed so far along the wonderful mystery of life?

But in that great advance to endorse our worthy separation from the restraints of evolution, we had got far too many things wrong. Not always in mistakes, because by good habit we played the hand we had been dealt. Unfortunately in being this fine, fine specie we are having developed to the same fine and good people we each are. We can rebuild first in expectation by setting the foundations to include the vast amount of others who do not celebrate the dawning of the day. Being constantly challenged by the daily rigors of finding food, or warmth or comfort or care, even of love, those things in some cases surround us all, but we have been to jaded to reach and know how to improve!

By knowing of the nature of our new foundation set by the prospects of the Fifth Dimension in overlay of our politics and religions, is another start to the end, this time even if to take up the suggestions that have been made time and time again, our success is guaranteed. Because never before have we been able and think in the true terms that we are this single specie we think we aren't?

'Out of Africa' has been our door and although we know it and have for years and years since Darwin's publication. No real bloom of flower to that seed has yet been seen, at first held back by our political ranting and raving. Which have been able and generate there own misery before any attempt at calm and unfortunately continue to do so? All generally in the name of commodity through the means of supply, demand and profit, even if encouraged by maintaining the matter of strife. From which by clever and devious means some banners of reaction to the matter of political strife are heralded to be of religious tones?

Making the first connection from what had been started half measured in nineteen forty five, carried on by our steps taken syndrome, thus being able and close needed response if any objections were made from any quarter since? Being prepared always and use military political force to settle all disputes by no other satisfaction than to protect market forces. Not least because by political terms, strife in conflict and war in part generated the idea to the spirit of economic strength, which when fought by internal reference has always been won by the foremost tentacle of political ambitions?

While at the same period of time, allowing the movement of religion have political voice, if suited to first world prominence and balanced unseen, yet again curtailing all religious tones if from second or third world sources even to be communistic!? Which again had found religious opinion from some communities even if associated to religious groups prepared to challenge the outcome of tenet if started politically at the Second World War ending? If any voice even to be heard through the United Nations was raised in opinion other than of the status quo, in posting reaction in the negative, the answer was given in silence never to be listened to what had been said. Leaving what alternative, if the voices of objection to be raised had there own case to be heard? Not in stated action but only by the same creed of what religious tolerance was supposed to mean?

There is no way ever that we can wipe the slate clean to cover the extent of how we have been blind to the obvious notion of equal living. Of all the ways tried throughout the world, even by the good heart of many small religious communities, but with our standard of political reasoning born from and in compliance to our steps taken syndrome! Because all action from that or those circumstances have flowed and ebbed to be political by outcome, which is not always to fail, although we are the same one group by specie, our edges have been roughened by no account reckoning of this fact?

What has not been available until now in this recent last modern era, has been the matter of direction, save what is done leaving the void of what to do now? But measured by the full acceptances that although we are one, we have all followed natural changes by the opened circumstances of geography, climate, diet and other reasons, but still to be one of specie, but only now realised again!

By having the one start point to the avenue of change needed by world ideals more so brought about because of the changing political standards at the end of world war two. In stating what was to be the intention of our route to stabilise those matters, by following into the trap of appeasement, the angle we first set to take was indeed half measured. Also the time without being able and rationalise that we moved only by those terms, when and why we took to the one standard of religious tolerance was a steps taken solution!

Although when in the first period of time in this last modern era, we tried and settle the matter of nationality and statehood, little was exact, which accounted for much, but not enough to seriously thwart what was supposedly the aim of good intentions. Even that in mix there had been political wars, battles and murder! Political to religious wars battles and murder, with religious political, to irreligious political wars battles and murder, only in this short period of time, how much was to be realised as right, just and fair and for all people?

Our one, two and three blind spots were that we hailed all by political standards from the supposed benefit of the political outfall of materialism. We conducted all political motivation to the one avenue of profit as a core to the full engine of how politics was to work on a one to one human basis. Second, not least by the introduction of statute, although we were all the same by civil liberties and human rights, we were also first, second or third world being the only way to diminish the first element of equality? Our third and what seems to have been written in stone has been of our political attitude to the set of religious tolerance, to be a block to further open and needed improvement!

All of which because of modus operandi, had created from some of us more and more reactive responses made by our own view of what the reflection really was, but had been crushed by our own and other reaction that might was always right, supported by the economics of politics! No matter of query or disagreement or dissent was heeded; if to challenge the great and fine standard that we have come to know in government, what voice could be used? Merely by establishment, establishment was made and from that the office of our steps taken syndrome was set.

To any of the above standards so cast, no reasonable suggestion of change was ever considered, because by style on offer. We, if through the matter of the United Nations, had cleansed our conscientious duty through the one office that was to be the conscience of politics. Not least if all times it could turn to close other matters with the standard of religious tolerance, at the same time addressing in completion, considered order by way of conduct in the future by all, even to include what would or had occurred in the Middle East! Because if known from the end of mandate there, some religions tolerance had taken new quarters, how better could that standard been reset?

If at instigation politically, supported in the ideals of democracy and on western bloc culture intention, when the new state of Israel emerged from the midst of blood and accusation. What had been achieved in best order was a step to be taken for one and all to be now promised nations? That by the incidence of coincidence she as a new nation in old lands was only supported politically if by standard from Americanism and Europeanism. What was done was thought of what was to be done else where throughout the world, even if standards were set first, second or third world!

When and even if not prompted, Israel stated in point she was to proclaim the means of religious fulfilment in care of the charges of the changes to have just happened, if then the act or acts were balanced nationally, they could not be internationally, because of our standards of religious tolerance to have been promoted politically!

Without cause there is no reason that the situation of today should not remain exactly as it is now, save the honest appraisal of minor adjustment and in cause and in concern and in honesty. What must change and is delighted to do so is the open acknowledgement to our first singularity of specie in being one, to what we all can only hope, is to do enough to raise Gods interest in us as specie by creation. Instead of for the measure of God continue in evolution for the next eight hundred million years, which is solely of Gods own time scale until the next correct voice in wanting is heard to hope that they are the ones to find Gods interest in the name of creation!

From the mist of all of our histories in association to how our many religions have been formed progressively. By court of reflection without the use of mirrors, without standardised accusation, without religious tolerance, but to look from face to face in Jew, Christian and Muslim, no difference can bee seen only if spoken! Because of the very matter in how we had all tried in the one way only to reach our known God, always acclaimed by Moses being Jewish with the greatest of respect. Always acclaimed by Jesus being Christian with the greatest of respect, then also always acclaimed by Mohammed to Islam with the greatest of respect!

Our pasts had different start times religiously, but never to forget Shanidar. We now understand the biggest reason of how and why we force the difference of our past through all the avenues of time. Even when we openly accept the same, totally the same expectation by and through the new common expectation in 'Natural Body Death', by terms of delivery in hope so that we can be part in the wheel of purpose!

By code, but alas somewhat unprepared at the time of our best steps taken forward, we have only found the path of discontent, not least because we have looked at the same, the same and the same issue from the one angle three different ways! Until the extent, that instead of what should have been our naturally selected steps to be taken to understand all. From the misfortune by political leads we had been wedged further apart, not least again by the force of religious tolerance before religious forbearance, which gives of the same light. But is far more in line to the dignity of dispute allowing that by all circumstances and conditions, we can stand our corners without the exchange of blows!

My better judgement to follow, in all cases puts our religious perception to the foremost, any standards set when followed are to be taken by the same caution I have laid out in these pages, not as my work in the way that Darwin's The Origin of Specie was his. But connected by the importance of what he started and left for us all in the best of good faith. By his work we were brought into the beginning of this last modern era because more than any other work before or since the vital points he raised had been enough for us to seriously be challenged in the rational in respect of life? Particularly in relationship with full regard to the question of what was religion in service and evolution in cause? His work without intention was a reflection of how far as specie we had travelled from evolution into creation!?

# CHAPTER 177

For us to take heed of our religious leaders is a means to self express all council, yet maintain the structure of society by political understanding also. To understand and follow that principle is to quote the full extent of our free will by listening in the first place and then following by our own judgement. I have set this realisation to be of now in this last modern era, because at no other time before did we have the circumstances or knowledge to proceed in that manner. Nor did we have the same index of our three taken and one active prospects of the Fifth Dimension in equal overlay upon all our religious and political propositions?

For the best code of balance I will use by example of what is to be our biggest task of reconstruction, stands to what our religious codes have meant to us or have been made to mean to us. I make that distinction because where ever we look in our civil calendar all to be seen is the matter of chaos driven by the outcome of sheer greed. Simply because so far, we have not been able and determine what or how to manage the recipe without note to the stimulation of supposed material reward counting as our only means of success politically. Unfortunately based from how in all cases, all political systems have only been able and operate half measured in response to any direction or proposed direction. Politics remains our errant child, but soon to be redirected into her teens by and in comparison to how our religious leaders meet the challenge to decipher the standing of our three taken and one active prospect of the Fifth Dimension!

Without the duty of conscience, because as an individual I have no mandate of judgement to what I have noticed to be the blade that has cut deep our original sensitivity to the unity of understanding or even the concept of unity. Politics is soon to manage her own affairs in a fitting manner to enhance what should have been started full measured in nineteen forty five!

My personal standard belonging to now, created these recent sixty to seventy past months, is to doubt nothing, but question everything to destruction if necessary, so that we might purge our ailments at the littlest cost. Like the overhang of religious tolerance to be upon all, but for me on our three monotheistic religions in particular. To have been laid side by side and filed by duty of our steps taken syndrome. For me to have awakened my own specific interest to the matter of full concern of how in place our religions might be used by the unscrupulous the devious and in cases the unconnected of mind. Has been put into the funnel of thought and been poured from the spout in the misused name of martyrdom.

From the nature of that sad and unmeasured fact, whereby people like you and I would seek only to murder, murder self by self in no name of courage or duty, only in voice to the ends of self gratification. Where I have been able and notice the cause to the true obscenity of ritual murder belongs only to how we were believed to have behaved in the new world up to some four hundred years ago. Other cases of ritual murder of similar intensity in all modern times have been conducted from the impetus of greed and wanting, more created in the misuse of how some of our political systems were made work. By centring the intention of interest away from the individual or society into the dark recesses of how any particular political motive had been misinterpreted by the stained rules of logic?

In general terms the only other case of reference of similar standard I mention is to those of us who are completely deranged of mind and follow to be serial murderers by the lack of any personal ability to rationalise the sophistication of the point in object of what is meant by anything! Being only mere mortals like us all, no case could ever have been made that we might kill of self by self if our plan had started outside the need to find creation!? Whereby a single human birth under conditions outside of the restraints of evolution is here and now to borrow in time from one to one and each other the intention of what life is for!? Then to be able and lead every minute of that merited birth into the realm of 'Natural Body Death', having been able at the very least hear of all or some of the options of what our individual creation is!

Without force, without direct connection to how we will operate by including the three taken and one active prospect of the Fifth Dimension. By example, so far I had centred to the one feature of how some of our imams, are to be accountable to have incited some of our young become honourable martyrs in the name of Islam, but under what format? No lack of judgement is offered here because the ground on which I walk trembles at the extent of how we have falsified our accounts; making only political points! By this last modern era no particular point of reference can be found to cite in modern times of how suicide bombing or for some bombers, became to be named as martyrs unnoticed? Without threading the mix but to maintain our best chance of proper reversal to such a trend on moral grounds, a look by reflection is needed, if only to see the obvious?

Without qualification to suggest the habit of religious significance has been accredited to have re-emerged in these modern times by events in the Iran/Iraq war which occurred in the last quarter of the last century. If reported that an act done by a young Iranian soldier who gave his own life to destroy an enemy tank and its crew or others also. Then the case ends there in all of its own fine but tragic consequence. Not least because the worse tragedy of that war, was imbedded to be wholly political or run from ideals outside of the fact that the antagonists were both of the Islamic faith. Without comparison and never to side step the main issue, the only reality to have emerged from that sad conflict is that at last in proof, we have a standard! That we no longer can fight religious war because four hundred years ago all of our sad and bad earlier decisions to the aim of human sacrifice have now been wasted!

What we have confused with the need of human death by religious tones has come from the misguided way we had entered into the matter of change post war by delivery of our half measured program, compounded with the error of statutory religious tolerance. Which by case and example have been exacerbated by political events in the Middle East since then, but carried on the one level of political disharmony?

My own Middle East reference is to stand clear on a world wide basis because throughout the world all acts of gratuitous violence to the outcome of murder are political, only political, and only motivated from the complete misunderstanding of what some of our political motivations have been? My personal focus to the Middle East in general and of some imams in particular gives no quarter on the matter of murder by self upon self under any circumstances. Except in the one political theme of crime being committed against the safety of an individual or a member of their family! Then to defend and protect stands! By the same token further exception is given to some of us who would act to prevent a crime leading to what might be murder and all of its avenues even if having to kill of self! But taken seriously in law, by the even and moral law of natural justice, beyond written law or the intention of unwritten law!

Above is the image even if by reflection of what text is and or can say in the void to understanding! If by comparison what is above can be used against the standards and standing of most unscientific text. Being a matter governed by the rules of experiment, where all the first standards have been set by rules in compliance with what have become fixed, balanced by what we have accumulated in and to the standards of all science balanced again by formula. Balanced in the temerity of how the fourth dimension can address its own standards to be better fixed than the ideals of unscientific text like for instance from what has been written in the name of many different religions! Because of how science delivers under the rules of supply while at the same time religions are to the future only?

From the standard of assumption that some of our imams have allowed religious text by understanding flow to be able and defend the faith of Islam. Then separate, but for the same reason explain that the faith also is to be protected and promoted by the same standards? Where and if such measures are permitted to be taken, do or can they be made and relate to the misguided uses mostly unseen, but to political reasoning? When at first if the measure of martyrdom was solely to the idea and ideals of only religious determinations and are still, how can they be turned to call upon there need of martyrdom for political balance? When politics cannot now be measured on the same scale we use for our religions, as we use for our religions from now on having to employ the three taken and one active prospect delivered from the Fifth Dimension overlay!

Unfortunately by reflective challenge much of the collective damage done by the class of imam some of us are, not to open our eyes and ears to the standard of progress indicated at by and from Mohammed in open acknowledgement to the business of all of our futures! By only looking over our shoulders some to have encouraged our young to the glory of martyrdom, have failed us all miserably, to the serious cost of turning better eyes and better harmony from our own, albeit to be acknowledged by later reactions if only from the reflection and not the primary image!

My connection with the lacking of some of our imams in relation to their role within our last monotheistic faith of Islam, has been in our own representation but to show a bigger case of danger than the worst of what politics can do? By direct statement that indeed some who would teach to the matter of conduct, yet fail to the matter of opinion, because of political opinion instead of religious belief? Have to face the charge that instead of helping the few to total redemption by the act of martyrdom. They further condemn all of mankind, all of this specie away from our attempt to claim Gods interest over any one of the millions of specie so far to have been through the one element of evolution in procreation. Before any could show the understanding of creation?

By that fixture I am obliged to continually chastise any class of person in you and I, who would and does take any form of religious change, even if newly manufactured, to promote the office of criminal murder to be a case for martyrdom. Even that others and I include all ages, from where the view of martyrdom is expressed as reason and cause. I have been deliberate in my choice of some imams who have lost track to the meaning of Islam by now taking it and entering it upon this world political arena as a countermeasure to political decisions. Knowing that to be fact, is enough cause in the case of warning that this deed, act or inclination will end, at first to comply with what Mohammed required when we were informed of the generational and global context of what Islam was supposed to bring. Not least already connected to Judaism and Christianity at a time when it was possible to be more laid back than when the edge of politics in the form of greed had been fully revealed.

Without judgement but to set the clock from the mid nineteen sixties by reference from where I on my own account had made the connection I have, bears upon us all in any case, simply because of the effect or outcome of local wars in the Middle East. Essentially to have been between the Israelis and the Palestine people if stateless by measure, but to have merged into the political standing with Arab countries by definition and not least in being also laid back? Fitting to one side of the bloc culture standoff without much in the way of any political conviction or perhaps not concerned enough of what a political standard was, when religion was to be the victor in any case!?

Without connection but of dour consequence what had occurred in the Middle East at or guided by any circumstances to have befallen us all, has been by slight of hand without any real players only supposition. From my own suggestions tied to the abstract of abstract reasoning, all said has been said before, but not seen by the way circumstances had led us make our judgements. My only certainty in all conflicts and or dealing and of intentions, when delivered half measured leading to the consequence of applying our steps taken syndrome, is that to some, all our yesterdays become important for tomorrow? Our only task is to properly connect with the important matters of the past and bring them forward as a means of escape and a means of change. Instead of operating by the suggestion of tomorrow in having to utterly change the past like when communism as a political agent was first introduced. With the cited intention to destroy all elements of religion!

How ever Karl Marx and Frederick Engels had set plan for the future benefit of all, their ideas were presented too early in time, although sound if found, they were unsuitable to be worked by any group who would seek to be revolutionary. Simply because in the design of building there ideal called for the destruction of the old to ring in the new, which on their intellectual level was a matter of shuffling a deck of cards, but by application was an open sore upon humanity. Although open rebellion was the key to their method of change when first applied the only outfall was in the name of murder of self by self. Rejecting religions at core to the new standards to the aim of political murder was further established.

Part of my charge to be upon some imams is to bring them in view of how some errors started half measured, followed by our natural tendency to react under the automatically induced steps taken syndrome? Marks and Engels had erroneously set up plans and by not making clear any statement to the value of an individual, but only for a group, who were to be the tools of change in allowing the outrageous forms of actual change to have followed adoption of plan or revolution its self was an error of thought!

From the first days of the Bolshevik communists in Russia in nineteen seventeen, under the illusion that religion was the opium of the masses holding the general population enslaved to the concept of political authoritarianism and being servants of the state and church! The only exercise of conscience shown to the saving of the populous was carried out to the political chime of murder and murder and murder. Of all those who would be dissident, who would try to refine the open intention of the regime to be workable, to further to humiliate the Christian aspect of the vast majority of the people leaving them only with the vacuum of hope, which was quickly replaced by the forced sanctity of the state!

Even from the next case of communistic control gathered by the Chinese in nineteen forty seven, at about the same time independence was a by word in the sub continent of India and in the Middle East proper? Even under the nose of a leading victorious force in the recent war in Europe, Great Britain, a smaller case of partition was taking place in the remote island country of Ireland. Who like India had part of her original emerald shore wrenched from the whole to be another country, Ireland was politically and religiously divided but under the one faith to be Christian!

Without count the small Irish matter was not because of religious concerns as could have been suggested, but fully economic under the unusual vagaries of the capitalistic system of democracy. So much so that the powers to be in Great Britain allowed all rules of politics in the name of statehood to be broken in that particular case. Although she had overseen the former colonies of Ireland and India become independent. Because of disbursement and if, the volatile nature of the Irish question in relation to Britain direct and the matter of the major partition of India drawing other conclusions. With Britain's own history of staunch Roman Catholicism now to be protestant mainly through the Churches of England, Scotland and Wales along with many other smaller Christian communities. Rules were made to be unique in modern history, that even both the countries of Ireland the north part to be British in king and country and the larger southern part of the same island Ireland to be an independent republic like India and Pakistan were also to become. It was only to the Southern Irish population who could travel into what was still classed as Great Britain without any type of border control. Because if for instance Great Britain was set to use the mainly barren and rural labour force of the Irish when needed without offering the same facility to Indian or Pakistani nationals! Falling into the trap of following steps taken, from half measured reasoning by other counts like finding one solution and by creating two problems?

It should not go unnoticed that perhaps in the back room study of such huge decisions to be made some were cast from different parameters. Like for instance with one group being first world and another second world politically, further, with one group being mainly Christian and the other Hindu and Muslim, which if gathered by religious tones that all should be religiously tolerant of all and all. Then the seed of apathy had been sown that religions will find there own level in any case, while all the rigors of politics are to be pursued in the greatest of needed haste rebuilding a shattered world that had endured the brutality of total war!

Unfortunately the brutality of war unavoidably was to lead to the brutality of political success, only this time without any perceived objective. Except only to fill what was to become our darkest hour leading without any form of connection to our darkest moment, when a suicide or a person intent on suicide of self only, of self only, was to murder and murder the innocence to the illusion that the hand of God might be upon them in the name of martyrdom, therefore that connection alone was endorsed!

Martyrdom was not a choice to the earlier Russian communists or to the latter day communists influenced by marks but not quite in the same way as the Russians first took up the banner. China as a huge fragmented country embraced the ideas of Marks in open acceptance even more readily than the Russians had. But in the sadness of using the same methods of terror which in all cases of attachment with any communistic system goes hand in hand as a means of coercion. Quite simply in both cases the level of equality was to be determined by the lowest common denominator, meaning that in order to show how new success was to be really shared those who had already succeeded were to be shot out of hand because of there overt greed and exploitation of the masses! They were openly shot and killed in sight of every villager, man women or child, every member of the population was shown examples of quick justice in the name of fairer and better change. No mask was heeded, this new rule was for all and what we had been waiting for and what we deserved without God!?

Communism in every corner in every infected country cut its own bloody furrow wherever it was raised to protect or defend the faithful, also as a good warning to everyone to toe the new line? All murder was done in the name of fair play and social justice and to show better what was to be produced in the name of the mother country! Of which in times of real conflict against others over political ideologies, in many cases the idea of being suicide bombers was implanted into the minds of young men and women to die for mother country! Which by reflection was a big ask, when considered, even as a national soldier in battle with others as national soldiers, the only objective in life was to kill ones enemy! What could that be for, when all done outside the image of God meant little unless by result such acts were nothing more than a misguided publicity stunt if only for mother country? Unless except to perhaps insure benefit for ones family, not to be shot as an example for others who would think not to toe the line?

Unfortunately without any comparison to what was done in the name of democracy which by example had set different parameters in any case. Just prior to this last modern era when from Europe leading to two world wars, conflicts to the matter of human slaughter amounted to the full art of mans suicide in the name of King, country or party or system which by refinement was to become to the matter and needs of commercialism?

Although supposedly fought beyond any reason at such times, when conflict began the plan of battle to include armed army's face to face, was a standard, but carried out on the above said whims of fighting for any cause like King or country. Alas what the time of our histories in progress had allowed was that the method of actual contact was one of attrition. Whereby under unwritten rules, in order to succeed, one side simply advanced into the ferocity of the other sides defending or making gains of protecting or defending what was ours, by reflection!

Column after column of men walked into certain death in duty, but above the terror of supposedly believing that they as individuals were not to fall. Weather buoyed by the courage of faith in there God or King or country or politics, slaughter was relentless always on either side. So much so to be debased as mass suicide, for the spoils to be fought over again were by those not involved in actual battle, only our armchair warriors, politically!? By connection of which we are to see in kind the waste of human life for reasons supposed to be noble and honourable of which there are virtually none, if we are to embrace the obvious of 'Natural Body Death' in the medium of now, but only as told of now, so must it be!

No future is certain by any means of forced death, leastways politically, when the odds can never be measured as to why we kill in the name of commodity! By the first events of this last modern era while settling new grounds for new grounds, although in theory the ideas of change in the name of settled peace were to cease in the pursuit of political murder. Even after the mayhem of the Arab/Israeli, India/Pakistan and in tone but to a lesser degree of activity the Christian equivalent of the north/south divide of Ireland. Proposals were that by political comment after the dust had settled in the name of religious tolerance, then the only ability for us to kill of self by self was to be of political motivation only. From which we had set in foundation of how the future was to be by standards we had not even been able and consider until now?

Unfortunately no such picture could have been seen to what might happen, what did happen, what is waiting to happen unless we open and change opinion led now by the true class of our religious leaders. To be constantly reminded by us all of how they as we must use the full intention in meaning of what is expected from us by taking to the cure of the Fifth Dimension in prospect!

# CHAPTER 178

By described plan and to my own reasoning, not classifying Islam to have a leadership structure in the same style that we can think of Judaism and Christianity holding through their priesthood. Has been done not to separate us by any motive, how could it, when we all accept the ideals of monotheism which are uncompromised? My opinion has been set in terms of correct representation so that we all might shout at the same time making only sound instead of sense? What is to be agreed without compromise is that beyond excuse we hold to the same, utterly the same God in Allah who in open acclimation is beyond our sphere. Save what we alone as specie have believed of ourselves if from Shanidar, if from ancient Egypt, at any time, ancient Egypt, or from our refinements of some four, two and one thousand four hundred years ago to also include in parallel our Eastern religions?

By claim but under the broad influence of many of our civil/political forms of control, God in concept as a singularity was carefully introduced to us all eventually. My own supposition in the respect of equal count is that we received the full knowledge of the situation left to us in different times to become aware of the wisdom of what had been discovered by us. Unfortunately capturing on three counts the full meaning of intent all ways, but missing much on the application, because the only bands of thought we could use were from the actual times in which we lived. From the factor that in being progressive by nature when we were told the same story three times, our reasoning had read the same signs differently.

With proof wanting, even delivered from the abstract, I have been able and measure much of the final differences between us all, relative to how we were then and are now. Leastways to be able and cast my first opinions to the matter of how Moses and Jesus and Mohammed represented us from their own same, but separate standard? Which I had pointed at the differences in two cases who had been able and accept a reading direct from God not to kill of self by self, followed by the third example of where we could kill of self by self!

I have never made any form of condemnation of that last connection, not least because I think in terms of today and Mohammed thought in writing to the terms of his day however delivered. But by direction the delivery was for that time and not the mistaken reading from six hundred or two thousand six hundred years earlier. From what I see of now no mistakes had been made then ago only errors, because of what was to happen on going, no body from any time of any of our pasts could ever have set the stage we are on today. Not least to take into consideration that a time would come when by token we would come to this new last modern era? When all the acts in the wings, were now formed in the right shape, but had each mixed to the wrong script, if not religiously then politically or both?

Without limit I have put one and only one point of reference from which our religions can regain the moral high ground to really force change. Without the need to apply the expediency of politics in destroying all not to be understood first time! My fated connection between religious leaders and imams has the vital link of religions, all religions. In knowing that indeed there are many high religious leaders in the one faith of Islam does not weaken or put into disarray my own standard. Not least because their code is first and foremost to what Islam already is, which in class without conflict, accepts other religions for what they are?

It has only been by the actuality of contact on the religious edge and the outer edges of politics we can attempt to assume the proper harmony desired! Although in real time most of our new attempts to move forward on block as we had tried in these past sixty plus years, have relied on the fixation that by the code of religious tolerance we had moved to close that area of discontent. But in reality from half measured political moves the name has come to roost by the one term of double handed self expression that I had put to be upon some imams, but by their control!?

I had set early appraisal of how from mainly Middle East activities at first between the Jews and Arabs then the Israelis and Palestinians before the political realisation had emerged that Jews and the Palestinian Israelis were one. Whereas the Arabs, Palestinians or other groups of tribal nationals even to include Egypt were first of Islam and second of politics. Which by outcome, because of how the bloc cultural standoff had resulted in the demise of the political edge of communism, no standing voice was raised to oppose the settlement of affairs from which the religious advantage of Israel through Americanism was bannered in triumph? To the reflection of a depleted political impetus that had befell several Arab Islamic countries in direct area, even when all religions there were major monotheistic faiths!

Having come along the same road as Israel these same years of this last modern and in not having had the same balance that might have been expected religiously and politically, even unnoticed? What voice might a less organised political orientated monarchical, semi democratic, semi dictatorial collection of Islamic countries raise in open query to the primary first world might of Americanism if that was the case? Of the time it is to be remembered that American political support was to an American political ally!

Without doubt only conjecture, supposition and abstract reasoning, can now be used to move all cases forward on the same level also without changing history! My standard about how I see some imams, stands because of the importance I personally attach to the full religion of Islam also being all of ours, but not by tolerant means, only by forbearance! I make no case against my counted first and second dimensional religions of Judaism or Christianity. Nor are charges to be made against any of our huge other collection of, what I have classed along with many other people to be our Eastern styled religions.

Hinduism and Buddhism in concept, because of there structure have been able to encompass the matter of death by self upon self through internal connection, which is not related by any count to monotheism, being no bad thing. But from which by reflection leads to what must now end instantly, whereby to the matter of human sacrifice religiously, there can never be any form of consideration. No religion ever in religious law holds to the matter of a lost human life through doctrine, leastways in this last modern era!

No charge can be made to history from the above save that where the matter of death was to be upon any person by doctrine, sanction was taken by use of our steps taken syndrome. Whereby in mix we had made the error of assuming our civil to be political reasoning had a place to be overlaid upon the physical strength of our new emerging religions! Our only emerging religion in those ancient times although God was our mentor, others in our king, emperor even our priests were of authority to make the wish of sacrifice in tenet apply to humans! Also as had been done before the light of supposed true understanding was able to shine so described in the bible at the cost to us all in example from Kane and Abel?

We had all been overlaid with the concept of death by self upon self in the subconscious which was first translated into civil then political law if by needs must? Again I state without being repetitive that the issue cannot be brought to religion as first show by Moses and Jesus when stating that we were not to kill of self by self. Then reviewed in the exact same manner by Mohammed then ago, allowing even if unrealised at the time for the extra two thousand six hundred years of added development. Still only showing that, yes, as we do kill of self and self in civil and political law, then to that feature while we are in transition to this final monotheistic religion if needed the same can apply? While at first to promote and protect the progression from what earlier prophets had introduced. Then at first again to defend the gains made so that the will of the written word was not to be compromised. Therefore no force was to be applied only by acceptance and in need if required, which is now, never!

Without confirmation what followed in times past, where through and then continued by outside influences, the charge of battle was able and be answered in religious terms. Then the full issue of whether or not to kill of self by self through Islamic reasoning still held in at first being sanctified, but not from being modernised. Making all relevance of this exact time of needed change because of how for the first real time in all of our histories by consent. We can study in open form to the danger of working half measured, then turning blind eyes to the rampant display of how we apply our steps taken syndrome?

Much of what is at first delivered outside of our direct concern can go unnoticed for the two counts that we genuinely are so concerned struggling with the rigors of our personal lifestyles yet to be drawn? Then the other might be that to any reaction caused, our standard has been to chart further reaction link by link, but in the end having no association with what the reflection was! Hence, that on an individual level of thought, some solutions, propositions and ideas multiply in response to support or condemn any so actions, but unguided, when seen from our own perspective, having been formed from multiplexed reasoning over unknown periods of real time??

Taking to issues on an international level in practically all cases from having always been programmed half measured. Equivalent steps taken in support of that medium can bloom in disarray. If for instance to have been reacted upon from political determining only as we naturally do in any case, perpetually turning the wheel of open discontent from the one aspect in all thinking, we are always right, but only half measured?

From where I have set one common standard to be upon some imams is a reference point to highlight the effect in reaction to what for some has become the open way of misreading all reflective impulses. I have set the standard of our general opinion, that in modern times we are more of a political society, who have graced all of our religions into religious tolerance. So much so that in order to buoy up that fixation we have set to overlay all aspects of our understanding in politics and religion, of which we have let them be equal in the way we have set it so for all people who might practice either. With such elements of invention in the likes of being politically correct, of more equality through civil liberties and human rights, then in order to make those features stand, we have made most of our civil/political legislation set the standard of most of our racial or ethnic laws. In attempt to rationalise our full equality by those very laws in referral to our mixture having to be taken the same, we create vast divisions from the attitude in working by our steps taken syndrome!

Although we are in such a situation having funnelled many mixes, we have always accepted what was delivered at the spout, which at last is in open review by now, when we can see some of our two biggest errors. Religious tolerance being one from a political standpoint and the political judgement that in equality under the idea of law even internationally, how can we operate first, second and third world economically, politically? Failing not in concept, but by example that to set division and remain unaware of it or blind to it, delivers the open account of hopelessness. Which if read outside the feelings of compassion can lead to any type of mix by any kind of view right or wrong, but in most cases always wrong again and again when violence is used!

From the viewpoint above I once again will turn to the matter of how some imams behave in our representation. Although not always acting alone in what they might encourage of some of our young to do in the end gain of martyrdom, because in some cases the families of who are essentially murderers by the way most have been known to die. Also reap benefit not from the offices of the religion but from some interested political leaderships who for different reasons supposedly, are against some other political standards even if to the assumption of being of international concern for all? When ever the incidence of martyrdom is raised and by whatever means, there is no standard to it by reflection or reaction or stipulation. Because the sphere when done, cannot be for any other reason but to defend, protect or promote all the effects of Islam only, all the effects of Islam only.

Some of our imams in us, but by there own endeavour, have tried to set what was seen to be unbalanced in reflection and of the reflection of the reflection, because it had been set to one peoples advantage over another's by perception. In any case, if the case is proved simply by emotion, any and all unbalanced matters of legislation risk the matter of personal response? How we should react now is to carry the same weight in all cases delivered through the Fifth Dimension. Then in class by our religious leaders whose open view even when expressed with our own religious bias stand to be heeded in forbearance and not by tolerant adherence. Making the breech between politics and religion having to be met in how politics must and will change at least to see the filling of the gap!

## CHAPTER 179

I have not made the full case of how I see how some of our imams have reacted to any number of the situations having developed world wide, but with particular note of this last modern era? References to the mid nineteen sixties from my own record, even if concerning on the one side the Palestinians and the Arabs mainly to the religion of Islam has been one of my own marks in reference? From where I would take the case that in real terms and of this era, Islam was to raise her voice on the world stage, but now politically if for the first time. What ever was to have occurred in or through or because of the tragic events of the Iraq/Iran war, Islam could never again step aside and let politics pick and choose all of our options. Because in voicing comment for the first time in this last modern era, Islam as we was now ready to approach the world stage, even if to find no listeners?

Unfortunately spread over, under, between and intermixed with the full volume of what was going on religiously and politically, with lots of concern to the Middle East. Much of her thunder was taken in step through or because of the bloc culture standoff situation at the time. Religions being overshadowed by the impetus of communism and democracy seemed to have to take second place. But in some cases if to be seen from the Israeli perspective that second place was fast becoming first place even nationally and internationally! Taking the twist that Israel in nationhood was a barrier to communist expansion only? No balance having been set that Israel like all other countries was tied to the matter of religious tolerance as were we all individually also? Allowing if necessary that politics in the one form of democracy was the voice against what irreligious communists stood for in the Middle East!

Even that the area was awash with hundreds of millions of Muslims. If that was the case we might ask in the same voice that which I have already answered in. How was any entry to be made into the reactive form of politics at any time when decisions had already been made but not perhaps fully displayed? Not least by and from a controlling body already thinking all was under concern and care through the United Nations? Whereby argument for or against could be voiced in the supposed knowledge that the balance had been set equal. Especially when the contention of religious matters in some cases had been confined to the reflective standard of religious tolerance!

How ever it is seen of our religious conduct over these past sixty plus years, by being unpartisan my case is ours, but without having to be upheld, only reviewed. I have always stressed the importance of religious matters to be our primary force of drive, which can only grow by influence, but now from the presentation by our religious leaders. Who are only to be our representatives and not to allow any and every objection or query be aired by opinion if not to have already put personal view points before our one active prospect in 'One for the Whole'. Because religions unlike politics is to the individual just like all separate religions should be and are intended to be throughout all of our histories uncompromised!

Otherwise our standard of religious tolerance takes us hand in hand back to a time past the four hundred years allowed when in a global sense for the first time, for the first time by the matter of discovery of the new world. It was found that we as specie practiced human sacrifice on a massive scale religiously? In relying on our own judgement on a personal level without our leadership structure who from now on are fully, fully approachable? Self opinion even if not from personal opinion, has been able and form whereby some of us have found it easy to operate outside the limits of piety, charity or concern or understanding for each other, yet remain aware of self to each other by whatever standard!

We know some of us have encouraged others who have not had enough time to have life experiences, yet throw away there own time and the time owed to others by the severity of brutal murder under the impression they were to God. Under the fed expression that in name there act was for God, ignoring or choosing not to point to the obvious fact that if in Gods will it was of Gods own concern, it was of Gods own concern? That is why we had been given or made our own free will. Not in the same manner that all other of Gods creatures had not done, but by that fact we were the only single specie of finches to have broken the code of evolution and entered the realm of creation with imagination and the will to use it?

Having now arrived at the height of all time from all perception, in mounting those barriers set by evolution and now released to be of creation. We have come to the last balance in that the formula to be exposed is for us all singly by our fifty plus or seventy plus year lifecycle, now continuing generationally on all sides. Without hindsight, without some elements of the past pre nineteen forty five even without trace for some of us, our place to be is here and now. Because for the first time in all of our histories, through technology it is possible for us to know what we each think continents apart, but of the same idea?

I have made the case to our standards being better from our religious base even if to have been managed by some of our civil and political attitudes. Simply because in all the other specie come and gone, here and now, all have had there civil and political means of control delivered as instinct. In the same manner that we were included, but ever under the same wings of development open to all creatures who like us had learnt to follow the pattern of the seasons for instance? Who like us had used in a haphazard way what a wildcat fire might have left in its wake, then in the same way that some other creatures apply measures of tool use, albeit at very simple levels?

Before the first split by what trigger, we had reviled in the aspect of specie selection in all cases save one? Unfortunately specie selection is a blind alley, it is also a wonderful concept by idea, but has led to the misreading of too many matters right up to our most important time of change post nineteen forty five. It, by concept has carried the wrong messages by equal measure to and for the wrong ideals. So that some of group in this one human specie have been able and forget or not even consider, that it was from one finch we had formed slightly different characteristics to suit our particular environment. Further carried to the extreme that because of natural selection making the case for excellence, the form so described on behalf of we humans, was simply claimed? Making the case that from results our differences were or had been appointed by the lords of evolution only, but carried into the aspect of creation by there own interested parties?

From the smallest amount of reasoning my partial claim of how some imams might be involved to further propagate the idea of natural selection to or through political reasoning is a hung over effect! More because in the same way that the errant nature of politics had been able and drive us humans forward, we who are the creators of all reasoning, into the state of unbalanced cross referral? Has been done in or from the unseen style of how we are all at the top of the natural selection choice. Meaning in all cases we are all of success through the counted style of how evolution works. Coupled with the other and later form of acceptance to the real theory of creation by or for any felt connection in any way, even to God or to God first?

By that code natural selection was to be a consideration having been taken first by Judaism, then denied by new Judaism, further denied by Islam. Then brought back into play by our misinterpretation of what Darwin had meant to highlight it in the simplicity of success as it was? Until some of us took the idea in hand and ran with it but only in the name of political reasoning, creating in a way the blind foundation of how we were to be satisfied by the naturally selected medium of religious tolerance, which to the proper grace of doctrine was settled for other reasons. Tied to the broad sense of the tolerant nature of politics expressed by behaving politically correct, beholding to the standard of civil liberties of human rights and other spoken commodities of token to the standing wisdom of what our politics was supposed to be for all? But managed instead to make further splits by the number of tolerant laws to our actual differences expressed through racial and or ethnic standards and or standing, which have only managed to split us more politically and religiously!?

Without connection, but if having indicated that at any time except from first connection, Israel took in charge the matter of faith in us for us her people being mostly Jewish as we were. No modern standards had been changed, but much had been altered which is why the Jewish religious leaders have a case to answer and work to do, not least to help straighten the bent pin of tolerance. They now need to set by deed without note, ideals that are linked to the past that had been managed with all sincerity even since nineteen forty five, but now carried ever forward. Like from this time with all others who need the penicillin of reason to be administered in reading through the Fifth Dimension overlaid upon the religion and politics of Israel. Coupled with all other political and religious attitudes throughout the rest of the world, even at this late date of necessity to the need of much change with little alteration, but carried in equal harmony!

My Middle East and our Middle East is much bigger than the matter of how

Israel might adjust to manage her own affairs. Just like everywhere in the three worlds we recognise economically, from where problems have to find there own solutions, but fortunately now in line with the aid of the four prospects of the Fifth Dimension. Without specifics, but not without the utmost of concern, in many places which do not make regular headlines in the same way that connected features do within the first world economic reasoning. Are committed some of the worst acts of human degradation by cruelty, exploitation and murder in the only point for personal gain, suited to the economics of profit politically. Then by religious terms socially, but aimed against what can only be delivered to the assumption that some religions are taken to be naturally selected, uncounted, but read? When without foundation we have only listened to the one voice we wanted to hear!?

By mention in regular brief to the rest of the world, what is gained is equal choice, is that under all circumstances, we all counting to be the rest of the world can make no viable case in acclamation from any quarter, that the matter of natural selection has any sway to any called or claimed separate tribe, clan group or society! Making the double claim in open review reflectively of what might be claimed ethnically as opposed to racially or monotheistic as opposed to any multi adjoined reincarnation ideals. Making the case in any case we are all of the one specie 'Out of Africa', embossed with changes from the circumstances of geography, diet, climate or other reasons only!

Although I accept the same reasoning of how and why we differ, but not through the connected avenue of natural selection, having been misread since Darwin published. My own style set out above, along with how I had added the ambiguous to the equation of us being progressive or laid back. Has been done without any claim of definition save what can be gathered by inclusion in overlay of the Fifth Dimension from now. By the same balance and for the same motives using the abridged version of abstract reasoning to make the case that less is more! I have deliberately directed a series of measures for our consideration to the matter of what we gain from some of our imams?

By personal choice the view is my own and led from the matters of my own study, even if to have drawn abstract conclusions based on new points of view. Those to have always been with us, unfortunately made dormant by other influences like from politics, which since nineteen forty five has grown in strength beyond its own capabilities because its food of reason has been drawn from our steps taken syndrome. Following in blind obedience of our first laid anew, half measured approach to tackle world matters only politically? Never knowing without preparation what might stir on any of our religious fronts if any was to voice a religious opinion in contrast to a made political statement?

When the events conducted on the sub continent of India with peoples as we becoming partisan religiously, instead of following what was to have been the same measure of continued half measured political decisions in any case. Results in outcome were left just so because Islam and Hinduism did not have the same political rating that first world choices of Christianity and by connection Judaism had in being both to be first world and of connected religious tolerance by political understanding? Unfortunately then ago what was directed to be a general standard of improvement for all, was allowed take form under that heading? Like for instance to call and settle the matter of religious taste, providing all matters were local and left no stain at the feet of the United Nations, our new reference point?

Islam against Hinduism on the sub continent was so localised that the issue was allowed run its course no matter the cost in human life. By the same token although to be set in standard for a political outcome the answer remained of the same consequence, that once found, the solution would always remain local? What was happening in the Middle East in parallel by the name of partition had a different connection religiously!?

Here although on a much smaller scale by the general count of population the emphasis was changed from how matters were on the political front. Once again I set the same balance that with one cup weighted with massive religious concern in being of Islam, plus the union of a non religious style of political communism. Our other cup filled with the first world rated group of Europeanism and Americanism already attuned to the matter of religious tolerance politically? When the balance was let swing by the process of time and circumstance, if the aspect of natural selection carried the pseudonym of might being right, then the first card again was set to be of the western political camp. Our balance automatically fell western, not necessarily in all cases by natural selection, but more so that the western case could be defended in having the measure of religious tolerance attuned to the standing for political interaction?

Communism in respect of, had no opinion religiously, therefore could make no representation of any aspect by any religious indication although heavily committed to the one standard of Islam. Which was a workable concept if allowing that to promote our bloc culture standoff each side was to accommodate some form of religious concern set from political perceptions? In the case whereby no set fixation by a religious stand could have been given from a political only representation through communism, by my own notice and in cross referral, in order for the western political machine disrupt the communistic ethic. In not being able and force issue through the matter of religious tolerance across the bloc divide? By adding more in the way of control or acceptance that we all should adhere to our western perception, was forced by the means of either international legislation which was thought to be the best medium of control. Human rights begot civil rights begot a profusion of race and or ethnic inclined approved laws whereby the supposed outcome was to how we were not to mix the matter of our differences in cover to be politically correct in the muddle of working steps taken!?

In another attempt if to suppress what was thought of communistic nations, all done, was a full mess whether to get people to say no more or less, than please or thank you at the edge of distraction? All done without vision, without hindsight, without conviction, certainly without foresight was for us to settle in acceptance of our law making by how we had made them, right or wrong or indifferent? Because now the full idea was created that whatever the law was to be it was to be from the first world perspective only, by taking the link that might is right yet not to consider the religious heart!

No more better proved than in notice of how communism was to shorten her ideas of expansion and turn to a type of political commercialism? Where she had previously trod throughout the world, all left was deprivation to the former championed mass of different nationals? Making for our worst collective failing caused by her misjudgements, pointing now the only available balance left, was that democratic commercialism worked? Making the parallel case that although she had held to no religious code, many of her former nationals returned to the aspect of religious sufferance, but also fell into the silence of how most of our religious leaders had done by the acceptance of religious tolerance?

Our worst legacy of burnt bridges was how many free of choice communist free countries, were to adapt to the matter of commercialism and religious tolerance, but did not mix into the happy state of affairs that was supposed to happen in hope? Which to its own way is open to be addressed by our political overlay of the Fifth Dimension to the matter of my own acceptance that the rules of conduct in this case are that the three taken and one active prospect are to be read politically first?

On the other foot in the measure of support by the Fifth Dimension overlay of now, cannot have a bearing on how we had previously acted, reacted or interacted by the means of reflection or not? Even by or from what I have set to have been a key matter how we this single specie have kept at bay of each other, not from or because of religious tolerance, but because of religious perception. From my own notice, even if generated by the religion of Islam in the open honest commitment for us to change for the better. But unfortunately to have been set in standard and style by my own interpretation of what some imams have delivered in the name of martyrdom. Also to be through the tools of suicide, murder, intimidation even outright brutality in the first matter of expression that this new might in the name of fear is right, is to oppose what is nothing more than our own reflection!

## CHAPTER 180

By using the term some imams, I have set no standard only that of supposition, because for now, some people without a defined objective or motive explained, cannot support what martyrdom means? Imams and Islam in the relationship I have set by association within these pages, from what we know or can imagine the role of the former within the latter, has been constant almost all of these one thousand four hundred years of Islam. By that consistency and from where imams best operate even across the broad range of their duties, in view of the close contact they in we have with many of the younger members of Islam. In view that they have been in a position to set all the standards of the one text for interpretation, puts them almost in class to be part of our structure in being religious leaders, leastways in terms of what real influence they as we might have upon some of our young?

In any case and measured from the political perspective, that running title holds if we look to the sad effect of what the dark side of martyrdom delivered means? Even if I personally feel fortunate to have been able and review some matters having first applied the overlay of the Fifth Dimension. No case can be made in support of what some imams in we have either said, implied at encouraged while dressed in the coat of one of our gigantic monotheistic religions. Because no case had been first made to address the imbalance of any religious situation, except by the impatience of reacting to political misuse?

What I can see without judgement is the creation of a situation with all the worst of our failings expressed beyond the control of reason and beginning to set a class and standard of its own making. I have laid the reactive core of blame without judgement at the open door of Islam, because of the closed view to danger some of our imams have now promoted. But only without classified or political judgement from how I personally have interpreted the reflective image of cause and reason or more precisely reason and cause?

In any time since nineteen forty five by any code, also not from any initiative, even if instigated by the United Nations, apart to the loose fit of religious tolerance, no further consideration had been given to any religion by name big or small. Even when no open display was given to the full meaning of what religious tolerance was to mean without the understanding of what forbearance really is! By that mix of pull or push if and when the Israelis on there own account were to express personal religious opinion, no case of objection could be set. Because without further undefined definition we had no score card for how we were able and check the excesses of claim if made. Because religious tolerance was a square fixture without definition, except that we were to be religiously tolerant, which was a blank cheque if any group decided to exercise there own form of what tolerance meant?

Between what was to happen, what went on or what was thought to have gone on in the Middle East on all of our account at any time since, I had set the mark from about the mid nineteen sixties. No other voice other than what had been aimed to be political was heard in any form of display than what had been the noted political exception from Israel. Who on her own account as well as taking all the political steps allowed, grew to switch the reflective image into taking religious steps first, which is certainly no bad thing, unless!

Except by religious tolerance, the standard had been set to create the same core balance for all, therefore by the unwritten rules of religious tolerance, the only effect was that we were all allowed maintain our own standards of belief genuinely held without the matter of outside interference by any degree! Unfortunately leading to the grand old mess we have found today whereby in order to substantiate the rules of religious tolerance. We have had to add unseen changes and charges every time we speak to the matter of God, in the cases of monotheism the one God in Allah for all. Which now at this time leads to the goal of what and how our religious leaders are to do in order to substantiate the full matter of religious forbearance of which all have a claim!?

I have not set any form of excuse of how or why some of our imams might have behaved in a certain way and still do? I will not set in charge the matter of how I have interpreted what I have seen of the conduct of Israel to have a bearing on other conduct by peoples who are not first Jewish of religion, and who were to have behaved as they were supposed to have behaved? Save that if opinion is fact, then the only case made is that what was done was done, without any ability to change the emphasis of from when it was done by later thinking to the reason or motive! No person as you or I are in any way sophisticated enough to balance such judgement, no person as you or I can change the matter of history only to improve the effect!

Form my own prolonged statements that events by example of partition in the sub continent of India, of the Middle East of many parts of Europe of Asia of Africa, were wholly political stands in all sincerity. That fixture is to be taken that by decision, such decisions were an open wound upon most societies throughout the world, but having to have been given credence, because for the first time ever we began to think whole world, one world! Naturally still keeping our eye to the first matter of gain, simply because that was how it worked, our choices although very tenuous were all that we could do led from the counted ignorance of victory in war, but without remidies1

When the first choice made behind closed doors was to root out the causes of all war once and for all, leastways from the thinking by first world perspectives. Whereby if examined the two major world conflicts of touchable time to the decision makers then, were done in the one name of economics and in more economics by what could be afforded and what was the profit margin, simply how first world thinking had developed. In later back room translation and from half study, when the first matters of religious perspectives were raised about the difference of style in the one country of colonial India and the one mandated country of Palestine. Our first world back room boys set all standards to the future seen by how their economic outcome in requirement might develop? But with the short stop of making only decided political decisions having set the principles of religious tolerance elsewhere! Leaving the supposed unalterable events of what has become history now having been done in all cases from political perspectives only, having been also done encased by religious tolerance!?

From whatever was or could have been, I have taken notice that what was done, cannot be measured in condemnation. But can now and only now be viewed in hindsight since we have begun to think of the future collectively, which is only of hindsight recognition. With the ideals of religious tolerance as a separate matter to how politics did work, in order for it to be a case, all individuals have been set with the exact same standing religiously! It is from about now we will need our religious leaders catch breath and count, because of the quandary I will set in our name, to be upon them all!

To the small point of recognition that a standard was set where a single Jewish state was brought into nationhood amid a vast sea of several Islamic countries, was to become a matter of political fact, but from political cause only? For me to have cited the situation of our bloc culture stand off, particularly in the Middle East was done to show the importance of the political perspective. From when Israel showed joy and rejoicing to a native land was purely a matter of how the standard of religious tolerance was supposed to work. From where some in Israel put the matter of the effect in outcome to the will of God anew, is not how or by any intention of the way religious tolerance was supposed to work! Which indicate the vast problems for our religious leaders in setting the table for further discussion, but now at last we have the bread of three taken prospects and the wine of the one active prospect of the Fifth Dimension as a centre piece to our new standards!

Although politics makes decisions it is our religions that from now on will make the rules, which is never to be a distraction, because it has been from our religious drive that has ever moved our progress past all the imaginings of evolution? We have come too far and too well on track since Shanidar for us to fall at the last hurdle by being blinded by the materialism of commercialism. Of which without the capacity to think in terms of tomorrow on an individual count, has long since lost the plot or where to look for purpose and stir hope in our own design as human beings! Politics is a great tool, but it is to be our individual servant, simply because she cannot ever or even plan to prepare us to purpose and set in us a self standard and style of collective acceptance to draw Gods interest in us. In separate form to the interest that might already be shown to other creatures of evolution not having touched the door of creation, but if to be part of Gods bigger plan? Politics although to lead can only cough and splutter at the idea of God, because inexorably she has been able and only show the means to make fast money, which on its own only benefits evolution?

As far as our politics and religions run we know that one or the other are to be our masters in one or another way. We also know that by personal choice on purely individual levels, in time we can extend that choice to one the other or both or none of the two in order for us to live and lead good blameless lives, which is a very good thing. No case ever will be successfully made against that proposition. I will not stamp my foot and shout foul to any person who would oppose either any form of our politics or of our religions, save the unconnected. Those of us unfortunate not to enjoy pleasant weather or the full ambience of life in general by the open show of human progress, we are the only ones to be able and value our progress, when considered the vast majority of us think in collective terms of specie instead if being tied to the limits of self only?

Not I add from the barrier of being a single person, those of us who have lost a loved partner. Those of us separated from our first circle of friends and companions by the march of time splitting us by our need to work wherever or by location for similar reasons. We in being single in most cases have been spread by the matter of uncontrollable circumstances to the matter of how time has run its course. Because we are very well connected with what we do be it at work or play or politics or pray. If some of us feel less so, it is by not realising that in fact we are covered with love and good wishes but do not know how to return them?

My own feeling of pleasantness and relief to the matter of our felt singularity is that indeed we are single but in the collective sense of being 'Out of Africa'. From three and a half million years ago, to Shanidar, to cave painting to Mesopotamia to Egypt the Middle East and now the past sixty plus years! Our bracing form of the singularity I stand for means us all singly in any case, becoming more and more important to be realised sooner rather than later so that we know that we individually wish not to have a place to hide. Because by the utmost expression of what our aims really are and how best to settle the one standard has been knocking on all of our doors these past sixty plus years at least, if not longer!

When suggested, not for the first time by myself and others better informed, if the primary aim of mankind is to be found and then to be workable by understanding? It is from us being mankind that all the moves in learning and understanding are to be made. Our first recognisable step in that direction having been taken from the blindness of normal living some sixty to one hundred thousand years ago displayed unknowingly at Shanidar, when we buried for tomorrow? When we buried for tomorrow!

By that realisation it is almost tempting to say, God only knows why! But once that step had been taken even as long as four thousand years ago, by our own conduct at delight in what we thought we knew. It has taken the same four thousand years for us to realise that by the excellence of mind we used to deliver ourselves to a result that could have been our first error, because of what we did not know what we had followed? Because in discovering the reality of God we had missed his reality of evolution. Which in place we did not realise, because from instigation we already had given God the credit of making the whole universe and everything in it? But anyhow by the joyous vanity of our simple pride we attuned all references to us in tribe to show in recognition that at last it was we who had been chosen, by finding?

Never to be an unreasonable standard because all tribes and clans at that time even to have intermixed in the strictest of formula by trade and commerce had developed the inbuilt human knack of being aware to our full difference. Even if at first to be shown by the bigger structure of our city states that had been spreading in theme for perhaps the previous six thousand years in area to be the Middle East, until we became locally global!?

## CHAPTER 181

Armed only with the thinking of the day, which at the time was to our best ability, any similar connected acts and there were many, were addressed to particular tribes in all cases. So if fringe of modern intelligence at such times would suggest that to the Jews as it was some four thousand years ago, to have called it they had been chosen, they were? They in parallel as all ancient tribal, clan, group or city state communities, were all chosen in the choice of there own Gods done since before we had found a way to count beyond the seasons, perhaps now until the Jews?

Without charge or claim or accusation it is not unreasonable ever again to suggest that we are all chosen but not so that the connection is to be one way only. Because choices or not from us or from or by God, choices were made three and a half million years ago 'Out of Africa'. We must settle in choice that for us to have come by that route to populate this world was a series of events measured in Gods own time span of evolution.

For me to have made a case whereby I have viewed what some imams might be responsible for today is done by the same reasoning of choice used in all time, why not? My own standard if to criticize some imams in we, is done only by result of what I would claim to have been their intention without realising the intensity of what they might really be doing? I personally wholly condemn some imams who of type have used there own ignorance to shadow ours, being all of us, to play their deadly game!?

For me to have taken the standard I now explain generated from actions or reactions in the Middle East, is done to our best value in the promotion of self, meaning all of us in type and kind who have always walked on the one path owing to self, but to the full and honest acceptance to the start of God! In these pages without compromise, I have set the ground of our path in future conduct, but only laced to what we all have to do if only to express the one sense of harmony. By laying our full case to the means of Gods own understanding in us creating the one standard to suit all Gods!?

This last modern era is our crossroads of decisions whether to turn left, right or carry strait on. For me to have taken note that our sense of behaviour, our physical behaviour and the conduct of that behaviour is to be best shown in example of our conduct in the Middle East representatively. Is core to how we will generally change by the same route never available before in its form, but now to be here because we are learning to move together.

To turn to the aspect of our religious conduct, once again to the utter importance of qualifying this last modern era by the stamp of those tolerant interactions. By the noted feature of how the failure has been manifest religiously we have at last now got the ingredient to flavour change favourably. Judaism and Islam in Palestine faltering all those years ago pre nineteen forty five, had both used the offices of politics to what was to have been considered their personal advantage!

For me to be able and express my own opinion, which in point might be agreed upon by numbers of either faction above, is done in the needed name of modernity and paid for with the standard to overlay our politics and all religions with the prospects of the Fifth Dimension. Until in the very near future when that necessity commences, my own personal view in expression fits into all the necessary requirements of what is to be our new modern era by word of common decency, C D. Which will give licence to be expressive as we see fit, but only to be positive, only to be positive, even if having to operate all the switches of criticism as seen? But only from the certainty of actions good or bad, right or wrong, balanced or unbalanced, but to have happened! Without having regard to what might be presented as need or reaction or cause, without having to defend the right of a group or a person or a tribe or a clan, be allowed to carry on with the practice of human sacrifice because that is how we have always done things, C D will close unwelcome doors!

My connection with the failing of what some imams had promoted, stems not from the matter of human sacrifice for spurious religious reasons, which fortunately is no longer to be practiced any where in the world. My association with the matter of religious standards of this one single specie is collected not from the tolerant nature of our supposed conduct. But by the reactive means of some who happen to be imams, who are prepared to steal the virtue of what martyrdom was and will always have been? To turn it into the human sacrifice of suicide and of murder by the inept view of falsely holding to the vital standards of Islam as portrayed by Mohammed delivered in verse from Gods own agent!

Some imams who have taken the fourteen hundred years of Islamic doctrine and ideals and turned all on its head in order to influence some of our young in our sons, cousins, brothers even sisters and daughters to die in the futile event to defend and protect and promote this fine religion of Islam. Then by the same inclination be able and promise those of us to be susceptible by the honesty of our youth, follow in belief that for the act of crushing the bodies of people at random into lifeless pulp, we will be honoured with the status of being martyrs. That the very status to be put upon us in our youth when by code we had nothing to defend or promote or protect save the value of murder on all sides and by all views except our own, because we only thought that to hear was to obey, but whom?

By my own personal code, which has no weight to any scale, save the one of specie, I had been able and value all life for what it is, I add, not without some cases suited to condemnation. From where I might seem to be selective and pick on the value of some imams is done under that brief. I do not wish to kill or have killed any imams for any reason what so ever, even if a line could be drawn to particular cruelties issued in idea or of vocal disarray. Criminal murder if to have included the tool or weapon of suicide bombers who are never, never, never to be martyrs, draw only the pity of sympathy from me, because in one case the criminal mind is generally half affected one way or the other!

If suicide bombers to be disaffected in some way, by the vicious nature they have been told to read the true words of God, which in the sympathy of time had been given to us all for the same reading, there is to be no let to their personal crime! Even with the whole secrets of Islam always having been there from day one some fourteen hundred years ago, martyrdom and murder were never to sit at the same table, therefore no case can now be made to use the former as a political remedy if for religious grievance?

With the reflective arm of our personal code to life having been first laid upon us pre Shanidar by the standard of how we managed that event of burial in the first place? Our own civil code giving us the time to waste upon one of kind to have died was a true step for mankind, who looking into the face of evolution, saw creation! Our own twist of monumental proportions was to leave grave goods with self. Because then ago and to what must have been similar events throughout the world, by that very act in turning the wheel so that our full concerns were no longer to procreation, we had opened the door of purpose to reveal the small matter in meaning of the reflection, creation!

A most elusive shell, but nevertheless the search of which has been the best news in the whole universe, because that very burial rite was also to reveal to us the matter of the universe! From which if we do not know the how by all the scientific studies in use from the fourth dimension, our conclusions are close to the why? Because of some of our closeness to the gods of explanation, we have needed to create in order that we become on route to the one God, by the many counts we have always found to be there? Some of which are so obvious we have made the obvious difficult to feel by locking our own emotions!

For any person for what ever reason, to talk in terms of what or how our imams might wish to encourage some of our young into taking drastic action is not a quantifiable event. I have made the personal observation because of the same reasons that any discussions are made without requiring to make gains. I have made the point as a good friend might with hand on shoulder, but not necessarily for the reasons of friendship when some of the assured acts I see being done are against all the principles of humanity. I will not allow any defence by any religious code, that it is, was or has been necessary to kill another human being, religiously, religiously or religiously!

By connection and in situation to the Middle East being a world example, in that one extended area, from political connotation than might be, depending on our mindset. There are several reasons to hold violent animosity from one group politically and even religiously but only from the direct emotions generated from an act, action or reaction instantly felt. In meaning, that to a particular singular act to the individual or a member of their direct family only. I put that charge to make the one claim beyond all religious law or political law, by all of the driven codes of progress the most important feature of specie distraction is the avoidance of extinction before due time! But this code can only be operated to the limits of family and more specifically the emotion of trust, because that is the fuel to the engine of how we must use our own free will in aid of our separation from all other specie.

I would wish to make no exceptions, but in our aid for total clarity and in terms of family, our connection is from all of us in that constant role who are to make better and wiser all those to follow us by that direct connection. This feature is not clan or tribally based, but to the strait forward link by any means whereby a person, male or female, mother or father, brother or sister, cousin or mentor are to lay the service of all of their experiences at the disposal of our young. So that the future, our total future can be carried into with less fear of the past, only the prospect of brighter discovery?

I put no charge to be upon our young, because they in turn carry all of our failings, all I seek is to put them on the one path at there own pace, at there own pace under the one feeling that they carry only our best consideration for them and for us. To the ideal that they carry none of our burden we had created by our own lacking in not at first assuming we might be right but always working to create change without having the means to replace what we wanted to change ready? For those young people by any group defined by the circumstances of climate, geography, diet and other reasons, if ever a death of self by self can be charged on account, it is to be upon those who would set to kill our young in waste, and the reasons are more than forthcoming. If time is taken in note that no religious code had ever been made except to us all being fully adult in death, which openly means no child sacrifice!

From where I can look and see what is in the reflection of the reflection, it is worth making the one standard for our young, but by my own misfortune I cannot lead in this expedition, I have readily left the case to our religious leaders. Of which amongst our monotheistic religions the three standing groups are our Rabbis to the Jewish faith, our Pope and cardinals, bishops and priests of the Catholic/Christian faiths and our Text of the Islamic faith. Our eastern styled religions are as fine and complete and honest as all and many other religions. But from where I had seen the complexity in relation to our western styled religions is from the generality of all life forms intermixed and included in Gods own plan filling all corners of the wheel? But cannot now be used in the same way we can think in terms of Gods time of evolution being separate to our time of creation in the same field, which is not a comparison, only a statement recognising our similarity before any of our differences.

Jewish Rabbis, Christian priests do not equate to Islamic text, but all three are of the exact same level of importance and always will be set in time to be four, two or one thousand four hundred years old to there respective start dates. By reaffirming those time scales I set always to make us aware of the fact that we do think differently because of the added extra knowledge between gaps? But we can now liken attitudes more by the equal overlay on the three religions with the three taken and one active prospect of the Fifth Dimension by today. Then without charge, without charge or without any attempt ever to alter the past, put those three and one active prospect to the imagination of overlaying our first three points of contact four, two, or fourteen hundred years ago when our religions were given first light.

I have made the case that the last of three in Islam, when brought to us had two vital or could be made to show of two main differences to previous religions of area. Allowing me use both of those anomalies in making my case by personal opinion in where those features could have led to the misjudgements of some imams breaking the first rules of all monotheistic religions? Even that the balance of my accusations are somewhat concealed, by the new circumstance of the Fifth Dimension there is no further hiding place for Judaism or Christianity. Which has always been my intention in how I set my plan to bring balance to all of our religions and will do so even if it appears at times I have my own foot to one cup?

I had set the case that from Moses and Jesus was given tenet for us not to kill of self by self religiously, but from time to time we do and continue so. I had set the case that from Islam the matter of active death by the hand of self upon self had been taken in principle to defend, protect or promote Gods one final religion relayed through the last prophet in man, Mohammed. Without condemnation I had set it so to accept that this principle was of its day but made from the human perspective of living through the changes to have occurred from the previous two thousand years six hundred years?

Coupled with the second divide I had put to belong to Islam more than to Judaism and Christianity, from where at outset this latest religious view had broken with the code of immanent conclusion and settled to be generational. More fitting if, to our length of time in the order of creation, but not realised then or since perhaps until now? What had not happened so far between all religions of the one God inclination, has been to the core that should and must be now given to the future. Because imminent in any case by Gods time scale of evolution, is openly generated by our time of creation in all realisations! Three and a half million years is imminent, therefore the haste of now must be generational which is to be shown not from the past only the future.

From Islam in showing us the first aspect of a generational concept by setting all text of then to now and further, was done in the opposition of change to continue which is now our task, even if uncommitted, but not to change the past. Islam was never to remove or deny Judaism or Christianity but run on the same tracks in the same vein of all? What she has done along with Judaism and Christianity has been to run in time with both, but not to have fallen into the trap that they in overlaying all outlooks to religious tolerance at the behest of politics had guaranteed?

Then having been classed in time from the mid nineteen sixties, for example by my own judgement; if to have been seen or felt classified or even cheated by political and or religious reasoning from/of/by Judaism and Christianity only. Islam in voice was muted by the misconceived view of how politics now regarded the standard of what religious tolerance was or is? From the western political outlook carried for the world by the United Nations, it was to be alright if democratic countries aligned to the western bloc carried to there own duties, but linked to the overall political plan if stimulated by the bloc culture standoff situation? Which would mean that more so from the Middle East in particular if relevant, that any sovereign nation by previous alignment could carry to their own religious standards of accepted tolerance even if to thwart other religions directly, but not yet having been made fully aware of linked situations, because then we did not have the three taken and one active prospect to assist further and future rational observations?

Like for example if Israel was to now promote her own religious standards, being no bad thing, but if done by tolerant means, where might the mandate be for her to ignore the events of two thousand and one thousand four hundred years ago? Under and justified by the principals of the mid nineteen sixties that might was right if divinely supported by the mightiest nation on earth, who just happened to have a full tolerant Christian background?

Although most of the principles of communism were questioned at that time over the main feature of commercial politics in comparison only, when moves were made and decisions were taken. All were supposed to be wholly political, making the case that in fact Israel had not acted by any religious mood or reason or connection. By some of us in closing our eyes to the assumption that she did, allowed that the principles of democracy were to stand as a reason of success against communism, If only communism! Who happened to be aligned with many Middle East countries, who happened to be of Islam by religion, having maintained a one thousand four hundred year continuous presence in their own lands mostly of the Middle East!

Unfortunately what was not available even in the mid nineteen sixties was the simple idea of overlaying all of our political concepts with the prospects of the Fifth Dimension. By fair choice and not in using hindsight, what that would have meant, is that in order to bring communism into the bearing to the consideration of people above state. We would have to have forced the political issues of human rights or civil rights or religious tolerance, to draw some form of compatibility, from what was a wasted effort of a style and system automatically regressing back to evolution! Having no aims other than instinct and procreation like forever owing to all other creatures including mother country and fatherland!

From the full standards of how communism actually worked by changing degrees especially in the Middle East proper, when she was in decline, the religious arm of the sides she supposedly had supported while in standoff. Had no collective voice that could have expressed any case from one source effectively. Because at no time, even to now, no comparison had been made between any forms of monotheism, save in us and how we individually had fallen to the first words of Moses seen through Abraham, later through Jesus and Mohammed, but not then seen to be one!

Our differences being that from very recent and modern interpretation although called, there is no active separate clause between religions and or politics because one is politics and the other is religion, held to be religiously tolerant. But or by or from the one tainted standard that politics had allowed, if to suit the primary needs of what original political standards were, to what they had now become commercially?

Without pointed direction but at first connected to what religious sufferance is and might yet try to become. By taking part notice of how Israel was allowed do, even by tolerant means, has laid the straightest path for us to better progress in the name of unity, but by direction of our religious leaders first, which is a matter to be drawn on page and has been so done but to wait for the refinement of natural justice!

My own view of the same matter, because of the deep disturbing times we might live in and wish to do so. I have made very pointed observations of how some imams behave and have done and still do. I put many in the practice of open deceit, of bending rules and in cases of self selection to exploit the made mayhem they had promoted. I do not support their particular type of ranting in tarnish to the true meaning of what Islam will always be, our last delivered monotheistic religion, born in standard to the future and the future in now?

Fortunately many of our imams in we, just get on with the job in hand, we see no further than we might by being able and set the path for our young in the faith as it was meant to be. Naturally recognising the standards of conduct throughout the world which are likely to have a serious bearing on many of our futures outside the important feature of leading the good life anyhow!

Unfortunately perhaps only looking from the outside to see the importance of having a defined religious leadership structure, has set undue problems to the matter in balance of how we should try and progress on the one level equally. Fortunately by the good advantage of how the generational picture actually works. Without charge and in full compliment of what can be taken as intention by any connected prophet from the past! What generational impulses were to actually mean above the matter of being imminent, was that in all conversation, we have the open right to translate all past present and future thinking into now. When done our reasoning will highlight the added importance to the separate overlay of the prospects of the Fifth Dimension upon all religions. Unfortunately taking for granted that we do need a direct role for the direction of another lead from Islam apart to the full writings of text, which have sufficed these years without the question being put?

I have made no claim that Islam is leaderless nor can anybody else. Many of us know in the best connection that indeed there is a fine body of individuals and groups to set the arrow of discontent strait. I make no case against or for or to those of us to be Sunni or Shiah, save we are of Islam. My injection of cure to that situation is for the utmost practice of religious tolerance on its own level between all factions within Islam.

For the best reasons, that to practically all religious standards of Islam, the one pull is to the religion first before civil order and or political control, which is no bad thing. Save that we forget that to spite what the past might wished to have projected of the future. Any cases were made before we were one world, that is before we had discovered the new world? Before we had discovered our burial at Shanidar, before Darwin's work was to miss-form emotion about our possible differences, then before nineteen forty five when for the first new time, we began again to try and set the triangle strait!?

By making this last modern era of vital importance, to have selected the one area of the Middle East and generally the one time from about the mid nineteen sixties, is done to world standards. Because the two main influences of control to be upon us in our religions and or politics, are ably expressed in parallel equivalent to almost anywhere else in the world going through similar upheaval. Although I have focused on one particular religion or more so on the taken role of some adherents of that religion, is done in open display to us all. Which can now be better asked of by using our one active prospect in 'One for the Whole' from the Fifth Dimension. It works both ways politically and religiously, but to be taken from different perspectives and of different emphasis!

I have taken note to the worst side of some imams who have turned what has been the most obvious misjudgement of our political half measured delivery into some sort of religious error? Political failing is political failing, even if to ignore some of the blindness of how politics might not see the conduct of religious tolerance! From that image, it is easy to imagine that even to profess to good intentions, actions if misguided in the first place can produce catastrophic countermeasures or even reaction if to take the reflected image as seen to be on its own account unmotivated?

I have given it that the steps taken syndrome by any choice comes from the unprepared, which is sometimes all that our half measured introduction of procedure is. So if from such success politically, that the matter of Americanism/Europeanism, Christianity, Judaism and our first world management was to fall by any means to the matter of attrition. Won on the single aspect to the growing expanding feature of materialism, commercialism being tentacles of democracy, also if that was the overbearing type of first world politics!? What role could the only futuristic, generational yet connected monotheistic religion have in the full game of new internationalism, if at first base its voice was silenced never having been heard internationally by any degree prior?

How could Islam enter the world stage if in only one case or only at one time, its main representative body in text, was given the only ear available, that of religious tolerance? Then if later, but still in this modern era by normal progression, her first silent partner to religious matters in communism was the other ear. Who had been forced to accept the features of human rights, civil liberties and politically correct directives and also to the matter of accepting religious tolerance superficially! How if, was any sovereign Islamic country associated too or close too or part of what was once Palestine, make the same political and or religious sounds made by the one Jewish inclined country new to that same area? Who in having come to fruition in her own right politically at the ending of the United Nations mandate to Great Britain, was heard to voice religious sentiment which is the right of us all ever? But not if to meet the intention of what religious tolerance was supposed to be and was acted upon in part, if only to those who had voice in the appropriate situation? If the only listening was to be done of ear to those we were best connected with who already held the attitude of might always being right!

From the situation arising, any number of combinations fall into view, what is of concern is how the jigsaw has been best suggested to be put together in the matter of might being right? For me to take a positive objection to some imams is not a negative act. I have taken a corner piece of the whole jigsaw and studied its importance by effect to what the expected picture should be. My direct reference to some imams, is born by the obvious in what many rant above the process of persuasion, of which I have been very liberal giving an account of how and why some of them in we had acted and continue to act!

My prime concern is to the full danger they create in the name of Islam and to what might be the reaction by those who only see the first reflection or the reflection of the reflection of their own intentions. I have made no defence of why some imams have behaved as they do even in the full name of Islam. My case has been far more pointed to the matter of our own failing to be open human beings by using the ambiguous in excuse of some of our excesses even to have been misread!

In as far as the matter of Israel has been put to view by my own examination, I have made no case of support or condemnation simply to reserve some of our misguided reactions set from the matter of our misinterpretation of the situation whatever? To balance that against how some imams in we have set there own case is now done by the better information that the Fifth Dimension can release for our equal benefit. Instead of playing the hand of politics in one game and periodically trumping the reactive force of some individuals with the ace of religious tolerance which is to become no longer viable by either division?

From the part or point of view or role our politics has put forward without listening or even understanding the concept of Islam. We, from our best efforts of not listening to what could be the voice of over one and a half billion people religiously inclined to Islam, is no less than a huge failing. By not listening to the tune of what religious tolerance actually is, is a further misjudgement, because at the same time the representative voice of politics has been silent to what has been done elsewhere?

From my own view even with added differences the balance that I have seen stems from the shortness of time we are all owed by the fact of our fifty plus or seventy plus years of actual living we are allowed under average example. What I can suggest without bias or let, to what some imams might do or have done or promote from all the wrong angles, is explain not the why they do what they do but the because, which is always loose but factual!

From where I had painted in the clear that in cases, perhaps because of no other reason than they are first world politically. Israel by representation has far too loud a voice on the world stage in respect of her main religion in being Jewish? She with us all had been given the same one standard to be religiously tolerant. Yet by code supposedly owed to us all, she in our name has led the field to her own religious matters first. Not even to consider that of a fine Jewish population of up to twenty million souls world wide, her religious influence from the one land of Israel with about five million souls, far outstripped the balanced voice of either Christianity or Islam.

Religious tolerance is a fine concept, but if it is to compromise the whole aspect of what democracy is supposed to represent, then the reflection has to be reviewed! By that very chart, in a small community having a larger say than the two much larger groups religiously, our review of that situation is to be urgent, but better balanced! Our care is to be that we do not tip the cup one way or the other, only set the scales to hold the same status for the individual but to the revised means of who the individual is now, having the full balance of all the prospects from the overlay of the Fifth Dimension anew? Even if to replace the one standard of human rights that we have come to use in supplement because when introduced we had no alternative to work with like common decency C D for that end until now!

Accepting as we must that religious tolerance is to be a gauge, then by that rule across the full range of intention, if to have allowed one camp its own head, we have to accept what all else are entitled to do! Fortunately no allowance has to be made to the matter of any religious group who advocate human sacrifice in doctrine, unless to be misconstrued by the fixers of all interactive matters, to whit, our politics. Not to allow that the range of human sacrifice from any of our religions, any of our religions, should be our standard! But in the case if, whereby for instance there is a spate of suicide bombers making there own standard of religious tolerance! If only by reaction to what the reflection of the reflection had shown, we now can and will use the prospects of the Fifth Dimension to address the one standard to all people post reflective countermeasure. Even if in using ideals from the mist of recollection whereby to die in religion was to become a martyr to the faith, no merit is given! Making that our open obligation for all intention is to set the triangle of birth, living and dying within its own limitations, but by no violence at any angle?

By any key to the future, because I have given the open ability for us to compare two small matters having huge consequence for us all, is done by myself not in the comparative note that is first shown. I have not, not made the case allowing for any sort of violent action or reaction to any situation, save reacting to personal crimes against the individual or family. All done is to show in my own terms how the reflections had appeared. So in order to show how balance is to be seen scientifically I have used our Fifth Dimension in overlay to all future events, but yet to be acted upon.

Irrespective to what others might do, which has never been blessed by the main force of religious standards even tolerantly. In part to the major, major mistakes that some of our imams have shown to think by? Much of my many points have been directed to what some have promoted only for our general improvement in the one form only of accepting the necessity to the possibility of change required? That is why in all cases throughout these pages, I have set to target some to our imams, so that all might be able and view our situation of now?

In the full light of what has been happening these past fourteen hundred years since the introduction of our last main monotheistic religion and these past sixty plus years of this last modern era. From the present situation in the world today to be expressed through the one avenue of religion, if not viewed from the aspect of generosity for whatever reason the matter of change has to be examined. Which on evidence is what we have done all the time, but from now we must be better attuned to the effect on others of what we might have to do!

By loose definition and only to be seen for reference without applying what the reflection might be imagined to produce. In first looking at Judaism, what has to be realised is to the nature of that religion and how it was formed, even if to have borrowed some pictures from earlier Egyptian religious thinking? Likewise of Christianity and now Islam, but always to remember that all were first to the thinking of their day, but under very extraordinary circumstances?

When in time Judaism was seriously questioned of theory by the prophet Jesus incarnate. For those who did not take to new ideas or ideals was to remain the original concept of what Judaism had started of to be and maintained in that form by an able priesthood. Having then worn the one coat for two thousand years in immanent desire until the time of Jesus, their standard had been set. If for the following two thousand years in the time of the vastly growing faith of Christianity the Jewish core was to remain with old habits, then that is a matter of concern to the individuals involved, which is to be asked but not questioned?

Although of depleted numbers because in first sweep many did come to the new Judaism as told by Jesus and from the fact that Judaism was originally a very local tribal religion, for some to have broadened there own horizons left a smaller core to maintain original standards! What followed in the name of Judaism and in parallel to every other sect or group to have continued in tribal form if and only to intermix to the purpose of self? Was to create a field of entrenchment by the means of disassociation to new approaches, supposedly of the same religion, which then had to have been delivered by a false prophet as had been done before? Which might again have been misunderstood and the new direction offered failed to impress, leaving in part a core of original adherents to the faith in how Moses had relayed his own messages from God not least not to kill of self by self!

Although our Jewish history is as diverse and diverted as any other over the full two thousand years of Christian history and the fourteen hundred years of Islam. By what ever means of association without having to change emphasis, those to have remained fixated to past doctrine remained so, but of false harmony. Not least because the first strands of originality were born from the thinking of two thousand years before Christ Jesus!

In owing to the nature of how events were to be involved in change generated by the growing need for nationality especially in Europe. When by no more than the flow of circumstances to have affected Europeans in a slightly different way than perhaps those encamped in the Middle East proper. Was in part determined on a similar scale whereby if from religious terms Europeans at a specific time were to be progressively inclined because of a more open acceptance of what the fourth dimension was to reveal. Balanced to how most of the Middle East was still making modern discoveries associated with the sciences. But were content to be laid back about such findings in any case, when in finality all our religious searching was to be achieved because of Islam without forfeit!?

Without the cost of compromise, but of costly outcome, by events in these past sixty plus years of this last modern era, our religions were to be counted for the first time on the world stage. Our findings were such to have now embarrassed all of our consciences because of what we had not seen but come to accept, whereby on the new world stage of economics our standards have been set to and by and from a first world perspective? Of which we have a huge Christian connection with stylised association of Judaism having been partly influenced in the similar field from past circumstances. From having been displaced numerically at first by Christianity, then Islam in time form?

Almost fourteen hundred years late, if Judaism was to have come home, it was by the coincidence of being first world and progressive at the only time in those fourteen hundred years that she could! By no more than the circumstances of modernity without the effort of construction, but due in cause to the advent of time only, timing only? Our open reality is to a very small religious sect born some four thousand years ago who have remained to those first standards of acceptance of creed without change. Yet have readily used and accepted all the expositions from the fourth dimension in the most up to date and modern uses, in order to set in standard what had been laid down those four thousand years ago religiously of which the results were imminent, less that span of years with the scientific advancement from them!?

Making the full opening of as many questions that might have to be answered about the morality of immanent desire, used generationally by only adopting the said advantages of the time produced bounty? Almost to compromise Gods own involvement fully, from whereby to have rejected other linked faiths like Christianity and Islam, yet embrace the modernity of science and commerce. In a range produced far in excess to the understanding of the first time of Abraham and of how Moses understood things to be. Keeping to the same religion for four thousand years without change is commendable, but not a worthy practice to close ones eyes to the advances of future study in God, while at the same time gather to self advantage what all in man had produced commercially through the course of history generationally?

Without comparison but of the same inclination, Christianity again by the turn of the weather found circumstances to be very similar. Leastways in one term of doctrine expressed not to kill of self by self but could, Christianity although of its own time two thousand years after Judaism, had taken to the more open fortune of opened doors to all men. Without notice at the time, this attitude was to create a broad base to the matter of change which had been ably demonstrated throughout modern history especially in Europe. Making that the religion in Gods name was for all men in their name, hence that there was no valid objection when Roman Catholicism became mixed to be also of European national Christianity!

I had previously indicated that for some reasons it was from the above combination that stimulated a certain drive from individuals to achieve national objectives, at the same time keeping under the umbrella of Christianity. Bringing eventually what I have called our progressive outlook, which can be viewed without cost even to conceit, which is not to be counted in the negative. Because all futurism was eventually to fall to us all in any case and by all time, which has brought us to now, but still at our own pace.

When mentioned of the crude and cruel European national interaction at the time of reformation at about the same time the new world was discovered. To have found the aspect of human sacrifice there and to obliterate the habit with as much fervour as it was originally practiced in. Further sign was taken by the European standard that their star was in the ascendancy stimulating the advances in discovery in revelation of many of our mysteries. Creating the new frontier of modernity in science to be able and guide most of our political decisions yet held in the check of knowledge by the church!

Creating in turn the new countries of the Americas with different perspectives whose outlook was now first political then because of such other circumstances championed by the American dream, was also to be religiously tolerant! Making the core of todays rational at a time of vital importance to what was happening in the independent America of some two hundred years ago. Even held against European attitudes to the people of the Middle East at about the same time from how England and France fought there! Although not linked at the precise time what was to eventually percolate to the people of the Middle East was the world standard of religious tolerance as a bi-product of commercial politics?

By review of Islam now in the depth of this last modern era, but only by my own acceptances and understanding, even by the misreading of concept of what might have occurred. My standard is ours, because like from all of our pasts we each have only fifty plus or seventy plus years to become aware of what purpose is, should be, and of how we are to bring one standard to all. Which is to be an occurrence in any case, because we have always known the answer is through the keeper of the mysteries, having received proof from the conduct of how science has continually worked and reworked the clay of creation!?

## CHAPTER 182

By my own terms only, and I make that statement in its form not to draw the wrong based judgement upon others if set with the wrong association that might have been drawn from what I have so far written! In this many pages so far, I have made but one statement of fact which now has to be balanced without being judged. In the same way for what has not been written no barriers can be made against anything said, because our new equal full licence to all is through the prospects of the Fifth Dimension!

With that in mind by any combination of past events to create what I personally consider to be the modern rational. I have concentrated my main efforts to bear upon what has become of our relationship with God, before the study of any type of our political or scientific endeavour. Politics has yet to be put into its proper perspective with regard to how it has grown from the one seed of instinct belonging to evolution to be able and distort the full study of creation!

By taking to our monotheistic religions in a different form of study owed to our Eastern religions, mainly Buddhism and Hinduism. I have been able and isolate all the standards that politics might put forward to lessen how any single religion might impact. Because also in theory politics from its first issue of civil control, only works to the secular which cannot include any form of transmigration of spirit if that is the case of any of our Eastern religions? With that aside, but not to be forgotten, by refocus to our monotheistic religions and or the political position of the Middle East, throughout. Any case made of how we have ignored the true impact of Islam, has been either totally forgotten or refined to the corner of religious tolerance?

Along with my personal concern to the fact that we have underplayed the world role of Islam if not politically mainly then by influence morally, my balance has been set against how some imams in the course of settlement to that issue, had overplayed their hand. Not least to the detriment of at first Islam and the good memory of Mohammed, then broader to all of us of any religion. When it was always Islam that led to the true spirit of what religious tolerance is by its proper maintenance of keeping to religious forbearance almost pointing to some prospects of the Fifth Dimension!

Using my own good office of abstract reasoning, if to forget the single word martyr and martyrdom in any context, also to discount for brevity, of how others might wish to resolve my own focus to what suicide in the name of religion means? I give my own account but enclosed with our learning, reasoning and understanding of events if not of these recent sixty plus years, then of this last modern era. The one standard of query I make is against the reason, cause and application of what new religious suicide means above what it has been thought to mean!?

I have put the name imams to account in accusation of leading a great misdeed we are all responsible for. I make no apology because my terms of reference are based on the supposition of fact and taken only through the religion of Islam. Not in condemnation, but because of the great wrong we have all set to Islam by the promotion of the matter of politics if not mainly democracy delivered only from a first world perspective!

I would add caution for us not to assume that I have allowed an excuse or a reason for the how and the why I have chosen the name imam to be a pivotal feature to our future progress. Because in the same manner of how I have chastised some of our political attitudes in not giving Islam its proper voice and status in representation? We as the one community have self compromised what the tone of our future outlook should be.

Imams as we are of the same skin and bone, of the same blood and heart, of the same desire even religiously? Which having been born is to live and then die in all consequence by 'Natural Body Death', completing the triangle in full, lasting fifty plus or seventy plus years. If without any connection to other religions or of other political conduct, Islam alone by my own account had maintained one of the seen differences I had focused upon between us. Again even at this close time when mentioned that our previous monotheistic religions set in claim not to kill of self by self and do. Is not weighted to our third religion of the same standard, not least because each were of their day and more specifically to the full thinking with the full ability to only think as they could, owing to the different times that we all lived in at such times?

Islam by well stated examples even if used out of context by myself, also hung by the strands of her day and time, but by impressing that feature whereby we could kill of self by self, given it was to protect, promote or defend this new concept at core? Was married to the fact that the original text of then, first entered some fourteen hundred years ago, has remained unchanged within that same time span. Therefore to be able and kill of self by self then ago, has now been distorted to allow that wanton murder can be used through the double murder of suicide of self and in squandering the real truth expressed from the open heart of our young, carried also to the innocence of other victims!

Where the mixture fails is to the obvious as I had indicated previously, not least from the first time I revealed our soon to be taken prospect of 'Natural Body Death' as the norm? Although the vital text of Islam has carried these fourteen hundred years, our use of the adage through martyrdom to kill or die in faith is an empty shell? There is no longer a case in Islam that we would have to defend, protect or promote the religion by name or in name to the other people other than the one and a half billion adherents there are already to the faith. They are testament that the religion is safe and has already passed all the precautions laid in abeyance to the hope of its first true success.

For some of our imams to address the matter that in order to stave off the invisible invasion of a mediocre political advance of a type called democracy, is to border on the edge of lost causes? For some to have encouraged as they might, some or our young to the sad and cruel acts of misadventure to take the option of martyrdom when it has no standard is an act of personal waste beyond recall. Especially when by these religious times there is no longer any requirement of duty to promote the faith of its obvious success. By swing, for any suggestion to be made that the faith is to be defended against for instance the onslaught of one small almost insignificant sect. Looses for it its own sense of dignity, first signed by its generous nature to all other religions by its own defined terms of religious tolerance in the name forbearance at outset almost fourteen hundred years ago.

Islam and imams go hand in hand almost like creed, but from the specific role to be able and guide some of our young. Imams can create what might be a somewhat different relationship with their charges in us. Without consent, even mention the vast, vast majority of those of us given the role or to have come to the role of imam. Have done so in the hushed care of duty and obligation to what it is and should be, irrespective of other circumstances save to the good health of this our faith, Islam!

No matter what provocation to the religion if any, we by definition continue in the success we have enjoyed after fourteen centuries. It is only by the sense of panic and worry and from the uncertain nature of our own standing that some of us have weakened and fallen to betray our spoken concept of open unity. By looking in reflection at the needless brutality of some political events, some of us have failed to review both the first image and the secondary images of further reflections, which had given us the wrong impressions at first take. But unfortunately had already sparked the means for us to incite reactions as we saw fit, but in reacting to what!?

For me to have stated in fact that at no time now, at no time now, Islam is ready to deploy the activity of martyrdom, is to be reality! If Islam was ever to use the maxim to kill of self by self for any reason or at any time, no call to faith had been made in this modern era save to politics, save to politics. Bringing dark and undefined results of what is meant, what has been done and what is yet to do?

I have brought imams to the fore for the two reasons which like other aspects of influence have several avenues. My first case is to the how it is from imams generally, that the standard of our conduct is checked in terms of how our young pay note to our original text? A quite natural offshoot to this, like in all societies, is how some of us can influence and are influenced by those in close contact by whatever connection? From which in this particular case some of our young in learning look to opinions from their mentors!

In the vast, vast majority of cases this relationship is a dormant exchange simply by familiarity. For centuries and centuries the standard was normal and full and truthful and uncomplicated, without sign of a spark or a match with which to light an idea or a negative thought of retribution. When arrived at this precise time of this last modern era, Islam by eyesight found herself without the voice owed her high standing to world events over the past millennium, even to have operated by a laid back style! If not to the first realisation that the new standard had been forced forward in the shaped form of politics, what was to be her opinion of now, at the time of that realisation?

From the opinion of nothing needed to be done because Islam on our account was well prepared to defend, protect and promote our objective in the form of peaceful progression progressively as it always had done, laid back? Then we have no story that needs to be reckoned with, except the one case of modernity generated from our exact times of now, but driven from the different angle of politics if at first to have made there own rules. Then how should we balance the equality of religious tolerance, if to have allowed some religious reactions from some of us to interact from our own religious standard without open consideration? Yet be covered by the might of politics to do so, not from the perspective of being different but only because we were first world?

I make no subtraction from how I had led that the new aura of sacrificial suicide by any terms had been revived in the Middle East, in particular in a certain style from about the mid nineteen sixties. In the same way that I have set the picture for us to kill of self by self in any cause, civil law is the one key only to that task! That I have now laid down our prospect of 'Natural Body Death' shows in reflection that the future does not have to be cast in line of human sacrifice to any end. Has not been done to tip any scale save only the true value of all religions and what they have always stood for, life unto 'Natural Body Death'.

Before the Fifth Dimension was to overlay the means and matter of all our politics and religions, I had gathered my own abstract rational to understand the why of our present situation in highlight through most of the events that had already occurred between us. From whatever the situation appeared to be, no fixed account had ever been given to what we should expect in death other than from our religious attitudes in all cases? For politics to have misled our true intentions was required for it to follow the creative pattern that we were something we are not! In doing so the picture although related to our religious choices was a wedge between intention, outcome and purpose. Which of the latter, purpose, politics was unable to enter the field, because purpose is a relative term to how the ambitions of God might yet be realised!

From its own standard politics unashamed has trod the one path only supposing any first intentions of what civil order was to be, she in effort by the mix of unseen circumstances has trod that path in the one name of commodity.

But worse, unseeing and on one leg only, which is a disease to have shown its symptoms in the most recent of times! Politics by all accounts has no master only the self perpetuating desire to show its own prowess, which having been done half measured in form from nineteen forty five, has propagated the imagination by terms of evolution only. To self generate on our behalf making and taking our steps taken syndrome on the one path that is inevitable?

Although attuned to the ideals of mankind, politics by choice has taken the one standard to be self serving, such in case to have put in class our religions to the silence of religious tolerance. Leaving the effect that choice in that form is the only rule running to the tune of expanded commodity fields of materialism and all of its servers, save religion. To boot, that the total style unread of how we manage in control, is the very single means whereby we do manage all things linked to the adage that might is right!

By the very fact that we have walked hand in hand since Shanidar, later connected through vital interactions in Mesopotamia, by the first classification of what were to be real religious connection if through the image of the Pharaohs and the heavens. Carried in line of succession if when through Abraham more viable links were made not between the king and the heavens, but man and the one God concept even if at first met through tribal associations?

If to have made those links with God, even to be tainted by the reasons of our excellence of mind but error of thought, that God had found us? Proof to the reality in point of fact it was we who had found God, that it was only we as the one same specie, we who were first to nail the ideals of hope and purpose to the mast of life and living unto death all for the same understanding, but perhaps much bigger?

Even unknowing that the mystery of Shanidar was uncovered to us all in succession from about four thousand years ago by the minute way we adopted the same theme differently. Our only single problem extended from then has the one core, whereby at no time have we been able and separate the fact that in line with the full knowledge of God and purpose. Our civil order of control had intervened so much so to give off its aura in similar style to what must lead, and what must be? With the vital difference that when civil control was to become politics, it by type expanded the standard of instinct only.

That is all it was able to do, because civil control is a feature of service. Politics by extension is part of that service which has always been taken so, until when? In making that case I have not changed any aspect of the past, but have only allied our present to then, in highlight of how when through the righteous knowledge of our religions, we joined our two forms of control matter of fact. Arriving to the one position of now, whereby we do what we do by who we still think we are?

For me to have built one case in the direction of reason and cause, then to introduce the reflective means of management expressed about or around how we have interacted through or because of religious tolerance, offers no excuse of how or of why some of us react or interact politically. Because of the shortcomings of our religious tolerance, which are supposed to take to the full political standard of not interfering, unless to be approved if allowing that the action or reference suited a particular political point of view? I make that pointed reference to highlight how many of the events of the Middle East have now been brought into the world arena.

Although to have used my own style of delivery even if to have been erected from the spirit of abstract reasoning, it is one of the best ways we might ask awkward questions and expect full answers. It is the best way to value how we interact and have done so in the past from the false premise that things in most cases were as they should be. Not least on the religious front in the Middle East in particular. I make no case for one voice over another on those terms which is standard to the ideals of religious tolerance, but better seen?

From where I had ventured at regular intervals in reference to some of our imams has been done in relation of how things actually are. But in all cases of reference I have made no separate standard for one against the other by any charge. In the same way I offer no excuse, no excuse for what is done by one to the other of us by us! My full and only intention has been to cut the grass underfoot to the one length so that we all stand on the one surface.

How ever what I say might be taken, my intentions are clear of which no further apologies need to be made, but before I close that door with the means of remedy to our present situation delivered first by our religious leaders. I enter the arena for near the last time when explaining what the matter of cause has been for some time of our imams in us to behave as we have?

How ever bad politics is or how ever good religions are, in all of our histories we have not been able and recognise their true relationship until now, because until now we have never expanded the attitude of concern for all, first taken if unseen from about nineteen forty five. Although our original plan was in the aftermath of two bloody world wars to the cost over eighty million deaths world wide. For those examples to have occurred in the first half of the twentieth century bringing us to nineteen forty five, new plans were first shown to be needed. At last our new first intention was to the desire to eliminate all the supposed reasons for world war and at the same time local wars!?

From the haste of want and desire even if through the offices of the United Nations, our plans although viable, worked from the one perspective of measured change, but created there own problems. By that, we must take heed of how we had misguided the original plan which has to be balanced? For me to have taken a major aspect of our political classification how we actually work on a first, second and third world basis to a commercial bias is fact. By the same standard for me to have made a point of reference about the misplaced management of what religious tolerance is or how it works, is also stained in fact. Allowing in recognition of the two main influences to be upon us, to be better examined from new perspectives by inclusion with the overlay of the Fifth Dimension, now!

Without any broad intervention, without using my plan, one of the first steps to be taken is the one I follow by intention. With the largest part of my own concern to the matter of specie unity which is a simple name for humanity. That it will be given in suitable time, from certain leads we will automatically change our political and civil attitudes to the edge of organised change. Because along with the original attitude from about nineteen forty five, even if only half measured, we now have the opportunity using new points of reference to expand the outcome of our steps taken syndrome in line with having taken corrective steps. But from now full measured in a new and better judged attitude to include all of us first hand and to be first world by our own standards of acceptance, but assuredly in the first world framework?

Before we might begin again I have set standards of cause to what could be our situation by how some events had developed over centuries and more pointed these past sixty plus years. At last, if at first our political path is set to change without having to alter the past, only the future. No matter how much we had tried before, we never had the strait forward simplicity of the three taken and one active prospects of the Fifth Dimension?

That we are to apply the same overlay to be upon our religions, I have used as a guide, the one and only blemish which could hold our ideals in ransom? By association one way or the other counted against all religions save one, I have taken the standard that we are not to kill of self by self but still do! Choosing not to labour that point in terms of direction my focus has been aimed only at Islam by example, but with the best of intentions?

For me to have raised in notice the standard that Islam only provides the means and agreement to kill of self by self if needs must. Stands to how I have placed the reasons to be in line to defend, protect or promote the faith in or at a time when peril and failure were possibilities! By style, Mohammed had enough of face to put forward the real issues of the day in order to progress in the open progression in our name and to our ability of understanding along with our limited wherewithal. Although planned for the future, Islam had first to attend to the attitudes of now, then ago!

Fourteen hundred years later for me to put in line how at first the tradition of defending, protecting and promoting the religion no longer exist. Is indeed reasonable in the light of how our religious conduct through Islam since Mohammed gave us direction has flourished. By any religious terms there is no longer ever to be a time when any Muslim might need to kill of self by self religiously. Unless in light that we had been misled to do so from the concept of instruction if to have been set in requirement from those of us who have lost the ability that within the same religion there is always a future unexposed?

From personal examination I have made the trip to study one reflective image from the bulk of our three monotheistic religions. I have made no attempt to steer Judaism or Christianity from close examination, but have left the field in the certainty that their own challenges are deep and dangerous with regard to the common ground they both hold. That is in terms from Gods law in equal acceptance whereby we must not kill of self by self under any circumstances even human sacrifice, but they do!

From Islam the standard is not the same but the intention is. Our only note of concern is to the one feature of what is civil murder as opposed to religious murder.

Being the open question to be asked every time a religiously minded devout young man or women straps on a vest of death and destruction in the name of what misdirected martyrdom is! My challenge has never been to the individual because we are all to be of the same inclination. I have made no case that suicide bombers if, are mentally defective. My concern from the personal view that I permit is drawn from the tedious in that two wrongs do not make a right. My concern by expansion covers the matter of waste to the efforts of all of us, who by the simple matter of living, we hold to the concept of God being first, not politics!

I even carry the same line to those of us cast in doubt to the reality of God. A doubt I believe generated by the very structure of all of our religions. That is why I have begun to set the task in hand to be determined by our religious leaders, being those of us better acquainted with how our separate concepts have allowed us to think apart. Yet with all having been lit from the same spark of wonder, if at first to the magic of life and its mysteries then to be drawn from the excitement of where death might lead!?

From how I have singled the one action of encouragement, if what some of our imams pass to our young are for the reasons already stated even if not understood? Our key to this viewpoint that I had formed, have honestly been generated by events of this last modern era in real form from about the mid nineteen sixties generally in the Middle East.

All of our reasons are to be there even that I had trod the path over and over. What is said from now is the open standard of how abstract, my abstract rational works. When, where and how I have continually put myself of the first person has been done in the case of a singletons relationship with God or of any expectation we might have! Unlike from the core of our best, even biggest religious communities, who in good faith have brought us all to be of Gods consideration, if the reflection had been read properly!

My pittance stands to the individual, which places us all to be Moses, Jesus and Mohammed. Which even if complicated by the structure of many of our Eastern styled religions, our eventual understanding is to be the same that it is we from the simplicity of our human form even to have undergone several transmigrations. Will be first or the one specie type, who are to attract Gods interest in us for what we have shown to be able and express understanding of! We are the only specie among our several transmigrations to have been able and break the shackles of evolution belonging to others and run the course of creation? Creation being the one concept throughout the universe that is our single triumph over all matters, all other creatures, even the fourth dimension!

When the concept was triggered at Shanidar some sixty to one hundred thousand years ago, it has taken us three and a half million years before in this last modern era of the past sixty plus years, before the warmth could be felt from the good glow of all of our best religions and from the understanding that we are right!

# CHAPTER 183

For you and I to have set our standards above the group is a false economy by any ruling or intention or inclination. Because in all cases from the unknown, whatever done is half measured and will always generate the worst type of our steps taken syndrome, so much so that in this last modern era particularly through our steps taken syndrome from any reflective code. Other hitherto unseen elements to the basis of how we are to interact are automatically raised, which will destroy the very essence of specie unity by promoting the worst wrongs that we had manufactured in the guise of natural selection. Especially if claimed to be part of any single group when by all examples we are the one single specie who in developing under the circumstances of geography, climate, diet and other reasons had set down differences?

With all the arrows of decision making pointing down from a position over each of our heads right now, the only way for us to answer that challenge is through or from our religious leaders. Who until now have never had the three taken and one active prospects of the Fifth Dimension exclusively to be a form of naturally selected guidance. When coupled with the already good and fine tenet, covenant, inclination and indicators from the past and the past and the past. Our one line of delivery in all cases is for us to use our natural inclination of real humanity wrapt with the observation of humility because we naturally care for our young, which is a following trait of our specie, because our young are incapable of looking after them selves! Therefore by natural selection we are bound in the community of family and specie. Until we are able and release what has to be the key to the answer of creation over the standard of evolution!

Always to be included is the matter that we are the one single specie whom by good and fortunate coincidence, have been able and bring substance to the universe. By no call other than from unmeasured time it is we who have also created the rational of existence. How done is by where we are and what we are, both in relation to all other life forms and then to the evidence of purpose. By the great, great good fortune that we had come to God on terms, remains the single and main total mystery of all measured time, I say measured time because that concept is wholly human found, above the limits of anything shown from the universe scientifically?

Before we had ever found the thought of science we had found the object of time in passing, not to the collection of answers by experiment. But to the continuing rational of purpose by counting the meaning of creation from amidst all that we did not know. Having arrived at the concept of God, progress has been the answer in all fields except true understanding which is where we stand at the door of right now?

Although our open standards are religions and politics the one feature to have caused us the most doubt has been what I had determined to be our fourth dimension? Being a feature of the utmost importance to our wellbeing, but also to be a feature whereby we must continually check the side effects?

Unlike at Shanidar when a new emotion was first felt if not realised. Our running key to the matter in revelation of our mysteries was already on the move that time ago. Because in parallel with the open mystery of a clan death and the unknown emotions that for some of us come with such an event, science in terms of some of our taken for granted acts already ran on the same track, but at a different speed if from the front? Making the now problem that we have treated the same aspects of politics and religion and science differently.

Without being appointed, science was already revealing the best stone, in the name of flint, for us to use to the growing number of tasks we had found for it. Knapped arrow heads for distance hunting, knapped blades for skinning and slicing our prey. Then even other uses of other materials, like reeds to make baskets or to form cup type receptacles made from soil to lift water? Enough to continue and find the best or right kind of reed or soil to use, sturdy marsh reeds or clay that when worked into shape and dried against the fire or sun were used until more were needed or better ideas brought newer innovation?

Although in the course of time in a hidden way yet to be understood, we had been automatically drawn to the habits and rules of the materials we used. No sign was given us that in type those implements and connected understanding, although mysteries just like the mystery surrounding the death of a clan or tribe member, would reveal themselves in time? Whereas the concept of death into religion was to take the drive in all of our aspirations even to cover better adaptations of the things and objects we now knew how to control, even civil order, but in paradox not yet religion?

By the limited steps we were to take through the advent of time, what had been realised in standard apart in owing to all other creatures? Was human intellect, being the full ability to rationalise our surroundings, not least by our adaptability, most to the one echo of our mysteries. Of which we had come to classify the main standard of living by our civil/political rational linked to the revelations from the sciences of our fourth dimensional aspect? Yet had and still remain locked to the one mystery of human death by its aftermath to be able and reveal far in excess of what the sciences can deliver. Because when we are honoured in 'Natural Body Death' we will have come to the time in understanding of what our emotions at Shanidar really meant!?

Providing that we have entered the spirit of what humanity is by any determination to show our actual separation in type or class or objective to any other life form to have endured any length of lifecycle in the expanse of what evolution is!? Therefore to contribute even in the smallest way that we can manage to the one objective showing our difference by having come through the equality to all life, evolution having come to the point of purpose, and what creation was for, it was our sophistication that aided us more than to be civilised!

Done by us alone this specie, who from now have little to do but show the honesty of what is required to attract Gods interest in us by our interest in what the majesty of the universe was aimed at? We are the one and only specie to have been ever able and stir God to what time is, because instead of wasting two hundred and fifty million years not looking for purpose like crocodiles are still doing. We have only fifty plus or seventy plus years to find the full extent of what purpose is beyond the assumption of evolution by being the only ones to have crossed the barrier into creation, which is not a reasonable reflex!

Bringing what I consider the trail of human existence to the small matter of how some of our young have been denied the reality of life through the importance of death? Is now the full extent of how politics has failed our mission, leastways through the standards of what the United Nations can deliver, albeit from some of the information it had received half measured! I make no case one way or the other because in operating as known, was to follow our steps taken syndrome road. Supporting whatever our half measured introduction was to have introduced, creating for us a very small circle of conduct?

By my own judgement I have taken one small number of events to create for us the best example I see, that I see, to be a fixed point of reference, whereby in all future issues we might judge what our steps taken syndrome really means. Not just to continue half measured but also to induce via the small circle, both attitudes to support a style for better improvement. Not to continue and patch up our differences by some first world individual being oversensitive to what they have always professed to support. But for us to have to introduced the new plan of action from where I have been fortunate to see the one major error we continue to make, in not giving our religious sentiments their correct status!

## CHAPTER 184

For me to have taken the many diversions I had felt necessary to take so far, is fair recognition to our human worth and worthiness. If I stop and I do at the way we have managed our religious tolerance by fitting to this last modern era, only one factual command emerges. That being in the way we were to tackle the concept of human rights, which is ever to be a noble ideal, and ever to be wanting?

When introduced, we were standing on the bones of the disgusting regime of fascism. From which was ample evidence of how far we had degenerated the flower of human intellect to be at a time when we began to think of new horizons socially and morally. Fortunately now with that stain partly removed some of our victors over fascism by taking there own ground of dictatorial communism. Were to compromise some of what was to be our first world half measured delivery of social and community welfare!

By my own score to have used our bloc culture standoff in example and to have classed one group to be religiously inclined with the other in opposition of holding religious sentiments, who instead gathered to the hollow concept of mother country only? Who by choice were also to support many Arab countries of the Middle East, who happened to be Muslim by heart? Without ever seeming to recognise this union, we in best behaviour continued at pace, even if misplaced, to set the idea of one human standard of which the clock has stopped at and remains, until the battery is changed which is now!

I had made previous comment to the effect that some of our first standardisations made to and through the United Nations were flawed by intention. My main pointer to the fact that our western styled political influences to have impacted on our eastern styled political influence of communism, stands. Our open plan was to force the ideals of communism by practice, to have better regard to the individual? At first they did not fully carry, but in calling to the aspect of human rights the reflection was to show the spirit of fair judgement was to be carried to all for all!

With both of our main communist groups in modern times Russia and China, who having grown only by the first use of the bullet before the ballot to remove any and all matters of discontent to the standard taken by dictatorial communism. The above ploy to the matter of human rights was to draw their integrity. Unfortunately in the time waiting the effort of being forestalled by clumsy and slow moving leaderships, millions upon millions of premature deaths in the name of that political system were to follow in both countries. Simply because in Russia and China before either government were able and measure the value of the individual, all pointers had been set to the state, to the very earth and grass we all walked upon in mother country!?

In the other cup the balance was almost overtly reversed that the matter of human rights was excessively expressed by giving the outlook which has done more too regress any form of progress we attempted to make over the value of the mediocre. By the standard and style to the adoption of what human/civil rights are supposed to mean we moved in half measured fashion because we had no idea that other examples explained in the death of millions above, were national and state rights! Nevertheless from our western societies in there own relentless way have caught the virus of compensation to the tune of what human rights are now considered to be! Spreading a similar message world wide in the proposition of equality, but not fully aware of how to manage the issue unless to follow our steps taken syndrome in any case because that is the norm? Now, instead of creating a platform on which we could all stand upon equally we have created a small room of which there is no escape unless we use the extreme?

In parallel form to the concept of human rights, and I say again, it was originally to thwart communism. Another error of thought was introduced via our delivery of what religious tolerance was thought to be and how it was supposed to work and to what standard? Although it was to be another leveller then, the idea was good, but set to be a bigger error than anything, because it sprung from some of the assumptions we had made of ourselves, which is almost arguable for ever, until the Fifth Dimension?

What has never been addressed at any time since nineteen forty five, certainly up to the mid nineteen sixties, is or was the question, how fits Islam into the picture? How fits Hinduism and Buddhism into the picture, save to be measured in concept to what is quiet by our first world perspective, is left quiet? If unseen then the actuality was real whereby although religious tolerance was to be our best standard for what ever reason, our reflective images were wrong? Because from outset, if too, and I say this again, but cast from my own different reflection of what was attempted, with religious tolerance also there to thwart communism, another wrong application was used!

In the fact of theory religious tolerance never cut across communistic doctrine because that was a vacant entity. Where it was to bear was on Hinduism and Buddhism, but far more against Islam without being counted? But on the count that Islam was monotheistic and high in number of adherents, but dormant to be able and offer political opinion, beacons for which have never been understood until now? Because no voice until now had been able and view our similarity far in excess than of our differences, which had been cast from the opinion that we are first, second or third world! When we had to succumb to that ruling of opinion, but to be denied voice, first because we are second or third world or second because of being first world religiously through monotheism. We, from the sign of being Muslim have never had a world voice in any name politically!

Islam until this day has never had a political voice under any circumstances, but has been addressed so of having one in the same name of death of self by self only. Meaning in open honesty, that Islam as our last closed visionary monotheistic religion, our intentions had been automatically left wanting in the name of humanity until we as a group, even larger than clan or tribe are able to come to terms with our specie singularity. Allowing in the best standards of expression how we as a unit might express singular opinions in a collective manner, being the one and only way whereby our heads are turned towards our first mystery at Shanidar!

Sixty to one hundred thousand years later, further expressed by fourteen hundred years of unheeded Islam in our open modern time span bringing us to now. Have placed one religious attitude in splinter, but not from the measure of doctrine, only the matter of opinion if expressed through the reflective image of religious tolerance!

For me personally to have put that image under threat stands in the factual reality of what has occurred in these most recent of modern times bordered by the past sixty plus years of this last modern era. In taking monotheism to be the largest collective religious attitude, we have seriously erred in not placing Islam to be more prominent by world standards of acceptance. Because in not doing so has allowed any measure of situations arise as they have without clear objectives or of no faithful purpose, save to use antiquity as a means of redress.

Islam has never had the same feel about politics that Judaism or Christianity had operated to. Our open misunderstanding to this fact has only been realised by us in these past forty plus years or so. When without the knowledge or understanding of why or how, we have allowed the mundane of what politics properly mean rule over the balance of what religions are? Meaning that we have readily allowed the political standard of now commercially, overrule the prospect of when, determined by the one, our one common singularity of 'Natural Body Death'?

'Natural Body Death' to be a factor, that on reflection Islam more than compromised our situation by the belief of expression cast from some of our imams, who by being prepared to defend or protect or promote the same aims of Islam that we all hold, but now to their own design? Have cut to the matter of human sacrifice and human murder in order to re-sanctify the relief of martyrdom upon some of us, even if to be young and vulnerable and not to really understand? To be a matter of prime concern which cannot now be justified or had never been justified either by the words of Mohammed or Gods agent in the angel Gabriel.

Fourteen hundred years of the same pace is the determining feature of the fact that Islam no longer needs to defend, protect or promote our ideals to the terms of yesterday. Because although the same in style, Islam has reached hundreds of extra millions of converts and adherents. Without ever having to refer the aspect of what I and others have determined a weakness of the faith. Simply by the case that many millions of Muslims do not understand the feature whereby we can kill of self by self in an open religion that has mass appeal!

My references to some imams in not being able and see the course of development is to the reflected shadow of those of us in doubt and more ready to use panic or jealousy instead of reality to reply to how we had seen the reflection! Some of us have been small of mind by attempting and use the dead from the designed act of murder, to speak our feelings of discontent? Some others of us have attempted to bring the ideals of Islam to other nationals in the old style of fear and intimidation, only assuming that the last fourteen hundred years of natural progress never happened? Being a matter of reflection more important to Islam than any other religion, because she of the three main monotheistic religions had set a future plan differently than others, where they were for the imminent, Islam planned tomorrow!?

So with one and a half billion Muslims worldwide there is no going back to the first necessary ways to originally convert or enlighten the faithful? That would be under the assumption that up to now all our successes were to have been only political from which the world needed conversion, which is almost what has happened because our present religions are served under political religious tolerance?

Although wrapt in style, it has not been the full fault of politics that has brought the situations of today to whatever form? But more from our own failings in having taken what Moses, Jesus and Mohammed had said in name to be a conversation to us direct first hand, even to have been misheard because we wanted it so. Which is no failing because religions by time were the first unseen signs that we had broken from evolution and entered creation, but never realised by some of us even now, because our spirits were so high then!

All that we were unable and do is recognise the one factor that God was to us we were not yet able to ever be to God? Because in the visionary exuberance of what had occurred to Moses, Jesus and Mohammed, was the why, not the how!? Each of three were carried by eureka, a wonderful state of awareness to the display of reality, each of the three acting as men acted as men and would never deny the issue. Each of the three carried the exact same message to the full care of self and to the full welfare of self and all from the three tributaries of four, two and one thousand four hundred years ago. Each of the three worked and lived in a time of separation, defined if not by our tribal standards and definitions, then to the further growth of our civil means of control! Happening in any event, but not noticed to be in competition with our newer and better discovered values in our young religions!

Forcing the misplaced judgement that the way forward was to be led in style politically, because at source we had already met the full expected standards of what life was about, by us having come to the understanding and purpose of what creation is, above what exuberance was?

Four thousand years of reasonable study and in part acceptance of the sublime has made for us the hazards of today fitted to the simple reality of questions posed most seriously over these past sixty plus years? Fortunately by our own tenacity and unshakable faith in the future we have been able and position ourselves to continue in spite of all the breakages we have piled against ourselves now realised to be in the name of politics.

By turning circles in the hope of first world stability, even to the protection of that idiom, we have only found new standards of loss or to have been originally wrong, from which in being in a circle we do not know where or how to turn to correct matters? Because in all realities politics as our servant in rule has lost all impetus having made most of the wrong decisions at the wrong time, but continued to support them all from and by the wrong motives. Politics has lost our way and until she begins to take the medicine of the Fifth Dimension her standard of health is in danger of becoming terminal!

Fortunately the best in antibiotics is at hand to begin treatment which surprisingly is to be measured by the high standard of our religious leaders. For it is from that line we will be able and set the direction we have always set. But this time they also are to be guided by what or how the three taken and one active prospect from the Fifth Dimension overlay are to chime when taken only to religious standards of interpretation, separate to how the influence is seen or absorbed cleared by our political overlay!?

Apart to the full complexity we have made about our means of interaction or how our intention of interaction has been determined of late in this last modern era. For me to have focused on one small area of total discontent is symptomatic of how difficult change will be unless we declare our openness to change, even without alteration, which by connection is less than difficult. From where I had indicated the main crossover in terms of reflective compatibility is denied us or has been denied us by the mistake of assumption, needs little done to be put right.

In standard that we do promote the concept of religious tolerance to include all religions has been badly displayed by past circumstances of fact. Not least to the charge I have made against us all, yet at the same time on our behalf, whereby because of our bloc culture standoff, particularly in the Middle East, even through the mid nineteen sixties and to include the Iran/Iraq war some twenty years later. By taking cover behind the matter of what religious tolerance was supposed to be, we allowed that feature stand, because it was done from the first world political standard of change or the considered justifiable wind of change under the present situation?

In line with that feature being put to a wounded philosophy of communism the concept of human rights ran with religious tolerance. Of area, with communism supporting much of the Arab cause, therefore to the misunderstanding that Islam was very much included? Islam was given no voice politically even through religious tolerance, because of her own settled religious countenance perhaps? Also in her case having a far more settled religious viewpoint than realised, allowed that she owing to the benchmark of political development now ongoing was second world!? Islam in representation of many Arab and or Middle East countries was only accepted to the fixation of what religious tolerance was meant to convey to communism, therefore being irreligious Islam was ignored?

Islam was held to one standard, to be controlled by a system of politics who in the name of convenience had been gathering on mass more and more uncounted necessities for the means of control before progress! Who instead of being aware of what religion was, what any religion was, had gathered so much dross in the name of society to be able and smother the concept of creation!? Therefore when and if failure or collapse is imminent the picture of blame has to endure in order to sanctify the case of error and if, so the mystical route of how to return to the status quo is protected! Creating the situation of now not ever being able and work to the theory of change because we had lost the ability to change, by the how and why we had gathered the clutter of convenience?

My case in study in the name of some imams in us, who have also taken the wrong path, is illustrative of how two different opinions have led to forms of entrenchment of which neither side would wish to relinquish. Because replacement or the necessity of it had never been covered when both factions had thought there actions were the right ones because of the others standard, who is automatically considered wrong!

For me to have held the one point of miss-contact or missed contact by understanding as an example, is to be put upon each cup in balance because once again, but this time more seriously. We must weigh the multiple errors we will endlessly pursue if we do not position ourselves to the working out of what the three taken and one active prospect of the Fifth Dimension mean politically and religiously. Remembering that in all cases and at all times of review they are looked at from two different perspectives?

With one cup of the scale in the name of political democracy carrying the standard of religious tolerance in the base form of unity, held against the other cup in line to follow in application a once vivid, but now defunct system and style of martyrdom held to the same balance! All we have to do is look into the whys and wherefores of either wrong case to find a fair and just settlement even if nothing is to change except us in the process of change by whatever key? But always from now to accept the one common standard of what our ending by death really means under any system or style but very much aligned to the time set of creation, only creation!

Our greatest travesty of any time has only just occurred, not least because of some of our political only failing? Being a matter of fact because for all the shoring up we have to do, nothing will work until we begin to operate the full theory of politics in line to our new thought religious ideals? To be delivered by all our religious leaders, including those who will emerge to have a sickened fixation to the wrong matters of right, for there own words will successfully condemn them to the obscurity they deserve, which is not my judgement. But an example of how we all have allowed the feature of our new form of self appraisal of common decency C D to have lapsed or to have been smothered by us pretending to be politically correct as the best option of communication!?

To the one simple and small reality how our main first world political influence had paid heed to Islam is a three side square. One, that up to nineteen forty five Islam as a basic religion to millions upon millions of people had no first line political voice, also more dangerously it was spread without a clerically defined religious leadership. Therefore in answer to the principle of the nature of what human conduct should be religiously! There was no ability to set to consideration any political input. Save to the distinction whereby Islam was left un-listened too, except in cover of reacting within the terms of what political religious tolerance was supposed to mean!

By token and in kind, when through the realisation if to include Eastern and Western styled religious interaction through the event of partition in the sub continent of India in parallel with other like events? Such as the smaller divisive features of settlement in Eastern Europe in Ireland and particularly in the Middle East. Although Islam was interacting religiously, she had not the same standing that was growing by the new political tempo if shown to previous standards? Islam in political terms was lagging behind what was thought to be the operation of new and genuine international progress!

By effect that we had never considered the full standard required to meet all the natural differences between us, has allowed the case in the basics and basis of what we have now settled for. Without having to sift through several layers of time with innumerable different types of filters in order to establish remedy. My standard in difference between the attention of how some imams have reacted and what I might recognise as Americanism worked to the prospect of the American dream, is purely representative!

By that case I carry no lesser opinion of some other events to have occurred throughout the world, which have truly been dreadful and in cases from much more sinister reasoning with greater final harm to more people. Without balance to those events, what emerges from the distinction I have made between the westernised political system of American/European democracy and the religious force of Islam, carried by one small feature of what is or is not doctrine. Stands to balance the full division of the secular held to be commodity, and the spiritual, held to be monotheistic!

In this instant I have not qualified Judaism or Christianity in representation, because they are the same in treatment through the westernised political wing of religious tolerance. Whereas I have not been able and make the same case for Islam, because as of yet there is no reflection of who can be addressed in the name of its religious leadership, save from my own interpretation of using the example from the outfall of how or why some imams have acted how I have seen them to act or react?

It has never been said within these pages that there is or will ever be reason whereby we might kill of self by self in name. Even that I have been generous to point to what could be exasperation for some holding to the faith of Islam, of how they might feel regarded to other religions and or political thinking? Gives no ground to how some of us as imams have doggedly held to the letter of past learning, which in all cases were only designed for short duration from the understanding of the day, even while being generational!

To have made the assertion that in Judaism and Christianity when in direct clause, God was to have passed the ideal in human kind not, not to kill of self by self, but from both faiths we do, is a wholly separate matter for the religious leaders of Christianity and Judaism to deal with, as the matter has never been looked at properly, ever by the code of doctrine! Balanced by our taken prospect of 'Natural Body Death', making it a separate matter from now for the same leaders to take that competitive edge from any type of religious standards new or old? Because for the first time ever, certainly within the bounds of this last modern era even leaning to the past, we can all adjust religious standards by what we now know?

Naturally the same applies to our Islamic religious leaders, but to be with those of us across all religious divides, to be able and set the matter of change without alteration? Of which the most prominent problem is to try and understand, so we might be given a brighter picture than I might cast on the subject, our query is how to quantify the personalisation of martyrdom at this time of now through Islam? Which is not a test or a trick to draw error, I, through my own level of study have set this task on our behalf, religious leaders or not when the answer is shown of the billion plus who do not!

For me to be able and charge imams to the matter of events to the extent of how some might actively encourage from a sense of duty, so that we might create the position of martyr for personal advantage and no other ends. Is a reflective observation, but done in the name of Islam in order to disarm the closed eye and plugged ear posturing from some political mediums! For I see most in the act of encouragement to the deed of martyrdom in the form of not knowing what else to do!?

My personal stand about if some of us as imams were to encourage others in our young to take extreme methods of communication by the supposed act of martyrdom instead of murder! Has to be formulated by how some of us have reacted to what has become a political beacon in the form of Americanism to the matter of world conduct leastways commercially. Of which I make no case in excuse of how some of our twisted opinions, have taken the one simple small anecdote of martyrdom and carried it as a vile tool of nothing more than murder to voice opinion!

I will not balance any situation like for like, kind for kind as a means to endorse the road some have taken by their own reason, cause, obligation or imagination? My standard had been set, even though my deafness of sound evokes blame and violence, as a voice is the only means to be heard, such reasoning is flawed! Not from any sense of duty or from disharmonies, nor ever to change things in retribution or by reaction or in comparison, but instead drawn from the ignorance of dogma however delivered, when no rhyme or reason can be seen in the mirror while reacting to the reflection only?

I have made much of how we have progressed being the one undivided specie we are. I have entered with many, many better informed people than my self to give self opinion of how I have seen our rights, wrongs or maybes develop? Mostly other opinions and understanding have brought us thus far on our journey. My small contribution is a credit to us all, because it is from us all. I have simply used what is strewn about for our wondering and using what I see and how I have seen it to put my own case. My positivity has been from how I have seen any particular situation from any time past, even to include this last modern era and put a balanced take on any situation without envy. I have made the important discovery of the realisation that we have little to do in life except survive in time to our fifty plus and seventy plus year lifecycle reaching the one equaliser for us all in 'Natural Body Death'.

Without being heavy handed which is never to be the case, when that taken prospect above is mentioned. I have been fortunate to naturally, not selected, but naturally be able and move by thought, supposition or just plain observation and query in the world of the abstract. Allowing me put different propositions to our same multiplexed problems. Be they in the two fields of influence to be upon us all in the political or civil drive and religious query leastways since Shanidar!

In taking note of what can be measured to be small consideration on the world stage, leastways if put to be balanced by the ideals of the United Nations. For me to place imams and what some may suggest or encourage or openly promote. Is done by the one standard only of what I see how we better need to react allied to the concept of specie unity. Which is owing to us all because by defining matters outside of refinement, I see the case made that the final meeting between our two main means of control in our politics and religions. Are illustrated by what democracy is and supposed to be and what religion is supposed to be expressed thus by our failure to understand our own ground rules.

Without any form of support to what some of us might encourage of our young in the name of religion opposed to politics or opposed to political ideals. Whereby in asking for our young to represent us so, is done first from within Islam beyond any other reasoning! I have drawn my own conclusions that the situations so described of the relationship between Islam and the democratic wing of politics, is a direct result from the misalignment of the assumption that politics had got all things right. Therefore religious tolerance in name was enough to cover any whimsical desires from any religion, even the uncounted or unexpected reaction from Islam?

By range that Islam had never been given credit to be of equal standing as other religions in the forefront politically, such as Judaism and Christianity, is a case to be reckoned? By being unrecognised in voice even through the United Nations, further denied by the condition of not having the one focus of a defined leadership, even to the similar standard of much smaller religions Eastern or Western. Even if attempting to have tried and put the same case as others religiously over the delicacy of some political decisions already made, how if seen or felt by desire in need could any Islamic voice be heard internationally?

I make no case in the positive for what was to have happened in the Middle East, especially when by my own fancy I have set events to the rational. From which is meant how I have turned some bloody and damning recent and past events to be counted now to be part of our personality!

For me to have offered my own opinions of how I have seen events occur and the reasons for, have come down to the simple and strait forward fact of how I see our two failings of the same magnitude to what politics is and what religions are!? Our human condition of politics has spun its own web at first in concern of civil control starting about those past sixty to one hundred thousand years ago. Loosely culminating in a style of government to be held as the best form of construction by all, all of our efforts ending to be the present form of westernised democracy, warts and all, but unfortunately taken too seriously by some! Many of us are already aware of its shortcomings, but allow that in type it is one of our best hopes to improve things generally and internationally, once we begin to measure its true value.

Unfortunately where it fails most exemplified in this last modern era has been from its overbearing view to our religious liberties? Which I myself have pointed too in relation to Islam in particular, but without being judgemental, only to have tried and show what is actually encouraged by some parts of modern Islam in its relationship with modern politics along with the whys and wherefores? Bringing me in case to the one other divisive issue I have seen preformed in answer to what our best political standards actually are?

There is never to be a way or reason, where any of us can offer any type of support to the supposition of what I have offered as I see it, to what has been encouraged by some of our imams. If democracy and or Americanism and or Europeanism have been lapse in record to our last of the first, second or third dimensional religion of Islam, I have been able and best measure where the vital standards of that relationship meet in this last modern era. But to be aware, my suggestion is not final only representative!

Our latest forms of government do work in this last modern era from the negativity of producing positive results? From our first world perspective we have been able and count the imagined effect of what we consider necessary by that first world availability being relative to the second and third worlds! Leastways that is how we guide the idea of policy if through the United Nations for all, but from that first world standard?

Without what others might see as overt arrogance, when I had referred to the matter of our religions in parallel or even to be more important than our political aspirations, but that we must always consider both seriously? That I have set my opinion to the fact of how we have ungenerously ignored the world place of where Islam is to figure politically, rather than being left in the same corner with all religions, even that some have the advantage of political weight, if that covers the aspect of might being right! Where can the twain meet in out true representation on equal terms unless our religious leaders promote the issue that all are to fit into all future plans?

Our case politically is to be settled by inclusion of the overlay of our three taken and one active prospect of the Fifth Dimension. But first again, I submit we have to settle the last conflict by the good intention of doing so by and with the honest heart we all already hold!

## CHAPTER 185

Without claim or condemnation, without judgement or settlement, I have set the clash of titans in the true modernity of now to be the balance through and between Americanism and Islam, but always to be open to the standard of correction at any time and from any source!

Politics through the eye of how I have determined an end product of Americanism is purely representative. Islam in value whereby I alone have made the connection if to what some of us as imams might encourage of our young? Is carried in the same consideration but not in comparison because one is purely of secular self concern with the other of spiritual significance, being far more important! Which is not an observation of my own making to antagonise such factions, but an effort to show how we as imams have reacted in some cases to the errant show of political output to whatever end?

No religion, no religion can offer now in settlement for any of its adherents the respect of what martyrdom was once supposed to have been or had been renewed as in these most recent of times. In any and all cases, martyrdom has only ever been an active reactive in the progressive, in order to account for the integrity of many religions! Where to the fact that because peoples could not kill of self by self in the defence of there faith, even against the most brutal of regimes? To die because of it, was justified by the title of martyr, which is a very honourable intention when entered into the unknown only by suggestion?

Martyrs who in many cases suffered before death could have been placed in the comfort of knowing that because of such suffering there place in Gods hand was assured. By break, those of us who were to suffer death in the face of adversity from others, would be rewarded for our resolve in the kingdom of heaven or paradise? Being a matter that was made more poignant when Jesus Christ was to die and from the idea of that death was able and proclaim that on the very day of death he would be at the right hand of God, specifically because the death had been foretold in the name of religion!

Without connection, but of theme, however I might have presented my own fix to what martyrdom is or is relative too. Any form of speculation since, like from me now, has to refer in standard of how the proposition was first changed by recognition, beginning at about the time of Mohammed!

For me to have used a style of modern reasoning has been done in example of how we might refer to past examples, without any aim to alter the past, but include its experiences because the past is very real and effectively cannot be changed in any case? By connection, for me to have put any kind of comparison to Jesus and or Mohammed is of the same cut, especially when both have been given status which by any case places them separate to normal men. But with both always showing self awareness and social awareness in expression for us all, but not really knowing how we, we to be their people, could honestly react to the wonders of belief they were to reveal!?

Not least because in the case of Jesus, his plan from his day in time was to extend the tribal religion of Judaism if, to a broader base, but from the ideals of Judaism! In so that even from his death and therefore his absence, we could carry on from a well established base in any case!

Mohammed of his day and inclined to the thinking of his own time some six hundred years after Christ, now had a different emphasis? But nevertheless his union with God was considered by himself to have been of the same class that was delivered to Jesus and Moses? Therefore any connection in religious terms were as one but in compliance to the supposed failings in practice of what or where Moses and Jesus had led, hence this newer more enlightened guide of approach he was to deliver on all of our behalves! Not as a means of direct replacement of what had gone before, but renewal. Because although Islam had its beginning from the need of force in the violent times it was first introduced, considerable forbearance was allowed to former similar religions!

For me to have used the expression in the name of Islam that human life was to suffer at times to the value of death in order to promote, protect or defend the new concept of the religion to be. Is a comparative note whereby we can understand our differences of now and then, even if accounted for by religious tolerance? Then ago the objective was for creation in the two natural terms owed to mankind, with the correct link to God, which even in the mire of associated confusion surrounding the full topic, still produced the wellbeing of what Islam was to be. But at the same juncture allowed the ideals of our religious forbearance continue when the first deeds of conversion or change had occurred if by conquest?

Fourteen hundred years later the charges to Islam are negligible, yet by what I have addressed as the re-emergence of what martyrdom is to be. Has been done from the field of our human conscience, which is now served by taking to the three taken and one active prospect of the Fifth Dimension! Knowing as I do that the same ingredients in overlay to our political perspectives has a less defined measurement because politics is only a commodity!

Nevertheless politics has benefitted us all immensely in those fields, but has also produced the false offering of purpose explained as our first world goal. So much so that from it being applied half measured because of our limited means of understanding to it, not least in these modern times. Some of us related to the religion of Islam have been carried back some fourteen hundred years because we can be by text? Because we should be, to count the cost of now or how we think of now in relation to then and set the correct balance before we pursue our own opinions in these modern times with old tools!

Where I have set the claim and need that any described inter consultations between all standards because of what we are trying to achieve in this last modern era, be done first hand by our religious leaders, is paramount. Not least because of how we have progressed without the input of what Islam might and can offer all, provided we are reminded of our own free will better than before!

By reading that we had all benefitted immensely from what politics could do and had done up to now, we have allowed the veil of silence fall upon the uneven way our mutual political success has presented itself, which is standard. When considered that as a process of success there is an automatic benefit structure. However, what has never been counted in human terms nor can be, is the relation between politics and what it does or serves or produces and the prospect of 'Natural Body Death', even that it operates in blindness 'Past Gods Words!

One feature left in the wake of politics acting to our steps taken syndrome through this last modern era, I have highlighted because of an effect on some of us imams? Even that we do not know what to do un-led, some of us feel it necessary to do something no matter what? Our only problem is to follow in furrow of what the devil or an enemy if, has done according to how we see the reflection of what had been done?

In the case of being imams who by standing are leaders, albeit to a distinct classification, but without a defined clause, our supposed reaction firsthand is to defend our faith by the style we recognise? Which in count is remembered by not being listened too at any time over the past two hundred years politically, nor counted again by division over these past sixty plus years of this last modern era? Therefore in the face of rejection if personified by Americanism and the feature of might always being right, our balance is set in aggressive martyrdom, not least to take the shine of what can be termed the glow of politics in any case.

I have not set any standard whereby excess can be offered by what I have written or how I have interpreted past events. My single direction is to the fact that we as specie are the one and only ever to be able and contemplate the ideals of purpose! Our story so far in general life, has been clouded by the actions of actual life and living. All that we can do is continue and move forward from now, now knowing what the offer of the Fifth Dimension holds?

For me to have set the standard of how we have and do interact at this time, has been further clouded by the generated imputes from how some of us, especially over our religious codes have also seen the same reflection, but reacted in different ways! Especially when we now know that apart to some other issues the one sense in reality of common ground to us all today, is the feature of us all having only one chance to get things right. Because in fifty plus or seventy plus years time, each of us will be no more?

So, from that perspective there are no grounds to invoke the ideals of martyrdom in relation to what the reflection was from the first image, or more vividly the last, when all were of the same line in any case. Nor can we invoke martyrdom even to be against religion, because from the time of Mohammed we had already come to the standard of religious forbearance? In full operation to those standards of the day, which automatically included the full applied learning of all previous times in the same way that previous examples had used the same key! We were able and set a difference by our understanding of our differences which were not apparent, only tribally!

Unfortunately to show that in fact we do understand and in order to prove the case some of us have taken the wrong charge to put the right moves in place. Bringing us to the ultimate of now and how we have improved our ability to misunderstand if pointed in direction by the fourth dimension, but we must recheck count from the one religious perspective of reusing martyrdom in the wrong place and now at the wrong time in all of our histories no matter who we are!

Using the Middle East for my own example only, I have set the full range of our differences which can be agreed upon, although somewhat misplaced from there own reasoning! For full effect by using the pre-fix I, I have set for us to be aware of how progress has been funnelled by charging the cup end of the funnel with multiple ideas and not knowing what the spout produces? Like for instance when going through our bloc culture political standoff in the Middle East, the one side aligned to Americanism and religious tolerance, did not really understand the relationship between communism and Islam? Therefore to thwart communism even if to apply religious tolerance along with human rights to those ends, Islam was left in shadow, and held in wanting! Because it had been assumed her voice was also of communism as there was no specific religious leader structure to implement understanding or improvement at any acceptable level if that method was ever tried!

By suggestion that I had stated the course politics had taken to lead us to the American dream under the umbrella of democracy, and taking acceptance that the collective standard of religious tolerance for the full code of our religious conduct has not produced the desired effect yet. Because our situation has turned to the importance of always being to now only, which by actual operation can only refer to the past and its supposed effects on the future or not! Making the open example of how I can assemble all of our past political efforts to point to the one achievement of reaching the American dream materialistically.

By that very declaration, for me to have put the conduct of some of us imams and what we might encourage in todays world fits in the abstract, but honed to the actions of reality. In the first instance, that we, from some of our political perspectives, have allowed the situations arise whereby it was easy to ignore what ever Islam had to say on the world stage, has made the substance of what the reflection of such matters were and or how to read them? Situations to have come upon us have arrived in the sadness of how we have failed on the political front to recognise what our politics was about or for?

Although politics has ranged from the application of civil law in the context of control and the progress of human endeavour indicating commercial success, which effectively is a naturally selected phenomenon? By turning us from the ready state of our held religious intentions, our compromise was to channel the importance of what religion is to be, into a small step in progress only. Instead of meeting the query of what purpose really is, leaving us at this very time of now, from which we have the two options standing in front of us in competition?

Unfortunately even if that is the most simple of cases, we have not been able and set sail until at first we display the full breath of our real failing, if only directed by myself from now! Politics in all cases is failing whereas religions in all cases are not progressing, because in both examples our only connection is with now or the time of now? We have no future prospects, or under that outcome, we had no future prospects until now, when we are able and adopt the three taken and one active prospect of the Fifth Dimension. In the meantime we have had no choice but to step into the past to see if we could put a better case or ply a better tack!

Until now we had considered the past was always to be the key to the future religiously, allowing the matter of our politics run its own course. But not realising the extent of its full objective and in believing it was the means to all liberty while at the same time working to its own ends which have never been clear? Made poignant in this last modern era by setting religious tolerance to have formed political standards and not being able and equate how if some religions were to act react or interact with the true dogma of politics!

I have seen and measured the full route of our religious reaction to what politics has ultimately achieved in its own sense, which from my point of view has brought us to now. In due course politics will receive its comeuppance, when its actual efforts of achievement will be exposed by my self under the second level of scrutiny it deserves. Putting it in its proper place of importance once and for all without destroying its better intentions driven from its own unknown conscience? So that we can connect to purpose explained by the best efforts of our religious leaders to be, or who already are!?

## CHAPTER 186

These past sixty plus years since wars end in nineteen forty five are to have been the most crucial in all of our histories, already proved by a plethora of world achievement in all fields. There is no standard of emphasis I can raise to promote this fact except through what I have encouragingly done in these pages. Our only focus is to now or has to be from now using all recent examples, because of our pasts and what we have already achieved.

From my personal note I have always admired our total connection with our histories in all of its forms, but particularly of its relationship to all of us now, before any comparison? I have even allowed opinions form in my self of what the meaning and or outcome of a particular event would be, unguided? Because in the making history by her charter, was not a series of events of any character, but only events in any case, it was we who have complicated matters by assuming the modern significance of all or most past occurrences? Because all that I write is ours, which is constructed from all of our efforts from all of our histories, but used now to reflect on the intentions of what or where our histories have brought us to!

My approval by definition to call this work the Fifth Dimension is a fair example of how I think in terms of us all. My effort of inclusion on religious terms is the most prominent because religions have always been the means/cause/reason we have been able and open the combination of what evolution is and will always remain. Religion points the arrow of purpose which is fair set towards what creation is and how it was meant to be understood.

By count, choosing the first three and mainly appointed monotheistic religions of Judaism, Christianity and Islam, to be my first, second and third dimensions are sited by expression of how important our religious countenance should be. Without balance my choice to me has a better historical connection to what all religions are about, also in these cases, because of there direct association with God? Promoted by standards of time, by distance, by area and more vital by inclination even if to have been laid down with slight differences by the line of prophets in Moses, Jesus and Mohammed!

From the unproved historical fact of Judaism and that some of our Eastern styled religions are of roughly the same age, mostly parts of Hinduism? My separate note to them had been previously cited of how I like many others, had determined them to be of a different agenda? Though I would never and have not made a clause, whereby the different base structure between Eastern or Western in style have any separate merit. Because all religions stem from the fact or from some of the reasoning born from the events at Shanidar!

In terms of any religion, no direct trace can be put between Shanidar and now, because that event was from any time between sixty to one hundred thousand years ago, when now we see the oldest of modern religion stem from about four thousand years ago? Without any perceived connection our Eastern group of religions started on the pretext to include all life forms in plan as an act of reverence to the understanding of what mysteries are. Still not realising what Shanidar was representative of, that discovery of inclusion if in context, is monumental to the collection of all of our philosophies.

Because in parallel to monotheism our Eastern religions had taken the like step of religion as a must, but on a different line, nevertheless to still terminate at the same junction. From that standard to the same fact that life past the restraints of evolution, was owing to a force far in advance of any other expression. Therefore in taking other life forms to share the split of how the mysteries were to be revealed or understood was the working plan to be able and understand the same fable by including all other life forms also, never to be a bad thing!

No distinction has been added to our monotheistic religions by my own dimensional titles. First, second and third remain and in taking the fourth to cover all the activities of science to be the other force in explaining by revelation other mysteries, I have still not made a case against our Eastern religions. What has been added in representation of all has been the Fifth Dimension to be the basis of our own lost reasoning without having to compare our example of specie excellence, which fits without neglect of our Eastern religions.

Because for some reasoning the answers required might need the mixture of experience of other creatures along with humans to fit how we can understand? Being a matter of judgement, because with other creatures working to the inbuilt habit of instinct only, how can/is the mix between that factor and politics to be met? Unless in us using that same thread of instinct in original overlay to our new formed habit of religion but to include every vestige of life, leastways by how we humans measure things even if to be in other species name?

Never to be used as an example unless only in the positive, which would be to turn the understanding of some events into an understandable commodity. For me to keep my line of inquiry on the same keel, but importantly in keeping with actual events of this last modern era post nineteen forty five. In using the example of partition in the sub continent of India where the one outcome was of huge human migration prompted by political reasoning, but instigated by religious desires. Some of our opinions are to be reset, even if from my own view which has only been a guide to some past events and why!

Two features for my own convenience of theory had emerged from that tumultuous and bloody event of partition. Without any value to be put on my judgements, because in the hindsight I have used my standard, has been generated from the abstract!

In taking the one connected incidence whereby it was from the decision of the inhabitants themselves even if politically led, who were to choose to decide there own religious style of government? Once stated, others who were to decide how the actual geographical split was to occur were influenced only by the level of European political activities in Europe after wars end in any case. Who if not to have realised the nature of the felt differences between the emotions of Hinduism and Islam on the sub continent then ago, is a charge against all of our consciences now.

By the same token, once done politically the open cruelty and devastation and murder done by self upon self being muslim or Hindu, should have remained as a beacon to us all. Because at outset from all parties involved, no credit was given or taken to our singularity of status, because we did not then know that we are the unique specie!

What had been done led from Europe or by let from the United Nations, what had been done in the name of Hinduism and the politics of India had been wrung or had been from a style of politics not to be familiar with the real concerns of the locality!? What had been pointed was the open face of Islamic desire when calling for political solutions to those religious desires. Under that combination of what had really been decided from the vast misconceptions of religious partition politically, was a standard to endorse the full result of our political modernity pre the mid nineteen sixties. Which unreservedly kept us at bay with each other along the two main influences to have always been upon us, those we are still trying to integrate, but at last in the correct time slot of now?

My own terms of reference via opinions about the sub continent of India again, are on the one level to point to the effect in respect of how Islam could and did interact religiously with our main Eastern religion. To endorse the feature of how Westernised religious tolerance had up to now misplaced its charge of concern to all religions was shown if intermixed by time and or intention. Because by that title of religious tolerance, Islam and Hinduism were continually left to there own devises leastways world wide. But strangely not blamed on the result of partition, only held by indifference, by a community who in thinking it had solved the matter of religious harmony, still did not understand the equality of being first, second or third world?

By reflective tones when we again look at Islam if from the mid nineteen sixties. Our total religious codes had now come into much the same review with different opinions generated from different perspectives. Now or since then the overall standard of opinion was led first world politically, which in turn allowed also that any religious voice was to be held in check by the supported ideals of religious tolerance!

If that is to say our present form of interaction today on religious terms had been drawn from past and recent experience only, then we have a dividable split which is to remain unarguable, but only in token. Because by statement we are self challenging our own united concept of what religious tolerance had become, instead of the clouded route of its original intention?

Abstract reasoning allows me use the true failure of those good intentions from when we thought we were to get things right, leastwise politically, in order to set the rigging? In this last modern era, with our new concepts delivered half measured from a situation of post war trauma. Unfortunately at the precise time of wanting, we did not yet have the correct formula or right lead in order to proceed in a meaningful way? Although victory was the first voice in choice and direction, even at that time, our half measured reasoning had been prominent enough for us to think we had most things under control. Not least when by extent at that time much of our religious prominence was waning in the light of what the ideals of my fourth dimension was revealing by replacing the icon of some of our mysteries into the reality of formula scientifically!

Not knowing what the future was to bring except progress, all of our steps taken syndrome reckoning was obliging in the sense that it complimented every first world reasoned act or reaction!? Even to the extent that when also in recognisable terms from about the mid nineteen sixties, Islam, was to voice world opinion for the first time, but without one leading voice internationally, nevertheless our actual future was beginning to form? Which was to set, taken in conjunction of how we rationalised anything to suit the way we had become entrenched to what following our steps taken syndrome is and still remains. Making the situation of now from which instead of promoting specie singularity we have entrenched into first, second and or third world tribalism politically and or religiously! Making new standards half measured?

Without example but in show to how that term works in modern times, when forced without any sort or sign or type of opposition, even in query above violence. Any further classification relays on how the fixture has been and is read from our first world perspective only. Unfortunately perpetuating the open myth that in or through our world order, if over seen by the United Nations, it for us also sets first world standards? Maintaining the unseen status quo of how we are now guided to interact, because of familiarity, which although covered has never included what Islam might hope to wish for us all beyond the aspect of religious tolerance!

By previous connection and under my own terms of modernity, for me to have chosen Americanism and how some imams from the religion of Islam have come to interact or more precisely react to one another if in group. Is my own take on of the level of extremities we have reached, leastways to judge by style that Americanism or the American dream personifies the total objective of what politics can/should/has or intends to deliver?

Although to think in terms of achieving the American dream, so far the ultimate of what we can expect from all the material effects of what politics is supposed to deliver. I would further submit in the same style I have used throughout these pages. Our American dream is all and the ultimate that politics can deliver, but now questioned by our three taken and one active prospect from our Fifth Dimensional political overlay!

On the other hand and comparatively of the same intensity, for me to have chosen how some imams in we have set to deal with modernity as the counterbalance to what the ultimate in politics is, fits. Because by that example I have been able to face the two most vital means of our control, head to head. Leaving both parties to ponder in the need for reflective change so that we can all hear our own thoughts individually and represented?

Americanism or to aspire to the American dream is no defence to the poor quality of how our politics actually delivers. Nor is it to be a concept from which we are entitled to raise our own hand in opposition to from deep and dark reasoning! In all cases my use of the term American dream is representative purely as an objective from within us all, to display the ultimate success of what politics could be to us all first, second or third world. Making the case that we do all aspire, but at now, we are not sure at what level?

By the same measure, but from a different cause, in using the full suggestion that through the religion of Islam only, some of us as imams have made a way to challenge something we have found to be objectionable. Alas to that fixture my own input is paramount, because in total balance I have used an aspect of how Islam has been interpreted by some, but to openly show our need to change some matters but without alteration!

With the due input of what our three taken and one active prospects of the Fifth Dimension really are, combined with the reasoning and rational of how our religious leaders, that is our religious leaders from the main intermixed combination of our Eastern and Western styled religions, from now, are to give us all the one inclined standard? Allows how I personally can and do put to example the cases whereby if, from some of our imams in pronouncing doctrine above intent rely on our young to deliver their tainted messages! Then the standard is to be made for the first time in this new modern era in compliance with all intentions, save intended reactions, even misinterpreted!

By open notice to be given in all cases by our religious leaders, by inclusion of the prospects from the Fifth Dimension in religious overlay. Common ground is to be the result of future thinking. That is why it is more than important for us as religious leaders and followers to take to the one standard of self monitoring, in order to improve what the understanding of 'Natural Body Death' is, whereby in shorter time than realised we bring the concept of equality to the fore!

With only two pointed features of deliberation in how I have classed Americanism and the American dream to the pinnacle of all that we can achieve from politics at this time. To be countered for whatever reason by some imams who openly promote the defiant and wrongly inclined lead of martyrdom in that cause of opposition? Have been made finally representative by myself, because in fact I have taken both aspects to be from the extreme in what politics is and or what religions are becoming!

By fact that politics is self perpetuating allows for separate treatment of how I might clip the wings of our server who was to assume leadership instead. On the other front which is of our religious connection, my standard in choosing the actions and opinions of what some imams have done or have said about what they/we do, even by my self, is representative in core to all of our religions? Because as any might and many have done, we still assume that the outcome of martyrdom is a guaranteed form of sanctity? It is not or never can be, because even before the ink is dry the feature of suicide quickly turns to the matter of murder by the qualified name of human sacrifice? Which when tried before, even in distant lands that now deliver the American dream, warts and all! Does not stand against an ideology that is honed and waits for the event of 'Natural Body Death' in theory to allow for Gods intervention to our community at any time yet to be?

For any imam to rant in the name of Islam by encouragement of our young to reach the heights of martyrdom is an open denial of the past fourteen hundred years, which by its self as a statement offers all the answers and arguments standing to my own point of view in aid to us all?

Self committed martyrdom is a denial of all religious concepts, it cannot be sanitised by the pretence that the stand had been taken in or for or to our own acceptances individually. Nor can it be taken in the defence, protection or promotion of any religious concept. Nor can the feature be spread out to include the vagaries of what politics might need, be or done to have caused the emotion to become inflamed?

Imams who if at first were to encourage the deed, in the belief of right, but always wrong, have been chosen by my own intention in order to magnify our different standing in attitude to the two governing influences to be upon us. With the type of tools I suggest are to be found with the use in overlay of the prospects of the Fifth Dimension upon both of our politics and religions, but quite naturally drawing different feelings about those same taken prospects, a different outlook will naturally occur?

With politics in a fluid state yet pretending to be solid and committed in cause, allows that she cannot self criticise. Because in offering any measure of doubt to any particular system assuredly desired, allows other systems promote there own cause by simply shifting the emphasis to another group, again vested in self interest?

With all religions leading to the final reason of purpose held in abeyance for the greater good, our only failings have always been human and not those of our religions. That in part is why our Eastern and Western religious philosophies will always remain obtainable, like when any of us can make any claim such as I already have. From where in this case I had particularly cited the relationship between Islam and the American dream making a play on the outcome of reality by that very expression?

Because even under and by our very many differences owed to the community of time, driven at first by the unseen circumstances of change in us generated by our own particulars of geography, climate, diet and other reasons. We had emerged to think we alone were special in class, but to the biggest mistake, that in order to highlight our assumed difference, our religions were the best means of separation! Not only from the other God made, evolved animals, but in our separate tribes, clans and later from our separate nationalities and or our separate political systems?

Unfortunately by the very longevity to the standard of those religions in relation to how parallel we had run our politics on a similar reading? Our way, our route had been led by the familiarity of covenants we had made with the best of our imaginings some thousands of years ago? Allowing that in all cases even if diminished by some of us, our religious fervour was to continue but not for its own sake. Now in cases it has been found to be in direct opposition to what and how some arms of politics had developed over the same periods of time. But aided in the confused political fervour of what and how the fourth dimension war ongoing was revealing about some of our hitherto religious mysteries, into the cold realities of the secular because of explanation?

Unfortunately further to the open display of some of our political/religious intentions all over the world since nineteen forty five, and me setting a further time slot from the mid nineteen sixties, and to the same value of our interaction by reflective cause, particularly in the Middle East. I have set it so that from those time periods up to the present moment, amid all of our religious/political turmoil throughout the world today. Our best example for total, even full reconciliation stems from that very area going back some four thousand years if necessary. But of the most vital concern in reference to the fact that we each will only live fifty plus or seventy plus years in full, showing the one opportunity of time only, where we have to get things right!

It is to be from that one simple open reality above which I have been able and set in jargon, where the standard is set in union in any case ex of religions or political influences for us all to die by 'Natural Body Death'? Because in open display it is the only period of time in which we as individuals have to display any ability of our wonder? Which in all cases is set from now to take first command from the events at Shanidar and why?

By the best good grace it is in open display that our religious leaders will from now be able and give us a more enlightened charter of recognition to God. Because at the moment we have slightly misaligned our own good will of thought by trying to act in how God might wish us to behave. Not least in display of how in the piety of hope we had previously grasped the flower of what purpose should be by our own imagination and then erred by excellence of mind but error of thought, taking the display of our mind to have been a full reckoning, which had never been a bad thing, but which was never the actuality!

Although in panic and in exuberance at the same time, the closer our own ideas had brought us to God! Allowing from the core of our specie excellence, the matter of impatience to grow from that flower, where we in full array to be able and fulfil our own measured terms of what Gods expectations might be? Which in fair understanding turned the matter in virtually all religions or through all religions, that we had come to force the one theme, although done, the matter has not yet been seen by us to really understand in all cases today? Nevertheless in order to force by imagination our understanding of what or who God really is, we created the impossible to be the factual theory of life after death from our own good!!

Unknowing that this as a philosophy was to drive the full imputes of human endeavour since Shanidar and over the past four thousand years in particular, which is no bad thing. But our theory of heaven and paradise or nirvana, flies in the face of our own interactive means and result of how our relationship with our religions and political institutions are? Not least because from some ideas of what politics really is, still unknown, because there is no end game? Our mention of the hereafter by whatever camp has to be denied by our committed act of separation in placing all of our religious aspirations to the corner of religious tolerance!

Although our political drive is to the intention for the good of all, its one laxity is to be exposed by how we mix the understanding of how we read the overlay of the three taken and one active prospect for the Fifth Dimension, against how we are to read the same overlay upon our religions, from which we now have a wider understanding of what real intentions are to be religiously? Because from them and in line to what might be revealed even by the element of the fourth dimension, it has been from all of our religious perspectives we hold the key to all of our futures!?

Even if opposed by the most stringent of means, the one and only unifying bond between all people, who are we of this one and single specie! Totally relies on the concept of the future, but only through the marvel of what has been achieved these past eight hundred million years. Oops I mean these past three and a half million years when at first we left foot steps track in volcanic dust on the planes of Africa never to know that as we walked in care if guiding our young also, we carried the opening key for all of our tomorrows!

Because from then we started to run in time to the one specie who were able and query the consideration of creation in separate reasoning than the millions of others who from within the former eight hundred million years by whatever skin. Have only run to the clock of instinct aimed only at procreation, named only by evolution? From when the great sea creatures who in cradle carried the time of hundreds of millions of years in age, from where the great crocodile carried its time of about two hundred million years of generational age to the present day, ever in abundance. Alone it was unable to master the theory of time unless to the call of instinct, which aided some changes, but unlike us, who fell to the whims of the same circumstances of geography, climate, diet and other reasons, they changed for changes sake. We remained the same if not through Shanidar then from the same wonder to what was to drive us all, humans, in the search to discover the meaning of our mysteries!?

All locked to the matter of evolution remain so, but still support our standard to find the fuel of creation and turn it into the matter of real purpose. Which is born beyond the reach of how/what science might ever discover, because our good evidence is that creation had to be stirred from within those that were created, from within any living creature! Who in place were able and notice an insight beyond the charge of instinct, like for instant any of any group who when in lament at the life loss of an offspring or a pack member? From where was it to come of mind that this still body was to be laid in the comfort of death and given tribute to enable further use of, at some later unseen time, save from us!

All our yesterdays are carried by how we wish to see them, which is no bad thing unless we take the habit of normal living and turn it into the reality of premature death. Which is bound to interrupt the forward flow of how we are to jointly progress on the same level and now from the one timescale if bound to be these past sixty plus years of this last modern era? Even that I have set the importance of it by magnitude to have began post the Second World War. From which in attempting to address our future prospects and in using the same equipment of habit, already fixed to the standard of how we automatically think politically first, second and third world, depending on the aspect of might being always right?

By turning into the wind in order to set our course we automatically set new standards of progressive change, but in using old charts when tacking left or right we sometimes took the wind from another's sail? Even if not intended, when done the matter in quite natural fashion, drew reactive sentiments from one quarter or another. How and what are still ongoing by such events can very readily be put right from wrong, when realised that in plan from nineteen forty five we had set sail under the wind of politics which has now taken us into the doldrums. Unfortunately instead of choosing the winds of real or intended religious fortitude, which are to be the real winds of change we can use without alteration? Our half measured first choice of political direction spread from the causes of war victory, have taken us to react by the steps taken syndrome even to at times ignore or override some of our best felt religious fortitude to sit in the doldrums politically and wait for change!

Of the one small feature we have set for the judgement of our religious leaders in task forthcoming, our beginning of the end will not change by one iota. Because in principle among all of our religious tones collectively, we had gotten all things right, how ever done we are still not sure of, but the best and most common of our expectations from all of us is to the future. Because when turning to the unanswerable, just like trying to understand the events at Shanidar at occurrence, when a group of frightened clansmen were to bury a tribesman in remembrance. Asking the one question, what next, yet daring answer it at the same time with the gift of grave goods, just in case!?

Fifty six to ninety six thousand years later even if to have taken a mixture of attitudes if at first from Mesopotamia of the Middle East and leading to Egypt of the same area. From where our fixation shown to the reasoning of death had come in theory to the philosophy set to the heavens, had given us a lead in link between the Egyptian God King and man? Translated to be from the standard of the sun by day to be the true giver of life to all through seasonal management and the one human lead, with the heavens at night charting the progress of such matters allowing us our measure of understanding? Making our own charted progress in any case that we in package as the one specie even at that time to be expressly expressed to be those from the tribe of Egypt, had stretched the mystery of Shanidar into a realised concept, even if a little wanting!

Although by the same token I had stated in my own loose fashion that the King God philosophy, if of Egypt first, was served by a dedicated priesthood? My picture is only to be related to one of their administration, because in line our standard was to the image of our King God structure tied to the measured habits of how the sun and stars interact with the means of human emotion by guiding human thought?

From which was to follow, but nevertheless from area our first monotheistic religion of Judaism, changing, even if borrowed, that the understanding to the creation of tribe was not from the king or tribal ruler. But had been done in benefit particularly to those who would first recognise the plan of mystery that we from tribe who saw so, were appointed so, when to have been able and realign our broad religious concepts of then ago from Egyptian thinking pre monotheism but of kind! To be able and reconstruct the element of purpose from God to man as a defined religion even if at time tribal, then to pre suppose what the past had been by Gods intention expressed in testament at first through the bible, a new era was first born some four thousand years ago!

Not allowing for human frailties, but accepting to the conditions of life and living of some four thousand years ago, if and when Judaism was created, our link between God and man was expressed in discovery through the opinion of tribe, which was normal. That the connection between God and the common man in we had to be managed, then by process it would not have been unusual for a similar priesthood structure to have taken shape if like to what had been in

Egypt? Without charge, this feature if, was of a go between nature, if copied, but with unseen differences?

Egyptian wise the link was between Gods agent in the king, the priesthood, and his ruled subjects, making the divide of having a subtly interwoven aspect of two governing bodies over the flock. Instigated at core or accepted as having been so, the Egyptian pattern when noted was interwoven because at the first time of realisation the aspect of religion actually ruled, because the political aspect of kingship was also the prime link with the God feature! Judaism being fully monotheistic through the recognition in image of our own similarity to God in tribal human form, had to settle the dilemma of how to act and react to the pattern of duty decided by our priesthood and at the same time our tribal leaders, who were first to be our civil governing body? Although also intertwined this pattern was to become more defined when in time the supposed lands of those to the Jewish religion only, were to be ruled by Roman influences who were not to be monotheistic religiously! But in time were full to the capable duty of controlling conquered peoples wholly to the civil means of secular management, which means politically?

In the crucial case of what happened to our first monotheistic religion under siege, without connection, religion was given its first second place reading by having to follow law over Gods law even if not admitted! Without connection because of two thousand and a further six hundred year passage of time. When Christianity and Islam came, they also were led second hand, because then if not only by the time lapse and the additional knowledge gained over those periods of time, we had become more inclined to the secular of now while still looking for the purpose of tomorrow? Although becoming confined to our own freedoms, our one element of surprise was always to remain in our individual control, that of our own free will, which is perhaps our ultimate connection with creation!? Making for the best situation like from today when so much doubt has crept into some of our religious ideologies from various new frontiers we can fortunately rationalise?

Controlled, our free will concept was last and best able to be applied from what we were to learn from Mohammed. Although in task he was to reset the clock of time by the process of realignment, religiously, because of the full intervention of our political/civil drive elsewhere. From which we were beginning to lose much of our real religious fortitude, if only by the advent of time, when if at first considered, our hope was to have been immanent, but now our planning was for tomorrow, but politically only, until Mohammed anew!

In addressing such matters in a similar way that had been done by Moses and Jesus, Mohammed included all the right ingredients, but quite naturally terms were unable to follow the full rational of what the general future was to bring? Because if being told of all the features of what the future was to bring, such as the discovery of the Americas or Australia, hitherto unknown continents. God would have exposed the open reality of God; and thus to have dismantled the aspect or our own free will; because from that moment, with open proof of the existence of the almighty, where might our own free will be in order that by personal choice we might make, if to choose our own destiny!? God would have exposed his hand in open display proving inevitability, which would have compromised the very essence of what the timescale of evolution was for or about, by never leading to creation!

At all times, Mohammed like Moses and Jesus delivered their wisdom in the most honest of ways, but to three distinctive evolutionary groups of the same specie in us, who at such times each group suffered only the full understanding of the day in which we lived. Dated to have been held in check by the thinking of four, two or one thousand four hundred years ago to the three different standards the same?

Our feature of free will has been the cause of our own impatience. We all carry it, so it is a real commodity, where it is frail is that it is from our own efforts by whatever charge that the concept is made work? Which first hand is a great and good ability, if not clouded by other later or better informed intention like from some of our political regimes or some of our religions or from some aspect of opinion from our religious guiders? But to be later realised, expressed from our religious leaders soon to set direction by our choice!

My own guide to the plus or negative of what our free will is, stems not to what Moses or Jesus or Mohammed or our politics or our priesthoods or our imams have said of us in duty. But what is/has been expressed by the following motives with which we always lived. Our first take has to be the principle that our free will is not a bodily organ, it is not skin, nor our liver of kidneys, nor our heart or our brain. It belongs to our emotion, leading to how we might think enough of a recently deceased comrade to at first lay them in comfort, because of death; and at the same time leave the additional comfort of grave goods because of life!

Free will is a human expression only, it is a step beyond instinct and procreation, if free will was not to exist then we are only to be like the great creatures of our oceans, we are to be part of what crocodilian are, a time worn specie group of much character but little substance. We are to know we have free will because of our very good histories, not least of religious concern expressed through Moses, Jesus and Mohammed by/from/through our monotheistic prospect. Further expressed in ally from our Eastern styled religions, because in all cases without exception all religions are set to a similar future prospect outlook? They in we, guarantee tomorrow in the only way possible, by first having broken from the restraints of what evolution is compounded to be by instinct and procreation!

Without fixed proof there is only one further and frail claim to the matter of what free will is and of its ultimate intention, which follows in the matter of reason? In stock, to take all other creatures no matter of what denomination save to evolution only. None have shown the ability to view the future in any degree. My point without force relies on classification, although all strive to procreate for the assured continuance of specie. No inclination to the habit or form of what might have been classed as natural selection is shown! All other specie simply strive to have babies no matter the consequences, their habits are only open to and unguided by the feature of instinct. Illustrated by how the vast majority of other life forms simply display in random and class or react to the matter of germination or fertilisation by seasonal preference. Highlighted by the sacrificial way most specie over produce an abundance of young to account for the vagaries of nature itself, instinctively driven by what the seasons might bring!

Sadly to be exposed in the habits of some of our more renowned specie of birds or mammals, who also display a faltering lack of concern to some of there offspring if times are hard? Instead of collecting to the ability of natural selection, by not allowing it take place, instead of promoting the aspect of care in the community for specie integrity, many of the higher creatures ably allow some or in many cases many of there offspring perish in order that at least one survives for the end game of procreation, instinctively!

With another look, but not of solid credentials, when viewed in the same light of other creatures also naturally selected because of the product they are. They in type propagate at an alarmingly high rate leaving the outcome of survival of their young to pure chance. Again without claim, but in having laid the above examples my picture for other creatures is only of now, none save we have given any inkling to what the future might bring or indeed what it is. Which is not to be a charge only a condition of comparison, like with we, and in the case of myself, can query into understanding the fact that we alone carry the ability of choice through our own free will, naturally selected or not!

My closing declaration of free will so far is for religious purposes only, the matter of free will stands? Because of the way that the rational leans to monotheistic terms, Moses, Jesus and Mohammed when being contacted by the assuredness of God, even through linking agents, were employed as mediums for our sensitivities. The full and final matter of choice to all three religions offered in addition from the thinking of the day, already expressed by Moses, Jesus and Mohammed. Even with or from or through the standard of thinking allowed by the advances of the sciences, in religious terms each religion although forthright. Could only allow for us individually to make the right choices by employing our free will of which all three were made to accommodate! Perhaps allowing from the unseen terms of contact the three main prophets, chosen to represent us, always offered free choice by the hidden terms of free will. Which in term if seen through Shanidar was the first proof that we as specie employed actions outside of Gods terms, to what the time span of evolution was!

In connection, perhaps a linked measure to understand our free will concept, had always been relayed by religious terms, leastways in respect of what Moses, Jesus and Mohammed was to set before us, is that each concept was in relay? At all times and to all three religions from their respective time set, we were always shown by comparison of what our best choice was to be of how we use our free will? Another matter in the same vein is related in question to or for Moses, Jesus and Mohammed. Not of each other comparatively or to be taken separately, but connected to times before the matter of monotheism.

When and if the relationship between the nature of man and God had been entwined to various similar creatures to man, leastways by construction? Whereby the God image was fixated to a man with a beasts head or any form of mixed combination of man or beast, simply to convey mystery, by our own understanding or the natural understanding of the day. Which although counted to be pre monotheistic, without forfeit, although some societies today still hold belief outside the scope of what modern Eastern of Western religions portray. They are carried by the thinking of the day to those societies. Who have loaded there own philosophies, although using their free will, have not had the opportunity to learn and express the better effects of other histories of where we now stand these recent sixty plus years of this last modern era, but by connection to old religions, old free will is used!

Fortunately aside to any previous fixations of how or why, when counting, we can stand alone as this single specie with the ability of choice and our best concern for the future. Never now to be challenged further by what might be presented as a new religion? Because in religious terms to be marked by the advances made over these past four thousand years our free will has survived the only challenge to deplete its reality, our time born religions!

If the image of God is ever to intervene on our behalf and appear more than a voice to us individually or collectively, then the actuality of creation over evolution is forever destroyed. Because at a single stroke Gods own timescale of evolution automatically replaces our collective discovery of creation! We, who by that direct contact would become no more as the single specie we are, because on that individual level we would now have lost the perspective of what purpose is or was or might be!

From that simple statement above can be derived our first tick to the matter of our specie singularity unquestioned even undoubted! Whereby in expression of our individual and in deep reverence to all our religions we might tweak the nose of politics and what it has become to set the path of reason into the possibility of purpose renewed!

Offering only from a religious perspective, no matter what religion we belong to, the one unifying acceptance however tied we are to our current religion, all are worthy in standard, save miscreants! Nevertheless to point that standard by collective reference and in aid from our religious leaders in keeping to the essence of our held religions, we as the earth bound people we are by specie. By stint of how we as one have emerged slightly changed by the circumstances of our climate, geography, diet or for other reasons. Can set the one claim in standard to also hold the open operation of pride that it is only we to have grown the aspect of free will! Being a unique tool to enable us self rationalise events as they were, are or might yet become, allowing us exercise our own personal choice always and under all circumstances, wherever!

From which we can draw back from our delighted pleasure of discovery in the one God or the full concept of the future. By openly not wishing to kill of self by self in case, that the harm done, belies God's trust in us. Not least if the person to become in choice of an early ending, even to fit to the/my/our standard of what 'Natural Body Death' is. Then it might be we as the same individual who compromise Gods intentions for any particular person, for whatever use? If God is ever to wish and intervene on our behalf at any time in the future before we have set our own show to what hope in purpose really is!? Because in case we erroneously kill one of self, who by searching throughout the three and a half million years of creation, was the one person able and offer mans insight to God, by Gods choice!

## CHAPTER 187

### REMEDY AND CURE!

Living to a ripe old age of fifty plus or seventy plus years in title to allow that we are to understand the meaning and purpose, is a very short time, but vital. Although done to the edge of following all the creatures of pure instinct, we have cast the one different shadow bringing us in entitlement to ask of what the future should be, will be or might be?

By our own good fortune and monumental effort to be able and stand atop of the hill, our view of the future far surpasses all others of our kindred life forms. It is not I who have self classified other life forms, but the slow fall of time relative to there own longevity, from which in passing no self drive or prime motivation had ever been seen in them wanting to know more? My own clause that all propositions have come from us the human element of specie only, stands on the one point of if! Whereby even that we have fully known and accepted what in time must be the full declaration to the view of purpose? We still question our own ability while in our own full search to discover all the cures to all of our mysteries through the sciences, now classified by myself as the fourth dimension!

By the nature of open study, I have already pointed to the inept standing of what the fourth dimension might yet reveal? Because as we know of the full reaches of science to have been set by a limited number of circumstances, by a limited number of elements, all acting together overseen by a limited number of conditions. So that we can at last read formula proved by experiment in its full and final array so that we know all? Alas without direction science from within the range of its own study for a very short time had done enough to swallow its own pride. But in the true sense to the meaning of what formula should be, it began to deny the basics of how we understood what science should be? Because from the secular perspective science realised it did not have enough information of what we already knew to be able and express what had already occurred in the universe without acceptance to the time span of evolution!

From which seen by myself with my own ability to understand, science had said that in order to be able and be in the true position to follow the natural progression of experiment. Without cause to humanity, which is us, science on our behalf had allowed it so that in order to understand the nature in or for or too the creation of the universe unaided. It is necessary to form by imagination, form by imagination extra dimensions, which in turn create by invention the need of extra features to follow our one standard. Moving the feature of us having to recalculate the limited number of known circumstances or elements or conditions into the uncalculable, which could mean we would require extra dimensions. Perhaps bringing us in full circle to when we thought we had discovered God!

But alas unashamedly moving the impetus of what science is supposed to be into the direct camp of civil control or government or politics, to be a self serving entity without any recourse to those who at first made her, use her, need her, but are not here to serve her. I mark the ultimate arrogance of what we have achieved by looking into the portholes of science. To reveal by the assumed matter of how we are to accept the feature of its discoveries as a means of explanation over the meaning/reasoning of how we have arrived at our own multiplexed religions. Which in turn are to the full expressions of what ever science might wish to understand, which will be still lost, because on delivery of extra dimensions, we will not be able and understand the reflection in any case, until 'Natural Body Death'?

If cause is to be ever explained through the invention of up to seven or so extra dimensions so that the full range of formula may be met, then the whole study has been put in jeopardy of what science really is. To place the standard of our religious excellence in its rightful place, one small mark we might use in full recognition of this is to be taken from the proposed style of what the Fifth Dimension is to be?

Although I have been able and borrow a theory of what and where science in conjunction with politics had proposed to lead us, my fixation of employing the three taken and one active prospect is never to be the final move because of our very nature. No matter what theories abound some part of us will always question, question and sometimes answer!

What has been done in the open and is to be further implemented in the setting of one standard to apply to us all of this single specie, which is no bad thing ever! My own view of how to use the Fifth Dimension in implementation is fairly basic and starts now. My cut to overlay the concept upon our two main influences in politics and our religion and let the water flow, of which for both examples our own levels of discovery in need will be found, is only part of our new experiment.

Without direction what is to emerge is the open need and requirement for our religious leaders to stand up and be counted for the purpose to highlight our specie singularity. This of course can be done without forfeit because for the first time ever, from all of our histories. we are gathered in time by the recent events of these past sixty plus years of this last modern era without having to compete for Gods favour!

Not least because by all religious channels, how ever previously delivered, we all now have the open link to Shanidar and our first concept of the future shown. Which allows the open expectation that we can accept the full spirit of organised religion owing to mankind by the measure of our own exuberant impatience?

From Shanidar to the Egyptian heavenly Gods, our Western styled to our Eastern styled religions, can be forever joined in our open plan for tomorrow first hand now, even to be from several previous takes before? That is the message we can expect from our religious leaders without the need to ever change the past and its meaning from then ago. But only to allow that in all of our present finery we have changed in attitude relayed by the changing circumstances to have met our challenge. Leastways from day one when some of us left frozen foot prints in volcanic ash over three and a half million years ago.

Politics as our second means of control has its own burden, when its own reflection seen has been of the wrong print? Because at all times even before reason, having had at first civil control as a collective lead, it entered our psyche giving us alone as specie, an added incentive to what? But in a way to make us become independent by forming our view, while at the same time also propagating as others but by a drive then unknown!

I had given notice of when or if or how we might have come to the general reasoning about religion, by the signs we left at Shanidar? But in order for that burial to have taken place, a measure of organisation in terms of control had to have occurred? Under the wind of natural selection this or the condition of civil control had to have pre-empted the possible duty of the event in any case, because of the prevailing circumstances driven by the physical and mental levels of awareness we had achieved.

Having turned sixty to one hundred millennium into the similar uncertainty of now, determined by this past sixty plus years of this last modern era. Much the same standard applies where now we have defined political systems to control us, albeit laid in parallel with our religions fixed by religious tolerance. Of which the latter now becomes the most important to be revealed how they are to be further read or led in conjunction of how our religious leaders in us see the future!

For politics to progress in a way hitherto unseen, but from the need for expansion shown us by its own stunted opinions generated by first, second or third world thinking. Is to reverse and re-cut the cloth to make a better cape in which we might better warm our wellbeing. Our need becomes more desperate in these recent of times, because when in due course our religious leaders honestly take the floor, to show what our political shortcomings are or how they were created in open error, then the cure desired might begin without comparison so far, but led from a new direction?.

Our future will always continue, because each day when the sun rises, and it rises every second somewhere in the world for someone. We would have been left the status quo save for our unchallenged query in all matters except those we have allowed ourselves take for granted. Because at any particular time when the idea of change made a dent, even a small one, our choice was to take the pill even if not to pronounce a cure?

Like for instance when before we knew almost anything, our first credit was given to the aspect of need, when, if noted by envy, we saw how others adapted better to certain environments. Like for instance if it came known, that at a certain time one group apart to others came to the use of flint! Eventually being able and knap the skill to a fine exportable art! Creating the two human elements of envy in the name of want and achievement in the name of effort! Again an added fixation to our separation and singularity because other specie did not show the same aptitude unless in the one field to be able and propagate. Although we in our tribal groups made collective, but in cases separate use of our particular environment and with what was at hand. My submission to the matter of how we did interact by trace comes together around the events when again one tribe, clan of group took time to bury one of our own dead! Opening the first mystery for us all by asking the question "what comes next"?

For balance, but not to be comparative, when our religious leaders stand up to be counted by expressing the matter of our true needs. In looking to one of our prospects 'Out of Africa', without pause, the explanation is to be of total acceptance without any reference to what has been already done by our particular religions. 'Out of Africa' is not a pre ordained requisite, is not a trophy to be claimed by the first bidder, is not even a challenge?

'Out of Africa' is a reference, but in stone, from which we can all trace a single line of descent channelled by the then only need to propagate? Which is to stand in fact without denial from any religious source, even our leaders of now, who in some cases from the haste of decision making, have taken the act of us finding the matter of creation to be a self appointed reality?

Although to remain a matter of controversy generated by some of our genuine held attitudes, our religions are already fully aware that in concept all run to the matter of supremacy through the one being to be God or Allah or nirvana or the Great Spirit! Then also we know that it is only some of our religious leaders who have failed to see this phenomenon until now held in balance to how we have used the same issue of 'Out of Africa' by or from our political overlay, but not really knowing the full connection!

From any of the questions borrowed because of the differences how 'Out of Africa' as a stamp of our true unity might be raised? We only need examine the means where we have been affected differently by the same occurrence, not least when realised by our lifecycle differential? Although the underlying means to our seen differences had been delivered by the circumstances of climate, geography, diet and other reasons, those features magnify how politics over the ages has always failed us! Not least by allowing a range of different lifecycles form over us, yet accepting the condition to have come by our own made standard of being first, second or third world commercially!

By any reckoning, because that fact had only been tackled half measured, and we have not yet read how to lay all propositions in their proper full state first hand, having always led that the plan is to be continually met under the non written rules of our steps taken syndrome? Now is to be the first real time that we have been able and review the nature of the prevailing situation, because of these past sixty plus years of intended change into our unknown future of the next four thousand years! Unless we bite the bullet of abounding discontent to settle the matter of what in reality will become our biggest concern through the use of the Fifth Dimensional overlay, in this instance, politically!

I have said that the best means of coordinated change will come from the application of the overlay of our three taken prospects with the one active remaining to act as a clutch and break to some of our proposed excesses. At first we must change in dramatic form the existing understanding of how we already apply politics through any form of government. Of which the signs are that in them we have already set the pecking order, that some forms hold to be more important than other forms? Allowing them have religious opinions or not, but enough to always surpass the required opposition of delivered religious chastisement by installing the fixture of religious tolerance so that they; some, might override the sensation of religious input!

In open display it is possible to agree with matters of what we have tried to do with our political leads, leastways over these past sixty plus years of this last modern era. One concern because of events and or our reflective reactions in the same time period, first, second or third hand. Makes for the crucial necessity to implement the idea of needed change, then to lay the means of remedy and cure!

Without much further pre-emption the time of our closest recognition is at hand when the steps to be taken, and left in print for the future, are to be backwards? I feel our religions are safe and in our own good hands when now viewed with the added implement of the Fifth Dimension. Which only allow for our religious leaders to be forward looking no matter of what existing religious favour?

Politics by previously casting its own shadow of self mastery needs to be slightly undressed before undergoing the same, but differently read treatment we are to take on religiously. At first and without ceremony the main loss of what had previously encased our confidence in politics has to be lessened then cast aside before we dress in its replacement. Fed in lead by the same three taken and one active prospect, but from the political wing of the Fifth Dimensional overlay, we will notice change in style, but without the need of any comparison with what has been claimed religiously!

In earlier pages I had set the task for us to question the aspect of what civilisation might be or might mean? Especially when from it in supposition, we can set the very standards whereby we alone can cast opinion to what is only seen in the first reflection of any mirror. With us making further judgements to what the reflection leads too, even if to have relayed its image to us through reflecting upon another mirror, but to always remember that when the right hand is raised the left responds at some time in the transaction?

In order that we might now stand on the same ground that we are expected to in this last modern era, then we must heed our own diagnosis from whatever corner. Because this last modern era in time is representative of our own lifecycle in general terms, therefore any remedial suggestion if offered by any individual, can be put forward, but only in the positive, which is not to be the end product only the suggestion of needed change!

Politics of these recent times has shown its own lacking, but as with many taken acceptances does not know the order of required change, because it has become set in working by our steps taken syndrome standard! Therefore in compliment to our own acceptances, when change or the need of change is put forward it has to be set in the tangible for the possible. Like for instance when we offer a new or better defined base structure, like our three taken and one active prospect of the Fifth Dimension, our method of delivery has to be one to lessen the claim of us already acting in ignorant disarray!

By first recognising the almost equal value of our politics in relation to the full value of our many religions, our true course is to be set from now in preparation to equal the best aspects of all religions by standard, with the changes owed our political progress to follow? Yet in order to tackle that task beyond the thought of it, bearing in mind that we are to treat both influences equally with the overlay of the Fifth Dimension, politics will need special consideration. Not least because in term it is the medium that had equated our religions to the box of religious tolerance which now deserves to be charged?

I had made the open connection between politics and the word civilisation which are party in the sense that the single word civil interposes on both expressions. Civil/political control; civilisation, meaning or connected to the feature of human management by conduct by effort by effect and by command are not always to be compatible. A further connection in being civilised can be made when looked to some of its products. Like for instance how we use the term to be the seer of all eventualities by papering over any cracks that appear in the final political philosophy with the invention of terms, like holding too or being religiously tolerant and without cause accepting that definition as sealed. While at the same time compromising its very essence in display to an extra adage of suffering human rights supported by behaving politically correct. To be balanced in concept not by relief but under a plethora of mixed laws to cover all eventualities of us being different and none for us being the same, save from our taken prospect 'Out of Africa', which when used to determine what the reflections really are, will show a fixed image!

It is never to be said that we should think the same or act the same. Because in far too many times past, the circumstances of our geography, climate, diet and other reasons have bore down on us all with the same mixed feelings those experiences offered. Even that some of us might not wish to recognise the importance of a better basic understanding of self in us all because of pointed self interest. No charge of morality is to be laid against any individual or type to have put in much personal effort for all, and quite naturally take benefit in the name of self, entering the realm of obtaining personal advantage? That feature of personal wealth is to be kept in the one field of self determination!

I would make no charge against the wealthy even if to have inherited there own benefit. I would make no charge that it would be more difficult for them individually to enter heaven or paradise than to thread the eye of a needle with a camel? By the same token and by how the factor of wealth was tackled in general terms by most religions, because of our personal bounty we are obliged to spread a measure of charity. Is not to be an active consideration by any religious standard, because in all religious cases the aspect of money or wealth spiritually, has no value by any form of inclusion, wealth, money, riches of the singular are purely of political interest!

By crossing the tees and dotting the I's, by doing all the sums in all corners about personal wealth, my show of opinion is a statement through our one active prospect of the Fifth Dimension? With 'One for the Whole' as our forum, my standards about some very lucky individuals, refers to the good of all in the simple form of a statement. If purpose or the intention to explain all the avenues of science to arrive at the door of understanding involves the aspect of civil management via our monetary system, then that is the case. But it is to be seen that the avenue of wealth is only a working means to purchase us the proper time of leisure so that all before us might take place.

By that meaning although first delivered from the means of our civil/political relationships in or to the control of how we bought the liberty of time so that we could pontificate religiously. Then we must run with that advantage before pretending that to tackle the matter of how some of us see greed spoils our best of good intentions religiously!

Personal wealth is to be a condition of politics only, because the nature of it can only be gathered by civil enterprise? My own view expressed to the matter, but from the aspect of abstract reasoning stems to the open view that through the study of wealth and personal wealth undefined. We will better be able and understand the matter of how much has gone wrong in society before the standard of success can be seen from our religions, who in standing in the shadow to the physical meaning of what the commodity of wealth is or means, have allowed our fervour be diverted!

Although we might refer to wealth in term, when looking at two of the anomalies related to it, our need of change can be expressed in the obvious without vicious intent. There will be no need to demand equality of resource when the first stand is to be taken from or through direction by us now delivered by our religious leaders. Because of there lack of concern to the element of power assumed by the material aspect of personal wealth, which in full light has no real place to any religious standard!

With the route of how to achieve our religious wellbeing having received the full explanation of what they believe by heaven or paradise or nirvana or of the end game by 'Natural Body Death' for all. The matter of personal wealth becomes a sharp realisation for all, especially politically, when viewed that it was only a status symbol ever. Seen through the biggest example shown how we were to bury our dead long after Shanidar?

In the gap between Shanidar and now we have had several great examples of how we arrived at our present condition, not least first shown some thirty to eight thousand years ago, even if at times undetected. When viewed from a spiritual perspective, in looking at various aspects of our cave paintings, from which others as my self had offered the study that some could have been tombs in memory of our dead. By first light of realisation that indeed there was to be an afterlife, then in chronicle I have supposed that the fine art work left on the walls of difficult to access caves, was indeed to have been grave goods. In other words and because of civil organisation the depictions on cold dark stone were an expression of our tribal wealth in deference to some of our dead!

By carrying that idea of thought to the time of the Egyptian Pharaohs it can be argued that by the now state of thinking in terms of life, death, God, Kingship, ruling and or with or through or from any religious perspective! Because the ruling element of our tribes had now become kings and did have or hold or feel religious affiliations, that the feature of death was to automatically carry reward. Then the religious interpretation of such a view was that the king at least by lead, was to carry his own emblem of power in accumulated wealth as a bastion in the afterlife? Whether the matter of commodity had now become of un-regular importance religiously, we carried that view until the split of wealth only was made to be of full political significance in these modern times.

My point made is to have a clearer picture for us all when we come to challenge the attached importance of what personal accumulated wealth can mean when reviewed from our joint but differently read political and religious overlay through the Fifth Dimension. More to show in highlight that the drift of opinion has remained morally the same from or to our religious determination, but in class, because of the full product of political materialism has become a tainted challenge by our acceptance to the supposed bounty of commodity? Our look into the divide is not even spread on moral grounds, but to highlight our own forms of misalignment from using our steps taken syndrome normally!

My open argument without quarrel stands to be answered not by myself, not by or from any religious perspective, but from the very heart of all of our political standards of today. By first putting these two examples in line, will be set the means whereby all future political come commercial come monetary arguments will now stem from! Not to forget our great pharaoh kings, but as shown time and time again the proof of death had always shown us that you can't take it with you?

Not to be a means of ridicule, but in the face of how we run our world economy today from which it is possible for one person to accumulate enough personal wealth to exceed the total value in worth of individual countries, has to be an alarm? With the whole concept of all features of modern living commercially, having the base structure of being a free market in terms of time relative to us all. Whereby in the broad sense we as individuals might ply our skill of time in order to support in a proper fashion those of close acquaintance! Spreading what we earn to the one aim of survival until the relay of 'Natural Body Death' if that is the case? From which even with all moral effects counted, can be seen the error of our ways that you can't take it with you no matter how much the desire or plan or wish. Becoming an unasked feature to question the meaning of political purpose balanced to religious purpose! My second example to our half measured reasoning in the first place, stands on the sound footing of negative reality? Whereby our whole stamp to how we are actually to survive has been given in true form to method, determined by our steps taken syndrome, which as a political factor automatically supports money!

Quite quickly in our time travel we were fortunate to grasp the idea of money as the crowning means of barter. Unfortunately from that stem came our ready support to the idea of money and of its own ability to work or lead as a focal point in the name of commodity. So much so that at pinnacle within these past sixty plus years of this last modern era, we have come to service the concept of money for no other reason than of its worth? Taking from us all the rational of what purpose is to be, because in concern of any philosophy most of our religions are beginning to bend under the weight of what money in the name of politics can deliver to us above honest morality!

By the simplest example of how we have cast the die in looking through or at any aspect of supply and demand or the refined form of market forces, we have adopted the false economy of applying a factor to be on money that is not there, or should never be! Money cannot propagate in any form save that in which we have uncontrollably allowed it in giving it voice to set opinion over necessity. Although unseen at any particular time, even to the time or reading this report the first, second or third time, that statement is constantly before us, yet so well hidden that most of the time we cannot see the wood for the trees.

My best example to explain the myth at one attempt comes from an aspect of outcome rather than direct input, by the good fortune of any of us holding a quantity of money above our direct need. In most cases that surplus is held in the safe hands of one of our banking institutions anywhere throughout the world, with the major advertised plus of extra interest accumulated on that held sum of whatever amount? Meaning by all accounts that money on its own account grows? Creating an unseen moral burden to many of our wealthy with some to have vast amounts of wealth to account, who in or from the moral turmoil created, blind or not, cannot even spend the extra interest they have earned for no effort. Because that amount exceeds need, which is not a condemnation!

My point to the effect of interest earned is an abstract observation when to talk just about currency. But in order that we might shelter the cared for money held in safe keeping by those who can afford it? In allowing that money to propagate in the name of return and in using the reflective charge of whatever interest was earned, our only option open is for us to borrow the same coin but at a higher rate of interest than what we give in dividend. Creating the double reflection whereby interest earned is paid for by higher interest charged. From which by taking the standard that those who borrow are the neediest in most cases, are left to pay the interest for those who have no need to borrow. Who are left look in glee at their own earnings without any understanding of how the system which is commercially generated by politics, actually works? By no form of criticism to them, but in due to how our steps taken syndrome works, we have allowed one of our best general forms of convenience in money, grow its own beard of interested deception.

Making the hard fact that somewhere in between the balance of the interest earned from investment and the interest charged to borrow, there is a further burden to be added to those that need to borrow money. All active premiums for such transactions are further challenged that those in the position to make or lead in such transactions. Draw there own costs from the borrower, giving an unwarranted political emphasis to the direct need to manage money as a source in commodity instead of the real convenience it was first supposed to be. Making the standard, that now instead of being the aid it should be, it has become the fuel of all enterprise and in cases to be above the aim of purpose, leastways to do enough to divert us from what should be?

Fortunately in style and in new time we will have better answers by way of direction from the heart of our religious leaders who without any perceived doubt will be able and offer the case of balance to what we already do. But I add in hot delight that their case is our case, because from now until; they can change emphasis in turning our modern steps taken syndrome practices into the proactive form of first setting all criteria full measured, that is full measured!

## CHAPTER 188

Once again we have the challenge to change without alteration. This can always be done if we move in a forward direction, but never to forget the past, which of course is never to be changed! All to be done in reference of the past and how it works, is remember in the most strait forward way how it was made? If not from the style generated by our own unseen circumstances of how our geography, climate, diet and other features might have also formed aspects of our thinking of the day. Of the day and time in which we lived, until we are here and now, but not yet fully accustomed with how so, not least because of the way some of us still propose to act and or react to the same in self but from other views!

'One for the Whole' is the lock which is to be undone for how we will be able and change for the better because we are here and now, but in order for us to find or shape the key that will unleash that avenue of our forum. Our must do or to be our must change acceptances have to come thick and fast, because I further propose that change does not necessarily mean alteration?

I put the tone that most of the ratchets to be engaged in the lock for it to spring, are of political definition. Because in the most important of instances it is from that aspect of our control we have been able and hide the importance of what our real religious attitudes should be. With the feature or standing of religious tolerance having been laid by political dexterity in order for its own progress, I now submit in formal fashion, but ever through our prospect in 'One for the Whole' which is soon to be our open medium. Because in term that prospect allows that if, I might support all the aspects of all religions I can use my argument 'Past Gods Words'. Which has to become our first means of communication from now, belying all of our political attitudes in all cases and turning heed to our religious leaders, who also from now in open speech speak to our understanding 'Past Gods Words' until the natural time of our 'Natural Body Death'!

For me to have put challenge to the claim of us being civilised by any section of our community or tribe or clan or political group or of any religious persuasion, stands on the grounds in the nature of what such supposed civilisation has brought us. Not least in terms of our regard to human rights or civil rights or in acting politically correct or introducing civil laws or of standardising religious tolerance!

Our acclaimed styles of what we assume to be civilised behaviour has to be realigned for the reasons of us all gaining any benefit of what we have tried to do these past sixty plus years of this last modern era. Without any form of negative charge, in reality the feature allows us our first, second and third world approach, but only heeded first world, which erases value. If also it had been the master of some of our attachments like from the above list, then human rights or the tone of or the intention of human rights has to make its own standing without having been supported from a background of being civilised!

Human rights in concept, has been abused by the very notion it purports to hold sacred, which I personally liken to the laxity of its use through our standard of religious tolerance. From my obscure case to highlight this acceptance we have only to look at many of the events during the past Second World War period from which the whole revamped idea was to be reinstated. In terms of it being a condition and not an event has aided our misuse of its full welfare of recent times. Our case for it in operation remains to follow our steps taken syndrome, naturally following the first impression of our original half measured introduction.

Although first introduced in a global sense to be a passion of the League of Nations pre world war two, without any objection it was totally laid down from the opinion of our first world standing, held at that time by America and Europe. But then lost in failure due to the events that seemed to drive a motive or some relented desire for us to enter a Second World War from which we were prepared to equate all the wrongful settlements due and passed by the victors of the First World War, who in case were still America but with greater European self involvement then!

When reintroduced again at wars end in this last modern era through the United Nations, remembering now that the world was thought of to be in a different sense of maturity even reacting while in our recent bloc culture standoff mentality. If the concept of human rights was to be played by both sides, then some forms of stability were achieved. But alas when the wheels began to fall of in part due to the political emphasis to the desire of us as our separate nationals seeking self determination. Our renewed concept of human rights was answered to by the overall powers that be, answering in solution to set the one standard by our first, second or third world reading. Using the arrogance of ignorance carry the flawed plan to fruition by the colossal assumption that might is always right!

There is no way ever that we are to disintegrate the element of what human rights was intended to be, what we are to do with all express, is bring the matter to its full intention by making all the attachments to it work from the full measured introduction we had always intended to deliver. Our best way to do so in aspect like other modern features of our working presentation to conduct, can only be tackled at this first chance, because we are all able and use instant communication to our best advantage from now, with the voice of 'One for the Whole'!

By working from the perspective of not, of not being fully civilised, in this case is the best way to realign the open concept of what human rights ever was intended to be. By taking out or taking us away from any standard of what or how any political system or style could have claimed of the matter at any time pre now is a step to be taken full measured. Which again is finding another piece of the jigsaw needed for rational understanding so that we might admit some of the errors of our ways from a set of previous ideals, first cut before we had reached the matter of international political reasoning? By anecdote, the feature of being civilised or not, has never been asked of by any of our main world religions, Eastern or Western, because in source civilisation is religion!

Born from the confusion in style of how many of us have tried to work what ever view we had to operate by, into our steps taken syndrome procedure? Was again tied to the matter of how we thought from our first world perspective. Leaving the stain of our reasoning to taint the afterthought of our second or third world equal necessities at least politically, certainly not fully religiously because the difference is not so obvious over these years of this last modern era!

My case for us to drop or alter or down grade what essentially was an unresearched application of who was to be civilised. Although to be in open forum by all codes from now through our one active prospect in 'One for the Whole', even that similar additions of communication will be forth coming, our over view is to centre us all to have been 'Out of Africa'. Shown by all of us who in walking backwards, we will all eventually leave the same mark in volcanic ash as it dusted its own environment some three and a half million years ago!

By taking note that one of our best collective achievement in full representation to us all had been pointed to by my self in term of us arriving to ponder what the American dream means. Falls to another piece of the jig-saw needed so that we are not to be afraid of some aspects of what the future or not reaching it might bring? My own standard is the future, but in order that our steps taken in that direction are the right ones in order to leave the matter of style in the hands of our religious leaders. We must learn how the past actually affected us before we realise the results of whatever we may have done?

From whatever time it was that we accepted the principle of human rights up to date, our perceptions drawn from the outlook of being civilised have erroneously made that feature what it is not. If first seen for my own argument that cracks began to appear in the principle of the American dream, because the base structure of general human rights in relation to all men had become self compromising. From then ago the study of this fact runs in parallel to what was happening in the Middle East in the mid nineteen sixties between the same people if first and second world! Spreading in attitude to the same people of the new emerging countries of Africa who were second or third world, but all entered to the spirit of hope in the future!

Although many transitional quirks had to be settled in or for the theory of right, our first big notice to the fact that we are not really civilised came from the land of the American dream and quickly spread to Europe, but was handled in a slightly different way?

Much of our tribal or national or social interaction to have occurred by reference in the late nineteen fifties and early to mid nineteen sixties of the just past century, were inbound with the principle of human rights for us all, which was meant to be a levelling feature for us. Without charge, but to take note that even if to have been of a shocking and damning nature, by very small signs at first, the cracks that were to occur in the basis of all of our principles came from deep within the land from where the American dream of personal achievement was most forthright.

Even by having the seat of our main United Nation forum in America at that time, it was first to be internationally realised that in point of fact America counted as two countries. One to be black skinned and one to be white skinned. Complicated by the overlay of being generally Christian by religion, which from the theory of doctrine was unable to notice the colour of any skin or of any status being first, second or third world? Even that the United Nation directive if of human rights was delivered in first world tones, which amounted to be a half measured delivery of the feature! Even if by American with heavy European involvement the feature was set again and again to thwart the principles of what communism was actually delivering to its people in the form of oppressive terror!

When the home standard of operating by a white skinned and a black skinned code came to head in the name of equality expressed by human rights in America herself. Our only option so as not to weaken the standard we were setting in our supposed equality for all first world ideals, even in communist countries. Was to address the position in note by our tried steps taken syndrome, which for settlement of, required that we were to review the original meaning of what human rights was ever to be? By and with the introduction of the new vision of civil rights, which if delivered from the confines of a wholly civilised point of view, had to wholly compromise any standard of what human rights are, was or became intended to be? We began to cloud what civilisation really meant not least when to introduce a mass of racial and or ethnic laws to cope with our political shortcomings, when in essence this later bout of law making only went to highlight our differences?

What was effectively laid down was a solution to a non problem, but by a standard of failing by taking the scientific lead as normal! We were only too ready to accept any leading connected tentacle of politics loud enough to push forward its own opinion/solution understood or not? By the same mix but from the perspective of already holding to be religiously tolerant politically, even that it was never fully understood that our religions were already deeply intertwined with the concept of human rights. Because in the case of all of our religions the end in all cases justified the means by the result of heaven or paradise or in learning nirvana or in connection with the Great Spirit, full measured!

By being classed from our political wing to the one standard of being religiously tolerant for all religions, our main religions by number and from the volume of there own numbers had been cut off from the immediate and directed means of contact with our politics. So much so that for politics not to hold to ear our exasperation, because of the fixture of religious tolerance, was to create our biggest human error over these past sixty plus years of this last modern era!

With religions held in ransom to a political code, none could comment like to what human rights was to mean religiously, when politics had already drawn up its charter? Also from the reflected image of the rising political standard of what civil rights was to mean and from where, this new standard did not have to hold the same meaning throughout the world. Which is to be a balanced argument when all the votes are in if delivered through the prospect of 'One for the Whole'!

Part of the standard, if to be, of the American dream, came into being in its altered and correct form from the early to mid nineteen sixties, when at first it was realised that instead of meaning human rights in general. All that we started through the old league of nations was first world human rights!

Allowing that if my prognosis is right, then and when it was realised that the standard of being first or second or even third world already operated in America and Europe, shown in context by our held attitudes to people we saw different than our self by whatever cut. Then the charge of being fully civilised failed, because in so or from so, no issue at that time in passing could ever have been in position for us to further compromise what had been tried in the name of human rights even through the United Nations!

Cracks that I have seen can only be noticed when the taken prospects of the Fifth Dimension are applied, because until now we have run with the flow over the value of human rights and or civil rights, even that in places they are contradictory. Without being able and rely on the single feature of 'Natural Body Death' as our totem of equality in all cases. Human rights and or civil rights are almost an expression to far, and have become the medium of our divide far more than to bring us together. By having stated, if with the maxim of human rights we are to use it as our penicillin to cure our ills is one thing. By having to introduce the later addition of civil rights to enhance that opinion has done little to strengthen our position of unity even if we try and blame our previous bloc culture standoff for setting us up in the first place, whatever done with both examples was in too much haste!

From how I have seen the reflection to the matter is only cleared by the good and gracious effort of people like me in us all, who at first were eager to accept the opinion of right over might by the process of discussion. Unfortunately what is becoming more into view than is necessary, is the full effect of us growing to the accepted standard of always using half measured delivery and complying by our steps taken syndrome, meaning that nothing needs doing because the lights still work?

Our needed change that I have seen revolves around the understanding of what being civilised means or of what behaving in a civilised manner means. Although the word its self sets standards, our time now is overdue to change our habits which are encased in the steps taken syndrome reactive mode because the lights still work. If civilisation had delivered the bounty of human right, civil rights, of behaving politically correct and of religious tolerance in accord. Then if any change or alteration is needed to any single aspect above, we are not really fully civilised either by reflection or from the reflection of the reflection.

With our actual standard of civilisation working only half measured, does not mean we have to discard our acceptance of it in order to change it by style. All that we are to do is improve how we are now to operate in class too or by or with our steps taken syndrome, which is to be a lot easier than what we might imagine, because we are who we are; the only specie ever to have seen the possible meaning of creation!!

## CHAPTER 189

With the wheel ever turning even after we close this book for the second last time; the matter of change required is by taking small steps but to be of huge consequence, not least when now realised that we really are the same people both in evolution and of creation!

From whatever trigger I could call in aid to my own presentation of remedy if needed, the two rather small issues I have chosen are only to be expressive of how we must set the balance in these times of this last modern era! With both issues I have selected having a longer history than first realised extending in time beyond the span of these past sixty plus years. One of over two hundred years, and the other of some fourteen hundred years, is to be no barrier of an excuse to be used against my own principle, especially when my tone is to set the precedent so far unseen, but relevant to us all. Because of how I have used these past sixty plus years of this last modern era to examine what I see to be the most representative examples of how our politics and religions are divided. Not least in the sense of how they each display an unreasonable amount of contempt to our values by all of class in this one specie we are!

Our American dream is to be my political form of what all politics, in the name of social objectives are aiming for from no matter what background we hold. In general terms we are naturally progressive in thinking, even when laid back by character, then that political image is a balanced assessment of our true condition? Because in all matters, most of us continually keep looking over the horizon of time into the material, commercial and in gainful hope for better, which is a fair condition when the manner of hope is to set the automatic standard of naturally selected improvement by all creeds!

Our American dream seen through the lens of reality translates to the fact that we simply want better, but as of yet we have never really looked to the full meaning of that concept unless through the medium of wanting more! Without the further use of protracted argument in using theory by my own interpretation of the event of what politics can ultimately deliver to us, my standard is only relative to our final political expectation on its own account. Because at best, in achievement to reach the American dream on whatever level, can only be done from the secular and in the name of politics? At no time has any suggestion been put forward that the position has anything to do with the fact that ultimately America herself in modern time had grown with mainly Christian influences?

From my self using that clause as a tenet to our achievements politically, has not been a reason to influence my own counterbalance of any religious situation, save over the mixture of events these past sixty plus years! By centring on one small supposed act and separate action by some individuals who are or might have been influenced to react to how our American dream of Americanism is perceived, especially in parts of the Middle East, where such ideals are represented by proxy? Is done to square all of our political and or religious arguments yet to be in the future, because from now, any supposition has to be measured by the three taken and one active prospects of the Fifth Dimension!

By taking what I see to be the misguided reactive countenance of some imams and how or on what they might wish to influence some of our young? Has been deliberate in link to this last modern era, because of the open ambiguity from which any half measured delivery has allowed other opinions form, even by other means to be considered false enough to excite some extreme reactions only! By my own deliberate move away from the convention of having to react in many cases even in not knowing what the deed was for. I have come to question the standard of what human rights are or of the additional need of civil rights separate from that context, vitally stitched to what we are to signal by behaving politically correct in our religiously tolerant first, second and third world societies. My balance is both reflective and also encounters what is to be the method to the understanding of what the reflection of the reflection tells us!

Imams who might face of the outcome of martyrdom in the form for acts of murder are not worthy to represent any of us in any cause, not because of why they have acted so, even if my own supposition is to be wrong! But because in the first influence of what they do, there is no manner or style or reason of how they might set standards for any of us in the name of Islam to act by reaction to any matter of pure politics!

Because of such action our aim is to be ever distorted, not by how we might have felt the future to become looked at from fourteen hundred years before, but because in the most obvious of cases even throughout the world, martyrdom in the name of Islam is taught to be used as a political tool! Belying all the features of what ever the future of Islam was to be, now that we can see it as directed, to already be a major world religion by religious terms. When all that can ever be achieved by some imams, or some voiced in the name of Islam who would lower the tone of Islam in response to what had been political activities in application from a perspective not to have been aligned to the full reaches of fair modernity now explained through the Fifth Dimension to show martyrdom as death only, not murder!

By so acting in a manner to have become more common as personal suicide, we involved those of us who are never to know the end, even if to be martyrs who now compromise all that was ever shown by Mohammed in our name? From such times that Mohammed was to the pen through Gods own agent; the picture of martyrdom was a vital concept. Because the start had to be maintained from within the violent times that we all lived in then ago.

In any case, but only of secular relevance, from then ago no signs were to have been seen of such a place as the Americas or the continents we know of today. No sign by any form of comparison was to be seen that the final concept of the American dream could or would be laid before us to be reflected upon being of vital importance under any explanation? No connection could have been made in the name America, because fourteen hundred years ago where was and who was America, therefore what was the American dream? With that as the case in point, where is the plan to challenge the American dream?

By the same progressive interlude used fully to the thinking by some of us in our Middle East monotheistic religions, even laid back? When that religious feature in its self was to be labelled second or third world, the angle of the triangle became more acute to the case of our religious divide. Making the one standard to be secular or spiritual in both cases, but as we now know to have been separated in class. So that the secular aspect of our thinking and actions are to be fully progressive, even if to be held by us being civilised, in operating our standard of human rights to civil rights of being politically correct and holding to religious tolerance. Then in concept by all modern times even by the standard of working from our own excellence of mind we continue with our error of thought!

Unless we can rely on one single principle other than evolution to be a focal point in the name of our conduct, we must run in mind to our starting point of Shanidar when in all of our names we first came to the matter of creation in the name of evolution. But only in representation of this single specie we are by the proof that naturally surrounds us, selected or not, but to be the only specie to hope beyond our death!

The second part of this book in section had the leader, Remedy and Cure. Without any form of judgement anywhere, I have laid the medicine on the table. What has been done is of us all, because in the vital lifecycle owed us individually has to be revealed the one standard of raising our self awareness. My style has been strait forward and of good heart, I have been able and know Moses, Jesus, Mohammed and our politics to all, who have been on our side. Of which there are no divisions, unless seen as I have done to look into the mirror of each separate enclave and take them to be a single entity of which none are!

For any individual in taking to or rejecting any political style, system or regime one way or the other is a personal matter. For any of the same individuals in us all to take on or reject any of our religions in being Eastern or Western or coded to another medium save the inclusion of personal or collective abuse of we the individual, is our challenge. Because without calling, without selection, even without choice, there is open evidence that of all specie throughout the world come and gone, only one voice has made a sound to be upon the universe. When we as specie wondered in awe at how one of our own when dead and quiet and unmoving of self, left the memory of reacting in our community in our own minds eye? So much so that we were compelled to leave with them in there loneliness, goods that could be further used when needed while on there own separate silent journey. And just in case, goods also to bring memories of us with them!?

Cure and Remedy is held in all cases by our own lifecycles, because again until we actually die to the true terms of 'Natural Body Death' the mist will always remain for those of us who close all lights in death, only then the case is there to remain closed. Because in all, the matter of creation is not to be explained in the way we have so far tried to raise the curtain? From which I see the standard of how our religious leaders can set the pattern of change without alteration, because from the ethic of most of our religious perspectives we have each done a tolerable job save excesses!

By keeping religions on side but not by the prefix of religious tolerance, each are to be held in the equality of what has occurred through the avenues of change. Even when generated by the circumstances of our geography, climate, diet and for other reasons which had been, but not to be held as final, but without excess if we acquire the openness of what religious forbearance really is, which will suffice in transition if necessary.

## CHAPTER 190
## OUR FIRST EPILOGUE

### THE FIFTH DIMENSION

Our first realisation of the Fifth Dimension, originally thought of because from the proposed study of the future by our best scientific means that additional dimensions would need to be created in order to understand reason before purpose? Has now become a reality not least by this our work from statement to be of the Fifth Dimension which forestalls any double dealing to be presented by the core standard of what our politics has actually become, a mishmash of uncertainty?

In as far that I had ventured the standing of politics and our religions to be the two main bodies of influence upon us in actuality, by how the matter of religious tolerance has now become standard has allowed politics become primary over us. If not on its own account then from our over reliance on some of our political tentacles in the form of commercialism followed in quick succession by science in devotion to the materialism of technology, now interpreted to be a social indicator. Allowing that politics in core seems better attuned to the sound of science and its other allies over what our religious leads have always given us to the full ideals of purpose, science has swept in with mystery busting.

For any of us to catch the speed of science and trim its feathers so that our pace of understanding is to be matched by our expectation is a real case. That I have held its wings of flight in order to rationalise its input for its own sake as well as ours is done to show that science is not a surrogate God or Allah or Nirvana or of the Great Spirit. She is only a small part of our memory to be used by our collective or individual desire, religiously!

Added to how I have seen the full value of science being the provider of formula only and ever was, even if to use the standard of experiment to those end results for record. Science has never had any depth because of her naturally selected state whereby her only possible aim is to supply information already there? That by our own hand we have let the study grow beyond its means of expectation of supply in all the answers humanity might ask or expect. Has allowed that some of us have attached too much importance to her and by link that she is a tentacle of politics, too much credit has been given to the ultimate aim of both in expectation!

By allowing or more precisely bowing to the fact that when science speaks we must listen, has given a false credibility to politics. Who by reflection, when seen the left hand of science raised has, taken the proposition right handed or in fashion to be steps taken, which has been a mirrored means whereby we have been able and put to silence the best value of all our religions. In doing so it is we in some of us who have also taken to the matter of science, who in being able and explain many of our mysteries have allowed it force our contentment by being religiously tolerant in case science questions our embarrassment by explaining something we might have thought of being divine until recently?

My effort to pour salt on the tail of science is ours, therefore not to be taken in malicious form, especially to those of us who use her to be first hand and first of belief. For me to have put the answer that the totality of science rests in formula is a small mathematical equation and is our way of accepting its immensity. Because even while through our religions we have worked from very few certainties to be with God in mind, and in part have closed some of our due study? Science in leading for her own answers has forcibly ignored our efforts by continuing in the field of formula before faith and belief to the misunderstanding that the same mystery of God is also to be proved by experiment!?

So much so, that to reiterate, my example of how and why I have helped end our chase by the introduction of the prospects used from the Fifth Dimension, has been done so that we can at least show science the full power of reasoning even if to use her own studies in a different manner or from different perspectives? Further, to the element of science alone, I had given her the prefix to be the fourth dimension by means to limit the expectation of total scope. Not to diminish scope, but show that even after the full run of all studies there will be more to be seen in God before formula. But again not by comparison, only to show that even with science being set by formula, there are vast studies for us to still undertake by imagination, which if first raised at Shanidar are to be wondrous and high over experiment!

From her study to the end of formula and in keeping to also remain within the limit of human understanding within the limit of scientific understanding reflectively. Science in taking her own steps to create or invent the avenue lined with more and more dimensions so that we might begin to understand some of what we already proposed or proved? Science in our name, and I add from her own conceit, but not blameful, has called that we accept in theory the use of extra dimensions so that the full jigsaw fits the picture of formula joined to evolution but not seen by creation!

For that very reason it is important we set her in mild rebuke because in us as a tentacle of politics she has cajoled the matter of commodity and materialism to a greater standing than its true worth against our sake. Not I add from spite or jealousy, for she does not know these emotions that we can easily run with. Her objective in maintaining formula is honest but shaded, because from her own revelations in moving only for formula, she had automatically criticised our own mysteries in God, Allah, nirvana, enlightenment or continuance by the way we have grown to follow the spiritual before the secular.

Having pointed to the edge of extra dimensions if, has led some of us read that incredible and fantastic visions are possible for us all. Unfortunately not allowing for the nature or how our own emotional fervour allows that our heart is more ready to reach the wishes of what we perceive of Gods need and wants before the ranting of where science might lead. Even so for science to deliver at best the thought of possible time travel if by its own imagination? Almost brings her into the spiritual aspect but without the acceptance or inclination to attempt and meet the final mystery, which can only ever be guessed at so far!

Having already reached the standard of thought to consider the fantastic, not of God, but to the mundane of how? Science in giving us thought to the wonder and nature of the universe, has allowed for our understanding the means to invent reasons, but only scientifically. Her charge also in that field is to invent ways of how something might occur or had occurred. Like for instance in reply to the literary thought of time travel already presented to us by many very well adapted authors writing in the field of science fiction? Whereby we in our human form, by show of our scientific achievement might traverse time in a forward or backward move even if only to observe!

Although an incredible thought, such a vision has been updated even if taken erroneously, to be able and mix all of our fantasies about such maters by scientific means, we will be able and manipulate such ideas by using the concept of extra dimensions! Mixing the realities of actual time relative to the individual in terms of the speed of light being a fixed marker! By further using the theoretical concept of features, extended to the exposed creations of such inventions as worm holes in space, of wave structures to count the reason and or structure of matter? Quantum physics, parallel worlds, of what and where dark matter might be, even to the means of finding the vast quantities of missing matter! Have allowed our awe grow to the full feature of what one side of science is, until we can now look to the reflection of the reflection of what science in our name has tried to present for our benefit. Even if to have been balanced in part to our political medium by being a part of people as my self, but not for us to ignore our religions!

In one of our best examples of how we as a single people, even through the pseudonym of specie can generate our own salvation, most of our gratitude is owed to this last modern era. Not necessarily because of what we have done or because what has occurred, but more precisely because it is only from now in this last modern era that the formula for the correct chemistry of change has been recognised by whatever plan save half measured!

Although on vast occasions I had intimated the vital importance of our religions in partnership with our political determinations. It is only now again that we can treat both the same by adding our three taken and one active prospect from the Fifth Dimension in overlay, but to allow that our reading of the same is to have different effects of outlook. Not to sectionalise us but create a broader understanding of what creation really means, which we are always entitled to query the staid outlook of what scientific evolution might suggest!

From the same opinion of what will naturally follow, even before I would go to our one active prospect in 'One for the Whole'. My case is made to the reality that we are all truly involved in the future! There is no standard to withdraw from the agreed or intended aim of using our Fifth Dimensional overlay in any case, save from the fact that some of us will not be able and rationalise normal intention? Not least for those of us who might be affected with some mental incapacity or some to have been poisoned of mind to the whims of any particular religious or political dogma of the worst kind taking us into closed opinion! With some allowing the nature of the fanatic extend their view by at first not accepting the generous spirit of what the Fifth Dimension has always intended! Fortunately our future although getting close, is not bound by or never was bound by the certainty of the fanatic, from which there is no case whereby any fanatic might turn to cast opinion to when or how anyone might die, other than by 'Natural Body Death'!

In all cases we are not generally disposed to think for each other because of the advent of free will. All that we might do from our own generosity are lay the foundations so that we each can react to the suggestions or proposals any might offer or not; delivered to and through our one active prospect in 'One for the Whole'.

Although mentioned again as above, with two more prospects already read in 'Out of Africa' and 'Past Gods Words'. What needs further opinion is to the obvious and less obvious, from which it is for us to decide without choice, but under our personal free will? Not least when I had already expressed the solid view that we are to be aware that when viewed or reviewed now, our Fifth Dimension can take natural opinion to suit personal choice as per free will. Leastways when in examination of the same overlay to our politics and religions, because now we are able to better value intent from those two separate determinations!

## CHAPTER 191

In using the overlay of the Fifth Dimension, what will be found, is that it is of its time and ours in this last modern era spanning these past sixty plus years of now. Not least in presentation, because we now know how to read the symptoms of our ills and we now know of what our medicine is, which is time delivered to the individual in us to compliment our existing actual, lifecycles? Fifty plus or seventy plus years is plenty long enough for us each to prepare our own comfort with all the best examples of how our joined histories should have been and yet can be!

For us to be aware that the overlay has two standards in how we behave religiously and or politically is to be a matter of self, but not of personal definition because of our singularity of specie, which in spiral is the single most recognisable acknowledgment to the future and what purpose has shown of its self. For any of us to be content to having a future in readable terms, then the key of purpose to the open desire of God has to be met within our ability to understand.

Allowing that under all conditions we started in the same standing owed all other specie throughout all time, even in owing to there/our longevity, be it counted in hundreds of millions of years like from crocodilian or three and a half million years from us condensed in gratitude to be fifty plus or seventy plus years long!

Having already proclaimed that it could only have come from our religious attitudes to give a basic standard of meaning to what purpose really is, stands to the single and only opinion of God. Which includes all standards by Eastern or Western religious definition? To be further endorsed now that we have been able and compress eight hundred million years of life on this our world to the relevance by one lifecycle owed to each of us.

By knowing full well that all histories come from the past, it is only we of this specie who have been able and give sign to what the future is for? Leastways to our expert ability to imagine it past the restraints of instinct set for procreation. We alone have drawn our own picture from what was at first a canvas of death owing to one of group. Who by unknown wisdom did enough to excite our emotions beyond clause, when we buried that new soul some sixty to one hundred thousand years ago in our own memory? Unfortunately that event had been hidden to reality, so that by the same drive when again looking for reality we drew different conclusions but still aimed to the future. We again come to the wonder of death perhaps in Egypt if just before real monotheism?

All done brought us by quick succession to hold the same query first felt at Shanidar, from where we now know was to be the cause of our study in God, relative in form by us, but apart and separate to all other life forms. But for us, making the primary reason of self control in conduct, unfortunately misused by self because we did not really know how to set religion alongside our taken means of civil control, then ago?

Although having come to where we are now, then, our two differences occurred from then until now by our own impatience. By hindsight, in knowing that our standards had already been set by needs must, if through what was to be normal tribal or clan or group interaction in the Mesopotamia area some ten to eight thousand years ago?

In all the times since, marked by how each group reacted or interacted, even when we carried all of our civil habits first hand. Our religions were squeezed between what politics and or civil control demanded or could deliver and what was owing from our refinement of how we had made God, of which the rest is history. But mainly carried in parallel by many of the same clans, tribes and expanded city states in the same general area of the Middle East! Which as the wheel turns, even after the introduction of what are considered some of our final religions, things are kept to the same format. Not least with or from our United Nations whom by accepting the standard of religious tolerance without question of our religions being held in lea of politics has compromised much of our real religious integrity!

We far too readily accepted standards we were not sure of and accepted the wrong reflection of what religions are, against what politics is supposed to be, leaving the gap until now, when again we might address the future!

It has always been difficult to make a true case for our religions while our politics now have ruled the roost, endorsed by how our politics have set recent standards in this last modern era marked by these past sixty plus years! By the same token, but from a different perspective, said again, it is to become clear that our previous standards cannot change until we have a catalyst which will enable rational change without alteration.

Our Fifth Dimension overlay taken in literal form of direction will automatically set a course better suited to how we must now think of the future by the strait forward use of taking the steps owed our three taken prospects followed with the aid to query all, through our one active prospect 'One for the Whole'?

'Out of Africa' in relation to this single specie is the most unifying of any other historical opinion. It is also a provable term by the modern means of genetic study which becomes a distraction when measured against our actual achievements. 'Out of Africa' has never been doubted by any of our serious religious concepts, simply because doctrine has never made any measurement of requirement leastways over these past fourteen hundred years? In open theory from all good religious concepts the picture of our differences percolated by the circumstances of geography, climate, diet and other reasons, has never been a religious issue! We are all to have been always one in God!

'Natural Body Death' is also to be of small religious concern, because even beyond the content of that expression, the matter of death religiously had always mostly been revered as the/a beginning? Although in coming to the conclusion of our death being of paramount importance religiously, our small but potent error in concern only seen now, was to mix that the haste of death by other means was to hold some special religious significance! Until now by our own dogged determination we can come to the one standard for all in accepting the open principle of 'Natural Body Death'. To be carried by all religions making no charge or claim against any individual, then further relayed to any of our adopted political principles which are also to be measured in how they might interpret that taken prospect on its own!

Our third taken prospect 'Past Gods Words' holds no form of barrier to how we might or should think of those people even if considered prophets that some of us revere in our own defined religions. For Moses and Jesus and Mohammed and many others fitted to the core thinking of Western or Eastern religious philosophies. To have given note of there own understanding in no way compromises how we are meant to think or behave by any standard save to the matter of our own free will which was earned before appointment!

'Past Gods Words' is a true feature of how politics works, and from so, because of its need we are to accept many delivered issues, but with the sole aim to alter most. With the sole aim to alter many, not least by how we at first can recognise a tentacle of politics being of the fourth dimension, being only a part of the whole?

After all there is no need to get excited about what science can tell us even about the universe and of its formation, because it is we who hold the key to creation within the boundaries of our religions!

By example only and in using our Western styled religions from concept, what can be seen in good view from the time of Moses, Jesus and Mohammed is change to our care in God! If, and the picture might always remain clouded, our monotheism was extracted from the political aspect of conduct in Egypt in or around or before the time of Moses. Then until that particular time when our choices were only civil or political, until by an unknown connection we regressed to think in the original full picture when at Shanidar, not just to lament, but at last begin to really understand?

If there was ever only one way for us to find out how the value of 'Natural Body Death' by thought was carried from Shanidar, was it not to be in similar fashion of how we might have constructed fresher organised thought to the cause of life, but still keeping the event of death paramount? Although I had put Egypt and her kings, if to have first made the break from full secular and by use of a defined priesthood offer a running plan for the spiritual to be in organised array. Although set to an Egyptian image whereby it was from the king if through the desire of the one sun God and or night time instructions from the heavens that priests were to help administer the state also!

Making the actual case that much of the Egyptian thinking or standard of religious conduct was of slight misalignment, if first proved of how it was so when through Moses, if the picture was made clearer so that the one sun was to become the one God? From the matter of Egyptian selection, if through that sun or king, was now to become the matter of selection to the now one God with the same type an organised priesthood? To administer the good conduct of all in plan coupled with the idea that the true discovery of God was now better organised! Not least if and when the matter was further sealed by the hearing of what was to be, whereby Moses was directly instructed by the one God to the conduct of self, but to be relayed to the chosen tribes of clan later to become the one religion of Judaism, which was misunderstood, because of tribal connection before specie acceptance!

From what ever spark or sign or image Moses was to feel and relay to us of tribe, was done from the effect of history up to then and form our ability of thought then ago? But in charge was the one tied concept that in order for all things to end by new expectation, there at first had to be a beginning? But now by refinement and again tied very much with our ability of projected thought governed by what we could know, Moses was to relay his message to us? Firstly by and from his own ability of delivery and second to how we might receive his words, which now although linked through him, were 'Past Gods Words' yet belonged to the realm of reason!

What ever done then ago, which has run for as many as four thousand years, continues? For me to have speculated that the term of monotheism in Gods name somehow replaced the standard of how and because the Egyptians might have thought of king and creation linked. Proves no case to the possibility of if there was also a priestly plus royal house led migration to the new world some four and a half to three and a half thousand years ago?

My direction to that matter is only by supposition and the possibility to indicate in a time set how the importance of when we first began to develop in the Middle East proper? From the time of our early group interaction in Mesopotamia, by result of, if the key brought by Moses was to be the lock, when undone it gave credence to the real study of God in humanity. While at much the same time, if an errant band of priests and a deposed king line, was to follow the sun in migration ever westward to carry with there own perceived faith in success. How then might they develop in new lands if found even across mighty oceans without the benefit of close interaction like from the best times in Mesopotamia? Unless to keep the standard what was brought by connection to the society who brought the one and full ability of holding doctrine to the imagined demands of the sun by day and the heavens by night? Until some four hundred years ago when we brought the world to its geographic entirety by the last discovery of the new world, only to find that of the full religions already there, all worked 'Past Gods Words'. By such steps of then ago, what we must reckon is how we now set the canvas to hold the concept of our one global world!

Although my standard is ours, what said is in the value of the probable before the actual but rounded by the possible. Because when by the meeting of giants in the discovery of the new world some four hundred years ago, although carrying the first words of Moses in form, but by the new Judaism of Christianity. All claimed in Gods name was 'Past Gods Words', for no ear had been put to gather the first commandment of God to man in not to kill of self by self, unless we would call self another name!?

By working in image 'Past Gods Words' we are at open liberty to kill of self by self in these modern times not least because that is how we had set the table! No study to the matter or content of what or of how our last and very poignant religious work in the Quran was recorded can ever be quantified until we draw the links of chain between fourteen hundred years ago and this last modern era together. Leastways by understanding of what humanity is, but not judged to be religiously tolerant!

Of all that is to be most relevant to us in these most recent of times and in balance to how the value of our religiously tolerant attitude had been most seriously compromised. Is in need of the utmost attention because on its own account without prompt, the concept has shown its biggest and most dangerous flaw ever! But more inclined to how the concept had/is being abused, but within the clouded terms of what religious tolerance is or means or has been allowed permit?

# CHAPTER 192

By reflection, for us to look to a time of our joined histories when the new world was discovered some four hundred years ago making us global for the first time? Without definition, I had laid the matter of that re-discovery to show us the full practice of human sacrifice found in parts of the new world but sadly conducted on a massive scale. Only to be halted by as much brutality that was allowed, even blessed by the same European conquerors, settlers or exploiters who saw the reach of gainful appointments there.

Of such matters, by the same reflection, I had eluded to the fact that it was from the European Roman Catholic, then Christian outlook that this settlement was made, but judged by the standards of the day! Without moral charge, what was done, although brutal and disgusting, by the same use of morality, there was to be a cleansing of what was considered the barbaric practice of human sacrifice, remembering that our young were also included to be objectives of sacrifice, without even understanding the morrow!

Although the world picture has changed since then, giving for my own purposes, that it was from America that the wheel was to turn full circle in the hidden aspect of interpretation? Even that I might have put the case that by other features America has left us the legacy of the American dream to be the political ultimate! Not least because we have adopted an empty promise by how religious tolerance is part of the support to that end? From which I have set the balance of what the ultimate in politics as the American dream is? Weighted to what has now become a primary religious practice how or in which elements might some of our imams promote their own excesses through our young, even though they were not to understand the morrow!!

My case to prove or disprove my own opinion, has always been in the positive, not least that the two features are both prominent in class and made so in this last modern era. If to take or ignore my assumption that America features so prominent in my own presentation is a matter of self choice for all. By the same token to accept in part or by inclination that my interpretation of how some of our imams in we, might have promoted events to end in martyrdom, is a matter of the same issue!

What cannot be questioned is a matter of such loose connection to have slipped unnoticed into the positive, which in effect is the prime example of how and why our open standard of religious tolerance needs the utmost attention first hand!

From when I had laid doubt in previous claim that the concept of accepting our judgements or accepting religious tolerance per se, are to be exposed in combination by events to have occurred and are now occurring to the worst flawed standards any of us might imagine again to be set world wide but no where more serious than in the Middle East! Need our utmost political attention, and in case will have to be first administered if through or from the United Nations, but to compliment the full aspect of the/our/my proposal in mention to what the American dream is supposed!

That I had stated in example if one side of modernity in the secular was to carry the explanation of what the American dream is to be counted as. Balanced in the spiritual sway by how some of our imams, would lead that we could apply murder and mayhem and suicide finally expressed as an act of self martyrdom. To set the picture of equals in what we might have to endure to bring the secular and spiritual of equal standing from different reasoning, is a measure of how hard some of us believe in the concept of the future as a beginning and not an ending?

Although I continue to move that our standing of accepting the standard of religious tolerance should now be changed to its full meaning of referral to us holding religious forbearance in place, stands, because urgent change is now needed. Even if the similarity is there, our intentions are to be far more profound than accepting any minor tentacle of politics making us supposedly religiously tolerant! From which in order that we might see the erred implications of what such minor dishevelment means by that fixation. We will have to regress and at the same time step back to the start of this last modern era, marked to be from about nineteen forty five and any time since over these past sixty plus years, but only to look at assumption!

In giving much to the fact that most deeds done since then in terms of developing nationhood had to be politically balanced or were led from the political perspective to fit into the associated plan of the United Nations as overseer? Remains in core settlement providing we can deal with all anomalies in or from the secular perspective, which I had advocated or not! By that token, in order not to excite ready made emotions to the contrary I had made the political connection of our standard for us all to be religiously tolerant. Further by my own sight, I had made the pointed reference that the terms of religious tolerance leading to religious forbearance, had first been truly generated from the religion of Islam at least some fourteen hundred years ago! I had made no case for that example, save that it was ultra modern and in part supposed the base thinking of what Islam should be, even from fourteen hundred years ago until this last modern era!

By how I had compared the foul deeds done by Palestinian to Palestinian in the mandated area of the Middle East just before Jewish independence into the nation state of Israel. Was in reflection to what had actually occurred between Arab and Jew or in fact to be the same people in Jew and Muslim! Was part of our joint conscience in operation to try and rationalise in any way the importance of nationhood even religiously, over the event of slaughter and murder first, second or third world?

Although I had given mention to most events until about the mid nineteen sixties, that interaction in count by both factions was of prime political importance supported by the fact of our bloc culture standoff at the time. By making that case it is easy and open to claim all features were of political importance first, being no bad thing by intention, but to be remembered it was always going to be questionable by outcome? All that we had forgotten until now was the manner of how we continually change?

Continuing, I had suggested to the fact that although the country of Israel had raised her own flag at about the same time of the religious partition of the sub continent of India. Her actions were in compliance to what her first world dealings were supposed to have produced, and because so, were wholly political! My further example relative to the mid nineteen sixties proper, is to set new standards of behaviour?

By allowing that we of the United Nations, we of westernisation, politically and religiously if counted. We of our first world designation, could now at this juncture close the door on politically based religious interaction irrespective of any previous happenings, and henceforth move to the one aspect of only judging matters politically. Thus allowing for any example in question, that if from now any objective religious voice was raised, if like in the case of Palestinians who had generationally lived for years in their own area of the Middle East, yet had been given no national standing! But wanted to reshuffle the issue, they in we can only move to political music. Which by effect although not to justify the past allows that we can make proper readjustment without recourse to God in Abraham or Allah! Such matters have to be now treated only from our new washed political perspective!

In the name of the United Nations, for example, even if to really mean American or European opinion, can any voice be attended to if having already been classed second world politically and therefore religiously by association proving any case to our first world rational? Who have already set incompatible standards, like the mix of what human rights again is, when meant to be spread through religious tolerance as one standard when we are first, second or third world politically, yet the same religiously? How are they to come together on this one point, without the use of our three taken and one active prospect from the Fifth Dimension?

My point by example is not to support my own introduction of when the overlay is to be set, even that parts have been introduced in the past, if like from the suggestion to behave politically correct, which of course cannot cross the void between the secular and the spiritual. Because it will show of its separate needs by introduction to what our religions require and what our politics deliver. Because from our habit of natural progression first world, by using the maxim of might to be always right we are further compromising human integrity. Especially when until we have fully changed our impetus by expectation of what the Fifth Dimension in prospect will or can do! Some of us in working outside of its delivery still manage to read all the volumes of how conduct should be, even to regress to react as we might some four hundred years ago when the feature of human sacrifice was ably conducted with full cover from a designated priesthood!

## CHAPTER 193

Post the mid nineteen sixties even if I had put the case of mixing our united time span and including the first years of this current third millennium. My proposition to the pinnacle of what our political image might be in representation only, where the full goal of what politics can deliver is held to be close to what the

American dream is. Stands as a statement on its own account when applied to the aim of politics in only being able and accommodate secular needs, which in class are to be materialistic.

By the same standard, for me to have put the balance by religious terms hat some of our imams might encourage in the name of martyrdom, stands in representation to how the spiritual and secular automatically differ, which is only part of the fable? Because in both examples of how I might have been able and draw those features to a particular conclusion if so described, falls short because both standards in coming from how we have reacted since, even before the end of the mid nineteen sixties. Had not been subject to the influences yet to come when overlaid by all the prospects of the Fifth Dimension!

By the same declaration I had alluded to, much of the Jewish claim to nationality was officially political, but carried religious sentiments? Has done all that was necessary to undo what ever the United Nations had intended by the reintroduction of religious tolerance if only to thwart communism! For me to have insinuated that virtually all nationality claims save the outstanding case of events in the sub continent of India were politically drawn only, stands! Not least because the plan made by both parties to be of Islam or Hinduism, had made choices before the standard of partition was made by the primary controlling influences at the time to also have been European and Christian by outlook. Results of which we now know are still unsettled, but carry a different complexion than other similar events in the Middle East at roughly the same time?

Without slight of hand and only born of or from the circumstances to have developed in India and the Middle East in parallel form. If in the main the dividing or the primary religion of Islam was in balance in both cases. What had been done is look to the relationship between Islam and Hinduism on one level, but not to be counted? Then look at Islam and Christendom represented by Americanism and Europeanism at such times, before the long awaited introduction of the Fifth Dimension in overlay in the name of modernity. Before we might dare and make rules apply before the thought, simply because then ago we began to think that whatever we done was always right?

Even before we block corridors with the intention of what the understanding of our three taken prospects from the Fifth Dimension are, all doors are opened with the key of 'One for the Whole' in standard to all the same parties. But first we must face the true reality of open events in the Middle East at this time more than ever before for our own and equal health check!

Without the image of repetition, because our reflections are always to be different when classing matters in the Middle East, having been done in the spirit of political judgement up to now, has been compromised by one of our most long standing forms of religious control in our priesthood!

Without balance, without balance or justification to what ever I might have suggested even in accusation to what some of our imams are responsible for, I have set for them in my own presentation their standard and standing to be at least second world? Because from my own study, Islam has not yet had a full political voice to spite what had first happened in the partition of India post world war two? When from about the same time if locked by the auspiciousness of our westernised political and or religious outlook, the formation of the new Jewish state of Israel was of a first world perspective. And I had delivered the statement that this fact was of political determination only, but with internal religious hopes.

It is from that standard that we can show the full failure of what religious tolerance in form has committed against earlier hoped for proposals to the understanding of what religious forbearance is. Although now to be a reflective image, until we gather all the meanings of religious forbearance in this new concept overlaid by the prospects of the Fifth Dimension. Our following of religious tolerance has the ability to lead us into the abyss of darkness covered by Armageddon!

Without balance without comparison, without accusation, but by association, not least because of area and of its significance, when said, the utmost importance of how we are prepared to try and solve any remaining problems if in the Middle East are now paramount. Not least because of recent developments which now turn our political propositions in aim, on their head? Because without open view or open acknowledgement even unnoticed through secular conduct. But only revealed from abstract study in the form of making the connection, we can now see how the feature of religious tolerance has truly been compromised so much so that vital and urgent change is needed no matter what the name is. But providing it is now run through our active prospect in 'One for the Whole' ever, our representations are to all and for all by whatever name or call!

Our vital charge, which in term is to be through our religious leaders, is to the matter of change, even by our political attitude which although not yet fully understood will need more help than just relying on our political fraternity in also using the Fifth Dimension? Because in entering this end of this last modern era if through the intention of what religious tolerance was ever supposed to mean, some of our priesthood are mixing the spiritual and secular only to make the clock turn back! Which has the very real danger of cutting the vitals of our new and open attempt of us trying to create, create our own position to stimulate Gods own interest in us who are the only specie ever to have shown that inclination. From which in using that word creation again, will be seen how no other specie has offered any term save instinct and evolution?

If ever there was to be an errant band of imams and notice of what they might or might not encourage from our young and do! Imams as we, act to what can only be the misunderstanding of what was ever meant to be our conduct, even if determined by us individually, then even from the hand of Mohammed if done some fourteen hundred years ago. Their error is ours and will be shown so in short reply like before, but always by its own accountability. But not to be compared to a similar but far more sinister happening now occurring in the full religious sector of or by some of the Jewish priesthood!

Although from that prefix of priesthood, there are similarities in the structure of how the Jewish and Christian religions are managed, but always to the spiritual.

Which by class or intention or inspiration had always accepted the one standard for all not to kill of self by self. Unless determined by the classed political structure of our continued political Jewish and or Christian societies, but to maintain the reflection to the fact that although we are all the same spiritually. Our secular set up was prevalent not least to be able and show at first our being led by princely sources, then in the name of materialism and commercialism because of the guiding structure of market forces. Fortunately pre Darwin and running up to the start of this last modern era, we have always been able and refer to how it is from our spiritual aspirations that we are to be treated equally in Gods name through 'Natural Body Death'!

That done and in being able and also outmode the practice of human sacrifice religiously, before we were able and justify the concept of religious tolerance. All of our other main religions have at last come to the standard that our godly influences are worldly, and to all who answer the call by the same case of asking the right questions to show of our ability to make the change from evolution to creation! The best that any and all religions have already done even when shown to have been chosen, is for all of us to stand in line for Gods own liberty, which is nothing less than what the Fifth Dimension makes clear to us all of whatever faith!

In making the case to be proved of what has happened recently by the behaviour of some of our priests of any faith, is no more than to repeat play history, but taken to be far more serious when we might turn the office of religious tolerance into the mill whereby the new flower milled is from old wheat. Which by the worst standards of modernity take us back by religious terms at best some four thousand years to have been specifically chosen? Or from how I might determine the actions of some of our Jewish rabbi, in setting the standard again to be at a time of only some four hundred years ago when the act of human sacrifice was a true count in another religious concept. But in case, then ago was separated to work or be offered in form for the sun by day and the heavens by night religiously. Being a past feature which in translation could only be described as an evolutionary event, because in term all deeds done from those sacrificed by the thinking of four hundred plus years ago were to the only aim of procreation!

How ever I might wish to translate the reflection of what I have seen to be the case, stands in terms of 'One for the Whole' which by device is the master of conversation allowing us all voice opinion! Which when translated even before I had made any un-final statement about the incidence whereby some of the Jewish rabbi are now actively willing to take up arms even if by reaction and are now ready to openly kill of self by self in Gods name without recourse to Moses effort in bringing us by hand the true word of God in not to kill of self by self! Some of our Jewish rabbi in we, have managed to far eclipse the wrong edge of how I had seen some of our imams in we behave by my own score? Which in all cases is to signal the end in standard of being religiously tolerant by whatever code, even if to make the case for religious forbearance! Making the better choice whereby our religious outlook only, instead of the previous examples of entrenchment, can turn all of our different religious perspectives into the rational of human adjustment only. So that if called by any religion that we are or had been chosen, the choice is always ours alone to be set by our religious leaders!

Even not unlike what or to how I had aimed charge to some of our un-priestly imams in setting standards ill of their charges in our young. There is to be no religious comparison to what some of our priests in the faith of Judaism are now prepared to offer of themselves!

Although of good heart and in good chime to the matter of there own standard of belief. Some rabbi have openly crossed the floor from the spiritual to the secular and are prepared to keep to the habits seen and felt post Moses but pre Jesus Christ of new Judaism, introduced some two thousand years into Judaism and also two thousand years ago?

How any of us might or can interpret the extent of the action some of the Jewish priesthood are prepared to endure even totally unsighted to reality, falls to the open proof in our need to relinquish the standard of how we have all abused the concept of what religious tolerance was ever supposed to be. Is the opening gambit before the application of what the taken and active prospects of the Fifth Dimension will deliver; not least to accept all religions as they are in the spiritual, without trying to balance them with the secular! In order that we might jointly if in both cases be able and excite Gods own interest in us as the unique specie of life so far so long!

Where and if in the same form of comparison we might compare what and how the terms of holy martyrdom endured in or at the time of the Iran/Iraq war of recent duration in this last modern era is held as an example to what some rabbis advocate, falls short of reasoning? None can be made to what or of how some Jewish rabbis being adult are now learning to partake in the act of battle as secular soldiers with all the mayhem that is prone to produce. Honing there military skills so that if in Gods name they might become expert in the art of murder of self by self, in Gods name?

By being prepared and turn time to stop, some rabbis as we are looking in the self same class used in the form of human sacrifice some four hundred years ago. Simply because in time, no sign had yet been given to the hope of imminent, from which will spring the trigger to be pulled in the name of being chosen!

Some of our Jewish rabbis have lost charge or found in name the wrong implications of what religious tolerance was ever meant to be? Not least because of the complicated nature of how time and our generational onslaught has now curried favour to the reintroduction of imminent thinking by our own made standards of impatience which can only compromise our known conscience, which is to be of little or less concern. When at last the true nature of our religious leaders show their ability to work the limited time owed us all by the generosity of finding the truth through the real understanding of what purpose is!

For those of our rabbis in we, who are prepared to die in the cause of battle in our name or to remain to their own tribal self classification, but for Gods sake! Might yet become our biggest error of thought without excellence of mind? Because in style the feature is even more bizarre than when we practiced human sacrifice!

Then ago the thoughts in deed were almost as simple in the thinking, because we had been given such bounty by the sun God, our duty in kind by associated obligation was to be earth wise and offer our highest regard to at best maintain the status quo, even to the cost of human life! Naturally such matters were to be overseen by our well established priesthood!

Now, even if driven by the pretence that such matters of recent concern are the way to counterbalance at least the effect of what some imams even from other similar sources might have contributed too? Wastes our ability to always improve or wastes our understanding of what the incidence of now means above the importance if shown to the imagination of imminent!

If and so I would make the case without comparison, that rabbis on there part or imams on there part, would both, in party or in part or individually, offer what they would consider to be important to the development of specie in self. Be it by or from first or second world perspectives, then the urgent case is for them in we to stop dancing on the souls of those of us encased in the spiritual, from their falsity of only acting to the political.

Fortunately with little encouragement, our intention with the idea and practice of the Fifth Dimension is the best means for us to see how we err in either case. Because of accepting reality ongoing, we have been discontent and have reacted to the first reflection of what was happening, then again to further react, seen or not to how the reflection of the reflection was received?

From where I had made the accusation of what some of our imams might encourage to be at first of a spiritual standard. If compared to how this later addition of why some of our Jewish priesthood, are coming to terms with modernity in a more sinister way and are also prepared to administer death wholesale, but only in the name of murder! Reflects one message only? That both religions and the new Judaism of Christianity included by the link of the same monotheism, have gotten all the sums wrong, and have set the standard to create multiples of our errors by only and ever being able and split our active concern of what politics delivers in fact, against what religious tolerance is to deliver in promise?

Which in translation by the only possible explanation, drawn from the probability of our ability of thought or how we thought then ago, with creating our religious concepts from within the bounds of civil order and or other social controls at such times. Instead of centring to the greater importance of what religions really are. At best we had only been able and squabble in argument, not because of our separate religions, but because we had never been able and recognise the importance of most of them in a truly global sense.

Although we have always sought answers it is only from this last modern era, sprung from the accident of our final attempt at specie unity, even if unclear to the meaning of that expression. Not least because all of our workable features of politics and or religion on entering the funnel of organisation and when leaving at the spout was seen only the spume of confusion to be acted upon? Leaving it so that in order for order, most of us have taken to what we understand, unfortunately more so to politics and its tentacles and run with their projections, man made?

Instead of following the real trail of progress, still man made, but cosseted by hope and purpose to be the light of understanding to what political science can never deliver although formula can be promoted, purpose cannot!

Without the necessity of creating new images, because what we already have by several different counts is well enough to begin with again. What we have, has already been put into the top of the funnel, all we require now is to pay open attention to what is all around us. Remembering, even if unseen by a large score, the full events of what happened at Shanidar some sixty to one hundred thousand years ago has always been at our disposal. Not unlike the three taken and one active prospect of the Fifth Dimension, always there but only now coded because we have had this opportunity of these past sixty plus years to re-examine our full potential. Which from now and for the first time since Shanidar we can gather all realities to at first re-examine the potential of 'Natural Body Death' owed us all!

## CHAPTER 194
## OUR LAST EPILOGUE

For any battles ever served in the name of religion, we are about to enter the last with any connection to the past. Our further steps to be taken from now aided from abstract reasoning can now begin, but to be carried by our religious leaders in our name, in our cause, in our society, all to quench our wonder! Even if that is to prepare us so that we reach the understanding by acceptance, that indeed there are up to eleven full dimensions. But only to be revealed to us at the time of 'Natural Body Death', because it has only ever been we in our human society who had asked the first question of what is tomorrow?

Our religious leaders in our name are fully entitled to offer the hope of our own purpose in their name because of their own religion, which in all cases is our only connection with God, unknown, but guessed at long before the wisdom of science had been squeezed by our inside desire of wanting to know? From which was turned the one wheel of life for our own humanity before we knew of, but made by us in unity 'Out of Africa', telling the one recorded story of hope in the throws of life for the first time even when we were to look into the face of purpose.

We and we alone without self acclimation have since been the axel of life to turn that wheel always for tomorrow. Proved by how we have been able and genuinely advance in bond all over this same planet yet unlike any other specie, we alone out of the shadow of all life had looked for the reason of why we must propagate? We of evolution in that name and by pulling at the query of the day dawning, had set the new standards of life to bear a great truth upon our universe!

Although almost losing track at the near point of our best understanding in the one generational sway of this last modern era. Once again we are lucky to be charged to tackle the wall of ignorance we have compounded into fact by creating the aspect of fear in the future. Which is to be dismantled by you and I giving charge to our religious leaders for the honour of settlement as far as our specie integrity is maintained until the meaning and understanding of all mysteries are to be known if necessary by our 'Natural Body Death'!

Fortunately much of the ammunition we will need for the struggle of such understanding already lies about in the form of our religions. But at this time because the force of our politics had been able and quash our first felt enthusiasm and exuberance at original instigation making us all now to be religiously tolerant. Because by reflection, when any of our religions were actually started, four two of fourteen hundred years ago; we already had forceful civil order or the princely prospect of it in control. From which we have not been able and rationalise what was to become the spiritual from the secular, unless only in following! Since shown to have been proved by the way that when we first showed our spiritual colours before our welfare, they were consigned to be secondary to the whims of what secular management should be. So much so that in the desperate need to understand the concept of God in any form or by any route, at the same time that we of our secular heart allowed religions there wings, we closeted them in style not to interrupt how the management of government was to work!

Not I add from superior wisdom, we only have to look at how when any society gathered to only be of the secular all have failed and ever will. Because in type virtually all of our political systems, even when proclaiming to hold to the spiritual in national religions. Have erroneously allowed their/our tentacle of science be the master of testing all of our mysteries. So much so, even when unproved, some of us look to science instead of God from the arrogant stand that because we understand the principle of one, two, three by experiment, therefore we will by able and understand Gods own matters above Gods own meaning!

Although we are proved formidable for what we have achieved when counted by three and a half million years against the two hundred and fifty million years of the crocodile, we really only know as little now than when we first queried our own motives when a group, clan or tribe buried at Shanidar, it is that simple!

Unfortunately in taking to accept our style to be content with accepting half measured deliveries of any secular policy, because we have never seen the alternative, we have set the situation to flounder by implementing the support in reacting half measured by our steps taken syndrome? Being most able not to question that situation because on the several counts we use it to maintain the status quo?

Firstly by loose connection, politics does work even when using flawed policies, second under its own wing of tentacle control through the hand of commodity and commerce, which only has a small sound, things do work! Thirdly although not fully understood we have come to assume that science in its designated form of being the fourth dimension will automatically supply the answers and solutions to all of our problems, because already it has given us the light bulb? From which by imagination when we hold a picture in minds eye of the light bulb above our own head, somebody will get a brighter idea which will be able and lead us in the correct direction of improvement this time!

Fourthly and the most complicated yet, in already holding all the answers of what science might deliver by attention to our religions Eastern or Western. Politics from our natural ability to grab and hold the one bird in hand instead of waiting to reach the two in the bush, has only been able and promote its own values in competition to our spiritual visionaries? Simply because science will never be able to prove the value of our individual imagination, yet attempts to placate the ear of politics with up to eleven unknown dimensional studies in order to be able and offer answers, by imagination! Even so by supporting the ideas of what science might bring in material advantage, politics has removed the true influence of our religions by the corner of religious tolerance, unless perhaps to suit the maxim of might always being right!

## CHAPTER 195

Politics to her code and in collecting most to be fed into the top of the funnel around nineteen forty five, accepting what ever was to spume from the spout as read, unchallenged! Instead of being able and manage intent even if with the supposed help of our United Nations has only managed to set the first standards to suit one of her master tentacles of finance!

Fitting all other needs to that principle and making for the one clear interpretation that wealth is health, but alas from the sinister that self wealth is the key to generate the industry of creating more self wealth, which has an already found stop, but for now to be ignored so as not to interrupt fashion! Which is by no means a bad philosophy, except when counted that in whole the fixture as standard was only ever designed to endorse what always had been done to suit a class, or leadership structure who already carried another form of wealth if only in self power! Which from past examples like when in some cases of our priesthood running the administration, even to maintain the principle of human sacrifice, such all standards became acceptable!

From now, with politics having to take charge of our outlook in this last modern era even running to the above principle of our steps taken wealth syndrome, what had been occurring in period since, has been the exposition of our shortcomings over the same generational time span owed us. If measured only on the least number of examples, my intention is to show how small steps taken from the original half measured introduction are sometimes responsible not only for big errors but missed opportunities, like for example much of what we have not yet done in this last modern era.

Further shown but without charge and using the example replicated all over the world. By index with the new creation from the old League of Nations which was definitely first world in all outlooks. When introducing its replacement through the United Nations, old habits remained to the first choice of first world legislation, which when viewed rationally had to be our starting point? As said along with other world examples, what was exposed in the lead by the American dream only a few years into term of our recent modernity, was that indeed America herself although primed first world, there were a substantial number of second class or second world citizens there!

Although to correct this situation by the counted terms of what modernity was supposed to mean both internally as a single nation state; and collectively as a prime member of the United Nations. America in sync with others drew up or added to the original concept of what human rights was supposed to be or mean or was taken for, by adding separate national laws around the fringe of civil rights, weakening both aspects. But by that adage civil, confused the supposed open singularity of how our political intentions were to lead and also of there chronicle!

Along with such necessary solutions and having already standardised some reactive attitudes by adopting further adage like behaving politically correct. Our United Nations aura of being in the right cause along with what and how we were finding how to act or react politically internationally set the standards we primarily use today. From which being totally necessary are also to be the full cause of the state of affairs we each now live under. Which in class is like that of a broken mirror strewn upon the floor with each piece only being able and act to or react to there own specific reflection!

What the Firth Dimension can now do on all of our behalf is set in proposal our desire not for change, but realignment. Although with little connection between Shanidar of some sixty to one hundred thousand years ago and its reintroduction by the time mark of when this book is published or first read. Both events bring us in order of what is to be done now, because then ago we were too new 'Out of Africa', and only able to think 'Past Gods Words', but had been driven by the query of when and of what will happen by 'Natural Body Death'?

At the same time we were also void to be able and set standards by discussion using our one active prospect in 'One for the Whole'. Leaving the intervening hundreds of generation to be better able and reflect upon our differences before seeing our similarity. Although politics in case has become our master, she in our name can show her own true ability of progress by now legislating to the proper meaning of what is to be gained from the overlay of the Fifth Dimension, not unlike what I expect from elsewhere through our religious charter! Won from all divisions of society and able to naturally surpass the aim of politics because no religion stands on any principle other than we are the same!

## CHAPTER 196

In previously using many key, lock or unlocking scenarios, has been a way to highlight how at various times in locking the door behind us because of our chosen certainty, because we have made them work! Has been only a means to paint ourselves into a corner, at time to be inescapable until the paint dries to reveal a different shade of colour from which we must find the correct path again? Uncounted by genuine terms, our religious aspirations have always been the only route for any full success for any specie so far to have graced life anywhere in this full universe!

On our behalf, our religious leaders in we are now entrusted to break the mould of eight hundred million years of primitive life forms until we, and turn the corner full to face tomorrow. Unlike at Shanidar when we only knew uncertainty. Now if from no more than the exertions of science we can fully face the prospect of a true religious beginning, by making ready to face the answer of purpose, but only to have been delivered in camera by the advent of 'Natural Body Death'.

All that our religious leaders have to do is take the hand of politics and show how we are to better behave all round by our specie design so that we might attract Gods own interest in us. Not only are we prepared to admit that much of our excellence of mind was in error of thought, because at one time we guessed that we might have been chosen into the realm of creation? Thus at the same time compromising Gods own time span of evolution which is far in excess to what we might ever have imagined when we had considered ourselves to have been chosen, but carried far beyond malice, fear or envy yet our show in reality was of specie separation. All to do now is reverse that proposition!

Without comparison, to have arrived now in this last modern era and be able and carry in mind the examples of martyrdom and priestly murder even in battle, is of the highest rebuke to God. Because in not to notice Gods own time span of evolution, we insult the spirit of all religions by having to be told in separate array than other specie, of Gods existence, which has forever been held in evolution! Proved when if only to look from eye to eye to eye of all other life forms and not to see a spark that might be turned to the hope of finding purpose instead to only follow instinct for the one aim of procreation.

As far as we might ever know, no other creature has given us Shanidar. As God did not give it to them, nor did we receive notice save by our own endeavour! We found God!

# SUMMERY

Our Fifth Dimensional overlay is not a new set of three or four of what some might consider commandments. All seen is to the fact that less is more to our needs of control, from which by the religious overlay will be given a better guide to how our politics are to react to the same medium!

'One for the Whole' is our joint or equal or continuing means of change if necessary, but to be far more serious, it is to be held up in demonstration that what ever is offered in choice through it, like martyrdom for instance or to murder for instance, is for the whole. Therefore those who would propose such an option have to take the option first and first hand like in what 'One for the Whole' is asked to portray? Meaning that the instigators of such violence would have to take to the same suggestion first, nothing less but to kill of self in demonstration of what 'One for the Whole' is to mean from the suggestion!

So as not to enhance the standard or standing to those of us sick of mind, unlike when some times we of the military in aim to defend and protect self in self nationally. But for those associated priesthoods like in the case of some who would oversee human sacrifice, there call is to die also first but not in the necessity of battle, but also at there own hand by suicide if to show the nature of death in murder, which is never to be of religious connection!

For such reasons 'Past Gods Words' has the double inclusion to be 'Past Gods Words', so that none can say it is Gods will that I act so, it is by Gods word I act so. Not least when any mans voice is a mere whisper to whatever the desires of God might be?

Also by 'Past Gods Words' we give God full grace to his own choosing, not least even beyond the imagination of what evolution ever was. Evolution is Gods own wait to make his own choosing of any who have or would or might find the path to creation and always has been the one fact of common ground between all religions. Therefore can never be our choice to kill of self by self lest we are left as the crocodile to ponder in another two hundred and fifty million years of evolution locked into instinct and driven only by the aim to procreate!

Gods choice done; not ours, unless to be 'Past Gods Words'!

With our religions having already taken to much of the prospects of the Fifth Dimension in less being more, like already accepting that although we differ because of the circumstances of geography, climate, our diet and other reasons. We are effectively one of specie and 'Out of Africa' to kind, which except to the above circumstances, we have to allow that at first we were already the same by being black skinned and from the one tribe. Being the reality account of what and where we began without motive almost like all other specie but were still able to break the mould of evolution!

Less is more if only to realise we are all 'Out of Africa'!

Politics is marked on two counts, not unlike our religions, but in the case of civil control, in order to walk with the prospects of the Fifth Dimension. Our link with the overlay has to be translated by what might be considered serious action, but is contained by a small and common feature owed us all, but in need of deep appraisal!

To have criticised the aspect of being politically civilised in term of reference to indicate advancement has to be now questioned, but in style from the abstract? Allowing that we do not have to qualify the quantity of how or for how long or to or from what criteria we have been able and measure the state of our relationship with each other. Always allows that the standard in any case is only another tentacle to our political understanding, because it has no value within any of our main religions. Their standard has been imposed to another ending which calls for no special signal to indicate our claimed differences from one another.

By following the supposed conduct of what civilisation is or brings, we are led to the thinking ability of how in claiming to be more civilised, like for instance being first world, in its self there is no challenge if that state is only considered to be a graded style of achievement. Where we abuse the principle of being civilised is how we prepare to protect the honour of being civilised if only theoretically. Which only allows that we react in the belief of being right because we are civilised enough to do so, which is only like accepting that the feature is there by appointment, when at all times ever, it had been a cold and unfeeling emotion to have accommodated any such brutish and brutal action any person might take because they/we are civilised?

Without gratifying the request until we can begin to read how to react to the reflection and or the reflection of the reflection of life through normal interaction. It will be better to wait and see how the Fifth Dimension will begin to affect our quality before we try to upstage each other by being super civilised?

Without testing the theory to the obvious of how some of us who claim to be civilised have and do behave towards each other even to kill of self by self. Then even without judgement in any case, if the matter is to stand, then the settlement if the same, will find its own standing under our new impetus?

What ever we see in the need of doing and of what cause, all lights will need to be focused to the one unifying fact of all life ever, in that to find God or not, when found, as in our case. We have only our own generational time to account for, the only account to settle ever, is of our own personal conduct spent in fifty plus years or seventy plus years of our, our own personal lifecycles, and in Gods hope until 'Natural Body Death'!?!?!?

## THE FIFTH DIMENSION

Named so has been done to imply in all honesty and grace that our religious aspirations are of far more importance than how we might behave under secular reasoning and civilised understanding!

By first choice and seriously connected, in choosing our former by time and if prime monotheistic religions of Judaism, Roman Catholicism and Islam to be my first, second and third dimensions was to set the balance against the fourth dimension of science. Who in turn is used to imply that our religions are only distractions of ignorance because we had allowed them hold the hidden answers to all of our mysteries!

Science in her coat, if tied to belong to civilisation along with the expositions of many mysteries has led some of us create a religion in her, needs the check of Gods own purpose, because science also was born at Shanidar. When we raised the first matter of mystery; be it in God or wonder; the wheel of life was first seen, only to be lost in our own divide of religions Eastern and Western and politics and all of its tentacles not lest through the voice of science.

I take up no challenge to science, save to mind its self challenge of being political. Science in taking any other route even to be maintained as a tentacle of politics, has made her own errors of mind by excellence of thought thinking the experiment of life was part of its own formula, when clearly not!

The Fifth Dimension is to be our chapel, church, temple, mosque or synagogue or our places where we can see new views balanced to what being politically civilised meant, even when supported by the tentacle of experiment into unfinished formula, for all of our un-killing of self by self tomorrows. Our Fifth Dimension is also to be of its own, expecting even before the asking, that already done, its taken and one active prospect are to be the fuel of further study into humanity. Not least because from now without pretentious acclimation, without forced indemnity, we are all charged to be fifth dimensional, because by being so no questions are to be asked, only answers given even unrequested?

By both counts and in permit, that our politics and religions, although allowed value the prospects of the Fifth Dimension at there own speed. None can now deny the validity of them even if not to cross the divide of what our politics and religions have become.

Our equality is to be assured in outlook by the full acknowledgement only in acceptance to the three taken 'Out of Africa', 'Past Gods Words' and 'Natural Body Death' prospects. Further endorsed that when spoken in the need of change it is to be through the one active prospect 'One for the Whole' we will ever speak the same language. To the concern that from all our best efforts we will be able and self deliver in understanding so that we might excite Gods interest in us as the only true members of creation by any terms, which will allow the future standing for all of our next generations also. Who in line with us, but by the hope of our own generosity can expect to obtain what we have to do in our fifty plus or seventy plus year lifecycle until 'Natural Body Death'.

That tomorrow can be achieved on better terms, whereby we all might live the full extent of our own lifecycles untouched by the fear of what the new day brings. Because in learning of what is truly expected, the best experiment we all might find is to follow the Fifth Dimension, if to have been read by both our politics and told also by our religious leaders!?

# THE FIFTH DIMENSION

## THE LAST EPILOGUE

By laying politics with all of its tentacles to one side of the balance, with the matter and means of creation to the other cup, our only aim of comparison between politics and religion, between science and God. Stands beyond the means of perfect understanding, because from both camps the means of invention have no provable case to answer?

Science in asking that we might need to create extra dimensions, even if to have set the number in total to be eleven, can do no more than speculate on what they might tell or reveal to us? While to the same end, we who might have found how to misuse some of our religious doctrine have set the same standard of understanding owed us, to be coherently answered when through the event of 'Natural Body Death'!

By those codes in the making, all that we might need to know stands by example in what or of where science might wish to lead us in the possible belief of actual time travel. Leastways in order that we could extend our puny fifty plus to seventy plus years lifecycle, in order to be able and look into the face of purpose? Of which the following is an example to end any bitter taste we might hold for each other! Without comparison, without balance and without understanding, our mission un-appointed was met before all ages of reason.

When unprompted, save by our unclaimed initiative, we discovered the mater of true time travel shown by our last record at Shanidar. Even before we opened what was to be the door to our mystery box, we had found a link never known to any other specie, expressed by our ability to time travel!?

Without any study of the elements, without any aim of experiment, without the need to mark the occasion, because it was now, when we came to bury one of group and leave grave goods for later use, we marked for one of the first times ever our split from evolution in what we did by preparing for a future guessed at?

There and then in record we could traverse the past by use of the unknown element of memory, being a key to have turned and split the present of now, at that time, so that we could genuinely time travel at first into the past because of any memory! Further strengthened in having been stirred by the loss of a partner of tribe or clan or group to death, but still to be able and relate such deeds and actions to have occurred when they were alive? Even when unknown, that any such memories were a certain break from the deed of instinct. What we were doing and had actually done was set in standard our ability to time travel?

Not only were we to become aware of the past by the use of un-appointed memory, but as shown at Shanidar by the leaving of grave goods, our journeys of time travel were unanimously declared to be for the future also. Which by extension we had left trace of in our cave paintings right up to the refinement of Egyptian burials which by now had become steeped in our ability to travel into the future? Having already left in understanding the motive of our past!

When the key was further turned in the same lock about four thousand years ago by us entering the era of monotheism, time travel was certified. Because from about then all of our histories had been written in testament in the bible by method of time travel to the past in order to account for the properties of now and in turn to project us to the future by outcome, even to the extent of where the future was, but in heaven!

When science is done, even with eleven dimensions, maybe more, if and when she might look to the ability of how she could ever deliver time travel to us? None will surpass what the experiments at Shanidar and since have brought by us using elements unknown to the statistics of what science might ever do!

If I take to the unrefined emotion, and I do, that we each hold to thought and reasoning, then that done eclipses any element of experiment. Because in experiment although using many different properties they are all bland, none can breath, unless mixed so by experiment, which bears to a planned ending anyhow, just like from evolution!

It will only be when science through her/our own reckoning balances our ability to think in terms of hope and purpose, before she like us can reach the annuls of God! Which when closing most doors, opens the one whereby we as specie, armed with all the experience of evolution were able to turn the corner into creation! To now make ready and excite Gods own interest in us as that specie, not separate to all others, because they in we were of the same standard, it is only we who can rationalise what the future might bring. Not in the name of how science might wish to present time travel, but in the name of God!

## THE FIFTH DIMENSION

To near reach the end of life is to near reach the beginning of purpose!

## NOTES

Dear reader, because of the nature, style and references used in the just read work and to account for the different people we really are in being the same! I had not added my interpretation of chapter content, because in showing that we are different by being the same, although the forces of circumstances such as climate, geograph, diet and other factors have made us what we are allows that we see things differently. Therefore I have allowed that in the course of time we can each settle in desire to our own view tempered with the immovable prospects just laid down.

'Out of Africa', 'Past Gods Words', 'Natural Body Death' and my one active prospect 'One for the Whole' are in all chapters as they are the open product of our time. All have only the one form of understandable reality because they are a product of our united time relayed to us these recent five to six years. My own personal definition in deference to you the reader, is to let us all be able and think differently, but for our individuality from now, ever to remember that we are in fact the same plan in time whether we see it so or not!

Joseph Donnelly

Dear Tom and Hannah

While on your journey through life together, I hope that from some pages within this book you might find a little support if ever needed.

All the very best to you both and family.

Love from Noelle and Sol